Interactive
Statistics

Interactive Statistics

Second Edition

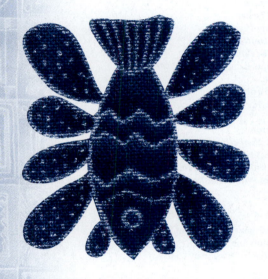

MARTHA ALIAGA
BRENDA GUNDERSON

University of Michigan

Prentice
Hall

Pearson Education, Inc., Upper Saddle River, New Jersey 07458

Library of Congress Cataloging-in-Publication Data

Aliaga, Martha.
  Interactive statistics / Martha Aliaga, Brenda Gunderson.--2nd ed.
      p.   cm
  Includes index.
  ISBN 0-13-065597-X (alk. paper)
  1. Statistics I. Gunderson, Brenda. II. Title.

  QA276.12.A3983 2003
  519.5--dc21   2002023109

Acquisition Editor: *Quincy McDonald*
Editor in Chief: *Sally Yagan*
Assistant Editor: *Vince Jansen*
Vice President/Director of Production and Manufacturing: *David W. Riccardi*
Executive Managing Editor: *Kathleen Schiaparelli*
Senior Managing Editor: *Linda Mihatov Behrens*
Production Editor: *Barbara Mack*
Manufacturing Buyer: *Alan Fischer*
Manufacturing Manager: *Trudy Pisciotti*
Marketing Manager: *Angela Battle*
Marketing Assistant: *Rachel Beckman*
Editorial Assistant/Supplements Editor: *Joanne Wendelken*
Art Director: *John Christiana*
Interior Designer: *Michelle White*
Cover Designer: *Michelle White*
Art Editor: *Thomas Benfatti*
Creative Director: *Carole Anson*
Director of Creative Services: *Paul Belfanti*
Manager of Electronic Composition: *Jim Sullivan*
Electronic Production Specialist: *Jacqueline Ambrosius*
Electronic Composition: *DonnaMarie Paukovits, Judith Wilkens, Clara Bartunek*
Art Studio: *Laserwords*

© 2003, 1999 by Pearson Education, Inc.
Pearson Education, Inc.
Upper Saddle River, New Jersey 07458

Printed in the United States of America

10 9 8 7 6 5 4 3

ISBN 0-13-065597-X

Pearson Education LTD., *London*
Pearson Education Australia PTY, Limited, *Sydney*
Pearson Education Singapore, Pte. Ltd
Pearson Education North Asia Ltd, *Hong Kong*
Pearson Education Canada, Ltd., *Toronto*
Pearson Educación de Mexico, S.A. de C.V.
Pearson Education—Japan, *Tokyo*
Pearson Education Malaysia, Pte. Ltd

CONTENTS

---

* = Optional sections.

v

PREFACE

Why Teach with Interactive Statistics?

We face many challenges when teaching an algebra-based introductory statistics course. Among the numerous obstacles is the perception of many students that statistics is boring and pointless. One of our principal goals in this text is to engage students in the subject and to teach them that statistics is full of ideas and methods that will make them more informed users of the information they encounter every day. This worktext encourages hands-on exploration of statistical concepts so that students take an active part in the learning process. With its strong emphasis on data analysis, this book seeks to make students more discerning consumers of statistics and to give them the skills to design and execute experiments in an undergraduate research class. We have tried to present statistical concepts economically and to reinforce them immediately with activities that will make the concepts clear and vivid.

Each chapter has cumulative material (exercises and examples). Students will encounter problems for which they must apply the knowledge learned from previous chapters. This shows students that statistics is not a collection of isolated techniques. Statistics and the scientific method provide a collection of principles and procedures for obtaining and summarizing data in order to make informed decisions.

Features

Interactive Exercises Are Built into the Text

Each chapter features many *Let's do it!* activities and *Think about it* questions that draw students into the text and reinforce statistical concepts.

- In many ways, the *Let's do it!* activities are the heart of the text. These activities are designed as individual or group projects to be completed in class. These activities reinforce the concepts just introduced or lead students to discover the next statistical concept. By working with the *Let's do it!* activities, students become engaged and active participants in the material. Students soon find themselves actually doing statistics—gathering data, analyzing the data they have collected, and discussing results with other members of their group.

- The *Think about it* questions ask students to reflect on a concept or technique just presented. *Think about it* boxes encourage students to make the leap to the next related statistical concept. The questions help students retain information and lead to new discoveries. These boxes also show students how to apply their new knowledge practically, rather than relying on rote memorization.

Real Data Used in Exercises and Examples

We attempt to pique interest in the examples and exercises by using data sets on current, timely topics that will engage students. The exercises and examples are drawn from newspapers, magazines, and journals with which most students will be familiar, thus underscoring the practicality of statistics.

Innovative First Chapter Highlighting Major Themes

Too many introductory statistics texts leave many of the most important ideas in statistics until the last few rushed class periods. We believe that students pay more attention to the path they will follow when they have some understanding of where it will lead. Consequently, in Chapter 1, "How to Make a Decision with Statistics," we introduce students to the major ideas and themes of statistics that they will use throughout the course. We show students how to work with data, make decisions in statistics, determine the chances of error, and assess the statistical significance of the results obtained at a simple level. This unique chapter gives students a grounding in statistical reasoning early on, allowing them to master the subject quickly.

Early Coverage of Sampling and Experimental Design

The first three chapters of this text deal with important issues in sampling. We cover practical topics, such as different sampling techniques, biases in the data, and the use of random samples. We provide a thorough introduction to factors for planning statistically valid experiments, including randomization, blinding, control groups, and the placebo effect.

Up-to-Date and Lucid Treatment of Probability

We have attempted to write a modern presentation of probability using examples and techniques that show how important probability is in understanding data and interpreting results. Coverage includes both estimation of probabilities through simulation and computation of probabilities through more formal results. The concepts of chance and likeliness are introduced as early as Chapter 1. This early introduction builds students' confidence in working with such concepts later.

Integrated Use of Graphing Calculator to Reinforce Concepts

We have found the graphing calculator to be a valuable addition to our course, allowing students to enter, plot, and summarize data quickly and conveniently. The TI-83 Plus graphing calculator has many enhanced statistical features that can be used for the more sophisticated and lengthy statistical analyses such as ANOVA, multiple comparisons, regression (simple and multiple), chi-square, and nonparametric. While the graphing calculator is not required, we rely on it in certain examples and exercises in which a calculator minimizes hand calculations and eases data plotting. More in-depth, keystroke instruction is placed in the ■TI Quick Steps sections that appear at the ends of selected chapters. These sections show students how to use their calculators efficiently.

Changes from the First Edition

Thousands of students have been successful in learning statistics from the first edition of this book, and we have made substantial improvements to the book based on feedback received from students, instructors, and reviewers. Substantive changes from the first edition include the following:

- New material has been added throughout the text to provide instructors the necessary materials to teach longer, more comprehensive courses. Additional content includes:

  Two-way ANOVA and the concept of *interaction* have been added to Chapter 12 on Comparing Many Treatments.

  The material on *regression* has been enhanced to include more details on *inference* in linear regression and a section on *multiple regression.* The chapter on regression has also been moved to a more appropriate place as Chapter 13, entitled "Is There a Relationship Between Two or More Quantitative Variables?"

  A chapter on *nonparametric procedures* has been added as Chapter 15, which presents the common rank tests: Sign Test, Wilcoxon Rank Sum Test, Wilcoxon Signed Rank Test, and the Kruskal–Wallis Test.

REAL DATA

- *Double the amount of exercises* and updated problems throughout the text. These exercises offer an abundance of material so students can apply and assess their knowledge of learned concepts. All exercises and examples that are based on real data and news articles have a *real data icon*, as shown here, to readily identify such material throughout the book.

- Chapter 10 on one-sample inference has been split into two chapters. Chapter 9 covers inference about a population proportion and Chapter 10 presents inference about a population mean. The smaller pieces help the students digest this material more easily.

Supplements for the Instructor

Comprehensive Instructor's Resource Manual (by Martha Aliaga and Brenda Gunderson)

To help prepare for teaching an interactive class with our text, we have developed an extensive set of materials to show the instructor how to get the most out of an interactive class. Detailed information is provided on how to set up effective student work groups, how to incorporate the graphing calculator into instruction, and how to prepare for the first day of class. Each chapter of the instructor's resource manual gives learning goals, ideas for teaching, solutions to the *Let's do it!* activities (including how long each activity takes, how to accomplish it, and its importance), solutions to the *Think about it* questions, and solutions to all of the exercises. (The ISBN for the instructor's resource manual is 0-13-065845-6.)

Printed Test Bank (by Brenda Gunderson, Martha Aliaga, and Kirsten Namesnik)

The printed test bank includes about 1000 additional problems for use on quizzes and tests. In addition to the print format, the test bank is available as Microsoft Word files from the publisher. (The ISBN for the printed test bank is 0-13-065849-9.)

Data Sets

The larger data sets used in problems and exercises in the book are available to download from the Aliaga/Gunderson Web site.

Web Site: http://www.prenhall.com/aliaga

Our Web site provides a central clearinghouse for information about the book for instructors and students. The Web site includes learning objectives, multiple-choice quizzes that provide immediate feedback to students, a syllabus builder for instructors, data files for the book, and a graphing calculator help function.

Supplements Available for Purchase by Students

Student Solutions Manual (by Brenda Gunderson and Martha Aliaga)

Fully worked out solutions to most of the odd-numbered exercises are provided in this manual. Careful attention has been paid to ensure that all methods of solution and notation are consistent with those used in the core text. (The ISBN for the solutions manual is 0-13-065846-4.)

Text and Student Version Software Packages

Interactive Statistics and SPSS 11.0 Student Version Integrated Package

A CD-ROM containing the SPSS 11.0 for Windows Student Version and the data files from the text may be purchased as a package with the textbook for a small additional charge. (ISBN 0-13-045396-X)

Interactive Statistics and MINITAB 12.0 Student Edition Integrated Package

A CD-ROM containing the MINITAB Release 12.0 Student Edition and the data files from the text may be purchased as a package with the textbook for a small additional charge. (ISBN 0-13-045398-6)

Technology Supplements

TI-83 Graphing Calculator Manual for Statistics (by Stephen Kelly)

This brief manual provides simple keystroke instructions for using the TI-83 graphing calculator in statistics. Class tested for many years, this manual is the perfect answer for frustrated students and professors. (The ISBN for the graphing calculator manual is 0-13-020911-2.)

An Introduction to Data Analysis Using MINITAB for Windows (by Dorothy Wakefield and Kathleen McLaughlin)

A hands-on guide to using MINITAB 12.0, this spiral-bound workbook provides step-by-step instruction for learning how to perform basic statistical analysis with MINITAB 12.0 for Windows. Each lesson is set up with an activity that is designed to be completed and handed in, making this manual ideal for lab sessions or independent study. (ISBN 0-13-012508-3)

Acknowledgments

We are thankful for the enthusiasm that everyone at Prentice Hall has brought to this project. Special thanks go to Quincy McDonald for his constant support and guidance. Barbara Mack has won our appreciation for managing the production of our book. It is a pleasure to acknowledge the encouragement we have received throughout the development of this worktext from our students and colleagues at the University of Michigan. A special thanks to the following professors, who used our text at Michigan and provided valuable feedback and suggestions: Julian Faraway, Janis Hardwick, Bruce Hill, P. Jeganathan, Bob Keener, Ed Rothman, Mary Meyer, Julie Berube, Ling Chen, Steve Coad, Dan Coleman, Jonathon Kuhn, Arden Miller, Rahul Murkerjee, Yingnian Wu, Sarat Dass, Bruno Sousa, and Yolande Tra. Our gratitude is also extended to the many graduate student teaching assistants for their dedication to teaching with this text. We thank Priscilla Gathoni for her diligent work as our accuracy checker. We express our appreciation to Mary Ann King and Tina Smith for their administrative support. Thanks to Robert Megginson and Mark Mikhael for their help in developing the programs for the TI graphing calculators. Finally, and most significant of all, thanks to our family and friends for their encouragement, support, and patience.

Reviewers

The following people served as reviewers for this edition: Trent Buskirk, University of Nebraska, Lincoln; Constance Eshbach Cutchins, University of Central Florida; Sarah Hardy, University of Maine at Farmington; Jim Harrington, Adirondack Community College; Thomas P. Kline, University of Northern Iowa; and Jim Robison-Cox, Montana State University.

Reviewers of Previous Editions

The following individuals provided user diaries based on their classroom experience in teaching from the preliminary edition. Their suggestions and detailed commentary were instrumental in shaping the first edition. We are indebted to: Vivian Anderson, Hamilton College; Alice Burstein, Middlesex Community—Technical College; Cathy Chu, Eastern Michigan University; Josaphine Hamer, Western Connecticut University; Debra Hydorn, Mary Washington College; Cary Moskovitz, University of New England—Westbrook College Campus; Robert Schaefer, Miami University of Ohio; Joan Smith, St. Petersburg Junior College; and Olaf Stackleberg, Kent State University. We also appreciate user comments from Paul O'Hara, Northeastern Illinois University; Tom Blackburn, Oakton Community College; and Yousceek Jeong, Calvin College, who class tested earlier drafts of the manuscript.

The following people served as reviewers of the first edition of this textbook: Patti Collings, Brigham Young University; Chris Franklin, University of Georgia; Carolyn Likens, Millikin University; Steve Rigdon, Southern Illinois University; and Yvonne Zubovic, Indiana-Purdue University at Fort Wayne.

Workshop Participants

We appreciate the help of the following workshop participants: Rhonda Alercia, Bio-Pharm Clinical Services; John Boyer, Kansas State University; Norman Bruvold, University of Cincinnati; Don Bullock, University of Illinois; Miguel A. Caceres, University of Houston; Beth Chance, University of the Pacific; Bob Cruise, Loma Linda University; Mike Frey, Bucknell University; Joe Fred Gonzales, University of Maryland; Irene Grohar, Loma Linda University; Bob Hogg, University of Iowa; Stacy Hoshaw, University of Missouri; Anita Hurder; Debbie Hydorn; Victoria Martinez; Jessica Lam; Tom Moore, Grinnell College; Nancy Priselac, Garrett Community College; Steve Priselac; Justine Ritchie, Grand Valley State University; Neal Rogness, Grand Valley State University; Tom Short, Villanova University; Paul Stephenson, Grand Valley State University; Sandra Stinnett, Duke University; Richard Swanson, Cincinnati State College; Elliot Tanis, Hope College; Jan van Schaik, Pharmacia & Upjohn, Inc.; Joan Weinstein, Pine Manor College—Harvard Extension School; Jeff Witmer, Oberlin College; and Roger Woodward, University of Missouri.

HOW TO USE THIS BOOK

Interactive Statistics, Second Edition is a hands-on worktext that shows students how to be informed consumers of everyday information. The following pages will give you an overview of how to use this book.

Interactive Exercises Are Built into the Text

Each chapter features many *Let's do it!* activities and *Think about it* questions that draw students into the text and reinforce statistical concepts. Designed as individual or group projects to be completed in class, these activities allow students to actually gather and analyze data.

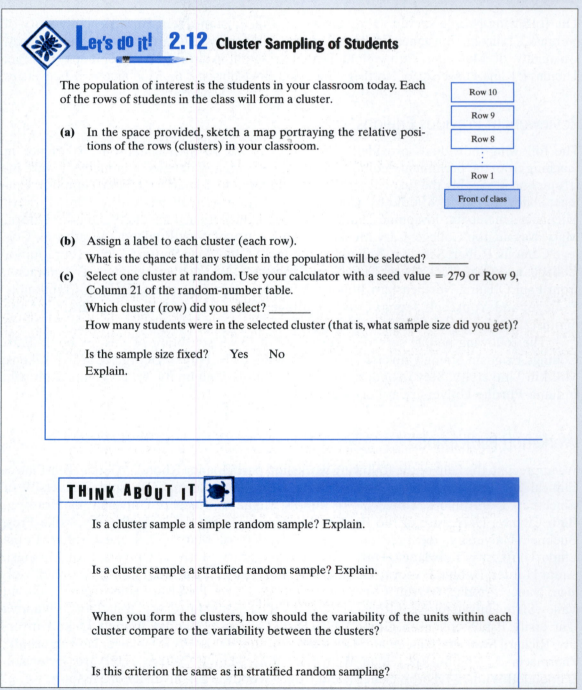

Let's do it! 2.12 Cluster Sampling of Students

The population of interest is the students in your classroom today. Each of the rows of students in the class will form a cluster.

| Row 10 |
| Row 9 |
| Row 8 |
| ⋮ |
| Row 1 |
| Front of class |

(a) In the space provided, sketch a map portraying the relative positions of the rows (clusters) in your classroom.

(b) Assign a label to each cluster (each row).
What is the chance that any student in the population will be selected? _____

(c) Select one cluster at random. Use your calculator with a seed value = 279 or Row 9, Column 21 of the random-number table.
Which cluster (row) did you select? _____
How many students were in the selected cluster (that is, what sample size did you get)?

Is the sample size fixed?   Yes   No
Explain.

THINK ABOUT IT

Is a cluster sample a simple random sample? Explain.

Is a cluster sample a stratified random sample? Explain.

When you form the clusters, how should the variability of the units within each cluster compare to the variability between the clusters?

Is this criterion the same as in stratified random sampling?

Real Data Used in Exercises and Examples

We attempt to raise student interest in the examples and exercises by using data sets on current, timely topics that will engage students. The exercises and examples are drawn from newspapers, magazines, and journals with which most students will be familiar, thus underscoring the practicality and real-world relevance of statistics.

REAL DATA

3.52 The November 26, 2000 edition of the *Orlando Sentinel* provided the following summary of a study under the section entitled "World News to Note":

Mobile phone dangers

LONDON -- Children who use mobile phones risk suffering memory loss, sleeping disorders and headaches, according to research published in the medical journal The Lancet. Physicist Gerard Hyland raised new fears over radiation caused by mobile phones and said that those younger than 18 were more vulnerable because their immune systems were less robust. "Radiation is known to affect the brain rhythms and children are particularly vulnerable, " Hyland said.

(a) Do you think this study was an experiment or an observational study? Explain.

(b) As best you can discern from the news report, give the explanatory variable and the response variable.

(c) A caveat is defined as a caution or warning. Give at least one caveat for this study.

Page 179

Early Coverage of Sampling and Experimental Design

The first three chapters of this text deal with important issues in sampling. We cover practical topics, such as different sampling techniques, biases in the data, and the use of random samples. We provide a thorough introduction to factors for planning statistically valid experiments, including randomization, blinding, control groups, and the placebo effect.

 3.5 Understanding Experiments

To establish a link between the explanatory variable and the response variable, we would like to hold everything constant except the explanatory variable. We would change the levels of the explanatory variable to see what happens to the response variable as a result. Although we can rarely achieve this ideal situation, we sometimes can come pretty close to it with an experiment rather than an observational study. In this section, we will study the basic terminology describing experiments and present some principles to strive for when planning an experiment. We shall see that, as in sampling, experiments also involve the planned use of chance.

3.5.1 Basic Terminology

Let us begin by recalling the vocabulary involved in studies in general, but presented in the context of experiments. The objects on which an experiment is performed and responses are measured are called the **experimental units**, or **subjects** if they are human beings. The intent of an experiment is to study the effect of changes in some variables on the outcome variables. The outcome variables are the **response variables**. The variables that the experimenter has control over are called **explanatory variables** or **factors**. The possible values of an explanatory variable are called the **levels** of that factor. A **treatment** is a particular combination of the levels of each of the explanatory variables. For most medical trials, the pa-

---

Integrated Use of Graphing Calculator to Reinforce Concepts

We have found the graphing calculator to be a valuable addition to our course. The TI graphing calculator lets students enter, manipulate, and plot data quickly and conveniently. While a graphing calculator is not required, we rely on it in certain examples and exercises in which a calculator minimizes hand calculations and eases data plotting. More in-depth, keystroke instruction is placed in the TI Quick Steps sections that appear at the ends of selected chapters. These sections show students how to use their calculators efficiently.

---

128 **CHAPTER 2** ▼ *Producing Data*

TI Quick Steps

Generating Random Integers

To be able to check your results in what follows you will need to specify a starting point by setting a seed value. The factory-set seed value for TI graphing calculators is 0. Suppose that you wish to generate a sequence of random integers between 1 and 50, using a seed value of 33. Using a TI graphing calculator to generate a list of random numbers between 1 and 50, starting with the seed value of 33, the keystroke steps are as follows:

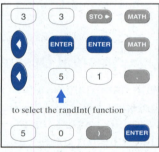

to select the randInt(function

Continue to push the ENTER button and your TI screen should look like this:

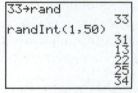

With the seed value of 33 and a population size of 50, the first value you get is 31, which corresponds to selecting the unit with the label of 31. By continuing to press the ENTER key four more times, the resulting simple random sample would consist of the units with labels 31, 13, 22, 25, and 34. You can request that more than one label be placed on a line. To have, say, two labels displayed on each line, you would use the command randInt(1, 50, 2). The third number in this command determines how many generated values to place on a line.

For further details of the TI number-generating functions, see your TI Graphing Calculator Guidebook under MATH operations.

Page 128

Available Supplements

Web Site: http://www.prenhall.com/aliaga

This companion Web site has multiple-choice quizzes, on-line graphing calculator help for the several versions of the TI graphing calculator, and data files for the textbook. For the instructors there will be lecture notes and a syllabus builder.

Student Solutions Manual *by Brenda Gunderson and Martha Aliaga* (0-13-065846-4)

Includes fully worked solutions to odd-numbered exercises.

Instructor's Resource Manual (0-13-065845-6)

To help prepare for teaching an interactive class with our text, we have developed an extensive set of materials to show the instructor how to get the most out of an interactive class. Detailed information is provided on how to set up effective student work groups, how to incorporate the graphing calculator into instruction, and how to prepare for the first day of class. Each chapter of the instructor's resource manual gives learning goals, ideas for teaching, solutions to the *Let's do it!* activities (including how long each activity takes, how to accomplish it, and its importance), solutions to the *Think about it* questions, and solutions to all of the exercises.

Printed Test Bank (0-13-065849-9)

Provides a full complement of additional problems that correlate to text exercises.

SPSS 11.0 Student Version (0-13-034846-5)

An affordable, easy-to-use student version of the professional statistical analysis tool. Available at a significant discount when packaged with the text.

Student Version of MINITAB® (0-13-026082-7)

A CD-ROM of MINITAB Release 12.0 Student Version may be packaged with the text at a significant discount.

Kelly TI Manual (0-13-020911-2)

This brief, spiral-bound manual provides simple keystroke instructions for using the TI-83 graphing calculator in statistics. Class tested for many years, this manual is the perfect answer for frustrated students and professors.

Dretzke *Statistics with Excel Manual* (0-13-022357-3)

A step-by-step introduction to using Excel to perform the statistical techniques commonly taught in introductory statistics courses. Appropriate for general, introductory statistics courses in disciplines such as psychology, sociology, mathematics, health care, and business.

PHStat2 *Microsoft® Excel Add-In* (0-13-035292-6)

This CD-ROM add-in is available as a stand-alone item, or can be packaged with the text at a discount. It features a custom menu of choices that lead to dialog boxes to help perform statistical analysis more quickly and easily than Excel alone permits.

Value Pack Options

Text and Student Solutions Manual (0-13-077988-1)

Text and TI Manual (0-13-045399-4)

Text and Excel Manual (0-13-045397-8)

Text and SPSS 11.0 (0-13-045396-X)

Text and Student Version of MINITAB® (0-13-045398-6)

Chapter 1

HOW TO MAKE A DECISION WITH STATISTICS

1.1 Introduction— Statistics and the Scientific Method

Statistics is the science of data. The word *science* comes from the Latin word "scientia," meaning knowledge. The scientific method is a procedure for systematically pursuing knowledge. Together with the scientific method, statistics provides us with a collection of principles and procedures for obtaining and summarizing information in order to make decisions. The scientific method is an iterative process for learning about the world around us.

We start with a *theory*—that is, a proposed, but unverified, statement. Suppose that we are making a product and recently

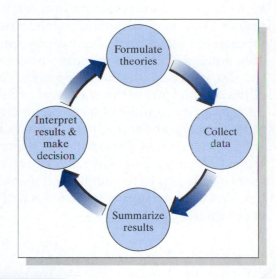

 The **scientific method** is composed of the following steps:
Step 1   Formulate a theory.
Step 2   Collect data to test the theory.
Step 3   Analyze the results.
Step 4   Interpret the results and make a decision.

some customers have returned the product, reporting that it did not work as they expected. We recognize this as an opportunity to improve. The feedback offered by the customers may give rise to a theory about

what is causing the product to malfunction. Since many customers reported a broken spring mechanism, it is hypothesized that the spring coil should be thicker. We wish to test this theory, so we begin to experiment. We *collect data* to help us check the theory. We might make a change in the production process of our product, and then measure the performance of the product

1

made under the change. These measurements are the *data*. We look at the data and *summarize* the results. We might summarize the percentage of product produced under this process change that still does not operate properly. We *interpret* the results and use the data to confirm or refute the theory. If the percentage of malfunctioning product has been sufficiently reduced, we might conclude that the theory has been supported. The process change is implemented and becomes the new standard for making the product. If the percentage of improperly operating product has not been sufficiently reduced, the theory may not be supported and a *new theory* should be developed and tested.

It would be nice if data could prove conclusively that a theory is either true or false, but in this uncertain world, that is generally not the case. Most theories are in a permanent state of uncertainty. There are always new observations about the world around us and new data coming in. Also, scientists are always thinking of new ways to test old theories or to interpret data, which may lead to exposing weaknesses of old theories, making it time to test them again.

If we cannot conclude whether or not a theory is absolutely true, at least it would be nice to be able to quantify how much "faith" we had in our decision—if we could say something like "We are 95% confident in our conclusion." This is where statistics and its collection of methods play a role. The ability to make such confidence statements stems from the use of statistics at every step of the scientific method. *A theory is rejected if it can be shown statistically that the data we observed would be very unlikely to occur if the theory were in fact true. A theory is accepted if it is not rejected by the data.*

The scientific method is an iterative process of learning. A conclusion may be that we need to update the theory and gather more data. The results do not necessarily give definite answers. Instead, the results may suggest new theories or decisions that are subject to further testing. The scientific method and the road map for studying statistics with this text are best represented with a circle. The various components in the circle are connected, and the circle does not end—just as learning is a never-ending process.

So, where do we start? We start here in Chapter 1 by providing you with an overview of the components of this statistical decision-making process. The goal of this chapter is to introduce you to some of its elements by involving you in the process. We will ask you to think critically about the information being presented. You will begin to learn how to examine assumptions, discern hidden values, evaluate evidence, and assess conclusions. The remaining chapters of this text reinforce and build on these various elements. Previous users of this text have commented that Chapter 1 presents some complex issues in statistics. However, mastering these ideas early on smoothes out the path through the remaining chapters.

 ## 1.2   Decisions, Decisions

Every day—in fact, almost constantly—we gather information to make decisions. Consider the following question: Can I cross the street at this intersection? What information, or **data**, would you need to answer this question? Without consciously thinking about it, your mind is processing the answers to a couple of questions: How many cars are approaching and at what speed are they traveling? Are there any obstacles or weather conditions that will impede my progress across the street? We wouldn't think of this simple, everyday decision as being a problem in statistics. However, what if you wanted to ask more complex questions, such as those in the following four examples:

- Suppose that you are a student at the University of Michigan and you are interested in gathering information about the student population registered full time during the winter term. Some of the questions you might be interested in are as follows: What percent of students are women? African-American? registered Democrats? vegetarian? married?

- Suppose that you are in the market for a new car. You have decided to purchase a General Motors car and will consider all such models produced in the current model year. The answers to the following questions will influence your purchase decision: What model is more efficient in average miles per gallon? What is the average price of all the models? How many inches of leg room are there for the driver?

- The makers of Advil® claim that "for pain, nothing is proven more effective or longer-lasting than Advil" (based on studies conducted by Whitehall Laboratories, the makers of Advil). As a consumer who does experience headaches, you might ask "What is meant by 'more effective'?" "Was Advil compared to all headache relievers?" "How were the comparisons carried out?" and "Does this mean Advil will work better for you?" "For every type of headache you have?"

- Many small-scale studies have found and reported that fish eaters live longer. Another study contradicts the popular belief that fish is good for the heart. What is a health-conscious person to do? How do you as a consumer weigh the information being cited daily? You learn to ask questions, such as "How was the study conducted?" and "What type of and how many subjects were used?" In the recent fish study, researchers followed the eating habits of 44,895 men and found that those who ate a lot of fish were just as likely to experience heart trouble as those who ate only small amounts. So, do these results extend to men of all ages? to women? What was the working definition of "experiencing heart trouble?"

In the first two examples, the group of individuals or objects under study is very large. Because it would be inefficient and expensive to question each student at the University of Michigan, or to obtain information on each car produced by General Motors in the current model year, we need to devise a reliable method for drawing conclusions based on a manageable amount of data. The last two examples illustrate how information can be a powerful and common tool of persuasion. We know how to discount some kinds of information. In the Advil example, are you surprised that the results of a study conducted by the makers of Advil are in support of Advil? If a company conducts a study comparing its product with a competitor's and the results show that the competitor's product is better, would these results be reported? If Advil is better, how can a competitor, such as Tylenol®, also claim to be the best and have studies supporting its claims? The example about eating fish illustrates that you should consider the weight of the evidence. You shouldn't base your decisions on every study that comes along. If there are, for example, three independent studies that have similar results, the weight of the evidence is stronger than one study that stands alone. Research is cumulative and no one study is definitive. *A study provides clues, not absolute answers.*

Over 100 years ago, H. G. Wells, author of such classics as *The Time Machine* and *The War of the Worlds*, stated a prediction based on the following theory: "Statistical thinking will one day be as necessary for efficient citizenship as the ability to read and write." Indeed, data collection and reporting of results are increasingly being used to confirm or refute theories. Individuals in nearly all job markets are finding the need to be able to evaluate data intelligently—that is, to use statistical reasoning to interpret the meaning of data and make decisions. We begin cycling through the scientific method by first establishing some standard terminology.

1.3 The Language of Statistical Decision Making

Learning statistics is like learning a new language whose new terminology involves special phrases, symbols, and definitions. The meaning of a word in everyday language can be different from how it is defined in statistics. In this section, we will discuss the statistical terminology involved in the decision-making process. There are many components in the decision-making process. The decision-making process involves theories, data, a measure of "likeliness," and the possibility of errors. Although we don't always think of these components from their statistical viewpoint, they are always present. In the following sections, we will discuss these components and introduce their corresponding more formal and accepted terminology.

1.3.1 Testing Theories

The group of objects or individuals under study is called the **population**. In the student example, the population consists of all University of Michigan students registered full time during the winter term. In the car example, the population consists of all car models produced by General Motors in the current model year. A part of the population is called a **sample**. We use the information from our sample to generalize and make decisions about the whole population—a process called **statistical inference**. The validity of the resulting generalizations will depend on the validity of the sample selected.

> *Definitions:* The **population** is the entire group of objects or individuals under study, about which information is desired.
>
> A **sample** is a part of the population that is actually used to get information.
>
> **Statistical inference** is the process of drawing conclusions about the population based on information from a sample of that population.

"Should I take Tylenol or Advil to relieve my headache fast?" One theory is that Tylenol is just as good as Advil, while another is that Advil is better than Tylenol. If we have in mind theories about a population that we want to test, those theories in statistics are called **hypotheses**. In statistics, we start with two hypotheses about the same population, each of which has a special name. We generally call the conventional belief—that is, the status quo, or prevailing viewpoint—the **null hypothesis** (denoted by H_0 and read "H naught"). The null hypothesis is generally a theory which states that there is nothing happening—no effect, no difference, or no change—in the population. The **alternative hypothesis** (denoted by H_1 and read "H one")* is an alternative to the null hypothesis. It is the theory that the researcher hopes to find evidence for—namely, the change in the population that the researcher is looking for. Let's look at some examples.

❖ EXAMPLE 1.1 Stating the Hypotheses _____

Aspirin Cuts Cancer Risk (SOURCE: The Associated Press, September 7, 1995)

Taking an aspirin every other day for 20 years can cut your risk of colon cancer nearly in half, a study suggests. However, the benefits may not kick in until at least a decade of use. According to the American Cancer Society, the lifetime risk of developing cancer is 1 in 16.

---

*Sometimes, the alternative hypothesis is denoted by H_a.

H_0: Taking an aspirin every other day will not change the risk of getting colon cancer, which is 1 in 16.

H_1: Taking an aspirin every other day will reduce the risk of getting colon cancer, so the risk will be less than 1 in 16.

Average Life Span

Suppose that you work for a company that produces cooking pots with an average life span of seven years. To gain a competitive advantage, you suggest using a new material that claims to extend the life span of the pots. You want to test the hypothesis that the average life span of the cooking pots made with this new material increases.

H_0: The average life span of the new cooking pots is seven years.

H_1: The average life span of the new cooking pots is greater than seven years.

Poll Results

Based on a previous poll, the percentage of people who said that they plan to vote for the Democratic candidate was 50%. The presidential candidates will have daily televised commercials and a final political debate during the week before the election. One hypothesis we might test is that the percentage of Democratic votes will change.

H_0: The percentage of the Democratic votes in the upcoming election will be 50%.

H_1: The percentage of the Democratic votes in the upcoming election will be different from 50%. ❖

Definitions: The **null hypothesis**, denoted by H_0, is the conventional belief—the status quo, or prevailing viewpoint, about a population.

The **alternative hypothesis**, denoted by H_1, is an alternative to the null hypothesis—the change in the population that the researcher hopes to find evidence for.

Tip: The null and alternative hypotheses should both be statements about the *same* population. In the previous average-life-span example, it would *not* be correct to have H_0 be a statement about pots made with the original material and H_1 be a statement about pots made with the new material. Both H_0 and H_1 are statements about the same population of pots made with the new material.

 **Let's do it! 1.1** Fair Die?

In a famous die experiment, out of 315,672 rolls, a total of 106,656 resulted in either a 5 or a 6. If the die is "fair"—that is, each of the six outcomes has the same chance of occurring—then the true proportion of 5's or 6's should be $\frac{1}{3}$. However, a close examination of a real die reveals that the "pips" are made by small indentations into the faces of the die. Sides 5 and 6 have more indentations than the other faces, and so these sides should be slightly lighter than the other faces. This suggests that the true proportion of 5's or 6's may be a bit higher than the "fair" value, $\frac{1}{3}$.

State the appropriate null and alternative hypotheses for assessing if the data provide compelling evidence for the competing theory.

H_0: The die is fair—that is, the indentations have no effect, and the proportion of 5's or 6's is _____.

H_1: The die is not fair—that is, the indentations do have an effect, and the proportion of 5's or 6's is _____.

 Let's **do it!** **1.2** **Stress Can Cause Sneezes**

REAL DATA

Excerpts from the article "Stress Can Cause Sneezes" (*New York Times*, January 21, 1997) are shown at the right. Studies suggest that stress doubles a person's risk of getting a cold. Acute stress, lasting maybe only a few minutes, can lead to colds. One mystery that is still prevalent in cold research is that while many individuals are infected with the cold virus, very few actually get the cold. On average, up to 90% of people exposed to a cold virus become infected, meaning that the virus multiplies in the body, but only 40% of those exposed actually become sick. One researcher thinks that the accumulation of stress predisposes an infected person to illness.

> ## Stress can cause sneezes
>
> FROM THE **NEW YORK TIMES**
>
> Winter can give you a cold because it forces you indoors with coughers, sneezers and wheezers. Toddlers can give you a cold because they are the original Germs "R" Us. But can going postal with the boss or fretting about marriage give a person a post-nasal drip?
>
> Yes, say a growing number of researchers. A psychology professor at Carnegie Mellon University in Pittsburgh, Dr. Sheldon Cohen, said his most recent studies suggest that stress doubles a person's risk of getting a cold.

The percentage of people exposed to a cold virus who actually get a cold is 40%. The researcher would like to assess if stress *increases* this percentage. So, the population of interest is people who are under (acute) stress. State the appropriate hypotheses for assessing the researcher's theory regarding this population.

H_0: _____

TIP: Having trouble determining the alternative hypothesis? Ask yourself "Why is the research being conducted?" The answer is generally the alternative hypothesis, which is why the alternative hypothesis is often referred to as the research hypothesis.

H_1: _____

 ### 1.3.2   How Do We Decide Which Theory to Support?

In statistics, the decision-making process is basically as follows: We have two competing theories or ideas about the same population of interest. To learn which of these two theories seems more reasonable, we gather some information, or data. We look at these data. Are the data more likely to be observed if the first theory is true or if the second theory is true? If the data are unlikely to occur if the first theory is true, then we reject that theory and support the second theory. When we support a theory, or hypothesis, it is customary to say that we *accept* or *favor* that theory. But do keep in mind that accepting a theory does not necessarily imply that the theory is true. Some texts prefer to avoid the term *accept* and state the decision as either rejecting the null hypothesis or *failing to reject* the null hypothesis.

The logic behind statistical decision making is based on the "rare event" concept. Since the null hypothesis is generally the status quo, **we start by assuming that the null hypothesis is true.** We then assess whether or not the observed result is extreme (that is, rare or very unlikely) under the null hypothesis. Example 1.2 provides an introduction to the idea of assessing how unusual the data are under the null hypothesis.

❖ EXAMPLE 1.2   Package of Balls

Suppose that you are shown a closed package containing five balls. The package is on sale because the label, which describes the colors of the balls in the package, is missing. The salesperson states that most of the packages of balls sold in this store contain one yellow ball and four blue balls, for a proportion of yellow balls of $\frac{1}{5}$.

You wish to test the following hypotheses about the contents of the package that is missing its label:

H_0:   The proportion of yellow balls in the package is $\frac{1}{5}$ (or 0.20).

H_1:   The proportion of yellow balls in the package is more than $\frac{1}{5}$.

You are allowed to collect some data. You are permitted to mix the balls in the package well, and then, without looking inside the package, reach in, select one ball, and record its color. After replacing the selected ball, you are allowed to repeat this procedure for a total of five observations. This is called selecting **with replacement**.

Scenario 1:   Suppose that the data are as follows: The first ball is yellow, the second ball is yellow, the third ball is yellow, the fourth ball is yellow, and the fifth ball is yellow.

Do you accept or reject the null hypothesis that the contents of the package are one yellow ball and four blue balls? Why?

The observed data are *possible* under the null hypothesis of one yellow ball and four blue balls. You could have picked the one yellow ball all five times! However, observing five yellow balls in a row is *very unlikely* to occur if the package indeed contains only one yellow ball and four blue balls. Thus, you are more inclined to reject this hypothesis based on the observed data.

Scenario 2:   Suppose that the data are as follows: The first ball is blue, the second ball is blue, the third ball is blue, the fourth ball is blue, and the fifth ball is blue.

Do you accept or reject the null hypothesis that the contents of the package are one yellow ball and four blue balls? Why?

The observed data are again *possible* under the null hypothesis of one yellow ball and four blue balls. Observing five blue balls in a row is *very likely* to occur if the package indeed contains only one yellow ball and four blue balls. Thus, you are more inclined to accept this hypothesis based on the observed data.

In each scenario, would the decision you made be a correct decision? You will know only if you purchase the package and look inside! You certainly could have made an error in either case.   ❖

In Example 1.2, the data presented in the two scenarios were somewhat extreme cases—observing five yellow balls in a row and observing five blue balls in a row. There are certainly other possible outcomes, each providing some evidence as to which hypothesis should be supported. Example 1.3 helps us to think about how much evidence is enough evidence to reject the null hypothesis.

❖ EXAMPLE 1.3   Is the New Drug Better?

We are faced with two competing theories, and we will gather some data that will be used to assess which theory should be supported. How will that decision be made? Suppose that you have developed a new and very expensive drug intended to cure some disease. You wish to assess how well your new drug performs compared to the standard drug by testing the following hypotheses:

H_0:   The new drug is as effective as the standard drug.

H_1:   The new drug is more effective than the standard drug.

A study is conducted in which the investigator administers the new drug to some number of patients suffering from the disease and the standard drug to another group of patients suffering from the disease. In Chapter 3, we will discuss why two groups are needed in studies like this and how to allocate the patients to the two treatment groups. The proportion of cures for both drugs is recorded. Based on this information, we have to decide which hypothesis to support.

Q1:   If the proportion of subjects cured with the new drug was exactly equal to the proportion of subjects cured with the standard drug, which hypothesis would you support?

Q2:   If 75% of the subjects were cured with the new drug while 55% of the subjects were cured with the standard drug, for a difference in cure rates of 20%, which hypothesis would you support? (Define the difference in cure rates as the percent of subjects cured with the new drug less the percent cured with the standard drug.)

Q3:   If the difference in cure rates was equal to 2%, which hypothesis would you support?

Q4:   How large of a difference in the cure rates is needed for you to feel confident in rejecting the null hypothesis?

The required difference will depend on a number of things: How much data were gathered—were 50 subjects assigned to each drug? 500? How were the subjects assigned to each group? How concerned are you about making a mistake?

Suppose that 75% of the subjects were cured with the new drug while 80% of the subjects were cured with the standard drug. Which hypothesis would you support? Certainly not the alternative hypothesis. If the cure rate for the new drug were the same or less than the cure rate for the standard drug, we would not have enough evidence to reject H_0. Although the null hypothesis states that the two drugs are equally effective, any evidence that the new drug is no better than the standard drug would not support the alternative hypothesis.

Suppose that 75% of the subjects who received the new drug were cured while 55% of the subjects who received the standard drug were cured. In this case, we might reject the null hypothesis and conclude that the new drug is more effective than the standard drug. However, a difference in cure rates of 20% is not proof that H_1 is true. It is possible that the observed difference in this study occurred just by chance even though the new drug is really not more effective.

Suppose that the difference in cure rates is 2%. In this case, we might not reject the null hypothesis and may conclude that the new drug is not more effective than the standard drug. However, the 2% difference in cure rates is not proof that H_1 is false. It is possible that the observed difference in this study occurred just by chance even though the new drug is really more effective. We must remember that a study provides clues, not absolute answers. ❖

Newspapers and articles often state such phrases as "but the results were not statistically significant" or "there was a statistically meaningful difference between the two groups." In general, the alternative hypothesis is the new theory, the researcher's claim. Thus, the researcher would like to have the data support his or her theory. The researcher would like to reject the null hypothesis. The data are said to be **statistically significant** if they are very unlikely to be observed under the null hypothesis. If the data are statistically significant, then our decision would be to reject H_0.

> *Definition:*   The data collected are said to be **statistically significant** if they are very unlikely to be observed under the assumption that H_0 is true. If the data are statistically significant, then our decision would be to reject H_0.

If the word *significant* is used in an article or a report, determine whether the word is being used in the statistical sense or is just being used in the usual sense to try to convince you that the result is important.

 **Let's do it! 1.3** **Complaints about Chips**

Last month, a large supermarket chain received many customer complaints about the quantity of chips in 16-ounce bags of a particular brand of potato chip. Wanting to assure its customers that they were getting their money's worth, the chain decided to test the following hypotheses concerning the true average weight (in ounces) of a bag of such potato chips in the next shipment received from the supplier:

H_0:   The average weight is at least 16 ounces.
H_1:   The average weight is less than 16 ounces.

If there is evidence in favor of the alternative hypothesis, the shipment would be refused and a complaint registered with the supplier. Some bags of chips were selected from the next shipment and the weight of each selected bag was measured. The researcher for the supermarket chain stated that the data were **statistically significant**.

What hypothesis was rejected?

Was a complaint registered with the supplier?

Could there have been a mistake? If so, describe it.

1.3.3 What Errors Could We Make?

One principle of the American justice system is that the defendant in a trial should be considered innocent until proven guilty. What would the null and alternative hypotheses be in the context of a criminal trial? The null hypothesis would be the status quo that the defendant is innocent. The alternative hypothesis would be the result that the prosecutor is trying to establish—namely, that the defendant is guilty.

H_0: The defendant is innocent.

H_1: The defendant is guilty.

The defendant and prosecutor present their cases. The jury must weigh the evidence presented, using it to assess whether or not there is enough doubt in the defendant's innocence to deliver a guilty verdict. The justice system is not perfect. If the jury delivers a guilty verdict, but the defendant is truly innocent, an error has occurred. If the jury determines that there is no reasonable doubt and delivers an innocent verdict, but the defendant is truly guilty, again there would be an error. In statistical terms, these errors have special names. If we reject the null hypothesis H_0 when in fact it is true, we have committed an error called a **Type I error**. If we accept the null hypothesis H_0 when in fact it is not true, we have also committed an error, called a **Type II error**.

> **Definition:** *Rejecting* the null hypothesis H_0 when in fact it *is true* is called a **Type I error**. *Accepting* the null hypothesis H_0 when in fact it *is not true* is called a **Type II error**.

A defendant is considered innocent until proven guilty. The burden of proof lies with the prosecutor. The jurors are instructed to find sufficient evidence, beyond a reasonable doubt, in order to return a guilty verdict. Convicting an innocent person is considered a serious consequence. In the trial setting, a Type I error would occur if an innocent person is falsely convicted and the guilty party remains free. A Type II error would occur if a guilty person is set free.

In hypothesis testing, we begin by assuming that the null hypothesis is true. The burden of the proof lies with the data. Generally, the null hypothesis is selected to be the hypothesis that may produce serious consequences to you if you erroneously reject it. Thus, rejecting the null hypothesis is a stronger statement than accepting it. *Accepting the null hypothesis means that we did not have sufficient evidence to reject that theory.*

We can summarize the two types of errors in hypothesis testing in the following table:

		The True Hypothesis	
		Null H_0 is true	Alternative H_1 is true
Your Decision	*Null H_0 is supported*	No error	Type II error
Based on the Data	*Alternative H_1 is supported*	Type I error	No error

The Type II error entry in the preceding table (upper right corner) is interpreted as follows: The true hypothesis (column heading) is H_1, but your decision (row heading) was to accept H_0—which is an error. Remember that a Type I error can *only* be made if the null hypothesis is true. A Type II error can only be made if the alternative hypothesis is true. If one of these errors occurs, it does not imply that we, in our decision-making process (i.e., the scientific method), did anything wrong. It only means that we were misled by the information collected.

❖ **EXAMPLE 1.4** **Rain, Rain, Go Away!** _____

You plan to walk to a party this evening. Are you going to carry an umbrella with you? You don't want to get wet if it should rain. So you wish to test the following hypotheses:

H_0: Tonight it is going to rain.
H_1: Tonight it is not going to rain.

The two types of errors that you could make when deciding between these two hypotheses are as follows:

Type I error: Decide that it is not going to rain, when in fact it is going to rain.
Type II error: Decide that it is going to rain, when in fact it is not going to rain.

   What are the consequences of making each type of error? The consequence of making a Type I error is that you are going to get wet, since you are not going to carry an umbrella. The consequence of making a Type II error is that you will carry around an umbrella you are not going to need. The conclusion that it is going to rain was selected to be the null hypothesis, since getting wet is considered a more serious consequence than carrying an umbrella that will not be used.
   You learn from the noon weather report that there is a 70% chance of rain tonight, which means that from records of days with similar atmospheric conditions, 70% of such days had rain and 30% did not. Therefore, it may rain tonight or it may not.   ❖

Let's do it! 1.4 **Which Error Is Worse?**

For each set of hypotheses that follows, decide whether a Type I error or a Type II error would be more serious. Recall that a Type I error occurs if you reject the null hypothesis H_0 when it is true. A Type II error occurs if you do not reject the null hypothesis H_0 when it is false. Answers may vary, but be prepared to discuss your answer.

(a) H_0: The water is contaminated.

   H_1: The water is not contaminated.

   A _____ error would be more serious because ...

(b) H_0: The parachute works.
   H_1: The parachute does not work.

   A _____ error would be more serious because ...

(c) H_0: The ship is unsinkable.
   H_1: The ship is sinkable.

   A _____ error would be more serious because ...

(d) H_0:   The shirt is not washable.

   H_1:   The shirt is washable.

   A _____ error would be more serious because ...

(e) H_0:   The value of my investment-stock portfolio is going to increase over the next six months.

   H_1:   The value of my investment-stock portfolio is not going to increase over the next six months.

   A _____ error would be more serious because ...

 **1.5 Testing a New Drug**

In the previous section, two drugs were being compared, a new drug and a standard drug, to treat a particular disease. The hypotheses were as follows:

   H_0:   The new drug is as effective as the standard drug.
   H_1:   The new drug is more effective than the standard drug.

A study is conducted in which the investigator administers the new drug to some number of patients suffering from the disease and the standard drug to another group of patients suffering from the disease. The proportion of cures for both drugs is recorded. What are the two types of errors that you could make when deciding between these two hypotheses?

A Type I error occurs if you reject the null hypothesis H_0 when it is true. A Type I error in this situation would be

_____.

A Type II error occurs if you do not reject the null hypothesis H_0 when it is false. A Type II error in this situation would be

_____.

What are the consequences of making a Type I error?

What are the consequences of making a Type II error?

Which error might be considered more severe from an ethical point of view?

To know the true proportion of patients suffering from the disease who would be cured using the new drug, we would need to administer the new drug to all such patients. However, this is not possible. Why not?

We must remember that the data are a sample, and depending on which sample we get, we could make a wrong decision. This is one reason why we continue to cycle through the scientific method as part of the process of learning. We usually want to protect the prevailing viewpoint by ensuring a small chance of committing a Type I error.

THINK ABOUT IT 🐢

If the consequences of making a Type I error are considered very serious, why not set the chance of making the Type I error to zero?

To achieve a value of zero for a Type I error, you would never reject the null hypothesis, and you would never support any new, alternative theory. We must be willing to accept some small chance of making an error.

In statistics, the chance of a Type I error occurring is called the **level of significance**, and it is denoted by the Greek letter alpha, α. The chance of a Type II error occurring is denoted by the Greek letter beta, β.

α = level of significance
  = the chance of a Type I error occurring
  = the chance of rejecting the null hypothesis H_0 when it is true.

β = the chance of a Type II error occurring
  = the chance of accepting the null hypothesis H_0 when the alternative hypothesis H_1 is true.

> *Definition:*   The **significance level** number α is the chance of committing a Type I error—that is, the chance of rejecting the null hypothesis when it is in fact true.

Since both α and β represent the chance of making an error, ideally, you want both to be small. The standard significance level used by most researchers is the value of $\alpha = 0.05$. However, there are situations for which a different value may be more appropriate. Sometimes this level is set at 0.01 and sometimes at 0.10. When do you use $\alpha = 0.01$ instead of $\alpha = 0.05$? The level of significance is related to the chance of making a mistake when deciding between two hypotheses. In particular, you select the level of significance to be extremely small to reduce the chance of rejecting the null hypothesis mistakenly.

We will see how α and β are related through examples shortly.* We will also learn how the significance level may be set in advance and used by statisticians to determine a rule for deciding whether to accept or reject H_0. The significance level can also be used as a statement regarding how much evidence against the null hypothesis is required in order to reject that hypothesis—that is, how unusual the data must be to reject the null hypothesis.

---

*The calculation of the chance of a Type II error, β, is a more advanced statistical concept in general and is not required for the statistical methods presented in the later chapters. In the examples and exercises of this chapter, the computation of β is quite easy and is included for completeness.

1.4 What's in the Bag?

In this section, we will discover many aspects of the decision-making process. We will have two competing hypotheses about the contents of a shown bag. We will be allowed to gather some data from the shown bag in order to make a decision about its contents. We will develop various rules for how that decision should be made and compare the rules in terms of the chances of the two types of errors occurring, α and β. We will see how α and β are related. Finally, we will focus on what the observed data tell us by assessing how extreme it would be to assume that the null hypothesis was true.

There are two bags—call them **Bag A** and **Bag B**. Each bag contains 20 vouchers of the same size and shape. The outsides of the two bags look alike. They differ in terms of the vouchers' face values and their frequencies as follows:

Bag A	
Face Value	Frequency
− $1000	1
$10	7
$20	6
$30	2
$40	2
$50	1
$60	1

Bag B	
Face Value	Frequency
$10	1
$20	1
$30	2
$40	2
$50	6
$60	7
$1000	1

It may be helpful to look at a graph of these data. This type of graph is called a **frequency plot**. These plots show the possible voucher values and the frequency of each voucher value.

Frequency Plot for Bag A

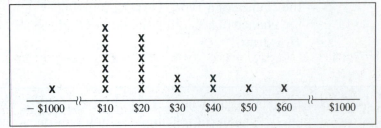

Bag A has a total value of −$560, while **Bag B** has a total value of $1890.

Frequency Plot for Bag B

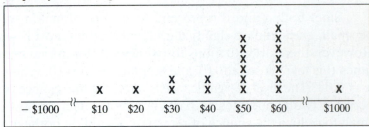

You will be **shown only one** of the bags. You will be allowed to gather some data on the basis of which you must decide whether to keep the shown bag or reject the shown bag and take the other bag. Once your decision is made and you have selected your bag, you are required to receive the **sum** of the face value of the vouchers in your bag (i.e., you win $1890 if the bag you pick is Bag B, or you pay $560 if the bag you pick is Bag A). Obviously, you do not want Bag A.

> *Definition:* A **frequency plot** displays a set of observations by representing each observation value with an X positioned along a horizontal scale. If there are two or more observations with the same value, the X's are stacked vertically.

The data will consist of selecting just one voucher from the shown bag (without looking in the bag) after mixing the contents of the bag well. In this case, we say the **sample size**, denoted by n, is equal to one, or $n = 1$.

> *Definition:*   The number n of observations in a sample is called the **sample size**.

Based on this one observation, you must decide between the following hypotheses about the shown bag:

H_0:   The shown bag is **Bag A**.
H_1:   The shown bag is **Bag B**.

Since **Bag A** is the undesirable bag, it is selected to be the null hypothesis. Note that both of these two hypotheses are statements about the same population—the shown bag. They are not statements about the bag that is not shown. What are the two types of errors that you could make when deciding between these two hypotheses?

Type I error = Reject H_0 when H_0 is true
= You decide the shown bag is Bag B when in fact it is Bag A
= You keep the shown bag, which is Bag A, and must pay $560.
Type II error = Accept H_0 when H_1 is true
= You decide the shown bag is Bag A when in fact it is Bag B
= You select the other bag, which is Bag A, and must pay $560.

How will you decide based on one observation from the shown bag whether to accept H_0 or to reject H_0?

Consider first the obvious choices: If the one voucher you select is $-\$1000$, then you will know that the shown bag is Bag A, and you will decide to accept H_0. Likewise, if the one voucher you select is $\$1000$, then you will know that the shown bag is Bag B and will decide to reject H_0 (that is, accept H_1). In each of these two cases, you will not have made an error.

THINK ABOUT IT 🐢

What if the voucher you select is $60? Would this observation lead you to think the shown bag is Bag A or Bag B?

Why?

How would you answer these questions if the voucher you select is $10?

In the preceding discussion, a **decision rule** is being formed. A decision rule is a formal rule that states, based on the data obtained, when you would reject the null hypothesis H_0 (and thus when you would accept H_1).

> *Definition:* A **decision rule** is a formal rule that states, based on the data obtained, when to reject the null hypothesis H_0. Generally, it specifies a set of values based on the data to be collected, which are contradictory to the null hypothesis H_0 and which favor the alternative hypothesis H_1.

 ### 1.4.1 Forming a Decision Rule

Let's develop a more formal and comprehensive decision rule based on the possible values that you could select from the shown bag. To do so, we need to examine the chances that a certain voucher value will be selected for each of the two possible bags. Since in Bag A there are 20 vouchers, exactly 7 of which have a value of $10, the chance that we select one of the $10 vouchers, *given* that the shown bag is Bag A, is $\frac{7}{20}$ (or 0.35). The other chances are computed similarly and are summarized as follows:

Face Value	Chance if Bag A	Chance if Bag B
−$1000	$\frac{1}{20}$	0
$10	$\frac{7}{20}$	$\frac{1}{20}$
$20	$\frac{6}{20}$	$\frac{1}{20}$
$30	$\frac{2}{20}$	$\frac{2}{20}$
$40	$\frac{2}{20}$	$\frac{2}{20}$
$50	$\frac{1}{20}$	$\frac{6}{20}$
$60	$\frac{1}{20}$	$\frac{7}{20}$
$1000	0	$\frac{1}{20}$

The voucher values of $30 and $40 have the same chance of occurring, whether from Bag A or Bag B. The voucher values of $10, $20, $50, and $60 give you some clue as to which bag you have been shown. If you select a $10 or a $20 voucher, you might conclude that the bag is Bag A, since such observations are more likely to occur from Bag A. If you select a $50 or $60 voucher, you might conclude that the bag is Bag B, since these observations are more likely to occur from Bag B. In other words, in this scenario, it is the larger voucher values that are extreme or unlikely under the null hypothesis H_0 and thus show the most support for the alternative hypothesis H_1. We thus define the **direction of extreme** for this scenario to be *to the right*, corresponding to values that are large.

> *Definition:* The **direction of extreme** corresponds to the position of the values that are more likely under the alternative hypothesis H_1 than under the null hypothesis H_0. If the larger values are more likely under H_1 than under H_0, then the direction of extreme is said to be **to the right**.

The direction of extreme may not always be to the right, which we shall see later in Section 1.4.2. We use the concept of determining values that are extreme in the development of a decision rule. We will first consider the **most extreme value**. (Note that in some cases it may not be possible to find the most extreme value, but determining the direction of extreme will be sufficient to establish a decision rule.)

> *Definition:* The value under the null hypothesis H_0 that is least likely, but at the same time is very likely under the alternative hypothesis H_1, is called the **most extreme value**.

Looking at the frequency plots for Bags A and B, of *all* the possible values specified under the null hypothesis (Bag A), the value that is both unlikely to come from Bag A and most likely to come from Bag B is the voucher value of $60. This most extreme value, indicated in the plots with an arrow, leads us to our first decision rule.

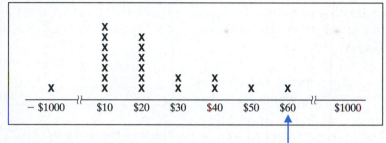

Frequency Plot for Bag A

Decision Rule 1:

Reject H_0 **if** you select a $60 or a $1000 voucher; otherwise, accept H_0; or **reject** H_0 **if** your selected voucher is $\geq$ $60.

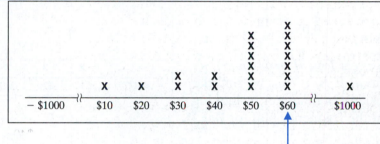

Frequency Plot for Bag B

Note that this decision rule is consistent with the direction of extreme being to the right—we reject the null hypothesis for values that are as large or larger than $60. In this case, the only voucher value larger than $60 is the $1000 voucher, which could only come from Bag B, so we certainly would reject H_0 if we selected it.

To any decision rule there corresponds a **rejection region**. A rejection region is the set of values for which you would reject H_0. For Decision Rule 1, the rejection region is $60 or larger. The extreme value of $60 in this decision rule is sometimes referred to as the **cutoff value**, or **critical value**.

You can think of the critical value as the trigger point that prompts you to change your decision from staying with H_0 to rejecting H_0.

> *Definitions:* A **rejection region** is the set of values for which you would reject the null hypothesis H_0. Such values are contradictory to the null hypothesis and favor the alternative hypothesis H_1.
>
> An **acceptance region** is the set of values for which you would accept the null hypothesis H_0.
>
> A **cutoff value**, or **critical value**, is a value that marks the starting point of a set of values that comprise the rejection region.

It is possible that we may make a mistake if we use Decision Rule 1. In particular, a voucher value of $60 is possible from both Bag A and Bag B. A Type I error is defined as rejecting H_0 when H_0 is true. For Decision Rule 1, a Type I error corresponds to selecting a $60 or larger voucher from Bag A. A Type II error is defined as accepting H_0 when H_1 is true. For Decision Rule 1, a Type II error corresponds to selecting a voucher less than $60— namely, a −$1000, $10, $20, $30, $40, or $50 voucher—from Bag B. Let's see what the chances of committing these errors, α and β, would be if this decision rule was used.

H_0:   The shown bag is **Bag A**.

H_1:   The shown bag is **Bag B**.

Since a *Type I error could only occur if H_0 is true,* we find the chance of this error by looking at the frequency plot for **Bag A**.

Frequency Plot for Bag A

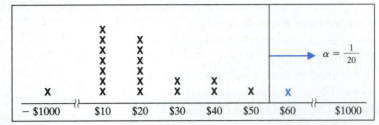

α = chance of rejecting H_0 when H_0 is true

= chance of selecting a "$60 or $1000 voucher from *Bag A*

$= \frac{1}{20}$

$= 0.05$.

Since a *Type II error could only occur if H_1 is true,* we find the chance of this error by looking at the frequency plot for **Bag B**.

Frequency Plot for Bag B

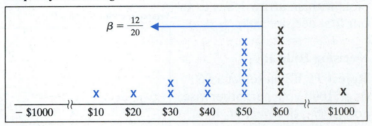

β = chance of accepting H_0 when H_1 is true

= chance of selecting a $-\$1000, \$10, \$20, \$30, \$40,$ or $\$50$ voucher from *Bag B*

$= \frac{12}{20}$

$= 0.60$.

The significance level α is 0.05. However, the value of β is 0.60. There would be a 60% chance of committing the error of accepting H_0 when H_1 is true. Are you satisfied with these levels for α and β? Would you prefer that either or both of them be lower? There are many possible decision rules, and the levels of α and β will depend on which decision rule is used. If the decision rule is changed, then the levels of α and β will generally change, and these levels are related to each other.

Decision Rule 1 resulted in a very large value of β because the rejection region was small, containing only the voucher values of $60 or $1000. So let's consider enlarging the rejection region. To enlarge the rejection region, we need to find the next most extreme value among the remaining possible values under the null hypothesis that shows the most support from the alternative hypothesis.

Among the remaining possible values under the null hypothesis, $-\$1000, \$10, \$20, \$30, \$40,$ and $\$50$, which value is least likely under the null hypothesis, but at the same time shows the most support for the alternative hypothesis? The answer is $50, with the largest chance from Bag B of $\frac{6}{20}$.

Face Value	Chance if Bag A	Chance if Bag B
$-\$1000$	$\frac{1}{20}$	0
$\$10$	$\frac{7}{20}$	$\frac{1}{20}$
$\$20$	$\frac{6}{20}$	$\frac{1}{20}$
$\$30$	$\frac{2}{20}$	$\frac{2}{20}$
$\$40$	$\frac{2}{20}$	$\frac{2}{20}$
$\$50$	$\frac{1}{20}$	$\frac{6}{20}$
$\$60$	$\frac{1}{20}$	$\frac{7}{20}$
$\$1000$	0	$\frac{1}{20}$

This next most extreme value gives us a new rejection region—namely, voucher values of $50 or more. The corresponding decision rule is as follows:

Decision Rule 2: Reject H_0 **if** your selected voucher is $\geq$ $50; otherwise, accept H_0.

Note that this decision rule is consistent with the direction of extreme being to the right—we reject the null hypothesis for values as large or larger than $50. The extreme value of $50 for this decision rule is the cutoff value, or critical value. Values larger than $50 are even more extreme than $50. The values for α and β corresponding to this decision rule are provided next.

Looking at the frequency plot for **Bag A**, we have

Frequency Plot for Bag A

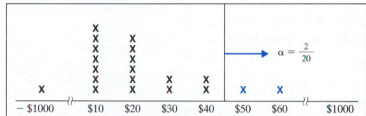

α = chance of rejecting H_0 when H_0 is true
= chance of selecting a $50, $60, or $1000 voucher from *Bag A*
= $\frac{2}{20}$
= 0.10.

Looking at the frequency plot for **Bag B**, we have

Frequency Plot for Bag B

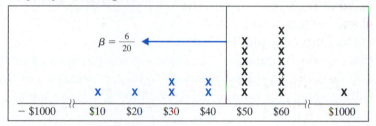

β = chance of accepting H_0 when H_1 is true
= chance of selecting a $-$1000, $10, $20, $30, or $40 voucher from *Bag B*
= $\frac{6}{20}$
= 0.30.

By enlarging the rejection region, the significance level α has increased to 0.10, but the value of β has decreased to 0.30. If we feel that the value of β is still too large, we must consider enlarging the rejection region again. What is the next most extreme value among the remaining possible values under the null hypothesis that shows the most support from the alternative hypothesis? Among the remaining possible values under the null hypothesis, $-$1000, $10, $20, $30, and $40, the values of $30 and $40 are both most likely, with a chance from Bag B of $\frac{2}{20}$. However, maintaining the direction of extreme for this scenario, we select the next largest value of $40.

Face Value	Chance if Bag A	Chance if Bag B
$-$1000	$\frac{1}{20}$	0
$10	$\frac{7}{20}$	$\frac{1}{20}$
$20	$\frac{6}{20}$	$\frac{1}{20}$
$30	$\frac{2}{20}$	$\frac{2}{20}$
$40	$\frac{2}{20}$	$\frac{2}{20}$
$50	$\frac{1}{20}$	$\frac{2}{20}$
$60	$\frac{1}{20}$	$\frac{7}{20}$
$1000	0	$\frac{1}{20}$

The corresponding decision rule is as follows:

Decision Rule 3: Reject H_0 **if** your selected voucher is $\geq$ $40; otherwise, accept H_0.

The values for α and β corresponding to this decision rule are provided next.

Looking at the frequency plot for **Bag A**, we have

α = chance of rejecting H_0 when H_0 is true

= chance of selecting a $40, $50, $60, or $1000 voucher from *Bag A*

$= \frac{4}{20}$

$= 0.20.$

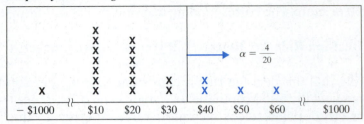

Frequency Plot for Bag A

Looking at the frequency plot for **Bag B**, we have

β = chance of accepting H_0 when H_1 is true

= chance of selecting a $-$1000, $10, $20, or $30 voucher from *Bag B*

$= \frac{4}{20}$

$= 0.20.$

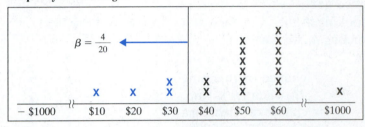

Frequency Plot for Bag B

Once again, by enlarging the rejection region, the significance level α has increased to 0.20, and the value of β has decreased to 0.20. A summary of the three decision rules and the resulting values for α and β is given in the following table:

Decision Rule	Rejection Region	We say . . .	α	β
1. Reject H_0 if the selected voucher is $60 or more.	$60 or more	. . . the critical cutoff value is $60 and larger values are even *more extreme.*	0.05	0.60
2. Reject H_0 if the selected voucher is $50 or more.	$50 or more	. . . the critical cutoff value is $50 and larger values are even *more extreme.*	0.10	0.30
3. Reject H_0 if the selected voucher is $40 or more.	$40 or more	. . . the critical cutoff value is $40 and larger values are even *more extreme.*	0.20	0.20

So far, we have started with various rejection regions and have compared the corresponding decision rules in terms of the chances of committing a Type I and Type II error, α and β, respectively. Given a particular decision rule, we can compute the levels of α and β exactly. The values for α and β were determined because we knew what vouchers were in Bag A and Bag B. The values for α and β depended on the contents of the bags and the applicable decision rule. We can also take the opposite approach. For a specified level of significance α, a corresponding decision rule can be determined. For example, for a 5% significance level we could determine the rule as Decision Rule 1. Therefore, if the selected voucher from the shown bag is a $60 voucher, we would reject H_0 and the result would be statistically significant. However, it may be that for some values of α, a decision rule with that exact significance level cannot be determined.

As we can see from the preceding summary table, there is no decision rule having a significance level of exactly $\alpha = 0.15$. In this case, finding a decision rule for a specified significance level α requires us to look for a rejection region that yields a significance level as close to α as possible without exceeding α. If the level of $\alpha = 0.15$ were specified, we would use Decision Rule 2.

The summary table also shows us that for a fixed sample size (here $n = 1$), there is a relationship between α and β. Enlarging the rejection region will increase the level of α and decrease the level of β. For a given level of α, is there any way we can reduce the level of β? You can see the answer to this question in Section 1.5, in which we are allowed to select two vouchers instead of a single voucher.

Summary of the Relationship between the Decision Rule and the Significance Level

If you start with a particular decision rule, with a given cutoff value, you can obtain the corresponding significance level α.

Decision rule $\rightarrow$ significance level α

For the Bag A–Bag B scenario with $n = 1$, if the decision rule is reject H_0 if the selected vouchers is \$60 or more, then the corresponding significance level is $\alpha = \frac{1}{20} = 0.05$.

You can also take the opposite approach. You can be given a specified significance level at which to perform the test, and for that level, determine the corresponding decision rule. That is, we can find the appropriate cutoff value that would yield a significance level equal to, or closest to without exceeding, α.

Significance level $\alpha \rightarrow$ decision rule

For the Bag A–Bag B scenario with $n = 1$, if the significance level is $\alpha = 0.10$, then the corresponding decision rule is as follows: Reject H_0 if the selected voucher is \$50 or more.

 ### 1.4.2 More on the Direction of Extreme

In the current Bag A–Bag B scenario, the direction of the extreme values was to the right. Since the rejection region contained values that were in just one direction, we say that the rejection region was **one-sided**. In the next example, we will also have a one-sided rejection region, but the direction of the extreme values will be to the left.

❖ EXAMPLE 1.5   One-Sided Rejection Region to the Left _____

We again have two bags, Bag C and Bag D, each containing 15 vouchers. The outsides of the two bags look alike. You will be shown **only one** of the bags. The null hypothesis will be that the shown bag is Bag C. The alternative hypothesis is that the shown bag is Bag D. Both of these hypotheses are statements about the same population—the one shown bag.

H_0:   The shown bag is **Bag C**.
H_1:   The shown bag is **Bag D**.

The contents of each bag, in terms of the face value, the frequency (or number) of vouchers, and the corresponding chance of selecting each voucher value, is described next. The corresponding graphical displays are also provided.

BAG C		
Face Value	**Frequency**	**Chance**
$1	1	$\frac{1}{15}$
$2	2	$\frac{2}{15}$
$3	3	$\frac{3}{15}$
$4	4	$\frac{4}{15}$
$5	5	$\frac{5}{15}$

Frequency Plot of Bag C

BAG D		
Face Value	**Frequency**	**Chance**
$1	5	$\frac{5}{15}$
$2	4	$\frac{4}{15}$
$3	3	$\frac{3}{15}$
$4	2	$\frac{2}{15}$
$5	1	$\frac{1}{15}$

Frequency Plot of Bag D

You are allowed to select just one voucher from the shown bag and must decide to accept H_0 or reject H_0. Looking at the frequency plots, what kind of values are contradictory to the null hypothesis (Bag C) and show the most support for the alternative hypothesis (Bag D)? It is the smaller voucher values. Thus, the direction of extreme for this scenario is *to the left*, corresponding to values that are small.

To develop a decision rule, we need to find the most extreme value. What is it? It is the value 1, because the voucher value of $1 is the least likely to come from Bag C $\left(\frac{1}{15}\right)$ and at the same time shows the most support from the alternative hypothesis of Bag D $\left(\frac{5}{15}\right)$. This most extreme value leads us to our first decision rule.

Decision Rule 1:   Reject H_0 **if** your selected voucher is ≤ $1; otherwise, accept H_0.

The rejection region is one-sided, containing only those values as small as or smaller than the cutoff value of $1. The values for α and β corresponding to this decision rule are as follows:

α = chance of rejecting H_0 when H_0 is true
  = chance of selecting a $1 voucher from *Bag C*
  = $\frac{1}{15}$
  = 0.067.

β = chance of accepting H_0 when H_1 is true

    = chance of selecting a \$2, \$3, \$4, or \$5 voucher from *Bag D*

    = $\frac{10}{15}$

    = 0.667.

With Decision Rule 1, the significance level is small (0.067), while the level of β is somewhat large (0.667). What should we do to reduce the level of β? We need to enlarge the rejection region, which you are asked to do in Let's do it! 1.6. ❖

Let's do it! 1.6 Enlarging the Rejection Region

Recall the **Bag C** and **Bag D** scenario of the previous example.

BAG C		
Face Value	Frequency	Chance
\$1	1	$\frac{1}{15}$
\$2	2	$\frac{2}{15}$
\$3	3	$\frac{3}{15}$
\$4	4	$\frac{4}{15}$
\$5	5	$\frac{5}{15}$

Frequency Plot of Bag C

```
                                    X
                          X         X
                X         X         X
      X         X         X         X
X     X         X         X         X
$1    $2        $3        $4        $5
```

BAG D		
Face Value	Frequency	Chance
\$1	5	$\frac{5}{15}$
\$2	4	$\frac{4}{15}$
\$3	3	$\frac{3}{15}$
\$4	2	$\frac{2}{15}$
\$5	1	$\frac{1}{15}$

Frequency Plot of Bag D

```
X
X
X         X
X         X         X
X         X         X         X
X         X         X         X         X
$1        $2        $3        $4        $5
```

H_0: The shown bag is **Bag C**.

H_1: The shown bag is **Bag D**.

Decision Rule 1 had a significance level α = 0.067 and a rather large level for β = 0.667.

Enlarge the rejection region and give a new decision rule that will have a smaller level for β. To do this, you need to find the next most extreme value among the values \$2, \$3, \$4, and \$5. What is it? _____

Decision Rule 2: Reject H_0 if your selected voucher is _____.

The rejection region for Decision Rule 2 is _____.

The values for α and β corresponding to Decision Rule 2 are as follows:

$\alpha =$

$\beta =$

How did these values compare to those for Decision Rule 1?

❖ EXAMPLE 1.6 Two-Sided Rejection Region to the Right and to the Left _____

We have two bags, Bag E and Bag F, each containing 30 vouchers. The outsides of the two bags look alike. You will be shown **only one** of the bags. The null hypothesis will be that the shown bag is Bag E. The alternative hypothesis is that the shown bag is Bag F. Both of these hypotheses are statements about the same population—the one shown bag.

H_0: The shown bag is ***Bag E***.
H_1: The shown bag is ***Bag F***.

The contents of each bag, in terms of the face value, the frequency (or number) of vouchers, and the corresponding chance of selecting each voucher value, is described next. The corresponding graphical displays are also provided.

BAG E			BAG F		
Face Value	Frequency	Chance	Face Value	Frequency	Chance
$1	1	$\frac{1}{30}$	$1	5	$\frac{5}{30}$
$2	2	$\frac{2}{30}$	$2	4	$\frac{4}{30}$
$3	3	$\frac{3}{30}$	$3	3	$\frac{3}{30}$
$4	4	$\frac{4}{30}$	$4	2	$\frac{2}{30}$
$5	5	$\frac{5}{30}$	$5	1	$\frac{1}{30}$
$6	5	$\frac{5}{30}$	$6	1	$\frac{1}{30}$
$7	4	$\frac{4}{30}$	$7	2	$\frac{2}{30}$
$8	3	$\frac{3}{30}$	$8	3	$\frac{3}{30}$
$9	2	$\frac{2}{30}$	$9	4	$\frac{4}{30}$
$10	1	$\frac{1}{30}$	$10	5	$\frac{5}{30}$

You are allowed to select just one voucher from the shown bag and must decide to accept H_0 or reject H_0. Looking at the frequency plots, what kind of values are contradictory to the null hypothesis (Bag E) and show the most support for the alternative hypothesis (Bag F)? Here, we have that both the smaller voucher values and the larger voucher values fit this description. The direction of extreme for this scenario is said to be to the left *and* to the right, corresponding to values that are very small and very large.

Frequency Plot of Bag E

				X	X				
			X	X	X	X			
		X	X	X	X	X	X		
	X	X	X	X	X	X	X	X	
X	X	X	X	X	X	X	X	X	X
$1	$2	$3	$4	$5	$6	$7	$8	$9	$10

Frequency Plot of Bag F

X									X
X	X							X	X
X	X	X					X	X	X
X	X	X	X			X	X	X	X
X	X	X	X	X	X	X	X	X	X
$1	$2	$3	$4	$5	$6	$7	$8	$9	$10

To develop decision rules, we need to find most extreme values. What voucher value is both the least likely to come from Bag E and at the same time most likely to come from Bag F? In this case, we have two values that are tied for the label of most extreme—namely, $1 and $10. We include them both in forming our first decision rule:

Decision Rule 1:   **Reject** H_0 **if** your selected voucher is either ≤ $1 or ≥ $10; otherwise, accept H_0.

The rejection region contains those values as small or smaller than $1 and as large or larger than $10. This type of rejection region, which includes values in both directions of extreme, is said to be **two-sided**. Here, we have two cutoff values, one for each of the two directions—namely, $1 and $10. Computing α and β and enlarging the rejection region are left as an exercise (Exercise 1.15) at the end of this chapter. Exercise 1.16 is a variation of this example in which there are two possible bags for the alternative hypothesis, and the computation of β is more complex.   ❖

> ***Definitions:***   A rejection region is called **one-sided** if its set of extreme values are all in one direction, either all to the right or all to the left.
>
> A rejection region is called **two-sided** if its set of extreme values are in two directions, both to the right and to the left.

1.4.3   How Unusual Are the Data? The *p*-value

In this section, we return to our original Bag A–Bag B scenario (page 14) and turn our attention away from the fixed decision rule, based on a fixed significance level. Instead we will focus on what the observed data tell us. Since H_0 is generally the prevailing viewpoint, we assume for the moment that H_0 is true—that is, that the bag you select your one voucher from is indeed Bag A. The idea is that we select our one voucher and assess whether the observed value would lead us to accept the null hypothesis or reject it.

Suppose that we select one voucher. We ask, "If H_0 is true (the bag is Bag A), how likely is it that we would get such a voucher value or a voucher that is even more extreme, which shows even more evidence against H_0?" Since in our example the larger voucher values show more evidence against H_0, we compute the chance of getting the observed voucher value or a larger voucher under the assumption that the bag is Bag A. That chance has a special name, called the *p-value*.

> *Definition:* The *p-value* is the chance, computed under the assumption that H_0 is true, of getting the observed value plus the chance of getting all of the *more extreme* values.

Since H_0 states that the bag is Bag A, and we are assuming it is true, we use only Bag A to determine the *p*-value. We find the *p*-value *after* we select and look at a voucher value from the shown bag. The *observed value* is the value of the voucher that we have selected. The *p*-value is a value between 0 and 1 that measures how likely the observed result is, or something even more extreme, if the null hypothesis H_0 is true. A small *p*-value indicates that the observed data, or data even more extreme, are unlikely if the null hypothesis is true. A *p*-value that is not small indicates that the observed data, or data more extreme, are somewhat likely if the null hypothesis H_0 is true. Therefore, a small *p*-value corresponds to the data showing stronger evidence against the null hypothesis or, equivalently, more support for the alternative hypothesis H_1.

> We should understand that the *smaller the p-value*, the stronger is the evidence provided by the data *against the null hypothesis* H_0.

Let's do it! 1.7 Chromium Supplements

According to a study in the *American College of Sports Medicine Journal* and *Science in Sports and Exercise*, men who trained while taking chromium supplements did not grow appreciably stronger than did men who trained without it. Sixteen men were monitored in the study. Half got the chromium supplements and the other half got a fake substitute. At the end of the training program, both groups had become stronger, but there was no statistically meaningful difference in strength gains, the study reported. Consider the following hypotheses:

H_0: No difference in strength gains between the two groups.

H_1: Chromium group had a higher strength gain.

Based on the stated results, which hypothesis was supported by the data?

Would the *p*-value for testing the preceding hypotheses have been somewhat small? Explain.

 **1.8** **Three Studies**

The following table summarizes the hypotheses and results for three different studies:

	Null Hypothesis	Alternative Hypothesis	*p*-value
Study A	The true average lifetime is ≥ 54 months.	The true average lifetime is < 54 months.	0.0251
Study B	The average time to relief for all Treatment I users is equal to the average time to relief for all Treatment II users.	The average time to relief for all Treatment I users is not equal to the average time to relief for all Treatment II users.	0.0018
Study C	The true proportion of adults who work two jobs is ≤ 0.33.	The true proportion of adults who work two jobs is > 0.33.	0.3590

(a) For which study do the results show the most support for the null hypothesis? Explain.

(b) Suppose that Study A concluded that the data supported the alternative hypothesis that the true average lifetime is less than 54 months, but in fact the true average lifetime is greater than or equal to 54 months. In our statistical language, would this be called a Type I error or a Type II error?

(c) If the results of Study C are "not statistically significant," which hypothesis would you accept?

(d) For each study, determine if the rejection region would have been one-sided to the right, one-sided to the left, or two-sided. (Circle your answer.)

Study A: one-sided to the right   one-sided to the left   two-sided

Study B: one-sided to the right   one-sided to the left   two-sided

Study C: one-sided to the right   one-sided to the left   two-sided

We may use a fixed level of significance α to make a decision. However, people may have different opinions as to which level is the appropriate one to use. The *p*-value is what provides a measure of the strength of the data against H_0. How small should the *p*-value be in order to reject the null hypothesis? The *p*-value is the observed significance level based on the data, so it is compared to the required significance level α to make a decision.

Let's return to the Bag A–Bag B scenario. Recall that you were asked to decide between the following two theories: H_0: The shown bag is Bag A, versus H_1: The shown bag is Bag B. You were allowed to select only one voucher. We will examine two approaches for making a decision: using a decision rule for a given significance level (called the *classical approach*) and determining the *p*-value to be compared to the significance level (called the *p-value approach*). As you will see, you will always make the same decision either way, so you only have to use one approach.

Suppose you select a voucher from the shown bag and it is a $30 voucher. If the significance level was set at 10%, what would you decide?

Classical Decision Rule Approach with $\alpha = 0.10$

For a 10% level of significance, we know that the corresponding decision rule would be "Reject H_0 if the selected voucher is $\geq$ $50." This was Decision Rule 2 given on page 19. Since the observed $30 voucher is smaller than $50, we accept H_0. Note that the observed $30 voucher is in the acceptance region.

p-value Approach with $\alpha = 0.10$

The *p*-value is the chance of getting the observed result or something even more extreme if the null hypothesis H_0 is true. In this case, the larger the value of the selected voucher, the more support for the alternative hypothesis—that is, that the bag is Bag B. So the *p*-value is the chance, assuming the shown bag is Bag A, of getting the observed $30 voucher or a voucher that is even more extreme—that is, a value larger than $30. Looking at the frequency plot for Bag A, we see that there are 6 vouchers out of the 20 that are valued at $30 or more. So the *p*-value is $\frac{6}{20}$ or 0.30, which is larger than the significance level of 0.10. So if the shown bag is Bag A, we would expect to select a $30 voucher, or something even larger, about 30% of the time. The *p*-value of 0.30 is larger than $\alpha = 0.10$ and our decision is to accept H_0.

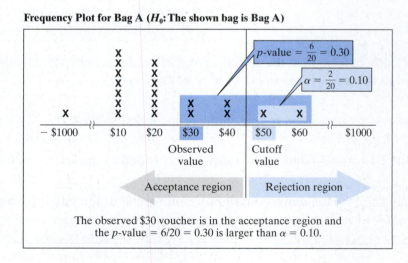

Frequency Plot for Bag A (H_0: The shown bag is Bag A)

$p\text{-value} = \frac{6}{20} = 0.30$

$\alpha = \frac{2}{20} = 0.10$

Observed value

Cutoff value

Acceptance region   Rejection region

The observed $30 voucher is in the acceptance region and the *p*-value = 6/20 = 0.30 is larger than $\alpha = 0.10$.

Suppose you select a voucher from the shown bag and it is a $60 voucher. If the significance level was set at 10%, what would you decide?

Classical Decision Rule Approach with $\alpha = 0.10$

For a 10% level of significance, we know that the corresponding decision rule would be "Reject H_0 if the selected voucher is $\geq \$50$." Since the observed $60 voucher is larger than $50, we reject H_0. Note that the observed $60 voucher is in the rejection region.

p-value Approach with $\alpha = 0.10$

The *p*-value is the chance, assuming the shown bag is Bag A, of getting the observed $60 voucher or a voucher that is even more extreme—that is, a value larger than $60. Looking at the frequency plot for Bag A, we see that there is exactly 1 such voucher out of the 20. So the *p*-value is $\frac{1}{20}$ or 0.05, which is smaller than the significance level of 0.10. So if the shown bag is Bag A, we would expect to select a $60 voucher, or something even larger, only 5% of the time. The *p*-value of 0.05 is smaller than $\alpha = 0.10$ and our decision is to reject H_0.

Frequency Plot for Bag A (H_0: The shown bag is Bag A)

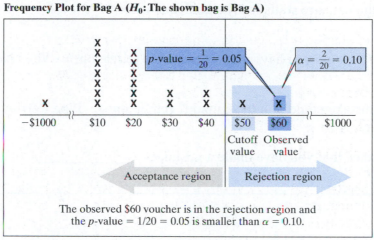

The observed $60 voucher is in the rejection region and the *p*-value = 1/20 = 0.05 is smaller than $\alpha = 0.10$.

The following table summarizes the decision that would be made for various observed voucher values:

Observed Voucher	Classical decision rule approach: $\alpha = 0.10$; Reject H_0 if observed voucher $\geq \$50$	*p*-value approach: Reject H_0 if the *p*-value $\leq \alpha = 0.10$
If observed voucher is $30	Since $30 < $50, we accept H_0.	Since the *p*-value = $\frac{6}{20}$ = 0.30 > 0.10, we accept H_0.
If observed voucher is $40	Since $40 < $50, we accept H_0.	Since the *p*-value = $\frac{4}{20}$ = 0.20 > 0.10, we accept H_0.
If observed voucher is $50	Since $50 $\geq$ $50, we reject H_0.	Since the *p*-value = $\frac{2}{20}$ = 0.10 $\leq$ 0.10, we reject H_0.
If observed voucher is $60	Since $60 $\geq$ $50, we reject H_0.	Since the *p*-value = $\frac{1}{20}$ = 0.05 $\leq$ 0.10, we reject H_0.

We will make the same decision whether we use the *classical decision rule approach* starting with a decision rule for the specified significance level and comparing the observed voucher value to the cutoff value in the decision rule or use the *p-value approach* and compare the *p*-value directly to the significance level. **In general, we will prefer to use the *p*-value approach.** Once the *p*-value is reported, a decision can quickly be made at any desired significance level.

The **relationship between the *p*-value and the significance level α** is as follows:

If the *p*-value is $\leq \alpha \rightarrow$ *the data are statistically significant at the given level α* and we reject H_0.

If the *p*-value is $> \alpha \rightarrow$ *the data are not statistically significant at the given level α* and we accept H_0.

THINK ABOUT IT

The significance level is $\alpha = 0.10$, the chance of committing a Type I error. The corresponding decision rule is: "Reject H_0 if the selected voucher is \$50 or more." A voucher is selected and it turns out to be \$60. Your decision is to reject the null hypothesis and conclude that the data are statistically significant at the 10% level.

You rejected H_0. Could you have made a mistake? (circle one)   Yes   No

What type of mistake could you have made? (circle one)   Type I error   Type II error

What is the chance that you have made a mistake?   _____

It is important to distinguish between "*before* we observe the data and state a decision rule" and "*after* we have observed the data and made a decision." Before we observe the data, we can state a decision rule and compute the corresponding chance of committing a Type I error—namely, α. For Decision Rule 2, we had $\alpha = 0.10$, which is the chance that you *will make a* Type I error. After we have looked at the data and have made our decision, our decision is either right or wrong. The shown bag is either Bag A or Bag B. If the observed voucher from the shown bag is \$60, the decision is to reject H_0. If the shown bag is actually Bag A, then we have made a mistake, a Type I error. If the shown bag is actually Bag B, then we have not made a mistake. If we have observed the data and decided to reject H_0, the chance that we have made a Type I error is either 1 (because the shown bag was Bag A) or 0 (because the shown bag was Bag B). The chance that we *have made* a Type I error is not equal to the significance level.

Key Idea   There is a difference between the chance that you *will make* a Type I error and the chance that you *have made* a Type I error. Once you have made a decision, the decision is either right or wrong.

❖ EXAMPLE 1.7 *p*-value for a One-Sided Rejection Region to the Left

Recall the two bags, Bag C and Bag D, from Example 1.5 on page 21. Each bag contains 15 vouchers and you will be shown only one of the bags. The null hypothesis will be that the shown bag is Bag C. The alternative hypothesis is that the shown bag is Bag D. The frequency plots that show the contents of each bag, in terms of the face value, are provided.

Frequency Plot of Bag C

				X
			X	X
		X	X	X
	X	X	X	X
X	X	X	X	X
\$1	\$2	\$3	\$4	\$5

You are allowed to select just one voucher from the shown bag and must decide to accept H_0 or reject H_0.

Frequency Plot of Bag D

X				
X				
X	X			
X	X	X		
X	X	X	X	
X	X	X	X	X
\$1	\$2	\$3	\$4	\$5

H_0: The shown bag is **Bag C**.

H_1: The shown bag is **Bag D**.

Recall that the smaller voucher values are less likely under the null hypothesis (Bag C) and show the most support for the alternative hypothesis (Bag D). Thus the direction of extreme for this scenario is to the left, corresponding to values that are small.

Suppose the observed voucher value is \$3. We wish to compute the corresponding *p*-value. First, we need to remember that the *p*-value is computed based on the assumption that H_0 is true. So, we need to find the chance of getting the observed voucher value of \$3 or something smaller under the H_0 frequency plot for Bag C. There are 6 vouchers (shown in blue) out of 15 that are \$3 or less, for a *p*-value of $\frac{6}{15} = 0.40$.

$$
\begin{aligned}
\textbf{\textit{p}-value} &= \text{the chance, computed under the assumption that } H_0 \text{ is true, of getting the} \\
&\quad\; \text{observed value plus the chance of getting all of the } \textit{more extreme} \text{ values} \\
&= \text{the chance of getting a voucher value of \$3 or less from Bag C} \\
&= \tfrac{6}{15} \\
&= 0.40.
\end{aligned}
$$

At the significance level of $\alpha = 0.10$, these data are not statistically significant since the *p*-value is larger than 0.10. These data are not statistically significant at $\alpha = 0.05$ or at $\alpha = 0.01$. The large *p*-value implies that we do not have enough evidence to reject the null hypothesis. We would support the conclusion that the shown bag is Bag C.   ❖

p-value for a One-Sided Rejection Region to the Left

Let's do it! 1.9

Consider again the two bags, Bag C and Bag D, from Example 1.5. You are allowed to select just one voucher from the shown bag and must decide to accept H_0 or reject H_0.

H_0:   The shown bag is **Bag C**. H_1:   The shown bag is **Bag D**.

Frequency Plot of Bag C

				X
		X	X	X
		X	X	X
	X	X	X	X
X	X	X	X	X
$1	$2	$3	$4	$5

Frequency Plot of Bag D

X				
X	X			
X	X	X		
X	X	X	X	
X	X	X	X	X
$1	$2	$3	$4	$5

(a) Suppose that the observed voucher value is $2. Find the corresponding *p*-value.

For the following significance levels, are the data statistically significant?

Significance level α	Circle one
0.10	Yes No
0.05	Yes No
0.01	Yes No

(b) Suppose that the observed voucher value is $1. Find the corresponding *p*-value.

For the following significance levels, are the data statistically significant?

Significance level α	Circle one
0.10	Yes No
0.05	Yes No
0.01	Yes No

❖ EXAMPLE 1.8   *p*-value for a Two-Sided Rejection Region

Recall the two bags, Bag E and Bag F, from Example 1.6 on page 24. Each bag contains 30 vouchers, and you will be shown only one of the bags. The null hypothesis will be that the shown bag is Bag E. The alternative hypothesis is that the shown bag is Bag F. The frequency plots that show the contents of each bag, in terms of voucher face values, are provided. You are allowed to select just one voucher from the shown bag and must decide to accept H_0 or reject H_0.

H_0:   The shown bag is **Bag E**.

H_1:   The shown bag is **Bag F**.

Recall that both the smaller voucher values and the larger voucher values are less likely under the null hypothesis (Bag E) and show the most support for the alternative hypothesis (Bag F). Thus, the direction of extreme for this scenario is to the left *and* to the right, corresponding to the values that are very small or very large. In Example 1.6, the values of $1 and $10 were tied for the label of most extreme. The rejection region consisted of two equal parts, one at each end of the possible voucher values, based on these two cutoff values.

Decision Rule 1:   Reject H_0 if your selected voucher is either ≤ $1 or ≥ $10.

Just as our rejection region included values in both directions, so will the *p*-value be computed using values in both directions. Suppose that the observed voucher value is $3. We wish to compute the corresponding *p*-value. A voucher value of $8 is just as extreme as a voucher value of $3. So, we need to find the chance of getting the observed voucher value of $3 or less or getting an observed voucher value of $8 or more under the H_0 frequency plot for Bag E. There are 12 vouchers (shown in blue) out of 30 that are $3 or less or $8 or more, for a *p*-value of 12/30 = 0.40.

Frequency Plot of Bag E

				X	X				
			X	X	X	X			
		X	X	X	X	X	X		
	X	X	X	X	X	X	X	X	
X	X	X	X	X	X	X	X	X	X
$1	$2	$3	$4	$5	$6	$7	$8	$9	$10

Frequency Plot of Bag F

X									X
X	X							X	X
X	X	X					X	X	X
X	X	X	X			X	X	X	X
X	X	X	X	X	X	X	X	X	X
$1	$2	$3	$4	$5	$6	$7	$8	$9	$10

p-value = the chance, computed under the assumption that H_0 is true, of getting the observed value plus the chance of getting all of the *more extreme* values;

= the chance of getting ≤ $3 or ≥ $8 from Bag E

= $\frac{12}{30}$

= 0.40.

At the significance level of $\alpha = 0.10$, these data are not statistically significant since the *p*-value is larger than 0.10. These data are not statistically significant at $\alpha = 0.05$ or at $\alpha = 0.01$. The large *p*-value implies that we do not have enough evidence to reject the null hypothesis. We would support the conclusion that the shown bag is Bag E.   ❖

Let's do it! 1.10   *p*-value for a Two-Sided Rejection Region

Consider again the two bags, Bag E and Bag F, from Example 1.6. You are allowed to select just one voucher from the shown bag and must decide to accept H_0 or reject H_0.

H_0:   The shown bag is **Bag E**.
H_1:   The shown bag is **Bag F**.

Frequency Plot of Bag E

				X	X				
			X	X	X	X			
		X	X	X	X	X	X		
	X	X	X	X	X	X	X	X	
X	X	X	X	X	X	X	X	X	X
\$1	\$2	\$3	\$4	\$5	\$6	\$7	\$8	\$9	\$10

Frequency Plot of Bag F

X									X
X	X							X	X
X	X	X					X	X	X
X	X	X	X			X	X	X	X
X	X	X	X	X	X	X	X	X	X
\$1	\$2	\$3	\$4	\$5	\$6	\$7	\$8	\$9	\$10

(a) Suppose that the observed voucher value is \$2. Find the corresponding *p*-value.

For the following significance levels, are the data statistically significant?

Significance level α	Circle one	
0.10	Yes	No
0.05	Yes	No
0.01	Yes	No

(b) Suppose that the observed voucher value is \$10. Find the corresponding *p*-value.

For the following significance levels, are the data statistically significant?

Significance level α	Circle one	
0.10	Yes	No
0.05	Yes	No
0.01	Yes	No

Let's recap the various directions of extreme. In our original Bag A–Bag B scenario (pages 14–21 and 25–30), it was the larger values that were unlikely under the null hypothesis H_0 and showed the most support for the alternative hypothesis H_1. So the direction of extreme was to the right, corresponding to values that are large. A p-value would be computed using this direction of extreme indicated by H_1, calculating the chance of getting the values that are equal to or larger than the observed value under the H_0 picture.

In our Bag C–Bag D scenario (Examples 1.5 and 1.7, pages 21 and 31), it was the smaller values that were unlikely under the null hypothesis H_0 and showed the most support for the alternative hypothesis H_1. So, the direction of extreme was to the left, corresponding to values that are small. A p-value would be computed using this direction of extreme indicated by H_1, calculating the chance of getting the values that are equal to or smaller than the observed value under the H_0 picture.

In our Bag E–Bag F scenario (Examples 1.6 and 1.8, pages 24 and 33), it was both the smaller values and the larger values that were unlikely under the null hypothesis H_0 and showed the most support for the alternative hypothesis H_1. So the direction of extreme was to the left *and* to the right, corresponding to values that are very small and very large. A p-value would be computed using this direction of extreme indicated by H_1, calculating the chance of getting the values from both ends of the H_0 picture.

In the next examples, we will have more practice at picturing the p-value. We will need to have the H_0 picture, know the value that was observed, and determine the direction of extreme by comparing the H_1 picture to the H_0 picture. Rather than working with more bags of vouchers and their corresponding frequency plots, we will simplify the H_0 and H_1 pictures by working with their smoothed versions. We will see more smoothed versions of frequency plots in Chapter 6. The frequency plot of the contents of Bag E and its smoothed version are shown here. In Example 1.8, we found the p-value when the observed voucher value was $3. The voucher values (X's) that are as extreme or more extreme under the H_0 picture are highlighted in both the frequency plot and the smoothed version.

Frequency Plot of Bag E

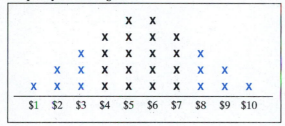

Smoothed Version of Frequency Plot of Bag E

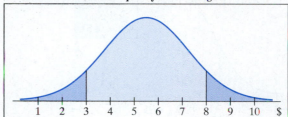

In Examples 1.9 through 1.11, we will depict the p-value by shading in the appropriate region under the smoothed H_0 picture. The following tips are helpful to think about as we go through the remaining p-value examples.

p-value Tips

- The p-value is always computed using the H_0 picture.
- The H_1 picture is used to determine the direction of extreme by comparing it with the H_0 picture.

❖ **EXAMPLE 1.9 Can You Picture the *p*-value?** _____

Consider two bags, each containing many vouchers. The outsides of the two bags look alike. You will be shown only one of the bags. The smoothed versions of the frequency plots depicting the contents of each bag, in terms of the face value, are provided. You are allowed to select just one voucher from the shown bag and must decide to accept or reject H_0.

Smoothed Version of Frequency Plot under H_0

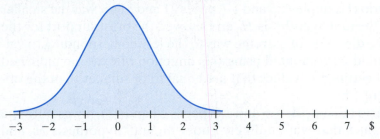

Smoothed Version of Frequency Plot under H_1

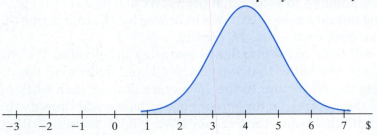

Since the picture for the alternative hypothesis is shifted to the right, to the larger values, the direction of extreme is to the right.

(a) If the selected voucher value is $2, then the picture of the *p*-value is given by the darker blue-shaded area.

Smoothed Version of Frequency Plot under H_0

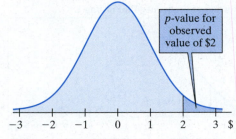

p-value for observed value of $2

(b) If the selected voucher value is −$1, then the picture of the *p*-value is given by the darker blue-shaded area.

Smoothed Version of Frequency Plot under H_0

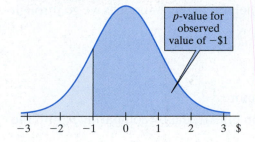

p-value for observed value of −$1

❖

❖ **EXAMPLE 1.10**   **Can You Picture the *p*-value?**

Consider two bags, each containing many vouchers. The outsides of the two bags look alike. You will be shown only one of the bags. The smoothed versions of the frequency plots depicting the contents of each bag, in terms of the face value, are provided. You are allowed to select just one voucher from the shown bag and must decide to accept or reject H_0.

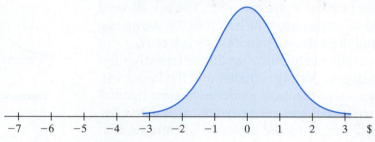

Smoothed Version of Frequency Plot under H_0

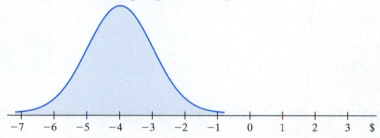

Smoothed Version of Frequency Plot under H_1

Since the picture for the alternative hypothesis is shifted to the left, to the smaller values, the direction of extreme is to the left.

(a)   If the selected voucher value is $2, then the picture of the *p*-value is given by the darker blue-shaded area.

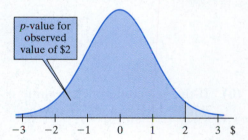

Smoothed Version of Frequency Plot under H_0

p-value for observed value of $2

(b)   If the selected voucher value is −$1, then the picture of the *p*-value is given by the darker blue-shaded area.

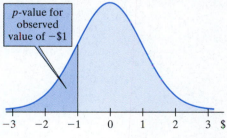

Smoothed Version of Frequency Plot under H_0

p-value for observed value of −$1

 ❖

❖ **EXAMPLE 1.11   Can You Picture the *p*-value?**

Consider two bags, each containing many vouchers. The outsides of the two bags look alike. You will be shown only one of the bags. The smoothed versions of the frequency plots depicting the contents of each bag, in terms of the face value, are provided. You are allowed to select just one voucher from the shown bag and must decide to accept or reject H_0.

In the picture for the alternative hypothesis, both the smaller and the larger values show the most evidence against the null hypothesis and in favor of the alternative hypothesis. Thus, the direction of extreme is to the left *and* to the right.

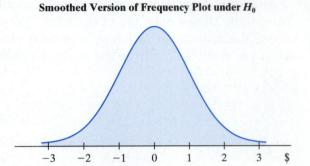

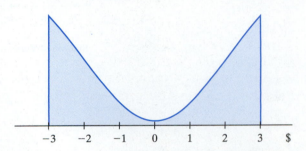

(a) If the selected voucher value is $2, then the picture of the *p*-value is given by the darker blue-shaded area.

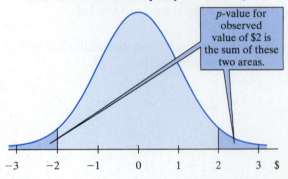

(b) If the selected voucher value is −$1, then the picture of the *p*-value is given by the darker blue-shaded area.

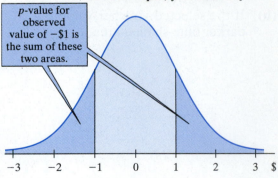

Machine A or Machine B?

Machine A makes parts whose lengths average around 4.6 mm. Machine B makes parts whose lengths vary similarly, but around a higher average of 4.9 mm. Suppose you have a box of parts that you believe are from Machine A, but you're not sure. You will test the following hypotheses by randomly selecting one part from the box and measuring it.

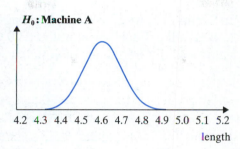

H_0: The parts are from Machine A.
H_1: The parts are from Machine B.

The hypotheses for the distribution of the length of the part (in mm) are presented at the right as the smoothed versions of the frequency plots.

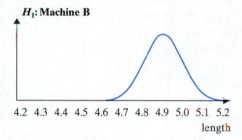

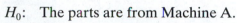

(a) The direction of extreme is (circle your answer)

one-sided to the right.
one-sided to the left.
two-sided.

(b) Consider the following decision rule: Reject H_0 if the selected part length is 4.8 mm or more extreme.
  (i) In the picture, shade in the region that corresponds to α, the significance level, and clearly label the region with α.
  (ii) In the picture, shade in the region that corresponds to β, the chance of a Type II error, and clearly label the region with β.

(c) The selected part length is 4.7 mm. In the picture, shade in the region that corresponds to the *p*-value and clearly label the region with the *p*-value.

(d) Is the result in (c) statistically significant? (circle your answer)   Yes   No
Explain your answer.

1.5   Selecting Two Vouchers*

In Section 1.4, we considered two bags—called **Bag A** and **Bag B**. The outsides of the two bags looked alike. Each bag contained 20 vouchers. The contents of each bag, in terms of the face value and the frequency of voucher values, is as follows:

Bag A				Bag B	
Face Value	Frequency			Face Value	Frequency
− $1000	1			$10	1
$10	7			$20	1
$20	6			$30	2
$30	2			$40	2
$40	2			$50	6
$50	1			$60	7
$60	1			$1000	1

We were to be shown only one of the bags. We were allowed to gather some data and, based on that data, had to decide whether to take the other bag, because we thought the shown bag was Bag A, or to keep the shown bag, because we thought it was Bag B. We had to decide between the following hypotheses about the one shown bag:

H_0:   The shown bag is ***Bag A***.

H_1:   The shown bag is ***Bag B***.

The data consisted of selecting just one voucher from the shown bag—that is, the sample size was $n = 1$. For this fixed sample size, there was a relationship between α and β: Enlarging the rejection region would increase the level of α and decrease the level of β. We then asked the question, "For a given level of α, is there any way we can reduce the level of β?" *The answer is to increase the sample size n.*

Let's see what happens when we increase the sample size—that is, instead of observing just one voucher, we are going to observe two vouchers, without replacement. **Without replacement** means that when you remove the first voucher from the bag, you keep it in your hand and then select another voucher from the bag. Alternatively, you might select two vouchers from the bag at the same time. Our decision will be based on the average of the two selected vouchers.

What are the possible averages that we could observe? It depends on whether we are selecting from Bag A or Bag B. As we did when we selected just one voucher, we will list all the possible combinations of two vouchers that can be selected, and then we will calculate the frequency that each combination can occur. We will prepare a table of the possible averages and their frequencies. With this information, we can develop decision rules and compute the corresponding levels, α and β.

Consider selecting the two vouchers without replacement from the shown bag. What pairs of values could we select?

---

*This section is optional.

THINK ABOUT IT

To calculate the number of different ways that a certain pair can occur—for example, −$1000 and $10—think of the seven $10's in Bag A as if they were of different colors, so you can distinguish them. The pair of −$1000 and $10 from Bag A can occur in seven ways, but cannot occur at all if the bag is Bag B.

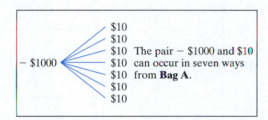

The pair of $10 and $20 can occur in 42 ways from Bag A, but in only one way from Bag B.

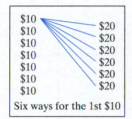 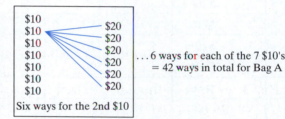

The pair of $20 and $20 can occur in 15 ways from Bag A, but cannot occur from Bag B (which has just one $20).

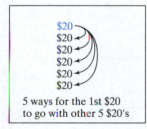 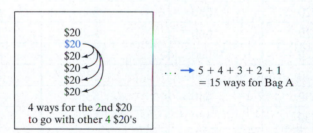

The pair of $30 and $50 can occur in two ways from Bag A and in 12 ways from Bag B. Why are these the number of occurrences?

Table 1.1 provides a listing of the possible pairs of values that could be selected (Columns 1 and 2), the corresponding average of the pair of values (Column 3), and the number of ways (frequency) of selecting that pair from Bag A (Column 4) and from Bag B (Column 5).

Notice that in Table 1.1, there are two entries in Column 3 corresponding to an average of $20 (entries shown in bold)—namely, when you select the vouchers ($10, $30) or ($20, $20). For Bag A, an average of $20 can occur a total of 14 + 15 = 29 ways, and for Bag B, an average of $20 can occur a total of 2 + 0 = 2 ways.

Table 1.1

Possible Two Values Selected		Average of the two selected values	BAG A Number of ways of selecting the two values	BAG B Number of ways of selecting the two values
−$1000	−$1000	−$1000	0	0
−$1000	$10	−$495	7	0
−$1000	$20	−$490	6	0
− $1000	$30	−$485	2	0
−$1000	$40	−$480	2	0
−$1000	$50	−$475	1	0
−$1000	$60	−$470	1	0
−$1000	$1000	0	0	0
$10	$10	$10	21	0
$10	$20	$15	42	1
$10	**$30**	**$20**	**14**	**2**
$10	$40	$25	14	2
$10	$50	$30	7	6
$10	$60	$35	7	7
$10	$1000	$505	0	1
$20	**$20**	**$20**	**15**	**0**
$20	$30	$25	12	2
$20	$40	$30	12	2
$20	$50	$35	6	6
$20	$60	$40	6	7
$20	$1000	$510	0	1
$30	$30	$30	1	1
$30	$40	$35	4	4
$30	$50	$40	2	12
$30	$60	$45	2	14
$30	$1000	$515	0	2
$40	$40	$40	1	1
$40	$50	$45	2	12
$40	$60	$50	2	14
$40	$1000	$520	0	2
$50	$50	$50	0	15
$50	$60	$55	1	42
$50	$1000	$525	0	6
$60	$60	$60	0	21
$60	$1000	$530	0	7
$1000	$1000	$1000	0	0

Table 1.2 presents a condensed version of Table 1.1, combining the entries corresponding to the same average. From Table 1.2, we are able to produce a frequency plot of the resulting possible averages for Bag A and for Bag B. The frequency plots for Bag A and Bag B show the possible averages and the frequency of each average when the sample size $n = 2$ for both bags. Examine the frequency plots that follow, and be sure you understand what they represent. Note that the number of ways to select two vouchers from a bag with 20 vouchers is 190. (See Exercise 1.57.)

Table 1.2

Average of the two selected values	BAG A Number of ways of selecting the two values	BAG B Number of ways of selecting the two values
−$1000	0	0
−$495	7	0
−$490	6	0
−$485	2	0
−$480	2	0
−$475	1	0
−$470	1	0
0	0	0
$10	21	0
$15	42	1
$20	29	2
$25	26	4
$30	20	9
$35	17	17
$40	9	20
$45	4	26
$50	2	29
$55	1	42
$60	0	21
$505	0	1
$510	0	1
$515	0	2
$520	0	2
$525	0	6
$530	0	7
$1000	0	0
Total	**190**	**190**

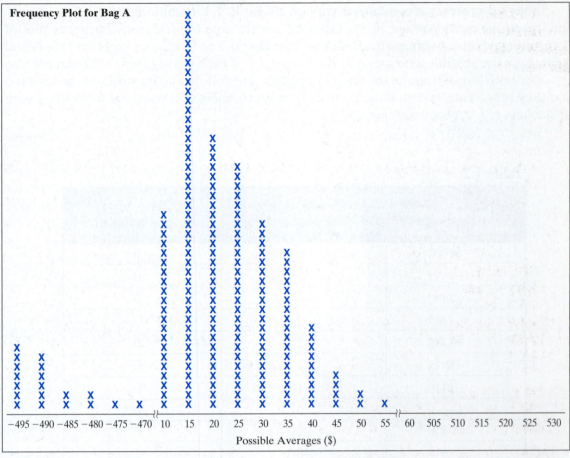

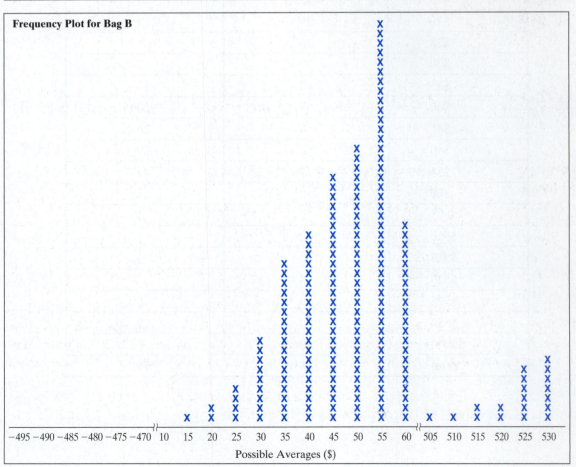

1.5.1 Forming a Decision Rule

The two competing theories are as follows:

H_0:   The shown bag is **Bag A**.

H_1:   The shown bag is **Bag B**.

How will you decide based on the average of two selected vouchers from the shown bag whether to accept H_0 or to reject H_0?

Consider first the obvious choices: If we obtain an average of −\$495, −\$490, −\$485, −\$480, −\$475, −\$470, or \$10, then we will know that the shown bag is Bag A, and we will decide to accept H_0. Likewise, if we obtain an average of \$60, \$505, \$510, \$515, \$520, \$525, or \$530, then we will know that the shown bag is Bag B, and we will decide to reject H_0 (that is, accept H_1). In each of these two cases, you will have not made an error. However, any of the other possible averages, \$15, \$20, \$25, \$30, \$35, \$40, \$45, \$50, or \$55, could result from either Bag A or Bag B.

THINK ABOUT IT

What if the average of the two selected vouchers is \$55?

Would this observation lead you to think the shown bag is Bag A or Bag B? Why?

How would you answer these questions if the average of the two selected vouchers is \$15?

An average of \$55 could result from either Bag A or Bag B. If the bag is Bag A, an average of \$55 could occur in just 1 out of 190 possibilities; but if the bag is Bag B, an average of \$55 could occur in 42 out of 190 possibilities. An average of \$55 is more likely to be obtained if the bag is Bag B. Likewise, an average of \$15 could result from either Bag A or Bag B. If the bag is Bag A, an average of \$15 could occur in 42 out of 190 possibilities; but if the bag is Bag B, an average of \$15 could occur in only 1 out of 190 possibilities. An average of \$15 is more likely to be obtained if the bag is Bag A. It is the larger voucher values that are extreme or contradictory to the null hypothesis H_0 and thus show the most support for the alternative hypothesis H_1. The direction of extreme is to the right, corresponding to values that are large, and we will use this direction to finding extreme values for developing possible decision rules.

We will first consider the most extreme value (in the direction of H_1) among the possible values under the null hypothesis that shows the most support from the alternative hypothesis. Looking at the frequency plots for Bag A and Bag B, of all the possible averages specified under the null hypothesis (Bag A), the average that is most likely to come from Bag B is the average of $55. This most extreme value leads us to our first decision rule.

Decision Rule 1:   Reject H_0 **if** your selected average of the two vouchers is $\geq \$55$.

We may make a mistake if we use Decision Rule 1. In particular, an average of $55 is possible from either Bag A or Bag B.

A Type I error could occur only if H_0 is true, so we find the chance of this error by looking at the frequencies for Bag A. Since under the null hypothesis (Bag A) there is only one average greater than or equal to $55, the significance level $\alpha = \frac{1}{190} = 0.0053$. A Type II error could only occur if H_1 is true, so we find the chance of this error by looking at the frequencies for Bag B. Under the alternative hypothesis (Bag B), there are a total of $29 + 26 + 20 + 17 + 9 + 4 + 2 + 1 = 108$ averages that are less than $55, so the chance of a Type II error is $\beta = \frac{108}{190} = 0.568$. The significance level α is quite small, only 0.0053; however, the value of β is a very large value, 0.568. So let's consider enlarging the rejection region. To enlarge the rejection region,

Average of the two selected values	Bag A Number of ways of selecting the 2 values	Bag B Number of ways of selecting the 2 values
− $1000	0	0
− $495	7	0
− $490	6	0
− $485	2	0
− $480	2	0
− $475	1	0
− $470	1	0
$0	0	0
$10	21	0
$15	42	1
$20	29	2 β
$25	26	4
$30	20	9
$35	17	17
$40	9	20
$45	4	26
$50	2	29
$55	1	42
$60	0	21
$505	0 α	1
$510	0	1
$515	0	2
$520	0	2
$525	0	6
$530	0	7
$1000	0	0

we need to find the next most extreme average among the remaining possible averages under the null hypothesis that shows the most support from the alternative hypothesis.

Among the remaining possible averages under the null hypothesis, which value shows the most support from the alternative hypothesis? The answer is $50, with the largest chance from Bag B of $\frac{29}{190}$. This next most extreme value gives us a new rejection region—namely, averages of $50 or more. The corresponding decision rule is as follows:

Decision Rule 2:   Reject H_0 **if** your selected average of the two vouchers is $\geq \$50$.

Since under the null hypothesis (Bag A) there are three averages greater than or equal to $50, the significance level $\alpha = \frac{3}{190} = 0.0158$. Under the alternative hypothesis (Bag B), there is a total of $26 + 20 + 17 + 9 + 4 + 2 + 1 = 79$ averages that are less than $50, so the chance of a Type II error is $\beta = \frac{79}{190} = 0.416$. Notice that enlarging the rejection region will increase the level for α and decrease the level for β.

❋ Let's do it! 1.12 Finding a 5% Level Decision Rule

Suppose that the specified significance level is $\alpha = 0.05$. Decision Rules 1 and 2 both resulted in actual significance levels that were less than 0.05. Continue to enlarge the rejection region to find a decision rule with a 5% significance level. Recall that if there is no possible cutoff value that will make the significance level exactly equal to 0.05, we use the value that yields a significance level that is as close to 0.05 as possible without exceeding 0.05.

The decision rule will be "Reject H_0 if the average of the two vouchers is _____".

This rule results in a significance level of $\alpha = $ _____.

The corresponding level for β is _____.

Summary of the Relationship Between the Decision Rule and the Significance Level

If you start with a particular decision rule with a given cutoff value, you can obtain the corresponding significance level α.

Decision rule → significance level α:

For the Bag A–Bag B scenario with $n = 2$, if the decision rule tells us to reject H_0 if the average of the two selected vouchers is \$55 or more, then the corresponding significance level is $\alpha = \frac{1}{190} = 0.0053$.

You can also take the opposite approach. You can be given a specified significance level for performing the test, and for that level, determine the corresponding decision rule. That is, we can find the appropriate cutoff value that would yield a significance level equal to, or closest to without exceeding, α.

Significance level α → decision rule:

For the Bag A–Bag B scenario with $n = 2$, if the significance level is $\alpha = \frac{1}{190} = 0.0053$, then the corresponding decision rule tells us to reject H_0 if the average of the two selected vouchers is \$55 or more.

Back on page 21, we asked whether, for a given significance level of α, there was any way we could decrease the level of β. We can answer this inquiry by comparing the results for $n = 1$ and $n = 2$. The following table shows that, for the same significance level of $\alpha = 0.05$, increasing the sample size from $n = 1$ to $n = 2$ reduces the level of β:

Significance level $\alpha = 0.05$

Sample size $n = 1$	Decision rule—reject H_0 if your result is $\geq$ \$60.	$\beta = \frac{12}{20} = 0.60$
Sample size $n = 2$	Decision rule—reject H_0 if your result is $\geq$ \$45.	$\beta = \frac{53}{190} = 0.278$

The following table shows that, for the same significance level $\alpha = 0.10$, increasing the sample size from $n = 1$ to $n = 2$ reduces the level of β:

Significance level $\alpha = 0.10$

Sample size $n = 1$	Decision rule—reject H_0 if your result is $\geq \$50$.	$\beta = \frac{6}{20} = 0.30$
Sample size $n = 2$	Decision rule—reject H_0 if your result is $\geq \$40$.	$\beta = \frac{33}{190} = 0.174$

If the sample size n is increased, more information is available on which to base the decision. For a fixed significance level, if the sample size n is increased, the value of β will decrease. We will return to the effect of increasing sample size in Chapters 2, 8, and 9.

1.5.2   What's in the Bag? *p*-value When the Sample Size Is 2

We again turn our attention to the observed data—to measuring how likely the observed result is if the null hypothesis is true—namely, the *p*-value. Suppose that we select two vouchers and observe their average. We ask, "If H_0 is really true, how likely would we be to observe an average of this magnitude or larger (in the direction supporting the alternative hypothesis) just by chance?"

Suppose that the observed average of the two selected vouchers is $30. What would you decide? What is the *p*-value? Looking at the frequencies for Bag A, we see that under the null hypothesis there are a total of $20 + 17 + 9 + 4 + 2 + 1 = 53$ ways to get an average of $30 or larger. Thus, the *p*-value is $\frac{53}{190} = 0.279$. Would you reject the null hypothesis H_0 at the 0.10 level of significance? The answer is no, because the *p*-value of 0.279 is larger than the significance level of 0.10.

Average of the two selected values	Bag A Number of ways of selecting the 2 values	Bag B Number of ways of selecting the 2 values
− $1000	0	0
− $495	7	0
− $490	6	0
− $485	2	0
− $480	2	0
− $475	1	0
− $470	1	0
$0	0	0
$10	21	0
$15	42	1
$20	29	2
$25	26	4
$30	20	9
$35	17	17
$40	9	20
$45	4	26
$50	2	29
$55	1	42
$60	0	21
$505	0	1
$510	0	1
$515	0	2
$520	0	2
$525	0	6
$530	0	7
$1000	0	0

We also see that the observed $30 voucher lies in the acceptance region for a 10% level decision rule.

Suppose that the observed average of the two selected vouchers is $50. What would you decide? What is your *p*-value? Looking at the frequencies for Bag A, we see that under the null hypothesis there are only $2 + 1 = 3$ ways to get an average of $50 or larger. Thus, the *p*-value is somewhat small, $\frac{3}{190} = 0.0158$. Would you reject the null hypothesis H_0 at the 0.10 level of significance? The answer is yes, because the *p*-value is very small, less than the significance level of 0.10. Note that the observed $50 voucher lies in the rejection region for a 10% level decision rule.

THINK ABOUT IT

Suppose that the significance level was set at $\alpha = 0.05$—that is, the chance of a Type I error was set at 0.05. Two vouchers were then selected and the observed average turned out to be \$50. Since the p-value for this observed result is 0.0158, which is less than the significance level of $\alpha = 0.05$, you would reject the null hypothesis and conclude that the data are statistically significant at the 5% level.

You rejected H_0. Could you have made a mistake? (circle one)   Yes   No

What type of mistake could you have made? (circle one)   Type I error   Type II error

What is the chance that you have made a mistake? Explain.

Once you have made a decision, the decision is either right or wrong, and the chance that you made the mistake is either 0 or 1. If your answer to the last question in the "Think about it" box was not 0 or 1, but was 0.05, go back and read page 30 on the distinction between *before* and *after* you look at the data.

Another common misinterpretation of the meaning of a p-value is to associate it with the chance that the null hypothesis is true. There are two competing hypotheses, and one of them really is true. The p-value is a chance that is computed assuming that the null hypothesis is true; it is the chance of observing a result as extreme (or even more so).

When you read the results of a study, you may be provided with a p-value, and the conclusion will be left up to you as the reader. You will need to consider the consequences of the two types of errors. It is also important to consider the sample size the study was based on. We must be careful—more data do not necessarily mean more understanding. Large and expensive data sets can be useless because there may have been no statistical design principles underlying the data collection, or those interpreting the results may have lacked the statistical expertise to make sense of the data. In Chapters 2 and 3, we will discuss the notion that not all data are good data and learn some guiding statistical principles behind the collection of good data.

 ## 1.6   Significant versus Important

When the observed data are very different from what would be expected under the null hypothesis (that is, the p-value is very small), we say the result is statistically significant. Such a difference between the observed data and the null hypothesis theory was unlikely to occur just by chance alone. However, the statistical term significant does not mean "important." A result may be statistically significant, but it may not be **practically significant**.

Consider a study conducted to compare two drugs for treating strep throat, a new drug and a standard drug. The hypotheses being tested were as follows:

H_0:   The new drug is as effective as the standard drug in terms of length of time to achieve a "cure."

H_1:   The new drug is more effective than the standard drug in terms of length of time to achieve "cure."

Suppose that the average time until "cure" for subjects treated with the new drug was 6.5 days, while the average for the subjects treated with the standard drug was 7 days. The new drug cured patients an average of 0.5 days sooner as compared to the standard drug.

Statistical analysis of these data indicated that this difference in time to cure was "statistically significant." The maker of this new drug launches a marketing campaign citing that a recent study has "proven" this new drug works faster than the standard drug. Doctors begin to prescribe the new drug to diagnosed patients.

Perhaps we should slow down a bit. The statistical analysis only tells us that the half-day difference is almost impossible to have occurred by chance if indeed the two drugs were equally effective. What other issues should be considered? Suppose that the side effects with the new drug, while tolerable, are generally more severe or more frequent. For patients who must pay for prescriptions, cost may also play a role. From a statistical viewpoint, a key factor to consider when assessing the results is the sample size. The *p*-value of a test depends on the sample size, as the following cases explain:

Case 1: With a large enough sample, even a small difference, a small amount of improvement, can be found to be statistically significant—that is, hard to explain by chance alone. This does not necessarily make it important.

Case 2: On the other hand, an important difference may not be statistically significant if the sample size is too small.

THINK ABOUT IT Case 1

The population is the 100 students in the Engineering 101 class. The hypotheses to be tested are as follows:

H_0: The class consists of all males (100% males).

H_1: The class does not consist of all males (< 100% males).

The decision rule is "Reject H_0 if you observe at least one female in your sample." Suppose that the class actually consists of 2 females and 98 males—that is, H_1 is actually true.

(a) If you take a sample of size $n = 2$, how likely would it be to obtain a female in your sample and thus reject the null hypothesis?

Not Very Likely   Very Likely

(b) If you take a sample of size $n = 80$, how likely would it be to obtain a female in your sample and thus reject the null hypothesis?

Not Very Likely   Very Likely

With a large enough sample, even this small difference (100% males under the null hypothesis versus the true percentage of 98% males in the population) can be found to be statistically significant.

THINK ABOUT IT Case 2

The population is the 100 students in the Engineering 101 class. The hypotheses to be tested are as follows:

H_0: The class consists of all males (100% males).

H_1: The class does not consist of all males ($< 100\%$ males).

The decision rule is "Reject H_0 if you observe at least one female in your sample." Suppose that the class actually consists of 30 females and 70 males, so again, H_1 is actually true.

(a) If you take a sample of size $n = 80$, how likely would it be to obtain a female in your sample and thus reject the null hypothesis?

Not Very Likely   Very Likely

(b) If you take a sample of size $n = 2$, how likely would it be to obtain a female in your sample and thus reject the null hypothesis?

Not Very Likely   Very Likely

If the sample size is too small, even a large difference (100% males under the null hypothesis versus the true percentage of 70% males in the population) may not be statistically significant.

❖ **EXAMPLE 1.12   Circuit Boards**_____

The introduction of printed circuit boards (PCBs) in the 1950s revolutionized the electronics industry. However, solder-joint defects on PCBs have plagued electronics manufacturers. A single PCB may contain thousands of solder joints. A particular company currently uses an X-ray-based inspection system that can inspect approximately 10 solder joints per second. The company manager is considering the purchase of the latest laser-based inspection equipment, which supposedly can inspect, on average, more than 10 solder joints per second. The manager wishes to test the following hypotheses:

H_0: The new laser equipment is as fast as the current X-ray equipment in terms of the average number of joints inspected per second; that is, the average is 10.

H_1: The new laser equipment is faster than the current X-ray equipment in terms of the average number of joints inspected per second; that is, the average is > 10.

Suppose that the true average number of solder joints inspected per second with the laser equipment is just 10.001. If enough observations are taken, that difference, however small, would be detected. The data would be statistically significant; the manager would decide to reject H_0 and conclude that the new equipment was indeed faster. The question that would remain is "Is that difference of any *practical* significance?"

The manager may need to weigh the cost of implementing the new equipment (purchasing, training of employees, etc.) against the actual increase in speed. The producer of the new equipment is interested in being able to advertise that this new product is indeed faster. Since the manager and producer have different needs, the level of significance α and the sample size selected will probably be different. ❖

❖ EXAMPLE 1.13   AIDS Affliction Rate

Consider the null hypothesis H_0: The country has a rate of AIDS affliction per population = 71.4 cases per 100,000 population. If the affliction rate is actually higher than 71.4 cases per 100,000 population, then the government will provide funding for the implementation of new programs. In this case, being able to detect a small difference is of practical significance. You want to gather many observations, to get all the possible information you can. If the rate is *actually* higher than 71.4 cases per 100,000 population, and this hypothesis is tested using a sample that is too small, you may not be able to detect the difference statistically, which is important. ❖

So, we have that if the sample size is too small, even a large difference may not be statistically significant. However, with a large enough sample, even this small difference can be found to be statistically significant. So what sample size should we use? In general, larger samples can give us a better idea about the population, but larger samples may mean more time and higher costs. In Chapter 2, we discuss methods for selecting a sample. In Chapters 9 and 10, we will discuss statistical techniques that help us determine the sample size needed.

The decision to accept H_0 does not mean that H_0 is true, but rather that the data failed to detect the difference. This can occur when not enough observations are taken, when the level of significance is small, or simply because the decision was based on just a sample from the population. The decision to reject H_0 does not mean that H_1 is true, but rather that the data were strong enough to detect the difference. Even though your data are statistically significant (H_1 is supported), you still might not act on H_1 because the difference is of no practical importance.

✳ Chapter Summary

In this chapter, we were introduced to some of the statistical elements that comprise the decision-making process. We have seen how data help us to make decisions, that errors are possible with any decision-making process, and that there is a difference between a result being statistically significant and practically significant.

The goal of this first chapter was to get you acquainted with the line of reasoning used in the statistical decision-making process. First, we determine the null and alternative hypotheses. We must take care in stating these hypotheses appropriately. Both hypotheses should be statements about the same population(s). In many of our examples, the hypotheses were represented pictorially with frequency plots. Since many of the computations in performing the test require that we assume that the null hypothesis is true, we need to state it correctly. The alternative hypothesis generally provides us with the direction of extreme that is needed to carry out the test (one-sided to the right, one-sided to the left, or two-sided).

Directions of Extreme:

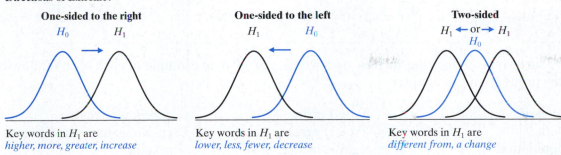

One-sided to the right	One-sided to the left	Two-sided
H_0 H_1	H_1 H_0	$H_1 \leftarrow$ or $\rightarrow H_1$ H_0
Key words in H_1 are *higher, more, greater, increase*	Key words in H_1 are *lower, less, fewer, decrease*	Key words in H_1 are *different from, a change*

We must understand that errors can be made in making a decision. A Type I error is rejecting a true null hypothesis, and a Type II error is failing to reject a false null hypothesis. The chance that we will make a Type I error is called the significance level, denoted by α. This level is specified in advance before the data are collected. Once we collect and summarize the data, we measure how unlikely the data are if the null hypothesis is true (the *p*-value). More formally, the *p*-value is the chance, computed under the assumption that H_0 is true, of getting the observed value plus the chance of getting all of the *more extreme* values. Based on this *p*-value, we can make a decision. The smaller the *p*-value, the stronger is the evidence provided by the data against the null hypothesis H_0. If the *p*-value is less than or equal to the significance level, we say the data are statistically significant and we reject H_0. Once a decision has been made, the decision is either right or wrong. Besides determining if the data are statistically significant, we should consider the role of the sample size in assessing practical significance. We will reinforce these decision-making concepts as we repeatedly apply, use, and discuss them in the remaining chapters.

Issues revolving around how to collect the data are addressed in Chapter 2. In Chapters 4, 5, and 6, you will learn various ways for summarizing data using graphs, numerical measures, and overall models. In many situations, the researcher's hypothesis is that some relationship exists—a relationship between aspirin intake and cancer risk, or the effect of owning a pet on the survival status of elderly patients. Studies are conducted to test such hypotheses. In the reporting of the results of such studies, often you are only told the final decision. Chapter 3 discusses various types of studies and the questions to ask to help us intelligently decide which results are worthy of our attention. Details on how to measure, present, and assess the significance of such relationships are discussed in Chapters 13 and 14. The *p*-value is the chance, computed under the assumption that H_0 is true, of getting the observed value plus the chance of getting all of the *more extreme* values. In Chapter 7, we study chance, more formally referred to as probability. In Chapter 2, we are introduced to the idea that samples taken from the same population using the same basic method will not all yield the same results. In Chapter 8, we experience what kind of dissimilarity we should expect to see among the various samples that could be obtained from the same population. The ability to quantify the amount by which sample results are likely to differ, from each other and from the population, is what allows us to use the statistical methods presented in Chapters 9 through 15. These last chapters bring us back again to statistical inference and decision making, discussing formal estimation and hypothesis testing. Your understanding of the material presented in this chapter provides you with a sound foundation for pursuing the knowledge presented in the remaining chapters.

Key Terms

Be sure you can describe, in your own words, and give an example of each of the following key terms from this chapter:

scientific method	Type I error	acceptance region
data	Type II error	critical value (cutoff value)
population	significance level (α)	p-value
sample	frequency plot	one-sided rejection region
statistical inference	sample size	two-sided rejection region
hypotheses	decision rule	smoothed frequency plot
null hypothesis	direction of extreme	practical significance
alternative hypothesis	most extreme value	relationship between the
statistically significant	rejection region	p-value and α

Exercises

1.1 In statistics, we use the symbols of H_0 and H_1. Write out an explanation of what these mean for a person who has not had statistics.

1.2 Decide whether each of the following statements is true or false:

(a) The first stage (step) in the scientific method is to formulate a theory (hypothesis).

(b) A theory (hypothesis) is rejected if it can be shown statistically that the data we observed would be very unlikely to occur if the theory were in fact true.

(c) A well-planned study will always provide proof for a theory.

(d) The null hypothesis and the alternative hypothesis are each a statement (sentence) about the resulting sample.

REAL DATA

1.3 The accompanying article suggests that "a vaccine treatment for malignant melanoma—the most deadly skin cancer—greatly improves survival with no serious side effects." The vaccine was administered to patients whose melanoma had spread. One measure of the effect of this vaccine is the percentage of such patients who survived five years since treatment. The article states that of those receiving other treatments, 10% survive five years. State the appropriate hypotheses for assessing whether this vaccine is just as effective as other

Vaccine battles melanoma

By Tim Friend
USA TODAY

SAN DIEGO—A vaccine-style treatment for malignant melanoma – the most deadly skin cancer – greatly improves survival with no serious side effects, researchers reported Sunday.

Results suggest it could be a major new treatment, says Dr. Donald Morton, John Wayne Cancer Institute in Santa Monica, Calif.

Morton says the vaccine, made from tumor cells, stimulates the immune system to attack the melanoma.

Researchers gave the vaccine to 355 patients whose melanoma had spread. Findings presented at the American Cancer Society's Science Writers Seminar show:

▶ 27% survived five years compared with 10% getting other treatments.

▶ 22% in the latest stage had tumor regression; tumors in three disappeared.

Dr. Laurence Meyer, University of Utah, thinks the vaccine may prevent melanoma in high-risk people or prevent recurrences.

treatments, or if this vaccine is more effective than other treatments, in terms of the five-year survival rates (SOURCE: *USA Today*, March 29, 1993).

1.4 Commercial fishermen working certain parts of the Atlantic Ocean sometimes have trouble with the presence of whales. They would like to scare away the whales without frightening the fish. In the past, sonar researchers have determined that 40% of all whales seen in an area do leave on their own, probably to get away from the noise of the fishing boat. The fishermen plan to try a new technique to increase that figure. The technique involves transmitting the sounds of a killer whale underwater. State the appropriate null and alternative hypotheses for testing whether or not the new technique works in terms of the percent of whales that leave the area.

1.5 For each set of hypotheses, decide whether a Type I error or a Type II error would be more serious. Explain your choice. Answers may vary.
   (a) H_0: The gun is loaded.
       H_1: The gun is not loaded.
   (b) H_0: The dog bites.
       H_1: The dog does not bite.
   (c) H_0: The mall is open.
       H_1: The mall is closed.
   (d) H_0: The watch is not waterproof.
       H_1: The watch is waterproof.

1.6 For each set of hypotheses, decide whether a Type I error or a Type II error would be more serious. Explain your choice. Answers may vary.
   (a) H_0: The electricity is turned on.
       H_1: The electricity is not turned on.
   (b) H_0: The brakes are not operational.
       H_1: The brakes are operational.
   (c) H_0: The snake is poisonous.
       H_1: The snake is not poisonous.
   (d) H_0: It is safe to cross the street.
       H_1: It is not safe to cross the street.

1.7 Suppose that you are an amateur gardener with a fondness for tomatoes and statistics. For the past few years you have always used one particular brand of fertilizer, Brand A, on your tomato plants, but now you think a new more expensive fertilizer, Brand B, may increase the yield you get from your tomato plants. Therefore, you decide that this year you will fertilize half your tomato plants with Brand A and half with Brand B and compare the average yields for the two types of fertilizer.
   (a) What is your null hypothesis and your alternative hypothesis?
   (b) Explain what Type I and Type II errors represent in this situation.

1.8 An international corporation plans to institute a drug-testing plan for its employees. An employee is assumed not to be a drug user, so the null hypothesis is "not a drug user," and the alternative hypothesis is "drug user." The test classifies a person as a drug user 4% of the time when the person is really not a drug user.
   (a) Explain what a Type I error and a Type II error are in this situation.
   (b) What is the chance of a Type I error in this situation?

1.9 The owner of a local nightclub has recently surveyed a sample of 100 patrons of the club. She would like to determine whether or not the mean age of her patrons is over 30 years. If so, she plans to remodel the club and alter the entertainment to appeal to the older crowd. If not, no changes will be made. Her hypotheses are

H_0: The average age for all patrons is 30 years.

H_1: The average age for all patrons is more than 30 years.

Explain the consequences of a Type I error. Are they serious?

1.10 Find an article in a recent newspaper in which a hypothesis is stated. State the corresponding null and alternative hypotheses and write down in words the meaning of both a Type I error and a Type II error.

1.11 A consumer protection agency received many complaints that the sodium content in a six-ounce can of Star Kist tuna is greater than the 250 mg that is stated on the label of each can. In response to these complaints, the protection agency tested the following hypotheses:

H_0: The average sodium content of all six-ounce cans is 250 mg.

H_1: The average sodium content of all six-ounce cans is greater than 250 mg.

The data from the study were *not statistically significant.*

(a) What hypothesis was accepted?

(b) Was a complaint registered with the provider, the Star Kist Co.?

(c) Could a mistake have been made? If so, describe the mistake.

1.12 Pharmaco, Inc., is a drug company that has come up with a new drug, Septaphine, to fight high blood pressure. They claim that their new drug is much better for reducing blood pressure compared to Cephaline, the drug that has been used until now to treat high blood pressure. They decide to test their claim.

(a) State clearly the null and alternative hypotheses.

(b) Pharmaco, Inc., found that Septaphine was significantly better than Cephaline using a 10% level of significance.

    (i) What hypothesis was accepted?

    (ii) Could a mistake have been made? If so, describe the mistake.

1.13 For a certain statistical test with a particular decision rule, $\alpha = 0.10$ and $\beta = 0.20$. Suppose that the sample size n will remain the same, but the decision rule is changed so that α is now 0.05. Determine which of the following is *possible* for the new value of β and explain your answer: 0, 0.15, 0.30.

1.14 If you do not reject the null hypothesis, you may be making... (select your answer)

(a) a Type I error.

(b) a Type II error.

(c) a correct decision.

(d) (a) or (c)

(e) (b) or (c)

1.15 Recall Example 1.6 on the two-sided rejection region (page 24). The first decision rule being considered was as follows: Decision Rule 1: reject H_0 if your selected voucher is either $\leq \$1$ or $\geq \$10$; otherwise, accept H_0. The rejection region contains those values as small or smaller than \$1 and as large or larger than \$10.

(a) Find the levels α and β corresponding to Decision Rule 1.

(b) Enlarge the rejection region to provide a Decision Rule 2. Compute the corresponding levels of α and β. How do they compare with those in (a)?

1.16 In this exercise, we extend Example 1.6 from having two identical bags, Bag E and Bag F, to having three identical bags, Bag E, Bag G, and Bag H. Each bag contains 30 vouchers, and you will be shown only one of the three bags. The null hypothesis will be that the shown bag is Bag E. The alternative hypothesis is that the shown bag is either, Bag G or Bag H. The frequency plots that show the contents of each bag, in terms of the vouchers face values, are provided. You are allowed to select just one voucher from the shown bag and must decide to accept H_0 or reject H_0.

H_0:   The shown bag is **Bag E**.
H_1:   The shown bag is **Bag G** or **Bag H**.

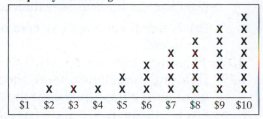

We still have that both the smaller voucher values and the larger voucher values are less likely under the null hypothesis (Bag E) and show the most support for the alternative hypothesis (either Bag G or Bag H). The direction of extreme for this scenario is to the left *and* to the right, corresponding to the values that are very small or very large.

(a) Consider the following decision rule: Reject H_0 if your selected voucher is either $\leq \$1$ or $\geq \$10$. What is the significance level, α, corresponding to this decision rule?

(b) Suppose that the observed voucher value is \$3. Compute the corresponding *p*-value.

(c) Recall that β is the chance of accepting H_0 when H_1 is true. If H_1 is true, the shown bag is either Bag G or Bag H. To compute the value for β, we need one more piece of information—namely, which of these two alternative bags is the shown bag.

   (i)   Compute β assuming that the shown bag is Bag G.

   (ii) Compute β assuming that the shown bag is Bag H.

1.17 Consider again the two identical bags, Bag C and Bag D, from Example 1.5 on page 21. You are allowed to select just one voucher from the shown bag and must decide to accept H_0 or reject H_0.

$$H_0:\ \text{The shown bag is } \textbf{\textit{Bag C}}. \qquad H_1:\ \text{The shown bag is } \textbf{\textit{Bag D}}.$$

```
Frequency Plot of Bag C                    Frequency Plot of Bag D

                          X                 X
                     X    X                 X    X
                X    X    X                 X    X    X
           X    X    X    X                 X    X    X    X
      X    X    X    X    X                 X    X    X    X    X
     $1   $2   $3   $4   $5                $1   $2   $3   $4   $5
```

Knowing only the contents of each bag, can you compute β, the chance of a Type II error? If yes, compute it. If no, explain what other information is needed to do so. (You may wish to review Example 1.5.)

1.18 Is a randomly chosen newborn baby equally likely to be a girl or a boy? Suppose that a researcher decides to investigate this question by checking the records of births for the past 10 years at a large metropolitan hospital. The null hypothesis can be stated as "The proportion of newborns that are girls is 0.50."

(a) Give the appropriate alternative hypothesis.

(b) Is the direction of extreme one-sided to the right, one-sided to the left, or two-sided?

1.19 Determine whether each of the following statements is true or false (a true statement must *always* be true; if the statement is false, explain why it is false):

(a) In hypothesis testing, if the significance level $\alpha = 0.05$, then $\beta = 1 - 0.05 = 0.95$.

(b) In hypothesis testing, if the null hypothesis is true, a Type II error could be made.

(c) In hypothesis testing, if the decision is to accept H_0, a Type II error could be made.

(d) In hypothesis testing, if the null hypothesis is rejected, it is because it is not true.

(e) A one-sided test is used whenever the sample size is small.

1.20 In hypothesis testing, if a level of significance of 0.05 is chosen, the 0.05 indicates (select your answer)

(a) the chance of a Type II error is 0.05.

(b) if H_0 is true, the chance of falsely rejecting it is 0.05.

(c) 95% of the time H_0 is true.

(d) 5% of the time H_0 is true.

1.21 Suppose that there are exactly two boxes, Box A and Box B. Each box contains 25 same-sized tokens. Each token is inscribed with a dollar value. The frequency table (distribution) for each box is as follows:

Box A

Value ($)	Frequency
0	1
5	1
10	2
15	3
20	4
25	6
30	8

Box B

Value ($)	Frequency
0	8
5	6
10	4
15	3
20	2
25	1
30	1

Suppose that you are shown just one of the boxes. A claim is made that the box is Box A. You must decide (judge) the probable truth of this claim *statistically* by selecting at random one token from the shown box.

(a) Write the two hypotheses for the statistical test.

(b) What is the direction of extreme?

(c) Write a decision rule so that the probability of a Type I error is as close to, but no more than, 10%.

(d) What is the numerical value for α for your decision rule in (c)?

(e) What is the chance of committing a Type II error using your decision rule?

(f) Suppose that you select a token inscribed with $5. Using the decision rule from (c), what is your decision?

1.22 Suppose that there are exactly two bags, Bag A and Bag B. Each bag contains 50 equal-sized tokens inscribed with a dollar value. The frequency table for each bag is as follows:

Bag A

Value ($)	Frequency
2	1
4	2
6	4
8	10
10	16
12	10
14	4
16	2
18	1

Bag B

Value ($)	Frequency
2	10
4	7
6	5
8	2
10	2
12	2
14	5
16	7
18	10

Furthermore, suppose that just one of these two bags is presented to you with the claim that the bag is Bag A. You must judge (decide) the probable truth of the claim *statistically* by randomly selecting one token from the given bag and using the *p*-value approach. Assume that you have selected a voucher inscribed with $14.

(a) Draw a frequency plot for the two hypotheses. What is the direction of extreme?

(b) Write the two hypotheses.

(c) Give a reasonable decision rule that could be used in the statistical test.

(d) What is the numerical value of α for your decision rule?

(e) What is the chance for committing a Type II error, β, if your rule is used?

(f) Suppose that a token inscribed with $8 is chosen.

   (i)   What would be your decision?

   (ii) What type of error (mistake) could you have committed?

(g) Suppose that a token inscribed with $16 is chosen.

   (i)   What would be your decision?

   (ii) What type of error (mistake) could you have committed?

1.23 Decide whether each of the following statements is true or false:

(a) If α is specified to be 0.10 for a statistical test, then β must be 0.90.

(b) A Type II error occurs whenever we accept the null hypothesis.

(c) If you suspect that the average life expectancy is not 76 years (as it has been accepted in the past), but is more than 76 years, then the rejection region for your decision rule would be one-sided to the right.

1.24 Suppose that there are two identical jars, Jar A and Jar B, that contain coins in the amounts shown by the following distributions:

Jar A					Jar B			
X	X						X	X
X	X						X	X
X	X						X	X
X	X						X	X
X	X						X	X
X	X	X				X	X	X
X	X	X				X	X	X
X	X	X	X		X	X	X	X
P	N	D	Q		P	N	D	Q

Note that P, N, D, and Q represent pennies, nickels, dimes, and quarters, respectively. Suppose that you are presented with one of these two jars with the claim that the jar is Jar A. The plan is to select **one coin** from the jar and decide which jar has been placed before you.

(a) What would be the two hypotheses?

(b) In order to conduct this test, each of the coins in the shown bag must have the same chance of being the selected coin. Is this the case here? Explain.

(c) If each of the coins in the shown bag did have the same chance of being selected, would it be appropriate to say the direction of extreme for this test is one-sided to the right? Explain.

1.25 Suppose that there are two bags each containing 25 same-sized candies. The distribution of the colors of the candies in each bag is given below. The hypotheses to be tested are as follows:

H_0:   The bag that is shown is ***Bag X***.

H_1:   The bag that is shown is ***Bag Y***.

Bag X

Value ($)	Frequency
Blue	1
Brown	7
Yellow	9
Green	7
Red	1

Bag Y

Value ($)	Frequency
Blue	9
Brown	3
Yellow	1
Green	3
Red	9

(a) Following is a graph of the contents found in Bag X:

Bag X

```
                  X
                  X
         X        X        X
         X        X        X
         X        X        X
         X        X        X
         X        X        X
         X        X        X
  X      X        X        X        X
 Blue  Brown   Yellow   Green     Red
```

Make a similar graph to display the contents found in Bag Y.

(b) Is it appropriate to say the direction of extreme for this test is two-sided? Explain.

1.26 What is a *p*-value? Write out an explanation of what this means for a person who has not had statistics.

1.27 Should the *p*-value be small or large to reject the null hypothesis, H_0? Explain your answer.

REAL DATA

1.28 A study was conducted to assess whether taking supplements of zinc might help certain women deliver larger babies—particularly those women who are thin when they get pregnant. Half of the women in the study took prenatal vitamins containing 25 mg of zinc. The other group received the same prenatal vitamins but without zinc. The results were reported in the *Journal of the American Medical Association*. The study compared average birth weight, premature delivery rates, and Caesarean delivery rates. There was no significant difference in the Caesarean delivery rates between the two groups. Consider the following hypotheses:

H_0: Caesarean delivery rates for the two groups are the same.

H_1: Caesarean delivery rates for the two groups are not the same.

(a) Based on the stated results, which hypothesis was supported by the data?

(b) Would the *p*-value for testing the hypotheses have been large or small? Explain.

1.29 You are on a game show and you must choose whether to accept or reject a bag presented to you. There are two possibilities for the contents of the shown bag.

The bag is the "winning bag" containing the following vouchers: $1, $1, $10, $10, $10, $100, $100, $100, $100, and $1000.

The bag is the "losing bag" containing the following vouchers: $-\$1000, \$1, \$1, \$1, \$1, \$10, \$10,$ and $\$100.$

Let the hypotheses be as follows:

H_0: The shown bag is the winning bag.

H_1: The shown bag is the losing bag.

(a) Make a frequency plot for H_0 and for H_1.

(b) Your decision rule is to reject H_0 if you pick a voucher of $1 or lower. Based on this decision rule, what is α?

(c) Your decision rule is to reject H_0 if you pick a voucher of $1 or lower. Based on this decision rule, what is β?

(d) Can you calculate a *p*-value?

If yes, calculate it. If no, explain in one sentence why not.

1.30 Machines A and B are the only two machines that produce steel rods at a steel manufacturing plant. The accompanying frequency plots give the distributions of the number of visual flaws in rods produced by Machines A and B. The frequency plots are based on a total of 15 rods produced by Machine A and 15 rods from Machine B. These plots will be used as the models for the number of visual flaws for rods produced by the two machines. Boxes of rods that were produced by Machine B are to be held back (not sent to suppliers). You have a box of steel rods that no longer have the label that tells you which of the two machines produced the rods. You will select a rod from this box and the number of visual flaws will be measured. We wish to test the following hypotheses:

Machine A:
```
                    X
                X   X
            X   X   X
        X   X   X   X
    X   X   X   X   X
    1   2   3   4   5
   Number of Flaws
```

Machine B:
```
X
X   X
X   X   X
X   X   X   X
X   X   X   X   X
1   2   3   4   5
Number of Flaws
```

H_0: The box of steel rods was produced by **Machine A**.

H_1: The box of steel rods was produced by **Machine B**.

(a) The direction of extreme for the test is (select one)

to the left   to the right   two-sided   can't tell

(b) It is decided to reject H_0 if the observed number of visual flaws of the selected steel rod is 2 or more extreme. Calculate the chance of a Type I error, α, and a Type II error, β.

(c) The observed number of visual flaws of the selected steel rod was 4. What is the *p*-value?

(d) Are the data statistically significant at the level α from (b)?

1.31 Jaeyun has two dice. Die A is a fair die; each of the six outcomes is equally likely. Die B is biased—that is, it is heavier on some sides so the opposite sides show up more often than others. The two dice look identical and are mixed so she cannot know which die is which. She decides to select one of the two die and roll that die one time. Based on the outcome, she must determine which die it is. That is, she must test the following hypotheses:

H_0: The selected die is **Die A** (fair).

H_1: The selected die is **Die B** (biased).

The following table provides the chances of each of the six possible outcomes for each die:

Outcome	Chance if Die A (fair)	Chance if Die B (biased)
1	1/6	3/10
2	1/6	2/10
3	1/6	2/10
4	1/6	1/10
5	1/6	1/10
6	1/6	1/10

(a) Since observing smaller outcomes (fewer dots) is more likely under the alternative hypothesis, the direction of extreme is (select one)

   to the left   to the right   in both directions   can't tell

(b) Jaeyun decides to use the following decision rule: Reject H_0 if she rolls a 1 (or less). Calculate α, the chance of a Type I error, and β, the chance of a Type II error.

(c) Jaeyun performs the experiment and she rolls a 2. Calculate the p-value and give her decision.

1.32 Suppose that there are two identical boxes, Box I and Box II, each containing 35 vouchers. The contents of each box are displayed at the right. You will be shown only one of the boxes and will be allowed to select just one voucher from the shown box at random to test the following hypotheses:

H_0:   The shown box is **Box I**.

H_1:   The shown box is **Box II**.

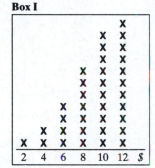

(a) The direction of extreme is (select one)

   to the right   to the left   two-sided   can't tell

(b) Consider the decision rule "Reject H_0 if the observed voucher value is \$4 or more extreme."

   (i) Circle, clearly label, and compute α, the significance level.

   (ii) Circle, clearly label, and compute β, the chance of a Type II error.

(c) The observed voucher value is \$6. In the above picture circle, clearly label, and compute the p-value.

(d) Is the observed result statistically significant? Explain.

(e) Give a new decision rule that will result in a larger significance level α as compared to the decision rule in (b).

   Complete the new decision rule:

   Reject H_0 if the observed voucher value is . . . _____

1.33 Suppose that there are exactly two bags, Bag A and Bag B. Each bag contains 40 equal-sized tokens inscribed with a dollar value. The frequency table for each bag is as follows:

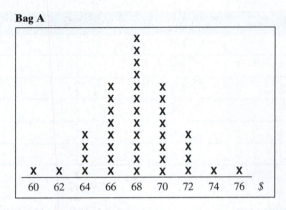

Furthermore, suppose that one of these two bags is presented to you with the claim that the bag is Bag A. You must judge (decide) the probable truth of the claim *statistically* by randomly selecting one token from the given bag and using the *p*-value approach.

(a) Write the two hypotheses.

(b) What is the direction of extreme?

(c) Suppose that you select a voucher inscribed with $74.

    (i)   What is the numerical value for the *p*-value?

    (ii) If the level of significance α is set at 10%, is the $74 voucher statistically significant?

    (iii) If the level of significance α is set at 5%, is the $74 voucher statistically significant?

(d) Suppose that you select a voucher inscribed with $68.

    (i)   What is the numerical value for the *p*-value?

    (ii) If the level of significance α is set at 10%, is the $68 voucher statistically significant?

    (iii) If the level of significance α is set at 5%, is the $68 voucher statistically significant?

1.34 Answer each of the following questions regarding the *p*-value of a test:

(a) Give two possible *p*-values for data that are statistically significant at 0.01.

(b) Give two possible *p*-values for data that are statistically significant at 0.05, but not statistically significant at 0.01.

(c) Give two possible *p*-values for data that are not statistically significant at 0.10.

1.35 Claude manages a plant that manufactures light bulbs. Claude is interested in finding out the average lifetime of a new model of a 100-watt light bulb that he is producing. The average lifetime for the population of 100-watt light bulbs produced by his competitor is about 40 hours. Claude wishes to test whether the average lifetime for his new-model bulbs is greater than that of his competition on average.

(a) What is the population under study?

(b) State the hypotheses to be tested.

(c) The manufacturer decides to test his theory at a level $\alpha = 0.10$. What is the chance of making a Type I error?

(d) He conducts a statistical test and the data were statistically significant at the $\alpha = 0.10$ level. Give a possible p-value for this test.

(e) He conducts a statistical test and the data were statistically significant at the $\alpha = 0.10$ level. Would the data be statistically significant at the $\alpha = 0.15$ level? at the 0.05 level?

1.36 Suppose that there are exactly two bags, Bag A and Bag B. Each bag contains 50 equal-sized tokens inscribed with a dollar value. The frequency table for each bag is as follows:

Bag A

Value ($)	Frequency
2	14
4	13
6	11
8	4
10	3
12	2
14	1
16	1
18	1

Bag B

Value ($)	Frequency
2	1
4	1
6	1
8	2
10	3
12	4
14	11
16	13
18	14

Furthermore, suppose that one of these two bags is presented to you with the claim that the bag is Bag A. You must judge (decide) the probable truth of the claim *statistically* by randomly selecting one token from the given bag and using the p-value approach. Assume that you have selected a voucher inscribed with $14.

(a) Write the two hypotheses.

(b) What is the direction of extreme?

(c) What is the numerical value for the p-value?

(d) If the level of significance α is set at 10%, is the $14 voucher statistically significant?

(e) If the level of significance α is set at 5%, is the $14 voucher statistically significant?

REAL
DATA

1.37 Examine the following information from the article "Children Today Watching More and More TV, Study Shows" (SOURCE: *The Associated Press,* December 30, 1997): A new study of children's viewing habits, conducted by the BJK&E Media Group of Manhattan, revealed that children aged 2–11 watch an average of 312.5 hours of kids' programs each year. That's up from 295.5 hours the year before. Boys watch an average of 53 more hours per year than girls. The study also found that 42% of children aged 6–11 have televisions in their bedroom. Nearly 25% of toddlers aged 2–5 also have their own TVs. Consider the following hypotheses to be tested:

H_0: The average number of hours of kids' programs watched this year by children between 6 and 11 years old is the same as that for last year.

H_1: The average number of hours of kids' programs watched this year by children between 6 and 11 years old has increased as compared to that for last year.

The preceding test was *not* statistically significant at a significance level of 0.10.

(a) Which hypothesis was supported?

(b) What can you say about the *p*-value for this test? (Be as specific as possible.)

(c) The data were observed and a decision was made. The decision could be right or wrong. Which type of error could have been made?

(select one)   Type I   Type II

(d) Is the rejection region one-sided to the right, one-sided to the left, or two-sided? Explain.

1.38 A statistical test was performed to test the following hypotheses, and the resulting data were statistically significant:

H_0: The average weight of a Bayer aspirin is 325 mg.

H_1: The average weight of a Bayer aspirin is not 325 mg.

(a) Which hypothesis was accepted?

(b) What is the direction of extreme for this statistical test?

(c) Suppose the significance level was 0.05. Give two possible values for the *p*-value.

1.39 A college newspaper would like to include an article on the cost of attending college. Part of the article discusses the cost of off-campus housing. A reference book about the cost of attending a college reports an average cost of $350 for monthly rental of a one-bedroom apartment within 1 mile of campus. A researcher decides to test the following hypotheses:

H_0: The true average cost is $350.

H_1: The true average cost is higher than $350.

Based on the data collected, the null hypothesis was rejected at $\alpha = 0.10$.

(a) An error could have been made. Could it have been a Type I error or a Type II error?

(b) Explain, using well-written sentences, the consequence of the error you chose in (a).

(c) What can you say about the *p*-value for this test?

(d) Are the results statistically significant?

1.40 Does exercise affect the birth weight of goats? A study was designed to address this question in which the pregnant goats were trained to walk on a treadmill. Suppose that the mean birth weight for goats born to mother goats that did not exercise was 1600 grams.

(a) Clearly write the hypotheses to be tested.

(b) The data were not statistically significant the 1% level. Which hypothesis was supported?

(c) What can you say about the *p*-value for this test? Be as specific as possible.

(d) Provide a possible *p*-value for this test.

(e) Suppose that the data were statistically significant at the 5% level, but not at the 1% level. Would you need to change the possible *p*-value you gave in (c)? If no, explain why not. If yes, give another value that will satisfy this statement on significance.

(f) According to (b), a decision was made at the 1% level. An error could have been committed. Which type of error could have been made?

1.41 The following table provides information regarding the hypotheses and results for three different studies:

	Null Hypothesis	Alternative Hypothesis	*p*-value
Study A	The true average lifetime is 73 years.	The true average lifetime is **more than** 73 years.	
Study B	The average time to relief for all Treatment A users is equal to the average time to relief for all Treatment B users.	The average time to relief for all Treatment A users is **not equal** to the average time to relief for all Treatment B users.	
Study C	The true proportion of students who work a part-time job is 0.40.	The true proportion of students who work a part-time job is **less than** 0.40.	

(a) Provide possible *p*-values for all three studies such that the following statements are true:

For Study A, the data were statistically significant at $\alpha = 0.05$ and at $\alpha = 0.01$.

For Study B, the data were *not* statistically significant at $\alpha = 0.05$.

For Study C, the data were statistically significant at $\alpha = 0.05$, but *not* at $\alpha = 0.01$.

(b) For which study do the results show the most support for the null hypothesis? Explain.

(c) Suppose that Study A concluded that the data supported the alternative hypothesis that the true average lifetime is more than 73 years, but in fact it is not. In our statistical language, would this be called a Type I error or a Type II error?

(d) For each study, determine if the rejection region would have been one-sided to the right, one-sided to the left, or two-sided.

1.42 Having a volatile nature may be hazardous to your health, say University of Michigan researchers, who found that hot-tempered men are more likely to suffer strokes than their cool-tempered counterparts. This study explored the relationship between level of anger (measured using Spielberger's Anger Expression Scale) and stroke risk (measured as the number of strokes over the past 10 years) in 2110 middle-aged men whose average age was 53. One set of hypotheses tested was as follows:

H_0:   Having a high anger level will not change the risk of having a stroke.

H_1:   Having a high anger level will increase the risk of having a stroke.

The researchers found that men with high anger levels were twice as likely to have a stroke than those less prone to explode. The preceding test was statistically significant at a 5% significance level.

(a) The rejection region for the preceding test would have been: (select one)

   one-sided to the right   one-sided to the left   two-sided

(b) Which hypothesis was supported?

(c) What can you say about the p-value for this test? (Be as specific as possible.)

(d) An error could have been made of which type? (select one)   Type I   Type II

1.43 The following table provides some information about hypotheses for three different studies:

	Null Hypothesis	Alternative Hypothesis	*p*-value
Study A	The true proportion of females is equal to 0.60.		
Study B	The average time to relief for all new users is equal to the average time to relief for all standard treatment users.		
Study C	The true average income of adults who work two jobs is equal to $70,000.		

(a) Give the appropriate alternative hypothesis for each study such that

   (i)   Study A is a two-sided test.

   (ii) Study B is a one-sided test to the left.

   (ii) Study C is a one-sided test to the right.

(b) Give a possible p-value for each study such that

   (i)   Study A results were statistically significant at 10%, but not at 1%.

   (ii) Study B results were statistically significant at 5% and at 1%.

   (iii) Study C results were not statistically significant at 10%.

(c) According to the p-values you entered in the last column of the table, which study results had the most support for the null hypothesis? Explain.

(d) Suppose that Study A concluded that the data supported the alternative hypothesis when in fact the true proportion of females is equal to 0.60. In our statistical language, would this be called a Type I error or a Type II error?

1.44 Proponents of a rehabilitation program at a local prison claim that the program has been successful because results of a study showed that the proportion of convicted persons who have been through the program and later reconvicted is lower than the national proportion of convicted persons who are released and later reconvicted.

(a) State the alternative hypothesis that was used in the study.

(b) If the results from the study were significant at the 10% level, but not at the 5% level, which of the following could have been the *p*-value? 0.04, 0.05, 0.06, or 0.11.

1.45 A fast-food store manager wishes to test hypotheses regarding the distribution of service time for customers using the drive-thru window. The hypotheses for the distribution of X = service time (length of time between order taken and order received in minutes) are presented at the right as the smoothed version of the frequency plots. Based on the service time for a randomly selected customer, you must decide whether or not to reject H_0.

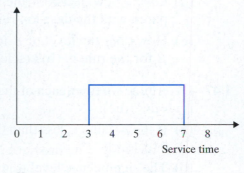

(a) The direction of extreme is (select one)
one-sided to the right,
one-sided to the left,
two-sided.

(b) Consider the following decision rule: Reject H_0 if the observed service time is 3.2 minutes or more extreme.

   (i) In the picture, shade in the region that corresponds to α, the significance level, and clearly label the region with α.

   (ii) In the picture, shade in the region that corresponds to β, the chance of a Type II error, and clearly label the region with β.

(c) The observed service time is 4.1 minutes. In the picture, shade in the region that corresponds to the *p*-value and clearly label the region with *p*-value.

(d) Is the observed result statistically significant? Explain.

1.46 The picture at the right presents the smoothed versions of the frequency plots depicting the contents of two bags, Bag A and Bag B, in terms of the face value of the vouchers within. One of the bags will be shown to you and you will be allowed to select just one voucher from that shown bag. Based on the one voucher value, you must decide whether or not to reject H_0.

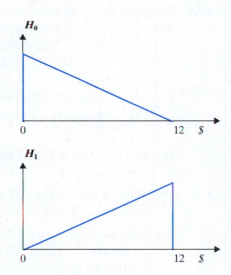

(a) The direction of extreme is (select one)
one-sided to the right,
one-sided to the left,
two-sided.

(b) Consider the following decision rule: Reject H_0 if the selected voucher is 10 or more extreme.

    (i)   In the picture, shade in the region that corresponds to α, the significance level, and clearly label the region with α.

    (ii) In the picture, shade in the region that corresponds to β, the chance of a Type II error, and clearly label the region with β.

(c) The observed voucher value is 3. In the picture, shade in the region that corresponds to the *p*-value and clearly label the region with *p*-value.

(d) Give a new decision rule that will result in a larger significance level α, compared with the decision rule in (b).

(e) How does the level of β for the new rule in (d), call it β^*, relate to the level of β for the rule in (b)? (select one) $\beta^* > \beta$ $\beta^* < \beta$ $\beta^* = \beta$

1.47 Determine whether each of the following statements is true or false a true statement is *always* true):

(a) If the *p*-value is less than 0.01, then the results are statistically significant at the 0.05 level.

(b) The significance level α is the chance that H_0 is true.

1.48 Suppose that the *p*-value for a statistical test is found to be 0.04 and the significance level was 0.05. Determine whether each of the following statements is true or false (a true answer is *always* true):

(a) The chance that H_0 is true is 0.04.

(b) The decision was to reject H_0.

(c) The results are statistically significant at 0.05.

(d) The results are statistically significant at 0.10.

1.49 A hypothesis test was performed, and the data were found to be statistically significant at the 5% level. Will the same data be statistically significant at the 1% level? Select one and explain your answer.

(a) Yes, all the time.

(b) No, never.

(c) Sometimes yes, sometimes no.

1.50 Suppose that you are asked to evaluate the abilities of an individual who claims to have perfect ESP (extrasensory perception). You decide to conduct an experiment to test this ability. You deal one card face down from a regular deck of 52 cards. The subject is then asked to say what the card is. Consider the following hypotheses:

    H_0:   The subject does not have ESP.

    H_1:   The subject does have ESP.

(a) What would a Type I error be in this context? (Give your answer in a nonstatistical manner.)

(b) What would a Type II error be in this context? (Give your answer in a nonstatistical manner.)

(c) Suppose that you decide to conclude that the individual has ESP if and only if he or she correctly identifies the card. What is the level of significance of this particular decision rule?

(d) What is the chance of a Type II error for the decision rule given in (c)?

(e) When the experiment is carried out, the individual *fails to identify correctly* the hidden card. What is the *p*-value?

(f) When the experiment is carried out, the individual *correctly identifies* the hidden card. The *p*-value is $\frac{1}{52}$. Is this the chance that the null hypothesis is correct? Explain.

1.51 In Example 1.2 (page 7), the following hypotheses were stated regarding the contents of a package of balls:

H_0: The proportion of yellow balls in the package is $\frac{1}{5}$.

H_1: The proportion of yellow balls in the package is more than $\frac{1}{5}$.

You are allowed to collect some data. You are permitted to mix the balls in the package well and then, without looking inside the package, reach in and select one ball and record its color.

(a) Suppose that you observed a yellow ball. What is the corresponding *p*-value for this result? Is your result statistically significant at a 10% significance level?

(b) You are allowed to take two observations—that is, after *replacing* the first selected ball and mixing the contents, you reach in and select a second ball and record its color. Suppose that the first ball is yellow and the second ball is blue. What is the corresponding *p*-value for this result? Is your result statistically significant at a 10% significance level? (*Hint:* List all possible outcomes and determine the number of ways each outcome could occur. Note that order will not be important—that is, getting a blue ball and then a yellow will be treated the same as getting a yellow ball and then a blue; both correspond to "one yellow and one blue.")

1.52 A social research scientist wants to test that the percentage of Republicans who favor the death penalty is greater than the percentage of Democrats who are in favor of the death penalty.

(a) State the corresponding null and alternative hypotheses.

(b) Will you need a one-sided or a two-sided rejection region? Explain.

Suppose that the data showed that the percentage of Republicans who are in favor of the death penalty is 42%, and the percentage of Democrats who are in favor of the death penalty is 40%.

(c) Do you think this result is statistically significant? Explain.

(d) Do you think this result is practically significant? Explain.

1.53 The population is the 50 students in a class. The hypotheses to be tested are as follows:

H_0: The class consists of all healthy students (100% healthy).

H_1: The class does not consist of all healthy students (<100% healthy).

The decision rule is "Reject H_0 if you observe at least one unhealthy student in your sample." Suppose that your class actually consists of just one unhealthy student. Which of the following statements about the sample size is correct?

(a) Your sample size should be large in order to find significance.

(b) Your sample size should be small in order to find significance.

1.54* Consider our Bag A–Bag B scenario in Section 1.5. We saw that for a fixed significance level of 0.10, increasing the sample size from $n = 1$ to $n = 2$ reduced the level of β (see page 48). Let's verify that this holds if we fix the significance level to be 0.05. Complete the following table by determining the decision rule and level of β for $n = 1$ and $n = 2$.

Significance level $\alpha = 0.05$

Sample size $n = 1$	Decision rule—reject H_0 if your result is	$\beta =$
Sample size $n = 2$	Decision rule—reject H_0 if your result is	$\beta =$

As the sample size is increased, what happened to the value of β?

1.55* The questions that follow are based on the Bag A–Bag B scenario when $n = 2$ vouchers are selected from the shown bag. The frequency plots for the possible outcomes are provided on page 44. The frequency table (so you won't have to count all of the little X's) is given on page 43. Consider again the following hypotheses:

H_0: The shown bag is **Bag A**.
H_1: The shown bag is **Bag B**.

(a) Suppose the decision rule is "Reject H_0 if the average of the two vouchers is at least $45."
  (i) What is the significance level for this decision rule?
  (ii) What is the cutoff value for this decision rule?
  (iii) What is the value of β for this decision rule?
  (iv) State very clearly what the value of β represents in this situation.

(b) Suppose that a new decision rule is given such that the significance level is 0.25.
  (i) What is the p-value if the two vouchers selected are $50 and $20?
  (ii) Based on your answer in (i), is the null hypothesis accepted or rejected? Explain.
  (iii) What is the p-value if the two vouchers selected are $30 and $30?
  (iv) Based on your answer in (iii), is the null hypothesis accepted or rejected? Explain.
  (v) What is the cutoff value for this decision rule? Explain how you know this is the correct cutoff value. [*Hint:* Think carefully about your answers to (i) through (iv).]

1.56* Recall Example 1.5 (page 21) on the one-sided rejection region to the left. Let's extend this scenario by increasing the sample size from $n = 1$ to $n = 2$.
(a) Construct frequency plots for Bag C–Bag D for the case in which the sample size is $n = 2$ and the average of the two selected vouchers (selected without replacement) is used to make the decision.

*These exercises follow from the optional Section 1.5.

(b) Consider the following decision rule: Reject H_0 if the average voucher value is $\leq$ \$2. Otherwise, accept H_0. Find the values of α and β corresponding to this decision rule.

(c) If the observed average voucher value is \$3, what is the corresponding p-value?

1.57* Many calculators can compute the number of ways of selecting 2 items (without replacement) from a total of 20 distinguishable items. In our Bag A–Bag B scenario with sample size of 2, we noted that the total number of ways of selecting 2 vouchers from a bag with 20 vouchers was 190 (see Table 1.2, page 43). More formally, this is called finding the number of *combinations* of 20 items taken 2 at a time. This is expressed as $\binom{20}{2}$ and is read "20 choose 2." The ■ TI graphing calculators have an operation called nCr for finding the number of combinations. This operation is found under the MATH PRB menu. The steps for finding the number of *combinations* of 20 items taken 2 at a time are as follows:

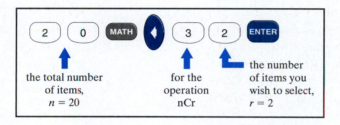

(a) Follow the above-mentioned steps to verify that there are 190 ways to select 2 vouchers from a total of 20 vouchers in a bag.

(b) Suppose that we were going to continue the Bag A–Bag B scenario to see what happens when we increase our sample size to 3. How many ways are there to select 3 vouchers (without replacement) from the total of 20 vouchers in a bag? That is, find $\binom{20}{3}$.

Chapter 2

PRODUCING DATA

2.1   Introduction

Chapter 1 gave us an overview of the statistical principles and procedures for obtaining and summarizing information in order to make decisions. We have a *theory* that we wish to test. We *collect data* to help us check the theory. We look at the data and *summarize* the results. We *interpret* the results and use the data to confirm or refute the theory. The decision made may be that we need to update the theory and gather more data. A decision may be made for now, but will be subject to test again at a later point in time.

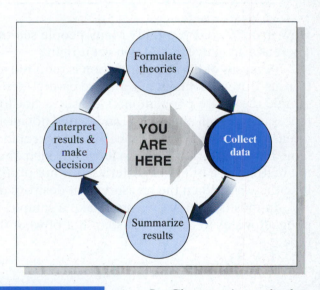

The **scientific method** is composed of the following steps:

Step 1   Formulate a theory.
Step 2   Collect data to test the theory.
Step 3   Analyze the results.
Step 4   Interpret the results and make a decision.

In Chapter 1, we had a quick tour all the way around our circle road map. In Chapters 2 and 3, we focus on Step 2—the way in which data are collected. You need data on some question of interest to you, to assess some theory. The data may have been collected by others or you may need to produce your own data. In either case, the quality of the decision you make will depend on the quality of the data you obtain. Not all data that are produced and presented are "good" data. By good data, we mean that we would like the data to be an accurate reflection of what is the truth. In this chapter, we will study various sampling methods—methods for selecting a sample. The idea of sampling is to gain information about the whole by examining only a part. We will begin by addressing the question, "Why sample?" We will turn to a discussion of the language of producing data and the basic terminology involved in sampling methods.

We will look at examples of data-collection schemes that may not lead to good data, and learn how to perform various sampling methods that help to produce good data.

2.2   Why Sample?

We want to learn about a population, so why not sample the whole population? Why not take a *census*?

> *Definition:*   A **census** is a sample consisting of the entire population.

Consider the process of proofreading. Using whatever method you choose, read the following sentence one time and determine how many times you see the letter "*f*."

> FINISHED FILES ARE THE RESULT OF YEARS OF SCIENTIFIC
> STUDY COMBINED WITH THE EXPERIENCE OF MANY YEARS*

How many *f*'s did you find? Many people see three *f*'s. Some people see four, five, or six. There are actually six. Did you get it right?

A census of the U.S. population is required every 10 years by the U.S. Constitution. The Census Bureau actually has a more careful definition of a census—namely, *an attempt* to sample the entire population. A census is not foolproof. We took a census of a short sentence, a very small population, and we did not all arrive at the truth. If the population is large, a census can be very expensive. Even if we could economically take a census, it may take so long that the information is no longer needed or useful. Suppose that you were interested in the lifetime of a brand of batteries. It is not profitable to test all of the batteries to see how long they last. When the measurement destroys the items, a census is not feasible.

In most situations, we must take a sample. With the world constantly changing, sampling allows us to get a quick snapshot of what the world looks like at a particular point in time.

2.3   The Language of Sampling

Sampling consists of selecting some part of a population to observe so that you can estimate something about the whole population of interest. In a national survey to gather information about the average income for the nation, only a sample of the people in the population is contacted, and the average income of those in the sample is used as an estimate of the true average income for the nation. In quality-control departments, only a sample of the parts from a large lot of parts is examined, and the proportion of defectives in the sample is used to assess if the lot can be labeled as acceptable. Let's recall our definitions of a population and sample and introduce the idea of a unit and a variable.

---

* SOURCE: *"The Deming Route to Quality and Productivity,"* William W. Scherkenbach, ASQC Quality Press, 1986.

> *Definitions:* The **population** is the entire group of objects or individuals under study, about which information is wanted.
>
> A **unit** is an individual object or person in the population. The units are often called **subjects** if the population consists of people.
>
> A **sample** is a part of the population that is actually used to get information.
>
> A **variable** is a characteristic of interest to be measured for each unit in the sample.
>
> The **size of the population** is denoted by the capital letter N.
>
> The **size of the sample** is denoted by the small letter n.

The population size $N = 16$.
The sample size $n = 4$.

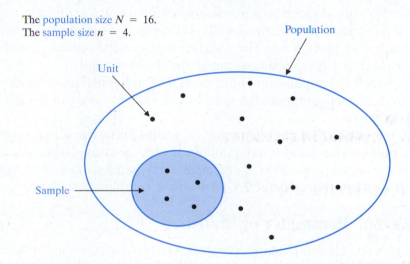

❖ EXAMPLE 2.1 Crossing the Street

Can I cross the street at this intersection? What data would you need to answer this question? How many cars are approaching and at what speed are they traveling? The population would be the list of oncoming cars. The variables of interest are the distance of the cars from the intersection and the speed of the cars. The number of cars is not so important as the distance and speed. ❖

Sometimes it is difficult to compile a list of the units of a population. How could we construct a list of all of the fish in a pond? How do we incorporate people who live in areas that are inaccessible or too costly to visit? Sometimes, the variable of interest is controversial. The population may be the students at a school. The question of interest may be "How many of the students have cheated on an exam?" Sometimes decisions are not easy to make because we do not know the complete situation. Suppose the president of a country wants to implement a program when the level of unemployment becomes too high. What level is "too high?" Can we know the actual unemployment rate?

We distinguish between a **population** and a **sample** from the population. When we want to learn something about the population—for example, the average income or the proportion who voted—the value that is computed from the whole population is called a **parameter**. The value of a parameter is a fixed number, which could be computed if we could examine the entire population. However, in general, we only have a sample from the population. When we calculate the average income for those in the sample, or the proportion in the sample who voted, each of these quantities is called a **statistic**. What if we want to know the value of the parameter for our population? If we cannot examine the entire population, we take a sample and use the information in the sample, the statistic, to estimate the corresponding information in the population, the parameter.

> *Definitions:* A **parameter** is a numerical value that would be calculated using all of the values of the units in the population.
>
> A **statistic** is a numerical value that is calculated using all of the values of the units in a sample.
>
> *Tip:* One way to remember this distinction is this: The letter **p** is for **p**opulation and **p**arameter, while the letter **s** is for **s**tatistic and **s**ample.

❖ EXAMPLE 2.2 Defective Parts

Parts from a supplier are shipped to a purchasing company in lots (that is, large shipments of 1000 parts). Upon receiving a lot, the purchaser examines a sample of 20 parts to check on the number of defective parts, if any. The contract between the purchaser and supplier usually states the procedure for determining the acceptability of a lot.

The lot of parts is the population under study. The individual parts are the units. The 20 parts selected from this lot comprises the sample. The variable being measured is whether or not the part is defective.

If all 1000 parts in the lot are examined, we would know the actual proportion of defective parts in the lot. This value is a parameter. If only 20 of the 1000 parts are examined, we would know the actual proportion of defective parts in this sample of 20 parts, but not for the entire lot. The proportion of defective parts in the sample is a statistic. ❖

❖ EXAMPLE 2.3 Parameter or Statistic?

The following is from "Health Update: Aerobics Are Hard on the Ears" (*Better Homes and Gardens,* February 1995):

REAL DATA

> Doctors at Henry Ford Hospital in Detroit studied 125 aerobic classes in five health clubs and found that sound levels in 60% of the classes exceeded safe noise limits.

These results are based on a study of 125 aerobic classes in five health clubs, not all aerobic classes in all health clubs. Thus, the 60% figure is a statistic and the sample size is $n = 125$. ❖

Let's do it! 2.1 Parameter or Statistic?

Nine percent of the U.S. population has Type B blood. In a sample of 400 individuals from the U.S. population, 12.5% were found to have Type B blood. Circle your answer.

(a) In this particular situation, the value of 9% is a (parameter, statistic).

(b) In this particular situation, the value of 12.5% is a (parameter, statistic).

❖ EXAMPLE 2.4 A Parameter Is Fixed and a Statistic Varies

Suppose our population consists of just five students. The variable of interest is the number of textbooks in their backpacks. The parameter of interest is the average number of textbooks in their backpacks. One student had 5 books, another had 3 books, the third student had 1 book, and the other two students had 2 and 4 books, respectively.

What is the value of the parameter? Since you can see all of the values in the population, you can easily compute it—the total of 15 divided by 5 is 3. The average number of textbooks in the population is equal to 3. If another person were asked to compute this parameter, he or she would also get the value of 3. The value of this parameter is fixed—it is 3.

Now, suppose that you can select only two students from this population. You still want to learn about the average number of textbooks. What would you do? You would average the two values for the two selected students. This average is a statistic because it is computed from a sample. The average that you get will depend on which two students you selected. If you selected the students with 1 and 2 textbooks, then your statistic would be 1.5. If you selected the students with 3 and 5 textbooks, then your statistic would be 4. So, will the value of this statistic necessarily be equal to the parameter of 3?

If another person were asked to select two students from this population and compute the average number of books in their backpacks, they would not necessarily get the same value that you just computed.

So a parameter is a fixed quantity. The value of the parameter is generally unknown since we do not examine the whole population. A statistic will vary from sample to sample. Once a statistic is computed from a given sample, it is known and it is used *to estimate* the corresponding unknown parameter value. How good is this estimate? In Chapter 8, we will discuss in detail how to measure the accuracy of using an average from a sample to estimate the true average for the population.   ❖

2.4   Good Data?

Producing good data does require some effort. We want to gather data effectively to help answer various questions, to try to produce accurate estimates. How? There are many different methods for taking a sample. At the start of this chapter, we discussed the fact that a census, sampling the entire population, may not necessarily lead to accurate or timely results.

Suppose that you are a vegetable wholesaler and a truckload of potatoes has just been brought in. It may be convenient to sample a few buckets of potatoes taken from the top of the truckload. But these potatoes may not be representative of the entire truckload since those on the bottom may be damaged in shipment. The potatoes that you would get in your sample would have a tendency to be better than the population of potatoes that the sample is meant to represent. *Bias* is defined as a systematic prejudice in one direction. A sampling method that produces results that systematically differ from the truth about the population is said to be **biased**. Although sampling potatoes from the top of the truckload may be convenient and easy to do, the sample results may be biased. Such **convenience samples** are generally biased—that is, not representative of the population—and can lead to inaccurate conclusions about the population.

❖ EXAMPLE 2.5   Biased Too High _____

Suppose that a food company would like to learn about the average monthly food expenditure for the households of a nearby city. The city has a mix of households, including a few areas of very low-income housing. Suppose that the study is conducted using a telephone survey by randomly selecting numbers from the most recent phone directory. Many of the households in the very low-income housing do not have phones and would not be able to be included in the sample. Also, the monthly food expenditures for these households typically would be lower than for households with phones that could be included in the sample. Thus, the responses obtained from the phone survey would tend to be larger. The systematic prejudice of this sampling method would be in one direction—with a tendency to include the larger responses. The resulting estimate of the average monthly food expenditure would be expected to be too large.   ❖

A magazine prints a survey in its latest issue requesting its readers to respond because "We want to know what you think." Will all of its readers respond to this survey? Most likely not. The results of such a survey would reflect only the opinions of those who elected to respond. What readers are likely to respond? Respondents tend to be those who have a strong opinion about the issues being addressed in the survey. The *voluntary responses* received would not be representative of any larger group.

> *Definitions:* A sampling method is **biased** if it produces results that systematically differ (in one direction) from the truth about the population.
>
> A **convenience sample** is a sample consisting of units of the population that are easily accessible.
>
> A **volunteer sample** is a sample consisting of units of the population that chose to respond.
>
> Convenience samples and volunteer samples are generally biased.

There are many other types of biases that can arise. We discuss some of the general types next and present examples of each.

> *Definitions:* **Selection bias** is the systematic tendency on the part of the sampling procedure to exclude or include a certain type of unit.
>
> **Nonresponse bias** is the distortion that can arise because a large number of units selected for the sample do not respond or refuse to respond, and these nonresponders have a tendency to be different from the responders.
>
> **Response bias** is the distortion that can arise because the wording of a question and the behavior of the interviewer can affect the responses received.

❖ EXAMPLE 2.6 The 1936 Literary Digest Poll

REAL DATA

The magazine *Literary Digest* had been predicting the U.S. presidential election results with great success, up until the 1936 election. The 1936 election was a contest between the Republican Alf Landon and the Democratic incumbent Franklin Delano Roosevelt. The magazine sent out its questionnaires to 10 million people taken from lists of car owners, magazine subscribers, telephone directories, and registered voters. In 1936, what type of people were most likely to be on these lists and thus eligible to respond? Those who were wealthy and not pleased with Roosevelt. Only 2.3 million responses were received, for a response rate of 23%. What type of people who received the survey would be most likely to respond? Those who felt strongly about the election outcome—the Landon voters who wanted a change. The *Literary Digest* predicted that Landon would win with a 57% to 43% victory. What went wrong? The results were spoiled by *selection bias* and *nonresponse bias*. They had a tendency to exclude the poor, who were Roosevelt supporters. They also relied on voluntary response, and the Roosevelt supporters were happy with the current situation and were likely to be nonresponders.

In 1935, George Gallup started up his own polling organization, the American Institute of Public Opinion. Gallup surveyed only 50,000 people and correctly predicted Roosevelt as the winner of the election. Gallup knew that a random sample, described in the next section, was likely to be quite representative. (SOURCE: *Freedman, Pisani, Purves, and Adhikari,* 1991, pp. 306–308.) ❖

❖ EXAMPLE 2.7 **Sampling University Students**

To select a sample of 50 university students, a person might go to the campus and interview various students walking around the campus. What are the dangers with this kind of convenience sample? The interviewers own personal biases can influence the sample that is obtained. If the interviewer is a female, she might feel more comfortable approaching female students as opposed to male students. Even if the interviewer made an effort to obtain a balanced or representative sample of students, the interviewer may not know the correct balance in the student population. If the interviewer selects every fifth person coming out of the university library, the sample would **overrepresent** those who use the library more often.   ❖

❖ EXAMPLE 2.8 **Separation Versus Attraction**

Give half the members of a group the following statement: "Psychologists have found that separation *weakens* romantic attraction. As the saying goes, *Out of sight, out of mind.* Can you imagine why this might be?" Most people can, and nearly all of those who read this statement will regard the finding as unsurprising. Give the other half the following, opposite, statement: "Psychologists have found that separation *strengthens* romantic attraction. As the saying goes, *Absence makes the heart grow fonder.* Can you imagine why this might be?" Most people can, and nearly all who read this statement will regard the finding as unsurprising.

There is a problem when a supposed finding and its opposite both seem like common sense. The type of bias portrayed in this example is **response bias** due to the wording of the statements.   ❖

❖ EXAMPLE 2.9 **Phone Surveys**

If a survey was conducted by phone and phone numbers were selected randomly from the phone book, those without phones and those with an unlisted phone number would have no chance of being selected. This type of selection bias is sometimes called **undercoverage**. Many polling organizations now use a method of sampling called **random digit dialing**. With the aid of a computer, the resulting sample is approximately a random sample of all households with telephones. Once a phone number is generated, it is important to make many attempts to contact someone at that household.   ❖

❖ EXAMPLE 2.10 **Mail Surveys**

In mail surveys, the lower and upper social classes tend not to respond, which implies that the views of the middle class are overrepresented. Many surveys promote prizes or gifts in exchange for a returned, completed survey in an attempt to boost the return rate; most mail surveys will at least provide a self-addressed stamped envelope for the return in an attempt to reduce **nonresponse bias**.   ❖

❖ EXAMPLE 2.11 **Concealing the Truth**

Even if the response rate for a mail survey is high, what questions are asked is another issue to contend with. Some questions may be irrelevant to some people. Most people do not know about China's foreign policy toward Albania. Some questions may be sensitive to some people. Most people would not admit to cheating on their taxes. Asking people such questions is likely to produce unreliable answers. People may have a tendency to lie about their income or their age. In Chapter 7, we will be introduced to a technique, Warner's randomized response model, that can help us obtain honest answers to sensitive questions.   ❖

❖ EXAMPLE 2.12   It's All in a Name!

It can be very difficult wording a question that is completely clear. It is important to be sensitive to the effect of question wording on the results that will be obtained. Simply identifying an attitude or position with a prestigious person or agency can bias responses. Starting out a question with "Do you agree or disagree with the President's proposal to . . . " would have this effect. Questions can be biased negatively as well as positively. "Do you agree or disagree with the communistic position that . . . " is an example of negative bias. Questions may be phrased such that they suggest the desired answer. The interviewer may, through tone of voice or facial expressions, suggest the desired response. In such cases, the responses will tend to cluster around or be anchored toward the suggested response. The following statement appeared on a course evaluation form: "Question 3: Comments regarding the Instructor (organized, enthusiastic, attentive to students' needs, etc.)." Approximately 75% of the responses emphasized at least one of the suggested characteristics, and one person simply responded "all of the above."

The wording of a question needs to be carefully examined. Interviewers should receive formal training with certain guidelines to be followed in order to reduce **response bias**.   ❖

❖ EXAMPLE 2.13   Prison Sentences

A study was conducted to estimate the average length of a prison sentence for prisoners at a correctional facility. A random sample of current prisoners was obtained on a particular day, and they were monitored to the completion of their sentence. The information from this sample was used to estimate the average length of a prison sentence. For this study,

the population is	all individuals who have or are currently serving a prison sentence at this facility;
the sample is	a few current prisoners serving a sentence; and
the variable is	the length of sentence served, in years.

Why is this estimate almost certainly biased? Do all individuals in the population have a chance to be included in the sample? Only those prisoners currently serving a sentence on that day had a chance to be sampled. Those prisoners serving a longer sentence are more likely to still be in the facility on that day compared to those serving a shorter sentence and may have already been released. If those serving a longer sentence are more likely to be included in the sample, and we are measuring prison sentence length, our estimate may be biased upward—it will tend to be too large. This is an example of **selection bias**, called **length-bias sampling**. If the units whose responses tend to be larger are more likely to be selected, the resulting average size being measured may be biased upward.   ❖

Let's do it!   2.2   Is It Biased?

A television show conducted the following opinion poll:

> Should gun control be tougher? Let us know what you, the public, think in a special call-in poll tonight.
> If yes, call 1-900-446-6444. If no, call 1-900-446-6445. The charge is 50 cents for the first minute.

Would you consider the results of this opinion poll to be trustworthy? Explain.

Let's do it! 2.3 Family Size

A study was conducted to estimate the average size of *households* in the United States. A total of 1000 *people* were randomly selected from the population and they were asked to report the number of people in their household. The average of these 1000 responses was found to be 4.6.

(a) What is the population of interest?

(b) What is the variable of interest?

(c) What is the parameter of interest?

(d) An average computed in this manner would tend to be larger than the true average size of households in the United States. Explain why this would be the case.

(e) To better estimate the average size of households in the United States, the *units* that should be labeled and thus sampled from are *not* the individual people, but rather the _____.

The manner in which data are obtained determines the nature of the data received. To assess whether a sample is any good, we need to ask questions. What was the population of interest? How does this population compare to the units from which the sample was selected? How was the sample selected? Was there selection bias? Was there nonresponse bias? What was the exact wording of the questions that were asked? In what order were the questions presented?

We know that convenience samples and voluntary samples generally result in bias. The primary reason for this bias is that people were allowed to make the choices, and it is very difficult for people to be impartial. Most people would agree that using a coin toss to select one person out of two people would be a *fair* selection method. Thus, a remedy for the partiality in human choice is to use chance to select the sample. The basic idea is to have every unit in the population have a known, nonzero (not necessarily the same) chance of being selected for the sample. Sampling methods having this property are called **probability sampling methods**. If proper sampling methods are used, even a relatively small sample is able to reflect accurately the responses of a large population. With such methods, we are able to quantify the accuracy of using a sample statistic to estimate the corresponding population parameter.

> **Definition:** A sampling method that gives each unit in the population a known non-zero chance of being selected is called a **probability sampling method**.

The simplest type of probability sampling is called a **simple random sample**.* With a simple random sample, every possible group of units of the required size has the same chance of being the selected sample. In the next section, we will see how to obtain a simple random sample. However, in practice, taking a simple random sample may not be very simple. Other probability sampling methods, which often have advantages over simple random sampling,

---

* Sometimes a simple random sample is abbreviated as SRS.

will be presented in subsequent sections. The probability sampling methods that we will discuss are as follows:

- Simple random sampling.
- Stratified random sampling.
- Systematic sampling.
- Cluster sampling.
- Multistage sampling.

Depending on the particular situation, samples obtained using one method may be easier to perform and may provide better information than samples obtained using another method. We will learn about these methods by actually doing them—you will be sampling from your own population of students.

One more note before you head out on your sampling journey. Many of the probability sampling methods contain the word *random*. The use of the word *random* here does not imply haphazard or without a definite system or plan. The appropriate synonym for *selecting at random* is *selecting indiscriminately*.

 ## 2.5 Simple Random Sampling

> ***Definition:***   A **simple random sample of size *n*** is a sample of *n* units selected in such a way that every possible sample of the **given size *n*** has the same chance of being selected as any other sample of size *n*. Samples of different sizes may have different chances of being selected.

We can think of simple random sampling as follows: Place a ball (same size, weight, etc.) in a basket for each unit in the population, with the variable of interest recorded on each ball. Mix up the balls in the basket. A simple random sample would be chosen by selecting, without looking, a few balls from the basket. Since we do not wish to select the same item twice, the balls are selected without replacement and the basket is mixed well before each selection is made.

In either case, with or without replacement, at each selection, each unit in the basket has the same chance of being selected—the method is impartial. Every unit has the same chance of getting in the sample. Bias arises in convenience and voluntary sampling due to human choice or the interviewer's discretion in selecting units. Simple random sampling removes the human-choice element in selection and thus is considered fair or *unbiased*.

The "physical" idea of taking a simple random sample can be accomplished instead by having a list of the N units in the population labeled from 1 to N and a source of random numbers. Random numbers can be found in published tables of random digits, or they can be generated by a computer or a calculator. Table I, provided at the back of this text, is a random number table. A portion of this table is presented as Table 2.1 in this chapter.

The random number table begins at Row 1 and Column 1 with the values 1 0 4 8 0 1 5 0 1 1 0 1 5 3 6. To make the table easier to read, the listing of the digits has been divided up into groups of five and presented in numbered rows. You can think of this table as being produced by repeatedly drawing a ball with replacement from a basket containing 10 balls numbered 0, 1, . . . , 9. After each selection, the result is recorded, the ball is returned to the basket, and the balls are mixed. Thus, a table of random numbers is just a long list of the 10 digits 0, 1, . . . , 9 such that any of the digits is equally likely to occur in any position in the list. This implies that any pair of digits has the same chance of being any of the 100 possible pairs 00, 01, . . . , 99; that any triple of digits has the same chance of being any of 1000 possible triples 000, 001, . . . , 999; and so on.

Table 2.1 Random Numbers

Column Row	1–5	6–10	11–15	16–20	21–25	26–30	31–35	36–40	41–45	46–50	51–55	56–60	61–65	66–70
1	10480	15011	01536	02011	81647	91646	69179	14194	62590	36207	20969	99570	91291	90700
2	22368	46573	25595	85393	30995	89198	27982	53402	93965	34095	52666	19174	39615	99505
3	24130	48360	22527	97265	76393	64809	15179	24830	49340	32081	30680	19655	63348	58629
4	42167	93093	06243	61680	07856	16376	39440	53537	71341	57004	00849	74917	97758	16379
5	37570	39975	81837	16656	06121	91782	60468	81305	49684	60672	14110	06927	01263	54613
6	77921	06907	11008	42751	27756	53498	18602	70659	90655	15053	21916	81825	44394	42880
7	99562	72905	56420	69994	98872	31016	71194	18738	44013	48840	63213	21069	10634	12952
8	96301	91977	05463	07972	18876	20922	94595	56869	69014	60045	18425	84903	42508	32307
9	89579	14342	63661	10281	17453	18103	57740	84378	25331	12566	58678	44947	05585	56941
10	85475	36857	53342	53988	53060	59533	38867	62300	08158	17983	16439	11458	18593	64952
11	28918	69578	88231	33276	70997	79936	56865	05859	90106	31595	01547	85590	91610	78188
12	63553	40961	48235	03427	49626	69445	18663	72695	52180	20847	12234	90511	33703	90322
13	09429	93969	52636	92737	88974	33488	36320	17617	30015	08272	84115	27156	30613	74952
14	10365	61129	87529	85689	48237	52267	67689	93394	01511	26358	85104	20285	29975	89868
15	07119	97336	71048	08178	77233	13916	47564	81056	97735	85977	29372	74461	28551	90707
16	51085	12765	51821	51259	77452	16308	60756	92144	49442	53900	70960	63990	75601	40719
17	02368	21382	52404	60268	89368	19885	55322	44819	01188	65255	64835	44919	05944	55157
18	01011	54092	33362	94904	31273	04146	18594	29852	71585	85030	51132	01915	92747	64951
19	52162	53916	46369	58586	23216	14513	83149	98736	23495	64350	94738	17752	35156	35749
20	07056	97628	33787	09998	42698	06691	76988	13602	51851	46104	88916	19509	25625	58104
21	48663	91245	85828	14346	09172	30168	90229	04734	59193	22178	30421	61666	99904	32812
22	54164	58492	22421	74103	47070	25306	76468	26384	58151	06646	21524	15227	96909	44592
23	32639	32363	05597	24200	13363	38005	94342	28728	35806	06912	17012	64161	18296	22851
24	29334	27001	87637	87308	58731	00256	45834	15398	46557	41135	10367	07684	36188	18510
25	02488	33062	28834	07351	19731	92420	60952	61280	50001	67658	32586	86679	50720	94953
26	81525	72295	04839	96423	24878	82651	66566	14778	76797	14780	13300	87074	79666	95725
27	29676	20591	68086	26432	46901	20849	89768	81536	86645	12659	92259	57102	80428	25280
28	00742	57392	39064	66432	84673	40027	32832	61362	98947	96067	64760	64584	96096	98253
29	05366	04213	25669	26422	44407	44048	37937	63904	45766	66134	75470	66520	34693	90449
30	91921	26418	64117	94305	26766	25940	39972	22209	71500	64568	91402	42416	07844	69618
31	00582	04711	87917	77341	42206	35126	74087	99547	81817	42607	43808	76655	62028	76630
32	00725	69884	62797	56170	86324	88072	76222	36086	84637	93161	76038	65855	77919	88006
33	69011	65795	95876	55293	18988	27354	26575	08625	40801	59920	29841	80150	12777	48501
34	25976	57948	29888	88604	67917	48708	18912	82271	65424	69774	33611	54262	85963	03547
35	09763	83473	73577	12908	30883	18317	28290	35797	05998	41688	34952	37888	38917	88050

Using the Random Number Table to Select a Simple Random Sample

Suppose you have $N = 50$ units in the population and you wish to take a simple random sample of size $n = 5$.

Step 1:   Assign labels—Give each unit in the population a numerical label.

   When assigning a label to each of the units in the population, be sure to assign a different label to each unit and use as few digits as possible for the labels (however, each label must consist of the same number of digits). If the population consisted of just $N = 10$ units, then we would use all of the single digits 0 through 9. If the population consisted of fewer than 10 units, then we would use some of the single digits 0 through 9. If the population consisted of more than 10, but no more than 100 units, then we would use double-digit labels 00, 01, and so on.

   Since we have $N = 50$ units, we must use double-digit labels. We could assign the first unit in the population the label 01, the second unit the label 02, and so on, until we get to the last unit, which would have the label 50. Any choice of 50 different double-digit labels would be a correct labeling scheme. The labels 01 through 50 are the most convenient for being able to identify quickly which units are selected. You do not need to try to scramble up the labels as you assign them, because labels will be selected at random in Step 2.

Unit	Label	Unit	Label	Unit	Label	Unit	Label	Unit	Label
unit 1	01	unit 11	11	unit 21	21	unit 31	31	unit 41	41
unit 2	02	unit 12	12	unit 22	22	unit 32	32	unit 42	42
unit 3	03	unit 13	13	unit 23	23	unit 33	33	unit 43	43
unit 4	04	unit 14	14	unit 24	24	unit 34	34	unit 44	44
unit 5	05	unit 15	15	unit 25	25	unit 35	35	unit 45	45
unit 6	06	unit 16	16	unit 26	26	unit 36	36	unit 46	46
unit 7	07	unit 17	17	unit 27	27	unit 37	37	unit 47	47
unit 8	08	unit 18	18	unit 28	28	unit 38	38	unit 48	48
unit 9	09	unit 19	19	unit 29	29	unit 39	39	unit 49	49
unit 10	10	unit 20	20	unit 30	30	unit 40	40	unit 50	50

Step 2:   Use the table—Go to the table and read off random labels.

   You select your sample of size n by selecting labels at random. You start anywhere in your random number table and you start reading off labels systematically (e.g., going across rows). If you come to a previously selected label or a label that does not correspond to a unit in your population, you simply skip over it and move on to the next label. You continue reading off labels until you have selected your sample of n units. Suppose that you start at Row 5, Column 1 and read across rows from left to right.

   Row 5: 37570 39975 81837 16656 06121 91782 60468 81305 49684 60672

The first double-digit label read off from the table is **37**, corresponding to the thirty-seventh unit in the population. The next double-digit label is 57, which was not used in our assignment, so we skip that number. Continuing with Row 5, we have **03**, 99 (skip), 75 (skip), 81 (skip), 83 (skip), 71 (skip), 66 (skip), 56 (skip), **06, 12**, and **19**. Thus, our sample consists of the units with labels 37, 03, 06, 12, and 19. If you were to reach the end of Row 5 and were not finished taking your sample, you would continue to Column 1 of the next row, Row 6.

Random numbers can also be generated by many calculators or computer packages. In general, calculators and computers will have stored in memory a very long list of random digits, just like a large random number table. The calculator or computer will go to a particular place in that list and produce a list of labels. To have consistent results, you can specify where to start in the list; the start point is often referred to as the "seed" value. Most calculators and computers can generate random numbers between 0 and 1, but the commands can be modified to have them directly generate a list of random numbers between 1 and any N. You can simply label the units in the population from 1 to N in any convenient way. The calculator or computer will do the work of assigning the appropriate number and length of labels to the units, read off the labels from its random number list, and report out the unit numbers that have been selected for the sample.

Using a Calculator or Computer to Select a Simple Random Sample

Suppose that you have $N = 50$ units in the population and you wish to take a simple random sample of size $n = 5$.

Step 1: Assign labels—Give each unit in the population a numerical label from 1 to N.

Step 2: Use the calculator or computer to produce a sequence of random labels between 1 and N.

If you used a TI graphing calculator to generate a list of random numbers between 1 and 50, starting with the seed value of 33, the keystroke steps are as follows:

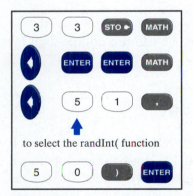

to select the randInt(function

Continue to push the ENTER button and your TI screen should look like this:

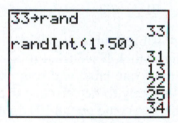

Details on how to generate random integers with TI graphing calculators are presented in the ■ TI Quick Steps appendix that follows the exercises at the end of this chapter.

If you used MINITAB, a statistical software package, to generate the random numbers, the commands and resulting output would look like this:

```
MTB > set c1
DATA > 1:50
DATA > end
MTB > sample 5 c1 c2
MTB > print c2
C2
     4 33 45 5 41
```

The main advantage of using a calculator or computer to generate the random labels is that your labeling scheme in Step 1 is very straightforward. Once the five units have been selected from the population, you still need to locate those five units and obtain their responses. If the units are people, and you gave up trying to reach someone after only one attempt, you would bias your results toward those people who were easily accessible.

❖ EXAMPLE 2.14   Simple Random Sampling of Forest Sites

As part of a study on the effect of forest fragmentation on the decline of the songbird population, a team of researchers need to gather information on nest predation. A large forest area is divided up into $N = 80$ sites, each of approximately the same size. Due to time and cost constraints, the researchers would like to select five of these sites at random and send out one team of researchers to each site for observation. We need to select a simple random sample of size $n = 5$ from a population of 80 sites.

■ Using the TI to Select Five Sites at Random

Step 1: Label the $N = 80$ sites from 1 to 80:

Site #	1	2	3	...	80
Label	1	2	3	...	80.

Step 2: Use the calculator with a seed value of 29 to produce a sequence of random labels between 1 and 80. The five labels, and thus sites, selected at random are as follows: Sites #50, #66, #43, #49, #74.

Using the Random Number Table to Select Five Sites at Random

Step 1: Label each site.

Since there are more than 10, but less than 100, sites, we can use double-digit labels. One possible labeling scheme is as follows:

Site #	1	2	3	...	80
Label	01	02	03	...	80.

Any assignment of 80 of the 100 possible double-digit labels to the 80 sites would work—each site receives a unique double-digit label. If we tried to use all 100 double-digit labels, we would have some sites receiving just one label and some receiving two labels. Those sites with two labels would be twice as likely to be selected as those sites with one label. If we only had 50 sites to select from, then we could use all 100 double-digit labels—each site would receive two distinct double-digit labels.

Step 2: Use the table—Go to the table and read off random labels.

We will use Row 10 of the table, starting at Column 1, and read off labels from left to right.

    Row 10: **85475 36857 53342** ... → **85, 47, 53, 68, 57, 53, 34, etc.**

The first double-digit label is 85, which does not correspond to a site, so we skip it. The next label is 47; thus, the first site selected is #47. The next label is 53, corresponding to site #53. The next two double-digit labels are 68 and 57, for sites #68 and #57. The next label is 53, which we have already selected, so we skip it. The final label to round out our list of sites is #34. So, the simple random sample of $n = 5$ sites consists of sites #47, #53, #68, #57, and #34. ❖

Let's **d**o **it!** **2.4** **A Simple Random Sample of Companies**

An investment magazine publishes data on sales, profits, assets, dividends, shares, and earnings per share for the nation's 500 most valuable companies. You are to select a simple random sample of 10 companies from the list of 500 companies. Explain how you would label the companies, and then use your calculator (with a seed value of 53) or the random number table (Row 26, Column 1, reading from left to right) to identify the labels of the 10 companies that would be selected from the list of 500 companies.

Let's **d**o **it!** **2.5** **Simple Random Sampling**

Form a group of 10 students. The population of interest is your group. Your task is to select a simple random sample of size $n = 3$ from your group.

Steps:

1. In the space provided, write the names of the people in your group.

2. Assign a different label to each of the names in your list. Be sure that everyone in your group assigns the same label to the same names.

3. Select your sample by selecting labels at random.

If you will be using a calculator, use a seed of 21 and your population size $N = 10$.

If you will be using the random number table, start at Row 13, Column 1 and read off labels going across rows, until three different labels have been selected. If you come to a previously selected label or one not in your list, then ignore it and move on to the next label.

What is the first label selected? = _____

Who is the first person selected from your group? = _____

What is the second label selected? = _____

Who is the second person selected from your group? = _____

What is the third label selected? = _____

Who is the third person selected from your group? = _____

Suppose that we wish to learn about the proportion of women in your population, denoted by p.

Count the number of people in your population: $N = $ _____.

Count the number of women in your population: COUNT = _____.

Compute the proportion of women in your population:

$$p = \frac{\text{COUNT}}{N} = \underline{\hspace{2cm}}.$$

Next, let's look at the results for your simple random sample of size $n = 3$. In many cases, the corresponding symbol computed for the sample is the same as that for the population, but a hat "∧" is written over the top, like this: $\hat{p}$.

Count the number of people in your sample: $n = $ _____.

Count the number of women in your sample: count = _____.

Compute the proportion of women in your sample, $\hat{p}$ (read p-hat):

$$\hat{p} = \frac{\text{count}}{n} = \underline{\hspace{2cm}}.$$

Recall that a parameter is a number describing the population, and a statistic is a number that can be computed from the sample.

In this example, p is a _____ (Select one—parameter or statistic.)

and $\hat{p}$ is a _____. (Select one—parameter or statistic.)

Does $\hat{p} = p$?_____ Will this always be the case? _____

THINK ABOUT IT 🐢

When will selecting a simple random sample be simple to do? Will it always be possible?

When will it be difficult to do? Why could it be difficult?

How would you label the units if the population size were 78? 292? 4000? Would it be simpler with the random number table or with a calculator (or computer)?

Now we know how to obtain a simple random sample—a sampling method that is objective or unbiased. As we saw in the previous section on bias, obtaining good data requires more than just using a probability sampling method. Even a simple random sample taken from a list of units that has excluded certain types of units will be biased.

A simple random sample is the best choice for certain kinds of surveys. However, there may be situations in which taking a simple random sample would be very time consuming and difficult to conduct. If a population is very large, obtaining a list of every unit in the population may not be easy or even possible. There are other probability sampling methods that can be used and have advantages over simple random sampling. Suppose that we feel it is important to obtain information from different regions, different subgroups of our population. We may want to be sure that both men and women are represented in our sample, or that each class level—freshmen, sophomore, junior, senior—is represented in our sample. If we were to take a simple random sample from the population, we might get mostly women in the sample or mostly sophomores in the sample. So how should we do it?

 ## 2.6   Stratified Random Sampling

Sometimes, a population of units falls into some natural subgroups, called *strata*. For example, if we wish to estimate the average cost for full-time infant childcare for a certain community, it might be convenient to view the population as being made up of three subgroups: (1) home daycare centers, (2) small licensed daycare centers, and (3) large licensed daycare centers. Rather than taking a simple random sample from the population of all daycare centers, we could take three separate random samples—one from each group of centers. These three samples would give us information about the costs for each of the three subgroups as well as about the costs for the overall population of centers.

> *Definition:*   A **stratified random sample** is selected by dividing or stratifying the population into mutually exclusive subgroups (strata) and taking a simple random sample of units from each stratum. The units sampled from each stratum are combined to form the complete sample.
>
> Mutually exclusive subgroups imply that each unit of the population belongs to only one stratum.

Stratified random sampling can be used if it is important to obtain information on the units within each stratum separately. Another advantage to stratified random sampling is that it sometimes provides more accurate inferences about a population than does simple random sampling. Stratified random sampling is most effective when the units within the individual stratum are very much alike with respect to the characteristic being measured. Why? Suppose that you know that males generally respond the same to a particular question (like, "What is your favorite type of movie?"), but females differ in their responses. You can only select a sample of size $n = 4$. How many females and how many males would you select? Since the male responses are much more homogeneous, as compared to the female responses, we would sample more females, perhaps three, and fewer males, just one. When the responses within a stratum are very much alike with respect to the characteristic we want to measure, we do not need to take a large sample from the stratum.

❖ EXAMPLE 2.15   Low-Fat, High-Carbohydrate Diet _____

Suppose a food expert wishes to determine the effects of a low-fat, high-carbohydrate diet on college students. The expert believes that the diet may have a different effect on students of different ages and gender, so the population of college students is subdivided into the following subpopulations: women over 21 years; women 21 years or under; men over 21 years; men 21 years or under. A simple random sample from *each* of these four nonoverlapping subpopulations is taken, yielding a stratified random sample of college students.   ❖

Let's do it! 2.6 Accounting Practices

Accountants often use stratified random sampling during audits to verify a company's records of such things as accounts receivable. The stratification is based on the dollar amount of the item and often includes 100% sampling of the largest items. One company reports 5000 accounts receivable. Of these, 200 are in amounts over $100,000, another 1000 are in amounts between $10,000 and $100,000, and the remaining 3800 are in amounts under $10,000. Using these groups as strata, you decide to verify all of the largest accounts (over $100,000) and to take a simple random sample of 5% of the midsize accounts ($10,000 to $100,000) and 1% of the small accounts (under $10,000).

(a) Based on this sampling design, how many accounts will be sampled?

# of large accounts to be sampled	=
# of midsize accounts to be sampled	=
# of small accounts to be sampled	=
total # of accounts to be sampled	=

(b) Describe how you will label the accounts in the *small* stratum to select some small accounts to be included in the sample. Use your calculator (seed value = 25) or the random number table (starting at Row 18, Column 1) to select only the first five small accounts to be verified. Think carefully about how many small accounts there are to label.

Let's do it! 2.7 Stratified Random Sampling

Form a group of about eight students. You need to have at least two females and two males in your group. As before, the population of interest is your entire group.

The question of interest is "How many times per year do you get a haircut?"

We want to learn about the average number of haircuts per year for your population. (*Note*: As a class, you may wish to come up with a different question of interest, such as, "How much money did you spend on lunch yesterday?")

In the space provided, write the names of the people in your group.

Ask the question of interest to each member of your population and record their responses next to their name. Compute the average response for your population. Add up all of the responses and then divide by the number of students in your population, N:

$$\text{Average} = \frac{\text{SUM}}{N} = \underline{\hspace{2in}}.$$

This number is a (parameter, statistic).

You are able to take a sample of size $n = 4$. Take a simple random sample of size $n = 4$.

Steps:

1. Assign a different label to each of the names in your list.

2. Select a place to start in your random number table (e.g., Row 10, Column 22) and read off labels until four different labels have been selected, or use your calculator (seed value $= 270$) to select your sample of size 4 from your population of N. Whom did you select from your group and what are their responses?

3. Compute the average response for your simple random sample of size $n = 4$. Add up the four responses and divide by 4:

$$\text{Average} = \frac{\text{sum}}{n} = \underline{\hspace{2in}}.$$

This number is a (parameter, statistic).

Now, you are able to take a sample of size 4, but you want to have two females and two males in your sample. How? STRATIFY!

Steps:

1. In the space provided, write a list of all the males and all of the females in your group (that is, form the strata). Also, include their response next to their name in parentheses—for example, Mary (2).

Females (Stratum 1)   **Males (Stratum 2)**

2. Assign a label to each unit in each stratum. Note that you can start with the same label for each stratum. For example, if there were four females and four males in your group, the females could be labeled 1 through 4 and the males could be labeled 1 through 4.

3. Select a simple random sample of size $n = 2$ females (start at Row 14, Column 1, or use your calculator with a seed value $= 24$) and a simple random sample of size $n = 2$ males (start at Row 23, Column 21, or use your calculator with a seed value $= 35$). Record the selected responses.

Responses for the two females selected: _____.

Responses for the two males selected: _____.

4. Compute the estimated average response for each stratum separately:

$$\text{Stratum 1, females:} \quad \text{Estimated average} = \frac{\text{sum}}{n} = \underline{\hspace{2cm}}.$$

$$\text{Stratum 2, males:} \quad \text{Estimated average} = \frac{\text{sum}}{n} = \underline{\hspace{2cm}}.$$

So now we have an estimated average for each stratum. How should we use these two estimates to obtain an overall estimate for the average of the whole population?

5. Compute the overall sample average response by pooling the estimated averages from each stratum together. Since the size of the strata may differ, we take a weighted average of the individual stratum's estimated averages. Each stratum's estimated average is weighted by the proportion of units in the population that make up that stratum.

Overall Sample Average

$$\left(\frac{\text{\# units in stratum 1}}{N}\right)\left(\begin{array}{c}\text{stratum 1}\\ \text{estimated average}\end{array}\right) + \left(\frac{\text{\# units in stratum 2}}{N}\right)\left(\begin{array}{c}\text{stratum 2}\\ \text{estimated average}\end{array}\right) =$$

This number is a (parameter, statistic). How does it compare to the average for the entire population?

❖ **EXAMPLE 2.16 Weighted Average**

Suppose that a population consists of four females with response values 1, 2, 3, and 4, and three males whose response values are 5, 6, and 7. The overall average for the population is

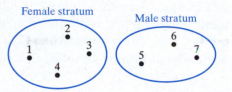

Female stratum   Male stratum

Population size = N = 7

$$\frac{1 + 2 + 3 + 4 + 5 + 6 + 7}{7} = \frac{28}{7} = 4.$$

- A simple random sample of size $n = 3$ females results in the sampled responses {1, 2, 3} and thus an estimated stratum average of 2.
- A simple random sample of size $n = 2$ males results in the sampled responses {6, 7}, for an estimated stratum average of 6.5.

The female stratum has four population units. The male stratum has three population units. In computing the overall weighted sample average, we replace each of the four female values by the estimated average for the female stratum, which is 2, and replace each of the three male values by the estimated average for the male stratum, which is 6.5.

Female stratum   Male stratum

Population size = N = 7

Overall Sample Average

$$\frac{(2 + 2 + 2 + 2) + (6.5 + 6.5 + 6.5)}{7}$$

$$= \frac{4(2) + 3(6.5)}{7} = \left(\frac{4}{7}\right)(2) + \left(\frac{3}{7}\right)(6.5) = 3.93$$

How did we do? This estimate, based on our stratified random sample, turns out to be quite close to the population average.   ❖

THINK ABOUT IT 🐢

When do you take a larger sample size from one stratum versus another?

When you form the strata, how should the variability of the units within each stratum compare to the variability between the strata?

Is a stratified random sample a simple random sample? Explain.

❖ **EXAMPLE 2.17** **Which Design?**

In a marketing study, a utility company wanted to estimate the average monthly bill of its residential customers. In the particular geographical area under study, there were 9050 customers. It was also known that 2210 of the customers received the minimum monthly bill of $8.65. Two different sampling designs were proposed:

   I. Select a simple random sample of 100 customers from the whole population.

   II. Select a simple random sample of 100 customers from the 6840 customers who received bills higher than $8.65.

Which design would you recommend?   Design II is preferred.
     The population can be stratified into two strata by the amount of the monthly bill:

     Stratum 1 = minimum charge of $8.65   Stratum 2 = more than $8.65.

By knowing that the actual monthly bill for those in Stratum 1 is $8.65, we have effectively sampled 100% from this stratum. We still need information on the monthly bills for those in Stratum 2. So, we sample as many as we can (for example, $n = 100$) from this stratum. The average obtained for each stratum would be combined using a weighted average.   ❖

Let's do it! 2.8 Daycare Compliance

A city has 40 licensed daycare facilities. A city inspector needs to check the compliance of these facilities with respect to a new city safety ordinance. Due to time constraints, the inspector will only inspect 12 facilities each month. She stratifies the facilities by size of facility. A facility with 15 or fewer full-time children is considered a small facility. The first stratum (I) consists of 10 large daycare facilities, and the second stratum (II) consists of the 30 small facilities. The city inspector plans to take a simple random sample from Stratum I, the large facilities, and a simple random sample from Stratum II, the small facilities. The variable "size of facility" is referred to as the stratification variable, the variable whose values determine which units are in which strata. How many large facilities would you recommend that the inspector sample from Stratum I? Thus, how many small facilities should be sampled from Stratum II? There is not just one solution here. If there is no prior information, you might elect to *sample in proportion to the size of the stratum relative to the size of the population.*

     Sample Size of Stratum I = _____   Sample Size of Stratum II = _____

The following list shows the population and their responses: the 40 daycare facilities, with "Yes" indicating that the facility is in compliance and "No" indicating that it is not in compliance. This list is not known to the inspector, of course.

Stratum I (Large facilities)		
Facility	1	Yes
	2	Yes
	3	No
	4	Yes
	5	No
	6	Yes
	7	Yes
	8	No
	9	Yes
	10	Yes

Stratum II (Small facilities)						
Facility	1	Yes	11	No	21	No
	2	No	12	No	22	Yes
	3	No	13	Yes	23	Yes
	4	Yes	14	No	24	No
	5	No	15	Yes	25	Yes
	6	No	16	Yes	26	No
	7	No	17	No	27	No
	8	Yes	18	Yes	28	Yes
	9	Yes	19	No	29	No
	10	No	20	No	30	Yes

Use your calculator or the random number table to choose a stratified random sample of size 12, as described previously. For Stratum I, use a seed value $= 21$ and $N = 10$, or Row 30, Column 6. For Stratum II, use a seed value $= 121$ and $N = 30$, or Row 29, Column 1. List the facilities in your sample along with their responses.

Based on your sample results, what is your estimate of the proportion of daycare facilities that are in compliance with the safety ordinance? Compute your overall weighted estimate:

$\hat{p}$ = estimate =

Based on the population listed, what is the true population proportion of daycare facilities that are in compliance with the safety ordinance?

p = true parameter =

How does your estimate, your observed sample statistic, compare with the true population parameter?

Stratified random sampling works well in situations in which the variability among the values within strata is small relative to the variability between strata. Information is obtained for each stratum separately. This information is combined through a weighted average to produce an overall estimate for the entire population. In the weighted average, the weights are the fraction of the population in each stratum, no matter what the sample size was from that stratum. The stratified random sample method, using the weighted average, is unbiased.

Sometimes simple random sampling or stratified random sampling can be costly to implement. In such cases, there are other sampling methods, such as systematic sampling and cluster sampling, that might be used. These methods are presented in the next two sections.

2.7   Systematic Sampling

Suppose that you had a list of 2000 names of students from which you wanted to select a sample of 100. You might consider selecting 1 name from every 20 names in the list. This is the idea behind a systematic sampling method.

> *Definition:*   For a **1-in-k systematic sample,** you order the units of the population in some way and randomly select one of the first k units in the ordered list. This selected unit is the first unit to be included in the sample. You continue through the list selecting every k^{th} unit from then on.

For our list of 2000 names, you would randomly select a starting point from among the first 20 names. Starting with this selected name, you would sample every 20th name after that on the list. The 1-in-20 systematic sample would consist of 1 person from every group of 20, spaced equally throughout the list. Once the starting point is determined, the sampling continues systematically. What makes this a probability sampling method is the feature that the starting point is selected at random. Before the starting point is selected, there are 20 different possible 1-in-20 systematic samples, and each is equally likely to be the actual sample taken.

An advantage is that the sample is taken from throughout the ordered population. This method is very fast and convenient when you already have a list of the units in your population. For example, you may have on file a list of patients with a certain type of ailment who visited a clinic. You may want to administer a certain treatment to every seventh patient on that list, starting with the patient selected at random from the first seven patients on that list.

A disadvantage is that this method can lead to a biased sample. A list of passengers on an airline flight was arranged by seat number, and there were 200 passengers sitting in 40 rows with five seats in each row. If you were to take a systematic sample of every fifth passenger, you would have passengers who sat in the same location in every row (for example, all on the aisle or all in a middle seat). This could lead to a biased sampling of opinions about the comfort of the flight.

❖ **EXAMPLE 2.18   A 1-in-4 Systematic Sample**_____

Suppose that you have 19 units in your population, represented by the labels A through S below, and you wish to take a 1-in-4 systematic sample.

A B C D E F G H I J K L M N O P Q R S
1 2 3 4

Only the first four units are labeled 1, 2, 3, and 4. You select a number at random from these 4 values using the random number table or your calculator. If the label selected is a "2," your systematic sample would consist of the 2nd, (2 + 4 =) 6th, (6 + 4 =) 10th, (10 + 4 =) 14th, and (14 + 4 =) 18th units in our ordered list—namely the ($n = 5$) units corresponding to B, F, J, N, and R (as shown with the arrows). If the label selected is a "4," your systematic sample would consist of the ($n = 4$) units corresponding to D, H, L, and P. So, for this example, the final sample size depends on which label is selected in the first group because the total number of units in the population ($N = 19$) does not divide evenly by $k = 4$. When the population size N divides evenly by k, then the sample size will always be

$$n = \frac{N}{k}.$$

Note: You do not need to divide the population into groups of size 4. You just need a list of the units in the population; then, randomly choose one of the first four units of the population, and select to include in the sample every fourth unit from then on. ❖

THINK ABOUT IT

Consider again the 19 units in a population represented by the labels A through S. You plan to take a 1-in-4 systematic sample.

A B C D E F G H I J K L M N O P Q R S
1 2 3 4

What is the chance that the letter A is selected? _____

What is the chance that the letter J is selected? _____

What is the chance that the letter P is selected? _____

What is the chance that the letter S is selected? _____

Does every unit have the same chance of being selected?   Yes   No

Does this imply that a systematic sample is a simple random sample? Explain.

Is a systematic sample a stratified random sample? Explain.

Is the sample size fixed in a systematic sample? Explain.

The accompanying figure is a map of census tract #5395 in Detroit, Michigan. Census tracts are small homogeneous areas with an average population of about 4000. Each block in the tract is marked with a Census Bureau identifying number.

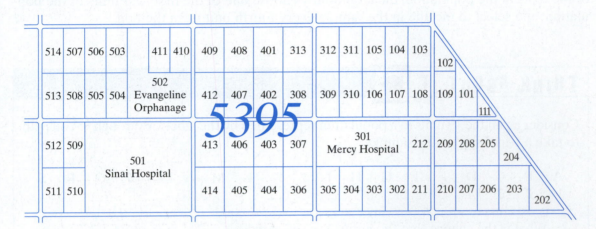

In the map, the ID numbers start at 101, which is along the right side, and end at 514 in the upper left corner. Notice that some of the numbers in between are skipped. For example, the numbers jump from 414, located in the middle of the bottom edge, to 501 for Sinai Hospital. Interviewers can be assigned to a *set* of blocks, such as the blocks with ID numbers in the 100s. The ID numbers are also assigned in a serpentine pattern so that the blocks that form a set are close to one another.

Your task is as follows: Take a 1-in-10 systematic sample of the blocks in this tract.

- Following the ID numbers for ordering the blocks in this population, use your pen to trace out this order on the map (that is, connect the blocks with a line in ascending order of ID number).

- For a 1-in-10 systematic sample, the first 10 blocks in the map form your first group. Label these 10 blocks and use your calculator (seed value = 39 and $N = 10$) or the random-number table (starting at Row 16, Column 6) to select your starting block, which is the first block in your sample.

  What is the ID number of the first block selected? _____

- Now, starting at the first block selected, count off every tenth block, in order, to be included in your sample. List the block ID numbers that form your sample:

  How many blocks are in your sample?
  Is the sample size fixed? Explain.

Let's do it! 2.10 Systematic Sampling of Presentation Attendees

An annual meeting for computer programmers was held in a convention center. A total of 1268 computer programmers were in attendance. At 10:00 A.M. on the last day of presentations, the 1268 programmers were attending exactly one of the 20 presentations that they had previously registered for. The 1268 programmers were given consecutive ID numbers based on the registration for the last presentation, as shown:

Presentation	1	2	3	4	5
ID Numbers	1–61	62–85	86–96	97–138	139–150

Presentation	6	7	8	9	10
ID Numbers	151–182	183–240	241–408	409–510	511–544

Presentation	11	12	13	14	15
ID Numbers	545–789	790–816	817–825	826–870	871–892

Presentation	16	17	18	19	20
ID Numbers	893–960	961–1017	1018–1120	1121–1249	1250–1268

The meeting organizers would like to survey the programmers to learn of their impressions regarding the annual meeting. It was decided to take a 1-in-50 systematic sample of the programmers, using their ID numbers.

(a) With this type of sampling plan, is each presentation guaranteed to be represented? That is, will the sample include at least one programmer from each presentation?

Circle one: Yes   No

Explain.

If yes, use your calculator with seed value = 18 (or starting at Row 33, Column 1) and give the ID numbers for the programmers that will be selected. If no, give the maximum value for k such that the 1-in-k systematic sample will always include at least one programmer from each presentation.

(b) With the correct k from (a), use your calculator with seed value = 18 (or Row 33, Column 1) and give the first 10 ID numbers for the programmers that will be selected.

(c) How many programmers will be included in your systematic sample?

(d) Suppose that the organizers would like to survey exactly two programmers from each presentation. Suggest a sampling plan to accomplish this.

2.8 Cluster Sampling

You want to learn about the television-viewing habits of households, in particular for children, adult females, and adult males. You expect that children's responses will be very similar, that adult female responses will be very similar, but not necessarily the same as the children, and that adult male responses will be very similar, but not necessarily the same as the children or the adult females. You want to obtain a sample that contains responses from all three groups.

Your first thought might be to do a stratified random sample with three strata: children, adult females, and adult males. However, such a sampling plan could be very costly, especially if the selected individuals must be interviewed and measured in person. For example, a child selected may live in the north region, an adult female selected may live in the west region, and an adult male selected may be from the south. In this case, a **cluster sampling** method may be used. Each family that has an adult male, an adult female, and a child would form a new unit called a **cluster**. We would select one or more families (that is, one or more clusters) and interview everyone in the selected cluster(s).

> *Definition:* In **cluster sampling,** the units of the population are grouped into clusters. One or more clusters are selected at random. If a cluster is selected, all of the units that form that cluster are included in the sample.

Cluster sampling is often confused with stratified random sampling. In both cases, the units in the population are divided into groups. In stratified sampling, the strata are to be formed such that the units within each stratum are homogenous, with each unit of the population belonging to one and only one stratum. We randomly select some units from within each stratum to form the sample. In cluster sampling, the clusters are to be formed so that they look like a "minipopulation"—that is, a smaller version of the larger population, with each unit of the population belonging to one and only one cluster. We randomly select one or more clusters of units, so the units of the population enter the sample in clusters, not individually.

The primary benefit of cluster sampling is a savings in cost and time. Less energy and money are needed if an interviewer stays within a specific geographical area as opposed to traveling across many areas. Sometimes, clusters are naturally defined, like families, hospitals, and so on. It may be that these naturally defined clusters do not reflect the variation that is present in the population. They may not look like a minipopulation. In this case, although there may be a savings in cost, there may be a loss in accuracy.

❖ **EXAMPLE 2.19** **A Cluster Sample of Students** _____

The president of a private, religiously affiliated university wants to learn about the opinions of the 820 freshmen students regarding the orientation training they received. At this university, all freshmen students are required to take one introductory religion class. A cluster sampling plan would be as follows:

- Write out a list of all of the introductory religion classes.
- Label each class in the list. Remember, if you use a random number table, all labels must have the same number of digits, so each class has the same chance of being selected.
- Take a simple random sample of five classes using a calculator or a random number table to select class labels at random.
- The sample would consist of all of the students in the five selected classes. ❖

2.11 **Cluster Sampling of Census Tract Blocks**

The following figure is a map of census tract #5375 in Detroit, Michigan:

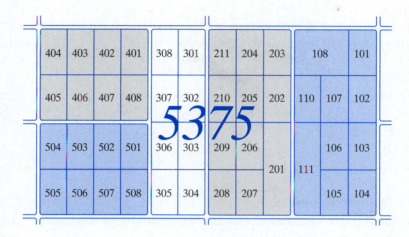

The blocks in the tract have been grouped to form clusters of blocks, corresponding to the Census Bureau identifying numbers.

In the map, there are five clusters of blocks based on the hundreds value for the ID number. Your task is as follows: Take a cluster sample by selecting two of the clusters of blocks at random.

(a) Label the five clusters.

(b) Take a simple random sample of two clusters. Use your calculator (with a seed value = 10 and $N = 5$) or the random number table starting at (Row 24, Column 31) to select two cluster labels at random.

What is the label of the first cluster selected?_____

What is the label of the second cluster selected? _____

(c) The sample would consist of all of the blocks in those two selected clusters. List the ID numbers of the blocks in your sample:

(d) What is the chance that a block will be selected for the sample?

Let's do it! 2.12 Cluster Sampling of Students

The population of interest is the students in your classroom today. Each of the rows of students in the class will form a cluster.

Row 10
Row 9
Row 8
⋮
Row 1
Front of class

(a) In the space provided, sketch a map portraying the relative positions of the rows (clusters) in your classroom.

(b) Assign a label to each cluster (each row).
What is the chance that any student in the population will be selected? _____

(c) Select one cluster at random. Use your calculator with a seed value = 279 or Row 9, Column 21 of the random number table.
Which cluster (row) did you select? _____
How many students were in the selected cluster (that is, what sample size did you get)?

Is the sample size fixed?   Yes   No
Explain.

THINK ABOUT IT

Is a cluster sample a simple random sample? Explain.

Is a cluster sample a stratified random sample? Explain.

When you form the clusters, how should the variability of the units within each cluster compare to the variability between the clusters?

Is this criterion the same as in stratified random sampling?

A population consists of 12 people, listed in the following table:

Population				
	Stratum I		*Stratum II*	
Cluster 1	Ann (1)	Bowei (2)	George (3)	Juan (4)
Cluster 2	Carrie (5)	Donna (6)	Bob (7)	Steve (8)
Cluster 3	Latisha (9)	Fran (10)	Paul (11)	Tom (12)

The two columns in the table divide up the population into two strata, labeled I and II. The population is also divided into three clusters by row. Below each name in the table is the person's corresponding ID number. Thus,

Cluster 1 consists of   Ann,   Bowei,   George, Juan.
Cluster 2 consists of   Carrie,   Donna,   Bob,   Steve.
Cluster 3 consists of   Latisha,   Fran,   Paul,   Tom.
Stratum I consists of   Ann,   Bowei,   Carrie,   Donna,   Latisha,   Fran.
Stratum II consists of   George,   Juan,   Bob,   Steve,   Paul,   Tom.

A sample of four people was obtained from this population. Listed next are three different samples. Consider the following sampling methods: (1) simple random sampling, (2) stratified random sampling, with equal sample sizes from each stratum, and (3) cluster sampling *by rows*. For each sample, determine which sampling method(s) could have generated that sample, by circling "Yes" or "No" for each. *Hint*: More than one method is possible.

	Sampling Method	(1) Simple Random Sample?	(2) Stratified?	(3) Cluster?
(a)	Carrie, Donna, Bob, Steve	Yes No	Yes No	Yes No
(b)	Ann, Fran, Carrie, Bowei	Yes No	Yes No	Yes No
(c)	Carrie, Donna, George, Tom	Yes No	Yes No	Yes No

(d) Take a systematic 1-in-3 sample from this population. Use the ID numbers as the ordered listing of the population items (that is, Ann = 1, Juan = 4, Steve = 8, etc.). Use your calculator with seed value = 78 and $N = 3$ or the random number table starting with Row 3, Column 3. List the names of the people in your sample.

Let's do it! 2.14 Name That Sampling Method

Read each scenario and identify the sampling method being described (simple random sample, convenience sampling, stratified random sampling, systematic sampling, or cluster sampling). Discuss your answers with your neighboring classmates.

(a) A shipment of 1000 three-ounce bottles of cologne has arrived to a merchant. These bottles were shipped together in 50 boxes with 20 bottles in each box. Of the 50 boxes, 5 boxes were randomly selected. The average content for these 100 selected bottles (that is, all 20 from each of the 5 selected boxes) was obtained.

Method_____

(b) A faculty member wishes to take a sample from the 1600 students in the school. Each student has an identification number. A list of all identification numbers is available. The faculty member selects an identification number at random from among the first 16 identification numbers in the list, and then every sixteenth identification number on the list from then on.

Method_____

(c) A faculty member wishes to take a sample from the 1600 students in the school. The faculty member decides to interview the first 100 students entering her class next Monday morning.

Method_____

2.9 Multistage Sampling

We have discussed a number of different sampling methods, each having various advantages and disadvantages. Many large, national sample surveys actually use a combination of these sampling methods. They might first divide the country into regions. Within each region, they stratify by the type of community (rural, urban, suburban). Within each of the selected communities, they select a few blocks at random. They systematically select some addresses within each selected block and interview a member at the selected address. This method of sampling in stages is called *multistage sampling*. At each stage, the type of sampling that is performed could be any of the types of sampling discussed so far. The items that are sampled at each stage become gradually smaller in size. The items selected at any given stage are chosen from within each item that was selected at the previous stage.

One advantage to multistage sampling is that a complete listing of all of the final units being selected is not needed. The selected units at the last stage were addresses, but we did not need to start with a very long list of all of the addresses in the country. We only needed a list of the units being sampled at each stage—that is, a list of communities within each region, a list of blocks within only the selected communities, and then a list of addresses only for the selected blocks. An interviewer could be sent to a selected block, update the list of addresses to make it current, and then sample addresses from the list. Another obvious benefit is a savings in cost and time.

❖ **EXAMPLE 2.20** **National Preelection Survey Procedure** _____

A separate study is conducted in *each* of four geographical regions of the United States (see the accompanying figure)—namely, Northeast, South, Midwest, and West. Within each region, towns are grouped together by size. One such grouping might be all towns in the South with a population between 50,000 and 250,000.

Step 1: Sample a few of these groups of towns.

Step 2: Sample a few towns from within each selected group.

Step 3: Sample a few wards within each selected town.

Step 4: Sample a few precincts within each selected ward.

Step 5: Sample a few households with each selected precinct.

Step 6: Some, or all, members of the selected households are interviewed.

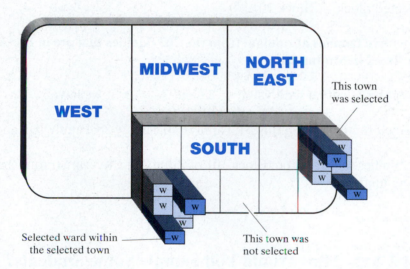

❖

❖ **EXAMPLE 2.21** **Selecting Families Using Multistage Sampling** _____

Suppose that we have eight cities of interest and we wish to select some families using a multistage sampling design. The eight cities are divided up by location into four strata with two (neighboring) cities in each stratum. The following procedure is conducted for *each* of the four strata. The procedure is demonstrated for Stratum 2 only.

Step 1: **Within Stratum 2, select either City 1 or City 2 at random.**

Using the TI with a seed value = 1015 and $N = 2$, we have

Selected city = City 2.

	Stratum 1		Stratum 2	
	City 1	City 2	City 1	City 2
	City 1	City 2	City 1	City 2
	Stratum 3		Stratum 4	

Step 2: **City 2 in Stratum 2 is divided into five zones. Select one zone at random.**

Using the TI with a seed value = 392 and $N = 5$, we have

Selected zone = Zone 2.

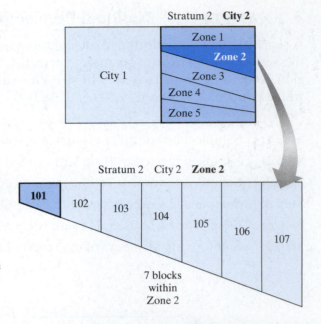

Step 3: **Zone 2 in City 2 in Stratum 2 is divided into seven blocks, numbered 101 through 107. Select one block at random.**

Using the TI with a seed value = 55 and $N = 7$, we have

Selected block = Block 101.

Step 4: **Select five families at random from the 203 families that are in Block 101 of Zone 2 in City 2 in Stratum 2.**

Using the TI with a seed value = 23 and $N = 203$, we have

Selected families = Family 31, Family 140, Family 38, Family 161, and Family 162.

These are the families selected from Block 101 of Zone 2 of City 2 in Stratum 2 using the multistage sampling method. ❖

❖ EXAMPLE 2.22 How Would You Sample Some Students?

A college has 15,000 students. Additional information about the students at this college is as follows:

	Women	Men	Total
Undergraduate	5683	5817	11,500
Graduate	2117	1383	3500
Total	7800	7200	15,000

We are to obtain a sample of students of size $n = 20$ from this population of students using each of the sampling methods listed in the examples. For each sample obtained, the number of undergraduate women (UW), graduate women (GW), undergraduate men (UM), and graduate men (GM) in the sample will be reported.

Simple Random Sampling All 15,000 need to receive a label. We label the undergraduate women from 1 to 5683, the graduate women from 5684 to 7800, the undergraduate men from 7801 to 13617, and the graduate men from 13618 to 15000. Using a seed value of 1024 and $N = 15000$, the following sample is obtained:

6814	GW	2562	UW	4034	UW	6948	GW	9569	UM
6633	GW	14434	GM	44	UW	11940	UM	3710	UW
11530	GM	14769	UM	12863	UM	3085	UW	1561	UW
7360	GW	12506	UM	11014	UM	1020	UW	9892	UM

Our simple random sample of $n = 20$ students consists of seven undergraduate women, four graduate women, seven undergraduate men, and two graduate men.

Stratified Random Sampling We first define the strata:

Stratum I: Undergraduate women.

Stratum II: Graduate women.

Stratum III: Undergraduate men.

Stratum IV: Graduate men.

We decide to take a simple random sample of five students from within each stratum. Suppose that we keep the labels that were assigned in the simple random sampling plan. We label the undergraduate women from 1 to 5683, the graduate women from 5684 to 7800, the undergraduate men from 7801 to 13617, and the graduate men from 13618 to 15000.

Stratum I: Using a seed value of 110 and randomly selecting integers between 1 and $N = 5683$, the five selected undergraduate women are 67, 848, 4488, 3497, 3046.

Stratum II: Using a seed value of 27 and randomly selecting integers from 5684 to 7800, the five selected graduate women are 5959, 5915, 5863, 7100, 6828.

Stratum III: Using a seed value of 25 and randomly selecting integers between 7801 and 13617, the five selected undergraduate men are 11518, 10113, 11488, 12051, 8773.

Stratum IV: Using a seed value of 297 and randomly selecting integers from 13618 to 15000, the five selected graduate men are 14214, 13898, 14906, 14117, 14927.

This stratified random sample of $n = 20$ students consists of five undergraduate women, five graduate women, five undergraduate men, and five graduate men.

Note that we could have decided to sample within each stratum in proportion to the size of that stratum relative to the size of the population. You might elect to take 10 women and 10 men overall (since 7800 is close to 7200). Of the 10 women, 7 would be undergraduate and 3 graduate. Likewise, of the 10 men, 7 would be undergraduate and 3 graduate.

Systematic Sampling We wish to sample $n = 20$ from the total of 15,000 students systematically. Since 20 divides into 15,000 evenly, for a value of 750, we should take a 1-in-750 systematic sample. If we use a seed value of 29 and select a random integer between 1 and 750, we obtain the number 466 as our first label. Selecting every 750th label from this point on results in the following sample of 20 students:

466	UW	3466	UW	6466	GW	9466	UM	12466	UM
1216	UW	4216	UW	7216	GW	10216	UM	13216	UM
1966	UW	4966	UW	7966	UM	10966	UM	13966	GM
2716	UW	5716	GW	8716	UM	11716	UM	14716	GM

Our systematic sample of $n = 20$ students consists of seven undergraduate women, three graduate women, eight undergraduate men, and two graduate men.

Cluster Sampling   Without any further information about the students, one possible idea is to form clusters of $n = 20$ students by listing all 15,000 students alphabetically by their last names. Cluster 1 would consist of the first 20 students in the list, Cluster 2 would consist of the next 20 students in the list, and so on. It is quite likely that each cluster would have at least one undergraduate woman, graduate woman, undergraduate man, and graduate man. If we use a seed value of 715, the first cluster to be selected is Cluster 57. All 20 students in that cluster would form the sample of $n = 20$ students. Without a more detailed listing of the students, we cannot know the exact number of undergraduate women, graduate women, undergraduate men, and graduate men.

Multistage Sampling   For a multistage sampling plan, we start with the list of 750 clusters that were formed in the cluster sampling plan.

Stage I: Take a simple random sample of 20 cluster units from the 750 clusters. Using a seed value of 1249, the selected clusters are listed.

Stage II: From each of the selected clusters of $n = 20$ students, select one student at random. For each selected cluster, the 20 students would be labeled from 1 to 20. Using a seed value of 10, random integers between 1 and 20 are obtained to indicate which one of the 20 students is selected from each selected cluster. The results are displayed in the following table.

Without a more detailed listing of the students, we cannot know the exact number of undergraduate women, graduate women, undergraduate men, and graduate men.

Stage I—The selected cluster is	Stage II—The label of the selected student within the cluster is
155	10
15	12
356	6
51	14
508	18
53	17
750	17
573	5
670	6
413	13
184	19
188	10
700	4
236	9
683	1
273	20
474	14
529	12
301	2
254	20

Chapter Summary

In this chapter, we have studied methods for producing data—namely, how we should obtain our sample(s). We discussed that we would like our sample to be unbiased and we learned that the results will vary from sample to sample. We learned how to perform the various sampling methods. Simple random sampling is the basic sampling method. Other methods can be quite complicated. But all of these methods, from the simple random sample to the multistage sampling design, have an important feature—there is a definite procedure for selecting the sample, and it involves the **planned use of chance**.

One threat to the validity of the results is bias. Certain sampling methods help reduce bias because of how the sample was selected. But there can be other sources of error, as cited in some papers reporting results. Such errors are not a result of which sampling method was used, but, for example, can be due to how the interview questions were phrased, the ordering of the questions, who asks the questions, the fact that a person's recall of past events often varies from the exact truth, and much more. Even in well-designed data-collection schemes, these other sources can be a significant source of total error. In large government surveys, measurement error, which arises from the respondent, the method of data collection, or the wording of the questionnaire, is likely to dominate total error, because coverage and nonresponse errors are relatively small. We need to keep such ideas in mind when evaluating the results of any sample.

We use data to make informed decisions. Sample information is used to gain insight about a population. Studies are conducted to test various hypotheses about a population. In the next chapter, we continue our discussion about producing data by presenting various types of studies and learning what questions to ask to help us intelligently decide which study results are worthy of our attention.

Sampling Design	Description
Simple Random Sampling	A sample selected in such a way that each sample of a given size n has the same chance of being selected as any other sample of the same size.
Stratified Random Sampling	The population is first divided into strata and a simple random sample is selected from each stratum.
1-in-k Systematic Sampling	The population items are ordered in some way. The sample is selected by randomly selecting the first item among the first k items (the starting point), and then selecting every kth item thereafter.
Cluster Sampling	The population is first divided into clusters and a simple random sample of clusters is selected. All items in the selected clusters make up the sample.
Multistage Sampling	Sampling is performed in various stages. The sample at each stage could be any one of the many sampling methods. The items selected at any given stage are selected from within each item that was selected at the previous stage.

Note: Samples obtained using stratified, systematic, cluster, or multistage sampling are probability samples, but they are *not* simple random samples.

Key Terms

Be sure you can describe, in your own words, and give an example of each of the following key terms from this chapter:

census	bias	random number table
population	convenience sampling	seed value
units	volunteer sampling	strata
sample	selection bias	stratified random
variable	nonresponse bias	sampling
population size N	length-bias sampling	response bias
sample size n	probability sampling	systematic sampling
parameter	methods	cluster sampling
statistic	simple random sampling	multistage sampling

Exercises

2.1 City records show that 35% of all residents are over the age of 60. To test a random-digit dialing device, you have the device randomly select 200 residential numbers. Of the 200 residents contacted, 28% were over the age of 60.

In the preceding paragraph, (select the best answer)

the 35% is a (parameter, statistic).

the 28% is a (parameter, statistic).

2.2 A public opinion poll in Ann Arbor wants to determine whether registered voters would vote "Yes" on Proposal A (regarding the use of public schools outside of school hours). They select a random sample of 50 registered voters and ask whether or not they would vote "Yes." The proportion of registered voters in Ann Arbor who would vote "Yes" on Proposal A is an example of (select one)

(a) a sample proportion.

(b) a statistic.

(c) a parameter.

(d) an unbiased statistic.

2.3 A parameter is to a population as a (select one)

(a) constant is to a variable.

(b) statistic is to a variable.

(c) statistic is to a sample.

(d) sample is to a statistic.

2.4 Consider a small population of five students. Their various ages are given in the following table:

Student	1	2	3	4	5
Age	18	20	27	21	26

(a) Compute the average or mean age of the five students in the population. This value is fixed for this population. For this population, there is only one value for the mean age. The value you have just computed is a parameter.

(b) One possible random sample of size $n = 2$ is Students 1 and 2 with ages 18 and 20, respectively. Compute the average or mean age of the two students in this sample. Does it equal the population mean in (a)?

(c) Another possible random sample of size $n = 2$ consists of Students 1 and 3 with ages 18 and 27, respectively. Compute the average or mean age of the two students in this sample. Does it equal the population mean in (a)? Was this sample mean the same as the sample mean from (b)?

(d) There are a total of 10 possible random samples of size $n = 2$ taken from a population with $N = 5$. Give the eight additional possible random samples of size $n = 2$ students. For each random sample, compute the sample mean age. The values of the sample mean vary from one sample to the next. There are many (possible) values for the sample mean age. The values you have computed in (b), (c), and (d) are statistics.

2.5 Your instructor will provide you with a copy of the obituary listing for a large (nationally distributed) city newspaper. The ages at death of the individuals in the listing will be your population.

(a) Using your calculator or the random number table, select a random sample of size $n = 3$ ages. Be sure to include which seed value or which row and column of the table you used. Compute the average age for this sample.

(b) Select another random sample of size $n = 3$ ages. Compute the average age for this sample. Was this sample mean the same as the sample mean from (a)?

(c) Compute the average age of all ages in your population.

(d) Which of the above values from (a), (b), and (c) are statistics? Which are parameters? Discuss the difference between a statistic and a parameter.

2.6 In a simple random sample of 200 items from a large population of items, five items were defective. It can be concluded that (select one)

(a) the percent of defective items in the population is 2.5%.

(b) out of every 200 items, there will be five defective items.

(c) the percent of defective items in the sample is 2.5%.

(d) the percent of defective items in the population must be larger than 2.5%, since the population is always larger than the sample.

(e) none of the above.

2.7 A particular gun-control bill is being sent to the U.S. Senate to be voted on. Each of the 50 states has two elected senators eligible to vote. To try to predict the outcome of the upcoming vote, a newspaper reporter randomly selected 10 Senators and asked each how he or she planned to vote.

(a) What is the population in this situation?

(b) What is the value of N, the population size?

(c) What is the sample in this situation?

(d) What is the value of n, the sample size?

2.8 For each of the following sampling situations, identify the population as accurately as possible, the sample, and whenever possible, give the value of the sample size n:

(a) Each week, the Gallup organization polls 1500 adult U.S. residents to learn about the national opinion on a variety of topics.

(b) The manager of a market is concerned that today's shipment of 1-gallon milk cartons might contain an unusually large amount of spoiled milk because of the recent hot weather. The manager randomly selects five 1-gallon cartons of milk from the shipment and assesses if they are sour before deciding whether to accept the shipment.

(c) A social researcher is interested in learning about the opinions of the members of a local women's business association regarding government funding of daycare facilities. She obtains a list of the 740 members of the business association and sends a questionnaire to a random sample of 100 of them. Only 38 questionnaires are returned.

2.9 To estimate the average age of all adult members of a club, a researcher randomly selects 100 adult members of the club, contacts them, and asks for their age. The average of the reported ages was 34 years.

(a) Is 34 the value of a parameter or a statistic?

(b) Examining the recorded ages from this survey the researcher found that there were a large number of adults reporting ages of 29, 35, and 39 years old and only a very few reporting an age of 30 or 40. Suppose that the reported 39-year-olds have been "39" for a good many years. This would be an example of what type of bias?

2.10 The Gallup Poll asks 1500 American residents age 18 or over, "If the election were held today, would you vote for the Republican candidate or the Democratic candidate for Congress in your district?" Based on the responses, 54% said "Republican," 34% said "Democratic," and 12% were undecided.

(a) Is the value of 54% a parameter or a statistic? Explain.

(b) If a large proportion of those polled refuse to answer the question, then this survey might be subject to _____ bias.

2.11 In a recent poll, 62% responded "Yes" when asked if they favored an amendment protecting the life of an unborn child. However, when asked whether they favored an amendment prohibiting abortions, 39% said "Yes." What type of bias is being observed? Explain.

2.12 Suppose that a gourmet food magazine wants to know how its readers feel about serving beer with various types of food. The magazine decides to send a survey to 1000 randomly selected readers (subscribers). Name a bias that this study is most likely to face and explain your answer.

2.13 A consumer interest agency will mail a questionnaire to addresses selected using a multistage probability sampling design. The agency decides that it will send a free trial issue of *Entertainment* magazine to all addresses that complete the survey. This free issue is being offered to minimize what type of bias?

2.14 The miles per gallon (mpg) were determined for five randomly chosen examples of four different luxury cars:

Rolls Royce	10.1	12.3	13.1	11.8	12.6
Porsche	14.8	13.1	13.7	14.5	13.9
Lexus	17.0	16.1	16.8	16.5	17.5
BMW	18.1	19.0	18.9	19.9	20.5

Assume that the true average mpg for all luxury cars (of these four types) is 17.4 mpg, and for each part, give your reason for your selected answer.

(a) If all the mpg information given came from one luxury car dealership in San Diego, the name of the bias that might occur in the data is

(i) selection bias.   (ii) nonresponse bias.   (iii) response bias.

(b) If the sampled mpg information was based on a sample survey of truthful, but somewhat lazy, car owners, of whom only 8% responded to the survey, then the name of the bias that might occur in this case is

(i) selection bias.   (ii) nonresponse bias.   (iii) response bias.

(c) If the sample of mpg data was based on a sample survey of randomly chosen dealerships, which generally believe lower mpg is a good attribute for a luxury car to possess, the name of the bias that might occur in this case is

(i) selection bias.   (ii) nonresponse bias.   (iii) response bias.

2.15 A department wishing to learn about the income earned by alumni sent out a survey to all of their alumni. One question in the survey was "What is your current income?" Approximately 30% of the alumni responded. Comment on the possible bias, if any, regarding the average income. Think about how people might answer this question, for those who do respond, and also how the responders might differ from the non-responders.

2.16 In a simple random sample (select one),

(a) nonresponse bias will be small.

(b) response bias will be small.

(c) any two samples of the same size will have the same chance of being drawn.

(d) each subject is asked easy questions and random responses are allowed.

2.17 A class consists of 100 students. Each student is assigned one and only one of the numbers $00, 01, 02, \ldots, 99$. Suppose that a box contains 100 tags, each tag having one of these numbers. Someone shakes the box well, and with a blindfold on, reaches in and withdraws 20 different tags simultaneously. The students assigned these numbers are taken as the sample.

(a) Is this a simple random sample from the class? Explain.

(b) If your answer in (a) is "Yes" is the sample drawn with or without replacement?

REAL DATA

2.18 The headline on Page 1 of an Illinois newspaper stated, "More people using drugs at work, survey reports." The article gave the following information: "The survey questioned 227 people, chosen at random, who called the national [cocaine] help line during a six-week period in February and March. . . . Ninety-two percent of the callers said they sometimes worked while under the influence of drugs" (*Rockford Register Star*, March 25, 1985).

(a) What kind of sampling method was used?

(b) What population do you think this sample was drawn from (that is, what is the sampled population)?

(c) Based on the information given, do you think the survey claim in the headline is justified? Explain.

2.19 A college newspaper would like to include an article on the cost of attending college. Part of the article discusses the cost of off-campus housing. There are exactly 4357 students in your college who live off-campus. The names of these students are available on a list, and they are given an ID number from 1 to 4357. A simple random sample of three students will be taken.

(a) Use your calculator with a seed value = 385 or the random number table (starting at Row 93, Column 1) and list the ID numbers for the students in this sample (in the order that they were selected).

(b) If the housing costs of the three selected students from (a) were averaged, would this number be a parameter or a statistic?

2.20 A business has a total of 100 employees. The business hires a statistician to conduct a survey to obtain information on the employees' attitudes toward the business. The statistician decides to take a simple random sample of 12 employees.

(a) What is the chance that a particular employee will be in the sample?

(b) Use your calculator with a seed value = 52 or the random number table (starting at Row 20, Column 1) and list the ID numbers for the employees in this sample (in the order that they were selected).

2.21 In the past, about 10% of customers were dissatisfied with the service received from automobile local dealers. This year, the manufacturer has reason to believe that the percentage of dissatisfied customers has decreased. He decides to test this theory.

(a) State the competing hypotheses H_0 and H_1 clearly.

(b) Out of a total of 40,000 customers, he decides to select a simple random sample of size 120. Use your calculator with a seed value of 32 or the random number table (starting at Row 18, Column 5) to give the labels of the first four selected customers.

(c) The data were found to be not statistically significant at $\alpha = 0.10$ level. What type of error could have been made?

(d) Give a possible p-value for this test.

(e) Would the same data in (c) be statistically significant at $\alpha = 0.05$ level? Explain.

(f) Suppose that in the selected sample, 9.5% of the customers were dissatisfied with the service received from automobile local dealers. The number 9.5% is a (select one)

   (i) parameter.

   (ii) statistic.

   (iii) simple random sample.

2.22 A class has a total of 100 students. Let H_0 state that the class of 100 students consists of all males, and let H_1 state that the class of 100 students has at least one female. The decision rule will be based upon a simple random sample of size 2 *without replacement* and will reject H_0 if there is at least one female in the sample. Determine whether each of the following statements is true or false (a true statement is *always* true):

(a) For this decision rule, the numerical value of α is exactly 0.

(b) For this decision rule, the numerical value of β is exactly 1.

(c) Suppose that the sample size is increased from 2 to 3, and again one rejects H_0 if there is at least one female in the sample. Then β will not increase from its value when the sample size was 2.

2.23 Jane has two purses. One contains two $1 bills and two $5 bills, whereas the other purse has four $1 bills. She decides to select a purse at random and reaches in to pull out two of the bills. She will use the selected bills to determine which of the follow-ing hypotheses to support:

H_0: The purse contains $1, $1, $5, and $5.

H_1: The purse contains $1, $1, $1, and $1.

Jane's *decision rule* is as follows: Reject H_0 and decide that the purse contains four $1 bills if the two selected bills are both $1. Suppose that both purses are sufficient-ly messy inside so that the selection of the two bills generates enough mixing of the contents to assume that we have a simple random sample of size $n = 2$.

Data   Jane selects two $1 dollar bills.

(a) What is Jane's decision?

(b) Were the data statistically significant? Explain.

(c) Can Jane be certain that she has the purse with the four $1 bills and that it is not the purse with two $1 bills and two $5 bills?

(d) What kind of error could Jane have committed?

(e) Suppose that each of the individual bills is distinguishable—for example,

Null purse: $1_1, $1_2, $5_1, $5_2.   Alternative purse: $1_1, $1_2, $1_3, $1_4.

List the possible simple random samples of size $n = 2$ that could arise if select-ing from the null purse.

(f) List the possible simple random samples of size $n = 2$ that could arise if select-ing from the alternative purse.

(g) Calculate the p-value for this observed result of two $1 bills. Remember, the p-value is based on the assumption that H_0 is true.

2.24 A zoning petition with a total of 4000 signatures was submitted to the city council. The council leader would like to find out the proportion of signatures that corre-spond to residents living on the east side of the city. It would take too much time to check all 4000 signatures on the petition, so it is decided to select a simple random sample of 400 signatures. You are asked to outline and conduct an appropriate sam-pling plan. Describe your plan, providing all details, including your seed value or ran-dom number table starting point and the resulting sampled signatures.

2.25 A company employs 1000 males and 200 females. A management task force officer polls a stratified random sample of 100 male and 100 female employees.

(a) What is the chance that a particular female employee will be polled?

(b) What is the chance that a particular male employee will be polled?

(c) If we took a stratified random sample of 100 male and 20 female employees at this company, each employee would have the same chance of being chosen. What is that chance? Is this a simple random sample? Explain.

2.26 A student organization wants to assess the attitudes of students toward a proposed change in the hours that the undergraduate library is open. They randomly select 100 freshmen, 100 sophomores, 100 juniors, and 100 seniors. What type of sampling design was used in this study?

2.27 A task force on fairness in employment wants to investigate allegations of gender dis-crimination in the promotion of bank tellers in a certain county. The study will focus on the 85 banks in this county. There are 300 female tellers and 100 male tellers across the 85 banks. Of the 300 female tellers, 75 are selected at random. Of the 100 male tellers, 25 are selected at random. What type of sampling design was used in this study?

2.28 The student services office of a college wishes to estimate the average amount of money that students spend on textbooks each semester. A stratified sampling method is being proposed. Suggested stratification variables (for forming the strata) are given. For each stratification variable, state whether you think it is a good variable for forming the strata and explain your answer.

(a) Stratify by declared major (i.e., field of study).

(b) Stratify by gender.

(c) Stratify by class rank: freshmen, sophomore, junior, senior.

2.29 A market research company draws a sample from the local residential phone directory by randomly selecting 10 people whose family names begin with A, then 10 people whose family names begin with B, and so on for every letter of the alphabet, for a total sample of 260 people.

(a) What kind of sampling method was used here?

(b) Does everyone in the phone book have an equal chance of being selected?

(c) Not all residential phone numbers are listed in the phone book. What kind of bias would this cause?

2.30 A class consists of 100 students. Suppose that we are interested in the heights of the people in the class. We could take a stratified random sample, using gender as the stratification variable. Suppose that it is known that 60 of the 100 students in the class are female. Someone shakes a box containing the 60 female tags, and with a blindfold on, reaches in and draws 10 tags. Then, this same person shakes a box containing the 40 male tags, and with a blindfold on, reaches in and draws 10 tags. Now we have two simple random samples, one for the males and one for the females, each of size 10. Suppose that the average height of the males in the sample is 70 inches, while the average height for the females in the sample is 63 inches. How should these two average heights be combined to produce an estimate of the average height of the people in the class?

2.31 A population of 100,000 people is stratified by age into three groups, those of age less than 21 years (Stratum 1), those whose age is at least 21 years but less than 60 years (Stratum 2), and those whose age is greater than or equal to 60 years (Stratum 3). It is known that 20% of the population are in Stratum 1 and 50% of the population are in Stratum 2. A stratified random sample is taken, with 300 from Stratum 1, 450 from Stratum 2, and 250 from Stratum 3. The average ages in the three samples are 16 for Stratum 1, 43 for Stratum 2, and 71 for Stratum 3. Give a numerical estimate of the average age for the entire population.

2.32 A sociologist wants to learn about the opinions of employed adult women on government funding for daycare. He has a list of 520 employed women, 20% of whom are members of a local business and 80% of whom are members of a professional club. He decides to randomly select 25 women members of the local business and 75 women members of the professional club to form a sample of 100 employed women.

(a) What type of sampling will he use to obtain the 100 employed women?

(b) Mrs. Smith is a member of the professional club. What is the chance that Mrs. Smith will be in the sample?

(c) Suppose that 15 of the 25 selected women of the local business reported favoring government funding for day care and 60 of the 75 selected women of the professional club also favor it. Give the overall estimate of the proportion of employed women that favor government funding for day care.

(d) The proportion obtained in (c) is a (select one)

    (i)   statistic.

    (ii) parameter.

    (iii) population.

    (iv) sample.

2.33 In an effort to survey the attitudes of former university graduate students, the registrar randomly selects 100 names from each of the four past graduating classes. There were 6050 students in all in the four past graduating classes.

(a) What is the population of interest?

(b) What sampling technique is being described?

(c) *True or false:* The chance that *any* past university graduate student is selected is 400/6050.

2.34 Suppose that a stratified random sample is to be used to survey alcohol consumption among teenagers in a certain small county. Three strata are used: 8th-, 10th-, and 12th-grade students. These have 450, 500, and 600 students, respectively. It is decided to take a simple random sample of 10% of the 8th-grade students, 5% of the 10th-grade students, and 8% of the 12th-grade students.

(a) Based on this sampling design, how many students will be sampled from each of the strata, and hence what is the total number of students in the sample?

(b) Use your calculator with a seed value = 26 or the random number table (starting at Row 3, Column 1) to select the first five students from the 10th-grade stratum for the stratified random sample.

2.35 The Gallup Poll gathered information about the 2000 election by telephone. The Gallup Poll divided each of the four time zones into three types of areas, according to the population density (heavy, medium, light). Thus, there were $4 \times 3 = 12$ sections. For example, one section consisted of lightly populated areas in the Eastern time zone, another consisted of heavily populated areas in the Central time zone. Within each section, the Gallup Poll took a simple random sample of 100 telephone numbers (excluding businesses).

(a) What type of sampling technique was used to obtain the 1200 selected phone numbers?

(b) Suppose that for each section, the phone numbers were sampled from telephone directories. Rich people and poor people are more likely to have unlisted numbers and therefore would not have a chance to be included in the sample. This is an example of _____ bias.

2.36 A company employs 800 people, 75% of whom are males and 25% are females. A management task force officer polls a stratified random sample of 90 males and 20 females.

(a) What is the chance that a particular female employee will be polled?

(b) Using your calculator with a seed value = 21 or the random number table (starting at Row 19, Column 1), list the first five females to be included in the sample.

(c) Suppose that the average response for the sampled females was 5, while the average response for the sampled males was 8. Give the overall estimate of the average response.

2.37 Suppose that you have four books on your shelf that you are planning to read in the near future. Two are fictional works, A and B, containing 212 and 379 pages, respectively. The other two are nonfiction works, C and D, with 350 and 575 pages, respectively. Suppose that you randomly select one of the two fiction books and also randomly select one of the two nonfiction books.

(a) What is the name of the sampling method described?

(b) What is the chance that *both* books A and B will be selected with this method?

(c) Give the name of two books that could be selected with this method.

(d) Give the name of another pair of books that could be selected with this method.

(e) What is the chance that the total number of pages in the two selected books will exceed 800 pages?

2.38 A study was conducted to estimate the average value of homes (in thousands of dollars) in a city. The city was divided up into three counties. There are 2000 homes in the city, with 50% of the homes in County I, 30% in County II, and 20% in County III. A list of the 2000 homes was constructed with those from County I listed first, followed by the homes in County II, and ending with the homes in County III. The homes on this list were then labeled from 1 to 2000. A random sample of 55 homes was selected from those in County I. A random sample of 15 homes was selected from those in County II. A random sample of 10 homes was selected from those in County III.

(a) What kind of sampling technique was used to select the homes that were included in the study?

(b) Use your calculator with a seed value = 75 or the random number table (starting at Row 8, Column 1) and give the labels (in order) of the first six homes selected from County I.

(c) The average value (in thousands of dollars) of the selected homes in Counties I, II, and III were 175, 200, and 195, respectively. Give the overall estimate of the average value of homes in the city based on this sample information. Include the units for your estimate.

2.39 An agency has prepared tax forms for 5000 clients. The agency's president would like to sample some of these clients and recheck their tax forms for accuracy. Of the 5000 clients, 1000 have a high gross income over $100,000; 2500 have a moderate gross income of $30,000 to $100,000; and the remaining 1500 have a low gross income of under $30,000. The president's plan is to sample 2% of the clients in the high category, 5% of the clients in the moderate category, and 3% of the clients in the low category. For each category, the clients will be numbered from 1 to the total number of clients in that category.

(a) What type of sampling technique is described for obtaining the selected clients?

(b) How many clients will be sampled from the high category? moderate category? low category?

(c) Using your calculator with a seed value of 98 (or starting with Row 8, Column 21 if using the random number table), give the labels for the first five clients selected from the moderate category.

2.40 A study was conducted to estimate the average size of homes (in square feet) in a community. The community was divided up into two sections. There are 700 homes in the community, with 60% of the homes in Section I and 40% in Section II. A list of the 700 homes was constructed with those from Section I listed first, followed by the homes in Section II. The homes on this list were then labeled from 1 to 700. A simple random sample of 30 homes was selected from those in Section I. A simple random sample of 10 homes was selected from those in Section II.

(a) What kind of sampling technique was used to select the 40 homes that were included in the study?

(b) Using your calculator with a seed value = 75 or the random number table (starting at Row 13, Column 1), give the labels (in order) of the first five homes selected from Section I.

(c) The average size (in square feet) of the selected homes in Section I was 2100 and in Section II was 2600. Give the overall estimate of the average size of homes in the community based on this sample information. Show all work and include the *units* for your estimate.

2.41 A class has 15 students with ID numbers from 1 to 15. You want to select a 1-in-4 systematic sample from all the students in the class with labels given by the ID numbers.

(a) Use your calculator with a seed value = 13 or the random number table (starting at Row 5, Column 1) to obtain the label of the first student to be included in your sample.

(b) What is the total number of students in your sample?

2.42 A class has 17 students with ID numbers from 1 to 17. You want to select a 1-in-3 systematic sample from all the students in the class with labels given by the ID numbers.

(a) Use your calculator with a seed value = 11 or the random number table (starting at Row 89, Column 1) to obtain the label of the first student to be included in your sample.

(b) What is the total number of students in your sample?

2.43 A school principal wants to take a survey of the 1500 students in the school. Each student has an identification number on the administrative computer. Identify the following methods as simple random sampling, convenience sampling, stratified random sampling, systematic sampling, or cluster sampling.

(a) The school groups students into 50 divisions for administrative purposes. Each of the 50 divisions contains 30 students. Students were assigned to divisions by listing them alphabetically by class rank. The division list started with freshmen listed alphabetically, then sophomores, juniors, and seniors. The school selected a random number between 1 and 30, obtaining number 23. Then they selected the twenty-third student from each division list.

What type of sampling method is being described?

Can you say exactly how many of the students selected will be freshmen?

If yes, give the number.

If no, explain why not.

(b) The school groups the students by class rank: freshmen, sophomores, juniors, and seniors. The computer's random number generator is used to select 25 identification numbers from each of the four class-rank lists. Interview the students corresponding to those selected identification numbers.

What type of sampling method is being described?

Can you say exactly how many of the students selected will be freshmen?

If yes, give the number.

If no, explain why not.

2.44 Suppose that a file contains 500 addresses, of which 5 must be selected. Systematic random sampling is to be used to select 5 addresses from the 500 addresses. The 500 addresses are already ordered and are identified as Address #1 through Address #500. You are to take a 1-in-100 systematic sample to select 5 addresses from the ordered list of 500 addresses. Use your calculator with a seed value $= 48$ and $N = 100$, or the random number table (starting at Row 55, Column 1).

(a) Describe your labeling procedure and list the addresses selected.

(b) What is the chance that any specific address will be chosen? Explain your answer.

2.45 The Michigan Department of State Police keeps track of the number of points received (for various traffic violations) by Michigan drivers. A county police department is interested in examining the relationship between the number of points received and the risk of being in a traffic crash. The following chart shows the number of points received by the drivers who reside in that county for the past year:

Number of Points Category	Number of Drivers
High (8 or more)	500
Mid (4–7 points)	1200
Low (0–3 points)	18,000

(a) The department will take a simple random sample of 2% of the drivers in the *High* category, 3% of the drivers in the *Mid* category, and 5% of the drivers in the *Low* category. The sample will consist of all drivers selected from each category.

  (i)   How many drivers will be included in the sample?

  (ii) What sampling method will be used to obtain the sample?

  (iii) Use your calculator with a seed value $= 87$ or the random number table (starting at Row 60, Column 1) to give the labels (in order) for the first five *High* category drivers to be included in the sample.

(b) An alternative sampling plan is proposed that would label all 19,700 drivers starting with those in the *High* category, continuing next with those in the *Mid* category, and ending with those in the *Low* category. Then, a 1-in-20 systematic sample would be obtained.

  (i) Use your calculator with a seed value $= 9$ or the random number table (starting at Row 50, Column 1) and give the labels (in order) for the first five *High* category drivers to be included in the sample.

  (ii) How many *High* category drivers will be included in the systematic sample?

2.46 There are 975 faculty employed at a college. The dean of the college wishes to conduct a survey to obtain faculty input on a new teaching method being proposed. The dean constructs a list of the 60 departments in the college. She plans to take a simple random sample of six departments. Her sample will consist of all faculty in those six departments. Each faculty member belongs to only one department.

(a) State what type of sampling method is being described.

(b) The 60 departments on the dean's list are identified as Department #1 through Department #60. You are to take a simple random sample of six departments from the list of 60 departments. Use your calculator with a seed value = 79 or the random number table (starting at Row 60, Column 1) to select your simple random sample of six departments. Be sure to describe your labeling procedure.

(c) Can you determine the chance that any specific professor will be selected?

    If yes, what is that chance?

    If no, explain why not.

2.47 *True or false.* (Explain your answer.) In cluster sampling, the chance that any unit is selected depends on the size of the clusters.

REAL DATA

2.48 Referring to the well-known President Clinton–Monica Lewinsky scandal, a sub-headline in the January 26, 1998, issue of the *Ann Arbor News* read, "Polls show asking former intern to lie would be cause for leaving." An ABC News/*Washington Post* poll stated that 59% of those surveyed said President Clinton should resign if he lied. This result was based on a poll of 1537 adults surveyed and had a margin of error of plus or minus 3%.

(a) How many of the adults surveyed feel President Clinton should resign if he lied?

(b) Suppose that a listing of the residential homes in the Washington, D.C. area was available. A simple random sample of homes was obtained and all of the adults residing at the selected homes were surveyed. What type of sampling technique is described for obtaining such a sample of adults?

(c) Adults residing in apartment complexes would not have a chance to be included in the survey. This is a type of bias called (select one)

(i) response bias.

(ii) nonresponse bias.

(iii) selection bias.

2.49 From the population consisting of all undergraduate classes taught at a particular university in a specific term, one class was selected at random. All of the students in the selected class form the sample.

(a) What kind of sampling method was used to select this class?

(b) Does every undergraduate student at the university have an equal chance of being selected? Explain.

(c) Suppose that we are interested in estimating the average height for students at this university. Most students on athletic scholarships at this university take courses in the school of education, and the randomly selected class described in this problem turned out to be a class from the school of education. Since all of the students in this class are included in the sample, the average height for this sample will probably overestimate the average height for the population. Is the procedure biased or is this a case of a poor design together with bad luck on the selection? Explain.

2.50 The College of Science at a certain university consists of five academic departments: (1) Physics, (2) Chemistry, (3) Biology, (4) Mathematics, and (5) Geology. There are

50, 70, 20, 10, and 40 faculty members in the departments of Physics, Chemistry, Biology, Mathematics, and Geology, respectively. It was decided to select a cluster sample of two departments.

(a) Using the above labels and your calculator with a seed value = 78 or the random number table (starting at Row 24, Column 1), which two departments are selected?

(b) What is the size of your cluster sample?

2.51 A school has exactly four dorms, one for freshmen, one for sophomores, one for juniors, and one for seniors. In each dorm, the rooms that are numbered 100 to 199 are for women, and 200 to 299 are for men. Every room is occupied by at least one student and no more than three students. A simple random sample of three rooms is taken in each dorm from its list of rooms. All the students in the selected rooms are interviewed.

(a) What type of sampling of students is performed in each dorm?

(b) What could be the minimum sample size of students from the four dorms? Explain.

(c) What could be the maximum sample size of students from the four dorms? Explain.

(d) How many women will be in the sample from the four dorms? Explain.

(e) Could all of the selected students be freshmen? Explain.

2.52 Consider the following study designed to determine which is the better of two submitted advertisements for an upcoming movie: The researcher wishes to get feedback from adults as to which ad is better. He randomly selects 20 city blocks from a large city containing 7820 blocks. An interviewer is sent to speak with all of the adults who live on each of these 20 selected city blocks. Each adult is asked to say which advertisement he or she prefers.

(a) What type of sampling is being used to select the adults interviewed in this study?

(b) Can you determine how many adults will be included in the sample? If yes, how many? If no, explain why not.

2.53 A college president wants to take a survey of 2250 students in the school. Each student has an identification number on the administrative computer. For each of the following descriptions, give the name of the sampling technique and the total sample size (if the sample size cannot be determined, write "cannot be determined"):

(a) Students are listed alphabetically and this list was broken into 50 sections of 45 students each. The administration selected a random number between 1 and 45, obtaining 33. Then they selected the 33rd student from each of the 50 sections.

(b) The school constructs a list of all of the possible undergraduate majors and takes a simple random sample of six majors, interviewing all the students in those six selected majors.

2.54 A maker of disposable diapers hired an investigator to conduct a survey in a big city about whether or not disposable diapers should be banned. It was desired to obtain a random sample of heads of households. In this big city, households are grouped geographically into 20 city blocks. Five city blocks were selected at random and all of the households in the five selected city blocks were included in the survey.

(a) What kind of sampling technique was used to select the households included in the survey?

(b) Suppose that each selected city block has 50 households. A list of all selected household was obtained. A 1-in-5 systematic sampling technique was used to obtain a sample of households. What is the chance that the household with label "9" was selected to be in the survey?

(c) The survey found that 80% of those sampled were *not* opposed to disposable diapers. Here was the actual wording of the question: "It is estimated that disposable diapers account for less than 2% of the trash in today's landfills. In contrast, beverage containers and other yard waste account for about 21% of the trash in landfills. Given this, in your opinion, would it be fair to ban disposable diapers?" What type of bias is portrayed in this question?

2.55 Determine whether each of the following statements is true or false (a true statement is *always* true):

(a) In cluster sampling, if all clusters contain the same number of units, then cluster sampling is the same as simple random sampling.

(b) In 1-in-3 systematic sampling, you can obtain a sample with three consecutive units of the population.

(c) If a population is divided into groups for which the units within a group are relatively homogeneous (i.e., similar or alike), then it is preferable to take a stratified sample (using these groups as the strata) rather than a cluster sample (using these groups as the clusters) from the population.

2.56 Suppose that the final area chosen in a multistage sampling design contains 200 addresses, of which 5 must be selected. The sample of 5 from the list of 200 addresses is chosen as follows:

Step 1: Choose 1 of the first 40 addresses in the list at random.

Step 2: The sample consists of the address chosen in Step 1 and the addresses that are 40, 80, 120, and 160 positions down the list from that first address chosen. For example, if the 10th address was chosen in Step 1, then the sample consists of the 10th, 50th, 90th, 130th, and 170th addresses on the list.

(a) What type of sampling method is described in Steps 1 and 2 (that is, in the last stage of the sampling design)?

(b) What is the chance that any specific address will be chosen among the 200 addresses in the final area chosen? Explain your answer.

2.57 A librarian wants to take data on the number of times fiction books are checked out during the year. The library has 1200 volumes of fiction, of which 400 are murder mysteries, 200 are science fiction, and the rest are "other." She decides to take a sample of 120 books.

(a) Suppose that she takes the next 120 books that are returned and calculates the average number of times these books were checked out in the last year.

    (i) What kind of sampling is this?

    (ii) Is there bias in this sampling scheme? What kind?

    (iii) Do you expect the average calculated for this sample to be higher or lower than the true average? Explain.

(b) Suppose that she randomly selects one digit from the list of 0, 1, 2, 3, 4, 5, 6, 7, 8, 9, and then chooses all books whose call number ends in that digit. What kind of sampling is this?

(c) Suppose that she takes a simple random sample of 40 books from each category.

    (i) What kind of sampling is this?

(ii) Suppose that of the sampled books selected in this way, the murder mysteries had been checked out an average of 20 times in the last year; the science fiction books had been checked out an average of 15 times; and the "other" books, an average of 10 times. Calculate an estimate of the average number of times checked out during the year for *all* fiction books at the library.

(d) Describe an example of a multistage sampling scheme the librarian might conduct.

2.58 The first sampling method introduced in this chapter was a census. The U.S. Bureau of the Census (the generator of the official statistics) is one government statistical agency that has a Web site, with an address of http://www.census.gov. The home page refers to this agency as "helping you make informed decisions." One piece of information that you can obtain is the current U.S. population count. For example, according the U.S. Bureau of the Census, the resident population of the United States, projected to 6/25/01 at 4:18:38 P.M. EST, is 284,513,279.

(a) Give the current projected resident population of the United States. Be sure to include the date, time, and even the second.

(b) Another link from this home page is called "State & County Quick Facts." Within this site, you can select a state and click on "GO!" to view a variety of facts about the people, businesses, and geography of your desired state. Check out some of the current information for your home state and then your home city. If this link is not available, check out the current feature(s) on the census Web page. Write a brief summary of your findings.

> **Note:** The parent agency for the Census Bureau is the U.S. Department of Commerce, with a Web site address of http://www.doc.gov. Should the census site change after the publishing of this text, a search for the agency's name should produce the current Web site.

2.59 A consumer protection agency has received many complaints that the sodium content in a six-ounce can of ABC Tuna is greater than the 250 mg stated on the can label. In response to these complaints, the protection agency will gather data for testing the following hypotheses:

H_0: The average sodium content of six-ounce cans produced by ABC Tuna is 250 mg.

H_1: The average sodium content of six-ounce cans produced by ABC Tuna is greater than 250 mg.

ABC Tuna is produced at one main plant and shipments are made to stores in five regions within Michigan. The protection agency decides to use a multistage sampling design. In the first stage, they randomly select one region from among the five regions: 1 = UP, 2 = Northwest, 3 = Southwest, 4 = Northeast, and 5 = Southeast.

(a) Use your calculator with a seed value = 116 or the random number table (starting at Row 55, Column 1) to determine the one selected region.

(b) In the second stage, the agency groups the stores in the selected region that sell the tuna into one of three types (big chain, local chain, mom & pop) and takes a simple random sample of two stores from each of these three types. What sampling technique is used in this second stage to obtain the selected stores?

(c) A total of 125 cans of ABC Tuna were recently shipped to the six stores selected in the second stage. The following table shows the labels of the cans sent to each of the six stores:

Type	Big Chain	Big Chain	Local Chain	Local Chain	Mom & Pop	Mom & Pop
Store	#1	#2	#3	#4	#5	#6
Cans	1–30	31–70	71–90	91–110	111–115	116–125

Of the first 10 cans (in this list of 125 labeled cans), one can will be selected at random. Then, starting with that first selected can, every 10th can in the list thereafter will be included in the final sample.

(i) What type of sampling technique is used to obtain the selected cans included in the final sample?

(ii) What is the chance that can #31 will be included in the final sample?

(iii) Use your calculator with a seed value = 182 or the random number table (starting at Row 53, Column 1) to give the labels for the first five selected cans.

(iv) How many cans will be included in the final sample?

(d) The data from the sampled cans were analyzed and the results were not statistically significant at the 0.10 level.

(i) Which hypothesis was accepted?

(ii) Give two values that could possibly have been the *p*-value for this test.

(iii) Could a mistake have been made? If yes, which type?

Generating Random Integers

To be able to check your results in what follows you will need to specify a starting point by setting a seed value. The factory-set seed value for TI graphing calculators is 0. Suppose that you wish to generate a sequence of random integers between 1 and 50, using a seed value of 33. Using a TI graphing calculator to generate a list of random numbers between 1 and 50, starting with the seed value of 33, the keystroke steps are as follows:

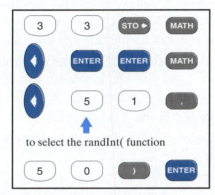

to select the randInt(function

Continue to push the ENTER button and your TI screen should look like this:

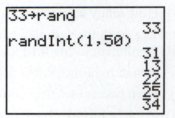

With the seed value of 33 and a population size of 50, the first value you get is 31, which corresponds to selecting the unit with the label of 31. By continuing to press the ENTER key four more times, the resulting simple random sample would consist of the units with labels 31, 13, 22, 25, and 34. You can request that more than one label be placed on a line. To have, say, two labels displayed on each line, you would use the command **randInt(1, 50, 2)**. The third number in this command determines how many generated values to place on a line.

For further details of the TI number-generating functions, see your TI Graphing Calculator Guidebook under MATH operations.

## Chapter 3

OBSERVATIONAL STUDIES AND EXPERIMENTS

3.1 Introduction

We have a *theory* that we wish to test. We *collect data* to help us check the theory. The data may have been collected by others or you may need to produce your own data. In either case, the quality of the decision you make will depend on the quality of the data you obtain. In Chapter 2, we studied various sampling methods—that is, methods for selecting a sample to gain information about the whole by examining only a part.

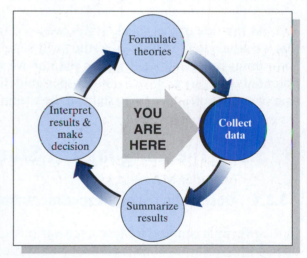

Studies are conducted to test various hypotheses about a population. In this chapter, we continue our discussion about producing data by presenting various types of studies and learning what questions to ask to help us intelligently decide which study results are worthy of our attention.

We will begin by addressing the question "Why study studies?" We will turn to a general discussion of the common types of studies, the basic terminology used in studies, and various problems that can arise in any study. We will then revisit each type of study in more detail. We will end with an overview of how you can be a more educated consumer about the data that you encounter every day. The goal of this chapter is to demonstrate that not all data are created equal—that it is important when analyzing data or making inference from the data analyses of others, to ask questions. The knowledge from your studies of statistics will help you to understand the methods and know what questions to ask.

3.2 Why Study Studies?

Pick up any newspaper and you will find an article containing conclusions based on some study. Don't always believe them. What you should believe depends on how the data were collected, measured, and summarized. Read the following excerpt from an article reported by the *Newhouse News Service*, May 1995:

Food for thought on what not to eat: Study the weight of the studies—

SOURCE: *Newhouse News Service*, May 17, 1995.

REAL
DATA

In the name of good health, you swear that butter will never find its way to your heart again. Beta carotene supplements are out of your life for good, too. Same goes for eggs. And no more dieting. But beer and wine can stay since a drink a day might keep the doctor away. Smart choices or rash decisions?

It's hard to tell. Just when researchers round a medical corner, another study emerges to challenge scientific dogma: Ban butter, they say; margarine is easier on the arteries. Wait, margarine is harmful too; stick with butter. Beta carotene, a vitamin A derivative, fights cancer. No, the antioxidant may increase your risk of cancer. Yo-yo dieting is bad for the body. Drastic weight loss and gain are not harmful. Actually, it may not matter. Your size may be predetermined by what's encoded on your DNA. Eggs aren't good for you. Eggs are OK for the cholesterol conscious. The saturated fat in the yolks poses a bigger problem for the arteries and is probably more of a problem than dietary cholesterol.

Surveys show that Americans are more mixed up than ever about what is good and bad for their health. Some people are so overloaded and frustrated by the conflicting data that they have tuned out. What's a health-conscious person to do?

As the title of this article states, we need to learn how to weigh the studies. The weight that we attach to the results of a study will depend on how well the information was obtained and summarized. In the rest of this chapter, we will learn about the common types of studies, potential difficulties of each, and proper guidelines for executing them. To make our discussion of studies clear, we first establish the common language that is used in statistical studies.

3.3 The Language of Studies

3.3.1 Observation versus Experimentation

In the previous chapter, we discussed various sampling techniques. In this chapter, we focus on how the variables of interest are actually measured. We obtain data by either *observation* or *experimentation*. For comparing the income of males versus the income of females, it would be sufficient to observe both the income and the gender of those individuals in our sample. If we simply observe the variables of interest for our sample, without controlling the process that is being observed, we are conducting an **observational study**. Observation is *passive*. Surveys are observational studies that measure various characteristics of the people in the sample.

Suppose that we are interested in whether or not a high-fiber diet reduces heart-attack risk in men. We could simply observe whether or not men ate a high-fiber diet and then whether or not they had a heart attack. It could be, however, that the men who ate a high-fiber diet were more health conscious overall—they may also have exercised more, did not smoke or smoked less, or consumed a diet lower in fat. To be able to establish support for a causal relationship between high-fiber diet and heart-attack risk, we would need to conduct an **experiment** by assigning the male subjects, at random, to one of two groups. Subjects who smoked, for example, would have the same chance to be assigned to either group. One group would eat a high-fiber diet, and the other group would be fed a standard diet. In an experiment, some treatment, such as a high-fiber diet versus a standard diet, is actively imposed on the units or subjects to observe the outcome. Experimentation is *active*.

Definitions: The **units** are the objects upon which measurements are made or observed. If the units are people, they are often referred to as **subjects.**

In a **designed experiment,** the researcher *actively* imposes some treatment on the units or subjects in order to observe the responses.

In an **observational study,** the researcher simply observes the subjects or units and records variables of interest. The researcher does not attempt to manipulate or influence the responses.

THINK ABOUT IT

In an experiment, the researcher assigns treatments to the subjects. In an observational study, the researcher simply observes what the subjects do naturally. Think about a situation in which an experiment would not be possible or feasible. Explain your reasons.

3.3.2   Relating Two Variables

Many studies, whether an observational study or a designed experiment, are conducted to try to detect relationships between two or more variables. In particular, they attempt to show that something *causes* something else. In a study to determine whether or not a high-fiber diet reduces heart-attack risk in men, whether or not the male has a heart attack is one variable and whether or not he ate a high-fiber diet is the other. A variable thought *to cause* something to occur is called the **explanatory variable** or **factor,** generally denoted by the variable X. The variable that is thought to be *caused by* the explanatory variable is called the **response variable** or **outcome variable**, generally denoted by the variable Y. Sometimes the variable Y is called the dependent variable and the variable X is called the independent or predictor variable.

The values of the explanatory variable are called the **levels**. When a study has just one explanatory variable, the levels of the explanatory variable are called the **treatments**. If there is more than one explanatory variable, a treatment is a combination of the levels of the explanatory variables.

Definitions: A **response variable (Y)** measures an outcome of the study. It is a variable that is thought to depend in some way on the explanatory variable.

An **explanatory variable or factor (X)** is a variable that is thought to explain or cause the observed outcomes. It is a variable that is thought to explain the changes in the response variable.

The possible values of the explanatory variable are called the **levels** of that explanatory variable. A **treatment** is a specific combination of the levels of the explanatory variables.

Let's restate the question of whether or not a high-fiber diet reduces heart attacks in men, using the language of studies. The response variable is *heart-attack status*, recorded as either the subject did have a heart attack or did not have a heart attack. The explanatory variable is *type of diet*. In this case, the levels of the explanatory variable are *high-fiber diet* and *standard diet*. If a second explanatory variable in the fiber study is amount of exercise, with three levels (High, Moderate, and Low) there would be a total of six treatments, as shown in the following table:

		Factor 2—Amount of Exercise		
		High	*Moderate*	*Low*
Factor 1—Type of Diet	*High Fiber*	Treatment 1	Treatment 2	Treatment 3
	Standard	Treatment 4	Treatment 5	Treatment 6

How can we determine which variable is the explanatory variable and which is the response? Suppose that you have two variables, $V1$ and $V2$, and you are trying to assess which is the response variable and which is the explanatory variable. Ask yourself

1. Are you trying to show that the values of $V1$ explain why you see certain values of $V2$? or
2. Are you trying to show that the values of V2 explain why you see certain values of $V1$?

By assessing which direction, $V1 \rightarrow V2$ or $V2 \rightarrow V1$, makes the most sense, you should be able to distinguish between the explanatory variable and the response variable. Let's look at two examples.

❖ EXAMPLE 3.1   Explanatory Variable and Response Variable

Baseball and the Divorce Rate

Do couples who go to baseball games have better relationships than those who don't? The Center for Marital and Family Studies at the University of Denver observed that in cities with major league baseball teams, the divorce rate is 23% lower than in cities without baseball teams. A spokesperson for the center stated, "I don't want to say having a team lowers the divorce rate, but it might help."

This is an *observational study*. The statement "in cities with major league baseball teams, the divorce rate is 23% lower than in cities without baseball teams," implies that *baseball-team status* is the explanatory variable, and *divorce rate* of the city is the response variable. The explanatory variable has two levels: The city does have a major league baseball team or the city does not have a major league baseball team. Thus, there are two treatments.

Production Time

A researcher wishes to study the effect of room temperature on the performance of a particular task. Three temperature levels are chosen for the study. Assuming that all other conditions remain the same, the number of times the subject performs the task in 15 minutes is recorded for each temperature level.

This is an *experiment*. The researcher actively imposes one of the three levels of the explanatory variable, room temperature, on a subject and measures the response, the number of times the subject performs the task in 15 minutes. There are some potential problems with this design. There might be a *learning effect* that will make the results difficult to interpret. A subject might perform the task faster on the second or third trial, which may not be due just to the different temperature levels: Having done it once or twice, the subject may have learned how to do it faster. Another problem is a *fatigue effect*. The task may require a lot of energy, and after performing it once or twice, a subject may get tired and take longer on subsequent trials. ❖

❖ EXAMPLE 3.2   Vaccine Battles Melanoma _____

REAL DATA

This article states, "A vaccine treatment for malignant melanoma—the most deadly skin cancer—greatly improves survival with no serious side effects." The vaccine was administered to 355 patients whose melanoma had spread. Measures of the effect of this vaccine were the percentage of patients who survived five years since treatment and the percentage of treated patients who had tumor regression or whose tumors disappeared (SOURCE: *USA Today,* March 29, 1993).

Vaccine battles melanoma

By Tim Friend
USA TODAY

SAN DIEGO — A vaccine-style treatment for malignant melanoma – the most deadly skin cancer – greatly improves survival with no serious side effects, researchers reported Sunday.

Results suggest it could be a major new treatment, says Dr. Donald Morton, John Wayne Cancer Institute in Santa Monica, Calif.

Morton says the vaccine, made from tumor cells, stimulates the immune system to attack the melanoma.

Researchers gave the vaccine to 355 patients whose melanoma had spread. Findings presented at the American Cancer Society's Science Writers Seminar show:

▶ 27% survived five years compared with 10% getting other treatments.

▶ 22% in the latest stage had tumor regression; tumors in three disappeared.

Dr. Laurence Meyer, University of Utah, thinks the vaccine may prevent melanoma in high-risk people or prevent recurrences.

Is this study an observational study or an experiment?

This article describes an experiment conducted to see if a vaccine is effective in the treatment of melanoma. The vaccine treatment was actively imposed on all of the subjects in this experiment.

What is the response variable(s) for this study?

Measures of the effect of this vaccine (that is, response variables) were the patient five-year survival status, tumor regression status, and tumor disappearance status.

What is the explanatory variable(s) for this study?

The explanatory variable is the melanoma vaccine.

How many treatments does this study have?

According to the article, the vaccine was given to (all) 355 patients. However, the rate of 27% was compared to rates for other treatments from previous studies. In this study, there was just one treatment: All subjects received the melanoma vaccine. ❖

Let's do it! 3.1 Explanatory Variable versus Response Variable

For each set of variables listed, identify which variable should be the response variable and which should be the explanatory variable.

(a) *V*1: The amount of money earned for a part-time job
   *V*2: The number of hours worked

(b) *V*1: The weight of a package
   *V*2: The first-class postage rate at the post office

(c) *V*1: The salary of a high school teacher
   *V*2: The number of years of teaching experience

Let's do it! 3.2 Food Myths

In a recent article entitled "Shed These Food Myths: Butter, Eggs, Starches Aren't Villains; Vitamins, Milk, Diet Pop Aren't Heroes," the following research study was presented:

> Research done at the Harvard School of Public Health showed that regular soda drinkers may have a tendency toward the weak and brittle bones associated with osteoporosis. The researchers surveyed 2622 women who were active athletes in college and classified them as to whether they *regularly* drank soft drinks or *rarely* drank soft drinks. The proportion of women in each group who suffered from bone fractures was determined. The findings: those who regularly drank soft drinks were twice as likely to suffer from bone fractures as those who rarely drank soft drinks. SOURCE: *Health and Fitness News Service*, August 21, 1996.

(a) Was this an observational study or an experiment? Explain.

(b) What was the population under study?

(c) What was the explanatory variable?

(d) What was the response variable?

(e) Suppose that the 2622 selected women were obtained as follows: From a list of all women's sports events available at Harvard (e.g., tennis, basketball, etc.), three events were selected at random. All alumni women athletes who participated in these three events were contacted for the study. What type of sampling technique is described?

3.3.3 A Problem Called Confounding

Any study has the potential to be wrongly interpreted because of **confounding variables.** Confounding variables are not of primary interest and may not even be measured, but are associated with the response of interest. For example, if we notice that men who eat a high-fiber diet have a lower incidence of heart attacks than men whose diet is low in fiber, it could be because men who eat a high-fiber diet also exercise more. In this case, exercise would be confounded with the high-fiber diet in how the diet influences the person's heart-attack risk.

In the debate about whether or not smoking is related to a higher risk of lung cancer, arguments were made that alcohol consumption and lung cancer incidence were associated. Since we might also expect that smoking and alcohol consumption are related, the observed increase in lung cancer among smokers might only reflect the association of lung cancer and alcohol consumption.

> *Definition* A **confounding variable** is a variable whose effect on the response variable cannot be separated from the effect of the explanatory variable on the response variable.

Confounding is generally more of a problem in observational studies than in experiments. Experiments allow the researcher to control for confounding variables. However, we cannot always perform an experiment. It may be impossible or unethical to assign subjects to receive a specific treatment, such as to smoke or to not smoke. It may be that certain explanatory variables are inherent traits or behaviors that cannot be assigned to subjects, such as having attached or unattached earlobes or being left versus right handed.

❖ **EXAMPLE 3.3** **Treating Multiple Sclerosis with Protein** _____

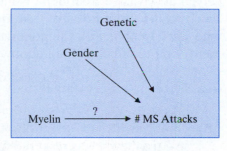

The first signs of promise in treating multiple sclerosis, MS, with simple oral doses of a common protein were reported back in 1993 (SOURCE: *Newsday*, February 26, 1993).

REAL DATA

The study involved 30 MS patients: Half were assigned to a treatment group that received daily doses of the protein myelin, and the other half received a *dummy treatment* that looked like the actual protein doses. Among the 15 persons given the myelin daily, six had no MS attacks during the year-long trial. Among those getting the dummy treatment, three had no MS attacks. The explanatory variable is protein myelin, with two levels—a daily dose or a dummy dose. The response variable was the number of multiple sclerosis attacks during the year-long trial.

Unfortunately, the results were unclear because men responded better than women, and it was observed that patients with a certain immune system genetic marker fared worse. The two groups did not contain equal numbers of men and women, nor were the groups balanced with respect to the presence or absence of this genetic marker. As a result, it was hard to tell whether the treatment itself, gender, genetic inheritance, or some combination of these helped the myelin-treated patients.

Gender and genetic inheritance are confounding variables. The effect of these variables could not be separated from the effect of the myelin treatment itself on the incidence of MS attacks. ❖

❖ EXAMPLE 3.4   Does Insurance Cause Whiplash?

In Lithuania, rear-end collisions happen much as they do in the rest of the world. Cars crash, bumpers crumple, and tempers flare. But unlike drivers in other nations, Lithuanians do not seem to suffer the long-term complaints of headaches and lingering neck pains, known as chronic whiplash or whiplash symptoms. They also do not necessarily carry personal-injury insurance, they have most medical bills paid by the government, and they are not in the habit of suing one another.

REAL DATA

Without disclosing the purpose of the study, the researchers gave health questionnaires to 202 drivers in Lithuania whose cars had been struck from behind one to three years earlier. The drivers were questioned about symptoms, and their answers were then compared with the answers of a control group made up of the same number of people, of similar ages and from the same town, who had not been in a car accident. The study found no difference between the groups. Thirty-five percent of the accident victims reported neck pain, but so did 33% of the controls. Similarly, 53% who had been in accidents had headaches, but so did 50% of the controls. The researchers concluded, "No one in the study group had disabling or persistent symptoms as a result of the car accident" (SOURCE: *The New York Times,* May 7, 1996).

The explanatory variable is car accident status, with two levels—whether or not the Lithuanian driver was in a car accident in which the car was struck from behind within a particular time period. The response variables were the various questions about symptoms.

A director of the head-and-facial-pain-management program at New York University Medical Center questioned the methods used in this study. Even though the people in the control group had not been in car accidents, they might have had other injuries. Falling down the stairs or slipping on the ice can set off a whiplash injury, and it can be longstanding. For controls, you should look at people without any injuries, even in childhood. The variable "previous injuries" could be confounded with the effect of car accidents on the prevalence of whiplash syndrome.   ❖

❖ Let's do it! 3.3 Invasive versus Noninvasive Cancer Treatments

Researchers *examined the records* of a large number of cancer patients. Some patients received an invasive treatment, while others received a noninvasive treatment. Patients who received the more invasive treatment survived much longer, on the average, than patients who were treated with the noninvasive treatment. The study concludes that invasive treatment is more effective than noninvasive treatment.

(a) Is this an observational study or an experiment? Explain.

(b) Upon further investigation of the data in the records, it was discovered that the invasive treatment was reserved for relatively healthy patients. Patients who were too ill to tolerate such treatment received the noninvasive treatment. Comment now on the validity of the conclusion of the study.

❋ Exercises

3.1 For each of the following situations, identify the explanatory variable and the response variable:

(a) Subjects were given 0, 4, or 8 ounces of coffee and exam performance was measured.

(b) Subjects low in self-esteem received 0 or 2 hours of counseling per week and were measured on grade point average six months later.

(c) Students living on campus were assigned one or two roommates and were measured for physical health at the end of the semester.

(d) Reaction time on a driving test was measured for people told that they were either a good or a bad driver.

REAL
DATA

3.2 Read the following excerpts from the *New York Times* (December 17, 1996) article "Chance of a Heart Attack Increases for Those Who Suffer from Depression."

A report in the December 16, 1996 issue of *Circulation* reports on a study that finds that those who are subject to depression are four times more likely to have a heart attack than those who are not depressed.

The study involved 1551 people in the Baltimore area who took part in a study in 1980. In this study, interviewers asked questions like "In your lifetime, have you ever had two weeks or more in which you lost all interest and pleasure in things you usually cared about and enjoyed?" They were also asked questions about unusual changes in their appetite and in their sleep. Based on this information, subjects were classified according to their *depression status* and their *heart attack status* was recorded.

Of the 444 who were classified as having been depressed, 27, or 6.1%, had a heart attack in the 13 years since the 1980 study while, of the 1107 who had not suffered from depression, only 37, or 3.3%, had a heart attack. The four-fold advantage mentioned earlier was obtained by controlling for other known risk factors for heart attacks such as age, sex, marital status, and high blood pressure.

(a) Is this study observational or experimental?

(b) Give the response variable for this study.

(c) Give the explanatory variable for this study.

(d) Is the value of 6.1% a statistic or a parameter? Explain.

(e) The last sentence refers to "controlling for other known risk factors for heart attacks such as age, sex, marital status, and high blood pressure." What is the statistical name for these types of variables?

3.3 A study of 5000 female nurses was conducted as follows: They were asked about their frequency of egg consumption, and their blood cholesterol level was measured. The study found that women who eat more eggs have higher cholesterol levels.

(a) Is the study observational or experimental?

(b) What is the response variable?

(c) What is the explanatory variable?

(d) Name a possible confounding factor and explain how this might impact the conclusions from the study.

3.4 A study will be conducted at a hospital to assess the relationship between the principal source of payment for hospital discharges (private insurance, government program, other) and the payment status (received on time versus incomplete or received late). Of the 2000 hospital discharge records from the past two months, 1200 had private insurance as the principal source of payment, 600 had a government program, and the remaining 200 had some other principal source of payment. Some records of each type will be selected, and the payment status will be recorded.

(a) Give the response variable for this study.

(b) Give the explanatory variable for this study.

(c) List the levels of the explanatory variable.

(d) Is this study observational or experimental? Explain.

(e) The study planned to select a simple random sample of 120 private insurance records, a simple random sample of 60 government records, and a simple random sample of 20 other records. Of the 120 selected private insurance records, 80% of the payments were received on time, while only 60% of the selected government records and 50% of the selected other records had on-time payments.

  (i) The percentage of 60% is a (select one): parameter statistic. Explain.

  (ii) What type of sampling technique was used to obtain the 200 selected discharge records?

  (iii) Use your calculator (seed value = 42) or the random-number table (starting with Row 20, Column 1) to list the labels for the first five private insurance records selected for this study.

(f) What is the population under study?

(g) What is the sample?

3.5 Homelessness linked to childhood years

SOURCE: *The Washington Post*

REAL
DATA

A large study of homeless people in Los Angeles has found that, for many, bad experiences as children rather than drug use or other problems as adults predisposed them to homelessness in adulthood. The Rand Corp. study found that many had been separated from their parents in childhood or raised in poverty. Many homeless women reported family problems or sexual or physical abuse. The researchers found in most cases the homeless adults had serious personal problems as kids. Nearly half had lived apart from their parents and 25% had been in foster or institutional care.

(a) Is this an observational study or an experiment? Explain.

(b) What is the response variable(s) for this study?

(c) What is the explanatory variable(s) for this study?

(d) What is the population under study?

(e) What is the sample?

(f) Is the 25% figure a parameter or a statistic? Explain.

3.4 Understanding Observational Studies

Observational studies generally try to assess how some factor or explanatory variable causes some response. When the response is whether a person develops some disease, the factors are often called *risk factors,* since they are thought to lead to an increased risk of the response. In observational studies, the explanatory variable is not actively imposed by the researcher, but simply observed. For example, the explanatory variable might be whether or not a person smokes and the response might be whether the person dies of lung cancer.

3.4.1   Types of Observational Studies*

There are two primary types of observational studies—retrospective and prospective. Each provides a different type of information.

Retrospective study of past events

A **retrospective study** identifies a group of people with a response of interest and tries to ascertain potential risk factors in retrospect. The basic approach is to search through records, examine those that have the response of interest, and see what proportion of these people have the risk factor. For example, researchers could look through records of those who have died of lung cancer and determine what proportion of these people smoked. This information provides an estimate of the likelihood of the risk factor—having smoked—given that a person had a certain response—died of lung cancer. However, to address the issue of causality, we really need to learn about the likelihood of a person having a certain response—dying of lung cancer—given that he or she has the risk factor—smoked.

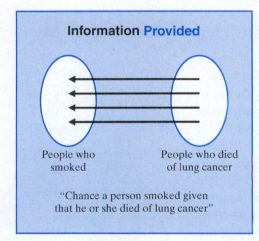

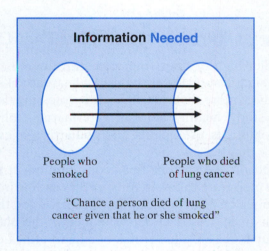

Although the likelihood of having smoked given that a person died of lung cancer could be pretty high, you would also have a high likelihood if you replaced the risk factor "smoke" with other events, like "have two arms" or "were alive at one point in time." A retrospective study works from the response, *Y,* to the potential explanatory variable, *X.* Many medical studies are of this type because they are relatively inexpensive to perform.

❖ EXAMPLE 3.5   The Study of Suicide

Suicide researchers are faced with a major problem—their subjects are no longer alive. How can investigators draw conclusions about the intentions, feelings, and circumstances of people who are no longer available to answer such questions?

   One strategy is to perform a retrospective analysis—a kind of psychological autopsy in which the researchers try to piece together data from the person's past. Retrospective data

---

*This section is not needed to read the rest of the book.

may be provided by relatives and friends who may remember past conversations and behaviors, from the notes taken in psychotherapy, or from the suicide notes that some victims leave behind. Such information would provide an estimate of the likelihood of having a particular behavior given that a person committed suicide. Unfortunately, this information is not always available, nor would it be necessarily valid. If researchers found that among suicide victims, 50% had experienced the death of a family member in the past three months, it does not mean that the death of a family member can lead to a suicide 50% of the time. ❖

Prospective studies

Prospective studies work from the potential explanatory variable to the response. The researcher might examine records of those who had smoked and records of those who had not smoked and learn what proportion of people in each group had the response of dying of lung cancer. The researcher could observe and follow a group of individuals who smoke and a group of individuals who do not smoke and record the number of individuals in each group who die of lung cancer. Such information provides an estimate of the likelihood of dying of lung cancer given that the person had smoked and the likelihood of dying of lung cancer given that the person did not smoke.

❖ **EXAMPLE 3.6** **High-fiber Diet Cuts Heart-attack Risk in Men**_____

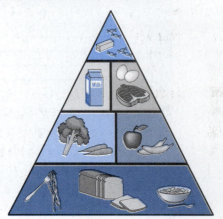

A diet high in fiber from cereal, fruit, and vegetables can significantly lower the risk of heart attacks in men, a study of more than 43,000 health professionals found. In the study, healthy dentists, veterinarians, pharmacists, optometrists, osteopaths, and podiatrists, ages 40 to 75, were divided into five groups, from highest to lowest, based on the current fiber consumption in their diet. The men were followed for six years ending in 1992. The number of heart attacks that occurred for the men in each group was recorded. The men in the highest-fiber group had a 36% lower risk of a heart attack during the study period than men in the lowest-fiber group.

So, the explanatory variable, X, is the amount of fiber in the diet, with five fiber levels for this study. The response or outcome variable, Y, is the number of heart attacks that occurred for the men in each group. This study examined individuals with the various levels of X, amount of fiber in their diet, and recorded the number of individuals in each group who experience the event Y, a heart attack. ❖

Definitions: Observational studies are often classified according to whether they are retrospective or prospective.

A **retrospective study** is a study of past events. Researchers identify subjects that have experienced certain responses and look back to see if the subjects also had various factors or explanatory variables. Subjects may be asked to recall past events. A retrospective study works from the responses to the potential explanatory variables.

A **prospective study** is a study of ongoing or future events. Researchers identify subjects which have various explanatory variables or factors and follow them into the future and record the responses. A prospective study works from the potential explanatory variables to the responses.

REAL
DATA

Are they right? Scientists say left-handers die sooner

SOURCE: *The Detroit News*, 1991.

Southpaws face more than their share of woe living in a world that is biased to the right. Now there's a new worry: Scientists say they have found that left-handers die nine years sooner than right-handers. "It is very scary," said Stanley Coren, a right-handed psychologist at the University of British Columbia in Vancouver.

Coren and fellow researchers examined randomly selected death records of 987 people in Southern California. The researchers said right-handers, about 94 percent of the group, died at an average age of 75. The left-handers died at an average age of 66.

Coren said the size of the group was statistically valid. He suspects left-handed people die sooner because of the way the world is built: to accommodate right-handers. The researchers found that left-handers were six times more likely to die from an accident and four times more likely to die while driving a vehicle.

Alan Searleman, a professor of psychology at St. Lawrence University in Canton, N.Y., and an expert on handedness, said the finding is "hard to believe," but the work might merit more analysis. "We are talking about a nine-year difference!" Searleman said.

(a) What type of observational study was conducted? Explain.

(b) How many left-handers were included in this study?

(c) Suppose that there may have been a disproportionate number of men—who are known to die earlier than women on average—in the left-handers group. In this case, gender would be an example of (select one)

(i) an extra variable

(ii) a control variable

(iii) a confounding variable

(iv) a response variable

Another issue is that years ago parents would try to force a child to be right-handed. It may be that some of the right-handers in the study would have actually been left-handers.

(d) Describe a better study to compare survival rates for left-handers versus right-handers.

3.4.2 Difficulties in Observational Studies

Observational studies have a disadvantage compared to experiments—the researcher can only observe, not control, the explanatory variable(s), nor can the researcher attempt to control for possible confounding variables. In observational studies, effort is usually made to identify all possible confounding variables and adjust for them in the analysis of the data. In the study of high-fiber diets and their effect on heart-attack risk for men, the reported results were said to have been "adjusted for other traits that could affect heart risk, such as smoking and fat consumption." In the whiplash study, the controls were selected to try to be as similar as possible to those in the accident-victim group.

One problem is that it is impossible to adjust for confounding variables that no one has considered. It can be argued that observational studies can never establish a causal relationship between two variables, such as smoking and lung cancer, and that causality can only be established by experimentation involving direct intervention. The science of epidemiology, which studies various aspects of public health, deals almost exclusively with observational studies, and, as a result, epidemiologists have identified a series of criteria for deciding that a causal relationship is very likely. The criteria for establishing causation in a nonexperimental setting given that a relationship between the explanatory variable X and the response Y has been established are as follows:

- The association between X and Y recurs in different circumstances; this reduces the chance that it is due to confounding.
- A plausible explanation is available showing how X could cause changes in Y.
- No equally plausible third factor exists that could cause changes in X and Y together.

Often an observational study is carried out and various questions or hypotheses are formulated based on what was observed. Then an experiment might be designed specifically to test these hypotheses. The results of that experiment may lead to new theories that are investigated with further experiments. For example, you observe that your friend drinks a glass of water before each meal. When asked why, she replies that it helps her to digest her food better. Is this valid for everybody? This is a possible research question or hypothesis to assess. An experiment is designed to assess if drinking a glass of water before each meal helps people better digest their food. Analysis of the experimental results showed that for women there was a significant improvement in digestion, while for men there was not a significant improvement. These new insights lead us to new hypotheses for further research—an evolving process.

In the study of high-fiber diets and their effect on heart-attack risk for men, the statement was made that the findings also most likely apply to women, who are now being studied. This is a new hypothesis being assessed. This study was based on men working in the health professions, such as dentists and veterinarians. One question to ask is whether or not these results can be extended to the general male population. Some observational studies use a convenience sample, which may not be representative of the population.

Of the two types of observational studies we have discussed, retrospective observational studies pose the greatest potential risks. A number of factors work against the researcher. First, he or she must work from the outcome or response to the explanatory variable using data that have been recorded in the past. When subjects are asked to recall various past events, they may not report accurate results. In general, these potential problems render the conclusions drawn from retrospective observational studies less reliable than those based on prospective studies.

 Exercises

3.6 **The Study of Suicide** In Example 3.5, the strategy of performing a retrospective analysis was discussed. Another strategy used is to study people who survive their suicide attempts and equate them with those who commit fatal suicides. Researchers often consider those who attempt suicide and those who commit suicide as more or less alike. Do you think this is valid? Give a potential problem with this substitution.

3.7 The elimination of the national 55-mph speed limit in 1995 provoked much debate about whether lowering highway speed limits reduces highway crash fatalities. The following questions focus on the association of speed limit with highway fatalities. The former national speed limit of 55 mph was imposed in the 1970s. A safety group gathered data on fatalities the year before and the year after the lower limit was imposed. The group compared fatality rates (deaths per one thousand people) for these two years and did a statistical test of the following hypotheses:

H_0: The incidence of a highway fatality is the same for different speed limits.

H_1: The incidence of a highway fatality increases with higher speed limits.

The data were significant at the 0.05 level.

(a) Which hypothesis did the data support?

(b) Was the *p*-value greater than 0.05 or less than or equal to 0.05?

(c) What type of study is being described? Select one of the following:

(i) Experiment.

(ii) Observational Study, Retrospective.

(iii) Observational Study, Prospective.

3.8 Margarine can stick it to your heart

SOURCE: *Rockland Journal-News*, March 1993.

REAL DATA

The bad news about margarine continues to mount. Eating stick margarine increases your risk of a heart attack, according to findings from the Harvard Nurses' Health Study, an ongoing analysis of the diets of 90,000 nurses. The women were divided into groups according to the amount of margarine in their diet. Over time, the incidence of heart disease was recorded. The study reported in the March 6, 1993, *Lancet* shows that women who frequently use margarine have more than a 50 percent higher risk of heart disease than those who infrequently use margarine. Those who ate a lot of foods high in trans-fatty acids—the fats that form when liquid vegetable oils are processed—have a 70 percent higher risk of heart disease than those who don't.

(a) Is this an observational study (and if so, which type) or is it an experiment? Explain.

(b) What would the response variable and the explanatory variable be for this study?

3.9 Milk may cut the risk of kidney stones

SOURCE: *USA Today*, 1993.

Doctors have always told patients with kidney stones—which contain calcium—to cut back on milk and other calcium-rich foods. But that advice may change. A new study shows men with diets high in calcium actually have a lower risk of developing kidney stones than men with low-calcium diets. Researchers at Harvard School of Public Health studied the diets of 45,619 men ages 40 to 75 who had no history of kidney stones. During the four-year study, 505 cases of kidney stones were reported.

(a) Is this an observational study (and if so, which type) or is it an experiment? Explain.

(b) What would the response variable and the explanatory variable be for this study?

(c) Give a new theory or hypothesis stemming from these results that might be further investigated.

3.10 **Study links breast cancer to pesticides, PCBs**

SOURCE: Dwight E. M. Angell, *The Detroit News.*

REAL DATA

PCBs and pesticides may be key factors in the development of breast cancer in women, according to a study released by a University of Michigan physician. "This is the first study that demonstrates a relationship between PCBs, pesticides and breast cancer," said the study's lead author, Dr. Frank Falck Jr. of The University of Michigan. The study, published in the *Archives of Environmental Health*, analyzed chemicals in breast tissue of women from Hartford, Connecticut. Half the women had been diagnosed with malignant breast cancer and half with benign breast disease.

Those with breast cancer had a 50- to 60-percent higher concentration of PCBs in their breast tissue. "That's an amazingly significant difference," Falck said. There was also a much higher level of the insecticide DDT. . . . The study's results are preliminary and Falck said more research must be done. This study, for example, didn't look at well-known factors associated with breast cancer, such as age, diet and family history.

(a) Is this an observational study (and if so, which type) or is it an experiment? Explain.

(b) The last sentence of this article refers to "well-known factors associated with breast cancer." What is the statistical name for these types of variables?

(c) Suppose that the likelihood of malignant breast cancer increases with age. Write a brief explanation as to how the variable age could confound the results of this study.

3.11 **Study: Alzheimer's may stalk victims early in life**

SOURCE: Brenda C. Coleman, *The Associated Press.*

REAL DATA

Alzheimer's disease may stalk its victims early in life, decades before it destroys the mind, a study of nuns who are donating their brains to science suggests. Alzheimer's may be like hardening of the arteries, resulting from a lifelong biological deterioration that becomes apparent only when people are older, authors of the study say.

The study analyzed nuns' youthful writings and found that those women who showed low linguistic ability when they were in their 20s had a much higher risk of Alzheimer's when they were elderly. The findings could indicate Alzheimer's impairs language ability when people are young, the researchers said. On the other hand, greater linguistic ability early in life might indicate a healthy brain resistant to Alzheimer's later on.

"It's a chicken-or-an-egg thing at this point," said the lead researcher, David A. Snowdon, an associate professor of preventive medicine at the University of Kentucky. The findings were published in the *Journal of the American Medical Association*.

The researchers studied the autobiographies of 104 nuns from the School Sisters of Notre Dame. The order's 678 nuns have agreed to donate their brains for the federally funded research. The women wrote one-page accounts of their lives for the order's archives just before taking their vows, at [an] average age of 22.

Scientists autopsied the brains of 25 nuns who died, 10 of whom had Alzheimer's. Those who had low linguistic ability when young had abundant neurofibrillary tangles, the lesions of Alzheimer's disease, when they were old. Nine of the 10 nuns who developed Alzheimer's disease, 90 percent, showed a low linguistic ability in their autobiographies, compared with only 13 percent among those who did not have Alzheimer's, the researchers said.

"That's what's most incredible about it—this relationship between what they wrote in their 20s and what their brains looked like 60 years later," Snowdon said. "It's a disease process that's underlying this."

(a) Is this an observational study (and if so, which type) or is it an experiment? Explain.

(b) What would the response variable and the explanatory variable be for this study?

(c) Is the "90 percent" stated in the article an estimate of the likelihood of the risk factor given a person has a certain response, or the likelihood of a person having a certain response given they have the risk factor?

3.5   Understanding Experiments

To establish a link between the explanatory variable and the response variable, we would like to hold everything constant except the explanatory variable. We would change the levels of the explanatory variable to see what happens to the response variable as a result. Although we can rarely achieve this ideal situation, we sometimes can come pretty close to it with an experiment rather than an observational study. In this section, we will study the basic terminology describing experiments and present some principles to strive for when planning an experiment. We shall see that, as in sampling, experiments also involve the planned use of chance.

3.5.1   Basic Terminology

Let us begin by recalling the vocabulary involved in studies in general, but presented in the context of experiments. The objects on which an experiment is performed and responses are measured are called the **experimental units**, or **subjects** if they are human beings. The intent of an experiment is to study the effect of changes in some variables on the outcome variables. The outcome variables are the **response variables**. The variables that the experimenter has control over are called **explanatory variables** or **factors**. The possible values of an explanatory variable are called the **levels** of that factor. A **treatment** is a particular combination of the levels of each of the explanatory variables. For most medical trials, the patients are the units, and the drugs are the treatments. In agricultural experiments, the experimental units are often plots in a field, and the treatments could be different plant varieties, fertilizers, or pesticides.

In an experiment, these treatments are not simply observed, but are actively imposed on the subjects. It is the active imposition of a treatment, not just the presence of explanatory and response variables, that makes a study an experiment. Many experiments have more than one explanatory variable. Here is an example of an experiment with two explanatory variables.

❖ EXAMPLE 3.7   **Producing Hard Gears** _____

A manufacturer of automatic car transmissions uses steel gears. An important characteristic of these gears is their hardness. An experiment is conducted to compare gears made from three different steel alloys (that is, three different compositions of steel). Gears made from each of these steel alloys were heat treated or hardened at two different temperature levels, 1600°F and 1800°F. Twelve gears of each of the three different steel alloys were available—6 were randomly assigned to each of the two temperature levels. Let's make a table to portray the layout of this experimental design. We will refer to such a table as the **design-layout table**.

| | | Factor 1:   Alloy Composition | | |
		Alloy 1	*Alloy 2*	*Alloy 3*
	1600°F	6 gears	6 gears	6 gears
Factor 2: Temperature	*1800°F*	6 gears	6 gears	6 gears

For this experimental situation, we find

(a) The *experimental units*: the gears

(b) The *response variable*: hardness of the gears

(c) The *explanatory variables*: factor 1 = alloy composition of steel,

factor 2 = heat-treatment temperature

(d) The *factor levels*: three levels of alloy composition—alloy 1, alloy 2, alloy 3

two levels of heat-treatment temperature—1600°F and 1800°F

(e) The *treatments*: There are 3 × 2 = 6 treatment combinations:

alloy 1 with 1600°F, alloy 2 with 1600°F, alloy 3 with 1600°F,

alloy 1 with 1800°F, alloy 2 with 1800°F, alloy 3 with 1800°F

(f) The *number of experimental units* required for this experiment: 36 gears, 12 of each alloy composition ❖

Definition: A **design-layout table** displays the various combinations of the levels of each of the explanatory variables in an experiment. The number of experimental units assigned to each treatment combination may also be presented.

If there is just one explanatory variable, the design-layout table would list the levels of that one variable along with the number of units assigned to each level.

If there are two explanatory variables, the levels of one variable would form the rows of the design-layout table, the levels of the second variable would form the columns of the table, resulting in a *two-way table*. The number of units assigned to each combination can be written inside each *cell* of the table representing that combination.

If there are more than two explanatory variables, we will need to use several two-way tables. For three explanatory variables, the design-layout table is a *three-way table*, presented as several two-way tables side by side, one for each level of the third variable.

❖ **EXAMPLE 3.8 Three Explanatory Variables** _____

An experiment is conducted to test the strength of a worsted yarn by repeatedly stretching a test specimen of yarn until it breaks, or fails. The number of repetitions until failure is used as a measure of the strength of the yarn and is assumed to depend on the *length* of test specimen, the *amplitude* (that is, how "long" the yarn is stretched in each repetition), and the *load* (that is, how "hard" the yarn is stretched in each repetition). The lengths of test specimens used in the experiment were 250 mm, 300 mm, and 350 mm. The amplitudes used were 8 mm, 9 mm, and 10 mm. The loads used were 40 g, 45 g, and 50 g.

(a) Name the experimental units: the test specimens of yarn

(b) Name the response variable(s): number of repetitions until failure

(c) Name the factor(s) and how many levels each factor has: length of specimen (3 levels), amplitude (3 levels), load (3 levels)

(d) Give the number of treatments: 3 lengths × 3 amplitudes × 3 loads = 27 treatments

(e) Give the number of experimental units: It is not clear how many yarn specimens were assigned to each of the treatments. If just one yarn specimen were assigned to each treatment, there would be 27 units—if two specimens were assigned to each treatment, there would be 54 units.

In general, it is best to have more than one unit assigned to each treatment. If you have one unit assigned to each treatment, and one of the units happens to have a defect, information about the response for that treatment will be wrong. Having at least two units for each treatment combination will also provide an idea of the variation in the responses between units receiving the same treatment.

Design-layout Table

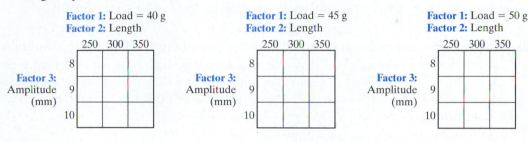

 **3.5** **Component Lifetime**

An experiment has been conducted to study the effect of temperature and type of oven on the lifetime of a metal clutch component. Four types of ovens and three temperature levels were used in the experiment. Two clutch components were assigned randomly to each combination of type of oven and temperature level.

(a) Why is this an experiment and not an observational study?

(b) For this experimental situation

  (i) The experimental units are _____.

  (ii) The response variable is _____.

  (iii) The factors or explanatory variables are _____.

    How many levels are there for each factor? _____

  (iv) How many treatments are there? _____

  (v) How many experimental units are required for the experiment? _____

(c) Make a design-layout table for this experiment.

 **Exercises**

3.12 In an experiment for improving the overall look (appeal) of pizzas made by a restaurant, the chef decides to investigate the effects of two factors: amount of cheese (with three levels) and type of crust (with two levels). Four pizzas at each treatment combination will be made and each will receive a score for its overall appearance. How many experimental units will be used in this experiment?

3.13 A textile researcher is interested in how the color of dye will affect the durability of fabrics. Because the effect of the dye color may be different for different types of cloths, he applies each of the four dye colors to five different cloth types. Six fabric specimens are dyed for each of the treatments. Each fabric specimen is then tested for durability.

(a) What is the response variable?

(b) What are the explanatory variables (or factors)?

(c) How many treatments combinations are there?

(d) How many total fabric specimens are required for this study?

3.14 An experiment was conducted to compare two different methods of teaching piano in individual lessons to beginning pupils of five different ages (6, 7, 8, 9, and 10 years). Four piano teachers took part in the experiment. Each teacher taught six pupils at each of the five age levels. For each group of six pupils at each age level, the first three randomly selected pupils were assigned to Method I and the remaining three were taught with Method II. An independent examiner rated each pupil's progress after one year of lessons. The examiner did not know which method had been used with each pupil.

(a) Identify the response variable(s).

(b) Identify the explanatory variable(s).

(c) How many experimental units (i.e., pupils) are needed for the experiment?

3.15 A chemical engineer is designing the production process for a new product. The chemical reaction that produces the product may have higher or lower yield, depending on the temperature and the stirring rate in the vessel in which the reaction takes place. The engineer decides to investigate the effects of the combination of two temperatures, 50°C and 70°C, and three stirring rates, 60 revolutions per minute (rpm), 100 rpm, and 140 rpm, on the yield of the process. Four batches of the feedstock will be processed at each combination of temperature and stirring rate. For this experiment,

(a) Name the experimental units.

(b) Name the response variable(s).

(c) Name the factor(s) or explanatory variable(s) and how many levels each factor has.

(d) Give the number of treatments.

(e) Give the number of experimental units.

(f) Make a design-layout table.

3.16 What are the effects of repeated exposure to an advertising message? An experiment investigated this question using a random sample of respondents to a local questionnaire that indicated they own other advanced electronic equipment at home. All subjects viewed a 30-minute television program that included ads for a new video camera. Some subjects saw a 30-second commercial, others a 90-second version. The sample commercial was repeated either one, three, or five times during the program. After viewing, all of the subjects answered questions about their recall of the ad, their attitude toward the camera, and their intention to purchase it. For this experiment, four subjects were randomly assigned to each of the treatment combinations of length of commercial and number of repetitions of the commercial. For this experiment,

(a) Name the experimental units.

(b) Name the response variable(s).

(c) Name the factor(s) or explanatory variable(s) and how many levels each factor has.

(d) Give the number of possible treatments.

(e) Make a design-layout table.

3.17 An experiment was conducted to find out what kind of diet was best for a weight-loss program. Three levels of fiber content were chosen: Low, Medium, and High; while two levels of carbohydrate content were chosen: Low and Medium. Twenty people were randomly assigned to each treatment combination of fiber and carbohydrate content. Their weights were measured two weeks after administering the diets.

(a) The response variable is _____.

(b) The explanatory variables are (list all of them) _____.

(c) List the corresponding levels for each explanatory variable given in (b).

(d) The total number of patients required for this experiment is _____.

3.18 An experiment was conducted to study the effect of fertilizers and moisture content on apples. Three types of fertilizers were selected and two levels of soil moisture (Dry and Moist) were used. Three apple trees were assigned to each treatment combination of fertilizer type and soil moisture. After two months, the number of apples in each of the trees was recorded.

(a) The response variable is _____.

(b) The explanatory variables are (list all of them) _____.

(c) List the corresponding levels for each explanatory variable given in (b).

(d) The total number of apple trees required for this experiment is _____.

 ### 3.5.2 Principles for Planning an Experiment

Experiments allow us to study factors that are of interest to us—in particular, to study the effects of the factors on various responses of interest. The validity of the results of an experiment depends to a great extent on the underlying structure of the experiment and especially on how the treatments are applied to the units: In this section, ideals to strive for in the planning of an experiment are discussed in the context of how an initially poor experimental design might be improved upon.

Experimental Design—Does Taking Vitamin C Lead to Fewer Colds?

Consider the following experiment: A researcher believes that taking a daily dose of vitamin C will help reduce the incidence of colds in elementary school children. To assess this claim, 100 elementary school children, who volunteered and for whom parental consent was received, were given a daily dose of vitamin C for a three-month period during winter. The number of colds during this season was recorded for each subject. For this experiment, we have 100 subjects, one response variable—the number of colds—and one explanatory variable—vitamin C—at one level—a daily dose.

Results					
Number of colds	0	1	2	3	4 or more
Frequency or count	45	28	12	7	8

Based on the data, does it appear that taking a daily dose of vitamin C helps reduce the incidence of colds?

THINK ABOUT IT

What proportion of subjects treated with vitamin C experienced at least one cold during the season? Does this seem like a low proportion? a high proportion?

Suppose that the proportion of people who had at least one cold over that season last year was 80%. Would that support the hypothesis that vitamin C leads to fewer colds? Is it appropriate to compare the incidence of colds for this season to the rate for last season?

Suppose that the proportion of people who did not take vitamin C over the period and had at least one cold was 80%. Would that support the hypothesis that vitamin C leads to fewer colds?

Suppose that the proportion of people who did not take vitamin C over that period and did not have a cold was 40%. Would that support the hypothesis that vitamin C leads to fewer colds?

The main problem is that we don't have anything to compare these data to—there is no basis for comparison.

How could this study design be improved? Perhaps the 100 subjects could be divided into two groups—one group would receive the treatment, vitamin C, and the other group would not. The group that does not receive the vitamin C would serve as a control and therefore is called the **control group**. Then we could compare the results—the distribution of the number of colds—for the two groups. The researcher would like to see that the treatment group has a higher proportion of no colds or one cold and the control group has a higher proportion of subjects with four or more colds.

Definitions: A **treatment group** is a group of subjects or experimental units that receive an actual treatment.

A **control group** is a group of subjects or experimental units that are treated identically in every way, except that they do not receive an actual treatment.

THINK ABOUT IT 🐢

How should the 100 subjects be divided into the two groups?

Should the experimenter decide which subjects should get the treatment?

Even with this improvement in the design, there are still some potential problems. We must be careful in how the subjects are assigned to the treatment and control groups. If the experimenter performs the assignment, he or she may, consciously or unconsciously, assign the "healthier" subjects to the treatment group. This could bias the results. Healthier subjects might be less likely to catch a cold. If the treatment group appears to have fewer colds, we could not assess whether this was due to the vitamin C or to the initial health of the subjects. The results would be **confounded** with the initial health of the subjects.

To reduce the effect of this and other bias, we use an impartial method for assignment—randomness. The subjects should be **randomly assigned** to one of the two treatment groups. How? Give each subject a label, outline the allocation scheme, then use the random-number table or a calculator to produce a random listing of the subjects. In our example, we could label the 100 subjects 1 through 100. The allocation scheme might be as follows: Assign the first 50 subjects randomly selected to the treatment group and the remaining 50 subjects to the control group. Using the TI calculator with a seed value of 41 and $N = 100$, the first 8 labels out of the 50 to be selected and assigned to the treatment group would be 57, 10, 24, 16, 49, 29, 50, 22. Notice that when reading off labels we skip over those that were previously selected or labels that are not possible. If we were to use the random-number table, we might label the 100 subjects 01 to 99, and end with 00.

Randomization is a simple and important principle in experimental design that helps to reduce bias in assigning the subjects to the treatments. Randomization also tends to make the groups similar with respect to other factors that might affect the experimental results, even if such factors are not known to the researcher. In our example, we would expect the two groups of subjects to be similar with respect to age, gender, various medical histories, such as asthmatic or not, and so on. Randomization tends to make the groups similar, but cannot guarantee similarity with respect to all such factors.

> *Definitions:*   **Random allocation** is a planned use of chance for assigning the units to the treatments. An experiment is **completely randomized** if the experimental units are randomly assigned to the treatment combinations.

Another concern in the design, even with a control group and random allocation, is that the subjects know whether they are in the treatment group or the control group based on whether or not they are being given the vitamin C. Subjects who receive the treatment may be less likely to "report" a cold ("I don't have a cold, just a little stuffy nose") because they know that the vitamin C is being tested as helping to reduce the number of colds. Research has shown that people respond not only to actually drugs, but also to **placebos**. A placebo is a dummy drug—it looks like the real drug, but has no active ingredients. The tendency of a subject to show the effect that the researcher is looking for is a *response bias*. Because the subject is receiving some treatment, he or she has a tendency to show improvement, even if the treatment is inert. This type of bias is called the **placebo effect**. To reduce the effect of this bias, we give both groups of subjects a "treatment"—one is the actual vitamin C and the other, which appears in every way to the subject the same, is a dummy treatment.

THINK ABOUT IT

A researcher has a treatment group and a control group. The control group will receive a placebo. The subjects will be randomly assigned to one of the two groups. Those in the treatment groups receive a bottle of medication marked "DRUG," while those in the control group receive a bottle of medication marked "PLACEBOS." What is wrong with the researcher's experimental design?

If the subjects know whether or not they are taking the real drug, it defeats the purpose of having a placebo in the first place. The subjects need to be **blinded** with respect to who is receiving the actual treatment. Who measures the response is also an important concern. If the researchers measure the response and they know whether or not the subject is in the control or treatment group, they may have a tendency to bias the results in their favor (for example, a borderline case may be classified in their favor). This type of bias is called **experimenter bias**. Thus, the experimenter or person measuring the response(s) should be blind to whether the subject is in the control or the treatment group. When both the experimenter and the subjects are blinded as to who is in which group, we say the experiment is **double blinded**.

> *Definitions:*   The **placebo effect** is a phenomenon in which receiving medical attention, even administration of an inert drug, improves the condition of the subjects.
>
> **Experimenter bias** is the distortion that can arise on the part of the experimenter due to how the subjects are assigned to the groups, which variables are measured and how they are measured, and how the results are interpreted. The bias is generally in the direction of the researcher's theory.
>
> A **single-blind** experiment is one in which only the subjects are ignorant of which treatment they receive. A **double-blind** experiment is one in which neither the subjects nor those working with the subjects knows who is receiving which treatment.

THINK ABOUT IT

Why not save money and assign just one subject to each of the two groups? Suppose that we only had two subjects and assigned one at random to the vitamin C treatment and the other to receive a placebo. The subject receiving the vitamin C was a male and had only one cold. The subject receiving the placebo was a female and had three colds. Would you trust these results and conclude that vitamin C leads to fewer colds?

Is it possible that the fewer colds of the vitamin C subject could have been due in part to being a male?

How might the female subject having a medical history of asthma affect the results?

We would not trust the results of a study with only one or two subjects in each treatment group. In general, it is best to have more than one unit assigned to each treatment. If you have one unit assigned to each treatment, and one of the units happens to have a defect or drops out of the study, information about the response for that treatment will be wrong or missing. Having at least two units for each treatment combination will also provide an idea of the variation in the responses between units receiving the same treatment. We prefer experiments with many subjects because with additional subjects, the effects of other factors not accounted for will tend to be about the same for each group, and thus wash out. We have **replication** in an experiment if we have more than one unit assigned to each treatment combination. An experiment is **balanced** if there is the same number of units assigned to each treatment combination.

> *Definitions:* If at least two units are assigned to each treatment combination, we have **replication** in an experiment. If there is the same number of units assigned to each treatment combination, the experimental design is **balanced**.

 **3.6** Chromium Supplements

REAL DATA

Chromium supplementation doesn't live up to its muscle-building, fat-shedding image—at least in normal doses in ordinary people, a study indicates. And contrary to many muscle builders' beliefs, the supplement may not make them stronger even if they use it in far larger doses, says one researcher. The finding is disputed by other researchers. . . .

Researcher Marc A. Rogers of the University of Maryland tested this over a 12-week exercise program involving 16 previously sedentary men with an average age of 23. Half got chromium supplements at 200 micrograms per day, the upper limit of what the federal government considers a healthy level. The others got a fake substitute. All trained three sessions a week for 45 minutes per session on resistive exercise equipment that had the same effect as lifting weights.

At the end of the training program, both groups had become stronger, but there was no statistically meaningful difference in strength gains, the study reported. This indicates that the exercise made the difference, and the chromium supplements did not help. The results indicate that the body simply cast aside the chromium it did not need without using it to build extra muscle or to decrease fat, Rogers said (SOURCE: *The Associated Press*).

(a) The article states that some received a "fake substitute." What would this be in statistical terms? What is the purpose of giving some a "fake substitute?"

(b) Suppose that the hypotheses being tested are as follows:

H_0: No difference in strength gains between the two groups.

H_1: The Chromium group has higher strength gain.

Which hypothesis was supported in the study?

If the researcher performed the test of these hypotheses at a significance level of 0.05, would the *p*-value have been more than 0.05 or less than or equal to 0.05? Explain.

❖ **EXAMPLE 3.9** **Random Allocation** _____

We want to test whether aspirin helps prevent repeated heart attacks. The following experiment is planned: twenty male patients who have had a heart attack will be randomly divided into two groups of 10 each. One group will receive 300 mg of aspirin each day for three years, the other group a placebo pill each day for three years. The 20 subjects are listed as follows:

Name	Label	Name	Label	Name	Label	Name	Label
Lee	01	Bill	06	Ed	11	Eric	16
Kyle	02	Matt	07	Robb	12	Ralph	17
Mark	03	Tim	08	Stan	13	Doug	18
Steve	04	Ryan	09	Phil	14	James	19
Pablo	05	Herb	10	Mike	15	Henry	20

Labels have been provided for performing the random allocation. Since there are 20 items that needed to be labeled, we labeled them from 01 to 20 ($N = 20$). With these labels, we must decide how we will allocate the subjects to the two groups and then use the generated random numbers to carry out the assignment of subjects to the two groups.

Here is one possible *allocation scheme* that we will use: The first 10 randomly selected subjects will be assigned to the aspirin group. The remaining 10, which do not have to be randomly selected because they are all that are left, will be assigned to the placebo group.

Another possible allocation scheme would be to alternate: The first randomly selected subject will be assigned to the aspirin group, the next randomly selected subject will be assigned to the placebo group, and so on, until all subjects are assigned to one of the two groups.

We will use the first allocation scheme—the first 10 selected will be allocated to the treatment group.

■ *Using the TI* with a seed value of 70 and $N = 20$, the first 10 selected labels are as follows:

4   19   16   17   2   19(skip)   13   11   19(skip)   11(skip)   10   9   5

Subjects in Group 1	*(treatment group)*	Steve	James	Eric	Ralph	Kyle
		Stan	Ed	Herb	Ryan	Pablo

So, the remaining subjects are assigned to the placebo group.

Subjects in Group 2	*(placebo group)*	Lee	Mark	Bill	Matt	Tim
		Robb	Phil	Mike	Doug	Henry

Using the random number table starting with Row 4, Column 11 and continuing through Row 5, the first 10 selected labels are as follows:

06 24(skip) 36(skip) 16 80(skip) 07 85(skip)...05...15...04...17...03...12...19

Subjects in Group 1 (treatment group)   Bill   Eric   Matt   Pablo   Mike
                                             Steve Ralph Mark   Robb   James

So the remaining subjects are assigned to the placebo group:

Subjects in Group 2 (placebo group)   Lee   Kyle   Tim   Ryan   Herb
                                           Ed   Stan   Phil   Doug   Henry   ❖

Let's do it! 3.7 Study Suggests Light to Back of the Knees Alters Master Biological Clock

REAL DATA

SOURCE: *New York Times*, January 16, 1998.

Authors report on a study that shows there are routes not involving the eye that allow the circadian rhythms in humans to be synchronized with the environment. The study involved 15 subjects who were assigned at random to either the control group or the active group. The active group subjects were subjected to a 3-hour pulse of light that was shined on the back of the knee. The subjects were not told whether they were in the control or the active group. Much care was taken to make sure that the subjects could not distinguish, by the presence of heat of light, whether or not they were in the active group.

(a) Explain why this is an experiment.

(b) The 15 subjects were labeled 1 to 15. The first 7 subjects selected at random were assigned to the control group. Using your calculator (with a seed value of 48) or the random number table (starting with Row 15, Column 1), give the labels for the subjects in the control group (in the order that they were selected).

(c) In this chapter, you learned about the following basic design principles: randomization, control, blinding, and replication. Based on the study details just provided (select your answer),

    (i)   was "randomization" used in the study?   Yes   No
    (ii)   was "control" used in the study?   Yes   No
    (iii) was "blinding" used in the study?   Yes   No
    (iv)   was "replication" used in the study?   Yes   No

Let's do it! 3.8 Controlling for Another Factor

To test whether aspirin helps to prevent repeated heart attacks, the following experiment is planned. Eighteen male heart-attack patients will be divided into two groups. One group, the treatment group, will receive 300 mg of aspirin each day for two years, and the other group, the control group, will receive nothing. The groups will be compared with respect to the number of additional heart attacks over the two-year period. The experimenters are concerned that the effect of the aspirin may differ depending on whether the patient has had only one previous heart attack or several previous heart attacks. So they decide to randomly assign half of the patients who have had only one previous heart attack to each group and half of the patients who have had several previous heart attacks to each group. Among the 18 subjects, 10 have had only one previous heart attack and 8 have had more than one. The 18 subjects are as follows:

One previous heart attack	Lee	McCane	Frakes	Arnold	Palmer
	Jackson	Girbach	Jones	Mitchell	Lewis

More than one previous heart attack	McConnell	Ihle		Daniels	Gernes
	Olson	Newman		Brock	Klein

(a) Label the subjects to carry out the random assignment as described previously.

(b) Use your calculator (with seed values of 27 and 39) or the random number table (starting with Row 10, Column 1 and Row 20, Column 1) to identify which labels, and thus which subjects, would be assigned to the aspirin group. List the names of subjects assigned to the aspirin group:

(c) The experiment as described has a serious flaw. What is it?

Note: The experimental design described here is an example of a **block design**. Block designs form blocks of similar subjects and assign treatments at random, separately within each block. Such designs are similar to the stratified samples in Section 2.6; however, we use two different names—blocks for experiments and strata for sampling. Blocking allows the experimenter to make conclusions about each block separately, and when the overall effects are studied across blocks, the conclusions are generally more precise than if blocking were not used. Exercise 3.47 is another example in which a block design is used.

To summarize, here are the basic principles for designing an experiment:

Basic Principles for Design of Experiments

■ **Randomization**—Use randomization to assign units to the groups. Randomization tends to produce groups of experimental units that are similar with respect to potential confounding factors. Randomization ensures that the experiment does not intentionally favor one treatment over another.

- **Control**—Control the effects of confounding variables by using a comparative design. If all subjects are treated exactly the same across the groups, except for the actual treatment they received, potential confounding variables should affect both groups equally and tend to cancel each other out when comparing the results from two groups.

- **Blinding**—If possible, give a placebo to the control group. Neither the subjects nor anyone working with the subjects should know who is receiving the treatment and who is getting the placebo.

- **Replication**—If possible, assign at least two units to each treatment combination to help assess the natural variation in the responses between units receiving the same treatment.

When selecting the design for an experiment, it is important to keep it simple, yet be statistically accurate and economical. The number of experimental units used should be determined so the results will be meaningful. You may have a particular response variable of interest, but there might be another response variable that would give you the same information and that is easier to measure or obtain.

Exercises

3.19 A vaccine has been developed for use against a virus. You have eight rats that will be exposed to the virus, named Alpha, Beta, Gamma, Delta, Epsilon, Theta, Omicron, and Omega. Design an experiment to test the effectiveness of the vaccine.

3.20 A researcher has a new method for training dogs. She would like to design an experiment to assess the effectiveness of the new training method as compared to the standard training method. The eight dogs available for the experiment have the following names:

Rover   Spot   Jasmin   Chelsea   Klaus   Charlie   Maggie   Bo

Design an experiment and use your calculator (seed value = 27) or the random number table (starting at Row 35, Column 6) to carry out any randomization.

3.21 Estrogen is taken by millions of American women to treat symptoms of menopause or prevent osteoporosis. A study was conducted to assess if using estrogen therapy can lower the risk of dying from heart disease. Researchers randomly assigned 100 women (who were all under 75 years of age) to one of two groups. Group I received estrogen therapy, while Group II received an inert substitute. The study lasted 10 years and found that the women under 75 who had estrogen therapy for at least 10 years had a lower risk of dying from heart disease than those who had not received the estrogen. This result was statistically significant at the 5% level.

(a) Is it an experimental or observational study?

(b) What is the response variable?

(c) What is the explanatory variable? How many levels does the explanatory variable have?

(d) It is stated that one group received an "inert substitute." What would this be in statistical terms?

(e) State the null and alternative hypotheses regarding the populations being compared.

(f) What can you say about the *p*-value for this study? Be as specific as possible.

(g) Could an error have been made? If yes, what type?

3.22 The key difference between a controlled experiment and an observational study is that in a controlled experiment (select one)

(a) treatments are assigned to subjects.

(b) response bias is smaller.

(c) inference is used when results are not observed.

(d) there are no confounding variables.

3.23 A group of 40 adults who suffer from severe migraines were asked to take a dose of vitamin B each day. After a period of two years, 28 of the adults reported that their migraines were less severe. From this information alone you can conclude (select one)

(a) vitamin B is effective in reducing the severity of migraines.

(b) nothing, because two years is too short of a period of time.

(c) nothing, because two years is too long of a period of time.

(d) nothing, because there is no control group for comparison.

REAL
DATA

3.24 The September 13, 2000 issue of the *Journal of the American Medical Association* (Web site www.jama.com) reported the results of a study in which three anti-inflammatory drugs were tested to compare the gastrointestinal (GI) side effects. A summary of this study was given in the December 5, 2000 edition of the *Orlando Sentinel* under the section entitled "Quick Study: Updates on Major Health Topics." The article reports that of the nearly 8000 arthritis patients, half took celecoxib (brand name Celebrex), a pain reliever called a COX-2 inhibitor. The other half took either ibuprofen or diclofenac—conventional nonsteroidal anti-inflammatory drugs (NSAIDs) commonly used for arthritis. NSAIDs can lead to gastrointestinal (GI) problems. In this study, patients taking celecoxib were less likely to get ulcers than those taking standard daily doses of the NSAIDs.

(a) Explain why this is an experiment.

(b) Suppose that there were exactly 8000 arthritis patients in the study and that the first half randomly selected were assigned to the Celebrex group, the next 25% randomly selected were given ibuprofen, and the remaining 25% took diclofenac. Use your calculator with a seed value = 38 or the random number table (starting at Row 77, Column 1) to give the first five patients who were given Celebrex.

(c) Suppose that only 6% of the Celebrex subjects reported ulcers, while the rates for the ibuprofen and diclofenac patients were 18% and 16%, respectively. How *many* Celebrex patients reported ulcers?

(d) It was reported that this study was double-blinded. Clearly explain the meaning of double-blinded.

(e) A caveat is defined as a caution or warning. The *Orlando Sentinel* article listed the following two caveats for this study:

(1) The study was funded by the manufacturer of Celebrex.

(2) The study lasted just six months and most arthritis patients take pain relievers for longer periods of time.

Comment on why these are caveats for this study.

3.25 A nutrition experimenter intends to compare the weight gain of male rats fed Diet A with that of male rats fed Diet B. Twenty rats are available for this experiment. She would like to randomly allocate 10 rats to Diet A and the remaining 10 rats to Diet B. She decides to label the 20 rats as follows: 01, 02, 03, ... , 20.

(a) Use the experimenter's labels and your calculator (seed value = 72) or the random number table (starting with Row 26, Column 1) to identify the labels of the first 10 rats selected that would be assigned to Diet A.

(b) The experimenter hopes Diet A is the better diet. To avoid possible bias in her results because of her favoritism, she blinds herself to her experiment. Explain how she would do this.

3.6   Reading with a Critical Eye

From the preceding sections, you see that certain problems can arise in both observational studies and experiments. What you should believe depends on how the data were collected, measured, and summarized. We will discuss what questions you should ask before you believe what you read. Even newspapers are beginning to give advice on how to critique the vast information available each day. The following excerpts are from an article reported by the *Newhouse News Service*:

REAL
DATA

Food for thought on what not to eat: Study the weight of the studies

SOURCE: *Newhouse News Service*, May 17, 1995.

In the name of good health, you swear that butter will never find its way to your heart again. Beta carotene supplements are out of your life for good, too. Same goes for eggs. And no more dieting. But beer and wine can stay since a drink a day might keep the doctor away. Smart choices or rash decisions? ... Surveys show that Americans are more mixed up than ever about what is good and bad for their health. Some people are so overloaded and frustrated by the conflicting data that they have tuned out. What's a health-conscious person to do?

Understand the methods and limitations of scientific research, experts advise. Research is cumulative and no one study is definitive. "You shouldn't base your decisions on every study that comes along," says Dr. Ted Mortimer, professor emeritus in epidemiology at Case Western Reserve University School of Medicine. "A study provides clues, not absolute answers."

If you want to evaluate those clues, get out the scale, says Janice Neville, a professor of nutrition at the medical school. "It's the weight of the evidence that's important," she says. "If you've got three studies that are similar, the weight of the evidence is stronger than the study that stands alone. I'm not saying that it's the premise that's true; I'm saying you can rely on it more."

But sometimes, reading the scales is not so easy. Consider the study released recently defying the popular belief that fish is good for the heart. Researchers followed the eating habits of 44,895 men and found that those who ate a lot of fish are just as likely to experience heart trouble as those who only ate small amounts. Smaller studies have found that fish eaters live longer. "I would say that fish is still beneficial if you're looking at it as an alternative source of protein," says Dr. Byron Hoogwerf, head of the lipid clinic at the Cleveland Clinic. "Fish is still lower in saturated fat than other protein sources like meat."

When enough evidence tips the scale, expect a non-profit authority, such as the American Heart Association, to make recommendations and adopt new guidelines. "There are experts on every medical issue who have spent their lives on heart disease, lung disease, etc., etc.," says Mortimer.

Connie Diekman, a spokeswoman for the American Dietetic Association, says medical wisdom may change, but the association's advice stays the same: Everything in moderation and you can't go wrong. "Moderation is not sexy," says Diekman, a registered dietitian. "But it works." The populace usually has trouble embracing moderation because it wants solutions that are easier to swallow. "Everyone wants the magic bullet," says Dr. Wayne Callaway, an internist in Washington who has assisted the government in revising nutritional guidelines. "What they need is a diet overhaul and exercise. No one wants to face that."

Still, when a study contradicts conventional wisdom, most doctors, like Edward White, are bombarded with questions. White, a family practitioner for 33 years, handles the calls by asking his own questions. Do you wear a seat belt? Do you exercise regularly? If the patient answers no to either one, White says a reality checkup is ordered. "Why are you so worried about breast cancer or any kind of cancer when your risk of dying in a car accident is far greater?" White says. "And if you don't exercise, it's nonsense to worry about eggs, butter and margarine." . . . "People are missing the big picture," says White. "It's not the one or two things that you do or don't do that'll make you a dead ringer for disease; it's a matter of your overall lifestyle." . . . Commit to the following in the listed order, experts say. Reduce overall fat intake, then concentrate on reducing saturated fats. Lower cholesterol. Exercise at each step.

Judging the Information

There is a lot of good research being done today, and progress is being made on many fronts. There are many ethical professionals who aspire to quality, objectivity, and accuracy. But there is no such thing as a perfect study, and judging the information presented is not always easy. To provide some structure to critiquing a news report, here are some questions to ask:

The Source: Who funded the study?

If you find that a study was funded by an organization that would have a strong preference for a particular result, and that turns out to be the result of the study, it is important to examine the study for sound procedures.

The Design: Which population is under study?
Who were the subjects of the study?
How many subjects were selected?
How were they chosen?
How long did the study go on?
How was the study controlled for bias?
Was there a control group?
How well were other variables controlled?
How were the responses being defined and measured?

The Results: How were the results analyzed?
What methods were used to summarize the results?
What are the underlying assumptions for performing the methods?
Are the assumptions valid?
Were the results "adjusted" in some way? If so, are they legitimate?
How were the results disseminated?
Through a peer-reviewed journal?
Presented at a conference?
In a marketing ad?

❖ **EXAMPLE 3.10** **How a Drug Firm Paid for a University Study and then Undermined it**

Full article appeared in the *Wall Street Journal*, April 25, 1996, A1, by Ralph T. King Jr.

The story involves a drug called Synthroid taken daily by about eight million Americans to control hypothyroidism. It is distributed in the United States by the drugstore chain Boots Co. There are cheaper drugs that their producers believe are "bioequivalent." However, these drugs have not been sufficiently tested, so doctors tend to stay with the more expensive Synthroid, which continues to capture about 84% of the market.

Feeling threatened by a drug called Levoxyl, which was claiming bioequivalence, Boots decided to support a new study hoping to show that the other drugs were not bioequivalent. They asked Dr. Betty Dong at the University of California at San Francisco to head a team to carry out the study, agreeing to provide significant financial support for the study (eventually they provided about $250,000). She accepted and the study was carried out. The conclusion of the study was that other drugs used in the study were bioequivalent to Synthroid. A paper providing the results of the study was submitted to the *Journal of the American Medical Association*. It was reviewed by five reviewers and accepted for publication in the January 25, 1995, issue of *JAMA*.

The article explains in great detail how the company tried to convince *JAMA* and administrators at UCSF that the study was seriously flawed and should not be published. All these attempts failed, and finally, Boots had to resort to brute force. It turns out the lengthy contract for the study included a clause that said that the results of the study were "not to be published or otherwise released without written consent" of the company. This clause actually violated UCSF policy, however, and Dr. Dong should not have signed the contract in the first place, as she naively did.

Finally, the company demanded that Dr. Dong withdraw the article, threatening to sue her if she did not. Pressured by Boots, the UCSF legal department said that they would refuse to support Dr. Dong if she would not withdraw the article. With no support to fight the case, Dr. Dong had little choice but to withdraw the article, making eight years of research rather useless. As added insurance against it ever being published, Dr. Mayor, a researcher for the company Boots hired to stop the *JAMA* publication, wrote a sixteen page analysis of Dong's data, aimed at showing it was too flawed to reach a definitive result. Dr. Mayor published this article in a peer-reviewed journal for which he was an editor. Does this seem too bad to be true?   ❖

❖ EXAMPLE 3.11   A Study on Bias

In 1983, T. C. Chalmers did a study on bias. He analyzed 145 studies involving treatments for heart attacks. He divided the studies into three groups. One group consisted of studies that were highly controlled for bias, meaning they were both randomized and blinded. Another group of studies was partly controlled for bias, meaning they were randomized but not blinded. The third group of studies was not controlled for bias, meaning they were neither randomized nor blinded.

Of the studies highly controlled for bias, 9% concluded the treatment was effective. Studies partly controlled for bias showed the treatment to be effective 24% of the time. Of the studies with no bias controls, 58% showed the treatment to be effective. As we might expect, studies with little or no control were more likely to conclude that the treatment was effective.   ❖

 **3.9** Study of HIV Patients

REAL DATA

A recent study that suggests that bodies infected by a weak form of HIV learn to protect themselves against AIDS gives hope that a vaccine against AIDS may be found. The study, summarized in the *Ann Arbor News* June 17, 1995 article entitled "Hope Rises on Vaccine for AIDS," tracked the health of 746 prostitutes in Dakar, Senegal. A weak type of the AIDS virus is called HIV-2, while the aggressive virus that causes AIDS is called HIV-1. At the beginning of the study, 618 subjects had no AIDS virus, while 128 sub-

jects were infected with the weaker virus, HIV-2. After nine years, subjects initially having HIV-2 had less than one third of the HIV-1 infections of subjects who were initially infection free. The study suggests that those with the HIV-2 virus have about a 70% less likelihood of acquiring HIV-1, or stated another way, a 70% protection from HIV-1. A statement of caution was also made that people should not try to get infected with HIV-2 as a means of protecting against HIV-1.

(a) What type of study is described—observational study or experiment? Explain.

(b) What is the response variable(s) for this study?

(c) What are the factors in this study and their corresponding levels?

(d) Consider the statement "After nine years, subjects initially having HIV-2 had less than one third of the HIV-1 infections of subjects who were initially infection free." Suppose that approximately 6% of the subjects who were initially infection free contracted the HIV-1 infection after nine years.
   (i) How many subjects who were initially infection free contracted the HIV-1 virus after nine years?
   (ii) How many subjects initially infected with HIV-2 contracted the HIV-1 virus after nine years?

(e) Why do you think prostitutes were used in this study? Why not a more general population of people with and without the HIV-2 virus?

(f) Would the results of this study extend to the general population in Senegal? to the general population in the United States?

(g) Are there any possible confounding effects?

3.7   What about Ethics?

> **Ethics:** *A principle of right or good behavior; a system of moral principles or values; a set of rules or standards of conduct that govern the members of a profession.*

All studies, both observational and experimental, should be guided by basic principles of ethics. In an observational study, researchers simply observe what the subjects do naturally. Before a study is conducted that gathers data on human subjects, the study plans should be reviewed by an institutional review board. The purpose of such review boards is to protect the subjects. The board reviews the study plans, the wording of the questions to be asked, and the consent forms to be signed by the subjects. Subjects must be informed before the start of the study as to the nature of the study and any potential risks. Subjects must give their consent in writing. It is also important that the data collected on the subjects be kept confidential.

In an experiment, researchers actively impose a treatment on the subjects. There are many situations in which an experiment would not be feasible for ethical reasons. We could not require some subjects to smoke and some not to smoke to compare the incidence of lung cancer. Even when an experiment may be feasible, there are other ethical issues. In any experiment in which half of the subjects receive the actual treatment while the remaining subjects receive a placebo treatment, there is a question whether it is ethical to give some patients a placebo. Medical researchers argue that there needs to be sufficient doubt in the treatment to warrant withholding the treatment to half of the subjects, while at the same time sufficient faith in the treatment to warrant administering the treatment to half of the subjects. There is a conflict between helping subjects now and getting good information to help others in the future. Sometimes, ethical issues in an experiment can be so controversial that even medical ethicists disagree, as described in the following article excerpt:

"Sham" surgery of clinical trials divides scientists

SOURCE: *The Boston Globe*, February 4, 1994.

REAL DATA

Scientists and medical ethicists are bitterly divided over a clinical trial in Colorado in which half of a group of 40 patients with Parkinson's disease will undergo a "sham" surgical procedure that involves cutting two holes in the skull, even though they will get no benefit from the surgery. The other 20 patients will get real surgery, in which needles will be inserted through the holes in the skull so that fetal tissue can be injected into the brain. The hope, partially supported by previous research, is that the transplanted tissue will replace a crucial brain chemical that is deficient in patients with Parkinson's disease.

The study, expected to begin soon, is so controversial that even medical ethicists who often see eye to eye hotly disagree. George Annas, director of the law, medicine and ethics program at the Boston University School of Medicine, says that as long as the surgery is relatively safe and the patients know they have a 50–50 chance of receiving the tissue, "It's not only OK. It's good they are doing it. Not only is it ethical to do it this way, it's probably unethical to do it any other way. I am a big believer in randomized, clinical trials," Annas said. "They are the only way to get valid conclusions as to whether a therapy is better."

But Arthur Caplan, director of the Center for Biomedical Ethics at the University of Minnesota, says he is skeptical about whether "it is necessary to go through the sham procedure to get adequate controls. . . . In this case, it seems to me that if we are going to bore large holes in people's heads, we have gone too far to pursue scientific accuracy and not far enough to respect the dignity of the subject."

In research with new drugs, it is normal practice to ask patients to agree in advance either to take the drug or a placebo, without their knowing which they will get. This allows researchers to assess the treatment's effectiveness. But human trials involving sham surgery are extremely rare because of ethical concerns about subjecting patients to invasive procedures that cannot possibly help them, even though research with clear-cut results might benefit society. Doctors interviewed yesterday could not recall any sham surgery studies in recent decades, noting that there are other ways to compare treatment involving surgery to non-surgical care. . . .

According to Dr. Curt Freed, a lead researcher and professor of medicine and pharmacology at the University of Colorado, none of the patients in his study will know who received the implants. Because all patients will be awake during the procedures, which will be done under local anesthesia, doctors will carry on similar conversations during all operations and will try to make sure all procedures take the same amount of time, he added. The patients will be followed by doctors in New York who will not know which patients got which surgery. Freed said that patients who get the sham surgery will be offered the real surgery a year later.

Freed defended the study design, noting this kind of study "is where progress in medical research comes from." He added that simply drilling holes in the skull, one on each side with each hole about the size of two postage stamps, is not very dangerous. The potential danger comes only with the insertion of the needles into the brain, which will not be done in the control patients.

Dr. J. William Langston, president of the Parkinson's Institute in Sunnyvale, California, praised the trial as "daring" and "first-class science." Noting that Freed's team will work with researchers from Columbia-Presbyterian Medical Center in New York and North Shore University Hospital on Long Island, Langston said "the double-blind study with neither patients nor doctors knowing who got the real surgery is the only way to get a good answer on this. I am not sure I would have had the nerve to do this myself. These people will fly 2,000 miles and won't know if they got the transplant or not." Langston said a larger concern is whether the Colorado study is premature. If its results are negative, he said, "People will conclude transplantation itself doesn't work."

Dr. Richard Penn, a neurosurgeon at Rush Presbyterian-St. Luke's Medical Center in Chicago, said in a telephone interview that the sham surgery is outrageous. He acknowledged that the government did not fund his own research, but said he found sham surgery objectionable "before we even put in a grant." He said he himself would not sign up for such a study.

THINK ABOUT IT 🐢

Do you think this is an ethical experiment? Why or why not?

Who do you think would be willing to sign up as participants in this study? Would you be willing to have two holes drilled into your head?

Can you think of any other way an experiment could be designed to determine if fetal tissue can be used to help patients with Parkinson's disease?

What information is not included in the article that would help you answer the preceding questions?

In both observational studies and experiments, issues of ethics arise in the reporting and presenting of the statistical results. The American Statistical Association Committee on Professional Ethics has produced *Ethical Guidelines for Statistical Practice*. These guidelines, summarized next, identify ethical relationships with the public, government, clients or employers, and other professionals.

Ethical Guidelines for Statistical Practice

Statisticians have a public duty to maintain integrity in their professional work, particularly in the application of statistical skills to problems in which private interests may inappropriately affect the development of applications of statistical knowledge. For these reasons, statisticians should

- present their findings and interpretations honestly and objectively
- avoid untrue, deceptive, or undocumented statements
- disclose any financial or other interests that may affect, or appear to affect, their professional statements

Since collecting data for a statistical inquiry may impose a burden on respondents, it may be viewed by some as an invasion of privacy, and it often involves legitimate confidentiality considerations, statisticians should

- collect only the data needed for the purpose of their inquiry
- inform respondents about the general nature and sponsorship of the inquiry and the intended uses of the data
- protect the confidentiality of information collected from respondents
- ensure that the means are adequate to protect confidentiality to the extent pledged or intended
- ensure that, whenever data are transferred to other persons or organizations, this transfer conforms with the established confidentiality pledges

Since statistical work must be visible and open to assessment with respect to quality and appropriateness in order to advance knowledge, and such assessment may involve an explanation of the assumptions, methodology, and data processing used, statisticians should

- establish boundaries of the inquiry and boundaries of the statistical inferences that can be derived from it
- emphasize that statistical analysis may be an essential component of an inquiry
- document data sources used in an inquiry, known inaccuracies in the data, steps taken to correct or to refine the data, statistical procedures applied to the data, and the assumptions required for their application
- make the data available for analysis by other responsible parties with safeguards for privacy concerns
- recognize that there may be alternative statistical procedures
- direct any criticism of a statistical inquiry to the inquiry itself and not to the individuals conducting it

Since a client or employer may be unfamiliar with statistical practice and may be dependent upon the statistician for expert advice, statisticians should

- make clear their qualifications to undertake the statistical inquiry at hand
- inform a client or employer of all factors that may affect or conflict with their impartiality
- not accept contingency-fee arrangements
- fulfill all commitments in any inquiry undertaken
- apply statistical procedures without concern for a favorable outcome
- state clearly, accurately, and completely to a client the alternate statistical procedures available
- not disclose any private information about any present or former client without the client's approval

Chapter Summary

Data can be obtained by either *observation or experimentation*. Most studies attempt to show that some explanatory variable "causes" the values of the response variable to occur. While we can never positively determine whether or not there is a distinct cause-and-effect relationship, we can assess if there appears to be such a relationship. In an observational study, we note differences in the explanatory variable and record whether or not they are associated with differences in the response variable. In an experiment, we create differences in the explanatory variable and examine the results.

Observational studies generally can be classified as being retrospective, looking back, or prospective, looking forward. Prospective studies are generally better procedures. Both observational studies and experiments can result in useless data due to confounding variables. It is harder to avoid confounding in observational studies. Experiments offer some features that help reduce the impact of confounding and various biases so as to understand more clearly the relationship between the explanatory variables and the responses.

Basic Principles for Design of Experiments

- **Randomization**—Use randomization to assign units to the groups. Randomization tends to produce groups of experimental units that are similar with respect to potential confounding factors. Randomization ensures that the experiment does not intentionally favor one treatment over another.

- **Control**—Control the effects of confounding variables by using a comparative design. If all subjects are treated exactly the same across the groups, except for the actual treatment they received, potential confounding variables should affect both groups equally and tend to cancel each other out when comparing the results from two groups.

- **Blinding**—If possible, give a placebo to the control group. Neither the subjects not anyone working with the subjects should know who is receiving the treatment and who is getting the placebo.

- **Replication**—If possible, assign at least two units to each treatment combination to help assess the natural variation in the responses between units receiving the same treatment.

In just about any newspaper, every day, there is some report of the results of a recently conducted study. Not all studies or experiments are well designed. Some studies are contradictory. How do we determine what results are good and which should be ignored? In this chapter, we have discussed common types of studies, potential difficulties of each, proper guidelines for executing them, and how to evaluate them critically. Statistics contributes a lot to our understanding of the world, to help us choose which studies we will listen to intelligently. In the next three chapters, you will learn what to do with data after it has been collected—how to summarize data in appropriate and useful ways.

Key Terms

Be sure you can describe, in your own words, and give an example of each of the following key terms from this chapter:

observational study	treatment	random allocation
experiment	confounding variable or	completely randomized
units	factor	placebo effect
subjects	retrospective study	experimenter bias
explanatory variable	prospective study	single-blind experiment
factor	experimental units	double-blind experiment
response variable	design-layout table	replication
outcome variable	treatment group	balanced design
levels of a factor	control group	ethics

 Exercises

3.26 Organic gardeners claim that planting marigolds nearby will reduce damage to vegetable crops by nematodes, small worms that live in the soil and attack plant roots. You want to test this claim. You have 12 plots that are similar with respect to type of soil, drainage, and amount of sun. You randomly select six plots and plant broccoli *with* marigolds, and, in the remaining six plots, you plant broccoli *without* marigolds. At the end of the growing season, you check for nematode damage.

(a) Is this an experiment or an observational study? Explain.

(b) What are the explanatory and response variables in this study?

(c) We could try to gather information about marigolds and nematodes by examining many gardens with and without marigolds. What are the advantages of conducting an experiment instead?

3.27 Have you ever waited in line for a public telephone while the person using the phone seemed to talk on forever? Do people tend to spend more time on a public telephone when someone is waiting than when they are alone? Researchers measured the length of time spent on the phone by 93 individuals at a shopping center. They recorded whether (a) they were alone, (b) a person was using an adjacent telephone, or (c) a person was using an adjacent phone and someone was waiting to use a phone. The data revealed that people talked less when alone than when a person was using an adjacent phone or when someone was waiting. Was this an experiment or an observational study? Explain.

3.28 A survey was conducted to learn about the possible influence of amount of work experience on the employers hiring decision. There are 252 city blocks and the interviewer decides to take a *1-in-5* systematic sample of these blocks. All of the eligible adults residing in these selected blocks will be surveyed.

(a) Is this an experiment or an observational study?

(b) What is the response variable?

(c) What is the explanatory variable?

(d) Use your calculator with seed value = 21 or the random number table (starting at Row 78, Column 1) to list the first five blocks to be included in the systematic sample.

(e) Can you determine the total number of blocks to be included in the systematic sample? If yes, how many? If no, explain why not.

3.29 The math department chair wanted to conduct a survey of the undergraduate students at a university to study the possible influence of gender on attitude toward mathematics. When students register, they are not allowed to sign up for a class if it meets at the same time with another class currently in their schedule. So the department chair generated a listing of all the undergraduate classes held on MWF from 10–11 A.M. There were 200 classes offered at this time. He selected 10 of these classes at random and sent his assistants to these 10 classes to survey all of the students attending those classes.

(a) Was this an observational study or an experiment? Explain.

(b) What was the response variable?

(c) What was the explanatory variable?

(d) What type of sampling technique was used to obtain the selected undergraduate students?

(e) A reviewer of this survey pointed out that some undergraduate students may not have a MWF 10–11 A.M. class and therefore did not have a chance to be included in the sample. This is an example of what type of bias?

REAL DATA

3.30 The September 27, 2000 issue of the *Journal of the American Medical Association* (Web site www.jama.com) reported the results of a study that examined the effects of whole-grain-food consumption on the risk of having a stroke. A summary of this study was given in the December 5, 2000 edition of the *Orlando Sentinel* under the section entitled "Quick Study: Updates on Major Health Topics." The article reports that researchers studied the diets of 75,521 predominately white women between the ages of 38 and 63 and divided them into five groups based on their whole-grain consumption. During the 12-year period of the study, 352 of these women had an ischemic stroke (one that is due to obstruction of blood vessels). The researchers found that the number of strokes decreased as the amount of whole-grain consumption increased. Those women who consumed the largest amount of whole grain (an average of 2.7 servings per day) had a 43% lower risk of stroke compared with those who did not consume any whole grain.

(a) Was this study an experiment or an observational study? If observational, was it retrospective or prospective?

(b) Give the explanatory variable and the response variable.

(c) A caveat is defined as a caution or warning. The *Orlando Sentinel* listed three caveats for this study:

   (1) The results may not be the same for men and other ethnic groups.

   (2) Some errors are possible because the study relies on self-report of diet through questionnaires.

   (3) Those who consumed more whole-grain foods also had a healthier lifestyle.

   (i) What is the name of the bias that Caveat (2) refers to?

   (ii) What is the statistical name for the variable "lifestyle status" mentioned in Caveat (3)?

3.31 High-fiber diet cuts heart-attack risk in men—Revisited

SOURCE: *Associated Press, CHICAGO.*

REAL DATA

A diet high in fiber from cereal, fruit, and vegetables can significantly lower the risk of heart attacks in men, a study of more than 43,000 health professionals found. While previous studies have shown the cholesterol-lowering benefits of oat bran and the various beneficial effects of grains such as wheat, "Our study confirms that fiber from all sources, but especially from cereal, significantly protects against coronary heart disease," said Eric B. Rimm, assistant professor of epidemiology and nutrition at Harvard School of Public Health.

The benefits of fiber appeared to be independent of how much fat the men ate, though experts outside the study said that remains to be proved. In the study, healthy dentists, veterinarians, pharmacists, optometrists, osteopaths, and podiatrists, ages 40 to 75 were divided into five groups from highest to lowest fiber consumption and were followed for six years ending in 1992.

Men in the highest-fiber group typically ate 29 grams of fiber a day the equivalent of about one cup of bran cereal, 1.5 cups of cooked pinto beans, or seven large apples. Those men had a 36 percent lower risk of a heart attack during the study period than men in the lowest-fiber group, who consumed about 12 grams of fiber a day, researchers found. That was after adjusting for other traits that could affect heart risk, such as smoking and fat consumption.

On average, white men in the United States eat about 13 grams of fiber a day, previous research indicates. "It's quite easy to get an extra 10 grams of fiber in the diet by eating an extra apple and banana or a bowl of beans or cereal," Rimm said. "It should be part of a healthy diet." The findings also probably apply to women, who are being studied now, Rimm said. Other studies have strongly linked high-fiber diets to a lower risk of colon cancer.

(a) Do you think this was an observational study (and if so, which type), or is it an experiment? Explain.

(b) Subjects in this study were "healthy dentists, veterinarians, pharmacists, optometrists, osteopaths, and podiatrists, ages 40 to 75." Comment on what population you think these results could be extended to. the general male adult population? females?

(c) Why were the results reported "after adjusting for other traits that could affect heart risk, such as smoking and fat consumption?" What is the statistical term for such traits or factors?

3.32 In a study about the effects of child abuse on crime, researchers recruited people convicted of violent crimes and interviewed them to determine their childhood abuse level (the level of abuse they were subjected to as children). The researchers recruited a comparable group of people who lived in the same neighborhoods the convicts had last lived in, and who were approximately the same ages as the convicts. The same interviews were performed with this second group.

(a) The variable "childhood abuse level" is the (select one)

   (i)   response variable.

   (ii) explanatory variable.

   (iii) confounding variable.

   (iv) control variable.

(b) The variable "conviction status" is the (select one)

   (i)   response variable.

   (ii) explanatory variable.

   (iii) confounding variable.

   (iv) control variable.

(c) This study is (select one)

   (i)   a controlled experiment.

   (ii) an experiment, but not a controlled experiment.

   (iii) a prospective observational study.

   (iv) a retrospective observational study.

   (v) a confounding study.

3.33 A food-products company is preparing a new pizza-crust mix for marketing. It is important that the taste of the pizza crust not change with small variations in baking time or temperature. An experiment is done in which batches of pizza-crust dough are baked at 410°F, 425°F, or 440°F for 20 minutes or 30 minutes. Six batches of pizza-crust dough will be assigned to each treatment.

(a) What are the explanatory variable(s) or factor(s)?

(b) What is the response variable(s)?

(c) How many treatments are there?

(d) How many experimental units does the experiment require?

3.34 A small locally owned pizza company is trying a new pizza crust. An experiment is conducted to find the optimal baking time (20, 25, or 30 minutes), baking temperature (400°F, 425°F, or 450°F), and amount of cheese (2 cups or 2.5 cups). Five batches of pizza-crust dough will be assigned to each treatment.

(a) How many treatments does this experiment have?

(b) How many experimental units are used in this experiment?

3.35 An experiment that was publicized as proof that yoga lowers stress was conducted as follows: The experimenter interviewed the subjects and rated their levels of stress. Then the subjects were randomly assigned to two groups. The experimenter taught one group how to meditate with yoga, and its members did yoga daily for a month. The other group was simply told to relax more. At the end of the month, the experimenter interviewed all the subjects again and rated their stress levels. The meditation group now had less stress. Why are the results suspect? Identify and explain the main source of bias in this experimental design.

3.36 The chemical element antimony is sometimes added to tin–lead solder to replace the more expensive tin and to reduce the cost of soldering. An experiment was conducted to determine how antimony affects the strength of the tin–lead solder joint. Tin–lead solder specimens were prepared using one of two possible cooling methods, water quenched or oil quenched, and with one of two possible amounts of antimony, 5% or 10%, added to the composition. Two solder joints are to be randomly assigned to each of the four treatments and the shear strength of each measured. The eight solder joints are labeled #1 through #8. Perform the assignment of the solder joints to the treatments. Use your calculator (seed value = 133 and $N = 8$) or the random number table (starting with Row 18, Column 1). Allocate the first two joints selected to Treatment 1, the next two joints selected to Treatment 2, and so on.

> Treatment 1: solder joints assigned to (water, 5%)
> Treatment 2: solder joints assigned to (oil, 5%)
> Treatment 3: solder joints assigned to (water, 10%)
> Treatment 4: solder joints assigned to (oil, 10%)

3.37 A New Zealand study that followed 1000 children through age 18 has found that those who were breast-fed as children fared better in school, both in teacher ratings and in performance on standardized tests. The authors of the study conjecture that fatty acids found in breast milk, but not in formula, may boost brain development.

(a) A criticism of the study is that breast-fed children tend to have mothers who were older, better educated, and wealthier. A trait that by itself could account for the differences in academic achievement is called a (select one)

(i) response variable.

(ii) continuous variable.

(iii) confounding variable.

(iv) control variable.

(b) The authors of the study stated that they adjusted for the traits mentioned in (a) and still found "small but consistent tendencies for duration of breastfeeding to be (positively) associated with IQ." Suppose that the hypotheses being tested were as follows:

H_0: There is no association between duration of breastfeeding and IQ.

H_1: There is an association between duration of breastfeeding and IQ.

Assume that the phrase "small but consistent" implied that the results were statistically significant at a 10% level. For each of the following statements, select True or False.

(i)   The explanatory variable is IQ.   True   False

(ii) The chance of a Type II error is 0.10.   True   False

(iii) The null hypothesis was rejected.   True   False

(iv) The p-value was larger than 0.10.   True   False

3.38 Seldane-D is a seasonal-nasal-allergy medicine produced by Marion Merrell Dow, Inc. This drug was marketed as "the first seasonal-nasal-allergy medicine that lets you stay alert." In an advertising brochure for Seldane-D, a table was given to summarize the frequently reported adverse events for Seldane-D in double-blind, controlled clinical trials. In these clinical trials, subjects were randomized to treatment with either Seldane-D, pseudoephedrine (a competing allergy medicine), or a placebo.

	Percent of Patients Reporting		
Adverse Event	Seldane-D ($n = 374$)	Pseudoephedrine ($n = 287$)	Placebo ($n = 193$)
Insomnia	25.9	26.8	6.2
Headache	17.4	17.1	22.3
Drowsiness	7.2	4.9	11.4
Nervousness	6.7	8.4	1.6
Anorexia	3.7	3.8	0.0
Fatigue	2.1	1.4	2.1
Restlessness	2.1	1.0	0.0
Irritability	1.1	0.0	1.0
Disoriented	1.1	0.0	0.5
Increased Energy	1.1	0.0	0.0

(a) One of the statements made in this brochure is, "In clinical studies involving (over) 300 patients, the reported incidence of drowsiness with Seldane-D did not differ significantly from placebo." What percent of Seldane-D patients reported drowsiness? What percent of placebo patients reported drowsiness?

(b) Why is it good to have a placebo group in such a study? That is, what is meant by the *placebo effect*?

(c) The brochure states that the clinical trials were *double-blind*. What does *double-blind* mean?

3.39 What is wrong with the following definition of placebo effect? "The placebo effect is the difference in outcome between a placebo-treated group and an untreated control group in an unbiased experiment."

REAL DATA

3.40 In the article "Special Tea Has Special Effects, Fans Say" (SOURCE: *Ann Arbor News*, April 17, 1995), enthusiastic Kombucha mushroom fans claim positive health effects from drinking the mushroom tea. Producers of the mushrooms state that the tea can be used to deter or cure almost any ailment, including some cases of cancer, diabetes, muscular dystrophy, and AIDS. Some doctors and scientists are skeptical of such a claim. A professor of health promotion and rehabilitation said, "There is no scientific basis to support home remedies such as the mushroom tea," and "a lot of the success of home remedies is psychological. If a person is convinced he will be cured by the remedy, he's likely to feel that way after taking it."

(a) What type of effect is the professor referring to?

(b) The article also states that FDA officials are interested in studying the tea and that there will be an upcoming study on the tea. You are asked to help design such a study, with the focus on the effect of the tea on treating immune disorders such as cancer or AIDS. Provide a brief write-up of your recommendations.

3.41 Consider a marketing study to evaluate a new type of juice drink. The drink is a mixture of seltzer water and juice, but the researchers wish to determine how much juice to use. They recruit a large number of volunteers, each of whom will try just one drink sample and rate it on a scale of 1 to 100 (100 being the best). For each volunteer, the proportion of juice in the drink is randomly selected to be either 5% or 10%. When the proportion of juice was 5%, the researchers added a bright food coloring (of no taste) to make it look just like the 10% level drink so the volunteers would not know which proportion they received. The researchers recorded the proportion of juice and the subject's rating of the drink. The results of the study are that the drink with 10% juice got the highest average rating of 82, compared with an average rating of 69 for the 5% juice. Upon further examination of the results, they discovered that the preference for the 10% juice was much stronger for younger subjects as compared to older subjects.

(a) Is this study an observational study or an experiment?

(b) What is the explanatory variable?

(c) What is the response variable?

(d) The variable age is a (select one)

    (i) explanatory variable.

    (ii) response variable.

    (iii) confounding variable.

(e) If the volunteer did not know what proportion of juice he or she was being given, but the experimenter did know, we say that the study is (select one)

    (i) controlled.

    (ii) single-blinded.

    (iii) randomized.

    (iv) double-blinded.

    (v) replicated.

(f) The volunteers were obtained over the course of a few days as the researcher stood in a grocery store with a table asking customers as they walked by to participate. Thus, all volunteers were not all available at one time, nor was the overall total number of volunteers known in advance. The researcher assigned Label 1 to represent the 5% juice drink and Label 2 to represent the 10% juice drink and randomly selected which drink each volunteer would receive. Use your calculator (seed value = 18) or the random-number table (starting with Row 20, Column 5) to determine which drink was given for the first five volunteers.

Volunteer Number	1	2	3	4	5
Drink Received (1 = 5%, 2 = 10%)					

(g) The average for the 10% juice was higher than the average rating for the 5% juice. This difference was found to be statistically significant at the 1% significance level. Is the following statement true or false? "The chance that the null hypothesis is true is equal to 0.01 (or 1%)." Explain.

3.42 Consider the following yields, in pounds, from equal-sized plots of wheat assigned in groups of three to four types of fertilizer and three different amounts of water:

Fertilizer	Water A	B	C
I	30, 25, 29	32, 34, 29	25, 22, 23
II	31, 37, 33	31, 32, 30	28, 24, 25
III	33, 29, 24	35, 41, 31	30, 29, 29
IV	31, 30, 32	33, 30, 31	26, 29, 25

(a) Name the explanatory variables(s) for this experiment.

(b) Name the response variable(s) for this experiment.

(c) How many treatments does this experiment have?

(d) What are the experimental units in this experiment?

(e) How many experimental units are used in this experiment?

(f) Compute the average yield for each of the treatment combinations. Which treatment combination resulted in the highest average yield?

(g) Compute the average yield for each level of water (combining across the types of fertilizer). Which water level resulted in the highest average yield?

(h) Compute the average yield for each type of fertilizer (combining across the water levels). Which type of fertilizer resulted in the highest average yield?

(i) It appears from the data that Amount B of water gives the greatest yield per plot of wheat. However, it is found later, after the experiment is over, that all plots given Amount B of water were also in the best-drained part of the field. It is very possible that these plots produced the greatest yield not only because they were given Amount B of water, but also because they were in the best-drained part of the field. How can randomization be used to prevent the problem described?

3.43 **Milk gets bad rap.** The results of a new study on lactose intolerance show that lactose intolerance is probably not responsible for bouts of intestinal mayhem that people blame on milk.

SOURCE: "Milk Gets Bad Rap, New Study on Lactose Intolerance Shows," *Ann Arbor News*, July 6, 1995.

Thirty volunteers who believed they were lactose intolerant were available for the study. The researchers tested all 30 volunteers to see if they were truly lactose intolerant using a simple breath test. Based on the test, 21 of the volunteers were truly intolerant. All 30 subjects were assigned to drink an 8-ounce glass of milk with breakfast. For one week, they got milk that was treated with the lactase enzyme to break down the milk sugar. For another week, they received ordinary milk that was slightly sweetened to taste like the treated milk. At the end of each week, the volunteers rated their intestinal discomfort on a scale of 0 to 5, with 0 representing the least discomfort and 5 representing the most discomfort. For both types of milk, the scores averaged less than 1. Analysis of the intestinal-discomfort scores for the two types of milk showed no difference. The study's senior author stated, "The final result is, there is virtually nobody out there who cannot tolerate a glass of milk a day."

(a) Is this an observational study or an experiment? Explain.

(b) What is the response variable of interest? What is being measured and how is it being measured?

(c) What is the explanatory variable? What treatments were given to the subjects?

(d) What is the population under study? Are these subjects a random sample from this population? Do you agree with the statement by the senior author? Explain.

Notes: A control was used in this study. Each subject received both types of milk, so the subjects were their own control. This is often referred to as a *paired design*. There was use of blinding in this study. The ordinary milk was sweetened to taste like the treated milk, so subjects should not be able to distinguish between the two types of milk. The study was conducted over two weeks. Suppose that all 30 subjects drank the treated milk for the first week and the sweetened ordinary milk the second week. It is possible that there could be some carry-over effect from the first week to the second week that could bias the results of the second week for all subjects. Randomization of the order of the milk presented to the subjects could be used to reduce the impact of this bias.

3.44 A researcher conducted a study on the effect of the amount of caffeine on anxiety. Of the 30 subjects available for the study, 10 were randomly selected and assigned to Treatment Group 1, the next 10 randomly selected subjects were assigned to Treatment Group 2, and the remaining 10 subjects were assigned to Treatment Group 3. Group 1 was given two cups of decaffeinated coffee (0 mg caffeine), Group 2 was given two cups of coffee with a small amount of caffeine (2 mg dose of caffeine), and Group 3 was given two cups of coffee with a large amount of caffeine (6 mg dose of caffeine). Subjects were then given a test that measures their amount of perceived anxiety. A statistical test of hypotheses was conducted to assess if the anxiety levels for the population of Group 3 coffee consumers were higher on average compared to those for the population of Group 1 coffee consumers. The results of this test were statistically significant at the 0.10 level.

(a) Is this study an experiment or an observational study?

(b) In this study, the amount of caffeine is (select one)

    (i)   a response variable.

    (ii) a confounding variable.

    (iii) a systematic variable.

    (iv) an explanatory variable.

(c) In this study, the perceived anxiety test score is (select one)

    (i)   a response variable.

    (ii) a confounding variable.

    (iii) a systematic variable.

    (iv) an explanatory variable.

(d) Using your calculator with seed value = 70 or the random number table starting with (Row 4, Column 1), give the labels for the subjects assigned to Treatment Group 1 (in the order selected).

3.45 A large study used records from Canada's national health care system to compare the effectiveness of two ways to treat prostate disease. The two treatments are traditional surgery and a new method that does not require surgery. The records described many patients whose doctors had chosen each method. The study found that patients treated by the new method were significantly more likely to die within eight years.

(a) Is this an observational study or a designed experiment?

(b) Give a plausible reason why the conclusion of the study might be suspect.

(c) Suppose that you have 100 volunteers who are prepared to participate in a new experiment to compare the effectiveness of these two methods. Describe a suitable experimental design, including what the explanatory and response variables are.

(d) Can this experiment be conducted double-blind? Explain.

3.46 Members of a high school language club believe that study of a foreign language improves a student's command of English. To see whether this is true, they randomly select three high schools in the city and obtain the scores on an English achievement test given to all the students in the selected schools. The average score of the selected students who had studied a foreign language for at least one year is 90 (out of 100), which is much higher than the average score of the selected students who studied no foreign language, which is 73. The difference is statistically significant at the 0.05 level.

(a) What sampling method is used to select the students?

(b) Is the study an experiment or an observational study?

(c) Give the explanatory variable(s) and response variable(s) of the study.

(d) Is the reported average score 90 a parameter or a statistic?

(e) How does the p-value compare to 0.05? (select one)

    (i)   It is greater than 0.05.

    (ii) It is less than or equal to 0.05.

    (iii) You cannot determine.

(f) How does the *p*-value compare to 0.01? (select one)

   (i) It is greater than 0.01.

   (ii) It is less than or equal to 0.01.

   (iii) You cannot determine.

(g) What is one problem with this study? Explain.

(h) How would you design an experiment to evaluate the effect of study of a foreign language on a student's command of English?

REAL DATA

3.47 The November 1997 issue of the *Journal of Personality and Social Psychology* reports research into greedy behavior—situations in which individuals take more than their fair share of limited resources. One goal was to determine whether greed is intentional or an instinctive part of human nature. The subjects were randomly divided into two equal-sized groups. Group 1 was told that they were sharing a box of sand and were instructed to take out what they considered a fair portion. They were also told they would receive $2 for every pound they scooped out. Group 2 was told that they would receive $10 if they accurately scooped a specified amount of sand. Group 1 results were to assess intentional greed. Group 2 results were to assess unintentional greed.

(a) Is this an experiment or an observational study?

(b) Suppose that there were 20 male subjects and 20 female subjects available for the study as listed in the following table:

| Male Subjects | | | | Female Subjects | | | |
Label	Initials	Label	Initials	Label	Initials	Label	Initials
	JD		BH		HN		SL
	MK		WB		BG		NL
	BJ		LG		MA		CJ
	JP		AB		JB		VC
	SK		KG		MM		EG
	MG		ZB		KS		CI
	JW		CF		SM		NI
	VN		TG		KM		KB
	RM		TD		AG		JG

To control for gender, half of the male subjects were assigned to Group 1 and half of the female subjects were assigned to Group 1. Provide appropriate labels in the table for performing the randomization. Then, using your calculator (with seed values of 23 and 41) or the random number table (starting with Row 12, Column 1 and Row 22, Column 1), list (in the order selected) the subjects (by their labels and corresponding initials) that were assigned to Group 1. Show *all* details of your work.

(c) In this study, the researchers were concerned about controlling for gender. What is the statistical term for gender?

3.48 Sixty subjects were randomly divided into two equal groups to study the effects of alcohol on reaction time to a stimulus. The members of one group consumed a specified amount of alcohol, and the members of the other group had a nonalcoholic beverage. The reaction times for all subjects were measured before and after drinking the beverage.

(a) What is the response variable?

(b) Was replication used in this experiment? Explain.

(c) The researcher used random allocation for assigning the subjects to either the alcoholic-beverage or nonalcoholic-beverage group. The first 30 subjects selected at random were assigned to the alcohol group. Use your calculator (with seed value = 34) or the random number table (starting with Row 20, Column 1) to list the labels for the first five subjects assigned to the alcohol group.

(d) What is the chance that any particular subject is assigned to the nonalcoholic group?

(e) The researcher wanted to test the following hypotheses:

H_0: Alcohol does not reduce reaction time.

H_1: Alcohol reduces reaction time.

After conducting his experiment, he came to the conclusion that the data were statistically significant at the 5% level. Which hypothesis was supported?

(f) What can you say about the numerical value of the p-value for this experiment? Be specific.

(g) Could an error have been made? If so, what type?

REAL DATA

3.49 Accolate (zafirlukast) is a type of asthma therapy that is now included in the National Institutes of Health Guidelines for Asthma. A study, reported in summary, was a double-blind randomized trial carried out in clinics nationwide. Of the 6090 subjects, the first 4058 selected received Accolate and the remaining 2032 were given placebo pills. Accolate was shown to be effective for controlling mild to moderate asthma. (SOURCE: *The Ann Arbor News*, September 23, 1997, p. D3 and pamphlet by Zeneca Pharmaceuticals.)

(a) Using your calculator with a seed value of 45, or starting with Row 10, Column 1 of the random number table, give the labels for the first five subjects selected who would have been assigned to the Accolate treatment group.

(b) The study gave a placebo pill to the second group of subjects to avoid the placebo effect. What is the placebo effect?

The incidence of various side effects was also studied. The table below provides a comparison of some of the side effects.

Side Effect	Accolate ($n = 4058$)	Placebo ($n = 2032$)
Headache	12.9%	11.7%
Infection	3.5%	3.4%
Nausea	3.1%	2.0%

(c) How many of the Accolate subjects experienced infection?

(d) Is the 11.7% figure for placebo subjects experiencing a headache a parameter or a statistic?

One set of hypotheses tested regarding side effects was

H_0: The percentage of Accolate users experiencing a headache is equal to the percentage of placebo users experiencing a headache.

H_1: The percentage of Accolate users experiencing a headache is greater than the percentage of placebo users experiencing a headache.

The preceding test was not statistically significant at a significance level of 0.05.

(e) Which hypothesis was supported?

(f) What can you say about the *p*-value for this test? Be as specific as possible.

(g) An error could have been made. Which type?

3.50 Two different methods of teaching first-graders to read are being compared. At the beginning of the school year, two first-grade teachers selected the students for their respective classes, and each teacher taught with one of the two methods. At the end of the school year, the students were given a reading test, and the two methods were evaluated on the basis of the test scores. It was found that there was no significant difference between the two methods at the 10% level.

(a) Is this an experiment or an observational study?

(b) What is the explanatory variable?

(c) Are there any confounding variables? If so, name one.

(d) Has randomization been used to select the samples?   Yes   No

(e) State the null and the alternative hypotheses in terms of the average test scores for the two methods.

(f) What can you say about the numerical value of the *p*-value for this study? (Be specific.)

(g) Would the results be significant at the 15% level? Select one and explain your answer.

Yes   No   Can't Tell

3.51 The November 26, 2000 edition of the *Orlando Sentinel* provided the following summary of a study under the section entitled "World News to Note":

Wine helps older brains

TOKYO -- Wisdom comes with age, but the odd glass of wine may also give senior citizens a boost in brain power. Those older than 40 who drink moderate amounts of wine or Japanese sake rice wine scored better in IQ tests than teetotaler and heavy drinkers, said a study conducted by Japan's national research on aging center and published in Saturday's (11/25/00) Tokyo Shimbun. "Alcohol protects certain brain functions against the aging process, which means it has some influence on intelligence as well," researchers were reported as saying.

(a) Do you think this study was an experiment or an observational study? Explain.

(b) As best you can discern from the news report, give the explanatory variable and the response variable.

(c) A caveat is defined as a caution or warning. Give at least one caveat for this study.

(d) Section 3.6 provided guidelines for critiquing a news report. Write a brief critique about this news report. Include questions regarding what additional information you would like to know about this study to be able to better evaluate its worth.

REAL DATA

3.52 The November 26, 2000 edition of the *Orlando Sentinel* provided the following summary of a study under the section entitled "World News to Note":

Mobile phone dangers

LONDON -- Children who use mobile phones risk suffering memory loss, sleeping disorders and headaches, according to research published in the medical journal The Lancet. Physicist Gerard Hyland raised new fears over radiation caused by mobile phones and said that those younger than 18 were more vulnerable because their immune systems were less robust. "Radiation is known to affect the brain rhythms and children are particularly vulnerable, " Hyland said.

(a) Do you think this study was an experiment or an observational study? Explain.

(b) As best you can discern from the news report, give the explanatory variable and the response variable.

(c) A caveat is defined as a caution or warning. Give at least one caveat for this study.

(d) Section 3.6 provided guidelines for critiquing a news report. Write a brief critique about this news report. Include questions regarding what additional information you would like to know about this study to be able to better evaluate its worth.

REAL DATA

3.53 The May 2, 2001 edition of *The Ann Arbor News* provided the following summary of a study under the section entitled "Health Briefs":

Study: Behavioral therapy helps improve sleep habits

Behavioral therapy can help people improve their sleep habits, a recent study indicates.

Researchers in North Carolina investigated the effectiveness of behavioral therapy for treating insomnia, which is often treated with medications. They randomly assigned 75 people with chronic insomnia into three treatment groups. One received cognitive behavioral therapy to help them come up with strategies to improve their sleep habits. A second group received instructions about muscle relaxation techniques. A placebo treatment group received instructions and practice mental exercise that were considered unlikely to have any impact on insomnia.

After six weeks, the behavior therapy-treated patients showed more improvements in their sleep than the others, based on logs they kept and on polysomnography measurements, which traced physiological activity during sleep.

Find this study in the April 11 (2001) issue of the Journal of the American Medical Association; abstract online at http://jama.ama-assn.org.

(a) Was this study an experiment or an observational study? Explain.

(b) As best you can discern from the news report, give the explanatory variable (or factor) and the response variable. Also give the levels of the explanatory variable.

(c) A caveat is defined as a caution or warning. Give at least one caveat for this study.

(d) Section 3.6 provided guidelines for critiquing a news report. Write a brief critique about this news report. Include questions regarding what additional information you would like to know about this study to be able to better evaluate its worth.

REAL
DATA

3.54 Who pays the most for new cars? A study conducted by automotive marketing specialists at Virginia Commonwealth University shows that older buyers, not women, pay more for new cars. (SOURCE: *Orlando Sentinel*, September 14, 2000, page F2.) The study sampled 4004 new-car purchases (from a J.D. Power & Associates dealer database) for one day in February, 2000. There were 40 different car models, and for a model to be included in the study, there had to be at least three sales for that model. Some of the variables recorded in the study were age, gender, car model, and price paid for the new car (purchase price).

(a) Was this study an experiment or an observational study? Explain.

(b) One of the goals of the study was to assess the effect of *gender* on *car purchase price*. In this case, which variable is the explanatory variable and which is the response variable?

(c) On average, respondents who were 70 to 89 years old paid 10.6% more for a new (2000-model-year) car than did respondents in the 20 to 29 year old age group. Is the rate of 10.6% a parameter or a statistic? Explain.

(d) A past theory has been that minorities end up paying more than other consumers when it comes to dealing with auto dealers. However, the database did not include information on the race of the purchasers, so this theory could not be put to the test. Suppose that the older respondents had been primarily minority and the younger respondents were not primarily minority. Fill in the blank. The variable minority would have been a possible _____ variable.

(e) The study found that on average, female respondents paid $29 more per vehicle than did male respondents. This increase, however, was statistically insignificant. Consider the following hypotheses that underlie the significance test that this last statement refers to:

H_0: The average price paid for a vehicle by all female vehicle purchasers is equal to the average price paid for a vehicle by all male vehicle purchasers.

H_1: The average price paid for a vehicle by all female vehicle purchasers is more than the average price paid for a vehicle by all male vehicle purchasers.

If the test were conducted using a 10% significance level, give a possible value for the *p*-value based on the provided conclusion.

(f) The figures used in this study were the ones the dealers reported to their state Department of Motor Vehicles. If the figures are falsely reported, the dealers could go to jail. Thus, the study is based on data that should be more credible than other past studies. Fill in the blank. The use of more reliable data, such as from the Department of Motor Vehicles, would help to reduce the effect of _____ bias.

3.55 The American Statistical Association (ASA) is a scientific and educational society founded in 1839 to foster excellence in the use and application of statistics to the biological, physical, social, and economic sciences. ASA's 19,000 members serve the industrial, governmental, and academic sectors. The ASA publishes six journals, conducts scientific conferences, fosters a continuing-education program, and is very active in bringing statistical education to elementary and secondary schools. The ASA has a World Wide Web site of http://www.amstat.org. At the date of publication of this text, this home page had links to various statistical committees, publications, and pamphlets. Check out some of the current features. Write a brief summary of your findings.

Chapter 4

SUMMARIZING DATA GRAPHICALLY

4.1 Introduction

We have a *theory* that we wish to test. We *collect data* to help us check the theory. We look at the data and *summarize* the results. In Chapters 2 and 3, we studied how the data are obtained. Once we have collected the data, we must "get acquainted" with the data. In Chapters 4, 5, and 6, we will learn how to take our data, our collection of measurements, and summarize them in useful ways. One of the most useful ways to look at data is through an appropriate picture of the data—a graph or a plot. Chapter 4 will focus on how to produce graphical displays of our data. We will enhance our graphical displays by presenting various

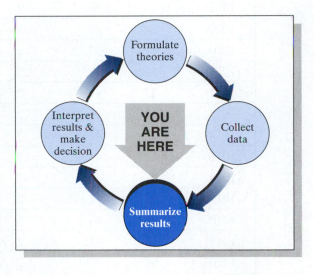

numerical summaries of the data in Chapter 5. A common numerical summary is the mean or average. In Chapter 6, we will take the graphical display of our data, which is based on a sample, and try to find a corresponding model to represent the results for the population.

We start out in this chapter by noting that there are different types of data. We might measure a subject's gender (say, male) and height (say, 69.5 inches). Sometimes the measurement may be a name; sometimes it will be a numerical value. Different types of data lead to different types of graphical displays. We will learn which graphical displays are appropriate for which types of data, and how to make them. We will include examples of common mistakes that are made in presenting the data pictorially, which can lead to misinterpretation of the results. The goal of this chapter is to demonstrate the usefulness of a plot or graph, when done well, in providing a visual summary of the data that have been collected.

4.2   What Are We Summarizing?

Data Set 1 contains various measurements on 20 individuals. These individuals were part of a medical study assessing both the ability of a new drug to reduce blood pressure and the relationship between the dosage of the drug and the amount of reduction in blood pressure. Important characteristics that could influence blood pressure are gender, age, and amount of blood-pressure-reducing medicine taken daily.

Data Set 1

Subject Number	Gender	Age	Dosage # Tablets Taken	Diastolic Blood Pressure Start of Study	End of Study
1001	M	45	2	100.2	100.1
1002	M	41	1	98.5	100.0
1003	F	51	2	100.8	101.1
1004	F	46	2	101.1	100.9
1005	F	47	3	100.0	99.8
1006	M	42	2	99.0	100.2
1007	M	43	4	100.7	100.7
1008	F	50	2	100.3	100.9
1009	M	39	1	100.6	101.0
1010	M	32	1	99.9	98.5
1011	M	41	2	101.0	101.4
1012	M	44	2	100.9	100.8
1013	F	47	2	97.4	96.2
1014	F	49	3	98.8	99.6
1015	M	45	3	100.9	100.0
1016	F	42	1	101.1	100.1
1017	M	41	2	100.7	100.3
1018	F	40	1	97.8	98.1
1019	M	45	2	100.0	100.4
1020	M	37	3	101.5	100.8

Data Set 2 gives the length in centimeters of 20 consecutive parts coming off an assembly line. The data are presented in order from left to right.

Data Set 2

Part #	1	2	3	4	5	6	7	8	9	10
Length (cm)	20.011	19.985	19.998	19.992	20.008	20.001	19.994	20.004	20.008	20.000

Part #	11	12	13	14	15	16	17	18	19	20
Length (cm)	20.007	20.004	20.001	19.997	19.984	19.975	19.969	19.984	20.004	20.002

What do you notice about these two data sets? Characteristics on individual people or things are variable; repeated measurements on the same individual are variable. *Data vary!* Because data vary, conclusions based on data are uncertain. Statistics helps us produce useful data that can be analyzed so that we may draw conclusions with a small degree of uncertainty. For example, a doctor considering prescribing a new drug for a patient would like to know how much reduction in blood pressure might be expected at various dosages. Carefully monitoring the dosage and blood pressure of the subjects in the study would provide information about the expected blood-pressure reduction for various dosage levels of the drug. Statistical reasoning allows us to quantify how uncertain our conclusions are.

> *Definitions:*   A **unit** is the item or object we observe. When the object is a person, we refer to the unit as a *subject.*
>
> An **observation** is the information or characteristic recorded for one unit.
>
> A characteristic that can vary from unit to unit is called a **variable.**
>
> A collection of observations on one or more variables is called a **data set.**

Data Set 2 consists of just one variable, length, which was measured for 20 units. Data Set 1 consists of five variables measured on each of 20 subjects. Two of the variables of interest were diastolic blood pressure and gender. How are these two variables different?

 ## 4.2.1   Types of Variables

Variables can be either **qualitative** or **quantitative**. Qualitative variables assume values that are not necessarily numerical, but can be categorized. Gender has two possible values: female and male. These two values can be arbitrarily coded numerically as 1 = female and 2 = male. We just need to remember the coding scheme. Adding, subtracting, or averaging such values has no meaning. Phone numbers are also qualitative with values that are numerical, but you would not add, subtract, or average a set of phone numbers. Type of job would also be a qualitative variable. Observations made on qualitative variables are often called **categorical data.**

Quantitative variables take on numerical values for which adding, subtracting, or averaging such values does have meaning. Examples of quantitative variables are weight, height, and number of children. Sometimes a quantitative variable can be changed to be qualitative. Income, recorded in dollars, is quantitative. However, we could record income as "less than $30,000, $30,000 to under $60,000, and $60,000 or more" and treat income as a qualitative variable.

There are two main types of quantitative variables: **discrete** and **continuous**. A quantitative variable is said to be discrete if its set of possible outcomes is finite or countable. The number of floors in an apartment building in a township with an ordinance that restricts such buildings to a maximum of 20 floors would be a discrete variable. The number of possible values is finite: 1, 2, 3, . . . , 18, 19, or 20. The number of floors could not be 2.1 or 7.6. The grade on your last quiz, which had five questions each worth 0 or 1 point, is discrete. The possible values would be 0, 1, 2, 3, 4, or 5. The number of phone calls made by a salesperson until she makes a sale is also discrete, with a countable number of possible values: 1, 2, 3, ... For a given set of salespeople who have made a sale, there would be a maximum value, but it would be possible that the next salesperson to be observed could have a value that exceeds that maximum. A quantitative variable is said to be continuous if its set of possible outcomes is an interval or a collection of intervals of real numbers.

Height of a newborn baby can be any value between the minimum possible height, h_{min}, and the maximum possible height, h_{max}.

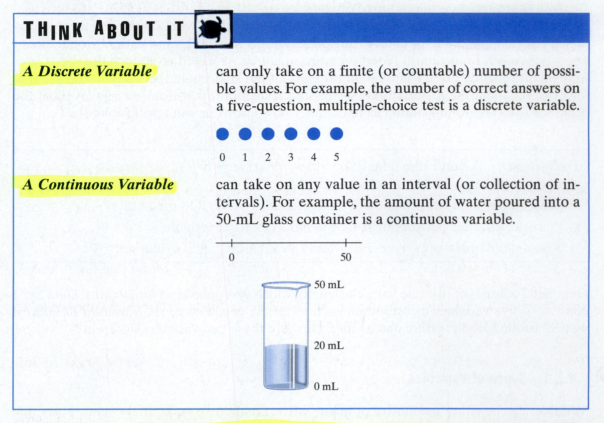

THINK ABOUT IT

A Discrete Variable

can only take on a finite (or countable) number of possible values. For example, the number of correct answers on a five-question, multiple-choice test is a discrete variable.

0 1 2 3 4 5

A Continuous Variable

can take on any value in an interval (or collection of intervals). For example, the amount of water poured into a 50-mL glass container is a continuous variable.

0   50

50 mL

20 mL

0 mL

Sometimes a variable can be treated as discrete *or* continuous. Consider the proportion of women in a population. If the population consists of only 10 people, then the possible proportions are 0.0, 0.1, 0.2,..., 0.9, or 1.0. There are a finite number of possible outcomes, so we have a discrete variable. However, if the population is very large, then the number of possible values for the proportion is also very large (a lot of the values between 0 and 1 are now possible), and for practical purposes, we would treat the proportion of women as a continuous variable. A continuous variable may seem to be discrete because of the way in which it was measured (to some nearest unit). Age is, in fact, continuous (we are constantly growing older with time). However, we often measure age discretely in years.

Data Set 2 consists of one continuous variable, length, measured on 20 units. A unit is a part from the production line. In Data Set 1, gender is a qualitative variable, while age and blood pressure are continuous, and the number of tablets is a discrete variable.

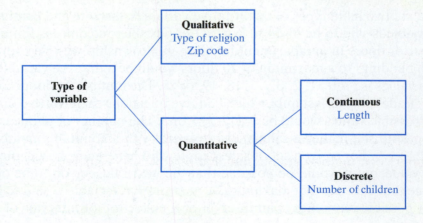

> *Definitions:* **Qualitative variables** are those that classify the units into categories. The categories may or may not have a natural ordering. Qualitative variables are also called categorical variables.
>
> **Quantitative variables** have numerical values that are measurements (length, weight, and so on) or counts (of how many). Arithmetic operations on such numerical values do have meaning. We further distinguish quantitative variables based on whether or not the values fall on a continuum. A **discrete** variable is one for which you can count the number of possible values. A **continuous** variable can take on any value within a given interval.

❖ EXAMPLE 4.1   What Type of Variable? _____

(a) The number of incoming people in the bank between 12:00 noon and 1:00 P.M. on Friday.
  quantitative—discrete

(b) Length of time to run a 50-yard dash.
  quantitative—continuous

(c) You roll two dice and record whether or not the resulting values on the two dice matched.
  qualitative

(d) The numbers on the uniforms of a basketball team. Since adding and subtracting of uniform or player numbers has no meaning—
  qualitative

(e) A children's meal at a fast-food restaurant contains one of five different prizes. A children's meal is purchased and you record which prize was received.
  qualitative

(f) A woman is selected at random from a city. You record whether or not the selected woman has breast cancer.
  qualitative   ❖

4.1 What Type of Variable?

Determine whether the variables that follow are qualitative or quantitative. If the variable is quantitative, state whether it is discrete or continuous.

(a) The weight of various postal letters

(b) The brand of car owned

(c) The total playing time of a CD

(d) The number of songs on a CD

(e) The temperature at noon in Duluth, Minnesota

(f) The amount of rainfall for a season in a city

(g) A person's religion

(h) The length of a rope

(i) The number of different birth dates of students in a class

(j) The number of undergraduate students at a college who suffered from the flu last season

THINK ABOUT IT 🐢

A number of packages are brought to a mailing center. The packages are weighed and the results are recorded as 9 pounds, 5 pounds, 4 pounds, 12 pounds, 20 pounds, and so on. These values are all whole numbers. Does this imply that the variable "weight" is discrete? The variable "weight" is continuous. We have just measured weight to the nearest pound. A package having a value for weight of 12 pounds could actually weigh 12.2 pounds, or 11.9975 pounds, or any value in the interval from 11.5 to 12.5.

Key Point: Don't let the appearance of the data after they are recorded be misleading as to their type.

Consider again the variable "weight." Packages weighing under 5 pounds are classified as *light* and cost a fixed amount to ship. Packages weighing over 20 pounds are classified as *heavy* and cost a fixed amount to ship. Packages weighing between 5 and 20 pounds are classified as *medium* and cost a fixed amount to ship. We record the variable "weight," which takes on the values *light*, *medium*, or *heavy*. Now the variable "weight" is qualitative.

Key Point: The type of variable depends mainly on the measuring process, not on the property being measured.

It is important to ask many questions about the data and how they were obtained, as discussed in the next section.

4.2.2   What? How? Who? When?

A data set is a set of observations or measurements. Whenever you are confronted with a set of data, it is important to ask some questions. Of course, you need to know *what* is being measured. But the *what* of measurement cannot be separated from the *how* of measurement. You need to know how the measurements were taken, as well as what instrument was used to make the measurement.

It is also important to record other information along with the actual variables of interest, if possible. For example, who made the measurement? What day and time was it made? What was the environment like when the measurement was made? Such information is very useful. When looking at the data, if an observation seems to stand apart from the rest (a potential outlier), then the extra information recorded may help you to understand why this observation was different and may warrant excluding it from further analyses.

Consider the following data on graduation rates, presented in the July 23, 1993 edition of the *New York Times*:

REAL DATA

Graduation rates at selected colleges and universities, as reported to the National Collegiate Athletic Association. Figures are for freshmen who entered in 1986 and had graduated by summer 1992. Athletes are those who had athletic scholarships. (SOURCE: *Chronicle of Higher Education*.)					
	All students	**Athletes**		**All students**	**Athletes**
Yale	96	*	**Univ. of S. California**	66	69
Princeton	95	*	**Indiana**	65	62
Notre Dame	94	84	**Syracuse**	64	69
Virginia	92	88	**Texas (Austin)**	63	55
Georgetown	92	95	**Alabama (Tuscaloosa)**	55	57
Stanford	92	86	**Miami (Fla.)**	55	52
Cornell	90	*	**Ohio State**	54	69
Rice	87	78	**Brigham Young**	48	42
Penn State	77	78	**Howard**	45	49
California (Berkeley)	77	61	**Houston**	31	20
U.S. Military Academy	76	*	**Morgan State**	16	40

* Institution does not award athletic scholarships.

Which college or institution listed had the highest graduation rate for all students? Which had the lowest graduation rate for all students?

The article "Seeking Data, Useful Data, on Graduates," by William Celis, starts out discussing the Student Right to Know and Campus Security Act passed by Congress. This law requires institutions to report campus crime and graduation rates. Graduation rates are often used as a marketing tool by colleges and universities to draw students to their institution. How useful do you find the data in this report?

The article raises questions regarding just what a "graduation rate" is. Is it the percentage of students who graduate in four years? Are there some programs that typically take longer than four years? How are transfer students handled? The primary problem in the reporting of these data is that there are no standards—no guidelines for what constitutes a "graduation rate" (that is, the *what* and *how* of measurement). Therefore, the data may not be comparable and not very useful.

Because of the variety of approaches to producing a graduation rate, it is possible for conclusions to change. For example, Stanford has higher graduation rates than Penn State in both categories. But perhaps if we break down the results by science versus nonscience degree students, Penn State may have higher graduation rates. Recording extra information, such as type of degree, may provide potential subgroups for which data should be examined separately to see if the overall results generalize across the subgroups.

This particular article recognizes the problems in reporting such data. Not all articles reporting data bring such questions to our attention. Asking such questions helps us to digest the information we receive so that we may use it to make good, sound decisions and conclusions.

Let's do it! 4.2 Missing Information

City records indicated that over the past month 12 car accidents occurred at a particular busy street corner. A man was the driver in 10 of the accidents and a woman was the driver in the other 2 accidents. You are asked to compare the car-accident results for men versus women drivers.

(a) What would be your conclusion if approximately 1000 male drivers and 20 female drivers had passed through that intersection over the past month?

(b) What would be your conclusion if approximately 100 male drivers and 200 female drivers had passed through that intersection over the past month?

(c) List various *what, how, who,* and *when* questions that you would want to ask before you compare the results for men versus women.

4.2.3 Distribution of a Variable

If we were to pick a variable and measure the value of this variable for one unit, then another unit, and another unit, and continue measuring for a large number of units, we would find that the values vary from one unit to the next. Variation is present in data (that is, we expect the values of a variable to vary). The **distribution** of a variable is just the collection of the possible values of that variable along with how often each possible value occurs.

The distribution of a variable can be summarized graphically, numerically, and with a model. There are many graphical ways to display a distribution. The method used will depend on the type of variable and the idea to be presented. In Section 4.3, we will examine **pie charts** and **bar graphs** for displaying qualitative data. In Section 4.4, we will see how to use **frequency plots, stem-and-leaf plots**, and **histograms** to display the distribution of a quantitative variable. We will learn how to use **time plots** to view quantitative data that have been gathered over time. Finally, we will use a graph called a **scatterplot** to examine the relationship between two quantitative variables.

> *Definition:*   The **distribution** of a variable provides the possible values that a variable can take on and how often these possible values occur. The distribution of a variable shows the pattern of variation of the variable.

 **Exercises**

4.1 State what type of variable each of the following is. If a variable is quantitative, say whether it is discrete or continuous.

(a) Religious preference.

(b) Amount of milk in a glass.

(c) Credit-card numbers.

(d) The number of students in a class of 35 who turn in a term paper on time.

(e) The brand of personal computer purchased by a customer.

(f) The amount of fluid dispensed by a machine used to fill cups with coffee.

4.2 Determine whether the following variables are qualitative, quantitative discrete, or quantitative continuous:

(a) The number of graduate applications in psychology each year at University of Michigan.

(b) The amount of time required to run 40 yards.

(c) The temperature recorded every half hour for a child with a fever.

(d) The number of traffic accidents each day in Detroit.

(e) The weight of baseball pitchers recorded every half inning.

4.3 A person's occupation, as classified by the Bureau of Labor Statistics, is coded by number. For example, the occupation class "managerial and professional" is assigned the number 1, the class "technical" is assigned the number 2, the class "sales" is assigned the number 3, and so on. These data are best described as (*select one*)

(i) discrete.

(ii) qualitative.

(iii) continuous.

(iv) quantitative.

(v) none of the above.

(vi) not enough information given.

4.4 Two government proposals are being considered for adoption. Proposal A is predicted to result in a 10% unemployment rate, while Proposal B is predicted to result in a 5% unemployment rate. Can you conclude that adoption of both proposals would result in a 15% unemployment rate? Explain.

4.5 A friend tells you that in his English class only 5 students failed the midterm exam, while in his chemistry class a total of 20 students failed the midterm exam. Is the total number of students failing the midterm a valid measure of the difficulty of the class (or perhaps the toughness of the professor)? If it is a valid measure, explain why it is. If it is not a valid measure, suggest a better one.

4.3 Displaying Distributions—Qualitative Variables

Qualitative variables are usually not measured on a numerical scale. They can be categorized. Two common types of graphs for qualitative variables are pie charts and bar graphs.

4.3.1 Pie Charts

To picture the distribution of a qualitative variable, we might use a pie chart. A circle or pie represents the whole (that is, all of the units). The pie is divided into slices or wedges, one for each category or possible value of the qualitative variable. The size of each pie slice (namely, the angle spanned by each wedge) is proportional to the percentage of units falling into each category. Figure 4.1(a) shows a pie chart with one shaded slice representing 25% of the whole. To represent that $\frac{7}{8}$ of a group of people are women, we would divide up the whole pie into eight equal parts and shade in seven of them, as shown in Figure 4.1(b). Pie charts show what percentage of the whole falls into each category. They are fairly simple to interpret and provide a visual picture of the size of each category relative to the whole.

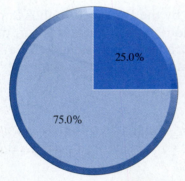
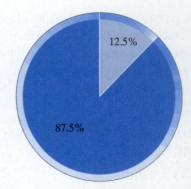

Figure 4.1 (a) Pie chart with one-fourth of the items having some property   (b) Pie chart with seven-eighths of the items having some property

❖ **EXAMPLE 4.2** **Pie Chart for Gender**_____

One of the potential confounding variables in the medical study on blood pressure was gender. In Data Set 1, we have the data on gender for the 20 subjects in this study. The corresponding pie chart for gender is shown here.

We see from the pie chart that 60% of the subjects were male. Since there were 20 subjects in all, we have that 60% of 20, or 12, subjects were male.

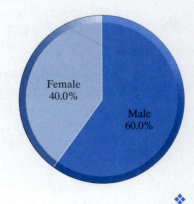

❖

Let's do it! 4.3 College Admissions

The following pie chart shows the breakdown of undergraduate enrollment by race at the University of Michigan for the fall term of 1996. The total number of undergraduates enrolled for that term was 22,604. (SOURCE: "Lawmakers Attack U-M Admissions," *The Ann Arbor News*, May 2, 1997, page A1.)

(a) What percentage of undergraduates enrolled were of nonwhite race?

(b) How many undergraduates enrolled had no racial category listed?

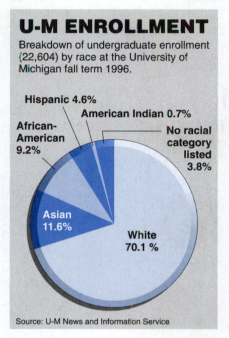

U-M ENROLLMENT
Breakdown of undergraduate enrollment (22,604) by race at the University of Michigan fall term 1996.

Hispanic 4.6%
American Indian 0.7%
African-American 9.2%
No racial category listed 3.8%
Asian 11.6%
White 70.1 %

Source: U-M News and Information Service

Let's do it! 4.4 Grade Distribution

In a summer-session short course at a community college, all students received a final grade of either A, B, or C. The distribution of the grades for the last class of 80 students is given in the following table:

Final Grade	A	B	C
Percentage of Students	25%	70%	5%

(a) How many students received an A in the class?

(b) Use the provided pie circle to produce an approximate pie chart of the grade distribution for this class. (Note: Light dashed lines have been provided that divide the pie into four equally sized pieces for your reference.) Clearly label the pie slices with the category name and the percentage of students in that category.

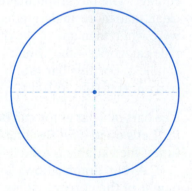

4.3.2 Bar Graphs

Bar graphs also show the percentage of items that fall into each category. A bar graph displays one bar for each category. ==The height of each bar is the proportion or percentage of items in each category. The width of the bars has no meaning, but should be the same for all categories.== Figure 4.2(a) is a bar graph of gender for the 20 subjects in the blood-pressure-reducing medical study. Figure 4.2(b) is a bar graph of the same data using count or frequency on the vertical axis.

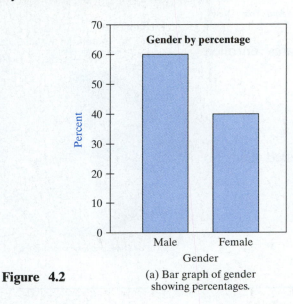 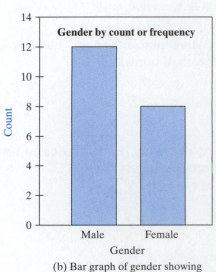

Figure 4.2

(a) Bar graph of gender showing percentages.

(b) Bar graph of gender showing counts or frequencies.

❖ EXAMPLE 4.3 Bar Graph for Race

REAL DATA

The accompanying bar graph displays the breakdown of undergraduate enrollment by race at the University of Michigan for the fall term of 1996.

There is no natural order to the race categories. In this bar graph, the categories are presented in order from the smallest percentage to the largest percentage. The bar graph makes it clear that more undergraduates were white than any of the other race categories—the "white" bar is tallest. For this bar graph the percentages sum to 100%.

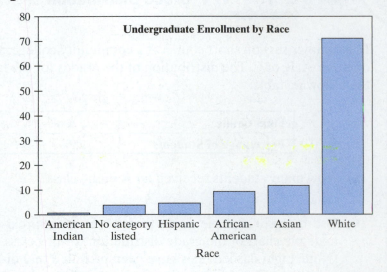

The bars of a bar graph can be either vertical, as in Figure 4.2 and Example 4.3, or horizontal as in "Let's do it! 4.5." ==Bar graphs can be used to represent two qualitative variables at the same time. One variable is used to label an axis and the other to label each of the bars in a group.==

A bar graph allows us to compare several groups or categories by comparing the height of the corresponding bars. Our eyes, however, tend to focus on not just the height of the bars, but their area. This is why it is important to keep the width of the bars the same.

Let's do it! 4.5 Nothing Really Matters

REAL
DATA

The bar graph shown here displays the percentage of respondents who think a particular problem is the most important problem facing America for two different years. (SOURCE: *The Economist*, March 30–April 5, 1996, page 33.)

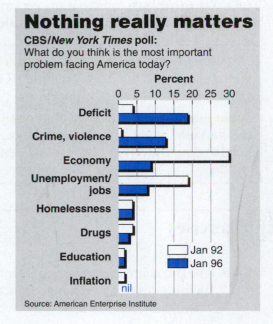

(a) In January 1992, which problem category had the highest percentage of responses?

Was this the same problem category that had the highest percentage of responses in January 1996?

(b) In January 1992, what percentage of respondents reported crime as the most important problem facing America?

In January 1996, what percentage of respondents reported crime as the most important problem facing America?

(c) What is the approximate sum of the percentage of respondents across all of the listed problem categories for January 1992? Is this sum approximately 100%? If not, give a possible reason why not.

❖ EXAMPLE 4.4   A Misleading Bar Graph

The bar graph that follows presents the total sales figures for three realtors. When the bars are replaced with pictures, often related to the topic of the graph, the graph is called a *pictogram*. It is easy to be misled by a pictogram.

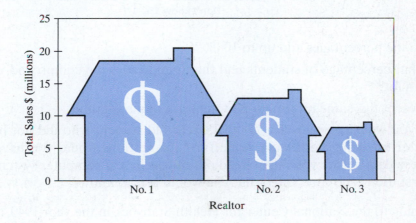

In this graph, the heights of the homes are at the correct levels. Realtor 1 is slightly over twice that of Realtor 3. However, to magnify the picture to avoid distortion, the area of the home for Realtor 1 is more than four times the area of the home for Realtor 3.

Tip: When you see a pictogram, be careful to interpret the results appropriately, and do not allow the area of the pictures to mislead you. ❖

> *Definitions:* A **pie chart** displays the distribution of a qualitative variable by dividing a circle into wedges corresponding to the categories of the variable such that the size (angle) of each wedge is proportional to the percentage of items in that category.
>
> A **bar graph** displays the distribution of a qualitative variable by listing the categories of the variable along one axis and drawing a bar over each category with a height (or length) equal to the percentage of items in that category. The bars should be of equal width.

✳ ## Exercises

4.6 The pie chart shown here presents information about how students have perceived their grades to have changed since they have started using a computer. (SOURCE: *USA Today*, March 4, 1997, page D1.)

REAL
DATA

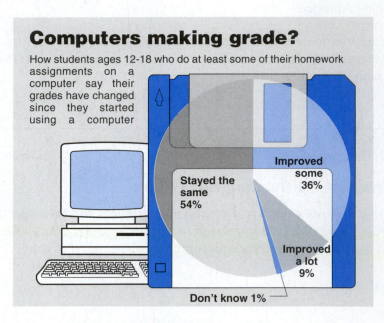

(a) Do the percentages add up to 100%?

(b) What percentage of students feel that their grades have improved, either some or a lot?

(c) Present this same information in the form of a bar graph.

(d) If you were a parent of a 12- to 18-year-old student who did not have a computer for doing homework, what further questions would you have regarding these data before you might head out to purchase a computer system? What information is not presented that you would like to know?

4.7 According to the National Center for Health Statistics, in the year 1940, the percentage of people age 65 who were expected to survive to the age of 90 was 7%. For the year 1960, that percentage was 14%. For the year 1980, that percentage was 25%. Based on projections, the percentages for the years 2000 and 2050 are 26% and 42%, respectively. (SOURCE: "The Old Get Older," *The Ann Arbor News*, February 25, 1997, page A1.) Present a bar graph to display the percentage of people age 65 who are expected to live to the age of 90 for the years 1940, 1960, 1980, and projections for the years 2000 and 2050.

4.8 Shown here is a re-creation of a graph that was presented in the *Country Sampler Decorating Ideas* magazine (October 2000). Is this an appropriate *statistical* presentation of the data? Explain.

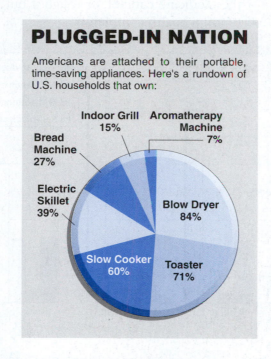

PLUGGED-IN NATION

Americans are attached to their portable, time-saving appliances. Here's a rundown of U.S. households that own:

Indoor Grill 15%
Aromatherapy Machine 7%
Bread Machine 27%
Electric Skillet 39%
Blow Dryer 84%
Slow Cooker 60%
Toaster 71%

4.9 **Thinking of You** An overwhelming number of cards are bought to celebrate Christmas and Valentine's Day. The accompanying graph shows how many cards were exchanged on various holidays. (SOURCE: *The New York Times*, May 15, 1995, page F9.)

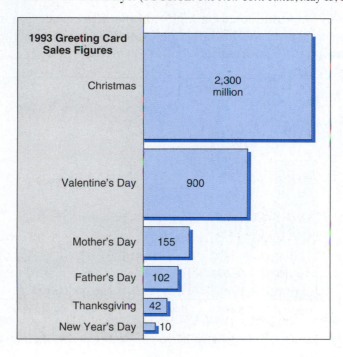

1993 Greeting Card Sales Figures

Christmas — 2,300 million
Valentine's Day — 900
Mother's Day — 155
Father's Day — 102
Thanksgiving — 42
New Year's Day — 10

(a) For Valentine's Day, there were 900 million in card sales, while for Christmas there were over 2.5 times that level (namely, 2300 million). However, the length of the Christmas bar is not 2.5 times the length of the Valentine's Day bar. Explain why this is the case (notice the width of the bars).

(b) Display the sales-figures data for greeting cards with a bar graph using bars of equal widths.

4.10 To learn about Americans' thoughts on health, *Parade Magazine* surveyed a nationally representative sample of 1752 men and women, ages 19 and older. Do we run to the doctor, reach for the medicine cabinet, or tough it out? Here's how the survey respondents treated common health complaints:

REAL
DATA

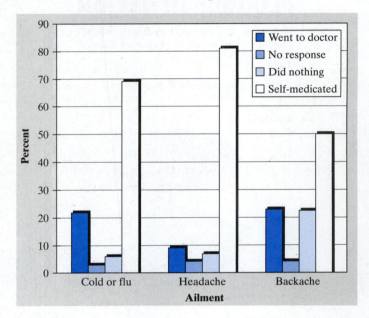

(a) Approximately _____ % of the respondents stated they self-medicate for a cold or flu.

(b) Approximately *how many* respondents did nothing for a backache? _____

4.11 A large survey was conducted on high-ranking officers serving in the army. One of the variables measured was his/her background before joining the army. The following graph displays the information on this variable:

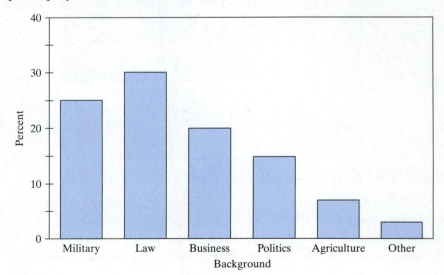

(a) If 620 high-ranking army officers responded to the survey, how many reported a background of business?

(b) In a draft of the written report summarizing the graph for this study, the following statement was given: "The distribution of background for high-ranking army officers is skewed to the right." Is this a correct statement to make? Explain.

 ### 4.3.3 Displaying Relationships between Two Qualitative Variables

Many studies are conducted in an attempt to learn about the relationship between two or more variables. Does having a pet increase the length of survival for patients with coronary heart disease? Is there a relationship between the nutritional status of elementary school children and their academic performance? Frequency tables or contingency tables are used to present the count data on two qualitative variables. As we shall see, the relationship between the two qualitative variables will be better seen if we compute appropriate percents and use a bar graph to display these percents.

   Is there a relationship between the nutritional status of elementary school children and their academic performance? The nutritional status of 1000 elementary school children was assessed and recorded as "poor," "adequate," or "excellent." The academic performance for these children was rated as "below average," "average," or "above average." The data are displayed.

| | | Nutritional Status | | | |
		Poor	*Adequate*	*Excellent*	**TOTALS**
Academic Performance	*Below Average*	70	95	35	200
	Average	130	450	30	610
	Above Average	90	30	70	190
	TOTALS	290	575	135	1000

This *frequency table* shows how many observations fall into the various categories. Each row and column combination is called a *cell* in the table. The value of 130 in the first column and second row of the table tells us that 130 of the 1000 elementary students, or 13%, were classified as having a poor nutritional status and average academic performance. In the preceding table there were 90 students with poor nutritional status who had above average performance, while there were only 70 students with excellent nutritional status who had above average performance. However, the 90 students were from a total of 290 students with poor nutritional status, for only 31%, while the 70 students were from a total of 135 students with excellent nutritional status, for 52%. The relationship between the two variables will be better seen if we compute appropriate percents. Let's take a closer look at the information that can be obtained from a frequency table.

What Information Can We Get from This Table?

From the two-way table for nutritional status and academic performance, we can get the **marginal distribution** for each variable. The marginal distribution of a variable summarizes the percentage of items having each of the possible values of that variable by looking at the row or column totals. To find the marginal distribution for nutritional status, the column variable, we take the column totals and divide by the overall total of 1000. Multiplying this result by 100 allows us to report the percentage for each nutritional level. We see that a little over half of the elementary school children have an adequate nutritional status and just shy of one-third have a poor nutritional status.

	Nutritional Status			TOTAL
	Poor	*Adequate*	*Excellent*	
Percent	29%	57.5%	13.5%	100%

The marginal distribution for academic performance, the row variable, can be found in a similar manner. The values are found by taking the row totals, dividing by the overall total of 1000, and then multiplying by 100 to report the percent.

Let's do it! 4.6 The Other Marginal Distribution

Give the marginal distribution for academic performance.

	Academic Performance			TOTAL
	Below Average	*Average*	*Above Average*	
Percent	_____%	_____%	_____%	100%

What percent of the elementary school children had an above average academic performance rating? an average rating? a below average rating?

From the two-way table for nutritional status and academic performance we can also get **the conditional distribution** of one variable given the other variable. The conditional distribution of the row variable, given the column variable, is found by expressing the entries in the original table as percentages of the column totals. The conditional distribution of the column variable, given the row variable, is found by expressing the entries in the original table as percentages of the row totals.

To find the conditional distribution of academic performance, the row variable, given nutritional status, the column variable, we go to the original two-way table and write out each column of values as a percent of that column's total. Starting with the first column, for poor nutrition, we would take the first value of 70, divide by 290 (which is the first column total), and multiply by 100 to give a percent:

$$\left(\frac{70}{290}\right) \times 100\% = 24.1\%$$

The second value in the first column, 130, would also be divided by 290 and expressed as a percent:

$$\left(\frac{130}{290}\right) \times 100\% = 44.8\%$$

For the third and last value in the first column, 90, we could divide it by 290 and express it as a percent:

$$\left(\frac{90}{290}\right) \times 100\% = 31.0\%$$

We could also obtain this last value by taking the sum of the previous two percentages and subtracting it from 100%:

$$100 - (24.1 + 44.8) = 100 - 68.9 = 31.1\%$$

The two numbers are off a bit due to rounding of the percentages to the nearest tenth. These percentages of 24.1%, 44.8%, and 31.1% give the conditional distribution of academic performance given a *poor* nutritional status. The conditional distribution of academic performance given an *adequate* nutritional status and given an excellent nutritional status would be found similarly. The divisors would be 575 and 135, respectively, instead of 290. The complete table summarizing the conditional distribution of academic performance given nutritional status is given:

		Nutritional Status		
		Poor	*Adequate*	*Excellent*
Academic Performance	*Below Average*	24.1	16.5	25.9
	Average	44.8	78.3	22.2
	Above Average	31.1	5.2	51.9
	TOTALS	100%	100%	100%

First, notice that when reporting the conditional distribution of the row variable given the column variable, the column totals are equal to 100%. What do we see when we examine this table? In general, as the nutritional status improves, we see a shift, a higher proportion of children, toward the higher academic performance class.

The following bar graph displays the conditional distribution of academic performance given nutritional status.

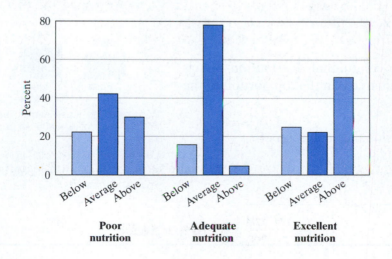

> *Definitions:*   Data on two qualitative variables are often presented in the form of a two-way frequency table. The **marginal distribution of the row variable** is found by computing the percentage of each row total based on the grand total (the entire sample size). The **marginal distribution of the column variable** is found by computing the percentage of each column total based on the grand total.
>
> The association between two qualitative variables is reported by computing some well-chosen percents in the form of a conditional distribution. The **conditional distribution** of the row variable, given the column variable, is found by expressing the entries in the original table as percentages of the column totals. The **conditional distribution** of the column variable, given the row variable, is found by expressing the entries in the original table as percentages of the row totals. If one variable is considered the explanatory variable and the other is the response variable, then the conditional distribution of the response variable, given the explanatory variable, should be examined.

 Let's **do** it!   **4.7** **Which Conditional Distribution?**

REAL
DATA

On March 19, 1995, the *New York Times* printed an article, "Abortion Foes Worry About Welfare Cutoffs," which contained the display shown at the right. The top portion of the display gives information on two variables: income (poor, low, higher) and unplanned pregnancy status (ended in abortion, did not end in abortion).

(a) Of the poor women (aged 15–44), what percent did not have their unplanned pregnancy end in abortion in 1988?

   Answer = _____ %

(b) Name the conditional distribution that is being summarized in the top portion of the display:

   The conditional distribution of

   _____ given

   _____ .

Poor Women & Abortion

Poor women are less likely than other women to terminate an unplanned pregnancy by abortion...

Women aged 15–44 whose unplanned pregnancies ended in abortion in 1988 by income level.

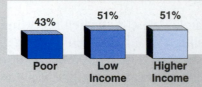

...especially if they are teenagers.

Women aged 15–19 whose unplanned pregnancies ended in abortion in 1988 by income level.

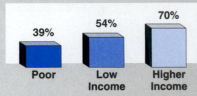

Poor refers to those earning below the Federal poverty level of $14,800 for a family of four in 1994.

Low income earning 100% to 199% of that level.

Higher income earning more than 200% of the Federal poverty level.

Let's do it! **4.8** **Beer Tastes**

The New Brew Pub manufactures and distributes three types of beers: a low-calorie light beer, a regular beer, and a dark beer. To investigate the relationship between gender and beer preference, a sample

		Beer Preference		
		Light	*Regular*	*Dark*
Gender	*Male*	60	120	60
	Female	90	90	30

of 450 beer drinkers was selected. After taste-testing each of the three beers, the individuals were asked to state their preference, defined as their first choice. The preceding table summarizes the responses.

(a) Describe briefly in words what the value of 30 in the table represents.

(b) Using the empty table at the right, give the conditional distribution of beer preference given gender.

		Beer Preference		
		Light	*Regular*	*Dark*
Gender	*Male*			
	Female			

(c) Based on your calculations in (b), should The New Brew Pub initiate a single advertising campaign for all of its beers, or tailor its promotions toward different target markets, such as male versus female? Explain.

(d) Present the conditional distribution in part (b) in the form of a bar graph.

Researchers are often interested in assessing whether or not the relationship between the two qualitative variables is significant. This can be achieved by testing the hypothesis of no association. The technique for actually doing the test involves computing a chi-squared statistic and reporting the corresponding *p*-value. The chi-squared statistic can be thought of as a measure of the distance between the observed counts present in the two-way table and the counts that we would expect to see if there really was no association in the population. Many statistical computer packages and even some calculators can perform this test for you. Details on conducting a chi-square test of independence (or association) are presented in Chapter 14.

❖ EXAMPLE 4.5 Having a Pet

Let's consider the data from a study involving patients with coronary heart disease. Each patient was classified as having a pet or not and whether the patient survived for one year. The data are given in the first table at the right. The objective of the study was to assess if having a pet increased the length of survival for such patients. Thus the explanatory variable is pet status and the response variable is survival status. The conditional distribution of survival status given pet status is displayed at the right. (SOURCE: E. Friedmann, "Animal Companies and 1-Year Survival of Patients after Discharge from Coronary Unit," *Public Health Reports*, 96 (1980) pp. 307–312.)

		Pet Status	
		No Pet	*Pet*
Survival	*Alive*	28	50
Status	*Dead*	11	3

Conditional Distribution

		Pet Status	
		No Pet	*Pet*
Survival	*Alive*	71.8%	94.3%
Status	*Dead*	28.2%	5.7%
	TOTAL	100%	100%

Does having a pet lengthen the lives of patients with coronary heart disease? Is there a relationship between pet status and survival status? From a descriptive standpoint there does seem to be an advantage to having a pet. Approximately 94.3% of the patients with a pet survived for one year, while only 71.8% of those without a pet survived for one year. Is this difference of 22.5% significant?

The output for assessing if having a pet is associated with a higher survival rate using the SPSS computer package is shown. The *p*-value for the test of no association is located under the heading of Significance. The *p*-value of 0.00293 is quite small, indicating support that there is a statistically significant relationship between pet status and survival status.

```
SURVIVAL  Survival Status  by  PETSTATU  Pet Status

                    PETSTATU
          Count  |
                 |No Pet |Pet
                 |       |       |  Row
                 |  1.00|   2.00| Total
SURVIVAL  -------+-------+-------+
          1.00 |   28  |   50  |   78
     Alive     |       |       |   84.8
          -----+-------+-------+
          2.00 |   11  |    3  |   14
     Dead      |       |       |   15.2
          -----+-------+-------+
          Column    39      53  |   92
          Total   42.4    57.6   100.0


Chi-Square              Value           DF     Significance
-----------------       --------        ----   -------------
Pearson                 8.85109          1       .00293
```

A final **cautionary note**: If the p-value for assessing the significance of a relationship is large so the decision is to accept the hypothesis of no association, it does not imply that a relationship was not observed. It means that the observed relationship was not strong enough to be statistically significant. It is possible that the lack of significance is partly attributed to the small sample size. With a small sample, a very strong relationship must exist in order for it to be detected statistically.

 **Let's do it! 4.9 About Your Class**

As a class, select two variables for which you would like to examine the association, if any, between the two variables. Be sure that at least one of the two variables is qualitative.

Variables:

Gather the count data and create a frequency table of the results.

Summarize the results by reporting the two *marginal* distributions and the *conditional distribution* of one variable given the other variable. Think about which conditional distribution you are most interested in.

Write a *brief summary* of your results in words. If you have a computer or calculator available, obtain the p-value for assessing if the relationship is statistically significant.

Exercises

4.12 As part of standard course evaluation, students are asked to rate the course as either poor, good, or excellent. The course evaluation form also asks students to indicate whether the course taken was a required part of their academic program or was taken as an elective. The data are summarized in the following table.

		Poor	Rating Good	Excellent
Reason	*Required*	16	38	16
	Elective	4	10	16

(a) Explain what the value of 4 in the preceding table represents.

(b) Give the conditional distribution of course rating given the reason for taking the course.

(c) Briefly describe your results in part (b)—that is, comment on how the distribution of course rating by students for whom the course is required compares to the distribution of course rating by students for whom the course is an elective.

4.13 Qualifications of male and female head and assistant college athletic coaches were compared in a recent paper. Each of 2225 male coaches and 1144 female coaches was classified according to the number of years of coaching experience. The data are given in the following table.

		Years of Experience 1–6	7–12	13 or more
Gender	*Male*	571	843	811
	Female	484	402	258

(a) Compute (in percents) the conditional distribution of years of coaching experience given gender and report it in a table, as shown.

		Years of Experience 1–6	7–12	13 or more
Gender	*Male*			
	Female			

(b) Write a brief description of the relationship between years of coaching experience and gender.

4.14 An insurance company has examined a large number of insurance claims and has classified them according to type of insurance and according to whether the claim was fraudulent. The data are shown.

		Type of Policy		
		Fire	*Auto*	*Other*
Category	*Fraudulent*	360	20	60
	Nonfraudulent	840	580	940

(a) Based on these data, what proportion of claims are fraudulent?

(b) The company would like to learn more about the relationship between claim category and type of policy. That is, it wishes to compare the fraudulent rates for each type of policy. What conditional distribution should the company compute? The conditional distribution of _____ given _____.

(c) Provide the conditional distribution stated in part (b).

4.15 More Americans take their own lives with guns than have their lives taken from them. In 1991, 18,526 of the nation's 38,317 gun deaths were suicides—outnumbering the 17,746 firearm homicides. Additional data are displayed below. (SOURCE: "Silent but Deadly Statistic in Gun-Control Debate: Suicide," *USA Today*, April 4, 1994.)

REAL DATA

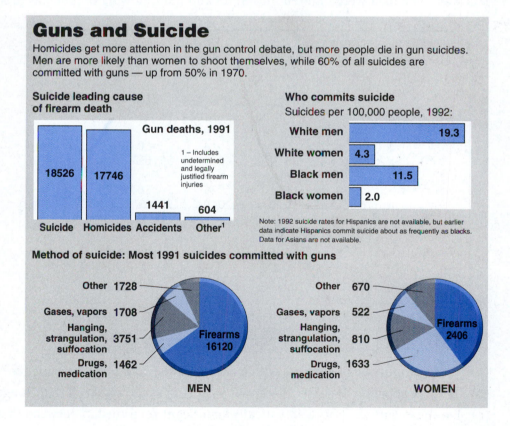

(a) What percentage of men who committed suicide used a gun? What percentage of women who committed suicide used a gun?

(b) Use the figures in the display to produce the conditional distribution of method of suicide given gender.

(c) Sketch the corresponding bar chart to display the conditional distribution in part (b).

(d) Comment on the conditional distribution.

4.16 Travelers who visited the Welcome Center restrooms showed the following results (at left).

Gender	Quality of Restroom Facilities Below Average	Average	Above Average
Female	60	30	10
Male	5	25	70

Gender	Quality of Restroom Facilities Below Average	Average	Above Average
Female			
Male			

In the empty table (at right), give the conditional distribution of quality of facilities given gender.

4.17 A geneticist wanted to learn whether the gene that could cause someone to have a Morton's toe (second toe bigger than the first toe) was related to the gene that allowed people to roll their tongues. A simple random sample of 100 people was obtained and for each person their toe status (BIG TOE: 1 = Morton's Toe, 2 = Not Morton's Toe) and tongue status (TONGUE: 1 = Can Roll, 2 = Can't Roll) were recorded. The data were entered into SPSS and the following output was produced.

```
BIGTOE  by  TONGUE
                      TONGUE          Page 1 of 1
             Count  |
             Exp Val|
                    |Can Roll Can't Roll  Row
                    |    1.00|    2.00| Total
BIGTOE      --------+--------+--------+
               1.00 |   36   |   14   |   50
             Morton |   30.0 |   20.0 | 50.0%
            --------+--------+--------+
               2.00 |   24   |   26   |   50
            Not Morton| 30.0 |   20.0 | 50.0%
            --------+--------+--------+
             Column     60       40      100
             Total    60.0%    40.0%   100.0%


Chi-Square              Value         DF    Significance
------------------      --------      ----  ------------
Pearson                 6.00000       1        .01431
```

(a) What proportion of subjects with a Morton's toe could roll their tongue?

(b) What proportion of subjects without a Morton's toe could roll their tongue?

(c) Do these data support a statistically significant relationship between these two genes? Use a significance level of 5%. Explain.

4.4 Displaying Distributions— Quantitative Variables

Quantitative variables are measured on a numerical scale. Common types of graphs for quantitative variables are frequency plots, stem-and-leaf plots, and histograms. If the data for a quantitative variable have been gathered over time, we might examine the values through a time plot.

 ### 4.4.1 Frequency Plots

A frequency plot is a quick way to display the distribution for the data on the real line. We made use of frequency plots in our Bag A and Bag B examples in Chapter 1. With most frequency plots, the actual values are retained. Each piece of data is represented by an X positioned along a scale. The scale is generally the horizontal scale, but could be the vertical scale. As the name implies, the frequency or count of the values is represented on the other scale.

> **Basic Steps for Creating a Frequency Plot**
>
> - Draw a real line.
> - On the real-line axis, mark the minimum and the maximum values of the data.
> - Fill in the scale for the numbers on the line in between the minimum and maximum in equal-spaced increments. The scale should cover the whole range of values in the data. If the data go from 1 to 29, you might make the scale start at 0 and go to 30.
> - Mark each data value with an X above the appropriate location on the scale. If there are two or more items with the same value, stack them vertically.

Some texts use a dot • instead of an X and refer to the frequency plot as a dot plot.

❖ EXAMPLE 4.6 Ice-Cream-Cone Prices

In September 1985, the prices (in cents) of a single-scoop ice cream cone at 17 Los Angeles stores were as follows:

25, 53, 70, 75, 90, 90, 91, 95, 95, 95, 95, 96, 100, 105, 110, 115, 120.

The minimum value is 25 and the maximum value is 120. The corresponding frequency plot is as follows:

Frequency Plot

The lowest price was 25 cents and the highest price was $1.20. We can immediately see how the data are "bunched" near the nineties. We also see a few values far removed from the rest of the data, the values of 25 and 51. These values stand out due to the gaps in the data between 25 and 51 and between 51 and 70. There is also a slight gap between 75 and 90. ❖

> *Definitions:* **If one or two observations are far removed from the rest of the data, these observations are called outliers.** A group of observations separated from the rest of the data form a **cluster.** A **gap** is a large distance between observations.

 **Let's do it!** **4.10** **Random Number Generator**

Many computers and calculators have the ability to produce a list of random integers. Using your calculator with a seed value of 22, generate 50 random integers between 1 and 10. If you do not have a calculator, use the random number table starting at Row 6, Column 1, and read off the first 50 digits between 0 and 9. Construct a frequency plot of the resulting integers. Note that all integers are used, including repeats.

Describe the distribution.

■TI Shortcut

With a TI, the following steps will generate a sorted L1 list of 50 randomly generated integers between 1 and 10. You can easily make a frequency plot from the sorted L1 list.

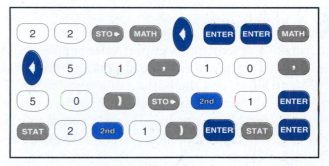

The TI screen should look like this:

```
22→rand
                22
randInt(1,10,50)
→L₁
{5 9 10 4 2 9 8...
SortA(L₁)
            Done
```

 Let's do it! **4.11** How Many Keys?

Question

How many keys do you have with you at this time?

Responses

Keys

_____

Students

Make a frequency plot of your results. Write a summary of the information included in your frequency plot. Be sure to discuss if there are any outliers, gaps, or clusters.

Exercises

4.18 A researcher is interested in learning about the exploratory behavior of a particular strain of rats. He has taken 20 such rats and deposited a rat individually (one at a time) on the bottom of a circular box. The floor of the box has been laid off with a grid, and the number of squares into which each rat stepped in a 30-second interval is recorded. This is a commonly used index of exploratory behavior called the "open-field test." The following data are the number of squares for each of the 20 rats:

Rat	1	2	3	4	5	6	7	8	9	10	11	12	13	14	15	16	17	18	19	20
# squares	3	2	4	4	6	5	6	7	7	6	8	8	7	9	7	8	9	9	11	10

(a) Make a frequency plot of the data.

(b) State an appropriate description for what the plot tells you about the open-field test responses for these rats. (To guide you in writing up this description, recall the paragraph description for the ice cream cone prices given in Example 4.6.)

4.19 The following is the number of goals scored during each game of a hockey team this past season:

3, 5, 2, 2, 1, 4, 0, 0, 3, 3, 2, 1, 8, 2, 4, 1, 1, 3, 2, 5.

Construct a frequency plot of these data.

4.20 Construct a frequency plot of the age data for the 20 subjects in the blood-pressure-reducing medical study. The age data are presented in Data Set 1 (page 184).

4.21 Obtain the scores for your class on a recent exam or quiz and present the distribution of those scores with a frequency plot.

 4.4.2 Shapes of Distributions

Based on most plots, we can comment on the *overall shape* of the distribution, the *approximate center* of the distribution, and any *deviations* from the overall shape. Next is a summary of the general shapes of distributions that often arise with real data. You can think of these curves arising from tracing out the overall pattern that is displayed in a plot such as a stem-and-leaf plot or histogram. One example of a characteristic whose distribution may have the specified shape is provided.

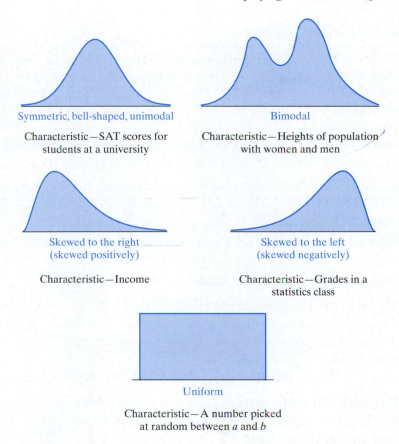

Symmetric, bell-shaped, unimodal

Characteristic—SAT scores for students at a university

Bimodal

Characteristic—Heights of population with women and men

Skewed to the right (skewed positively)

Characteristic—Income

Skewed to the left (skewed negatively)

Characteristic—Grades in a statistics class

Uniform

Characteristic—A number picked at random between *a* and *b*

Shapes of Distributions

- **Symmetric** The distribution can be divided into two parts around a central value, and each part is the reflection of the other.
- **Unimodal** The distribution has a single peak that shows the most common value(s) in the data.
- **Bimodal** The distribution has two peaks. This often results when two populations are being sampled.
- **Uniform** The possible values appear with equal frequency.
- **Skewed** One side of the distribution is stretched out longer than the other side. The direction of the skewness is the direction of the longer side.

 ### 4.4.3 Stem-and-Leaf Plots*

A stem-and-leaf plot, also called a stemplot, is a quick way to display the distribution for a data set with a relatively small number of units. It has the added benefit of retaining the actual values of the variables.

*This section is optional.

Basic Steps for a Stem-and-Leaf Plot

- Separate each measurement into a stem and a leaf—generally the leaf consists of exactly one digit (the last one) and the stem consists of one or more digits. For example,

$$734 \rightarrow \text{stem} = 73, \text{leaf} = 4 \quad \text{or} \quad 2.345 \rightarrow \text{stem} = 2.34, \text{leaf} = 5.$$

Sometimes the decimal is left out of the stem, but a note is added on how to read each value. For the 2.345 example, we would state that 234|5 should be read as 2.345.

Sometimes, when the observed values have many digits, it may be helpful either to *round* the numbers (round 2.345 to 2.35, with stem = 2.3 and leaf = 5) or *truncate* (or drop) digits (truncate 2.345 to 2.34).

- Write out the stems, in equal-spaced increments, in order increasing vertically (from top to bottom) and draw a line to the right of the stems.
- Attach each leaf to the appropriate stem.
- Arrange the leaves in increasing order (from left to right).

❖ EXAMPLE 4.7   Basic Stem-and-Leaf Plot for Age

Consider the ages of the 20 subjects from Data Set 1 (page 184). They range from 32 to 51; so for a basic stem-and-leaf plot, we would use the last digit as the leaf and the first digit as the stem. The basic stemplot is given at the right.

```
3|279
4|011122345556779
5|01
```
Note: 3|2 represents 32 years

Note that the smallest value represented by 3|2 should be read as 32 years.

From this basic stem-and-leaf plot we can see that most of the subjects were in their forties. However, with only three stems and a large number of leaves on a stem, the variation and shape of the distribution is not very well displayed.

One useful modification to the basic stem-and-leaf plot is to use **split stems**. Instead of having all of the digits 0 through 9 represented on the same stem, we can have two stems that are the same. One of the stems is for the digits 0 to 4 and the other (second) stem is for the digits 5 to 9, shown as follows:

Stem-and-Leaf Plot for Age with Split Stems

```
3|2                             3|2
3|79         For some statistical  *|79
4|01112234   packages, the stem-  4|01112234
4|5556779    and-leaf plot output  *|5556779
5|01         looks like this.     5|01
5|                                 *|
```
Note: 3|2 represents 32 years.

Now we can better see that the distribution of the ages of the subjects is roughly symmetric, centered at approximately 43–44, with no apparent outliers (observations that fall outside the overall pattern of variation).   ❖

❖ EXAMPLE 4.8 Length of Parts

Consider the lengths of the 20 consecutive parts from Data
Set 2 in Section 4.2. The parts were targeted to have a length
of 20 centimeters. An important criterion in the manufacturing of parts is to maintain consistent output as close to the
target as possible. These measurements show small deviations from 20 cm, with the smallest value being 19.969 and
the largest value being 20.011. Using the last digit as the leaf
and the remaining digits to determine the stem, we have the
stem-and-leaf plot at the right.

```
1996 | 9
1997 | 5
1998 | 4 4 5
1999 | 2 4 7 8
2000 | 0 1 1 2 4 4 4 7 8 8
2001 | 1
```
Note: 1996|9 represents 19.969 cm.

Here we have a nonsymmetric distribution, with more variation below the targeted
20 cm than above it. The tail of the distribution trails off to the lower values. If we turn
the stem-and-leaf plot on its side, the tail appears to trail off to the left. Such a distribution is skewed to the left, or negatively skewed. For the center of the distribution, we
might actually find the center value—that is, the value that divides the distribution in
half. With $n = 20$ observations, we might simply take the tenth ($n/2 = 20/2 = 10$) observation (when listed in ascending order), which is 20.000. The average of the tenth and
eleventh observations, which is 20.0005, is the center value, in that exactly half of the observations fall below it and the other half fall above it. (*Note:* This measure of center is
called the *median*. We will study measures of center more formally in Section 5.2.) We
will examine Data Set 2 using another useful graphical technique, called a boxplot, in
Chapter 5.

Suppose that we want to compare a second production process to the one that generated the Data Set 2 results. The lengths of 20 consecutive parts produced by the second
process were obtained. The results for both processes are represented as follows, using a
back-to-back stem-and-leaf plot:

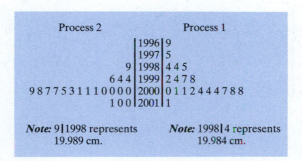

```
            Process 2              Process 1
                        | 1996 | 9
                        | 1997 | 5
                      9 | 1998 | 4 4 5
                  6 4 4 | 1999 | 2 4 7 8
      9 8 7 7 5 3 1 1 1 0 0 0 0 | 2000 | 0 1 1 2 4 4 4 7 8 8
                  1 0 0 | 2001 | 1
```

Note: 9|1998 represents   *Note:* 1998|4 represents
19.989 cm.   19.984 cm.

A back-to-back stem-and-leaf plot is useful for comparing two distributions. Note
that the leaves for the second process are arranged in increasing order away from the
stem (from right to left). In this case, it appears that the second process produces a more
symmetric, less variable distribution (as compared to the first process), and 20.001 is the
center value, for which exactly half of the observations fall below it and the other half
fall above it.   ❖

4.12 **Increase in Income**

The accompanying data are the percentage gain in per capita personal income by state (from 1993 to 1994). (Reported in the article "What Do People Earn?" *Parade Magazine*, June 18, 1995. SOURCE: Bureau of Economic Analysis, U.S. Department of Commerce.)

AL	5.1	HI	2.4	MA	4.9	NM	4.6	SD	9.5
AK	3.1	ID	4.1	MI	8.5	NY	4.7	TN	5.7
AZ	5.1	IL	5.4	MN	7.0	NC	5.4	TX	3.7
AR	5.6	IN	6.1	MS	7.4	ND	8.6	UT	5.6
CA	2.7	IA	10.9	MO	5.9	OH	6.3	VT	4.0
CO	3.9	KS	5.3	MT	2.8	OK	4.2	VA	4.3
CT	4.4	KY	5.4	NE	4.1	OR	5.1	WA	3.8
DE	4.5	LA	6.3	NV	4.9	PA	4.9	WV	6.4
FL	5.0	ME	4.7	NH	4.8	RI	4.7	WI	6.1
GA	5.2	MD	4.3	NJ	4.3	SC	4.9	WY	3.6

(a) Make a stem-and-leaf plot of these data. Be sure to include a note on how to represent each value.

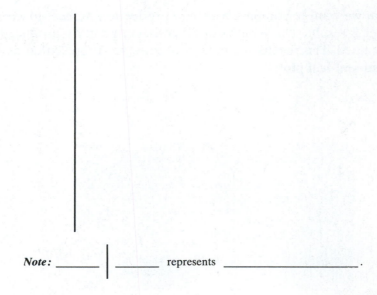

Note: _____ | _____ represents _____ .

(b) Describe the shape of the distribution. Is it roughly symmetric or distinctly skewed? What is the approximate center of this distribution? Are there any clear outliers?

(c) Which state had highest percentage gain? Is this state necessarily the best state to be in with respect to personal income? *Note:* The 1994 per capita income ranged from $16,898 for Arkansas to $29,402 for Connecticut. For Iowa, it was $20,265.

Let's do it! 4.13 Length of Names

Question How many total letters and spaces make up your full name? (e.g., for the name *Fernando Morney*, the number of letter spaces is 15.)

Data Gather this response for 20 people and record the results below.

Full Name	Number of Letter Spaces	Full Name	Number of Letter Spaces
1.		11.	
2.		12.	
3.		13.	
4.		14.	
5.		15.	
6.		16.	
7.		17.	
8.		18.	
9.		19.	
10.		20.	

(a) Make a stem-and-leaf plot of your results. You may wish to try split stems.

Note: _____ | _____ represents _____.

(b) Write a brief summary of the distribution name length based on your plot.

THıNK AʙOᴜᴛ ıᵀ **What Is Wrong?**

Explain why each of the following stem-and-leaf plots does not provide a very good display of the distribution:

Stem-and-leaf Plot 1	Stem-and-leaf Plot 2	Stem-and-leaf Plot 3
27|9	2|1 1 2 2 2 3 4 4 5 5 6 7 7 8 9	18|1
32|0 1 1 7 8	3|0 2 2 3 3 4 6 7 8	19|0
33|1 2 2 5 9	4|0 1 1	20|
34|0 3 4		21|1 2 8
35|1 1		22|0
41|0		23|7
58|6		24|
		25|5 8
		26|2 3
		27|0 5
		28|1 2 9
		29|2
		30|7
		31|6
		32|
		33|0
		34|
		35|0

Note: 27\|9 represents 279.   *Note:* 2\|1 represents 21.   *Note:* 18\|1 represents 181.

Tips: Be sure your increments are equally spaced. Use split stems, rounding, or truncating if you have too few stems with too many leaves or if you have too many stems with too few leaves.

 Exercises

4.22 For each shape presented on page 213 provide another real-life example of a characteristic whose distribution may look like the specified shape.

4.23 The following data on octane ratings for various gasoline blends were gathered by a consumers' organization:

90.5, 88.0, 86.7, 83.4, 87.7, 89.1, 88.8, 88.4, 87.5, 88.5.

Construct a stem-and-leaf plot such that the leaves correspond to the first digit after the decimal point. Be sure to include a note on how to represent each value.

(*Note*: _____\|_____ represents _____.)

4.24 The following data are the percentage gain in per capita personal income by state (from 1996 to 1997). (Reported in the article "What Do People Earn?" *Parade Magazine*, June 14, 1998. SOURCE: Bureau of Economic Analysis, U.S. Department of Commerce.)

REAL DATA

AL	3.9	HI	2.4	MA	5.8	NM	4.1	SD	3.4
AK	2.9	ID	3.1	MI	4.0	NY	5.2	TN	4.5
AZ	4.8	IL	5.0	MN	4.3	NC	4.9	TX	6.0
AR	3.3	IN	4.3	MS	4.0	ND	−1.0	UT	5.4
CA	4.7	IA	3.5	MO	4.4	OH	5.0	VT	3.8
CO	5.1	KS	5.4	MT	4.0	OK	5.0	VA	4.7
CT	6.1	KY	4.5	NE	3.6	OR	5.5	WA	5.7
DE	4.5	LA	4.9	NV	2.8	PA	4.9	WV	4.0
FL	4.4	ME	4.7	NH	4.8	RI	4.7	WI	4.6
GA	4.5	MD	4.7	NJ	4.4	SC	4.3	WY	4.9

(a) Make a stem-and-leaf plot. Be sure to include a note on how to represent each value. (*Note*: _____|_____ represents _____.)

(b) Describe the shape of the distribution. Is it roughly symmetric or distinctly skewed? What is the approximate center of this distribution? Are there clear outliers?

(c) Which state had the highest percentage gain?

(d) Compare the income distribution for 1993–94 (in Let's do it! 4.12) to the distribution for 1996–97 by making a back-to-back stem-and-leaf plot using split stems.

4.25 The following is a list of homework scores for two students:

Student A: 80,   52,   86,   94,   76,   48,   92,   69,   79,   45.

Student B: 73,   87,   81,   75,   78,   82,   84,   74,   80,   76.

(a) Construct a back-to-back stem-and-leaf plot of the data.

(b) Which student do you think has done better work? Explain your answer.

4.26 **Seafood: Fit or Fat?** Believe it or not, some types of seafood are high in fat. The following chart provides the fat, saturated-fat, cholesterol, and caloric content for various 3-ounce servings of seafood.

Type (3 oz)	Fat (g)	Saturated Fat (g)	Cholesterol (mg)	Calories
Salmon (Chinook)	11.4	2.7	72	196
Salmon (Atlantic)	6.9	1.1	60	155
Tuna (Bluefin)	5.3	1.4	42	157
Swordfish	4.4	1.2	43	132
Oysters (Eastern) (12 medium)	4.2	1.1	93	117
Salmon (Pink)	3.8	0.6	57	127
Halibut	2.5	0.4	35	119
Clams (19 small)	1.7	0.2	57	126
Snapper	1.5	0.3	40	109
Crab (Alaska king)	1.3	0.1	45	82
Shrimp	0.9	0.2	166	84
Orange Roughy	0.8	0.0	22	75
Lobster	0.5	0.1	61	83

(a) Which of the four properties are most important to you?

(b) Make a stem-and-leaf plot for the property you listed in (a). Describe its distribution.

(c) Which type of seafood performed the best according to your choice in (a)? the worst?

(d) Pick another of the four properties and make a stem-and-leaf plot for it. Describe its distribution.

(e) Was the best type in (c) still the best? Was the worst type in (c) still the worst?

4.27 The following stem-and-leaf plot shows the lifetime in hours of 15 Longlife light bulbs and 15 Everbright light bulbs:

Longlife		Everbright
	71	
	72	5
6 3 1 0	73	1
9 9 8 7 6 2	74	6
9 7 3 0	75	2
	76	0 2 2 5 7
	77	1 1 3 4 6 8
1	78	
	79	

Note: 72|5 represents 725 hours and 0|73 represents 730 hours.

(a) What is the longest that any light bulb lasted?

(b) Give a good reason why someone might prefer a Longlife light bulb. Give supporting details.

(c) Give a good reason why someone might prefer an Everbright light bulb. Give supporting details.

4.28 A study was conducted to assess how well a new brand of pacemakers are performing. The study involved 20 heart patients who received the new pacemaker. The time (in months) to the first electrical malfunction of the pacemaker was recorded. The data are as follows:

22	14	6	21	24	12	18	16	28	18
16	24	26	28	13	16	23	20	3	22

(a) Create a stem-and-leaf plot for the time-to-malfunction data.

(b) Which of the following description(s) are appropriate for the distribution of time to the first electrical malfunction? Note: Select all that apply.

  (i)   symmetric.

  (ii)   unimodal.

  (iii) bimodal.

  (iv) skewed left.

  (v)   skewed right.

  (vi) uniform.

(c) What proportion of subjects had their first malfunction before five months?

4.29 The table shown here reports percentages of on-time arrivals and on-time departures at 22 major U.S. airports for the fourth quarter of 1991.

   (a) Make a back-to-back stem-and-leaf plot of these data.

   (b) Compare the distributions.

   (c) Do airports perform better overall with respect to on-time arrivals or on-time departures? Explain.

Airport	% On-Time Arrivals	% On-Time Departures
Boston	75.4	83.3
Seattle–Tacoma	75.6	86.6
Los Angeles	76.0	83.9
Chicago	77.0	80.8
Denver	77.0	83.1
Dallas	77.1	84.3
Miami	77.8	88.6
San Francisco	79.6	85.7
Newark	80.1	87.0
Atlanta	81.1	88.1
Minneapolis	81.1	83.8
St. Louis	81.2	85.7
Philadelphia	81.5	86.1
San Diego	82.5	86.5
New York (LaGuardia)	83.1	88.9
Orlando	84.1	91.3
Washington	84.9	89.9
Dulles Int.	85.8	88.8
Las Vegas	86.0	89.0
Charlotte	86.4	87.6
Detroit	87.9	89.8
Raleigh	90.3	92.6

4.4.4   Histograms

A histogram is another way to display the distribution of a quantitative variable. A histogram is better than a stem-and-leaf plot for larger data sets, but does not retain the actual numerical values. It displays the distribution of a variable by showing the frequency (count) or percent of the values that are in various ranges.

Basic Steps for Creating a Histogram

- Find the minimum value, maximum value, and overall range of the data.
- Divide the range of the data (smallest to largest) into classes of equal width. The classes should cover the whole range of values, but they should not overlap. If the data go from 0 to 29, you might make your classes start at 0 and go to 30 with a width of 5.
- Count the number of observations that fall into each class. Recall that the counts are also called frequencies.
- Draw a horizontal axis and mark off the classes along this axis.
- The vertical axis can be the count, the proportion, or the percentage.
- Draw a rectangle (a vertical bar) above each class with the height equal to the count, the proportion, or the percentage.

❖ EXAMPLE 4.9   Histogram of Age

Let's go back to the ages of the 20 subjects from Data Set 1. They range from 32 to 51, so for a histogram we might create classes starting at 30 with increments of 5 through 55. You may have to try different numbers of classes with different widths until you get a good picture. Sometimes a frequency table can help to obtain the cell counts.

Class	Tally	# Observations	Percentage
[30,35)	/	1	1/20 = 0.05 → 5%
[35,40)	//	2	2/20 = 0.10 → 10%
[40,45)	////////	8	8/20 = 0.40 → 40%
[45,50)	///////	7	7/20 = 0.35 → 35%
[50,55)	//	2	2/20 = 0.10 → 10%

The notation [30,35) represents the interval $30 \leq x < 35$—that is, the observations whose values are 30 or greater, but strictly less than 35. The bracket "[" on the left means to include the endpoint 30, and the parenthesis ")" on the right means to consider observations up to, but not including, the endpoint 35. Shown here are two histograms for the age data—one with count as the vertical-axis scale and the other with percent as the vertical-axis scale.

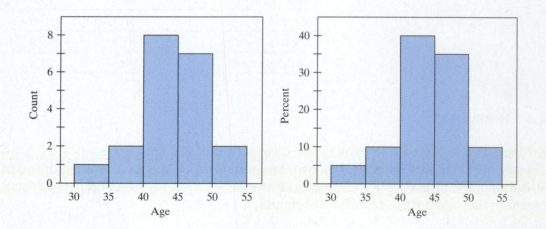

Again we can see that the distribution of the ages of the subjects is roughly symmetric with no apparent outliers. Many of the ages are between 40 and 45 years.   ❖

   In the preceding histograms, note the visual impact of area. In particular, the area of each rectangle is proportional to the frequency or count for each class. The concept of area will come up again when we talk about models for continuous variables or density curves in the next chapter.

The visual perception of the distribution could change dramatically if you change the width of the classes. For example, suppose that data were recorded on the time that accidents at an intersection took place. A histogram with the data grouped by hours might tell you that there were 10 accidents that took place between 9:00 and 10:00 A.M. However, with a histogram in which the data are grouped by the half hour, you might discover that all of the accidents occurred between 9:00 and 9:30 A.M. and none between 9:30 and 10:00 A.M.

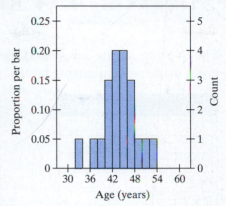

Shown here is a histogram of the age data from Data Set 1 in which the width of the class is two years instead of five years. Does your perception of the distribution change? Is there any new information portrayed in this histogram?

The visual perception of the distribution could also change if you change the scaling on one or both axes. The following two histograms are displaying *the same data*. The histogram on the left portrays a tightly clustered distribution with little spread, while the histogram on the right portrays a distribution with greater variation.

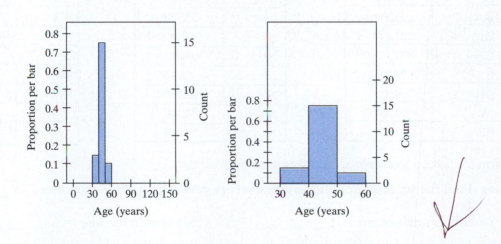

Finally, if we wanted to compare two distributions based on two sets of data, we might prefer to use percents instead of counts when constructing the histograms, especially if the data sets have different numbers of observations. Also, both histograms need to have the same scaling on the horizontal axis.

Back-to-back stem-and-leaf plots are useful for comparing two distributions. Back-to-back histograms can also be used. A common horizontal *x*-axis is used, and the bars for one group go up (as in a regular histogram), while the bars for the other group go down.

Back-to-back histograms for the data on the lifetime of light bulbs from Exercise 4.27 are shown here. The vertical axis of count is appropriate here since there were 15 light bulbs of each brand.

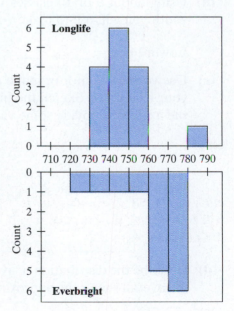

Let's do it! 4.14 Histogram of Household Income

REAL DATA

The average yearly household income in each state and Washington, D.C., for 1995–96 is given in the accompanying tables. (SOURCE: U.S. Bureau of the Census, March 1997, Current Population Survey.)

State	Income($)
AL	28,530
AK	51,074
AZ	31,706
AR	26,850
CA	38,457
CO	41,429
CN	41,775
DE	37,634
Wash, DC	31,811
FL	30,632
GA	33,801
HI	42,944
ID	34,175
IL	39,375
IN	34,759
IA	34,888
KS	31,911

State	Income ($)
KY	31,552
LA	29,518
ME	34,777
MD	43,123
MA	39,604
MI	38,364
MN	40,022
MS	27,000
MO	35,059
MT	28,631
NE	33,958
NV	37,845
NH	39,868
NJ	46,345
NM	25,922
NY	34,707
NC	34,262

State	Income($)
ND	30,709
OH	35,022
OK	27,263
OR	36,470
PA	35,221
RI	36,695
SC	32,297
SD	29,989
TN	30,331
TX	33,029
UT	37,298
VT	33,591
VA	38,252
WA	36,647
WV	25,431
WI	41,082
WY	31,707

You are asked to construct a histogram of these data.

(a) Find the smallest and the largest average yearly household income.

Minimum income = $_____ Maximum income = $_____

(b) Using about 9 or 10 classes, determine the lower limit, the upper limit, and the class width you will use to make your histogram.

Lower limit = _____ Upper limit = _____ Class width = _____

(c) Use your calculator or make a frequency table (like the one on page 222) to construct your histogram. Sketch the histogram using the axes provided below. Be sure to label the axes and provide some values on each.

(d) Describe the distribution of average yearly household income. Include a comment about the overall shape, the approximate center, and any deviations from the overall shape.

Let's do it! 4.15 Money, Money, Money!

Question

How much money in coins, expressed in cents, do you have with you today?

Data

Make a histogram of your group's results. You might need to try different options for the width of each class. Write a summary of the information included in your histogram.

Consider the following four distributions and four characteristics listed. Your task is to determine which distribution best matches which characteristic.

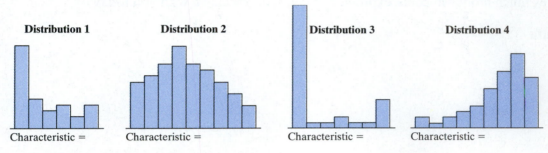

Distribution 1	Distribution 2	Distribution 3	Distribution 4
Characteristic =	Characteristic =	Characteristic =	Characteristic =

Characteristics

1. The age for the population of the United States in the year 1980. Describe and explain the shape of the distribution.

2. The miles of coastline for the 50 United States. Describe and explain the shape of the distribution. Which states do you think would be in the last class furthest to the right?

3. The number of miles traveled to work (that is, commuting distance) for employed adults in a city. Describe and explain the shape of the distribution.

4. The age at death for the population of the United States in the year 1980. Describe and explain the shape of the distribution.

❖ **EXAMPLE 4.10** **Histogram versus Bar Graph** _____

Graph I
Count

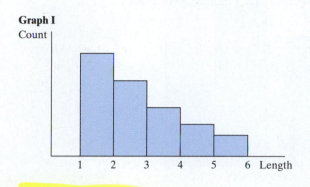

This graph is a histogram.
The variable on the *x*-axis is quantitative and continuous.

Graph II
Count

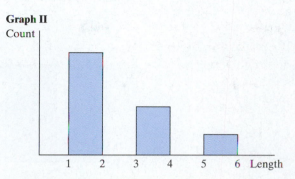

This graph is also a histogram, not a bar graph.
The variable on the *x*-axis is quantitative and continuous.
Some classes just happen to have no observations,
so there is no bar over those classes.

Graph III
Count

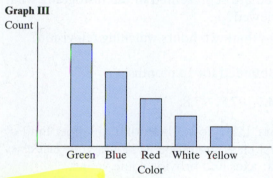

This graph is a bar graph.
The variable on the *x*-axis is qualitative.
The categories on the *x*-axis could be listed in any order.

Graph IV
Count

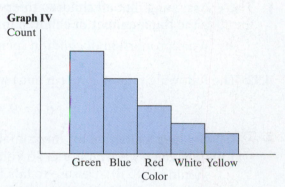

This graph is also a bar graph, not a histogram.
The variable on the *x*-axis is qualitative, but the
bars have no space in between. We should *not*
discuss shape (like skewness) for a bar graph.

 Exercises

4.30 Give a realistic example of a quantitative variable whose distribution is left skewed. Briefly explain your answer.

4.31 A study was conducted to find out the number of hours children aged 8 to 12 years spent watching television. Out of all of the 100 households in a certain housing area identified as having children aged 8 to 12 years, 20 households were to be selected at random, and all children aged 8 to 12 years in the selected households were interviewed. The following histogram was obtained for all the children aged 8 to 12 years interviewed:

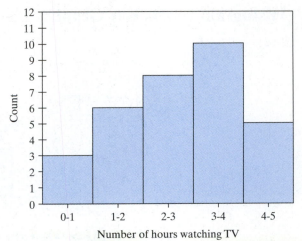

Note: The first class 0–1 represents [0,1), the second class 1–2 represents [1,2), and so on.

(a) Assuming that all children interviewed are represented in the histogram, what is the total number of children interviewed?

(b) What proportion of children spent less than two hours watching television?

4.32 The following rainfall data (in mm) was obtained for 10 months:

1.5, 3.5, 4.4, 5.8, 5.2, 6.7, 9.7, 4.1, 7.8, 7.8.

(a) Create a histogram for the rainfall data. Use a lower limit of 0, an upper limit of 10, and a class width of 2. So your first class should be [0,2) and your last class should be [8,10). Be sure to label your axes and provide some values.

(b) How many months had less than 6 mm of rain?

(c) Describe the distribution of amount of rainfall.

4.33 Data on the amount of time (rounded to the nearest second) that 31 randomly selected customers spent in line before being served at a branch of the First County Bank were entered into a statistical package called SPSS to produce the output shown at the right.

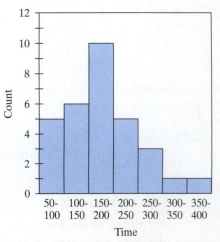

Note: 50−100 represents [50,100).

(a) Describe this distribution of length of time spent in line.

(b) Approximately what percentage of the customers spent less than 100 seconds in line?

(c) How many customers spent five minutes or longer in line?

(d) Based on this histogram, can you determine the maximum length of time spent in line by one of the 31 customers? Explain.

4.34 A study of merit raises at 16 U.S. corporations was conducted to discover the extent to which merit-pay policies for employees are actually tied to performance. One phase of the study focused on the distribution of merit raises (measured as percentage increase in salary) awarded at the companies. The histogram of the 3990 merit increases awarded at one of the largest firms in the study is given. ***Note:*** The first class represents [0, 1).

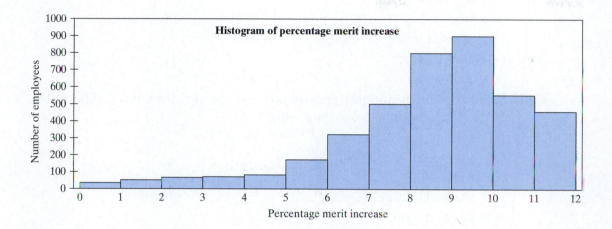

(a) Approximately what proportion of the 3990 merit increases were 10% or higher?

(b) Is the distribution skewed? If so, is it skewed to the right or to the left?

4.35 Consider the data in Exercise 4.29 on percentages of on-time arrivals and on-time departures at 22 major U.S. airports for the fourth quarter of 1991. Make back-to-back histograms of these data.

4.36 Manufacturing Plants A and B produce sports shoes. The histogram of all 20 shoe prices for the 20 different shoes made at Plant A is given by the following figure:

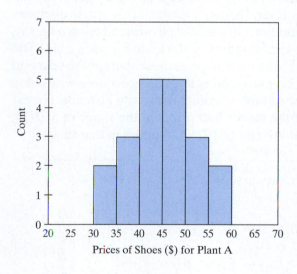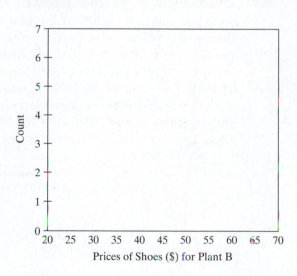

(a) How many different shoes produced by Plant A were priced at $40 or higher?

(b) The following are prices of all 20 sports shoes produced by Plant B:

27, 37, 29, 44, 31, 33, 34, 46, 58, 59, 32, 32, 56, 57, 63, 33, 56, 58, 62, 51.

Make a histogram of shoe prices for Plant B. (Use a lower limit = 20, an upper limit = 65, and a class width = 5.) Sketch the histogram in the space provided.

(c) Select the description(s) appropriate for the distribution of shoe prices for Plant A (*hint*: can be more than one):

(i) symmetric.

(ii) unimodal.

(iii) bimodal.

(iv) skewed left.

(v) skewed right.

(vi) uniform.

(d) Select the description(s) appropriate for the distribution of shoe prices for Plant B (*hint*: can be more than one):

(i) symmetric.

(ii) unimodal.

(iii) bimodal.

(iv) skewed left.

(v) skewed right.

(vi) uniform.

(e) An unknown shoe (that is, it is not known whether it was produced at Plant A or Plant B) has a price tag of $55. We wish to test the following hypotheses:

H_0: Unknown shoe comes from Plant A.

H_1: Unknown shoe comes from Plant B.

(i) The direction of extreme is (select one)   one sided   two sided   can't tell.

(ii) Find the *p*-value corresponding to the observed price of $55.

(iii) At $\alpha = 0.15$, the decision is (select one)   reject H_0 accept H_0 can't tell.

4.37 **Distributions in the stock market**   The *Investor's Business Daily* (July 10, 1995) ran an article entitled "Splits As Popular As Ever, Despite Debate." This article discusses the question of why stock splits are so popular. (In a stock split, instead of owning, say, one $80 share of stock, you might own two $40 shares of stock after a stock split.) The distribution of NYSE (New York Stock Exchange) prices is presented for the year end of 1925, 1960, and 1994 (SOURCE: *The Puzzle of Stock Splits*, by Roger Huang and Martin Weingartner). All three histograms are presented on one graph so comparisons can be made. For example, at the end of 1994, over 20% of the stocks had prices in the range of $10–15, while approximately 6% of the stocks at the end of 1925 had prices in that range.

REAL DATA

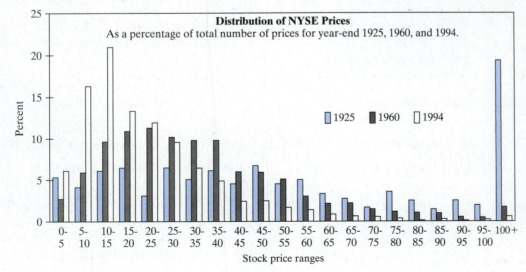

Distribution of NYSE Prices
As a percentage of total number of prices for year-end 1925, 1960, and 1994.

Stock price ranges

(a) Describe the distribution of stock prices for each year (shape, most frequent price class, gaps, clusters).

(b) How has the distribution of stock prices changed over these years?

(c) Does this change or shift in the distribution agree with the idea that stock splits have been so popular lately? Think about what happens to the price of a stock when it splits.

Note: Sometimes, plots can be more confusing than clear. There may be too much information being displayed in the preceding histogram.

REAL DATA

4.38 More distributions in the stock market. In the same *Investor's Business Daily* article on stock splits, two additional distributions are presented—the distribution of presplit stock prices and the distribution of postsplit stock prices.

(a) Compare these two distributions: Do the two distributions look very different? Does it look like there has been a significant change in the distribution? It is difficult to compare these two distributions accurately since they are not on the same scale. A change in the scale of a graph can drastically distort the way information is presented and lead to misrepresentations. The price-range (x) axis ranges from $10–70 for presplit, but $0–40+ for postsplit. The vertical (y) axis reports the number of stocks and not the proportion (or percent) of stocks. After a stock split, there will be more stocks than before the split.

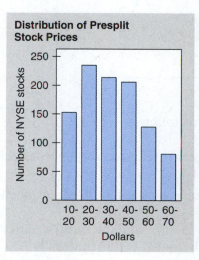

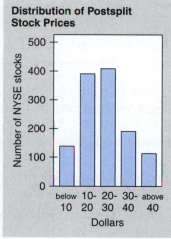

(b) Produce histograms of these two distributions, adjusting for these two factors and removing the space between the consecutive classes. Summarize how the distribution has changed.

4.4.5 Time Plots

When data on a variable are gathered over time, it may be informative to plot the data against either time or the order in which the data were obtained.

A **time plot**, also called a **time series plot**, is a plot of observations against time, or the order in which they were obtained. The consecutive points are connected with lines to help us determine whether the distribution appears to be

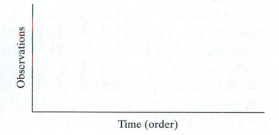

changing over time. Figure 4.3 shows a distribution of the lengths of incoming parts that appears to be changing with respect to the location of the center. Perhaps the shift in the distribution's center is the result of a change in the incoming material.

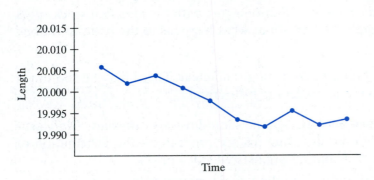

Figure 4.3
An example of a time series plot showing a decreasing trend

Figure 4.4 shows a distribution that appears to be changing with respect to spread or variation. The increasing variation could be the result of tool or machine wear.

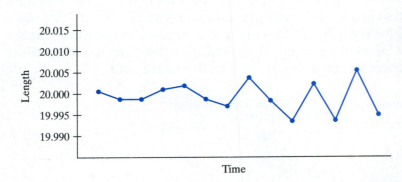

Figure 4.4
An example of a time series plot showing an increasing variation

Patterns that we might see in a time plot are as follows:

- **Trends**—increasing or decreasing, changes in location of the center, changes in variation or spread.
- **Seasonal variation or cycles**—fairly regular increasing or decreasing movements.

Remember, we expect to see some variation in a set of data. But if a pattern is apparent, it is important to ask questions. What may be the cause of that pattern?

In many industries, enhanced versions of the basic time plot are used to monitor the various processes. These enhanced plots are called *control charts*. They are used to study the *stability* of a process. The basic idea is that an interval is identified, using probability and statistics, within which all or most of the plotted observations should fall; points falling outside the limits are highly improbable and therefore suggest that the process may be unstable (out of control). Such unstable patterns or points may lead to a search for a special cause of the instability; and then appropriate action, such as changing the tool or adjusting some settings, may be taken to try to bring the process back under control.

It should be noted that a process that is out of control is not necessarily bad. It may indicate that the process is better. Being in control implies that the distribution of the process is not changing. These series plots will be useful for checking assumptions for some of our inference techniques. Most techniques require that the data be a *random sample* (that is, that the data look like they came from one population with one distribution, not a series of populations with the distribution changing). So a plot of the data against the order in which they were obtained will provide a check that the random-sample assumption is plausible.

Lᴇᴛ's ᴅᴏ ɪᴛ! 4.17 Lateness Patterns?

The following data are the number of students who were tardy (late in arriving) to school during a three-week period:

	Monday	Tuesday	Wednesday	Thursday	Friday
Week 1	10	7	6	8	11
Week 2	14	5	10	8	7
Week 3	9	3	6	4	6

Examine the data through a time plot and comment on what you see. Is there a relationship between the day of the week and the number of late arrivals? Explain.

❖ **EXAMPLE 4.11 Two Ways to "See" the Data**

Frequency plots, stem-and-leaf plots, and histograms display the distribution of a quantitative variable. Since data are collected over time, a time plot can also provide useful information about a set of data. The information from both of these plots can be displayed on the same graph. The accompanying graph

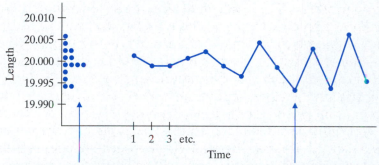

Here, you have some information—namely, that the distribution is roughly symmetric.

The time plot also gives you some information—namely, that the variability appears to be increasing with time.

shows the distribution of the data (see the •'s on the *y*-axis) and the series plot of the data over the time that the data were collected.   ❖

Let's do it! 4.18 Tension Data

The data shown here are measurements of the tension on a wire grid behind the screen in a computer display. The targeted tension level is about 310, with some variation around this level being tolerable. However, if the tension is very low or very high, the display may not work correctly. Thus, the manufacturer measures the tension of a small group of displays sampled each hour from the production line. Here are the hourly averages (means) for 20 consecutive hours of production (starting at the first row and reading from left to right):

269.5	297.0	269.6	283.3	304.8	280.4	283.5	257.4	317.5	327.4
264.7	307.7	310.0	343.3	328.1	342.6	338.8	340.1	374.6	336.1

(a) Make a stem-and-leaf plot or a histogram of the tension data.

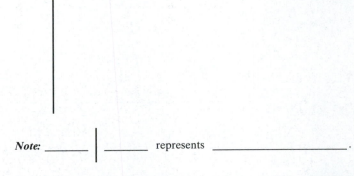

Note: _____ | _____ represents _____ .

(b) Describe the shape of the distribution based on your plot in (a). What might the manufacturer conclude based on this plot alone? (Note: There will be some variation in the data; this is expected.)

(c) Make a time plot of the tension data.

(d) Describe the pattern apparent from this time plot. Based on this plot, what might you recommend that the manufacturer do?

✳ Exercises

4.39 Consider the accompanying two time series plots and two descriptions listed. Your task is to determine which time series best depicts which characteristic. Give an explanation for your match.

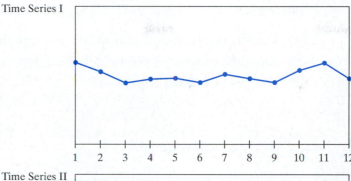

Descriptions

1. A plot of the proportion of girls born each month over the past year for the three hospitals in a city.

2. A plot of the cumulative number of failed batteries each day over the past 12 days.

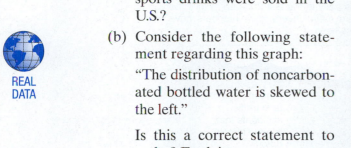

4.40 The graph shown here is from a recent article in the *Ann Arbor News* (May 13, 2001) that discussed whether the bottle-returns law should be expanded to other bottled drinks, such as water, tea, and sports drinks.

(a) Based on the graph, in 1996, approximately how many units of sports drinks were sold in the U.S.?

(b) Consider the following statement regarding this graph:

"The distribution of noncarbonated bottled water is skewed to the left."

Is this a correct statement to make? Explain.

REAL DATA

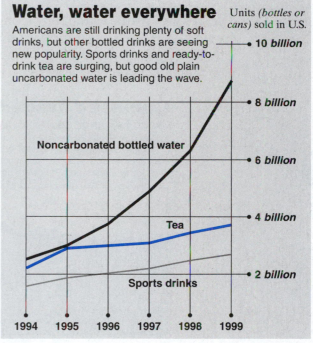

Water, water everywhere Units *(bottles or cans)* sold in U.S.

Americans are still drinking plenty of soft drinks, but other bottled drinks are seeing new popularity. Sports drinks and ready-to-drink tea are surging, but good old plain uncarbonated water is leading the wave.

4.41 The following data are the sales figures (in $1000s) for a grocery store for 12 consecutive months:

140, 131, 136, 139, 141, 145, 149, 148, 151, 152, 151, 154.

(a) Construct a time plot of the data. Be sure that you label your axes.

(b) The time plot in (a) shows (select all that are correct)

    (i)   an increasing trend.

    (ii) a decreasing trend.

    (iii) increasing cycles.

    (iv) decreasing cycles.

4.42 The following data are the times (in seconds) for a person to attach a component inside a machine in an assembly line for 15 consecutive trials:

150, 152, 148, 140, 141, 136, 134, 130, 125, 128, 125, 118, 120, 110, 112.

(a) Construct a time plot of the data. Be sure that you label your axes.

(b) The time plot in (a) shows (select all that are correct)

    (i)   an increasing trend.

    (ii) a decreasing trend.

    (iii) increasing cycles.

    (iv) decreasing cycles.

4.43 The article "Traffic Fatalities among Women on Rise" (SOURCE: *The Ann Arbor News*, February 12, 1995) presents information about the increasing trend in traffic fatalities among women. A time series of the percentage of fatal traffic accidents that involve female drivers is presented.

REAL DATA

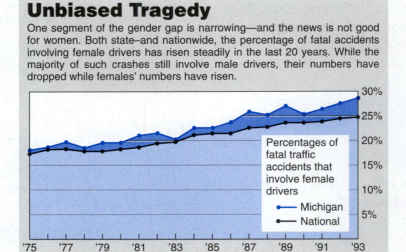

Unbiased Tragedy

One segment of the gender gap is narrowing—and the news is not good for women. Both state–and nationwide, the percentage of fatal accidents involving female drivers has risen steadily in the last 20 years. While the majority of such crashes still involve male drivers, their numbers have dropped while females' numbers have risen.

Percentages of fatal traffic accidents that involve female drivers

— Michigan
— National

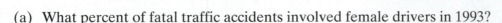

(a) What percent of fatal traffic accidents involved female drivers in 1993?

(b) What percent of fatal traffic accidents involved female drivers in 1975?

(c) What percent of fatal traffic accidents involved male drivers in 1993?

(d) Comment on the results for 1983.

(e) What could be some confounding factors as to why more women are involved in fatal traffic accidents?

4.44 The following data represent the average score on a standardized exam for students at a local school, over the last 20 years, in order of year (from left to right starting with the first row):

40.4 39.2 38.7 40.0 39.1 40.6 38.6 38.7 38.1 38.8
38.5 36.4 36.8 36.9 36.5 36.1 37.1 37.9 35.9 36.4

(a) Make a stem-and-leaf plot of the data. Don't forget a note that reports on how the data are represented.

(b) Make a time plot of the data.

(c) What information is apparent in the time plot, but not in the stem-and-leaf plot?

4.4.6 Scatterplots

As we discussed back in Chapter 3, many statistical studies are conducted to try to detect the relationships or associations between variables, that is, to attempt to show that some explanatory variable predicts the values of some response variable to occur. The data used to study the relationship between two variables is called bivariate data. Bivariate data are obtained by measuring both variables on the same individual or unit. Suppose that we wish to examine the relationship between midterm-exam scores and final-exam scores. Our purpose may be simply to explore the nature of the association, or we may be interested in developing a model that could be used to predict the final-exam score for a student having a given midterm score. In the latter case, the two variables have special names. The **response** variable, also called the **dependent** variable, is the response of interest—that is, the variable we want to predict—and is usually denoted by y. The **explanatory** variable, also called the **independent** variable, attempts to explain the response and is usually denoted by x. So our response variable would be "final-exam score," because we think it may depend on the midterm score, which is our explanatory variable. In the following Let's do it! exercise, we are asked to think of some possible explanatory variables for some specified response of interest.

Let's do it! **4.19** **Possible Explanations**

For each of the following response or dependent variables, list one or more possible explanatory or independent variables (the first one is done for you):

Response Variable	**Possible Explanatory Variable(s)**
Height of a son	*Height of the father, height of the mother, age*
Weight	
GPA at end of a semester	
Exam 2 score	
Quantity demand for a product	

One graphical display of the relationship between two quantitative variables is *called* a **scatterplot**. In Section 4.4.5, we actually constructed a certain type of scatterplot—namely, a series or time plot, which is a scatterplot of the response versus the order or time. In a scatterplot, the values of one variable, generally the independent variable, are marked on the horizontal axis. The values of the other variable, generally the dependent variable, are marked on the vertical axis. Each point on the scatterplot represents a single pair of observations measured on a particular unit (that is, a single **data point** or **case**), and is generically represented as (x_i, y_i), for $i = 1, 2, \ldots, n$, where n is the total number of pairs of observations or the total number of units in our sample. The midterm and final exam scores for a sample of 25 students in a statistics class are listed in the accompanying table and the corresponding scatterplot is shown.

Student Number	x Midterm Score	y Final Score
1	39	62
2	44	69
3	32	68
4	40	86
5	45	88.5
6	46	88.5
7	33	76
8	39	66.5
9	32.5	75
10	**21**	**38**
11	30	71
12	39	88
13	44	96.5
14	28.5	71.5
15	38	96
16	43	82.5
17	42	85
18	25.5	28
19	47	95
20	36	39
21	31.5	58
22	32	49
23	42	62
24	21	59
25	41	90

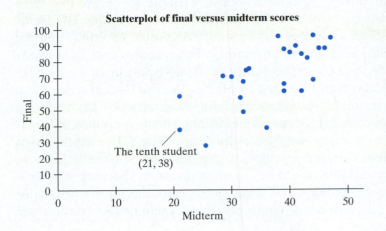

Scatterplot of final versus midterm scores

The tenth student (21, 38)

Let's **d**o **it!**   **4.20**   **Some Data Points**

(a) The point of the scatterplot that corresponds to the scores for Student 10 has been marked. Circle the point that corresponds to the scores for Student 4.

(b) Suppose that a new student scored 30 on the midterm and 40 on the final. Add this point to the scatterplot.

(c) Using your calculator or computer, enter the data and reproduce the above scatterplot.

As with histograms, stem-and-leaf plots, and other graphical techniques, we first want to look for the overall pattern and then look for deviations from the overall pattern. In assessing the overall pattern in a scatterplot, we might comment on the **direction, form**, and

strength. For **direction**, we say the two variables are **positively associated** when larger values of one variable tend to go with larger values of the other, as shown in Figure A. The two variables are said to be **negatively associated** when larger values of one variable tend to go with smaller values of the other, as shown in Figure B.

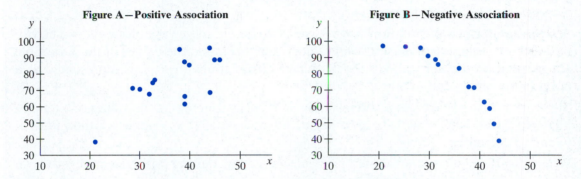

The underlying **form** of the relationship might be linear, curved, quadratic, seasonal, or cyclical, or perhaps there is no clear underlying structure. In Figure A, we might say the underlying form is linear, while there is some curvature in Figure B. When a scatterplot does not show a particular direction, neither positive nor negative, such as Figures C and D, we say that there is **no linear association**. Although there is an association between the two variables in Figure D, the association is not a linear one.

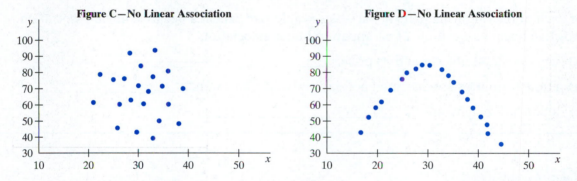

By **strength** we mean how tightly clustered the points are about the underlying form. Here, we have to be a bit careful, because a change in the scale of the axes can "change" the picture to our eye. Since Figures A and B are on the same scales, we can say that the relationship in Figure B is stronger than in Figure A. Later, we will discuss a measure of the strength of the linear relationship, called the correlation coefficient. Observations that appear to deviate from the underlying form may also become evident in a scatterplot. In Figure A, the data point (21, 39), although following the overall linear pattern, does stand apart from the rest. To summarize the relationships shown in the figures, we have the following:

■ Figure A shows a moderately strong, positive, linear relationship.
■ Figure B shows a strong, negative, slightly curved to linear relationship.
■ Figure C shows no association between the two variables.
■ Figure D shows a very strong association, not a linear one, but rather more quadratic.

How might we interpret the scatterplot of the final exam scores versus the midterm-exam scores? The association is certainly positive. Higher midterm scores generally go with higher final exam scores. But there are some exceptions. The pattern is generally linear and of moderate strength.

> ***Definitions:*** Studies are often conducted to attempt to show that some explanatory variable "causes" the values of some response variable to occur. The **response** or **dependent** variable is the response of interest—that is, the variable we want to predict—and is usually denoted by y. The **explanatory** or **independent** variable attempts to explain the response and is usually denoted by x.
>
> A **scatterplot** shows the relationship between two quantitative variables x and y. The values of the x variable are marked on the horizontal axis, and the values of the y variable are marked on the vertical axis. Each pair of observations (x_i, y_i), is represented as a point in the plot.
>
> Two variables are said to be **positively associated** if, as x increases, the values of y tend to increase. Two variables are said to be **negatively associated** if, as x increases, the values of y tend to decrease.
>
> When a scatterplot does not show a particular direction, neither positive nor negative, we say that there is **no linear association**.

Let's do it! 4.21 What Direction?

For each of the following pairs of variables, would you expect a negative linear association, a positive linear association, or no apparent linear association?

(a) age (in years) of a car and its price _____

(b) number of calories consumed per day and a person's weight _____

(c) a person's height and IQ _____

Let's do it! 4.22 Oil-Change Data

The following table presents data on x = the number of oil changes per year and y = the cost of repairs for a random sample of 10 cars of a certain make and model, from a given region:

Number oil changes per year (x)	3	5	2	3	1	4	6	4	3	0
Cost of repairs (\$) (y)	300	300	500	400	700	400	100	250	450	600

(a) Make a scatterplot of the data. Be sure to label your axes and include a few values on each axis.

(b) Comment on the relationship between the two variables with respect to the overall form, direction, strength, and any unusual values.

Let's **d**o **it!** **4.23** **Right versus Left**

When the instructor gives you the signal, begin X'ing in the boxes with your pencil or pen. Use your **right hand**. The instructor will ask you to stop after one minute. Cross out as many boxes as you can during that time.

Number of boxes crossed out = _____.

Now at the signal, do the same, but with your **left hand**.

Number of boxes crossed out = _____.

Let's examine the relationship between right-hand counts and left-hand counts.

Record your class data on x = the number of boxes crossed out with the right hand.

y = the number of boxes crossed out with the left hand.

Data	
x **(right count)**	
y **(left count)**	
x **(right count)**	
y **(left count)**	

(a) Construct a scatterplot of the data.

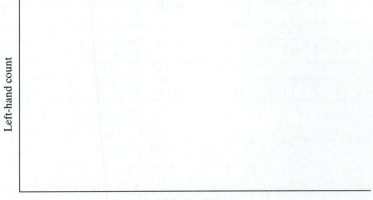

(b) Draw the 45-degree line where x would equal y on your scatterplot. This line would generally divide the observations into two groups. What are those two groups?

(c) If you needed to hire a person who could write quickly and well with both hands, which person would you select?

(d) If you needed to hire a person who could write quickly and well with just their right hand, which person would you select?

✳ Exercises

4.45 (a) Which of the following scatterplots best resembles a positive linear association between the variables x and y?

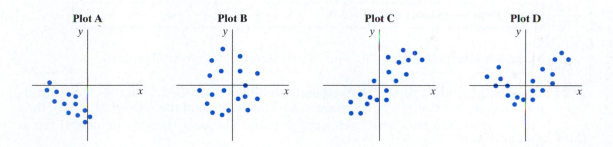

(b) Write a description of the relationship between x and y that is shown in Plot A.

4.46 A student was writing a paper about the income levels of people and religions in which people believed. The student wanted to know if there was any association between these two variables. Religious denomination was selected as the explanatory variable (x) and income as the response (or predicted) variable (y). Data were collected on these two variables, and the scatterplot at the right was obtained. The student concluded that a strong negative association exists between religion and income. Do you agree with the student's conclusion? Explain.

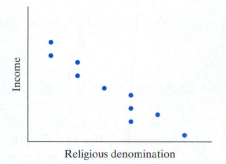

4.47 For each of the descriptions given,

(a) Identify the two characteristics that vary and decide which should be the independent variable and which should be the dependent variable.

(b) Sketch a graph that you think best represents the relationship between the two variables. (You don't have to plot individual points, but do label your axes and draw the general shape.)

(c) Write a sentence that explains the shape and behavior shown in your graph.

  (i) The population of the United States for each census since 1800 (census taken every 10 years)

  (ii) The temperature of an ice-cold drink left in a warm room for a period of time

4.48 Give a realistic example of two quantitative variables that would have a positive association.

4.49 Give a realistic example of two quantitative variables that would have a negative association.

4.50 A random sample of seven households was obtained, and information on income and food expenditures for the past month was collected. The data (in $100s) are given.

Income ($100s) (x)	22	32	16	37	12	27	17
Food Expense ($100s) (y)	7	8	5	10	4	6	6

Make a scatterplot of the data and briefly describe the overall pattern.

4.51 A student wonders if people of similar heights tend to date each other. She measures herself, her dormitory roommate, and the women in the adjoining rooms; then she measures the next man each woman dates. The data (heights in inches) are as follows:

Height of Woman (x)	66	64	66	65	70	65
Height of Man (y)	72	68	70	68	71	65

Make a scatterplot of these data and briefly describe the overall pattern.

4.52 The table at the right reports percentages of on-time arrivals and on-time departures at 22 major U.S. airports for the fourth quarter of 1991. Make a scatterplot of these data and briefly describe the overall pattern.

Airport	% On-Time Arrivals	% On-Time Departures
Boston	75.4	83.3
Seattle–Tacoma	75.6	86.6
Los Angeles	76.0	83.9
Chicago	77.0	80.8
Denver	77.0	83.1
Dallas	77.1	84.3
Miami	77.8	88.6
San Francisco	79.6	85.7
Newark	80.1	87.0
Atlanta	81.1	88.1
Minneapolis	81.1	83.8
St. Louis	81.2	85.7
Philadelphia	81.5	86.1
San Diego	82.5	86.5
New York (LaGuardia)	83.1	88.9
Orlando	84.1	91.3
Washington	84.9	89.9
Dulles Int.	85.8	88.8
Las Vegas	86.0	89.0
Charlotte	86.4	87.6
Detroit	87.9	89.8
Raleigh	90.3	92.6

4.53 The following table lists distances (the explanatory variable, in miles) and cheapest airline fares (the response variable, in dollars) to certain destinations for passengers flying out of Baltimore, Maryland (for January 8, 1995):

Observation	Destination	Distance (x)	Airfare (y)
1	Atlanta	576	178
2	Boston	370	138
3	Chicago	612	94
4	Dallas/Fort Worth	1216	278
5	Detroit	409	158
6	Denver	1502	258
7	Miami	946	198
8	New Orleans	998	188
9	New York	189	98
10	Orlando	787	179
11	Pittsburgh	210	138
12	St. Louis	737	98

(a) Make a scatterplot of these data. Be sure to label your axes.

(b) Is airfare linearly associated with distance? If so, is the association positive or negative?

(c) Would you characterize the association as weak, moderate, or strong? Explain.

4.54 The golf scores of 12 members of a women's golf team in two rounds of tournament play are shown in the following table (a golf score is the number of strokes required to complete the course, so low scores are better). Construct a scatterplot of the scores from round 2 versus the scores from round 1 and briefly describe the overall pattern.

Player	1	2	3	4	5	6	7	8	9	10	11	12
Round 1 (x)	79	91	88	83	106	101	81	85	96	88	91	90
Round 2 (y)	81	89	90	86	90	105	78	81	89	90	96	93

4.5   Guidelines for Plots, Graphs, and Pictures

There are a number of common errors that appear in graphs and plots. Sometimes these errors lead to misinterpretation of the information that is summarized. Here are some guidelines to check before you interpret a plot or as you construct your own graphs:

- Provide an appropriate title of the picture.
- Include the source of the data and relevant details of the sample size and how the data were collected.
- Be sure to label the axes appropriately.
- Check to see if the frequency, proportion, or percentage axis starts at zero.
- Check to see if the axes maintain a constant scale.
- Include the measuring units of the variables being summarized.

THINK ABOUT IT

Time plot 1 shows the percentage of incoming students at a college who have had a statistics course in high school over a five-year period.

Time Plot 1:

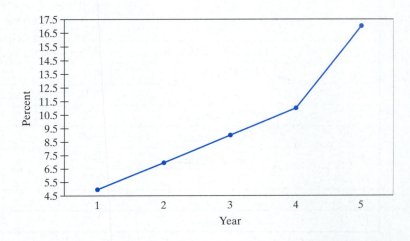

Time plot 1 gives the appearance that over time the percentage has (circle one)

dramatically increased.   *moderately increased.*   *slightly increased.*

Time plot 2 gives the percentage of incoming students at a college who have had a statistics course in high school over a five-year period.

Time Plot 2:

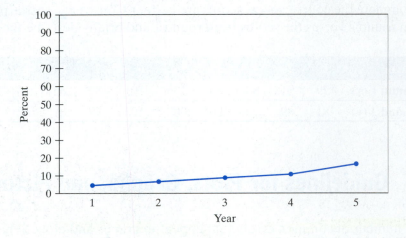

Time plot 2 gives the appearance that over time the percentage has (circle one)

dramatically increased   *moderately increased.*   *slightly increased.*

Look carefully at the two time plots. They display the same data. Notice in Time plot 1 that the vertical axis starts at 4.5 instead of 0. This suppressed zero axis shows a more dramatic increase in the percentage of incoming students than if we had started the axis at zero. By stretching the vertical axis in Time plot 2, we have the appearance of a slightly increasing trend.

The proper way to display these data is shown in Figure 4.5. The vertical axis starts out at zero but is not stretched out so much that it distorts the graph.

Tip: Check out the scaling of the axis!

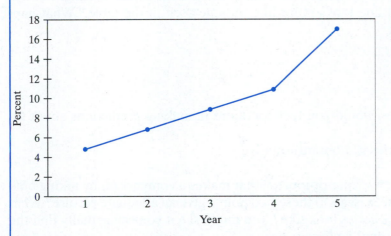

Figure 4.5
Proper time plot of the percentage of incoming students who have had a statistics course in high school over a 5-year period

❖ EXAMPLE 4.12 Where Have the Axes Gone?

A transportation department published its annual report, which included the graph of the number of users of the city bus system over the past five years, shown here. The graph shows an increase in the number of users. However, the graph did not include a horizontal axis, which is needed to interpret the data presented. There was also no mention of how the term *user* was defined.

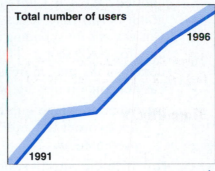

✳ Exercises

REAL DATA

4.55 A letter to the editor entitled "What Wealth Gap?" (SOURCE: *The Wall Street Journal*, July 11, 1995) presents some statistics on wealth in America. Often, information is reported and a comparison is made between *where we are now* and *where we were.* The present figures are compared to one single data point from the past rather than reporting the trend or the time series over the past to the present. (*Note:* The median is the center value such that half of the observations are at or below that value and half of the observations are at or above that value. The median will be

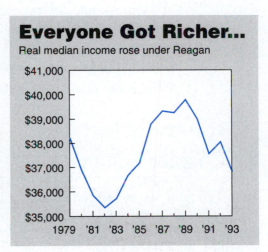

discussed in detail in Chapter 5.) Consider the following statement from the letter: "Regarding income, [the labor secretary] . . . often repeats the line, 'for a decade and a half, ordinary families have been working harder and getting less.' But that is only partly true. True, real median family income was $1300 higher in 1979 ($38,248) than it was in 1993 ($36,959). Taking only the endpoints, that looks like a straight downhill line. But the actual line looks more like the one seen in the table." What happened during the years

(a) 1979–1981?

(b) 1982–1990?

(c) 1990–1993?

(d) Based on this series, would you feel confident in making predictions about future observations?

(The complete time series is more informative.)

4.56 Find an example in a newspaper or journal that makes a comparison by taking only two points from the time series. Do these two points give an adequate picture? What might the complete time series look like? You might see if you can actually find the rest of the data in the complete time series.

REAL DATA

4.57 The time series plot shown here is from the *Investor's Business Daily* (July 10, 1995) article entitled "What Is Clinton Fighting For?" This graph is two time series in one, with the left vertical axis for spending and the right vertical axis for SAT scores. The graph is shown to demonstrate that more spending on schools has not paid off in student achievement.

(a) What are some questions that you would ask in order to better understand the data presented here?

(b) With the current scales, there does appear to be a dramatic increase in spending and pronounced decrease in SAT scores. What is the actual range for SAT scores? Provide a sketch of the data using the actual SAT-score range for the right axis and a range from 0 to 10 (in thousands of 1993 constant dollars) for the left axis.

How do things appear now?

What is the effect of this change in scale?

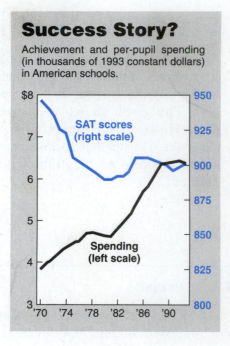

Success Story?

Achievement and per-pupil spending (in thousands of 1993 constant dollars) in American schools.

✳ Chapter Summary

In this chapter, we have been introduced to some of the more common techniques for summarizing data graphically. Graphing our data helps us to "get acquainted" with our data, and graphs provide a visual summary of the data. Many graphical summaries display the distribution of the variable of interest—that is, the pattern of variation. Looking at our data with a visual summary is often our first step toward drawing conclusions from the data. By having an idea of the distribution of a variable, other observations can be compared to it to assess whether the observation is consistent with the distribution.

The distribution of a qualitative variable can be displayed using a pie chart or a bar graph. The relationship between two qualitative variables is best seen by computing the conditional distribution of one variable given the other variable. A bar graph can also be used to display the conditional distribution.

To display the distribution of a quantitative variable, we might use a frequency plot, a stem-and-leaf plot, or a histogram. If the quantitative data have been collected over time, it may be useful to examine the data over time by constructing a time plot. A scatterplot can be used to display and examine the relationship or association between two quantitative variables.

When looking at a graph, we should focus on the overall pattern being displayed and not get lost in the details. Once we have captured the overall pattern, we can focus on any apparent departures from the overall pattern. Graphs are a very powerful way to communicate the data. A visual impression is much stronger than data presented in numerical form. But we must also remember that there are good graphs and graphs that can be misleading and lead to false impressions. Graphs can be used to distort the picture. Changing the scale or class intervals can change the visual outcome.

 Key Terms

Be sure that you can describe, in your own words, and give an example of each of the following key terms from this chapter:

unit	stem-and-leaf plot	conditional distribution
data set	skewed (right or left)	chi-squared test
discrete variable	back-to-back stem-and-leaf	scatterplots
distribution	plot	dependent variable
frequency plot	trends	independent variable
uniform	quantitative variable	response variable
bimodal	bar graph	explanatory variable
time plot	symmetric	positive association
cycles	unimodal	negative association
observation	histogram	direction of a relationship
qualitative variable	seasonal variation	form of a relationship
continuous variable	frequency table	strength of a relationship
pie chart	marginal distribution	

 Exercises

4.58 For each of the variables that follow, state whether it is qualitative or quantitative. If a variable is quantitative, say whether it is discrete or continuous.

(a) Marital status.

(b) The diameter of pebbles from a stream.

(c) The number of automobile accidents per year in Virginia.

(d) The length of time to play 18 holes of golf.

(e) The number of building permits issued each month in a particular city.

4.59 Determine whether the following variables are qualitative, quantitative discrete, or quantitative continuous:

(a) The height of senior students at a college.

(b) The amount of time (measured by a stop watch) it takes an adult to chug a beer.

(c) The temperature at noon during the month of June in International Falls, Minnesota.

(d) The number of countries a student has visited in his or her lifetime.

(e) Whether or not the student has been to Australia.

(f) The number of visitors per day to Walt Disney World in Florida during the month of December.

(g) The amount of gas in a 1-gallon gas can.

4.60 A study was conducted to examine the differences in job performance of white-collar workers with Type A (active) and Type B (passive) behavior. Data were collected for several workers on the variables listed. For each variable, state whether it is qualitative or quantitative. If quantitative, state whether it is discrete or continuous.

(a) Behavior type (A or B).

(b) Age (in years).

(c) Number of employees supervised.

4.61 Pie charts are often used to summarize qualitative data. The accompanying picture shows three pie charts for race, one chart for each of the three possible ways a student may be selected for admission. (SOURCE: University of California at Berkeley, from "Gains in Diversity Facing Attack at University of California," *The New York Times*, June 4, 1995.)

REAL
DATA

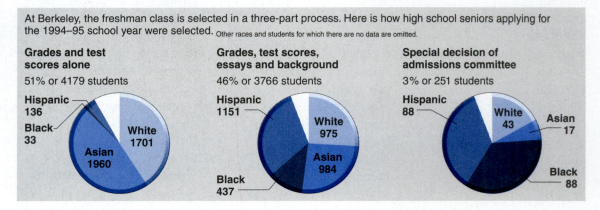

Of all of the students, 51%, or 4179, were selected based on grades and test scores alone. Of these students, 1701 were white, 1960 Asian, 136 Hispanic, 33 Black, and the remaining 349 were of other races or no data available.

(a) How many students were there in all?

(b) What proportion of students selected by a special decision of the admissions committee were Hispanic?

(c) Sketch a pie chart for the students selected by a special decision of the admissions committee that portrays the percent of students in each race category rather than the count.

4.62 On March 25, 1995, *The New York Times* printed an article "HMO's on Rise in New York State" that contained the display shown at the right.

(a) Using the information presented for 1994, fill in the appropriate percentages in the following table.

		Insurance Program		
		Traditional	*H.M.O.*	*Other*
	Northeast			
	Midwest			
Region	*South*			
	West			

(b) Name the conditional distribution that is being summarized in the display and preceding table:

The conditional distribution of _____ given _____.

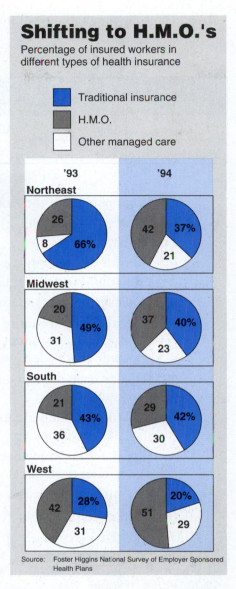

Shifting to H.M.O.'s

Percentage of insured workers in different types of health insurance

- ■ Traditional insurance
- ■ H.M.O.
- □ Other managed care

'93 — '94

Northeast: 26, 8, 66% | 42, 37%, 21

Midwest: 20, 31, 49% | 37, 40%, 23

South: 21, 36, 43% | 29, 42%, 30

West: 42, 28%, 31 | 51, 20%, 29

Source: Foster Higgins National Survey of Employer Sponsored Health Plans

4.63 An American Red Cross brochure provided the following information about the blood of its donors.

		Blood Type			
		Type O	*Type A*	*Type B*	*Type AB*
Rh Factor	*Positive (+)*	45.1%	37.8%	12.2%	4.9%
	Negative (−)	44.4%	38.9%	11.1%	5.6%

This table displays: (select one)

(a) The marginal distribution of blood type.

(b) The marginal distribution of Rh factor.

(c) The overall distribution of blood type and Rh factor.

(d) The conditional distribution of blood type given Rh factor.

(e) The conditional distribution of Rh factor given blood type.

4.64 The following table gives the number of students of a certain college in various categories.

| Gender | Undergraduate | | | | Graduate |
	First Year	Second Year	Third Year	Fourth Year	
Male	1520	1035	876	950	1210
Female	600	610	1465	1328	2120

(a) What percent of the students in this college are undergraduate students?

(b) What percent of the students in this college are male and undergraduate students?

(c) Considering only the female students, what percent of them are either first-year or second-year undergraduate students?

(d) Considering only the graduate students, what percent of them are male?

(e) Overall there are more undergraduate students than graduate students at this college. If each student belongs to one and only one department, then we can conclude that (select one)

  (i)   for each department, there are more undergraduate students than graduate students.

  (ii) for each department, there are more graduate students than undergraduate students.

  (iii) there are some departments that have more undergraduate students than graduate students.

  (iv) none of the above can be concluded.

(f) The table at the right gives (select one)

  (i)   the percentage of males out of the third-year undergraduate students and the percentage of females out of the fourth-year undergraduate students.

	Third Year	Fourth Year
Male	37.4%	41.7%
Female	62.6%	58.3%

  (ii) the conditional distribution of the third-year and fourth-year undergraduate students given they are male.

  (iii) the overall distribution of gender and number of years in school.

(g) The table at the right gives (select one)

  (i)   the percentage of female students.

  (ii) the percentage of undergraduate students and the percentage of graduate students.

	Undergraduate	Graduate
Female	65.4%	34.6%

  (iii) the percentage of undergraduate and graduate students among all female students.

4.65 A study was conducted in a small town on the possible influence of gender on color blindness. A simple random sample of size 200 was taken from among the 1158 female adult residents of the town. Another simple random sample of 200 was taken from among the 1005 male adult residents of the town. Among the 200 selected female residents, 4 had color-blindness, for a proportion of 0.02. Among the 200 selected male residents, 10 had color-blindness, for a proportion of 0.05. The results are presented below.

		Gender Male	Gender Female	TOTAL
Color Blindness Status	Yes	10	4	14
	No	190	196	386
TOTAL		200	200	400

(a) This is an example of (select one)   an observational study   an experiment.

(b) What was the response variable?

(c) What was the explanatory variable?

(d) The proportion of 0.02 is a (select one):   parameter   statistic.

(e) What type of sampling technique was used to obtain the 400 selected adult residents?

(f) One set of hypotheses tested was as follows:

H_0:   There is no association between gender and color blindness status.

H_1:   There is an association between gender and color blindness status.

Based on the results, the null hypothesis H_0 was accepted at a level of 0.05.

For each of the following questions, select your answers:

(i)   Would H_0 also be accepted at a level of 0.02?   Yes   No   Can't Tell

(ii)   Would H_0 also be accepted at a level of 0.10?   Yes   No   Can't Tell

(iii)   Are the results statistically significant at 0.05?   Yes   No   Can't Tell

(g) The table at the right summarizes the (select one)

(i)   joint distribution of color blindness status and gender

(ii)   conditional distribution of color blindness status given gender

(iii) conditional distribution of gender given color blindness status

		Gender Male	Gender Female
Color Blindness Status	Yes	5%	2%
	No	95%	98%

4.66 A new high school in Ann Arbor, Michigan, called the New School, has a loose educational system with an emphasis on student-driven projects. It provides one-on-one attention and a sense of belonging to many students who feel lost in a large public high school. A review was conducted to evaluate the school and provide information to assess the future fate of this alternative high school. The graph given below was part of the evaluation report. (SOURCE: *The Ann Arbor News*, January 18, 2000.)

REAL DATA

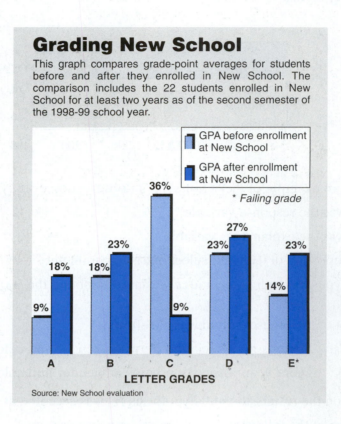

Grading New School

This graph compares grade-point averages for students before and after they enrolled in New School. The comparison includes the 22 students enrolled in New School for at least two years as of the second semester of the 1998-99 school year.

◻ GPA before enrollment at New School
◼ GPA after enrollment at New School

* Failing grade

LETTER GRADES

Source: New School evaluation

(a) The information in the graph can also be presented in a table. Complete the table by providing the missing percentages.

Letter Grade	A	B	C	D	E
Before			36%		
After		23%			

(b) What percent of students had a GPA equivalent to a letter grade of A before they attended the New School? What percent of students had a GPA equivalent to a letter grade of A after they attended the New School?

(c) What percent of students had a GPA equivalent to a letter grade of E before they attended the New School? What percent of students had a GPA equivalent to a letter grade of E after they attended the New School?

(d) Can it be concluded that performance has improved? Is this an appropriate type of summary to assess if the New School is better? What if the grading policy (how grades are determined) at the New School is different from that at the public high school? What other information would you want to know before comparing this before-and-after grade data?

4.67 One of the problems with which health service administrators must deal is patient dissatisfaction. One common complaint focuses on the amount of time that a patient must wait to see a doctor. In a survey to investigate waiting times, medical-clinic secretaries were asked to record patient waiting times (measured from the time of their arrival at the clinic until they saw the doctor) for a sample of patients. The data for one day (30 patients) are summarized here with a stem-and-leaf plot.

Waiting Time (recorded to the nearest five minute interval)

```
 0 | 5
 1 | 0 5 5 5
 2 | 0 0 0 5 5 5 5
 3 | 0 0 0 5 5 5
 4 | 0 5 5 5
 5 | 0 5
 6 | 0 5
 7 | 0
 8 | 5 5
 9 |
10 | 5
```

Note: 0|5 represents 5 minutes.

(a) Is waiting time discrete or continuous? Explain.

(b) Describe this distribution.

(c) What is the approximate center of this distribution?

(d) How do you think the sample was selected? Was it a simple random sample? Explain.

4.68 Health-care issues are receiving much attention in both academic and political arenas. A sociologist recently conducted a survey of citizens over 60 years of age whose net worth is too high to qualify for Medicaid, but who have no private health insurance. The ages of 25 uninsured senior citizens were entered into SPSS, and the histogram shown here was produced.

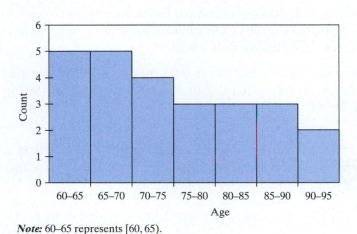

Note: 60–65 represents [60, 65).

(a) What percent of uninsured senior citizens were at least 70, but less than 80, years old?

(b) Are the ages relatively uniformly distributed, or are they predominately the younger or older segment of the senior-citizen population? If they are skewed, in which direction is the skewness?

4.69 In department stores, the amount of radiation emitted by a display of color television sets in a relatively confined area may pose a health problem. The following histogram summarizes readings of radiation (in milliroentgens per hour, i.e., mR/hr) readings for 20 different stores, each having at least five sets in their display areas:

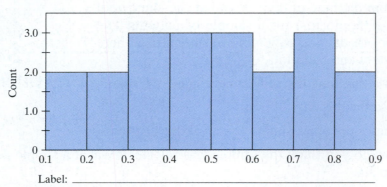

Label: _____

Note: Left endpoint of each class is included; for example [0.1, 0.2).

(a) Give a label to the bottom axis of the graph (i.e., what variable is represented there).

(b) Based on the histogram, can you give the exact range of measurements? If yes, give the exact range. If no, explain why not.

(c) The recommended safety limit for radiation exposure has been set at 0.5 mR/hr by the National Council on Radiation protection. What percentage of the stores have values at or above that limit?

(d) If a small television set emits 0.08 mR/hr, what is the maximum number of sets that you may put in your room (and turn on) before you exceed the safety limit?

4.70 A long-term study of the environmental conditions in a bay gave the following annual average salinity readings for 12 consecutive years (given in order from left to right):

9, 12, 14, 15, 13, 15, 15, 15, 13, 17, 17, 16.

(a) Obtain a time plot for the average salinity readings. Be sure that you label the axes and provide the minimum and maximum values on each axis.

(b) Which of the following is (are) an appropriate description of the plot in (a):

(i)   increasing trend.

(ii) decreasing variability.

(iii) multimodal.

(iv) skewed right.

(v)   decreasing trend.

4.71 A graduate student kept track of the amount of sleep that she has had over the three weeks leading up to midterm exams. The sleeping times in hours are given in the following table:

	Sun	Mon	Tue	Wed	Thu	Fri	Sat
Week 1	9	9.5	9	8	8	9	9
Week 2	8.5	7.5	8	7	7.5	8	7.5
Week 3	7.5	7	7	6	6.5	6	6

(a) Obtain a time plot for the sleeping times shown above. Label the axes and provide the minimum and maximum values on each axis.

(b) Which of the following is an appropriate description of the plot in (a):

   (i)   increasing trend.

   (ii) decreasing variability.

   (iii) in control.

   (iv) skewed right.

   (v)   decreasing trend.

4.72 A local daycare business has six separate daycare centers in Ann Arbor, Michigan. The owner has asked a statistician to conduct a survey to obtain information on the parents' opinions about various aspects of the care their child receives. The ID numbers for the families who have a child attending one of the centers are as follows:

Type of care	Infant	Toddler	Toddler	Toddler	Infant	Toddler
Center	#1	#2	#3	#4	#5	#6
ID numbers	1 through 20	21 through 70	71 through 110	111 through 158	159 through 180	181 through 240

(a) The statistician decides that it is important to have opinions from parents of infants and parents of toddlers. So in the first stage, she divides the centers into two groups by type of care (infant versus toddler) and randomly selects one of the infant-care centers and two of the toddler-care centers. For the next three Mondays, she goes to one of the selected centers at the end of the day and interviews the first eight parents who arrive to pick up their child on that Monday.

   (i)   What type of sampling technique is described in the second stage for selecting the parents to be interviewed at the selected centers?

   (ii) If a family does not have their child signed up for daycare on Mondays (e.g., only Tuesdays and Thursdays), then they would not have a chance to be included in the survey. This is an example of what type of bias?

(b) The rankings by the 24 parents that were interviewed are given in the accompanying table. A higher ranking implies a more favorable opinion about the care received.

81	86	84	41	87	69	76	61	77	79	73	77
77	80	52	78	67	71	64	56	72	65	71	82

(i)   Make a histogram of the rankings by these 24 parents. (Use a lower limit = 40, an upper limit = 90, and a width = 5.) Be sure to label your axes and include a value on the y-axis also.

(ii)   Which description(s) is (are) appropriate for the distribution of rankings? (Select all that apply.) *symmetric, unimodal, bimodal, skewed left, skewed right, uniform*

(c)   The statistician also produced the following time plot of the monthly costs for the daycare business over the past 12 months:

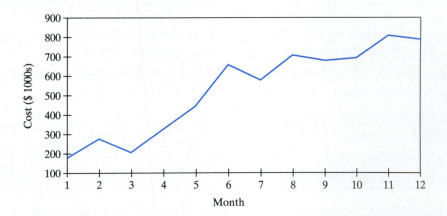

Which description(s) is (are) appropriate for describing this cost time plot? (Select all that apply.) *skewed left, more than one mode, increasing trend, increasing variability*

4.73   Recall the lengths of the 20 consecutive parts from Data Set 2. The parts were targeted to have a length of 20 cm. We have already examined a stem-and-leaf plot and a histogram for these data which showed some left skewness and more variation below 20 cm than above it.

Part #	1	2	3	4	5	6	7	8	9	10
Length (cm)	20.011	19.985	19.998	19.992	20.008	20.001	19.994	20.004	20.008	20.000

Part #	11	12	13	14	15	16	17	18	19	20
Length (cm)	20.007	20.004	20.001	19.997	19.984	19.975	19.969	19.984	20.004	20.002

(a)   Make a time plot of these data.

(b)   What do the data tell you?

(c)   What questions would you want to ask regarding what you see?

4.74   A math teacher is interested in studying the relationship between the level of stress a student had before taking a final math exam and the score on the exam. In particular, she wishes to develop a model for predicting the final math exam score from the stress test score. The following table gives the stress scores before the exam (a higher stress score indicates a higher level of stress) and the final math exam score for 12 students:

Stress test score (x)	6.5	4	2.5	7.2	8.1	3.4	5.5	9.1	7.5	3.8	6.9	2.9
Final math exam score (y)	81	96	93	68	63	84	71	57	70	86	64	90

(a) Make a scatterplot for examining the relationship between math score and stress score. Be sure to label your axes and include values on the axes for the last student with a stress score of 2.9 points and a math score of 90 points.

(b) Based on the scatterplot, does the relationship between math score and stress score appear to be linear?

(c) Based on the scatterplot, does the association between math score and stress score appear to be positive or negative?

(d) Based on the scatterplot, are there any apparent unusual values or outliers?

4.75 The PJH&D Company is in the process of deciding whether or not to purchase a maintenance contract for its new word-processing system. They feel that maintenance expense should be related to usage and have collected the following information on weekly usage (in hours) and annual maintenance expense:

Observation #	1	2	3	4	5	6	7	8	9	10
x = Weekly Usage (in hours)	13	10	20	28	32	17	24	31	40	38
y = Annual Expense (in \$100s)	17	22	30	37	47	30.5	32.5	39	51.5	40

(a) Make a scatterplot for examining the relationship between weekly usage and annual maintenance expense. Be sure to label your axes and include some values on the axes.

(b) Based on the scatterplot, does the relationship between weekly usage and annual maintenance expense appear to be linear?

(c) Based on the scatterplot, does the association between weekly usage and annual maintenance expense appear to be positive or negative?

(d) Based on the scatterplot, are there any apparent unusual values or outliers?

AL
ATA

4.76 In 1929, Edwin Hubble investigated the relationship between distance and radial velocity of extragalactic nebulae (celestial objects). It was hoped that some knowledge of this relationship would give clues as to the way in which the universe was formed and what may happen later. His findings revolutionized astronomy and are the source of much research today. Given here are the data that Hubble used for 24 nebulae. The two variables recorded are the distance (in megaparsecs) from earth and the recession velocity (in km/sec). (SOURCE: Hubble, E. (1929), "RA Relationship between Distance and Radial Velocity among Extra-galactic Nebulae." *Proceedings of the National Academy of Science,* 168.)

Distance	Recession Velocity
0.032	170
0.034	290
0.214	−130
0.263	−70
0.275	−185
0.275	−220
0.45	200
0.5	290
0.5	270
0.63	200
0.8	300
0.9	−30
0.9	650
0.9	150
0.9	500
1.0	920
1.1	450
1.1	500
1.4	500
1.7	960
2.0	500
2.0	850
2.0	800
2.0	1090

(a) Draw a scatterplot of these data.

(b) Comment on the direction and the shape of the relationship (if any).

(c) Are there any observations that stand apart from the rest?

4.77 An article from the November 9, 1998 edition of *The Ann Arbor News* (page A1) included the following graphs to show the distribution of population age for southeast Michigan for the years 1965, 1995, and 2025 (projected):

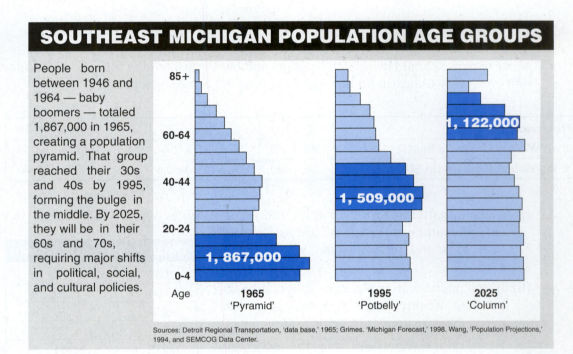

(a) The graphic summary states that "people born between 1946 and 1964—baby boomers—totaled 1,867,000 in 1965." Why does the first graph for 1965 show these 1,867,000 people covering the ages of 0 to 19 years of age?

(b) The graphic reports that a highlighted section totals only 1,509,000 for 1995. Why is this count less than the 1,867,000 count for 1965, but represents the same baby boomers?

(c) Based on the graphs as provided, can you determine the approximate number of people who were 20 to 24 years old in the year 1965? If yes, give the approximate number. If not, explain what information is missing in the graphs.

(d) Describe the shape of the age distribution in 1965, in 1995, and for the projected year 2025.

4.78 An article entitled "Lucky to Be Alive" from the June 12, 2001 edition of *The Ann Arbor News* (page D1) included the following graph to provide information on the automobile fatalities for young drivers:

REAL
DATA

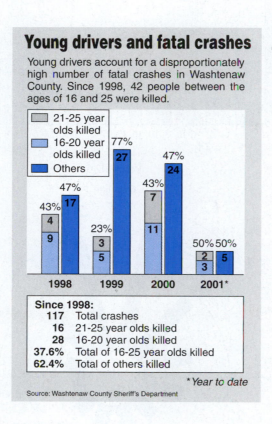

(a) How many people between the ages of 16 and 25 were killed in a fatal car crash in the year 1998? 1999? 2000?

(b) What percentage of the fatal car crashes were of people between the ages of 16 and 25 for the year 1998? 1999? 2000?

(c) What percentage of the fatal car crashes were of people over 25 (other) for the year 1998? 1999? 2000?

(d) Comment on the readability of this graph. Is there any information provided in the graph that is not clear? Explain.

(e) Provide an alternative graph to display the information from (b) and (c). Be creative.

4.79 **What time of the day are crimes most likely to happen?**
Many states summarize their crime statistics each year. In the article "Expanding the Role of the Bar Chart in Representing Crime Data" (*Chance* magazine, Vol. 5, No. 3–4, 1992), data from Utah crime reports were used to produce the following two figures:

R AL
D TA

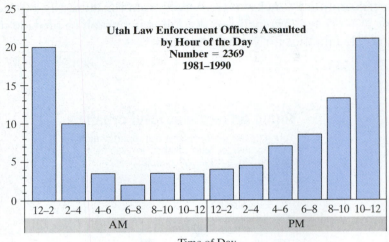

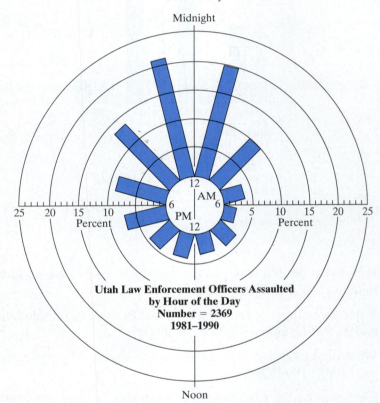

(a) Give a brief summary of what the figures show regarding the time of day that Utah law enforcement officers were assaulted.

(b) Discuss the effectiveness of the two figures in displaying this crime data. Include the pros and cons for each figure. Do you prefer one figure over the other? Why?

(c) What questions arise when you attempt to interpret these data?

4.80 What percent of murders are solved?
The following table summarizes the number of Utah murders and the percentage of those that were solved by the police for each year from 1981 through 1990 (values are approximate):

REAL
DATA

Year	1981	1982	1983	1984	1985	1986	1987	1988	1989	1990
# Murders	53	53	54	48	50	50	54	47	45	51
% Solved	88	67	92	92	80	76	80	96	75	80

Develop a method to display these two levels of information (number of murders and percent solved by year) on one graph or chart. Be creative. Include a brief description about your chart and summarize the crime information.

TI Quick Steps

Producing a Histogram

Consider the 20 ages from Data Set 1 in Section 4.2. You want to make a histogram of that data.

Step 1: **Clear data and plots.**
First, you need to clear out any previous data and graphs by using the following sequence of buttons:

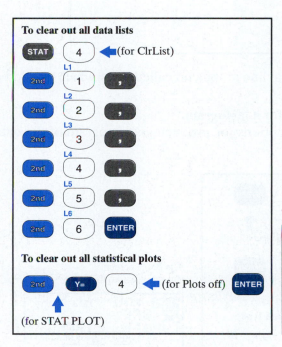

The screen should look like this:

```
ClrList L₁,L₂,L₃
,L₄,L₅,L₆
              Done
PlotsOff
              Done
```

Step 2: **Enter data to be plotted.**
Next, you need to enter your data in Column L1 using the following sequence of buttons:

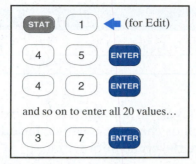

The screen should look like this:

L1	L2	L3	1
45			
42			
41			
40			
45			
37			

L1(21) =

Step 3: **Setting the data range and class width.**

To generate a histogram, you need to specify the range of the values that you want to display and the size of the classes. For our age data, we will start classes at 30 (Xmin) with increments of 5 (Xscl) through 55 (Xmax). We want the *y*-scale to reflect the count so that the scaling of this axis is set at 1 (Yscl = 1). We will start the axis at a negative value, say, −2 (if it were set at 0, we would not be able to see the *x*-axis, which would be hidden on the bottom edge of the screen), and the maximum value must be at least the largest number of observations in a given class (you may need to try various values for Ymax until you can see the complete histogram in your display).

> Press the WINDOW button and change the entries to read
> Xmin = 30
> Xmax = 55
> Xscl = 5
> Ymin = −2
> Ymax = 9
> Yscl = 1
> Xres = 1

Note that there is an additional line at the end called Xres, which should be set to 1, Xres = 1.

Step 4: **Set the STAT PLOT options for a histogram.**

Finally, set the STAT PLOT options for producing a histogram of the data in L1 as Plot 1.

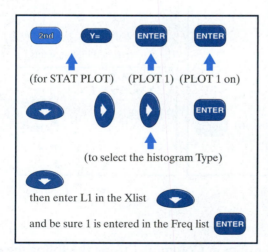

The Plot 1 screen should look like this:

Press the GRAPH button to have the histogram displayed. Use the TRACE button and the right and left arrow keys to see the endpoints for each class and the number of observations in each class.

Producing a Time Plot

Consider the 20 lengths of parts from Data Set 2 in Section 4.2. You want to make a time plot of that data.

Step 1:   **Clear data and plots.**
First you need to clear out any previous data and graphs.

Step 2:   **Enter data to be plotted.**
You need to enter the data and a variable to represent the time or order variable. Enter into L1 the values 1, 2, 3, . . . to 20. Enter into L2 the length data for the 20 consecutive parts.

Step 3:   **Set the STAT PLOT options for a time plot.**
The sequence of steps is as follows:

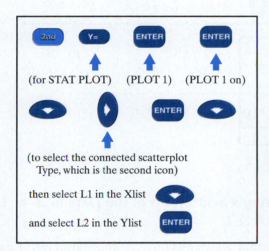

The Plot 1 screen should look like this:

Press the ZOOM button and then "9" to have the time plot displayed. Use the TRACE button and the right and left arrow keys to see values of each point plotted.

Producing a Scatterplot

The TI graphing calculator can be used to make a scatterplot. The steps are given using a small set of data on Test 1 and Test 2 scores.

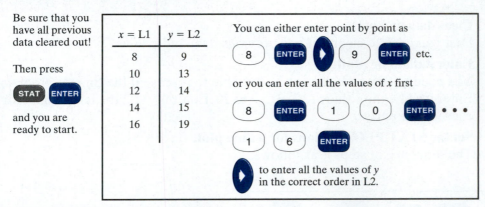

Be sure that you have all previous data cleared out!

Then press

STAT ENTER

and you are ready to start.

x = L1	y = L2
8	9
10	13
12	14
14	15
16	19

You can either enter point by point as

8 ENTER ▶ 9 ENTER etc.

or you can enter all the values of x first

8 ENTER 1 0 ENTER • • •

1 6 ENTER

▶ to enter all the values of y in the correct order in L2.

Your screen should look like this:

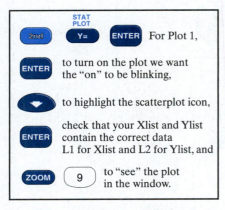

Obtaining a Scatterplot of the Data in $X = $ **L1** and $Y = $ **L2**

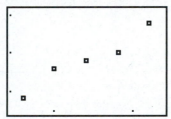

2nd Y= ENTER For Plot 1,

ENTER to turn on the plot we want the "on" to be blinking,

▼ to highlight the scatterplot icon,

ENTER check that your Xlist and Ylist contain the correct data L1 for Xlist and L2 for Ylist, and

ZOOM 9 to "see" the plot in the window.

Be sure that all graphs have been cleared out! Once the scatterplot is produced, we can comment on the overall form, direction, and strength of the association. We can also assess if there are any observations that seem unusual.

Your screen should look like this:

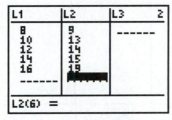

Chapter 5

SUMMARIZING DATA NUMERICALLY

5.1   Introduction

In Chapter 4, we began to summarize and organize our data in order to get information from our data. We focused on how to produce graphical displays of our data Graphical displays allow us to answer questions, such as, "Are most of the values clumped together in the middle with a few values trailing off at each end?" "Which value(s) occurred most often?" and "are there any values that seem unusual in comparison to the rest of the data?"

In this chapter, we continue to learn how to take our data—our collection of measurements—and summarize them in useful ways. We will enhance our graphical displays by presenting various numerical summaries of the data. We begin with the idea of the "center" of the data. What is a typical or average value? We will discuss two common measures of center—the mean and the median—and present another measure called the mode. We will see that these different measures of center can lead to different interpretations of the same data. As we saw in Chapter 4, data vary from unit to unit. Another aspect of a set of data is the spread in the values. How spread out are the values? Do they vary over a wide range or are they close together?

Just as different types of data lead to different types of graphical displays, so do they lead to different types of numerical summaries. We will focus on numerical summaries for quantitative data and will briefly discuss some alternatives if the data are qualitative. The goal of this chapter is to demonstrate the usefulness of a few well-chosen numbers in providing a summary of the data that have been collected.

Formulate theories

YOU ARE HERE

Collect data

Interpret results & make decision

Summarize results

5.2 Measuring Center

In Chapter 4, we examined the characteristic age thought to influence blood pressure. The age data, from Data Set 1, for the 20 individuals who were part of a medical study assessing the ability of a new drug to reduce blood pressure are shown here.

Suppose that you had to give a single number that would represent the most typical age for the 20 subjects. What number would you choose? You would probably pick a number near the center of the age distribution. Just about any presentation of data uses a **measure of center**—average gas mileage, median income, grade point average (GPA). Measures of center are numerical values that tend to report in some sense the middle of a set of data. There are a few measures of center to choose from. The two that we will focus on are the **mean** and the **median**. The mean and median are measures of center for numerical data. Even though gender could be numerically coded with 1 = female and 2 = male, obtaining the average is not meaningful in this case.

If the data are a sample, the mean and median would be called *statistics*. If the data form an entire population, then these measures of center would be called *parameters*.

Data Set 1

Subject #	Gender	Age
1001	M	45
1002	M	41
1003	F	51
1004	F	46
1005	F	47
1006	F	42
1007	M	43
1008	F	50
1009	M	39
1010	M	32
1011	M	41
1012	F	44
1013	F	47
1014	F	49
1015	M	45
1016	F	42
1017	M	41
1018	F	40
1019	M	45
1020	M	37

5.2.1 Mean

> **Definition:** The **mean** of a set of n observations of a quantitative variable is simply the sum of the observation values divided by the number of observations, n.

To find the *mean*, we simply add all of the observation values and divide the total by the number of observations. To find the mean age of the 20 subjects in the medical study, add up the 20 ages and divide by 20:

$$\text{mean} = \frac{45 + 41 + 51 + 46 + 47 + \cdots + 45 + 37}{20} = 43.35 \text{ years.}$$

This most common measure of center has some special notation that we should be familiar with. If $x_1, x_2, \ldots, x_n$ denote a sample of n observations, then the *mean of the sample* is called x-bar and is denoted by

$$\bar{x} = \frac{\sum x}{n} = \frac{x_1 + x_2 + \cdots + x_n}{n}.$$

The Σ is summation notation that basically means "add them all up." The *x*-bar notation is primarily used to represent the mean of *n* observations that are a sample (not the whole population). Any statistical package and most calculators will compute the mean of a set of observations. The TI has a built-in function called 1-Var Stats under the STAT CALC menu. You must first enter the data into a list. With the 20 ages entered into list L1, the steps for finding the sample mean (and many other summary statistics) are as follows.

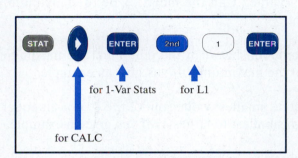

The corresponding result screen should look like this:

```
1-Var Stats
x̄=43.35
Σx=867
Σx²=37981
Sx=4.568484719
σx=4.452808103
↓n=20
```

The 1-Var Stats are now displayed in the window. The only mean provided is $\bar{x}$. If, instead of a sample, you actually have all of the population values, you could compute the mean of the population in the same way: Add up all of the values and divide by how many there are. The *mean of a population* is denoted by the Greek letter μ.

❖ EXAMPLE 5.1   Mean Number of Children per Household _____

The following data are the number of children in a household for a simple random sample of 10 households in a neighborhood: 2, 3, 0, 2, 1, 0, 3, 0, 1, 4. The mean of these 10 observations is

$$\bar{x} = \frac{2 + 3 + 0 + 2 + 1 + 0 + 3 + 0 + 1 + 4}{10} = \frac{16}{10} = 1.6.$$

We report 1.6 even though it is not possible to have 1.6 children in any given household; that is, the 1.6 is not rounded to 2. We are reporting an *average.* Now suppose that the observation for the last household in the previous list was incorrectly recorded as 40 instead of 4. What would happen to the mean?

$$\bar{x} = \frac{2 + 3 + 0 + 2 + 1 + 0 + 3 + 0 + 1 + 40}{10} = \frac{52}{10} = 5.2$$

Note that 9 of the 10 observations are less than the mean. The mean is *sensitive to extreme observations.* Most graphical displays would have detected this observation as an outlier. ❖

THINK ABOUT IT Is the Mean Always the Center? Be Careful!

Suppose that a sample of size $n = 10$ observations is obtained.

Can the mean, $\bar{x}$, be larger than the maximum value or less than the minimum value? If yes, give an example.

Can the mean, $\bar{x}$, be the minimum value? Can the mean, $\bar{x}$, be the maximum value? If yes, give an example.

Can the mean, $\bar{x}$, be exactly the midpoint between the minimum and maximum value (when the minimum does not equal the maximum)? If yes, give an example.

Can the mean, $\bar{x}$, be exactly the second smallest value (out of the 10 not-all-equal observations, when they are ordered from smallest to largest)? If yes, give an example.

Can the mean, $\bar{x}$, not be equal to any value in the sample? If yes, give an example.

What you are discovering with this exercise is that the *mean can be any percentile*. Percentiles are presented in the next section.

Let's do it! 5.1 A Mean Is Not Always Representative

Kim's test scores are 7, 98, 25, 19, and 26. Calculate Kim's mean test score. Explain why the mean does not do a very good job at summarizing Kim's test scores.

Let's do it! 5.2 Combining Means

We have seven students. The mean score for three of these students is 54 and the mean score for the four other students is 76. What is the mean score for all seven students?

❖ **EXAMPLE 5.2** **Mean Salary**

(I) A sample of size $n = 10$ monthly salaries is obtained. Suppose that nine of the values are equal to $1000 and one value is equal to $10,000. Is the mean a "good" measure of average? The mean is

$$\bar{x} = \frac{9(1000) + 10,000}{10} = \frac{19,000}{10} = 1900,$$

which does not measure the center since 90% of the salaries are below the mean.

(II) A sample of size $n = 10$ monthly salaries is obtained. Suppose that nine of the values are equal to $1000 and one value is equal to $100,000. How does this mean compare to the mean in Part I? The mean is

$$\bar{x} = \frac{9(1000) + 100,000}{10} = \frac{109,000}{10} = 10,900,$$

which is much larger than the mean in Part I.

The mean shifted not toward the values that are all the same, but rather toward the extreme value. This can be seen visually through a physical representation of the mean. ❖

The mean is also defined as the point of equilibrium—the point at which the distribution would balance. If the distribution is symmetric, as in the first picture, the mean would be exactly at the center of the distribution. As the largest observation is moved further to the right, making this observation somewhat extreme, the mean shifts towards the extreme observation. If a distribution appears to be skewed, we may wish also to report a more *resistant* measure of center. A measure of center that is more resistant to extreme values is the *median*.

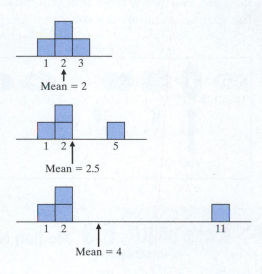

5.2.2 **Median**

> *Definition:* The **median** of a set of n observations, ordered from smallest to largest, is a value such that at least half of the observations are less than or equal to that value and at least half the observations are greater than or equal to that value.

To find the *median* of the 5 numbers 4, 7, 3, 9, 5 we first need to arrange the observations *in order* from smallest to largest: 3, 4, 5, 7, 9. Since the number of observations is *odd*, the median is the middle observation, which is 5. Note that two observations fall below 5 and two fall above 5. If the number of observations is *even*, the median is any number between the two middle observations, including either of the two middle observations. To be consistent, we will define the median as the mean or average of the two middle observations. So the median of 3, 4, 5, 7, 9, 11 is the mean of 5 and 7, which equals 6.

In general, the location of the median can be found by computing $(n + 1)/2$, where n is the number of observations. If $(n + 1)/2$ is a whole number, count up that many observations from the smallest, and the value of that observation is the median. If $(n + 1)/2$ is not a whole number (that is, it looks like xx.5), then count up $(n/2)$ observations from the smallest and average that observation with the next higher observation.

The ages of the $n = 20$ subjects in the medical study have been ordered from smallest to largest and are listed. Since the number of observations is even, we need to find the two middle observations and average them. Calculating $(n + 1)/2$, we get $(20 + 1)/2 = 10.5$. So the two middle observations are the 10th and 11th observations—namely, 43 and 44. The median is the mean of these two middle observations, $(43 + 44)/2 = 43.5$ years.

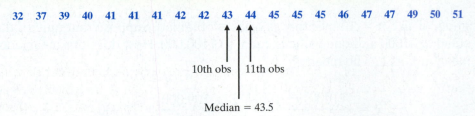

32 37 39 40 41 41 41 42 42 **43** **44** 45 45 45 46 47 47 49 50 51

10th obs | 11th obs

Median = 43.5

Note: The median would be fairly easy to find from an ordered stem-and-leaf plot.

For the TI, the 1-Var Stats option under the STAT CALC menu produces many summary measures, including the median. You must first enter the data into a list. With the 20 ages entered into list L1, the steps for finding the sample median (and many other summary statistics) are as follows. Note that you must use the up and down arrow keys to be able to view the various summary statistics that are provided in the result screen.

The corresponding result screen should look like this:

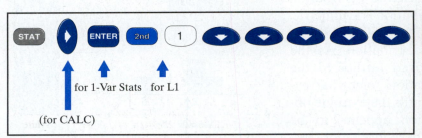

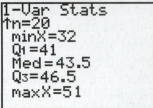

```
1-Var Stats
↑n=20
  minX=32
  Q₁=41
  Med=43.5
  Q₃=46.5
  maxX=51
```

STAT ▶ ENTER 2nd 1 ▼ ▼ ▼ ▼ ▼

for 1-Var Stats   for L1

(for CALC)

✳ Let's do it! 5.3 Median Number of Children per Household

Find the median number of children in a household from this sample of 10 households—that is, find the median of the following:

Observation Number	1	2	3	4	5	6	7	8	9	10
Number of Children	2	3	0	1	4	0	3	0	1	2

(a) Order the observations from smallest to largest.

(b) Calculate $(n + 1)/2 = $ _____.

(c) Median = _____.

(d) What happens to the median if the fifth observation in the list was incorrectly recorded as 40 instead of 4? Note that the value 4 was the largest value.

e) What happens to the median if the third observation in the list was incorrectly record-ed as -20 instead of 0? Note that the value 0 was the smallest value.

Note: The median is resistant—that is, it does not change, or changes very little, in re-sponse to extreme observations.

5.2.3   Another Measure—The Mode

> *Definition:*   The **mode** of a set of observations is the most frequently occurring value; it is the value having the highest frequency among the observations.

To find the mode, it may be helpful to arrange the observations in order so you can see how often each value occurs. The mode of the values $\{0, 0, 0, 0, 1, 1, 2, 2, 3, 4\}$ is 0 be-cause this value occurs most often, a total of four times. For the observations $\{0, 0, 0, 1, 1, 2, 2, 2, 3, 4\}$, there are two modes, 0 and 2, because they are the most fre-quent. This set is said to be *bimodal*. What would be the mode for the set of values $\{0, 1, 2, 4, 5, 8\}$? Here, each value occurs only once. Instead of referring to all of the ob-servations as a mode, it is the practice to say the data have no mode. For the set of values $\{0, 0, 0, 0, 0, 1, 2, 3, 4, 4, 4, 4, 4, 5\}$, we might say there are two modes, 0 and 4, because these values are the most frequent values among their neighboring values.

REAL DATA

The mode is not often used as a measure of center for quantitative data since the most frequent value may be far from the center of the dis-tribution. However, it is the one measure that can be com-puted for qualitative data. Consider the accompanying bar graph, which displays the breakdown of undergraduate enrollment by race at the Uni-versity of Michigan for the fall

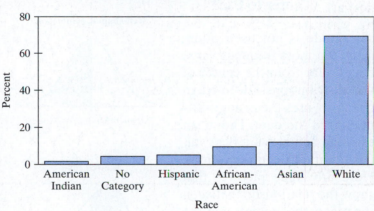

term of 1996. In a bar graph, the *modal category* corresponds to the category with the high-est (longest) bar. We can report that the modal race category is "white." If the various cate-gories were coded as

    1 = White,
    2 = Asian,
    3 = African-American,
    4 = Hispanic,
    5 = American Indian,
    6 = No category listed,

then the mode would be the value 1. However, it is not reasonable to compute the mean race category nor the median value for race since the coded values were arbitrary. The nu-merical values only serve to distinguish the categories, and they cannot be ordered or aver-aged. Sometimes, the categories are presented in order from the smallest percentage to the largest percentage.

❖ **EXAMPLE 5.3   Different Measures Can Give Different Impressions**

The famous trio—the mean, the median, and the mode—represents three different methods for finding a so-called typical value. These three values usually are not the same. When they are different, they can lead to different interpretations of the data being summarized. Consider the annual incomes of five families in a neighborhood:

$$\$12,000, \$12,000, \$30,000, \$90,000, \$100,000.$$

What is the typical income for this group? The mean income is

$$\overline{x} = \frac{100,000 + 90,000 + 30,000 + 12,000 + 12,000}{5} = \$48,800,$$

the median income is $30,000, and the modal income is $12,000.

If you were trying to promote that this is an affluent neighborhood, you might prefer to report the mean income. If you were trying to argue against a tax increase, you might argue that income is too low to afford a tax increase and report the mode. If you want to represent these values with the income that is in the middle, you would report the median. The three different measures are each valid and informative in their own way.   ❖

5.2.4   Which Measure of Center to Use?

The mode is not used often, since the most frequent value may be far from the center of the distribution. The median primarily uses the order information in the data. The mean uses the actual numerical values for all of the observations. Here are some pictures that show the relationship between the mean, the median, and the mode.

The mode is the value at which the distribution peaks. Because areas in a histogram represent frequencies, the median is the value such that half of the area under the distribution lies to the left and half to the right. The mean is the center of gravity of the distribution—that is, the point at which the distribution would balance.

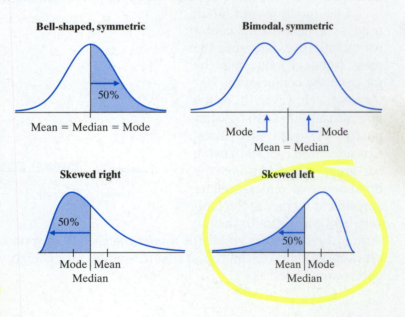

As you can see, if the distribution is approximately symmetric, the mean and median will be quite close to each other. If you have a distribution that is skewed, the median may be preferred as a measure of center. The median is a resistant measure of center, resistant to the effect of extreme observations.

THINK ABOUT IT

Suppose that you compute the mean, median, and mode for a list of numbers. Which must always appear as one of the numbers in the list?

If the distribution is symmetric, which measure of center do you calculate, the mean or the median? Why?

Let's do it! 5.4 The Usefulness of Randomization

Recall that in Chapter 3 we learned about the importance of designing experiments carefully. Random allocation of the subjects to the various treatment groups was one method to reduce the impact of bias. There are many other variables that are not controlled for directly in the experiment. Randomization helps to ensure that at least the treatment groups may be *balanced* with respect to these variables. If the groups are balanced, then any confounding variable should affect the responses somewhat equally across the groups. Thus, when the groups' responses are compared, the effect of a confounding variable has a tendency to cancel out.

Consider a study to compare two antibiotics for treating strep throat in children, amoxicillin and cefadroxil. At one clinic for this study, 23 children (who met the study entrance criteria and for whom consent was given) were randomly assigned to one of the two treatment groups. One concern is that the age of the child might influence the effectiveness of the antibiotics. The ages of the children in each treatment group are given. Calculate the mean, median, and mode of the ages for each of the two treatment groups.

Amoxicillin Group($n = 11$)   14   17   11   10   11   14   9   12   8   10   9

Mean

Median

Mode

Cefadroxil Group($n = 12$)   9   14   8   10   13   7   9   11   16   10   12   9

Mean

Median

Mode

How do the two groups compare with respect to age?

Let's do it! 5.5 **Attend Graduate School?**

When do undergraduates make the decision to continue their education and attend graduate school? An undergraduate attending a four-year college with a semester system (versus a quarter system) would have a total of eight semesters of classes (excluding any summer sessions). A sample of 18 senior undergraduates who would be graduating and attending graduate school was asked the following question: "In which semester {1, 2, 3, 4, 5, 6, 7, or 8} did you decide you would continue your education and attend graduate school?" The responses are given below:

2	4	3	1	1	6	5	5
4							
4	1	2	2	1	4	5	1
4							

(a) Construct a frequency plot of these data.

_____

(b) Obtain the following sample statistics for these data.

Minimum = _____   Maximum = _____

Median = _____   Mean = _____

(c) How do the two measures of center, the median and the mean, compare? Select one:

(i)   Median > Mean
(ii) Median < Mean
(iii) Median = Mean

 Let's do it! 5.6 **A Different Distribution**

The distribution for a variable is given.

(a) Is this distribution symmetric?   Yes   No

(b) What is the numerical value of the median of this distribution? Circle one:

larger than 4.   equal to 4.   less than 4.

Why?

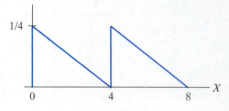

(c) What is the numerical value of the mean of this distribution? Circle one:

larger than 4.   equal to 4.   less than 4.

Why?

Mean, Median, and Mode

The most common measure of center is the **mean**, which locates the balancing point of the distribution. The mean equals the sum of the observations, divided by how many observations there are. The mean is also affected by extreme observations (outliers and values that are far in the tail of a distribution that is skewed). So the mean tends to be a good choice for locating the center of a distribution that is roughly symmetric.

The **median** is a more *robust* measure of center (that is, it is not influenced by extreme values). The median is the middle observation when the data are ordered from smallest to largest. If you have an odd number of values, the median is the one in the middle. If you have an even number of values, the median is the mean of the two middle values and falls exactly half way between them. If you have n observations, then $(n + 1)/2$ tells you the location or position of the median. For skewed distributions or distributions with outliers, the median tends to be the better choice for locating the center.

The **mode** is the value(s) that occurs most often. For a distribution, the mode is the value associated with the highest peak. The most frequent value can be far from the center of the distribution, so the mode is not really a measure of center. However, the mode is the only measure of the three that can be used for qualitative data.

Tips

- When you see or hear an "average" reported, ask which average was really computed—the mean or the median.

- Think about or examine the distribution of values to assess if the measure of center used is appropriate.

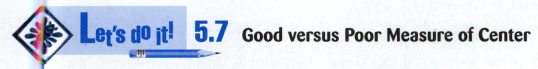

Let's do it! 5.7 Good versus Poor Measure of Center

Draw a distribution (a smoothed version of a histogram) for which

(a) the **mean** would be a **good** measure of the center of the distribution.

(b) the **mean** would be a **poor** measure of the center of the distribution.

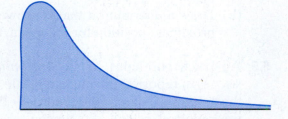

(c) the **median** would be a **good** measure of the center of the distribution.

(d) the **median** would be a **poor** measure of the center of the distribution.

(e) the **mode** would be a **good** measure of the center of the distribution.

(f) the **mode** would be a **poor** measure of the center of the distribution.

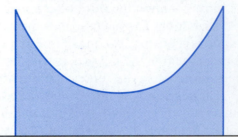

 Exercises

5.1 Which measure of center is most sensitive to extreme values? Select one:

   (a) mode.

   (b) median.

   (c) mean.

5.2 Which measure of center is located so that at least half of the values are less than or equal to that value and at least half of the values are greater than or equal to that value? Select one:

   (a) mode.

   (b) median.

   (c) mean.

5.3 A professor teaches two statistics classes. The morning class has 25 students, and their average on the first test was 82. The evening class has 15 students, and their average on the same test was 74. What is the average on this test if the professor combines the scores for both classes?

5.4 (a) Draw a distribution that is skewed to the right, and on the *x*-axis mark the approximate position for the mean, the median, and the mode.

   (b) Draw a distribution that is skewed to the left, and on the *x*-axis mark the approximate position for the mean, the median, and the mode.

REAL DATA

5.5 An article entitled *A New Home on the Ever-Costlier Range* (SOURCE: *New York Times*, May 7, 1995) reported the median resale house prices for the nation and by region for the months of March 1995, February 1995, and March 1994. Explain why the median was reported instead of the mean (or average) resale house prices.

5.6 The salaries of superstar professional athletes receive much attention in the media. The million-dollar annual contract is becoming more commonplace among this elite group with each passing year. Nevertheless, rarely does a year pass without one or more of the players' associations negotiating with team owners for additional salary and fringe-benefit considerations for all players in their particular sports.

   (a) If the players' association wanted to support its argument for higher "average" salaries, which measure of center do you think it should use? Why?

   (b) To refute the argument, which measure of center should the owners apply to the players' salaries? Why?

5.7 (a) The average (or mean) age for 10 adults in a room is 35 years. A 32-year-old adult now enters the room. Can you find the new average age for the 11 adults? If so, find it. If not, explain why not.

   (b) The median age for 10 adults in a room is 35 years. A 32-year-old adult now enters the room. Can you find the new median age for the 11 adults? If so, find it. If not, explain why not.

5.8 Where are the kids' cereals located on the shelves at a grocery store? at the bottom (Shelf 1), at the very top (Shelf 3), or in the middle (Shelf 2), where kids can see them? The following data provide the sugar content of 76 cereals and the display shelf upon which the cereal was found in the store:

Shelf 1 (lowest) 10, 6, 1, 3, 2, 11, 15, 10, 11, 6, 2, 3, 0, 0, 0, 3, 3, 3, 8.

Shelf 2 (middle) 14, 12, 9, 13, 12, 13, 0, 13, 7, 12, 9, 11, 3, 6, 12, 3, 9, 12, 15, 5, 12.

Shelf 3 (highest) 6, 8, 5, 0, 8, 8, 5, 7, 7, 3, 10, 5, 10, 5, 3, 4, 6, 9, 11, 11, 13, 7, 2, 10, 14, 3, 0, 0, 6, 8, 6, 3, 14, 3, 3, 12.

   (a) For each shelf location, find the mean and median sugar content.
   Note: For the Shelf 1 data, we have $\Sigma x = 97$; for the Shelf 2 data, we have $\Sigma x = 202$; and for the Shelf 3 data, we have $\Sigma x = 235$.

   (b) For each shelf location, produce a histogram or frequency plot of the sugar content. Describe the shape of each distribution, and comment on how the values for the mean and median relate to the shape.

REAL DATA

5.9 The "Ask Marilyn" column (SOURCE: *Parade Magazine*, June 11, 1995) discusses the three measures of center in an attempt to learn which value corresponds to a reported "typical" value.

   Reader's Question: I read that a sex survey said the typical male has six sexual partners in his life and the typical female has two. Assuming the typical male is heterosexual, and since the number of males and females is approximately equal, how can this be true?

   Marilyn's Response: You've assumed that "typical" refers to the arithmetical average of the numbers. But "average" also means "middle" and "most common." (Statisticians call these three kinds of averages the *mean*, the *median*, and the *mode*, respectively.)

   Here's how the three are used: Say you're having five guests at a dinner party. Their ages are 100, 99, 17, 2, and 2. You tell the butler that their average age is 44 [(100 + 99 + 17 + 2 + 2)/5 = 220/5 = 44]. Just to be safe, you tell the footman their average age is 17 (the age right in the middle). And to be sure everything is right, you tell the cook their average age is 2 (the most common age). Voila! Everyone is treated to pureed peas accompanied by Michael Jackson's latest CD, followed by a fine cognac. In the case of the sex survey, "typical" may have referred to "most common," which would fit right in with all the stereotypes (that is, if you believe sex surveys).

(a) Do you agree with Marilyn's conclusion that the survey is reporting the mode as "typical"?

(b) Marilyn's five ages yield very different measures of what is so-called typical. Construct an example of a set of five observations for which 80% of the observations fall below the mean and for which the median is equal to the minimum.

5.3 Measuring Variation or Spread

Measures of center are useful, but often give an incomplete interpretation of the data. Consider the following two lists of numbers and their corresponding frequency plots:

List 1: 55, 56, 57, 58, 59, 60, 60, 60, 61, 62, 63, 64, 65   **mean = median = mode = 60**

```
                                        X
                                        X
                              X X X X X X X X X X X
   |_____|_____|_____|_____|_____|_____|_____|_____|_____|_____|
   35    40    45    50    55    60    65    70    75    80    85
```

List 2: 35, 40, 45, 50, 55, 60, 60, 60, 65, 70, 75, 80, 85   **mean = median = mode = 60**

```
                                        X
                                        X
   X     X     X     X     X     X     X     X     X     X     X
   |_____|_____|_____|_____|_____|_____|_____|_____|_____|_____|
   35    40    45    50    55    60    65    70    75    80    85
```

Both sets of data have the same mean, median, and mode, but the values obviously differ in another respect—the variation or spread of the values. The values in List 1 are much more tightly clustered around the center value of 60. The values in List 2 are much more dispersed or spread out.

Measures of variation include the range, interquartile range, and standard deviation. These numerical summaries describe the amount of spread that is found among the data, with larger values indicating more variation. If the data are a sample, these measures of variation would be called *statistics*. If the data form an entire population, then these measures of variation would be called *parameters*. As with the mean, the notation to represent the standard deviation of a sample will be different from that for the standard deviation of a population.

5.3.1 Range

The **range** is the simplest measure of variability or spread. Range is just the difference between the largest value and the smallest value. The ages of the 20 subjects in the medical study were from 32 years to 51 years, for a range of 51 − 32 = 19 years. The range is often used in weather reports, which provide the high and low temperature of the day. Since the range is computed from the two most extreme values, it can give a distorted picture of the actual pattern of variation. The following two distributions have the same range, but very different patterns of variation—the distribution on the left has most of its values far from the center, while the distribution on the right has most of its values closer to the center:

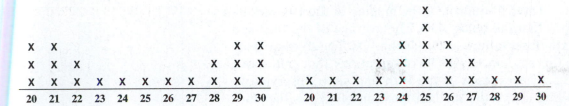

 ## 5.3.2 Interquartile Range

A measure of spread that still maintains the idea of a range, but is not influenced by the extreme values, is the **interquartile range**. The interquartile range measures the spread of the middle 50% of the data. You first find the median (represented by Q2—the value that divides the data into two halves), and then find the median for each half. The three values that divide the data into four parts are called the *quartiles*, represented by Q1, Q2, and Q3. The difference between the third quartile and the first quartile is called the *interquartile range*, denoted by IQR = Q3 − Q1.

Finding the Quartiles

1. Find the median of all of the observations.
2. First quartile = Q1 = the median of the observations that fall below the median.
3. Third quartile = Q3 = the median of the observations that fall above the median.

Notes
- When the number of observations is odd, the middle observation is the median. This observation is not included in either of the two halves when computing Q1 and Q3.
- Although different books, calculators, and computers may use slightly different ways to compute the quartiles, they are all based on the same idea.
- In a left-skewed distribution, the first quartile will be farther from the median than the third quartile is. If the distribution is symmetric, the first and third quartiles should be at the same distance from the median.

❖ EXAMPLE 5.4 Quartiles for Age

The ages of the 20 subjects in the medical study are listed in order. The histogram is also provided. We found the median to be 43.5 years.

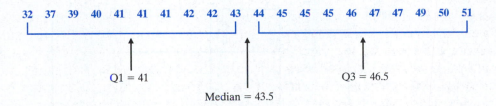

The first quartile is the median of the 10 observations that fall below 43.5. The average of the fifth and sixth observations is 41, which is Q1. The third quartile is the median of the 10 observations that fall above 43.5. The average of the fifth and sixth observations falling above the median location of 43.5 is the third quartile Q3 of 46.5. The range for this set of data is 19 years. The interquartile range for this data is $46.5 - 41 = 5.5$ years. With the TI, you would use the 1-Var Stats option under the STAT CALC menu to find the median and quartiles for a set of data that have been entered into a list.

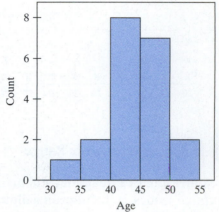

We see that the distribution of age is approximately symmetric and that the first and third quartiles are about the same distance from the median. ❖

The first quartile, Q1, is the value such that 25% of the observations fall at or below that value (and 75% fall at or above it). The second quartile, Q2 or the median, is the value such that 50% of the observations fall at or below that value (and 50% fall at or above it). The third quartile, Q3, is the value such that 75% of the observations fall at or below that value (and 25% fall at or above it). The quartiles are actually the 25th, 50th, and 75th **percentiles**.

> *Definition:* The *p*th **percentile** is the value such that $p\%$ of the observations fall at or below that value and $(100 - p)\%$ of the observations fall at or above that value.

 ### 5.3.3 Five-number Summary

When the median is used as a measure of centrality, the IQR (interquartile range) is often used as a measure of spread. If we take the quartiles and add the minimum and maximum values, we have a **five-number summary** of the data. You can show the five-number summary in a graph called a **boxplot**. The five-number summary is a set of five numbers that provide a good summary of the entire set of values. It provides a measure of center through the median. It provides measures of spread through both the interquartile range and the overall range. The distance of the first and third quartiles from the median can provide an indication of the skewness, which is best checked by examining a histogram or a stem-and-leaf plot of the values.

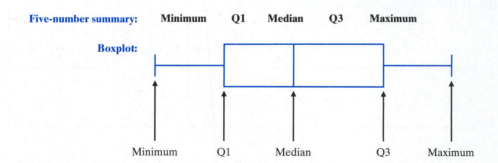

To Build a Basic Boxplot

- List the data values in order from smallest to largest.
- Find the five-number summary: minimum, Q1, median, Q3, and maximum.
- Locate the values for Q1, the median, and Q3 on the scale. These values determine the "box" part of the boxplot. The quartiles determine the ends of the box, and a line is drawn inside the box to mark the value of the median.
- Draw lines (called whiskers) from the midpoints of the ends of the box out to the minimum and the maximum values.

❖ **EXAMPLE 5.5   Five-number Summary and Boxplot for Age** _____

In Example 5.4 we ordered the age data and found the median and quartiles. So the five-number summary for the age data is

$$\text{min} = 32, \quad Q1 = 41, \quad \text{median} = 43.5, \quad Q3 = 46.5, \quad \text{and max} = 51.$$

A basic boxplot for the age data set is shown. The distance between the median and the first and third quartiles is roughly the same, supporting the rough symmetry of the distribution as seen previously from the histogram. The STAT PLOT options in the TI can be used to create a box-

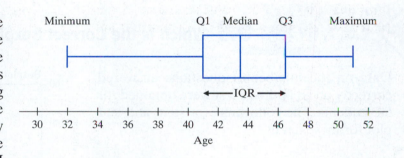

plot for a set of data that have been entered into a list. With the 20 ages entered into the list L1, the steps for creating a basic boxplot are as follows. Note that here we select the second boxplot icon in the Type list. The first boxplot icon, which shows some points plotted separately, is called a modified boxplot and is discussed later in this section.

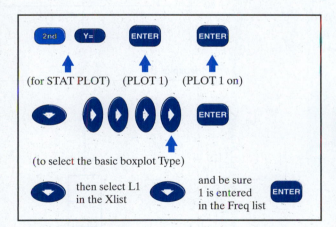

The Plot 1 screen should look like this:

Press the ZOOM button and then "9" to have the boxplot displayed. Use the TRACE button and the right and left arrow keys to see the values for the five-number summary.   ❖

A boxplot gives information about five aspects of a distribution. Side-by-side boxplots are helpful for comparing *two or more* distributions with respect to the five-number summary. The accompanying figure shows the boxplots for the length of parts for two processes. You can compare these plots to the back-to-back stem-and-leaf plot provided earlier.

Although the median of the first process is closer to the target value of 20.000 cm, the second process produces a less variable distribution.

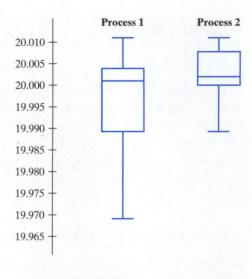

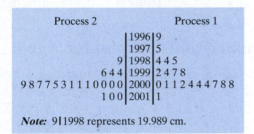

Note: 9|1998 represents 19.989 cm.

 **5.8 Which Is the Correct Boxplot?**

Data on the number of questions answered correctly on an aptitude test were obtained for 50 students. The following is a stem-and-leaf plot for these data:

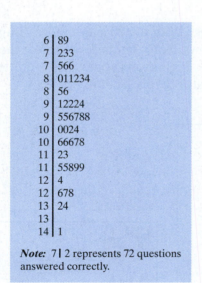

Note: 7|2 represents 72 questions answered correctly.

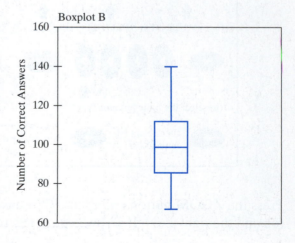

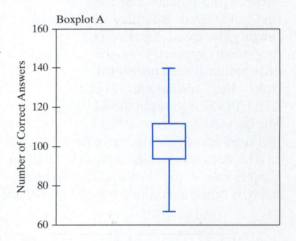

Based on the stem-and-leaf plot, which of the three boxplots is the correct boxplot corresponding to these data?

Answer = Boxplot _____

Explain how you made your selection (what features did you look for, etc.).

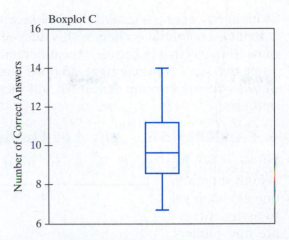

Boxplot C

Many statistical computing packages incorporate some modifications to the basic boxplot. One simple enhancement places an asterisk in the box at the location of the mean. Thus, the mean and median are both presented and may be compared. Another modification incorporates a rule of thumb for identifying potential outliers, called the **1.5 × IQR rule**. These potential outliers are plotted separately, and the resulting plot is called a **modified boxplot**.

Using the 1.5 × IQR Rule to Identify Outliers and Build a Modified Boxplot

- List the data values in order from smallest to largest.
- Find the five-number summary: minimum, Q1, median, Q3, and maximum.
- Locate the values for Q1, the median, and Q3 on the scale. These values determine the "box" part of the boxplot. The quartiles determine the ends of the box and a line is drawn inside the box to mark the value of median.
- Find the IQR = Q3 − Q1.
- Compute the quantity STEP = 1.5 × (IQR).
- Find the location of the inner fences by taking one step out from each of the quartiles:

  lower inner fence = Q1 − STEP;   upper inner fence = Q3 + STEP.

- Draw the lines (whiskers) from the midpoints to the ends of the box out to the smallest and largest values WITHIN the inner fences.
- Observations whose values fall OUTSIDE the inner fences are considered potential outliers. If there are any outliers, plot them individually along the scale using a solid dot.

The boxplot shown here was produced using a statistical package. The five-number summary for these data is   min = 1,   Q1 = 21, median = 32, Q3 = 66, and max = 325. So with an IQR = 45,   one   step   is 1.5(45) = 67.5. Going one step from the Q1 and Q3 (not from the median!) gives

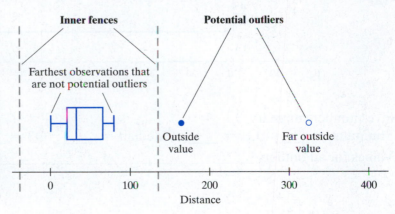

us inner fences at −46.5 and 133.5. Note the solid dot and clear dot on the plot. These represent two potential outliers with values of approximately 165 and 325, with 165 being an outside value (that is, between the inner and outer fence, where outer fences are defined as being two steps from the first and third quartiles) and 325 being a far-outside value sitting outside the upper outer fence. We will focus on using only the inner fences for detecting outliers.

❖ **EXAMPLE 5.6**   **Any Age Outliers?**

Let's apply the rule of thumb to our age data set to assess if there are any outliers.

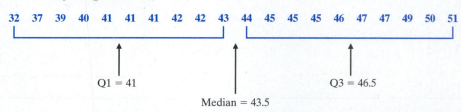

We have that the IQR = 46.5 − 41 = 5.5. A *step* is then equal to 1.5 × (5.5) = 8.25. The inner fences are located at 41 − 8.25 = 32.75 and 46.5 + 8.25 = 54.75. These inner fences are shown as dashed lines. There is exactly one observation that falls outside the (lower) inner fence—namely, 32, the minimum. Thus, there is one potential outlier based on this rule. This may be important information when we analyze the responses of interest (namely, blood pressures). The modified boxplot for age is shown. Note that the left whisker is drawn

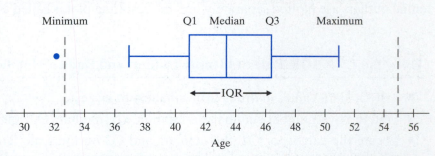

out to 37, the smallest observation that is not considered an outlier. The STAT PLOT options in the TI can be used to create a modified boxplot for a set of data that have been entered into a list. The first boxplot icon in the Type list shows some points plotted separately and is the modified boxplot option. ❖

Let's do it! 5.9 Five-number Summary and Outliers

For each of the following modified boxplots, report the corresponding five-number summary and list the values for all outliers (if any):

(a)

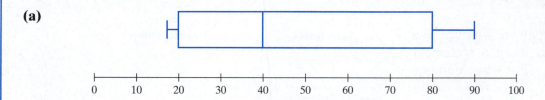

Five-number summary
Minimum = ____;   Q1 = ____;   Median = ____;   Q3 = ____;   Maximum = ____.
Values for all outliers _____

(b)

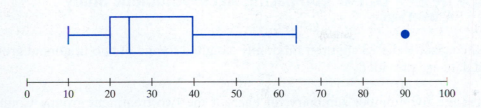

Five-number summary
Minimum = ___;   Q1 = ___;   Median = ___;   Q3 = ___;   Maximum = ___.
Values for all outliers: _____

(c)

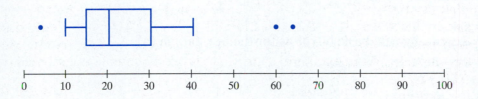

Five-number summary
Minimum = ___;   Q1 = ___;   Median = ___;   Q3 = ___;   Maximum = ___.
Values for all outliers _____

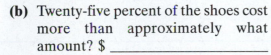 **5.10** **Cost of Running Shoes**

The prices for 12 comparable pairs of running shoes produced the following modified boxplot.

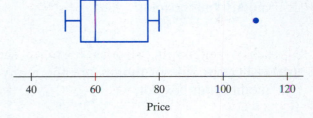

(a) What was the approximate range of prices for such running shoes?
Range = _____

(b) Twenty-five percent of the shoes cost more than approximately what amount? $ _____

5.11 **Comparing Ages—Antibiotic Study**

Recall the ages for the 23 children randomly assigned to one of two treatment groups. The ordered data are provided.

(a) Give the five-number summary for each of the two treatment groups. Comment on your results.

Amoxicillin Group ($n = 11$) 8   9   9   10 10 11 11 12 14 14 17

Five-number summary

Cefadroxil Group ($n = 12$) 7   8   9   9   9   10 10 11 12 13 14 16

Five-number summary

(b) Make side-by-side basic boxplots for the age data in (a).

(c) Using our rule of thumb, are there any outliers for the Amoxicillin group? If so, modify your boxplot.

(d) Using our rule of thumb, are there any outliers for the Cefadroxil group? If so, modify your boxplot.

To recap, the five-number summary and corresponding boxplot provide a nice summary of a set of data. The median provides a measure of the center of the distribution. The length of the box—the IQR—provides a measure of spread, and the distances of Q1 and Q3 from the median provide an indication of the skewness. However, when examining boxplots, you should be aware that they can hide gaps and multiple peaks; that is, **boxplots don't fully show the shape of the distribution**. Symmetry of a distribution does not imply that it is unimodal.

THINK ABOUT IT

If the boxplot is symmetric, can we conclude that the distribution is symmetric?

❖ **EXAMPLE 5.7**   **Symmetric Distribution Implies Symmetric Boxplot**

1. Consider the following (ordered) observations:

−1 1 2 2 3 3 3 4 4 4 4 5 5 5 6 6 7 9

The five-number summary for these observations is as follows:

$$min = -1$$
$$Q_1 = 3$$
$$median = 4$$
$$Q_3 = 5$$
$$max = 9$$

The boxplot and histogram for these observations is shown here:

2. Consider the following (ordered) observations:

1 1 1 1 1 1 3 3 4 6 7 7 9 9 9 9 9 9

The five-number summary for these observations is as follows:

$$min = 1$$
$$Q_1 = 1$$
$$median = 5$$
$$Q_3 = 9$$
$$max = 9$$

The boxplot and histogram for these observations is shown here:

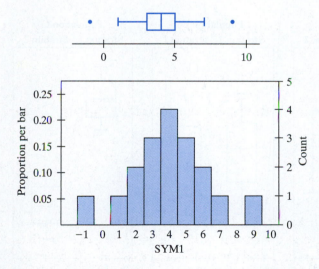

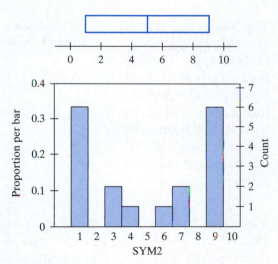

Key: If the distribution is symmetric, the boxplot will be symmetric.   ❖

❖ **EXAMPLE 5.8** **Symmetric Boxplot Does Not Imply Symmetric Distribution**

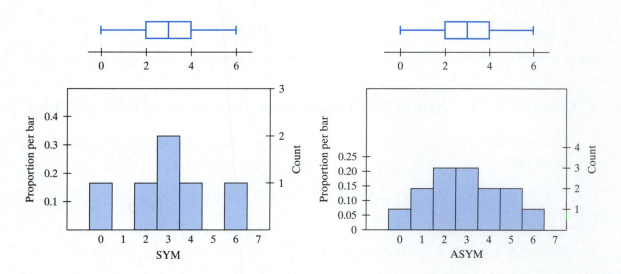

Symmetric Boxplot and Symmetric Distribution   *Symmetric Boxplot and Asymmetric Distribution*

The two boxplots are exactly the same, but the two distributions are not the same. The answer to the previous "Think about it" question is *no*.

Key: From a symmetric boxplot, you cannot conclude whether or not its distribution is symmetric. ❖

Let's do it! 5.12 Effective Sampling Design

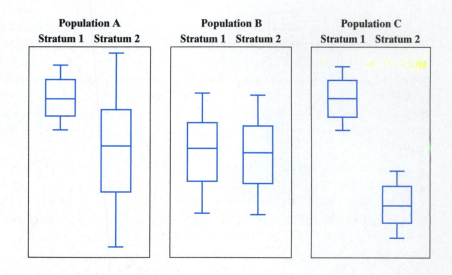

The following graphs are side-by-side box-plots of some variable for two strata in three hypothetical populations, A, B, and C. In each population, the units are evenly divided between the two strata.

Population A
Stratum 1   Stratum 2

Population B
Stratum 1   Stratum 2

Population C
Stratum 1   Stratum 2

Consider three sampling designs to estimate the true population mean:

(i)　simple random sampling.

(ii)　stratified random sampling taking equal sample sizes from the two strata.

(iii)　stratified random sampling taking most units from one strata, but sampling a few unifrom the other strata.

Assume that the total sample size is the same for all three designs.

(a)　For which population (A, B, or C) will Designs (i) and (ii) be comparably effective? Explain.

(b)　For which population (A, B, or C) will Design (ii) be the best design? Explain.

(c)　For which population (A, B, or C) will Design (iii) be the best design? Explain.

Which stratum in this population should have the higher sample size? Explain.

5.3.4　Standard Deviation

When the mean is used to measure center, the most common measure of spread is the **standard deviation**. Like the mean, standard deviation makes use of the information contained in all of the observations. The standard deviation is a measure of the *spread of the observations from the mean*. It is actually the square root of an average of the squared deviations of the observations from the mean. Since this is a bit cumbersome, we can think of the standard deviation as an average (or standard) distance of the observations from the mean.

To find the standard deviation, we first compute the **variance**, which is an average of the squared deviations of the observations from their mean. The standard deviation is the positive square root of the variance. It is not easy to compute the standard deviation, but many calculators will perform that task for you. In the next two examples, we show how to obtain the standard deviation by hand. Once you understand how the standard deviation works as a measure of spread, and how to interpret it, it is better to let your calculator compute the standard deviation for you (see the ■ TI Quick Steps at the end of this chapter).

❖ EXAMPLE 5.9 Standard Deviation—What Is It?

Suppose that we have just three values in our sample: $x_1 = 0$, $x_2 = 5$, and $x_3 = 7$. We wish to compute the standard deviation of the sample.

First note that the sample mean is equal to $\bar{x} = 4$ (check it out!). The picture shows the *deviations* of the observations from their mean.

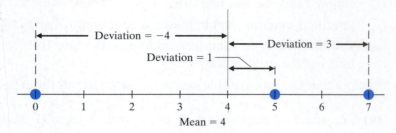

$$\text{Deviations} \quad x_1 - \bar{x} = -4, \quad x_2 - \bar{x} = 1, \quad x_3 - \bar{x} = 3$$

These deviations are $-4, 1$, and 3, which add up to 0. This will always be the case. Since we cannot use the sum of the deviations to measure spread, we use the squared deviations.

$$\text{Squared Deviations} \quad 16, \quad\quad\quad 1, \quad\quad\quad 9$$

The following table summarizes the calculations that lead up to computing the variance, and thus the standard deviation:

Observation x	Deviation $x - \bar{x}$	Squared Deviation $(x - \bar{x})^2$
0	$0 - 4 = -4$	16
5	$5 - 4 = 1$	1
7	$7 - 4 = 3$	9
mean = 4	**sum always = 0**	**sum = 26**

The sample variance is defined as the sum of these squared deviations divided by the sample size less 1—that is, divided by $n - 1$. (Further discussion about $n - 1$ is presented later.)

$$\text{sample variance} = \frac{(-4)^2 + (1)^2 + (3)^2}{3 - 1} = \frac{16 + 1 + 9}{2} = \frac{26}{2} = 13.$$

The value of 13 represents the size of the average squared deviation. The sample standard deviation is the square root of the variance and represents approximately the size of the average deviation of the observations from their mean. Thus,

$$\text{sample standard deviation} = \sqrt{13} \cong 3.6.$$

❖

❖ EXAMPLE 5.10   **Standard Deviation for the Number of Children per Household**

Recall the data on the number of children in a household for the 10 households in a neighborhood:

$$2, \quad 3, \quad 0, \quad 2, \quad 1, \quad 0, \quad 3, \quad 0, \quad 1, \quad 4.$$

(a) First compute the mean. We previously found the mean to be 1.6.

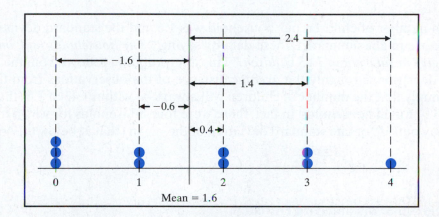

(b) Compute the squared deviations of the observations from the mean.

Observation x	Deviation $x - \bar{x}$	Squared Deviation $(x - \bar{x})^2$
2	0.4	0.16
3	1.4	1.96
0	−1.6	2.56
2	0.4	0.16
1	−0.6	0.36
0	−1.6	2.56
3	1.4	1.96
0	−1.6	2.56
1	−0.6	0.36
4	2.4	5.76
mean = 1.6	**sum always** = 0	**sum** = 18.40

(c) The sample variance is then

$$\text{sample variance} = \frac{\text{sum of squared deviations of the observations from their mean}}{(\text{number of observations}) - 1}$$

$$= \frac{18.4}{9} \approx 2.044.$$

(d) The sample standard deviation is the square root of the variance:

$$\text{sample standard deviation} = \sqrt{2.044} \approx 1.43.$$

The mean number of children per household was 1.6 and the standard deviation was 1.43. Therefore, we might summarize these data by saying, "*The households had, on average, 1.6 children, give or take about 1.43 children.*" This interpretation follows from the idea that the standard deviation is roughly the *average* distance of the observations from their mean. It does *not* imply that the number of children will always be within 1.43 of 1.6 (that is, between 0.17 and 3.0) for all households. In fact, there were four households for which the number of children lay outside of one standard deviation of the mean (that is, below 0.17 or above 3.0).

❖

Interpretation of the Standard Deviation

Think of the standard deviation as roughly an *average distance* of the observations from their mean. If all of the observations are the same, then the standard deviation will be 0 (i.e., no spread). Otherwise, the standard deviation is positive and the more spread out the observations are about their mean, the larger the value of the standard deviation.

Since the standard deviation is the most common measure of spread, it also has some special notation with which we should be familiar. If $x_1, x_2, \ldots, x_n$ denote a sample of n observations, the **sample variance** is denoted by

$$s^2 = \frac{\Sigma(x_i - \bar{x})^2}{n - 1} = \frac{(x_1 - \bar{x})^2 + (x_2 - \bar{x})^2 + \cdots + (x_n - \bar{x})^2}{n - 1} = \frac{n\Sigma(x_i^2) - (\Sigma x_i)^2}{n(n - 1)}.$$

Recall that the Σ is summation notation that basically means "add them all up." The last expression for the sample variance is a formula that is sometimes easier to use. However, when possible, have a calculator do the computational work for you.

The **sample standard deviation**, denoted by s, is the square root of the variance: $s = \sqrt{s^2}$. If, instead of a sample, you actually have all of the population observations, where N is the total number of observations in the population, the population variance is the average of the squared deviations of the observations from the population mean μ. The **population standard deviation**, denoted by the Greek letter σ (sigma), is the square root of the population variance and is computed as

$$\sigma = \sqrt{\sigma^2} = \sqrt{\frac{\Sigma(x_i - \mu)^2}{N}}.$$

For the TI, the 1-Var Stats option under the STAT CALC menu produces many summary measures, including the standard deviation. You must first enter the data into a list. In the result screen for the 1-Var Stats both the sample standard deviation s and the population standard deviation σ are provided. If the values in the list are a sample, you would use s. If the values in the list represent a population of values, you would use σ.

Unless noted otherwise, we will assume that our data sets are samples and do not constitute entire populations, and will use the appropriate equation and notation for the mean, $\bar{x}$, and the standard deviation, s.

Remarks

- The variance is measured in squared units. If age is measured in years and the variance is 1.44, we would report the variance as 1.44 (years)^2. By taking the square root of the variance, we bring this measure of spread back into the *original units*. The standard deviation for age is 1.2 years, or we might say that on average the observations were 1.2 years from the mean age.

- Just as the mean is not a resistant measure of center, since the standard deviation used the mean in its definition, it is not a resistant measure of spread. It is heavily influenced by extreme values.

- There are statistical arguments that support *why we divide by $n - 1$ instead of n in the denominator of the sample standard deviation*. If we wish to use the sample standard deviation as an estimate of the population standard deviation, the $n - 1$ version generally gives a better estimate. We will examine this idea in Chapter 9 and will see the quantity $n - 1$ referred to as the *degrees of freedom*. Briefly, since the sum of the deviations of the observations from their mean is always 0, we only need to know $n - 1$ of them to be able to determine the remaining one. Thus, $n - 1$ of the deviations are *free to vary*.

 **Let's do it! 5.13** **Increasing Spread**

Consider the following three data sets:

I: 20 20 20   **II:** 18 20 22   **III:** 17 20 23.

(a) Which data set will have the smallest standard deviation? _____

(b) Which data set will have the largest standard deviation? _____

(c) Find the standard deviation for each data set and check your answers to (a) and (b).

THINK ABOUT IT

Given that two (or more) sets of n observations yield the same standard deviation, will these sets show the same amount of *variability*? Just what is *variability* anyway?

❖ EXAMPLE 5.11   What Is Variability?

Consider the following four data sets along with the histograms displaying the distributions:

Data Set I

2 3 3 3 4 4 4 4 5 5 5 5 5

Data Set II

3 3 3 3 3 4 4 4 4 5 5 5 6

Data Set III

2 3 3 4 4 4 4 4 4 5 5 6

Data Set IV

3 3 3 3 3 3 4 5 5 5 5 5 5

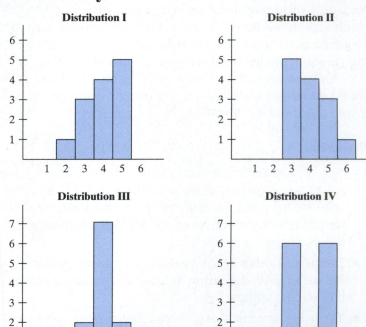

The mean for all four data sets is $\bar{x} = 4$. The table presents three measures of variability for each of the four data sets.

If we look at the range, Data Set III is *most variable*; if we look at the IQR, Data Set III is *least variable*; while all four data sets have the same standard deviation.

Some people associate variability with range, while others associate variability with how values differ from the mean. There are many measures of variability, with the standard deviation being the most widely used measure. But keep in mind, **a distribution with the smallest standard deviation is not necessarily the distribution that is least variable with respect to other definitions** or to your own definition of variability. ❖

Measure of Variability	Data Set			
	I	II	III	IV
Range	3	3	4	2
IQR	2	2	1	2
Std dev	1	1	1	1

SOURCE: A. J. Nitko, 1983, *Educational Tests and Measurement: An Introduction.*

THINK ABOUT IT

What would happen if the last value in all four of the preceding data sets were changed to 16? Recompute the three measures of variability for the four data sets and comment.

Let's do it! 5.14 Variation in Scores

Three instructors are comparing scores on their finals. Each instructor has 21 students.

- In Class A, one student received a score of 30 points, one received a score of 70 points, and the rest received a score of 50 points.
- In Class B, one student received a score of 30 points, one student received a score of 32 points, one student received a score of 34 points, and so forth, all the way through 70 points.
- In Class C, 10 students received a score of 30 points, 1 student received a score of 50 points, and 10 students received a score of 70 points.

The distributions are as follows:

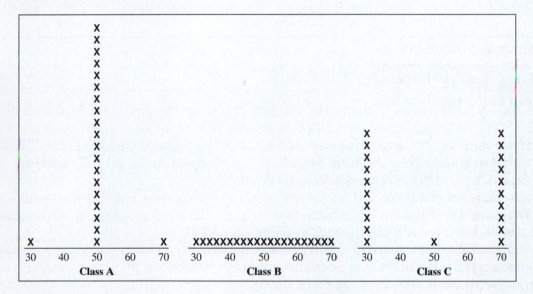

(a) Measuring variation by the range, which class has the largest variation? (Select your answer.)

Class A   Class B   Class C   All three are equal

(b) Measuring variation by the standard deviation, which class has the largest variation? (No calculations are required; select your answer.)

Class A   Class B   Class C   All three are equal

(c) Report the five-number summary for each class.

Let's do it! 5.15 Standard Deviation for Age

The ages of the 20 subjects in the medical study are listed in order. We found the mean to be 43.35 years.

   32 37 39 40 41 41 41 42 42 43 44 45 45 45 46 47 47 49 50 51

(a) Find the standard deviation for these data.

(b) Complete this sentence: On average, the ages of these subjects are about _____ years from their mean of _____ years.

(c) How many of the 20 subjects had ages within one standard deviation of the mean?

(d) How many of the 20 subjects had ages within two standard deviations of the mean?

IQR and Standard Deviation

The **interquartile range**, IQR, is the distance between the first and third quartiles (Q3 − Q1), and measures the spread of the middle 50% of the data. When the median is used as a measure of center, the IQR is often used as a measure of spread. For skewed distributions or distributions with outliers, the IQR tends to be a better measure of spread if your goal is to summarize that distribution.

Adding the minimum and maximum values to the median and quartiles results in the **five-number summary**. A graphical display of the five-number summary is a **boxplot**, and the length of the box corresponds to the IQR.

The **standard deviation** is roughly the average distance of the observed values from their mean. The mean and the standard deviation are most useful for approximately symmetric distributions with no outliers. In the next chapter, we will discuss an important family of symmetric distributions, called the normal distributions, for which the standard deviation is a very useful summary.

Tip: The numerical summaries presented in this chapter provide information about the center and spread of a distribution, but a graph, such as a histogram or stem-and-leaf plot, provides the best picture of the overall shape of the distribution. *Graph your data first!*

 ## Exercises

5.10 Consider the following sample of four observations: 20, 30, 40, 50. Give another set of 4 observations that will have the same sample mean but a larger sample standard deviation.

5.11 (a) For the following data sets of scores, calculate the sample standard deviation:

    (i)   2, 4, 7, 3, 9, 1, 27.

    (ii) 7, 7, 7, 7, 7, 7, 7, 7, 7, 7.

    (iii) 3, 5, 6, 7, 8, 9, 10, 11.

  (b) Why is the sample standard deviation for Data Set (i) so large? Describe the effects of extreme deviations on the standard deviation.

  (c) Determine the range for each of the data sets in (a). For which of these data sets is the range a misleading measure of variability and why?

5.12 The *five-number summary* for the distribution of test scores is
340,   460,   580,   780,   950.

  (a) Draw a basic boxplot for the test score distribution.

  (b) Suppose that you scored 470 on the test. What can you say about the percentage of the students who scored higher than you?

  (c) If the top 25 % of the students received an "A" on the test, what was the minimum score needed to get an "A" on the test?

5.13 Suppose that you are a budding entrepreneur. The start-up costs, in thousands of dollars, for *Entrepreneur Magazine*'s 25 hottest franchises are summarized here.

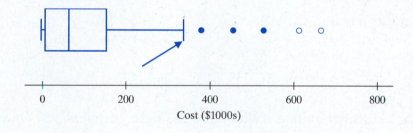

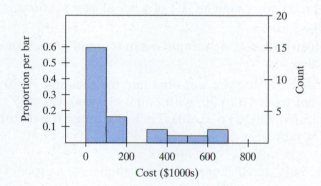

  (a) Hardees had the highest start-up costs. What is the approximate start-up cost for a Hardees restaurant?

  (b) Describe the shape of this distribution.

  (c) It was reported that for these 25 franchises, the typical start-up cost was $154,700. Which measure of *center* does this value represent? Explain.

  (d) What does the value marked with an arrow on the boxplot correspond to?

5.14 Determine whether the following statements are true or false (a true statement must be *always* true):

(a) The interquartile range (IQR) is half of the range.

(b) The mean is always strictly between the first and third quartiles.

(c) The median is always between the first and the third quartiles (including these endpoints).

(d) The standard deviation of a symmetric distribution is always equal to the interquartile range.

5.15 Determine whether each of the following questions is true or false (a true statement is always true):

(a) The distribution

```
            X
            X
 X          X          X
 X    X     X    X     X
 ───────────────────────
 1    2     3    4     5
```

has a smaller variance than the distribution

```
 X                     X
 X                     X
 X                     X
 X                     X
 X                     X
 ───────────────────────
 1    2     3    4     5
```

(b) For a distribution that is skewed to the right, the mean will always be greater than or equal to the median.

(c) If the interquartile range (IQR) of a set of observations is zero, then the range is also zero.

(d) If for a given data set, the sample mean is equal to the minimum value, then the range is zero.

(e) Suppose that a histogram was obtained for a sample of 20 observations, and its classes went from 10 to 30, with equal class widths of 5 (i.e., the classes were [10,15), [15,20), [20,25), [25,30)). Then the range for the observations is always $30 - 10 = 20$.

5.16 Where are the kids' cereals located on the shelves at a grocery store? at the bottom (Shelf 1), at the very top (Shelf 3), or in the middle (Shelf 2), where kids can see them? The following data provide the sugar content of 76 cereals and the display shelf upon which the cereal was found in the store:

Shelf 1 (lowest): 10, 6, 1, 3, 2, 11, 15, 10, 11, 6, 2, 3, 0, 0, 0, 3, 3, 3, 8.

Shelf 2 (middle): 14, 12, 9, 13, 12, 13, 0, 13, 7, 12, 9, 11, 3, 6, 12, 3, 9, 12, 15, 5, 12.

Shelf 3 (highest): 6, 8, 5, 0, 8, 8, 5, 7, 7, 3, 10, 5, 10, 5, 3, 4, 6, 9, 11, 11, 13, 7, 2, 10, 14, 3, 0, 0, 6, 8, 6, 3, 14, 3, 3, 12.

Note: For the Shelf 1 data, we have $\sum x = 97$ and $\sum x^2 = 857$; for the Shelf 2 data, we have $\sum x = 202$ and $\sum x^2 = 2284$; and for the Shelf 3 data, we have $\sum x = 235$ and $\sum x^2 = 2049$.

(a) For each shelf location, find the IQR and sample standard deviation for sugar content.

(b) Based on the graphs from (b) of Exercise 5.8, which summary measures would you prefer to describe sugar content: the mean and standard deviation or the five-number summary? Explain.

5.17 Suppose that you are about to purchase a car and have narrowed your choices to two models—Model A and Model B. All things are about equal, such as price, options, even the average annual maintenance costs. You find an owner survey in an auto magazine that indicates that the standard deviation of maintenance costs is smaller for Model B. Based on this information, which of the following statements is most appropriate?

(a) Model A with the larger standard deviation is preferable, because the larger value implies a smaller amount of variation in the data.

(b) Model B with the smaller standard deviation is preferable, because the smaller value implies that the mean is a more reliable representation of maintenance costs.

(c) They are equally acceptable, because standard deviations are not useful for comparisons of data sets.

5.18 Consider the distribution for prices of running shoes given in the first figure.

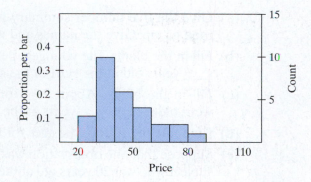

(a) Which of the following values, 40 or 45, would best represent the mean for this distribution and which would best represent the median for this distribution? Explain.

(b) Suppose that the largest observation was incorrectly entered, as shown in the second figure. Select the best answer:

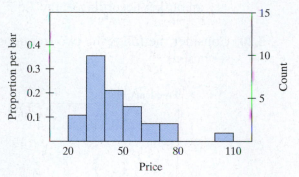

  (i) The mean would (increase, decrease, stay the same).

  (ii) The median would (increase, decrease, stay the same).

  (iii) The standard deviation would (increase, decrease, stay the same).

5.19 A study was conducted to assess whether adding whole-grain foods to one's diet significantly reduces the risk of having a stroke. A total of 450 women were available for the study. These subjects were divided into three groups based on their whole-grain consumption (low, regular, high). Over the period of the study (12 years), the stroke status (whether or not the subject had a stroke) was recorded for each subject for comparison between the groups. Those women who consumed the high amount of whole-grain foods in their diet had a 23 % lower risk of stroke compared with those who consumed a *low* amount of whole-grain foods. The age at the start of the study for each subject was also recorded. The side-by-side boxplot for age for the three whole-grain-consumption subject groups is shown here:

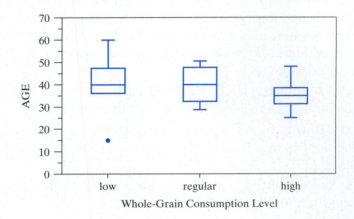

(a) Give the five-number summary for the age data for the regular whole-grain food group. Give the names and values.

(b) Fill in the blank. The youngest person in the low whole-grain food group was _____ years old at the start of the study.

(c) Fill in the blank. About 75 % of the high-whole-grain-food-group subjects were older than _____ years at the start of the study.

(d) Compute the IQR of the ages for the regular-whole-grain-food-group subjects.

(e) Suppose that one observation was inadvertently left out. It was a regular-group subject who was 20 years old at the start of the study. Would this observation be an outlier according to the 1.5 × IQR rule? Show all supporting work.

5.20 Consider the following two frequency plots based on two sets of data with 10 observations:

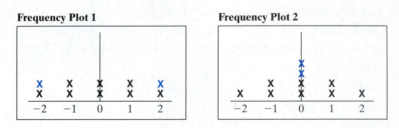

Note that Plot 1 can be converted to Plot 2 by moving the two blue X's.

(a) Do you think the range for Plot 1 is smaller than, equal to, or larger than the range for Plot 2?

(b) Do you think the mean for Plot 1 is smaller than, equal to, or larger than the mean for Plot 2?

(c) Do you think the standard deviation for Plot 1 is smaller than, equal to, or larger than the standard deviation for Plot 2? Explain.

(d) Check your answers to (a), (b), and (c) by computing the range, mean, and standard deviation for each set of data.

5.21 Consider the following two frequency plots based on two sets of data with 15 observations:

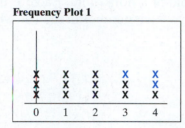
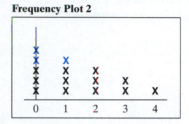

Note that Plot 1 can be converted to Plot 2 by moving the three blue X's.

(a) Do you think the range for Plot 1 is smaller than, equal to, or larger than the range for Plot 2?

(b) Do you think the mean for Plot 1 is smaller than, equal to, or larger than the mean for Plot 2?

(c) Do you think the standard deviation for Plot 1 is smaller than, equal to, or larger than the standard deviation for Plot 2? Explain.

(d) Check your answers to (a), (b), and (c) by computing the range, mean, and standard deviation for each set of data.

5.4 Linear Transformations and Standardization[*]

A worker has the task of measuring the length of parts coming off a production line. The three selected parts were measured and recorded as 2 feet, 2.5 feet, and 3 feet, respectively. What would be the lengths of these parts if the unit of measurement were inches instead of feet? Since there are 12 inches in 1 foot, we have 24 inches, 30 inches, and 36 inches. A conversion between two measurement systems can often be expressed as a linear transformation. In the case of converting from feet to inches, we multiply the values by 12: value in inches = 12 × (value in feet). In this section, we will explore how such transformations affect the various numerical summaries. The ■ TI Quick steps at the end of this chapter provide the steps for performing a transformation of a set of data.

---

*This section is optional.

Let's do it! 5.16 A Transformation

Recall the data on the number of children in a household for the 10 households in a neighborhood:

2, 3, 0, 2, 1, 0, 3, 0, 1, 4.

We found the sample mean to be 1.6 and the sample standard deviation to be 1.43. Suppose that we wish to summarize the number of people in a household. Each of the households has two adults so we can simply add the value 2 to each number in the list, which gives

4, 5, 2, 4, 3, 2, 5, 2, 3, 6.

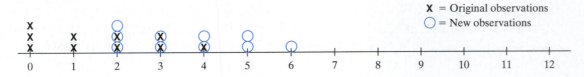

X = Original observations
◯ = New observations

(a) Find the sample mean and the sample standard deviation of this new set of observations and compare them to those for the original observations. How did the mean change? How did the standard deviation change?

(b) Summarize how adding the same constant to each observation affects the mean and standard deviation of the observations. Knowing how the standard deviation is computed, does this make sense?

Suppose that each child receives a weekly allowance of $3. The total allowance expense in a household can be obtained by multiplying every number in the original list by 3:

6, 9, 0, 6, 3, 0, 9, 0, 3, 12.

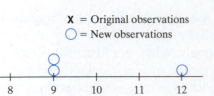

X = Original observations
◯ = New observations

(c) Find the sample mean and the sample standard deviation of this new set of observations and compare them to those for the original observations. How did the mean change? How did the standard deviation change?

(d) Summarize how multiplying each observation by the same constant affects the mean and standard deviation of the observations. Knowing how the standard deviation is computed, does this make sense?

(e) Suppose that for a local recreation program, 3 credits are deducted for each child in the household. The adjustment in credit hours can be obtained by multiplying every number in the original list (the number of children in the household) by -3. Note that the multiplier is now negative. Determine the new values and find the sample mean and the sample standard deviation for these new values.

New values _____

Mean for the new values _____ Standard deviation of the new values _____

Note: Even though the multiplier was negative, the sign for the new standard deviation is positive.

(f) Suppose that the cost is $5 for each child to enter an indoor play park. Adults are free. Each household also has a coupon to save $2. So the total cost for a household to enter the park, Y, can be represented by a linear transformation of the number of children in the household, X, as follows: $Y = 5(X) - 2$. Without actually transforming the number of children to the total cost separately for each household, can you determine what the sample mean and the sample standard deviation would be for the total cost?

In part (f) of the previous exercise, the equation was $5X - 2$. This is called a **linear transformation** of the variable X and, in general, is denoted by $aX + b$. A linear transformation is a particular type of transformation, conversion, or recoding. Recoding will be very useful when we work with a particular family of distributions, called the normal distributions, in the next chapter.

The rules developed in "Let's do it! 5.16" can be summarized as follows:

Linear Transformation Rules

If X represents the original values, $\overline{x}$ is the average of the original values, and s_X is the standard deviation of the original values, and if the new values (represented by Y) are a **linear transformation of X, $Y = aX + b$**, then

the **mean for Y** is given by $y \rightarrow \overline{y} = a\overline{x} + b$,

and the **standard deviation for Y** is given by $s_Y = |a|s_X$.

Note: Recall that the standard deviation is never negative. However, the multiplying coefficient a in a linear transformation could be positive or negative. So we must take the absolute value of the multiplier, $|a|$, in computing the standard deviation of a linear transformation, namely $s_Y = |a|s_X$.

❖ **EXAMPLE 5.12** **Temperature Transformations** _____

In a recent letter from one of your cousins in Europe, he stated that this past summer had been very hot. In particular, the high temperature of each day for a week was

$X =$ **temp** (Celsius)	Monday	Tuesday	Wednesday	Thursday	Friday	Saturday	Sunday
	40	41	39	41	41	40	38

Based on these data, the mean and standard deviation for temperature in the Celsius scale are

$$\bar{x} = 40°C, \quad \text{and} \quad s_X = 1.15°C.$$

You are not familiar with the Celsius scale. You remember that temperature in the Fahrenheit scale, Y = temperature in Fahrenheit, is related to temperature in the Celsius scale, X, by the following linear transformation:

$$F = \frac{9}{5}C + 32 \quad \text{or, in terms of } X \text{ and } Y, \quad Y = \frac{9}{5}X + 32.$$

So the mean and standard deviation for temperature in the Fahrenheit scale are, respectively,

$$\bar{y} = \frac{9}{5}(40) + 32 = 72 + 32 = 104°F, \quad \text{and} \quad s_Y = \left|\frac{9}{5}\right|1.15 = 2.07°F.$$

Now understand just how hot it was! Better than that, you did not need to transform each value. You just used what you learned in statistics.   ❖

Let's do it!   5.17   Standardization: A Special Transformation

Let's perform a special transformation of the original data on the number of children in a household:

   2,   3,   0,   2,   1,   0,   3,   0,   1,   4.

(a) The first step in the transformation is to subtract the mean, $\bar{x}$, from each number in the list: $x - \bar{x}$.

(b) The second step in the transformation is to divide the difference $(x - \bar{x})$ by the standard deviation, s_X.

(c) Calculate the mean and standard deviation of the resulting values.

         Mean = _____   Standard Deviation = _____

This special transformation is used to **standardize a variable**.

A variable X is said to be **standardized** if the variable has a ***mean of 0*** and a ***standard deviation of 1***. To standardize a variable, you first subtract its mean and then divide by its standard deviation. Note that the **standardized variable**, $(x - \bar{x})/s_X$, can be expressed in the form of a linear transformation:

$$\frac{x - \bar{x}}{s_X} = \left(\frac{1}{s_X}\right)x - \left(\frac{\bar{x}}{s_X}\right), \quad \text{with } a = \left(\frac{1}{s_X}\right) \text{ and } b = \left(\frac{\bar{x}}{s_X}\right).$$

The value of a standardized variable is often called a **standard score**.

❖ **EXAMPLE 5.13**   **Standard IQ Score** _____

Jessica is 12 years old and her IQ is 132. The mean IQ score for 12-year-olds is 100 and the standard deviation is 16. Let's find Jessica's standardized score.

$$\text{Jessica's standard score} = \frac{132 - 100}{16} = +2.$$

Jessica scored 2 standard deviations or average distances above the mean IQ score for the 12-year-old age group.

Suppose that Jessica has an older brother, Mike, who is 20 years old and has an IQ score of 144. It wouldn't make sense to directly compare Mike's score of 144 to Jessica's score of 132. The two scores come from different distributions due to the age difference. Assume that the mean IQ score for 20-year-olds is 120 and the standard deviation is 20. Standard scores allow direct comparison of scores from different distributions.

$$\text{Mike's standard score} = \frac{144 - 120}{20} = +1.2.$$

Thus, relative to their respective age groups, Jessica has a higher IQ score than Mike. Jessica's and Mike's IQ scores will be examined further in the discussion of normal distributions in Chapter 6.   ❖

Let's do it!   **5.18**   **What Will He Earn?**

X = the number of products sold per week by a salesperson
(based on the past six months of data)
mean = 18   standard deviation = 6

Five-number summary	Min	Q1	Median	Q3	Max
	4	12	20	24	37

Suppose that this salesperson earns $150 per week and $5 for each product sold.

(a) Give an expression that relates weekly earnings (E) to the number of products sold per week (X).

(b) What is the salesperson's mean weekly earnings?

(c) Find the standard deviation for weekly earnings.

(d) What would be his or her median weekly earnings?
Note: A median is a measure of center and will be transformed in the same way as a mean.

(e) Find the IQR for weekly earnings.
Note: The IQR is a measure of spread and will be transformed in the same way as a standard deviation.

 **Exercises**

5.22 The heights of four male students in inches are 70, 72, 69, and 74, respectively.

(a) Find the mean and standard deviation for heights in inches.

(b) Find the mean and standard deviation for heights in feet. (Do not transform each individual height to feet.)

5.23 A student who currently lives in Argentina plans to study at a university in Michigan during the winter. She would like to know about the average winter temperature and variation in winter temperatures. She has learned that the mean winter temperature is 28°F and the standard deviation is 10°F. She would like to have the mean and standard deviation for winter temperature in °C. Find them for her.

5.24 The mean score for all students taking Exam 1 was 60. Since the exam was so hard, the instructor decided to add five points to all Exam 1 scores.

(a) What is the new mean for the transformed Exam 1 scores? Select one:

  (i)   55.

  (ii) 60.

  (iii) 65.

  (iv) can't be determined with the information given

(b) How would the standard deviation for the transformed Exam 1 scores change? Select one:

  (i)   increase by 5.

  (ii) decrease by 5.

  (iii) increase by the square of 5.

  (iv) remain unchanged.

5.25 The 10 measurements that follow are the furnace temperatures recorded on *successive (ordered)* batches in a semiconductor-manufacturing process. The temperature unit is degrees Fahrenheit (°F).

Observation #	1	2	3	4	5	6	7	8	9	10
Temperature (°F)	949	949	951	950	954	953	955	955	959	957

(a) On average, the temperatures were about _____ °F from their mean of _____ °F.

(b) Give the five-number summary of this data. (Clearly label each of the five numbers.)

(c) How much could the largest temperature measurement increase *without* changing the sample median?

(d) Sketch the time plot (series plot) of these data. Be sure to label your axes. What information is obtained from this graph?

(e) On average, the temperatures were about _____°C from their mean of _____°C.

(f) Give the five-number summary of these data in °C. (Clearly label each of the five numbers.)

5.26 Julie is in Lecture 001, and her statistics midterm test score was 75 out of 100. John is in Lecture 002, and his score was 68 out of 100. For Julie's lecture, the mean midterm test score was 70 and the standard deviation was 5. For John's lecture, the mean midterm test score was 62 and the standard deviation was 4.

(a) Who had the higher test score overall?

(b) Find the standard score for Julie based on the scores in her lecture. Did she score above the mean? Complete this sentence: Julie scored _____ standard deviation(s) above the mean test score of _____.

(c) Find the standard score for John based on the scores in his lecture. Did he score above the mean? Complete this sentence: John scored _____ standard deviation(s) above the mean test score of _____.

(d) Relative to their respective lectures, who had the higher test score?

(e) Can you determine the mean of the test scores for the two lectures combined? If yes, find it. If no, explain what additional information you need to compute it.

5.27 The test scores of a student and the population mean and standard deviation on each of three tests are as follows.

Test	Mean	Standard deviation	Student's score
Quantitative (Stats)	67.2	10.3	63
Analytical (Logic)	56.3	6.7	61
Verbal Comprehension	50.9	3.2	57

(a) Convert each test score to its corresponding standard score.

(b) Based on the standard scores, on which test did the student score the highest?

5.28 (a) A subject gets a score of 45 on a test with a mean of 63 and a standard deviation of 12. What is the subject's standard score?

(b) A subject's standard score is -1 on a test with a mean of 30 and a standard deviation of 5. What is the subject's raw score on the test?

(c) Subject A gets a standard score of -2 on a test. Subject B gets a standard score of 2 on the same test. How many standard-deviation units separate the two subjects?

Chapter Summary

In Chapter 4, we saw how graphs are a powerful way to present data. In this chapter, we learned how to summarize specific aspects of a distribution by using some numerical summaries. We have undoubtedly seen and heard the "average" of something reported recently. We now know to ask "which average?" since there is more than one single-number summary of the center of a distribution. We should also think about what kind of values are being averaged to assess if the measure of center being used is appropriate. Although the mean is perhaps the most common measure of center, it is most appropriate for symmetric data sets with no outliers. Since the mean can be distorted by extreme values, the median is a better choice if the distribution is skewed or has outliers.

Often what is missing when the average of something is reported is a corresponding measure of spread or variability. To have an adequate summary of a distribution, we need both measures of center and measures of spread. When the median is used as a measure of center, the inter quartile range (IQR) is often used as a measure of spread. Adding the minimum and maximum values to both the median and the quartiles results in the five-number summary, which can be graphically displayed in a boxplot. When the mean is used as a measure of center, the standard deviation is often used as a measure of spread. The standard deviation is roughly the average distance of the observed values from their mean. The standard deviation is a useful yardstick for assessing whether a result is unusual or not. We will use this idea of a yardstick in our more formal inference techniques in later chapters.

Linear transformations are very useful when we change the units of measurement (for example, from height in feet to height in inches or centimeters). We learned how to convert various numerical summaries to the new measurement units without having to convert each individual observation to the new units. One of the most useful transformations is called standardization, which subtracts the mean and then divides by the standard deviation. Observations that have been standardized, also called standard scores, will have a mean of 0 and a standard deviation of 1. We saw that standard scores allow direct comparison of values from different distributions. Standard scores will play an important role in our discussion in Chapter 6 of a particular type of distribution called a normal distribution.

The numerical summaries presented in this chapter provide information about the center and spread of a distribution, but a graph, such as a histogram or stem-and-leaf plot, provides the best picture of the overall shape of the distribution.

Key Terms

Be sure that you can describe, in your own words, and give an example of each of the following key terms from this chapter:

measures of center	interquartile range (IQR)	standard deviation
mean	quartiles	variance
median	percentiles	linear transformation
mode	five-number summary	standardizing
measures of variation or spread	boxplot	standard score
	potential outliers	
range	modified boxplot	

Exercises

5.29 A class of 100 students consists of 70 students who attend Lab Section 1 and 30 students who attend Lab Section 2. The mean score of the students in Lab Section 1 is 52. If the mean score for the entire class is 64, then what is the mean score of the 30 students in Lab Section 2?

5.30 (a) Find the mean, median, and mode for each of the following data sets of scores:

   Data Set I: 2, 3, 4, 6, 6, 6, 7, 7, −9, 9.

   Data Set II: 6, 8, 4, 8, 9, 6, 42.

(b) For which of the data sets of scores, (I) or (II), is the mean a poor measure of central tendency? Why?

5.31 Consider the salaries in a company that employs 100 factory workers and two high-paid executives. How would the mean and the median compare for such a collection of salaries? Select your answer.

(a) The mean would be the same as the median.

(b) The mean would be higher than the median.

(c) The mean would be lower than the median.

5.32 The following are the weights (in pounds) of six randomly selected infants:

11, 10, 13, 9, 10, 7.

(a) Determine each of the following values for this sample:

   (i)   The sample median.

   (ii) The sample mean, $\bar{x}$.

   (iii) The range.

   (iv) The sample standard deviation, s.

   (v)   The 80th percentile.

(b) Which quantities in (a) are measures of variability?

5.33 For a list of positive numbers, can the standard deviation be bigger than the mean? If yes, give an example. If no, explain why not.

5.34 For a list of positive numbers, can the interquartile range be bigger than the median? If yes, give an example. If no, explain why not.

5.35 What about a 0? Here are the midterm scores (out of 25 points) for the $n = 40$ students in a computer lab section:

   19 19 20 20 22 20 17 19 20 19 14 21 21 12 17 23 17 23 16 22
   22 22 17 18 13 14 20 19 23 20 23 21 0 24 23 21 19 22 20 21

Note: For the midterm-score data, we have $\sum x = 763$ and $\sum x^2 = 15{,}253$.

(a) Summarize this data set graphically and describe the distribution of midterm scores.

(b) Based on (a), would you prefer to use the five-number summary or the mean and standard deviation to summarize these data numerically? Compute the summary you choose.

(c) Going back to the records, it is noted that the 0 score is for a student who did not take the midterm due to hospitalization. A note was added that the final exam would be weighted accordingly to account for the missing points. Thus, it may be appropriate to remove this observation. Redo part (b) with this observation removed. How do things change?

5.36 Determine whether the following statements are true or false (a true statement must be always true):

(a) The standard deviation of $-1, -4, -3$ is a positive number.

(b) The median of a symmetric distribution is the midpoint between the minimum and the maximum.

(c) Given the observations $0, -1, 1, -2,$ and $2,$ the median is 1.

(d) If you remove the largest observation from a data set, the average of the remaining values always changes.

(e) The range is always greater than the IQR.

5.37 A medical study was performed to assess whether or not the amount of calcium in one's diet can lower blood pressure. One group of 10 men was given a daily calcium supplement, while a control group of 11 men received a placebo or dummy pill. The seated systolic blood pressure of all men was measured before the treatment began and again after 12 weeks of treatment. Listed are the differences (before minus after) in the blood pressures for the 10 calcium group men.

$$7 \quad -4 \quad 18 \quad 17 \quad -3 \quad -5 \quad 1 \quad 10 \quad 11 \quad -2$$

(a) What does a negative value correspond to?

(b) Find the mean difference and standard deviation for the change in blood pressure.

5.38 Data were obtained on the number of aircraft owned by the 12 largest U.S. airlines for the two years 1990 and 1995 (estimated). The output shown here summarizes the difference in the number of aircraft (number in 1995 less the number in 1990) for the 12 airlines. For example, Delta Airlines had 444 aircraft in 1990 and is expected to have 583 aircraft at the end of 1995, for a difference (increase) of 139 aircraft. Use the given output to answer the following questions:

(a) Approximately what percentage of the airlines are expected to have *fewer* aircraft at the end of 1995 as compared to 1990?

(b) American Airlines is projected to have the largest increase in the number of aircraft from 1990 to 1995. What is the projected increase?

(c) Based on these data, 75% of the airlines had a difference that fell *below* _____ aircraft.

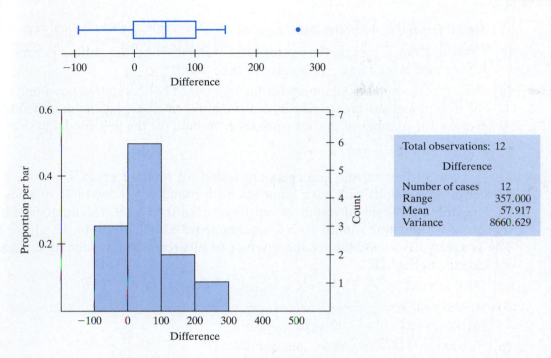

Total observations: 12

Difference

Number of cases	12
Range	357.000
Mean	57.917
Variance	8660.629

5.39 Machines are often used for inserting cotton into bottles of vitamins. A new-model machine has been developed that is designed to stuff more bottles per minute, on average, as compared to an older model. Data on the number of bottles stuffed per minute for 20 minutes has been obtained for both the new and the old model machines. The boxplot for the 20 observations with the older model is provided.

(a) The best performance by the old model resulted in a total of _____ bottles stuffed in a minute.

(b) (Approximately) 25% of the time, the old model stuffed at least _____ bottles per minute.

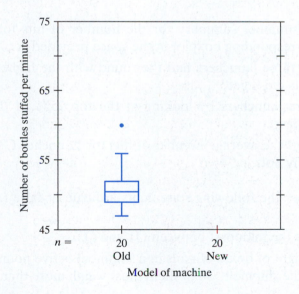

Data for the New Model

62	65	66	67	62	61	60	62	58	63
64	65	69	61	64	62	68	70	65	67

Note: For the new-model data, we have $\sum x = 1281$ and $\sum x^2 = 82{,}237$.

(c) Complete this sentence: On average, the number of bottles stuffed per minute for the new model was _____, give or take about_____.

(d) Give the five-number summary for the number of bottles stuffed per minute for the new model and create side-by-side boxplots by sketching the modified boxplot for the number of bottles stuffed per minute for the new model in the display.

5.40 Three air-to-surface missile launchers are tested for their accuracy. The same gun crew fires nine rounds with each launcher, each round consisting of 20 missiles. A "hit" is scored if the missile lands within 10 yards of the target. The number of hits registered for the nine rounds with Launcher A are 13, 11, 10, 14, 14, 12, 11, 10, 15. The boxplots below summarize the number of hits registered for the nine rounds for Launchers B and C.

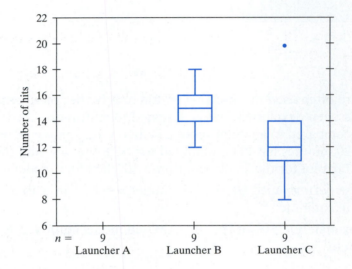

(a) Give the five-number summary for the number of hits for Launcher A and sketch the corresponding boxplot in the space provided.

(b) Which of the three launchers had the round with the highest number of hits? How many hits did it get?

(c) If we compare launchers by looking at the top 25% of their rounds, which launcher did best?

(d) Can you report the average number of hits for Launcher C? If yes, report it. If no, explain why not.

5.41 Determine whether the following statements are true or false (a true statement is always true):

(a) The median is the midpoint between Q1 and Q3.

(b) The mean weight of boxes of nails in a shipment is five pounds. Since there are 100 boxes in the shipment, 50 of them must weigh more than five pounds.

(c) If two data sets have the same mean and standard deviation, their histograms must be the same.

5.42 There are many different product lines of frozen meals marketed as being "light." Shown here are boxplots of the sodium content in all of the different single-serving frozen meals from two product lines, B and C.

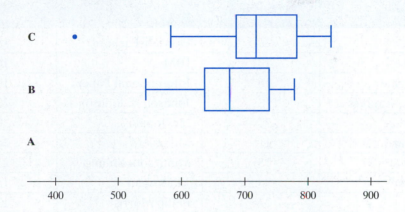

(a) The following sodium-content data were collected on the single-serving "light" frozen meals from Product Line A:

730, 560, 750, 500, 440, 610, 880, 850, 650, 810, 680, 780.

Give the five-number summary and sketch the corresponding modified box-plot in the space provided.

(b) If we compare product lines by looking at just the frozen meal with the very lowest sodium content, which product line does the best?

(c) If we compare product lines by looking at just the lowest 25 % of all their frozen meals, which product line does the best?

5.43 The manager of a bank is reviewing the performance of its tellers for possible salary increases and promotions. One variable that is measured is customer traffic—that is, the number of customers served each day. The traffic data over 35 business days for two of the tellers are obtained. The boxplots are shown. Based on these boxplots, what can you say about the distributions of customer traffic? Select one:

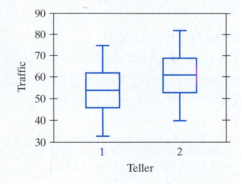

(a) Both distributions appear approximately symmetric.

(b) The shapes of the distributions are very similar.

(c) On the average, Teller 2 serves a higher number of customers each day as compared to Teller 1.

(d) All of the above.

(e) None of the above.

5.44 *Barron's* published the 1998 earnings forecasts of the price/earnings (P/E) ratios for the companies listed in the Dow Jones Industrial Average (*Barron's*, December 8, 1997).

REAL
DATA

Company	1998 P/E Forecast	Company	1998 P/E Forecast
AT&T	20	Hewlett-Packard	18
Alcioa	10	IBM	16
Allied Signal	16	International Paper	17
American Express	18	Johnson & Johnson	23
Boeing	21	McDonald's	18
Caterpillar	11	Merck	24
Chevron	18	Minnesota Mining	21
Coca-Cola	38	J.P.Morgan	15
Disney	27	Philip Morris	13
Dupont	16	Procter & Gamble	27
Eastman Kodak	15	Sears	13
Exxon	20	Travelers	17
General Electric	26	Union Carbide	12
General Motors	8	United Technologies	17
Goodyear	13	Wal-Mart	24

(a) Use split stems to construct a stem-and-leaf plot.

(b) Construct a histogram. Use a lower limit = 5, an upper limit = 80, and a class width = 5.

(c) Construct a modified boxplot.

(d) What did you learn about this data by graphing it in various ways?

5.45 An instructor distributes graded midterms back to her students and announces that the average score on this exam was 76 (out of 100). You get your test back to see that you received a score of 88. How should you feel? Perhaps you are just happy to be above average.

(a) Could your score be the top score?

(b) Could 50% of the students have scored higher than you?

(c) If the instructor also reports the standard deviation of the test scores, for which standard deviation would you feel happier about your score, 4 points or 16 points? Explain.

5.46 A study on college students found that men had an average weight of 66 kg with a standard deviation of 9 kg. The women had an average weight of 55 kg with a standard deviation of 9 kg.

(a) Can you tell whether the heaviest student was a man or a woman? Explain.

(b) Find the averages and standard deviations in pounds (1 kg = 2.2 lb).

(c) If you took the men and the women together, would the standard deviation of their weights be less than, about equal to, or greater than 9 kg? Explain.

5.47 Best Foods, Inc., operates a chain of grocery stores in the Midwest. Recently, the chain's stores have offered customers the option of paying for their purchases with credit cards in addition to the usual options of paying with cash or personal check. This new option has been implemented on a trial basis with the hope that the credit-card option will encourage customers to make larger purchases. After the first two weeks of providing this option, a random sample of 100 customers was selected over the third week. For each customer sampled, their purchase amount ($) and method of payment (cash, personal check, or credit card) was recorded. The data for cash- and check-paying customers are summarized:

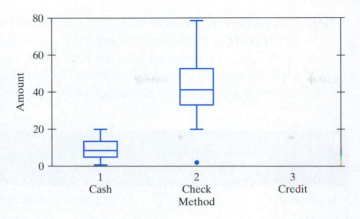

METHOD = 1 (Cash)	
	Amount
No. of cases	38
Mean	8.837
Standard dev.	5.284

METHOD = 2 (Check)	
	Amount
No. of cases	40
Mean	42.444
Standard dev.	15.333

The (ordered) data for those sampled customers who paid with a credit card are as follows:

16.55 19.92 22.54 25.60 26.88 26.95 29.85 33.85 43.58 44.53 44.58
46.59 48.26 48.59 49.52 51.24 51.90 52.64 52.96 54.29 55.59 69.85

Note: For the credit-card data, we have $\sum x = 916.26$ and $\sum x^2 = 42{,}313.74$.

(a) Give the five-number summary for the credit-card data and then sketch the corresponding basic boxplot in the space provided in the figure.

(b) The mean and standard deviation were provided for the purchases made by customers paying with cash and check. Provide these summary measures (round to two places after the decimal) for the purchases made by the customers paying with a credit card.

(c) Before the new credit-card option, approximately 50% of Best Foods' customers paid with cash, and approximately 50% paid with a personal check. Some of Best Foods' customers who previously paid with cash or by check may now be paying with a credit card. What proportion of the sampled customers paid with cash? with a check? with a credit card?

Comment on the shift in the distribution of the method of payment.

(d) Approximately what was the smallest purchase amount made by a customer who paid with a check?

(e) Does it appear that this credit-card option is encouraging customers to make larger purchases? Explain.

5.48 In a clinical trial, subjects enter the experiment over time and are assigned to either the treatment or the control group. Data collection continues until the termination of the trial. Variables such as the subject's weight may be important to the outcome, so these might be recorded. Consider the following information about the subjects entering a clinical trial for a particular drug:

Hospital 1				Hospital 2			
	Sample Size	Average Weight	Standard Deviation		Sample Size	Average Weight	Standard Deviation
Treatment	123	162 lb	23 lb	**Treatment**	162	143 lb	25 lb
Control	131	164 lb	24 lb	**Control**	167	178 lb	23 lb

Which hospital do you think failed to use randomization? Explain.

REAL DATA

5.49 The mutual-fund world can be a very confusing mess. There are many different types of funds. In an article entitled "Do You Own the Right Mutual Funds?" (SOURCE: *Smart Money*, July 1995), *Smart Money* conducted an extensive statistical analysis of the six major equity-fund groups rated by Morningstar to examine just how different these six categories really are. The two criteria were long-term-risk performance patterns as measured by average total return, and long-term-risk patterns as measured by the standard deviation. The authors of the article found that, on average, several of the categories were more alike than they were different. Based on their findings, they have formed three distinct groups: lower risk, steady earners, and high octane. To identify which of their categories an existing fund falls into, they suggest finding out two statistics:

1. The fund's yield over the past year.

2. The fund's monthly standard deviation measured over the past three years.

Here is their explanation of the standard deviation:

> *Standard deviation indicates the degree to which a fund's return is likely to stray in any single month from the fund's average performance. For instance, a fund with a standard deviation of 10 on an annualized basis will perform at best twice that amount, or 20 percentage points, above average in a good market and at worst 20 points below average in a bad one.*

What does the 20 correspond to?

5.50 A study was conducted to assess whether added calcium in the diet significantly reduces blood pressure. Thirty-two men with high blood pressure were willing to serve as subjects. Each subject took a white tablet that contains either calcium pill or a dummy pill that looks and tastes like calcium pill but has no active ingredient. The men did not know which pill they took but the researchers knew. The diastolic blood pressure of each subject was measured after taking the tablet daily for three months. The diastolic blood pressure reductions (in millimeters of mercury) for the group of men who took calcium pill are given below.

3.1	2.7	3.5	3.3	3.2	2.2	2.6	2.4
0.4	2.6	2.1	2.2	2.8	1.9	4.0	1.4

(a) This study is (select your answer)

    (i)   an observational study.

    (ii) a single-blinded experiment.

    (iii) an experiment, but not a single-blinded experiment.

    (iv) a confounding study.

(b) The variable "blood-pressure reduction" is (select all correct answers)

    (i)   the explanatory variable.

    (ii) a discrete variable.

    (iii) a confounding variable.

    (iv) a quantitative variable.

    (v) the response variable.

    (vi) a continuous variable.

    (vii) a qualitative variable.

(c) To assess overall compliance, the number of tablets taken over the 30-day study was recorded for each subject. This variable *"total number of tablets taken"* is (select all correct answers)

    (i)   the explanatory variable.

    (ii) a discrete variable.

    (iii) a confounding variable.

    (iv) a quantitative variable.

    (v) the response variable.

    (vi) a continuous variable.

    (vii) a qualitative variable.

(d) Give the sample mean and sample standard deviation (carried out to two places after the decimal) for the blood-pressure-reduction data.

(e) Use the following steps to determine how many men had their blood pressure reduced within one standard deviation of the mean:

    Step 1:   Compute the interval $[\bar{x} - s, \bar{x} + s]$.

    Step 2:   Count the number of observations in the interval $[\bar{x} - s, \bar{x} + s]$.

(f) Identify what each of the five-number summary is and give its value for the blood-pressure data.

(g) Determine whether there are any outliers, using the $1.5 \times$ IQR rule. List the values of all outliers if any.

(h) Draw a modified boxplot for the blood-pressure-reduction data. Be sure to include the values and labels for the five-number summary.

5.51 The following data on the amount of rainfall (in mm) for Ann Arbor, Michigan were obtained for 15 consecutive days in September 1999 (presented in order from left to right):

REAL DATA

3.5, 3.7, 2.8, 2.4, 2.9, 3.5, 3.9, 4.3, 4.5, 4.3, 4.4, 4.7, 4.8, 4.9, 5.3.

(a) Make a time plot (series plot) of the data.

(b) Select the appropriate description for what the plot tells you about the daily rainfall in Ann Arbor over this time.

  (i)   increasing variation

  (ii) increasing trend

  (iii) skewed to the left

  (iv) both (i) and (ii)

(c) Give the five-number summary for the rainfall data (include your units).

(d) Give the interquartile range, IQR, for the rainfall data (include your units).

(e) You find out later that the amount of rainfall on Day 15 was recorded incorrectly. The correct value should have been 6.0 mm. With this corrected observation, (select one)

  (i)   the new median > the old median.

  (ii) the new median < the old median.

  (iii) the new median = the old median.

  (iv) can't tell.

(f) The new sample mean for the corrected rainfall data (circle one)

  (i)   will be smaller than the old sample mean.

  (ii) will be larger than the old sample mean.

  (iii) will be the same as the old sample mean.

  (iv) can't tell.

5.52 A nutritionist wants to study the effect of age on muscle mass (in milligrams per ml) for women. Suppose that the following graph displays the boxplots for the muscle-mass measurements for the population of all women, divided by age into four groups:

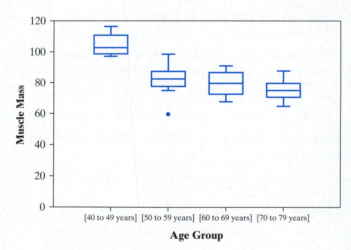

(a) What is the response variable?

(b) What is the explanatory variable?

(c) Based on the graph, would the best design for sampling from this population be simple random sampling, stratified random sampling, or cluster sampling? Explain your answer.

(d) Based on the graph, what percentage of women in the oldest age group have a muscle mass of 80 or less?

(e) Complete this sentence: Based on the graph, 25% of the women in the youngest age group have a muscle mass of at least _____.

(f) The distribution of muscle mass for women in the 60s age group is symmetric.
  Select one: true false can't tell

(g) The 50s age group had a woman with the lowest muscle mass among all women.
  Select one: true false can't tell

(h) The IQR for the youngest age group is 50%.
  Select one: true false can't tell

(i) What does the plot suggest about the relationship of muscle mass and age?

  (i) There is no relationship between the two variables.

  (ii) A woman's muscle mass decreases with age.

  (iii) The relationship is skewed to the right.

5.53 What is the preferred treatment for breast cancer when it is detected in its early stages? The most common treatment has been removal of the breast (Treatment 1). It is now more typical to remove only the tumor and the nearby lymph nodes and follow up with radiation (Treatment 2). To study whether these two treatments differ in their effectiveness, a medical team examines the hospital *records* of women who had either treatment and compares the survival times after surgery (in years). Boxplots for the survival times of women in the two treatment groups are as follows:

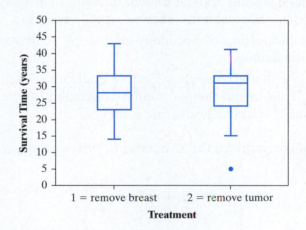

(a) This is an example of (select one)

  (i) an experiment.

  (ii) a prospective observational study.

  (iii) a retrospective observational study.

(b) Give the five-number summary for the survival times of women who used Treatment 1.

(c) Determine whether each of the following statements is true, false, or can't tell (a true statement is always true):

  (i) Treatment 1 had the patient with the lowest survival time among all patients in this study.

  (ii) Treatment 2 had the smaller number of patients.

  (iii) The distribution of scores for Treatment 1 is symmetric.

  (iv) About 75% of the Treatment 2 patients in the study survived at least 24 years.

5.54 Application forms for graduate schools can be quite lengthy and costly to complete. A researcher was interested in learning about the number of graduate schools to which students apply. Thinking that this response may vary from one department to another, she decides to obtain a listing of all departments at a university and take a random sample of two departments to include in her study. All of the students graduating within the next year for these two departments will be included in the study.

(a) What type of sampling method was used to obtain the students?

(b) One of the selected departments was Sociology. The number of graduate schools applied to by the 25 graduating students in this department are as follows:

9	6	2	4	12	7	4	16	8	6	2	11	4
3	7	10	12	5	8	5	10	5	8	15	6	

(i) Give the sample mean and sample standard deviation of these data (up to two places after the decimal) on the number of graduate schools applied to.

(ii) Obtain a histogram of these data. (Use a lower limit = 0, an upper limit = 20, and a class width = 2.)

(iii) Select the description(s) appropriate for the distribution of number of graduate schools applied to (can be more than 1): symmetric. unimodal. bimodal. skewed left. skewed right. uniform.

(iv) What proportion of Sociology-graduating students had applied to fewer than six schools?

(v) Suppose that each application to a graduate school costs $50. What is the sample mean and sample standard deviation for the total application cost for the 25 selected graduating students?

5.55 Data on rainfall (in mm) for the 12 months in 1990 was collected by three stations in a city.

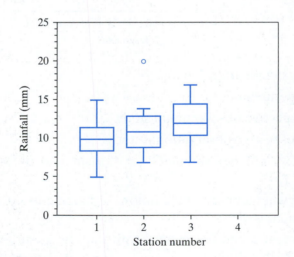

(a) Which station (1, 2, or 3) showed the largest range in the amount of rainfall collected in 12 months?

(b) Since the boxplot for Station 1 is symmetric, the distribution of the rainfall values must be symmetric. Select one: yes   no   can't tell

(c) Here are the rainfall data (in mm) for Station 4:

3, 5, 9, 9, 9, 10, 11, 11, 11, 11, 12, 15.

(i)   Give the sample mean and sample standard deviation for this data.

(ii) Give the five-number summary of the rainfall data for Station 4. Identify what each of the five numbers is and give its value.

(iii) Are there any outliers for Station 4 using the 1.5 × IQR rule? If yes, give the values of the outliers. Show all work.

(iv) Draw the modified boxplot for Station 4.

(v)   The rain gauge used to record the Station 4 data was not read correctly by the researcher. Each measurement should have an additional 1 mm added to it. What will be the sample mean and sample standard deviation of the correct rainfall data for Station 4?

5.56   Recall Examples 5.7 and 5.8 from pages 289 and 290. Create two sets of data (or distributions), one that is symmetric and the other not symmetric, such that they both have the same boxplot.

TI Quick Steps

Obtaining Summary Measures

Consider the 20 ages from Data Set 1 in Section 5.2. You want to obtain numerical summaries of that data.

Step 1   **Clear data.**

First, you need to clear out any previous data. See the TI Quick Steps at the end of Chapter 4 for a review.

Step 2   **Enter data to be summarized.**

Next, you need to enter your data in Column L1. See the TI Quick Steps at the end of Chapter 4 for a review.

Step 3   **Obtain the summary measures for the data in L1.**

Summary measures are obtained by requesting the 1-Var Stats from under the STAT CALC menu list. The sequence of buttons is as follows:

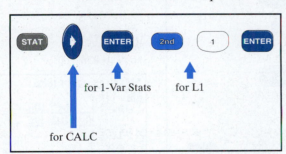

The corresponding result screen should look like this:

```
1-Var Stats
 x̄=43.35
 Σx=867
 Σx²=37981
 Sx=4.568484719
 σx=4.452808103
↓n=20
```

The 1-Var Stats are now displayed in the window. Notice that both the sample standard deviation s and the population standard deviation σ are provided, depending on whether the values in L1 are a sample or the entire population. The only mean provided is $\bar{x}$, but this would be the population mean μ if the values are the entire population. To find more information, in particular the five-number summary, press the down arrow button.

Performing a Transformation of the Data in L1

We would like to create a data set in L2 that multiplies the data in L1 by 4 and subtracts the result from 7—that is, it performs the linear transformation L2 = 7 − 4 (L1). The sequence of buttons is as shown here.

As for L1, we can obtain the summary measures for the transformed data in L2.

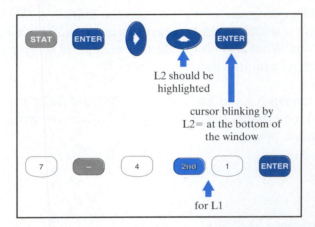

Producing a Boxplot

Consider the 20 ages from Data Set 1 in Section 5.2. You want to make a boxplot of that data.

Step 1   **Clear data and plots.**

First, you need to clear out any previous data and graphs.

Step 2   **Enter data to be plotted.**

Next, you need to enter your data in Column L1.

Step 3   **Set the STAT PLOT options for a boxplot.**

Finally, set the STAT PLOT options for producing a boxplot of the data in L1 as Plot 1.

The sequence of steps is as follows:

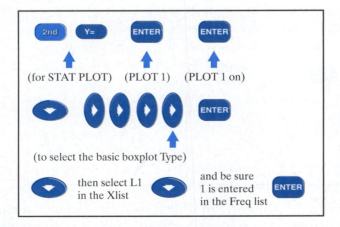

Press the ZOOM button and then "9" to have the boxplot displayed. Use the TRACE button and the right and left arrow keys to see values for the five-number summary. Note that the modified boxplot type is the fourth graph icon in the Type list.

Chapter 6

USING MODELS TO MAKE DECISIONS

6.1   Introduction

In Chapters 4 and 5, we began to summarize and organize our data in order to get information from our data. We first studied how to draw a picture of a set of data. Certain pictures provided information about the shape of the data. Next, we enhanced our graphical displays by presenting various numerical summaries of the data. We focused on numerical summaries that describe the "center" of the data and numerical summaries that characterize the "spread" of the data. In this chapter, we will learn how to take the graphical display of our data, which is based on a sample, and try to find a corresponding model to represent results for the population.

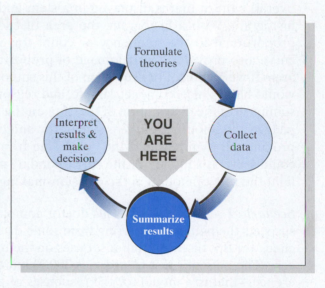

A *model* is a representation of a real-world object or phenomenon. Models are generally oversimplified and represent only prominent features of the object or phenomenon. A road map of the state of Minnesota is a model. It is oversimplified, but it can be used to measure the driving distance between cities, which can be used to predict the actual driving time.

Statisticians design experiments, gather data, summarize the results, and make predictions and decisions about populations. Some of the statistical decision-making techniques require us to make assumptions about the population from which the observations (the data) were obtained. The assumptions are often stated in terms of a *model for the population*. Suppose that we have a population of interest and each unit in the population has a value for a variable of interest. A statistical model for this population would summarize how the values of the variable are distributed in the population. The summary would be in the form of a mathematical model (a mathematical formula) that describes the relationship between the possible values of the variable of interest and the proportion of items in the population having the various values. In Chapter 7, we will continue discussing models for variables. As we shall see, the primary difference between the models presented in this chapter and the next is the replacing of the word **proportion** with **probability**. Models incorporate the inherent *variability* among people or units. People vary with respect to height, IQ, salary, and so on. Even objects such as one-gallon milk containers will vary in terms of the exact amount of milk with which they are filled.

Although the data in various disciplines may differ substantially with regard to the type of units observed and the variables measured, it is possible to present a collection of statistical models that are useful in a wide variety of situations. In this chapter, we will study some of these statistical models. We can use these models to provide a frame of reference that represents the variability of units in a population. Models will help us understand what kind of results to expect and how often they might occur. Models will help us make sound decisions between competing theories.

6.2   Why Do We Need to Know Models?

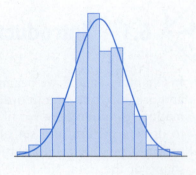

In Chapter 4, we summarized the general shapes of distributions that often arise with real data. The shape of a distribution was found by drawing a smooth curve that traces out the overall pattern that is displayed in a stem-and-leaf plot or a histogram. With a histogram, the area of each rectangle is proportional to the frequency or count for each class. The curve also provides a visual image of proportion through its area. If we could get the equation of this smoothed curve, we would have a nice, compact, simple, and reasonably accurate summary of the distribution of the data. In the accompanying picture, the smoothed curve is symmetric and bell shaped, even though the data are only approximately symmetric. If the data came from a representative sample, the smooth curve could serve as a model for the corresponding population. The following two scenarios highlight the role of models in the decision-making process:

Scenario I   A patient visits his doctor complaining of a number of symptoms. The doctor suspects the patient is suffering from some disease. The doctor performs a diagnostic test to check for this disease. High responses on the test support that the patient may have the disease. The patient's test response is 200. What does this say? The doctor has a frame of reference—that is, a model for the responses of the diagnostic test for "healthy subjects"—as shown in the accompanying figure:

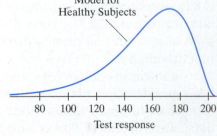

Model for Healthy Subjects

Test response

Based on this model, it is very unlikely that a test response of 200 or greater would have occurred if the subject were actually healthy. Thus, either the patient has this disease or a very unlikely event has occurred.

Scenario II   Suppose that we wish to compare two drugs, Drug A and Drug B, for relieving arthritis pain. Subjects suitable for the study are randomized to one of the two drug groups and are given instructions for dosage and how to measure their "time to relief." Results of the study are summarized by presenting the models for the time to relief for the two drugs.

Which drug is better overall? Consider any point in time, say time = T as indicated on the axis in the figure. A higher proportion of subjects treated with Drug A have felt "relief" by this time point as compared to those treated with Drug B. If the study design was sound and the models based on the study results adequately portray the models for the populations, then we might conclude that Drug A appears to be better than Drug B in terms of having a quicker time to relief.

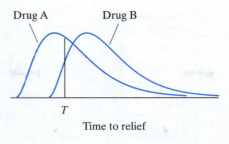

We may wish to assess whether the difference between these two drugs is statistically significant by conducting a more formal statistical test.

THINK ABOUT IT

Suppose that Marie and Mollie were subjects in the aforementioned study and that Marie received Drug A and Mollie received Drug B. Does the conclusion of Drug A being better then Drug B imply that Marie will have her arthritis relieved more quickly than will Mollie?

On the time axis, can you indicate locations for Marie and Mollie's time of relief such that Marie takes longer to have relief than does Mollie?

The conclusion is a statement about the two populations. It may not necessarily hold for particular individuals in the populations.

Statistical models bring order and understanding to the overwhelming flow of data. Models serve as a useful frame of reference—for comparison, to determine if an observation seems unusual or not. Models help us make sound decisions in the face of uncertainty. Statistics helps us evaluate and improve the quality of information in the face of uncertainty, to present and clarify options and to model available alternatives and their consequences.

6.3   Modeling Continuous Variables

Recall that, in constructing a histogram, the vertical axis can be the count or the proportion, and we draw a rectangle (a vertical bar) above each class with the height equal to the count or the proportion of units in the class. Whether the vertical axis represents the count or proportion, our eyes respond to the area of the bars. One feature of histograms is that the area of the bars is proportional to the count or proportion in the classes. In fact, we can rescale the vertical axis such that the area of a bar actually equals the proportion of observations in the class. Then the total area of all of the bars would have to be 1. The histogram would look the same, and only the scaling of the vertical axis would change. In Chapter 4, we constructed a histogram for the ages of the 20 subjects in a medical study. Consider the class of 40 years of age to just under 45 years. Histogram 6.1(a) tells us there are eight subjects in this class. From Histogram 6.1(b) we see that the proportion of subjects in this class is 0.40 (from $\frac{8}{20}$). With Histogram 6.1(c), the proportion of subjects in this class is equal to the area of the bar for this class (area is equal to base times height = $5 \times 0.08 = 0.40$).

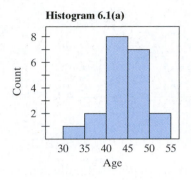

Histogram 6.1(a)

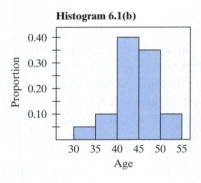

Histogram 6.1(b)

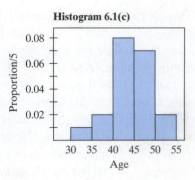

Histogram 6.1(c)

If we draw a curve through the tops of the bars in Histogram 6.1(c) and require the smoothed curve to have total area under it equal to 1, we would have what is called a **density function**, also called a **density curve**. The key idea when working with density functions is that the *area* under the curve, above an interval, corresponds to the *proportion* of units with values in the interval.

In the accompanying figure, we have a model, specifically, a density function, for the distribution of scores, *X*, out of 100 possible points. Based on the model, what percentage of students had scores of 80 or below? Since area corresponds to proportion, we would find the area under the density curve to the left of 80. This area is equal to 0.75, so we would say that 75% of the students had scores of 80 or below or, equivalently, 80 is the 75th percentile or third quartile. This is an approximation since our model is an approx-

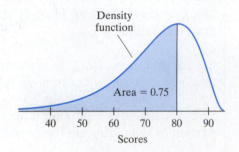

mation of reality, but it is a useful approximation. Note that the area under the density curve to the left of a value *x* (area under curve < *x*) and the area under the density curve at or below a value *x* (area under the curve ≤ *x*) are the same since there is no area under a curve at a single point or value.

Notation   We will continue to use $\overline{x}$ to denote the sample mean and *s* to denote the sample standard deviation. We will use μ (mu) to denote the population mean and σ (sigma) to denote the population standard deviation. Since we will be discussing models for populations, the mean and standard deviation for a density curve or model will be represented by μ (mu) and σ (sigma), respectively.

Definition: A **density function** is a (nonnegative) function or curve that describes the overall shape of a distribution. The total area under the entire curve is equal to 1, and proportions are measured as areas under the density function.

There are many different types of density functions, corresponding to the various shapes of distributions, where they are centered, and how much variation is present. In the next section, we will focus on a particular family of density functions called *normal distributions*. Normal distributions are important because many continuous characteristics tend to have this type of distribution, and the distribution of many statistics may be described by normal curves.

6.3.1 Normal Distributions

In this section, we will focus on one very common family of density functions called the **family of normal distributions**. A normal density function is symmetric, bell-shaped, and centered at the mean μ. Because normal densities are symmetric and unimodal, the mean, the median, and the mode are all identical. The amount of spread of a normal density is determined by the standard deviation σ. Note that the normal density function is concave downward around the mean and then becomes concave upward toward the tails. The point of transition from curving downward to curving upward is called a *point of inflection*. For the normal curve, there is a point of inflection on each side of the mean. The distance from the mean to the projection of the point of inflection is one standard deviation.

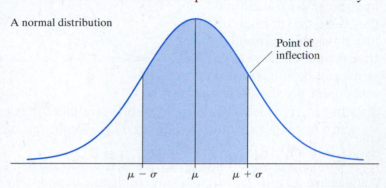

The accompanying picture shows three members of the family of normal distributions. Looking at Distributions 1 and 2, we see that changing the mean μ while the standard deviation σ is unchanged shifts the curve to either the right or the left, leaving the shape unchanged. Looking at Distributions 2 and 3, we see that making the standard deviation σ smaller while the mean μ is unchanged squeezes in the curve toward the mean, resulting in a higher peak. The values are more concentrated around the mean if the standard deviation is smaller.

The general notation X is $N(\mu, \sigma)$ means that the variable or characteristic X is normally distributed with mean μ and standard deviation σ. The first number in the $N(\_\_, \_\_)$ notation can be any number, and the second number must be at least zero. So X is $N(-3, 4)$ is a possible normal distribution, but X is $N(3, -4)$ is not.

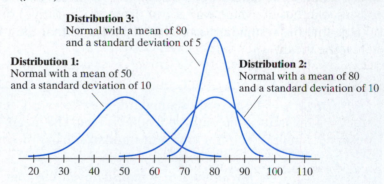

General Notation

X is $N(\mu, \sigma)$ means that the variable or characteristic X is normally distributed with mean μ and standard deviation σ.

❖ EXAMPLE 6.1 IQ Scores

Let the variable X represent IQ scores of 12-year-olds. Suppose that the distribution of X is normal with a mean of 100 and a standard deviation of 16—that is, X is $N(100, 16)$. Jessica is a 12-year-old and has an IQ score of 132. What proportion of 12-year-olds have IQ scores less than Jessica's score of 132? Since the area under the density curve corresponds to proportion, we want to find the area to the left of 132 under an $N(100, 16)$ curve.

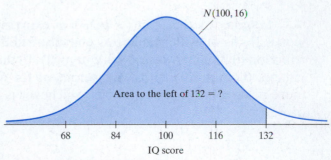

❖

How to Calculate Areas under a Normal Distribution

A normal distribution with mean 0 and standard deviation 1 is called the **standard normal distribution**. Any normal distribution can be transformed into the standard normal distribution by using a particular linear transformation. Any area of interest can be transformed to an area under the standard normal

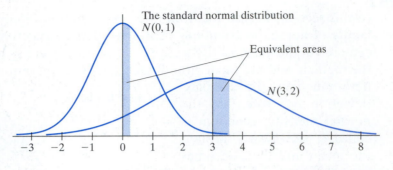

distribution. Fortunately, there are tables available that provide various areas under the standard normal distribution. (See pp. 372–373 for one such table.) Many calculators also have the ability to provide such areas.

We saw a preview to the process of standardization in Chapter 5 ("Let's do it! 5.17"). To standardize a variable, you first subtract its mean and then divide by its standard deviation.

> ***Definition:*** If X is $N(\mu, \sigma)$, **the standardized normal variable** $Z = \dfrac{X - \mu}{\sigma}$ is $N(0, 1)$.

An observed value of Z is often referred to as a **z-score** or a **standard score**. Values from a normal distribution are often expressed as standard scores (that is, in terms of how many standard-deviation units the value is from its mean). The sign of the standard score tells us whether the value was above the mean (positive) or below the mean (negative). So a value that has a standard score of -1 indicates that the value lies one standard deviation below the mean.

> ***Definition:*** The **z-score** or **standard score** for an observed value tells us how many standard deviations the observed value is from the mean—that is, it tells us how far the observed value is from the mean in standard-deviation units. It is computed as follows:
>
> $$Z = \frac{X - \mu}{\sigma} = \textbf{the number of standard deviations that } X \textbf{ differs from the mean } \boldsymbol{\mu}.$$
>
> If $Z > 0$, then the value of X is above (greater than) its mean.
> If $Z < 0$, then the value of X is below (less than) its mean.
> If $Z = 0$, then the value of X is equal to its mean.

Looking at the $Z = (X - \mu)/\sigma$ expression, we can see how this standardization works. In the first step, the mean μ is subtracted from every value. This shifts the whole distribution to be centered at 0 ($\mu - \mu = 0$). In the second step, we divide by the standard deviation, the natural unit of measurement for normal curves. This makes the distribution more peaked or more spread out, whichever is appropriate, so that the standard deviation of Z is 1.

❖ **EXAMPLE 6.2** **Standard IQ Score**

Going back to the distribution of IQ scores, let's find Jessica's standardized score.

$$\text{Jessica's standard score} = \frac{132 - 100}{16} = +2.$$

Jessica scored two standard deviations, or average distances, above the mean IQ score for the 12-year-old age group. Suppose that Jessica has an older brother, Mike, who is 20 years old and has an IQ score of 144. It wouldn't make sense to directly compare Mike's score of 144 to Jessica's score of 132. The two scores come from different distributions due to the age difference. Assume that the distribution of IQ scores for 20-year-olds is normal with a mean of 120 and a standard deviation of 20. Standard scores allow direct comparison of scores from different normal distributions.

$$\text{Mike's standard score} = \frac{144 - 120}{20} = +1.2.$$

Thus, relative to their respective age groups, Jessica has a higher IQ score than Mike. ❖

❖ **EXAMPLE 6.3** **Finding Proportions for the Standard Normal Distribution**

(a) Find the area under the standard normal distribution to the left of $z = 1.22$.

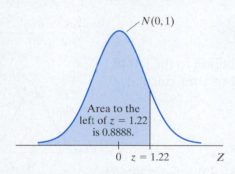

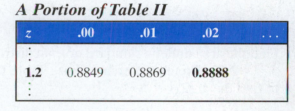

A Portion of Table II

z	.00	.01	.02	. . .
⋮				
1.2	0.8849	0.8869	**0.8888**	
⋮				

Using Table II Table II, presented at the end of this chapter, summarizes areas under the standard normal distribution to the left of various z-values. The z-values are found in the margins: The left margin has the units digit and the tenths digit; the top has the hundredths digit. Let's look up a z-value of 1.22. We find that the area to the left of $z = 1.22$ is 0.8888.

Using TI To find the area to the left of $z = 1.22$, the keystroke steps are shown in the accompanying diagram.

For further details on finding proportions for a normal distribution, see the ■ TI Quick Steps at the end of this chapter.

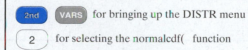

2nd **VARS** for bringing up the DISTR menu

2 for selecting the normalcdf(function

normalcdf(lowerbound, upperbound, μ, σ)
computes the area between lowerbound and upperbound for the specified normal distribution with mean μ and standard deviation σ; the defaults are $\mu = 0$ and $\sigma = 1$ so if you are finding areas under the standard normal, the values for the mean and standard deviation do not need to be specified.

Note $-$E99 represents minus infinity, which is the lowerbound when finding areas to the left; if you want an area to the right you would have E99 as the upperbound to represent infinity.

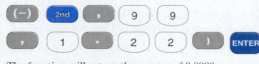

The function will return the answer of 0.8888.

(b) Find the area under the standard normal distribution to the right of $z = 1.22$. We already know the area to the left of $z = 1.22$, and we know that the total area under any density is equal to 1. So the area to the right of $z = 1.22$ must be $1 - 0.8888 = 0.1112$. We could also find the area to the left of $z = -1.22$. By the symmetry of the standard normal distribution, the area to the right of $z = 1.22$ is equal to the area to the left of $z = -1.22$.

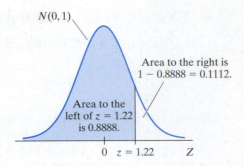

Tips It can be very helpful to picture the area you are asked to find by drawing a sketch and shading the corresponding region. Table II and some calculators provide only the area to the left of various z-values. Remember that the normal density is symmetric and that the total area under any density is equal to 1. ❖

L̲et's do it! **6.1** More Standard Normal Areas

(a) Find the area under a standard normal distribution between $z = 0$ and $z = 1.22$. Sketch the area and use Table II or your calculator to find the area.

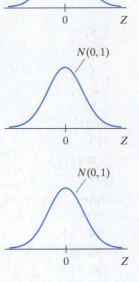

(b) Find the area under a standard normal distribution to the left of $z = -2.55$. Sketch the area and use Table II or your calculator to find the area.

(c) Find the area under a standard normal distribution between $z = -1.22$ and $z = 1.22$. Sketch the area and use Table II or your calculator to find the area.

❖ EXAMPLE 6.4 Finding Proportions for the IQ Distribution

Our original question was, "What proportion of 12-year-olds have IQ scores less than Jessica's score of 132?" Since Jessica's score of 132 is equal to a standard score of $z = 2$, we need to find the area under the standard normal distribution to the left of $z = 2.00$.

The area to the left of $z = 2.00$ is 0.9772. We would say that 97.72% of 12-year-olds had an IQ score less than or equal to 132, or, equivalently, an IQ score of 132 is the 97.72th percentile. So Jessica scored better than or equal to 97.72% of the 12-year-olds based on this model. ❖

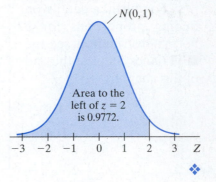

Let's do it! 6.2 IQ Scores

We will continue with the model for IQ score of 12-year-olds. In answering the following questions, remember to use the symmetry of the normal distribution and the fact that the total area under the curve is 1. It may also be very useful to draw a picture of the area you are trying to find so you can establish a frame of reference (for example, should it be larger or smaller than 50%?) and see the way to approach getting the answer with your calculator or using Table II.

$$X = \text{IQ score (12-year-olds) has an } N(100, 16) \text{ distribution.}$$

(a) What proportion of 12-year-olds has IQ scores below 84?

z-score = _____

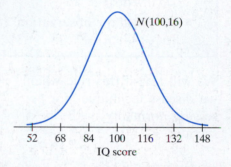

Note: Since the distribution is symmetric, we have that the area below 84 is the same as the area above 116.

(b) What proportion of 12-year-olds has IQ scores of 84 or more?

z-score = _____

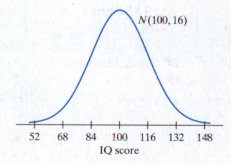

(c) What proportion of 12-year-olds has IQ scores between 84 and 116?

z-score (for 84) = _____

z-score (for 116) = _____

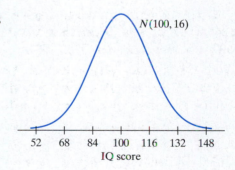

In (c) of "Let's do it! 6.2," you computed the fraction of observations that would fall within one standard deviation of the mean (i.e., between $\mu - \sigma$ and $\mu + \sigma$). The answer is about 68%. If you go out two standard deviations from the mean, you would have approximately 95% of the observations. And if you go out three standard deviations from the mean, you would have nearly all of the observations, 99.7%. These three statements are summarized next.

Definition: The **68–95–99.7 rule** for *any* normal distribution $N(\mu, \sigma)$

- 68% of the observations fall within one standard deviation of the mean $[\mu - \sigma, \mu + \sigma]$.
- 95% of the observations fall within two standard deviations of the mean $[\mu - 2\sigma, \mu + 2\sigma]$.
- 99.7% of the observations fall within three standard deviations of the mean $[\mu - 3\sigma, \mu + 3\sigma]$.

We will revisit this rule when we discuss techniques for assessing normality in the next section.

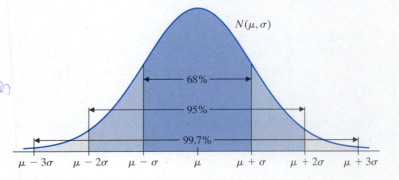

Let's do it! 6.3 A More Accurate Sketch

Let the variable Y represent the lifetime in years of a computer component. Suppose that an appropriate model for Y is $N(15, 2)$ in years. Draw the density curve for this distribution, and include a scale on the horizontal axis.

THINK ABOUT IT 🐢

Suppose that cholesterol measures for healthy individuals have a normal distribution. Kyle's standardized cholesterol measure was $z = -2$. Using only the 68–95–99.7 rule, what percentile does Kyle's measure represent?

Lee's standardized cholesterol measure was $z = +3.2$. Using only the 68–95–99.7 rule, does Lee's cholesterol seem unusually high?

 **Let's do it! 6.4 Pine Needles**

Different species of pine trees are grown at a Christmas-tree farm. It is known that the length of needles on a Species A pine tree follows a normal distribution. About 68% of such needles have lengths centered around the mean between 5.9 and 6.9 inches.

(a) What are the mean and standard deviation of the model for Species A pine-needle lengths?

(b) A 5.2-inch pine needle is found that looks like a Species A needle, but is somewhat shorter than expected. Is it likely that this needle is from a Species A pine tree?

> *Hint:* Calculate the proportion of pine needles that are 5.2 inches or shorter.

Let's do it! 6.5 Last Longer?

Time to failure of Brand A light bulbs follows an approximate normal distribution with a mean of 400 hours and a standard deviation of 20 hours. Scientists have developed a new filament that they claim will make the bulbs last longer.

(a) Set up the null and alternative hypotheses to test the scientists' claim.

H_0: _____. H_1: _____.

(b) It was observed that a randomly selected bulb with the new filament lasted 438 hours. Based on this observed value, what is the p-value for the test in (a)? Recall that the p-value is the chance, assuming that H_0 is true, of getting the observed value plus the chance of getting all of the *more extreme* values.

(c) Test the hypotheses in (a) using a 5% significance level. Give your decision and state your conclusion in a well-written sentence.

(d) Sketch the distribution of time to failure for Brand A light bulbs assuming that the null hypothesis is true. Suppose that there is a Brand B light bulb whose time to failure has a mean less than the mean of Brand A light bulbs, but a Brand B light bulb is more likely to last at least 480 hours than is a Brand A light bulb. Sketch such a possible distribution for Brand B on the same scale as the Brand A sketch.

Finding Percentiles for a Normal Distribution

Recall that the *pth percentile* is the value such that $p\%$ of the observations fall at or below that value and $(100 - p)\%$ of the observations fall at or above that value. The median of any distribution is the 50th percentile, and the first and third quartiles are the 25th and 75th percentiles, respectively. In any normal distribution, the value that is two standard deviations below the mean (that is, the value that has a standard score of $z = -2$) is the 2.5th percentile. From the 68–95–99.7 rule, 95% of the values are within two standard deviations of the mean, leaving 5% outside of two standard deviations on each side of the mean. Half of the 5%, or 2.5%, are in the lower tail. In fact, every z-score corresponds to a certain percentile. In the next example, we learn the steps for finding percentiles for a normal distribution.

❖ EXAMPLE 6.5   The Top 1% of the IQ Distribution

Continuing with the $N(100, 16)$ model for IQ score of 12-year-olds, consider the following question: What IQ score must a 12-year-old have to place in the top 1% of the distribution of IQ scores? Again, it may be helpful to draw a picture.

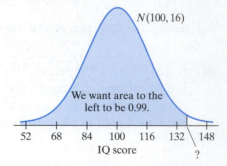

    We want to find the IQ score such that the area to the right of that score is 0.01 (that is, we want to find the 99th percentile of this IQ-score distribution). The first step is to find the 99th percentile of the standard normal distribution (that is, the value of z such that the area to the left of z is 0.99).

Many calculators have the ability to find various percentiles of a normal distribution. The TI has a built-in function called invNorm under the DIST menu. You must first specify the desired area to the left, then the mean and standard deviation for the normal distribution. The steps for finding the 99th percentile of our $N(100, 16)$ distribution are as follows:

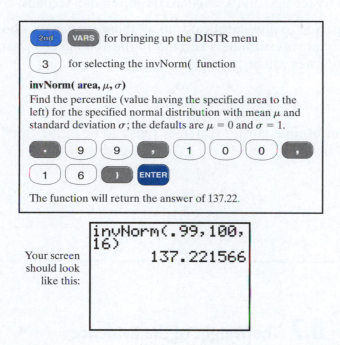

The function will return the answer of 137.22.

Your screen should look like this:

```
invNorm(.99,100,
16)
          137.221566
```

Table III, at the end of this chapter, provides various percentiles of the standard normal distribution. We look under the area column for the value 0.99 and see that the corresponding z-value is 2.326. The final step is to convert this standard score back to its value for the $N(100, 16)$ distribution:

$$z = \frac{x - 100}{16} = 2.326 \rightarrow x = 100 + (2.326) \times 16 = 137.216.$$

Therefore, a 12-year-old must have an IQ score of at least 137.216 to place in the top 1%.

❖

Percentiles from the Normal Table

Many common percentiles are already given in the normal table. If the exact percentile is not given in Table III, take the closest value from either Table II or Table III. In general, the unstandardized x-value is found from the corresponding z-score as $x = \mu + (z)\sigma$.

Let's do it! 6.6 Finishing Times

The finishing times for swimmers performing the 100-meter butterfly are normally distributed with a mean of 55 seconds and a standard deviation of 5 seconds.

(a) The sponsors decide to give certificates to all those swimmers who finish in under 49 seconds. If there are 50 swimmers entered in the 100-meter butterfly, approximately how many certificates will be needed?

(b) In what amount of time must a swimmer finish to be in the "top" fastest 2% of the distribution of finishing times?

Let's do it! 6.7 The Weight of the Evidence

A man is shot and killed while hunting in North Forest. It looks like an accident, but police question another hunter who is known to have had a grudge against the victim. They find a pine needle stuck to the suspect's coat. The suspect claims to have been hunting in South Forest on the day of the incident.

North Forest has only Species A pine trees, and South Forest has only Species B pine trees. Species A pine trees have needle lengths that are approximately normally distributed with a mean of 5.4 cm and a standard deviation of 0.4 cm. Species B pine trees have needle lengths that are approximately normally distributed with a mean of 6.4 cm and a standard deviation of 0.5 cm. The prosecution hires *you* to be the expert witness statistician. You are to test

H_0:   The needle is from a Species B pine tree.
H_1:   The needle is from a Species A pine tree.

(a) Sketch the two distributions on the same axis. Be sure to label the distributions.

(b) Suppose that the pine needle found on the suspect's coat measures 5.42 cm. What is the corresponding *p*-value?

(c) Give your decision and state your conclusion using a well-written sentence. Use $\alpha = 0.05$.

Assessing Normality

You have gathered some data on a characteristic of interest. You would like to see whether or not a normal distribution might serve as a good model. You could

- Look at a **histogram or a stem-and-leaf plot** and check for nonnormal features such as gaps, outliers, or pronounced skewness.
- Compare your observations to the **68–95–99.7 rule** for normal distributions. Calculate the sample mean, $\bar{x}$, and sample standard deviation, s, and then see whether approximately 68% of the observations fall in the interval $[\bar{x} - s, \bar{x} + s]$, approximately 95% of the observations fall in the interval $[\bar{x} - 2s, \bar{x} + 2s]$, and approximately 99.7% of the observations fall in the interval $[\bar{x} - 3s, \bar{x} + 3s]$.

❖ **EXAMPLE 6.6   Approximately Normal?** _____

The following data are the number of home runs hit by American League home-run leaders in the 39 years from 1949 to 1988 (note that the data have been ordered by magnitude, but they are not listed in order of time):

22 32 32 32 32 32 32 33 33 36 37 37 37 39 39 39 40 40 41 42
42 42 42 43 43 43 44 44 44 45 45 46 48 49 49 49 49 52 61

A histogram of the data reveals an approximately symmetric, unimodal distribution. But can we conclude that it is approximately normal? The sample mean and sample standard deviation of the data are 40.7 and 7.2, respectively. The following are the intervals that represent one, two, and three standard deviations from the mean and the percentage of observations within each interval:

$$[\bar{x} - s, \bar{x} + s] = [33.5, 47.9]; \qquad \textit{Percentage: } \tfrac{23}{39} \rightarrow 59\%.$$

$$[\bar{x} - 2s, \bar{x} + 2s] = [26.3, 55.1]; \qquad \textit{Percentage: } \tfrac{37}{39} \rightarrow 94.9\%.$$

$$[\bar{x} - 3s, \bar{x} + 3s] = [19.1, 62.3]; \qquad \textit{Percentage: } \tfrac{39}{39} \rightarrow 100\%.$$

Since these observed percentages are somewhat close to those expected for a normal distribution, we conclude that the distribution appears to resemble a normal one.   ❖

Let's do it! **6.8** Check for Nonnormal Features

The following frequency plots display distributions of samples of hypothetical exam scores, two of which were drawn from populations that are normally distributed.

Exam #1

```
                                    X   X
                      X       X   X   X   X   X
                      X  XX  X X X X X X X X X X      X  X X    X
        50          60              70          80          90          100
```

Exam #2

```
              X   X
              X   X
              X X X
              X X X X      X X                          X
              X X X X      X X      X X X   X X          X X      X
        50          60              70          80          90          100
```

Exam #3

```
                      X           X
                      X           X           X   X           X X
                      X X       X           X X   X           X X
                      X X X X X      X   X X X       X           X X
        50          60              70          80          90          100
```

Exam #4

```
                                                              X
                                                              X
                                              X               X X
                                  X           X X X       X   X X
                      X           X       X   X           X X X       X X X X
        50          60              70          80          90          100
```

Exam #5

```
                                          X       X
                                          X X X   X
                                  X       X X X   X   X X
                              X   X X   X X X X X X X X X X X X
        50          60              70          80          90          100
```

Identify the three samples that do not come from normal populations, and in each case, explain why the sample is clearly nonnormal.

(a) Distribution for Exam _____ is nonnormal because

(b) Distribution for Exam _____ is nonnormal because

(c) Distribution for Exam _____ is nonnormal because

Exercises

6.1 When a variable is normally distributed, what is the relationship between the mean, the median, and the mode?

6.2 Given a normal distribution, a standard score of 0 ($z = 0$) represents (select one)
  (a) a score at the mean.
  (b) a score below the mean.
  (c) a score above the mean.
  (d) cannot answer without knowing the standard deviation.

6.3 When a group of scores from a nonnormal distribution is transformed to standard scores, (select one)
  (a) they become normalized.
  (b) the mean of the distribution becomes 0 and the standard deviation becomes 1.
  (c) the distribution becomes symmetrical.
  (d) the scores become bimodal.

6.4 Based on the 1990 census, the number of hours per day adults spend watching television varies with a mean of 5 hours and a standard deviation of 2 hours. Based on these results alone, can you conclude that about 95% of the adult population spends between 1 and 9 hours per day watching television? Explain.

6.5 Math SAT scores follow a normal distribution $N(500, 100)$. Math ACT scores follow a normal distribution $N(18, 6)$. Eleanor scores 680 on the Math SAT and Gerald scores 27 on the Math ACT.
  (a) What is Eleanor's standard score?
  (b) What is Gerald's standard score?
  (c) Who has the higher score relative to the corresponding distribution of scores?

6.6 The test scores of a student and the population mean and standard deviation on each of three normally distributed tests are given in the following table:

Test	Mean	Standard Deviation	Student's Score
Quantitative (Stats)	67.2	10.3	63
Analytical (Logic)	56.3	6.7	61
Verbal Comprehension	50.9	3.2	57

  (a) Convert each test score to its corresponding standard score.
  (b) On which test did the student score the highest?
  (c) The student's score in statistics was surpassed by what percentage of the population?
  (d) The student's score in logic was surpassed by what percentage of the population?
  (e) The student's score in verbal comprehension was surpassed by what percentage of the population?

6.7 Sketch a picture of the following areas under the standard normal density, and then find the corresponding area:

(a) the area to the left of $z = 3.49$.

(b) the area to the left of $z = 4.00$. Think about how it should relate to your answer to (a).

(c) the area to the right of $z = -0.84$.

(d) the area to the left of $z = 0.84$.

(e) the area to the right of $z = 2.31$.

(f) the area between $z = -3.17$ and $z = -1.84$.

(g) the area between $z = 0.17$ and $z = 2.12$.

6.8 Sketch a picture of the following descriptions regarding the standard normal density, and then find the corresponding z-value(s):

(a) the z-value representing the 95th percentile.

(b) the z-value representing the 5th percentile.

(c) the z-values (symmetric about 0) whose area between is equal to 0.90.

(d) the (approximate) z-value representing the 78th percentile.

6.9 Mathematics aptitude test scores for males are normally distributed with $\mu = 67$ and $\sigma = 4$. Scores for females also have a normal model but with $\mu = 63$ and $\sigma = 5$.

(a) Meagan obtained a score of 70. What is her percentile rank on the female population? (i.e., she scored higher than what percentage of the female population?)

(b) Simon also obtained a score of 70. What is his percentile rank on the male population? (i.e., he scored higher than what percentage of the male population?)

6.10 Homes in Chicago, Illinois, in the price range of $95,000 to $130,000 are on the market for an average of 70 days before selling. Assume that the distribution of days on the market in Chicago for homes in this price range is approximately normal with a standard deviation of 25 days. Homes in Fargo, North Dakota, in the same price range are on the market for an average of 150 days. Also assume that the distribution of days on the market in Fargo for homes in this price range is approximately normal with a standard deviation of 50 days.

(a) Sketch the distribution of days on the market for both homes in Fargo and in Chicago (in this price range) on the same axis. Be sure to label your distributions and values on the axis.

(b) Indicate on your sketch what area will correspond to the proportion of homes in Chicago (in this price range) that sell in the first month (that is, in less than 30 days); then find the proportion.

6.11 The variable Y has a normal distribution with mean μ and a variance of 4 [i.e., Y is $N(\mu, 2)$]. The distribution of Y is shown, in which the values of 5 and 9 are equidistant from the mean μ.

(a) Find the value for μ, the mean of Y.

(b) Approximately what percentage of the values of Y are between 5 and 9?

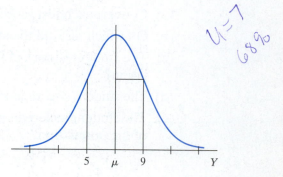

6.12 Your professor returns a set of exams to the class and announces that the distribution of scores followed an approximate normal distribution with a mean of 62 points and a standard deviation of 11 points. He also announces that the lowest 10% of the test scores will correspond to a failing grade. You get your exam back and you received 55 points. Did you fail?

6.13 A young woman needs a 15-ampere fuse for the electrical system in her apartment and has decided to buy either Brand A or Brand B. The length of life for Brand A is approximately normal with a mean of 1000 days and a standard deviation of 30 days. The length of life for Brand B is approximately normal with a mean of 990 days and a standard deviation of 10 days.

   (a) Sketch a picture of the two distributions for the lifetime of fuses. Use the same axis and be sure to label the two distributions appropriately.

   (b) The woman would be completely satisfied if the fuse she buys lasts longer than 980 days. Which brand of fuse should she buy? Explain.

REAL DATA

6.14 The article "Men Outnumber Women at Top and Bottom IQ Levels, Study Says" (SOURCE: *The Ann Arbor News*, July 7, 1995) states that the average man and the average woman share about the same level of intelligence, but men account for a higher proportion of both geniuses and the mentally deficient, according to a new study of IQ test results. If, for both men and women, the distribution of IQ scores is normal, what does the preceding statement infer regarding how the mean and the standard deviation for the two distributions compare? Provide a sketch of how the distribution of IQ scores for men would look relative to the distribution of IQ scores for women.

6.15 In a conference meeting, the duration of 100 presentations is normally distributed with a mean of 30 minutes and a standard deviation of 5 minutes. What is the maximum duration of the quickest 5% of the presentations?

6.16 According to the Opinion Research Corporation, the time men spend in the shower follows a normal distribution with a mean of 11.4 minutes and a standard deviation of 1.8 minutes.

   (a) Based on this model, what proportion of men spend less than 10 minutes in the shower?

   (b) Calculate Q_1, the first quartile.

   (c) What is the minimum time the slowest 5% spend in the shower?

6.17 The useful life of a computer terminal at a university computer center is known to be normally distributed with a mean of 3.25 years and a standard deviation of 0.5 years.

   (a) What proportion of computer terminals will have a useful life of at least 3 years?

   (b) Historically, 20% of the terminals have had a useful life less than the manufacturer's advertised life. What is the manufacturer's advertised life for the computer terminals?

6.18 The Bank of Connecticut issues VISA credit cards. Assume that the distribution for account balances on all VISA cards issued by this bank follows a normal distribution with a mean of $800 and a standard deviation of $200.

(a) Draw a distribution for the account balances. Be sure to include *all* important features and labels.

(b) What proportion of accounts have a balance below $500?

(c) The bank decides to extend a special offer to the accounts with balances in the top 10%. What is the minimum balance needed to receive this offer?

6.19 The distribution of Professor Anderson's final scores is normal with a mean of 70 and a standard deviation of 10. The distribution of Professor Johnson's final scores is normal with a mean of 80 and a standard deviation of 5. Both give "A's" if the final score is above 90.

(a) Using the same axis (note the values on the axis) shown here, draw both test-score distributions. Clearly label each curve with the professor's name.

(b) Which professor gives a higher proportion of "A's"?

(c) Which professor gives a higher proportion of "E's" (below 50)?

(d) Another professor on campus has said he will only give "A's" to the top 10% of scores on the final exam, regardless of the actual scores. If the distribution of scores on the final is normal with a mean of 69 and a standard deviation of 9, how high would a score need to be to earn an "A"?

6.20 If your score on a statistics exam was 76 and the professor gave you the distribution for the exam scores for your Section 001, you could find your percentile to understand where you stand in comparison to your fellow students. Assume that the distribution for exam scores for Section 001 is normally distributed with a mean of 80 and a standard deviation of 8.

(a) Draw the distribution for the exam scores for Section 001. Label the *x*-axis from 50 to 110 by 10. Mark the value of your score on the axis. Be sure to include all important features.

(b) What proportion of students in Section 001 scored 76 or lower?

(c) The professor announces that the top 10% of scores will receive a grade of "A." What is the minimum score needed to receive an "A"?

(d) In which section would you be better off—your Section 001, or Section 002, in which the exam scores were normally distributed with a mean of 78 and a standard deviation of 8?

6.21 Varivax, a vaccine for chicken pox, was approved by the FDA. One injection of the vaccine is recommended for children 12 months to 12 years old. Two injections, 4 to 8 weeks apart, are recommended for those 13 or older who have not had chicken pox. It was reported that 2.8% of vaccinated children still develop chicken pox. However, such cases are generally mild, with an average of 50 lesions, compared with an average of 300 lesions that develop from natural chicken-pox infection.

Let X = the number of lesions for vaccinated children who have developed chicken pox.

Let Y = the number of lesions for nonvaccinated children who have developed the natural chicken-pox infection.

Suppose that the number of lesions for each population can be modeled normally, in particular,

$$X \text{ is } N(50, 10) \text{ and } Y \text{ is } N(300, 100).$$

Suppose that a case is classified as "very mild" if the number of lesions is below 40, and classified as "very extreme" if the number of lesions is above 500.

(a) What proportion of *nonvaccinated* children with chicken pox have a "very mild" case?

(b) Fill in the blank. Show corresponding work.
25% of the *vaccinated* children with chicken pox have *at least* _____ lesions.

6.22 Suppose that the weight of tomato juice in mechanically filled cans follows a normal distribution with a mean of 464 grams and a standard deviation of 8 grams. The mean of the distribution can be set by adjusting the filling machinery, while the standard deviation reflects the precision of the filling machinery.

(a) What proportion of cans contain between 454 and 474 grams of tomato juice?

(b) How much tomato juice must a can contain in order for it to be in the top 10% of the distribution of weights?

(c) Without performing any calculations, explain how your answer to (a) will be affected if the standard deviation were halved.

6.23 Machine A makes parts whose lengths are approximately normally distributed with a mean of 4.6 mm and a standard deviation of 0.1 mm. Machine B makes parts whose lengths are approximately normally distributed with a mean of 4.9 mm and a standard deviation of 0.1 mm. Suppose that you have a box of parts which you believe are from Machine A, but you're not sure. You decide to test the hypotheses H_0: The parts are from Machine A versus H_1: The parts are from Machine B, by randomly selecting one part from the box and measuring it.

(a) Draw the distributions for the lengths of parts under H_0 and under H_1. For both sketches, label the *x*-axis from 4.2 to 5.2 by 0.1. Be sure to include all important features.

(b) Suppose that you get a length of 4.8 mm.
(i) In your sketch for (a), shade in the region that corresponds to this *p*-value and clearly label the region as such.
(ii) Compute the *p*-value for your test.

(c) What is your decision at the 0.01 level?

6.24 A five-year-old study claims that selling prices of new houses in Ann Arbor, Michigan, have a normal distribution with mean $\mu = \$180,000$ and standard deviation $\sigma = \$20,000$.

(a) A person looking to buy a house in the city claims that the average price has gone up. State the appropriate null and alternate hypotheses to test her claim in terms of μ, the population mean selling price for new homes today.

(b) What is the direction of extreme for testing H_0 against H_1?

(c) A randomly selected house has a selling price of $225,000. What is the corresponding *p*-value of her test?

6.25 The height of pine trees, X, in a forest is believed to be normally distributed. We wish to test the following hypotheses:

H_0: X is $N(15, 3)$.

H_1: X is $N(10, 3)$.

All heights are measured in meters. We decide to reject H_0 if the height of a randomly selected pine tree from the forest is less than 8 meters.

(a) Calculate the chance of a Type I error, α.

(b) Calculate the chance of a Type II error, β.

(c) Calculate the p-value if the height of the selected pine tree from the forest was measured to be 8.5 meters.

6.26 (a) The distribution for selling price, in thousands of dollars, of a sample of $n = 50$ houses that sold over the last month is summarized with a frequency plot and a frequency table.

(i) If you could afford a house at the 40th percentile of this distribution, how much would that house cost you, in dollars?

(ii) If you could afford a $122,000 house, that would mean you could afford a house that cost more than _____% of the other houses in this distribution.

(b) Now suppose that the distribution for selling price (in $1000s) of houses that sold over the last month is summarized approximately by a normal distribution with a mean of 116 and a standard deviation of 34. So X = selling price of homes follows a $N(116, 34)$ distribution.

Price (in $1000s)	Number of Homes
60	1
70	3
80	6
90	3
100	8
110	7
120	5
130	5
140	2
150	3
160	2
170	2
180	1
190	1
200	0
210	0
220	1

(i) If you could afford a house at the 40th percentile of this normal distribution, how much would that house cost you, in dollars?

(ii) If you could afford a $122,000 house, that would mean you could afford a house that cost more than _____% of the other houses in this normal distribution.

(c) Compare your answers to the problems in (a) to the corresponding problems in (b). Are the answers similar or different? Explain why you might expect this similarity or difference.

6.27 Many companies conduct annual assessment of their managers and professional workers. For one company, the performance scores follow a normal distribution and they assign an "A" to the upper 10% of scores, a "C" to the lower 10% of scores and a "B" to the middle 80% of scores.

(a) How many standard deviations above and below the mean do the A–B and B–C cutoffs lie?

(b) Suppose that, for this past year, employees with a score less than 50 received a "C" and those with a score more than 250 received an "A". What are the mean and standard deviation of the scores?

6.28 For a population under study, the variable "weight" follows a normal distribution. If the variance in weights is 100 and the 20th percentile corresponds to 130 pounds, what is the mean weight?

6.29 A product is packaged with a label that states a net weight of 250 grams. The product manager would like the packages to be filled with at least 250 grams, but at most 258 grams. To check the performance on this objective, a random sample of 60 packages was selected and weighed. The results are presented in the following list (rounded to the nearest gram):

251	258	256	260	255	255	261	257	258	257
257	257	255	257	254	257	255	256	249	257
255	255	255	254	255	256	259	257	251	253
256	257	257	253	256	255	256	251	254	260
250	253	253	259	250	252	258	261	257	259
252	256	258	252	254	254	252	258	256	253

(a) Find the sample mean and sample standard deviation of these data.

(b) Make a frequency plot, a stem-and-leaf plot, or a histogram of these data. Describe the distribution.

(c) Find the percentage of observations within one, two, and three standard deviations of the mean.

Percentage of observations in $[\bar{x} - s, \bar{x} + s]$ = _____.

Percentage of observations in $[\bar{x} - 2s, \bar{x} + 2s]$ = _____.

Percentage of observations in $[\bar{x} - 3s, \bar{x} + 3s]$ = _____.

(d) Does the distribution for these data appear to resemble a normal distribution based on the observed percentages? Explain.

(e) Report your results to the product manager and state your assessment about meeting the goal of filling the packages with at least 250, but at most 258, grams, of product.

6.30 A study was conducted by researchers at the University of Maryland to assess whether the mean body temperature of humans is indeed 98.6 degrees Fahrenheit. The researchers obtained the body temperatures of 93 healthy humans, and these data are provided in the list that follows. Further details of the study by P. Mackowiak, S. Wasserman, and M. Levive appeared in the article "A Critical Appraisal of 98.6 deg. F, the Upper Limit of the Normal Body Temperature, and Other Legacies of Carl Reinhold August Wunderlich" (*Journal of the American Medical Association*, 268, pages 1578–1580).

REAL
DATA

98.0	97.6	98.8	98.0	98.8	98.8	97.6	98.6	98.6
98.8	98.0	98.2	98.0	98.0	97.0	97.2	98.2	98.1
98.2	98.5	98.5	99.0	98.0	97.0	97.3	97.3	98.1
97.8	99.0	97.6	97.4	98.0	97.4	98.0	98.6	98.6
98.4	97.0	98.4	99.0	98.0	99.4	97.8	98.2	99.2
99.0	97.7	98.2	98.2	98.8	98.1	98.5	97.2	98.5
99.2	98.3	98.7	98.8	98.6	98.0	99.1	97.2	97.6
97.9	98.8	98.6	98.6	99.3	97.8	98.7	99.3	97.8
98.4	97.7	98.3	97.7	97.1	98.4	98.6	97.4	96.7
96.9	98.4	98.2	98.6	97.0	97.4	98.4	97.4	96.8
98.2	97.4	98.0						

(a) Find the sample mean and sample standard deviation of these temperature data.

(b) Make a frequency plot, a stem-and-leaf plot, or a histogram of these temperature data. Describe the distribution.

(c) Find the percentage of temperatures within one, two, and three standard deviations of the mean.

Percentage of observations in $[\bar{x} - s, \bar{x} + s]$ = _____.

Percentage of observations in $[\bar{x} - 2s, \bar{x} + 2s]$ = _____.

Percentage of observations in $[\bar{x} - 3s, \bar{x} + 3s]$ = _____.

(d) Does the distribution for these temperature data appear to resemble a normal distribution based on the observed percentages in (c)? Explain.

(e) Find the standard score for the value 98.6. Comment on the statement "Is the mean body temperature of humans 98.6 degrees Fahrenheit?"

6.31 **Selecting a jury** In 1969, Dr. Benjamin Spock came to trial before Judge Ford in Boston's federal court. The charge was conspiracy to violate the Military Service Act. Of all defendants, Dr. Spock, who had given wise and welcome advice on child-rearing to millions of mothers, would have liked women on his jury. The jury was drawn from a "venire," or panel of 350 persons selected by the clerk. This venire included only 102 women, although the majority of eligible jurors in the district was female. At the next stage in selecting the jury to hear the case, Judge Ford chose 100 potential jurors out of these 350 persons. His choices included only nine women.

(a) Under a null hypothesis, which states that 350 people are selected by simple random sample from a large population that is 50% female, the distribution of the number of females selected is approximately normal with a mean of 175 and a standard deviation of 9.35. What is the chance that 102 or fewer females would be selected if the null hypothesis were true? Can this chance be called a *p*-value? Explain.

(b) Under a null hypothesis, which states that 100 people are selected by simple random sample of 350 people that is 29.1% female, the distribution of the number of females selected is approximately normal with a mean of 29.1 and a standard deviation of 4.54. What is the chance that nine or fewer females would be selected if the null hypothesis were true? Would you reject the null hypothesis? Explain. Is the result statistically significant? Explain.

6.3.2 Uniform Distributions

Another set of continuous distributions is called the **uniform distributions**. A uniform density curve is a flat, rectangular-shaped curve that covers a given interval. The accompanying picture displays a general uniform distribution, denoted as X is $U(a, b)$. The total area under any density curve is 1. We have a rectangle whose area is equal to the length of the base times the height. Since the length of the base is $(b - a)$, the height must be $1/(b - a)$. A uniform distribution is also symmetric, so the mean and the median are the same value—namely, the midpoint of the interval, computed as $(a + b)/2$.

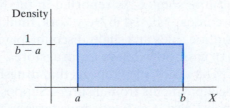

Note: Contrary to the notation for a normal distribution, a is not the mean, but rather the minimum possible value, and b is not the standard deviation, but rather the maximum possible value.

❖ EXAMPLE 6.7   A Uniform Density

Consider the uniform density X is $U(-1, 3)$. This density curve could also be written as a function:

$$f(x) = \begin{cases} \frac{1}{4} & \text{if } -1 \leq x \leq 3 \\ 0 & \text{otherwise.} \end{cases}$$

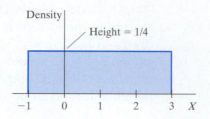

Density

Height = 1/4

−1   0   1   2   3 X

Q1 What is the mean, and thus the median, of this distribution?

A1 The mean is equal to the median and is the midpoint of the interval—namely, $(-1 + 3)/2 = 1$.

Q2 What proportion of the values is greater than or equal to 1.3 ($X \geq 1.3$)? Shade in the area under the density curve. Will the proportion be greater than 0.5 or less than 0.5? Why?

A2 The proportion will be less than 0.5 since the value 1.3 is larger than the median of 1 and we want the area to the right. We find the area under the curve to the right of $1.3 = (\text{base}) \times (\text{height}) = (1.7) \times (1/4) = 0.425$ or 42.5%.

Q3 What percentile does the value of $x = -0.1$ correspond to ($X \leq -0.1$)? Shade in the area under the density curve. Will the proportion be greater than 0.5 or less than 0.5? Why?

A3 The proportion will be less than 0.5 since the value -0.1 is smaller than the median of 1 and we want the area to the left. We find the area under the curve to the left of $-0.1 = (\text{base}) \times (\text{height}) = (0.9) \times (1/4) = 0.225$. So $x = -0.1$ is the 22.5th percentile. ❖

Let's do it! 6.9 A Uniform Model for Waiting Time

REAL DATA

Even drugstores have their own headaches, due to chronic late and incomplete payments from insurers and other medical personnel. As reported in the March 25, 1995 *New York Times*, "Pharmacy Fund, a new company based in New York, thinks it has a cure. It buys bills from the previous day's prescription sales at a slight discount, paying the pharmacist immediately and eliminating weeks of cash-flow delays for drugstores. The Pharmacy Fund then collects from the health plans." The article also reports that drugstores typically had to wait between 30 and 45 days to collect payments from insurers. Suppose that the waiting time, in days, to collect payment for non-Pharmacy Fund drugstores is uniformly distributed between 30 and 45 days.

(a) Sketch the distribution for the variable X = waiting time in days to collect payment for non-Pharmacy Fund drugstores. Be sure to label and give appropriate values on the axes.

(b) What is the mean waiting time to collect payment for non-Pharmacy Fund drugstores?

(c) What proportion of time would a non-Pharmacy Fund drugstore have to wait at least 6 *weeks*?

 Let's do it! 6.10 **Testing a Bank's Claim**

The waiting time, in minutes, for customers at a bank is known to follow the uniform distribution $U(0, b)$. The bank claims that the mean waiting time is 3 minutes. A consumer organization doubts the validity of this claim and wishes to test

H_0: The mean waiting time is 3 minutes—that is, $\mu = 3$.
H_1: The mean waiting time is greater than 3 minutes—that is, $\mu > 3$.

If it is decided to select one customer at random and reject H_0 provided that his or her waiting time is 5 minutes or longer, then what will be the significance level of this test?

Hint   Think about what the value of b is if the null hypothesis is true.

 Exercises

6.32 The travel time for Viviana, a college student, between her home and college is uniformly distributed between 40 and 70 minutes.

   (a) Sketch the distribution for Viviana's travel time. Be sure to label and give values on the axes.

   (b) Viviana leaves home each school morning to go to campus exactly one hour before her class starts. On what proportion of school days will Viviana be late to class?

6.33 Suppose that a commuter arrives at a bus stop at 10:00 A.M., knowing that the bus will arrive some time uniformly distributed between 10:00 A.M. and 10:30 A.M.

(a) For what proportion of time does the commuter have to wait longer than 10 minutes?

(b) If at 10:15 A.M. the bus has not yet arrived, for what proportion of time does the commuter have to wait at least an additional 10 minutes? (Questions of this type will be more formally addressed in the next chapter. For now, see if you can picture and reason out the answer.)

6.34 The distribution for the exact weight of 50-pound bags of Idaho potatoes is uniform over the interval 49 pounds to 51 pounds $U(49, 51)$.

(a) Sketch the distribution. Be sure to include all relevant details.

(b) Based on the distribution, what proportion of 50-pound bags actually weigh more than 50 pounds?

(c) Find the IQR for this distribution.

6.35 We have two variables X and U that have the following distributions:

$$X \text{ is } N(10, 4) \text{ and } U \text{ is } U(10, 16).$$

(a) Sketch the distribution for X and the distribution for U.

(b) Determine whether each of the following statements is true or false, and explain your answer:

(i)   The percentage of values of X that exceed 11 is more than 50%.

(ii) The percentage of values of U that exceed 11 is more than 50%.

6.36 Many calculators and computers have a function that is designed to produce random numbers from the $U(0, 1)$ distribution.

(a) What is the chance that a given number generated using this function is between 0.3 and 0.7?

(b) Use your calculator (with a seed value of 33) or the random number table (starting at Row 18, Column 1, reading from left to right) and generate 10 random numbers from the $U(0, 1)$ distribution. What fraction was between 0.3 and 0.7? Does the fraction match your answer to (a)? If there is a difference, how would you explain it?

6.37 No one likes to wait at a bus stop for very long. A city bus line states that the length of time a bus is late is uniformly distributed between 0 and 10 minutes.

(a) Sketch the distribution for the variable X = time late (in minutes). Be sure to provide all important features.

(b) For the distribution in (a), what proportion of the time will a bus arrive no more than 3 minutes late?

(c) The mean time a bus will be late is 5 minutes according to the distribution in (a). The director has been receiving a number of complaints from its riders regarding how late buses have been running. He obtains the help of a statistician to test the alternative theory that the mean time a bus will be late is greater than 5 minutes.

(i)   What is the direction of extreme for this test?

(ii) Suppose that a randomly selected bus stop results in the bus arriving 9 minutes late. What is the corresponding p-value for this observation?

(iii) At $\alpha = 0.15$, what is the decision?

6.38 A supermarket advertises that the waiting time spent by customers waiting in line at the cash register is uniform between 0 and 10 minutes (with an average waiting time of 5 minutes). A number of customers complain that they wait in line too long for this to be true. Sebastian, the manager, decides to test whether the complaints are justified by testing the following hypotheses:

H_0: $\mu = 5$.

H_1: $\mu > 5$.

(a) Sketch the distribution of waiting time when H_0 is true. Be sure to label and give appropriate values on the axes.

(b) The waiting time for a randomly selected customer will be obtained and Sebastian will test these hypotheses using the following decision rule: Reject H_0 if the observed waiting time is 8.5 minutes or longer.

   (i) In your sketch for (a), shade in the region that corresponds to α, the significance level, and clearly label the region with α.

   (ii) Compute α, the significance level.

(c) Suppose that the next randomly chosen customer waits in line 5.5 minutes.

   (i) In your sketch for (a), shade in the region that corresponds to the p-value and clearly label the region as such.

   (ii) Compute the p-value.

(d) For the customer in (c) who waits in line 5.5 minutes, are the results statistically significant at the 10% level? Explain.

6.39 A fast-food-store manager wishes to test hypotheses regarding the distribution of service time for customers using the drive-thru window. The competing hypotheses for the distribution of X = service time (length of time between order taken and order received, in minutes) are

H_0: The distribution of X is $U(3, 7)$.

H_1: The distribution of X is $U(0, 4)$.

(a) Draw the distributions for X under H_0 and under H_1. For both distributions, label the x-axis from 0 to 8 by 1. Be sure to include all important features.

(b) The service time for a randomly selected customer will be obtained and the manager will test these hypotheses using the following decision rule: Reject H_0 if the observed service time is 3.2 minutes or less.

   (i)   In your sketch for (a), shade in the region that corresponds to α, the significance level, and clearly label the region with α.

   (ii) Compute α, the significance level.

   (iii) In your sketch for (a), shade in the region that corresponds to β, the chance of a Type II error, and clearly label the region with β.

   (iv) Compute β.

(c) Suppose that the observed service time was 4.1 minutes.

   (i) In your sketch for (a), shade in the region that corresponds to the p-value and clearly label the region as such.

   (ii) Compute the p-value.

(d) Are the data in (c) statistically significant at the level α found in (b)? Explain.

6.40 Any nonnegative function with a total area under the function of 1 could be considered as a density function model for some continuous characteristic. Consider the following density function:

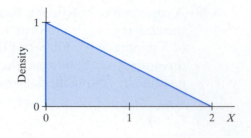

$$f(x) = \begin{cases} -\frac{1}{2}x + 1 & \text{if } 0 \leq x \leq 2 \\ 0 & \text{otherwise.} \end{cases}$$

(a) Is it a density function?

(b) Do you think the median for this distribution is less than 1, equal to 1, or greater than 1?

(c) Find the proportion of values that are less than or equal to 1.2.

6.41 A priority-mail rate is available for packages that weigh under 2 pounds. A mailing company collected data in order to develop a model for the distribution of weight of priority-mail packages. The shape of the shown histogram suggests the triangular density as an adequate model.

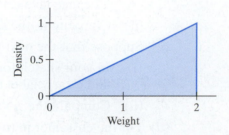

(a) Based on this model, what proportion of packages weighed less than 1 pound?

(b) Will the median weight be more than 1 pound, equal to 1 pound, or less than 1 pound? Explain.

(c) Will the mean weight be more than 1 pound, equal to 1 pound, or less than 1 pound? Explain.

6.4 Modeling Discrete Variables

The normal distribution and the uniform distribution are just two of the many types of density functions. In both the normal and the uniform case, we are working with continuous functions (that is, these functions are used to model continuous variables). If the variable is discrete, the number of possible values is either finite or infinite, but countable.

Consider the number of typing errors found on a page of a manuscript. For a given manuscript, the possible number of errors on a page is 0, 1, 2, 3, 4, 5, or 6. We could make a histogram to show how often each of these values occurs. Suppose that we smooth out the histogram and derive a corresponding density function as a model. With a density function, we would find the proportion of pages having between 1.5 and 1.8 errors on a page by computing the area under the density between 1.5 and 1.8. This area would not be 0, although it is not possible to have between 1.5 and 1.8 errors on any given page. A density function would not serve as a very useful model.

If the variable is discrete, a much simpler description of the distribution is to list the values that are possible along with the proportion of times they are expected to occur. This is exactly how to display a discrete model.

❖ EXAMPLE 6.8 Discrete Model for the Number of Books in a Backpack

Let X represent the number of books in a backpack for students enrolled at a particular college. We can summarize the distribution of such a discrete characteristic using a table. The accompanying discrete model tells us that 50% of the students have two books in their backpack, 30% of the students have three books, and 20% of the students have four books. Based on this model, no other values for X are possible. The percentages of any discrete model must add up to 100%. Note that 50% + 30% + 20% = 100%.

Value of X	2	3	4
Proportion	0.5	0.3	0.2

The distribution of a discrete variable X is sometimes called a **mass function**. The values of a discrete function must be between 0 and 1 (inclusive) and add up to 1.

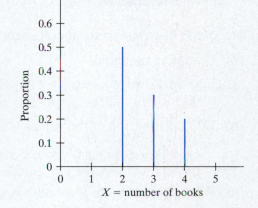

X = number of books

Q1 What proportion of the students have 3 or fewer books ($X \le 3$)?

A1 $0.5 + 0.3 = 0.8$, or 80%.

Q2 What proportion of the students have more than 3 books ($X > 3$)?

A2 0.2, or 20%.

Q3 What proportion of the students have between 2.1 and 2.8 books ($2.1 < X < 2.8$)?

A3 0.

Q4 What proportion of the students have 4 or fewer books ($X \le 4$)?

A4 $0.5 + 0.3 + 0.2 = 1$, or 100%.

Q5 What proportion of the students have less than 2 books ($X < 2$)?

A5 0.

Q6 Describe the general shape of this distribution.

A6 Skewed to the right. ❖

THINK ABOUT IT Not All Tables Present a Discrete Model

A discrete model can be used to describe the distribution of a qualitative or categorical variable, but be careful. Consider the variable X, which represents the type of pet owned, and the accompanying table regarding this variable.

Pet	Cat	Dog	Other
Proportion	0.40	0.70	0.30

Is this a legitimate discrete distribution? What do you get if you add up these proportions? What happened here? Obviously, some people own more than one type of pet.

Tip: When defining a model for a discrete variable, be sure that the possible values of the variable divide up the items into disjoint or nonoverlapping groups. Each element can take on one and only one of the possible values of the variable.

> *Definition:* A **mass function** is used as a model for a discrete variable. For each possible value, the mass function gives the proportion of units in the population having that value. Thus, the values of the mass function must be between 0 and 1 and add up to 1. Proportions are measured directly as the values of the function, not as areas under the function.

Let's do it! 6.11 A Discrete Model

Part of the distribution for the discrete variable X is given in the following table:

(a) Complete the table.

Value of X	−1	2	3	4
Proportion	0.1		0.6	0.2

(b) What proportion of the values is less than 3.1 ($X < 3.1$)?

(c) What proportion of the values is greater than or equal to −1.1 ($X \geq -1.1$)?

(d) What proportion of the values is between 2 and 3 ($2 < X < 3$)?

(e) What proportion of the values is less than or equal to 3.1, but more than 2 ($2 < X \leq 3.1$)?

Let's do it! 6.12 Zip Codes

Develop the distribution of local residential zip codes for the members of your class. Determine the main zip codes and enter them in the space provided. You may wish to include an "other" category.

Zip Code _____

Count _____

Proportion _____

Determine which zip codes correspond to those living farthest away from your university or college. What proportion of the members of your class lives in this (these) farthest away zip-code area(s)?

As we have seen in this chapter, there are many types of models, corresponding to the type of variable, the various shapes of distributions, their centers, and their variations. We conclude this section with two final "Let's do it!" exercises that include both discrete and continuous distributions.

Let's do it! 6.13 Three Distributions

Let X, Y, and L be three variables whose distributions are as follows.

X is $N(1, 2)$,   Sketch the distribution for X.

Y is $U(1, 2)$,   Sketch the distribution for Y.

L is discrete with the following distribution:

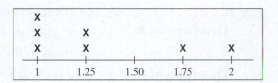

(a) Find
the mean of X_____.
the mean of Y_____.
the mean of L_____.

(b) Find
the interquartile range (IQR) of X_____.
the IQR of Y_____.
the IQR of L_____.

(c) Find
the standard deviation of X_____.
the standard deviation of L_____.

(d) Find
the proportion $(X > 1)$_____.
the proportion $(Y > 1)$_____.
the proportion $(L > 1)$_____.
the proportion $(X = 1)$_____.
the proportion $(L = 1)$_____.

Let's do it!   6.14 Comparing Distributions

Consider the following three distributions—the mean and standard deviation for each distribution are also provided:

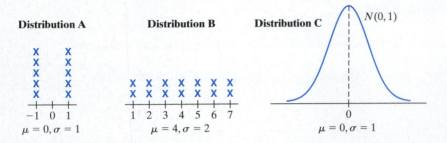

(a) Is Distribution A symmetric?   Yes   No

Is Distribution B symmetric?   Yes   No

Is Distribution C symmetric?   Yes   No

(b) For each distribution, find the proportion of values that are within one standard deviation of the mean.

Distribution A $[\mu - \sigma, \mu + \sigma] = $ _____.

Proportion in this range $= $ _____.

Distribution B $[\mu - \sigma, \mu + \sigma] = $ _____.

Proportion in this range $= $ _____.

Distribution C $[\mu - \sigma, \mu + \sigma] = $ _____.

Proportion in this range $= $ _____.

(c) Which distributions have 68% of the values within one standard deviation of the mean?

Note: There are many types of distributions, and not all distributions are normal. Knowing that a distribution is symmetric does not imply that it will follow the 68–95–99.7 rule for normal distributions.

✳ **Exercises**

6.42 Let Y be the number of televisions in a randomly selected Zone City household. The mass function of Y is given in the following table:

Number of televisions	0	1	2	3	4
Proportion of households	0.05	0.30	0.40	0.20	

(a) Complete the mass function and sketch the graph of the mass function. Be sure to label all important features.

(b) What proportion of households have at least one TV?

6.43 Let X represent the number of cars sold (per day) by a particular salesperson. The following table summarizes part of the distribution of X:

Value of X	0	1	2	3	4
Proportion	0.1	0.4	0.3		

On what proportion of days would this salesperson sell more than two cars?

6.44 Let the variable X represent the number of children per family for the hypothetical Lima City. The distribution of X is

Number of children	0	1	2	3	4
Proportion of families	0.40		0.20	0.10	0.05

(a) What proportion of families have one child?

(b) What proportion of families have at least one child?

(c) The shape of the distribution of X, the number of children per family for Lima City, is (select one)

   (i)   symmetric.

   (ii) uniform.

   (iii) normal.

   (iii) skewed to the left.

   (iv) none of the above.

6.45 M&M's were first produced in 1941 in six colors (brown, yellow, red, violet, green, orange). In 1949, tan replaced violet. In early 1995, Mars, Inc., concluded that people didn't like tan, and they conducted a survey to pick a new color. The winning color was royal blue. In the *Detroit Free Press* (March 31, 1995), an article entitled "No Sweet Secret: M&M's Go Blue" contained the following information about M&M's:

■ "The candy company makes 300,000 pieces a day. In the plain bags, 30 percent should be brown. You also get 20 percent each of red and yellow; 10 percent each of green, orange and now, blue."

■ "In the peanut variety, they added blue, so now they've got six colors: 20 percent each of brown, red, yellow and blue; 10 percent each of green and orange."

■ "In almond and peanut butter, it's five colors, evenly divided."

(a) Let Y = color for plain M&M's. Write out the distribution for Y and give a corresponding sketch of the distribution.

(b) Can you comment on the shape of the distribution for Y? If yes, describe the shape. If no, explain why not.

(c) If red is one of the five colors in almond M&M's, what proportion of almond M&M's should be red?

(d) Describe how you might go about assessing whether or not the above stated distributions for the various varieties of M&M's are plausible.

(e) (optional) What does M&M stand for?

6.46 For *each* of the following distributions, compute the proportion of the values that exceed 2:

(a) Suppose that X has a $N(1.8, 2)$ distribution.

(b) Suppose that T has the following discrete model:

$(T = t)$	1.8	1.9	2.0	2.1
Proportion$(T = t)$		$\frac{7}{20}$	$\frac{5}{20}$	

It is also known that $P(T = 1) = P(T = 4)$.

(c) Suppose that L has a $U(1.8, 4)$ distribution.

Chapter Summary

In this chapter, we have taken the display of the distribution of a variable one step further. We have discussed how to model variables. We started out in the continuous case with density curves that are smoothed histograms with the property that the total area under the curve equals 1. We studied perhaps the most important family of density curves, called the normal distributions. These distributions will be very useful in Chapter 8, which discusses sampling distributions. Another family of density curves is the family of uniform distributions. One key idea regarding any density curve is that the area under the curve corresponds to the proportion of items falling in that range (area = proportion). Models for discrete characteristics or qualitative data were also presented, sometimes called mass functions. For each possible value of the discrete variable, the mass function gives the proportion of units in the population having that value. Thus, the values of a mass function must be proportions— that is, between 0 and 1 (inclusive)— and the sum of all the values of a mass function must equal 1.

Remember that models will serve as a frame of reference. We will be able to compare observations to a potential model and assess whether it is likely or unlikely to have observed the response under that model. If the result is somewhat likely, we might support that model or theory. If the result is somewhat unusual, we might reject that model or theory.

In the next chapter, entitled "Probability," we will return to discussing models for variables. The primary difference between the models presented in this chapter and the next is the replacing of the word **proportion** with **probability**. So far, we have used models to find the proportion of units in the population taking on certain values. In Section 7.5, the variables will be called random variables, their models will be called probability distributions, and these models will be used to find the probability that a randomly selected unit from the population will take on certain values.

Key Terms

Be sure you can describe, in your own words, and give an example of each of the following key terms from this chapter.

model	68–95–99.7 rule for	points of inflection
normal distributions	normal distributions	standardization
standard normal	discrete distributions	z-score
distribution	density curve or density	uniform distributions
standard score	function	mass function

Exercises

6.47 The life span of a calculator manufactured by the Calc Company has a normal distribution with a mean of 60 months and a standard deviation of 8 months. The company guarantees that any calculator that malfunctions within 36 months of purchase will be replaced with a new calculator. What proportion of such calculators made by this company will be replaced?

6.48 Determine which of the following statements are true and which are false (a true statement must be always true):

(a) The median of a normal distribution is 0.

(b) The median is the midpoint between the first and the third quartile.

(c) The mean of a normal distribution is the midpoint between the first and the third quartile.

(d) The standard deviation of a normal distribution is always equal to the interquartile range.

6.49 Let X represent the height of American women, which is normally distributed with a mean of 63.6 inches and a standard deviation of 2.5 inches. (This model is based on data from the National Health Survey.) To fit into a Russian *Soyuz* spacecraft, an astronaut must have a height between 64.5 inches and 72 inches.

(a) Find the percentage of American women who meet the *Soyuz* height requirement.

(b) Among 1000 randomly selected American women, how many would you expect to meet the *Soyuz* height requirement?

(c) There is a club in the United States exclusively for the tallest 5% of American women. What is the minimum height for an American woman to be admitted to this club?

6.50 Suppose that weights of elite distance runners are approximately normally distributed with a mean of 145.1 pounds and a standard deviation of 9.4 pounds. If 90% of elite distance runners weigh less than Jack, how much does Jack weigh (approximately)?

6.51 Let the variable X represent the breaking strength of a 15-foot length of Proline fishing line. The distribution of X is normal with a mean of 30 pounds and a standard deviation of 3 pounds—that is, $X \sim N(30, 3)$.

(a) What proportion of 15-foot Proline fishing lines will have a breaking strength more than 31.5 pounds?

(b) Complete the following sentence: 5% of the 15-foot Proline fishing lines will have a breaking strength of more than_____ pounds.

6.52 A set of 25 test scores has a mean of 70 and a variance of 100. If from each of the original scores we subtract 70 and divide the difference by 100, the new set of scores has (select one)

(a) mean 0 and variance 1.
(b) a normal distribution.
(c) none of the above.
(d) both (a) and (b) are true.

6.53 The distribution of the demand for a product can often be approximated by a normal distribution. A bread company has determined that the number of loaves of its multigrain bread demanded daily has a normal distribution with a mean of 480 loaves and a standard deviation of 60 loaves. This was determined by examining the distribution of the number of loaves demanded each day for a large number of days. Based on cost considerations, the company has decided that its best strategy is to produce a sufficient number of loaves so that it will fully supply demand on 90% of all days.

(a) How many loaves of bread should the company produce daily?

(b) Based on the production in (a), on what percentage of the days will the company be left with more than 50 loaves of unsold bread?

6.54 For a certain law school, the entering students had an average LSAT score of about 700 with a standard deviation of about 40. The histogram of the LSAT scores followed a normal distribution.

(a) Approximately what percent of the entering students scored below 750 on the LSAT?

(b) Joe's LSAT score was 0.5 standard deviations above the average LSAT score. Approximately what percent of the entering students had an LSAT score lower than Joe's?

(c) This law school has a program to help entering students who have low LSAT scores. Students with an LSAT score in the lowest 15% of the distribution of LSAT scores are required to attend this program. What is the minimum score needed for a student to be exempt from this requirement?

6.55 The distribution of resistance for a certain type of resistors is known to be normal; 2.5% of these resistors have a resistance exceeding 10.5 ohms and 2.5% have a resistance smaller than 9 ohms. What are the mean and the standard deviation of this resistance distribution?

6.56 The length of human pregnancies from conception to birth varies according to a distribution that is approximately normal with a mean of 266 days and a standard deviation of 16 days.

(a) What percentage of pregnancies lasts less than eight months (240 days)?

(b) What is the interquartile range for length of pregnancy?

6.57 Personnel tests are designed to test a job applicant's cognitive or physical abilities. A particular dexterity test is administered nationwide by a private testing service. It is known that for all tests administered the year before, the distribution of scores was approximately normal, with mean 45.5 and standard deviation 1.5.

(a) A particular employer requires job candidates to score at least 50 on the dexterity test. What percentage of the test scores during the last year was at least 50?

(b) Using the 68–95–99.7 rule, approximately what percentage of the test scores during the last year exceeded 50?

(c) The testing service reported to a particular employer that one of its job candidate's scores fell at the 98th percentile of the distribution—that is, approximately 98% of the scores were lower than the candidate's and only 2% were higher. What was the candidate's score?

6.58 A study is conducted to test the following hypotheses:

H_0: The mean score on the exam is 30; that is, $\mu = 30$.

H_1: The mean score on the exam is more than 30; that is, $\mu > 30$.

The experimenter will select one exam randomly and look at the score. She will reject the null hypothesis if the score is 34 or greater. Suppose that the real distribution of scores on the exam is approximately normal with a standard deviation of 2.

(a) The experimenter has a decision rule: Reject H_0 if the score is 34 or greater. What is the significance level corresponding to this rule?

(b) The randomly selected exam had a score of 35. What is the p-value?

6.59 A researcher wishes to test the following hypotheses about the distribution for a variable X:

H_0: The distribution for X is $N(10, 2)$.

H_1: The distribution for X is $N(12, 2)$.

(a) The researcher will use the following decision rule: Reject H_0 if the observed value is 13.2 or more.

(i) Calculate α, the significance level.

(ii) Calculate β, the chance of a Type II error.

(b) Suppose that the observed value is 11.2. Calculate the corresponding p-value.

6.60 The height, X, of students in a class is believed to be normally distributed. We wish to test

H_0: X is $N(170, 4)$.

H_1: X is $N(180, 4)$.

All heights are measured in centimeters. We will reject H_0 if a randomly selected student from the class has a height more than 190 cm.

(a) Find the probability of a Type II error, β.

(b) Find the p-value if the observed height of the selected student is 185 cm.

6.61 Consider two competing models to describe a population of values. You will be allowed to select one value from the population at random and must decide which of the two models to support.

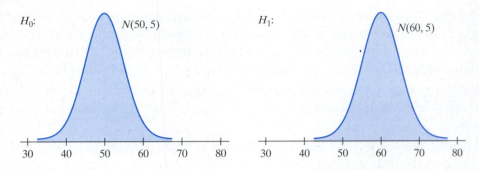

You select a number at random from the population of values, and the resulting number is 62.5.

(a) Calculate the *p*-value for this test.

(b) Do you accept or reject H_0 at a significance level 0.10? Explain your answer.

6.62 Determine whether each of the following statements is true or false (a true statement is always true):

(a) A normal curve $N(100, 25)$ is more spread out (that is, has more variability) than a normal curve $N(120, 10)$.

(b) The mode of a symmetric distribution centered at 0 is always equal to 0.

(c) A uniform distribution is symmetric.

6.63 A doctor commutes daily from his home to his city office. On average, the one-way trip takes 19 minutes, with a standard deviation of 1.7 minutes. Assume that the distribution of trip time is normally distributed. If the doctor leaves his home each day at 6:40 A.M. and a continental breakfast is served each day at his office from 6:50 A.M. until 7:00 A.M., what proportion of the time will he miss the breakfast?

6.64 For every 100 births in the United States, the distribution for the number of boys is approximately normal with a mean of 51 and a standard deviation of 5.

(a) Complete this sentence: Approximately 95% of the time, the number of boys for every 100 births will range between_____ and_____.

(b) Suppose that of the next 100 births at a hospital, there were a total of 64 boys (and thus 36 girls). Do you think that this result is unusual? Explain your answer.

6.65 Let X be the waiting time at a local oil-changing station. The variable X is assumed to follow a normal distribution with a mean of 15 minutes and a standard deviation of 3 minutes— that is, $X \sim N(15, 3)$.

(a) The manager has decided that all customers whose waiting time places them in the top 5% of the waiting-time distribution will receive a $5 discount. What is the minimum time that a customer would have to wait to receive that discount?

(b) The manager of the station has hired a new technician who will change the oil with a waiting time Y defined as $Y = 1.2X - 4$. What are the mean and the standard deviation for the new waiting time Y?

(c) What is the minimum waiting time for a person to get a discount with the new technician?

6.66 San Diego has an average high temperature in January of 65.9 degrees Fahrenheit. Suppose that the high temperatures on January days in San Diego over the years have followed a normal distribution with a mean of 65.9. Suppose further that 22% of these high temperatures exceed 75 degrees. Use this information to determine the value of the standard deviation of daily high temperatures in January in San Diego.

6.67 Suppose that the variable X has a uniform model over the interval from -3 to 3.
  (a) Sketch the density for X. Be sure to mark all important features.
  (b) Is the interquartile range for this distribution equal to 50%? Explain.
  (c) Sketch the boxplot for the distribution of X. Be sure to mark all important features.

6.68 The waiting time at the local oil-changing station is uniformly distributed between 15 and 22 minutes.
  (a) What is the corresponding density function? (Sketch it.)
  (b) What is the mean waiting time?
  (c) If the station policy is to take 20% off the price of an oil change if the wait is longer than 20 minutes, what proportion of the customers get a reduction in price?

6.69 Let the variable X represent the length of life, in years, for an electrical component. The following figure is the density curve for the distribution of X.

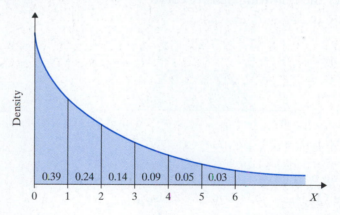

  (a) What proportion of electrical components lasts longer than 6 years?
  (b) What proportion of electrical components lasts longer than 1 year?
  (c) Describe the shape of the distribution.
  (d) What can you say about the value of the median of this distribution? Give an approximate value for the median. Do you think the mean of this distribution is larger than, smaller than, or equal to this value? Explain.

6.70 Consider the following two variables and their distributions:
$$X \text{ is } U(-2, 2) \text{ and } Y \text{ is } N(-2, 2).$$

  (a) Draw the distributions for the variables X and Y. Be sure to include all important features.

(b) Give the following numerical quantities for X and Y:

    (i)   mean of X.

    (ii)   mean of Y.

    (iii) Q1 for X.

    (iv) Q3 for X.

    (v)   Q1 for Y.

    (vi) Q3 for Y.

    (vii) proportion $(X > 0)$.

    (viii) proportion $(Y > 0)$.

6.71 For *each* of the following distributions, compute the proportion of the values that exceed 3:

(a) Suppose that $X \sim N(1, 2)$.

(b) Suppose that T has the following discrete model:

$T = t$	1	2	3	4	5
Proportion$(T = t)$		0.3	0.4	0.2	

It is also known that $P(T = 1) = P(T = 4)$.

(c) Suppose that $L \sim U(1, 4)$.

6.72 Consider the following density curve:

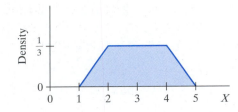

Select the best answer for the following statements:

(a) The Q1 is 2.   True   False

(b) The range is 5.   True   False

(c) The shape is   symmetric.   skewed.   uniform.

6.73 For each of the following distributions, find the proportion of X values that exceed 3. It may be helpful to sketch the distribution first.

(a) when X is normally distributed with a mean of 2 and a variance of 4, X is $N(2, 2)$.

(b) when X is uniformly distributed over the range of 2 to 4, X is $U(2, 4)$.

(c) when X has the distribution

Value of X	2	3	4
Proportion	1/25	7/25	

6.74 A student was given the distribution of a continuous variable (i.e., a density curve). He found the mean to be 20 and the standard deviation to be 3. He then concluded that approximately 95% of the values are between 14 and 26. Do you agree with the student's conclusion? Explain.

6.75 Consider the following density function for the variable X:

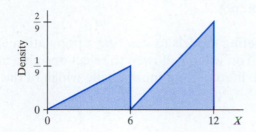

(a) The median is (select one)
   (i)   equal to 6.
   (ii) less than 6.
   (iii) more than 6.
(b) Explain your answer to (a) using one well-written sentence.
(c) We are interested in learning about the proportion of units with values between 9 and 12. Explain in words or show on a graph what corresponds to this proportion, then calculate the proportion.

6.76 The distribution of a variable X is symmetric with a mean of 20 and a variance of 25. You are asked to provide two numbers such that about 95% of the values of X are between the two numbers. The two numbers are (select one)
(a) $[20 - 5, 20 + 5] = [15, 25]$.
(b) $[20 - 10, 20 + 10] = [10, 30]$.
(c) can't tell.

6.77 Consider two competing models to describe a population of values for the lifetime of a computer chip. You will be allowed to select one value from the population at random and must decide which of the two models to support.

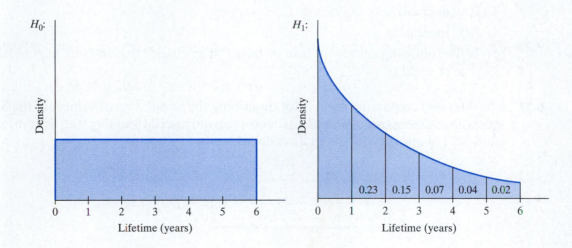

(a) What is the height of the density specified under the null hypothesis?

(b) What proportion of chips last less than one year under the alternative hypothesis?

(c) You select a chip at random from the population, and the resulting lifetime is 0.5 years. Calculate the *p*-value for this test.

(d) Give your decision at a 10% significance level and state your conclusion using a well-written sentence.

6.78 Consider two competing models to describe a population of values for the lifetime of a computer chip. You will be allowed to select one chip at random from the population, measure the lifetime, and must decide which of the two models to support.

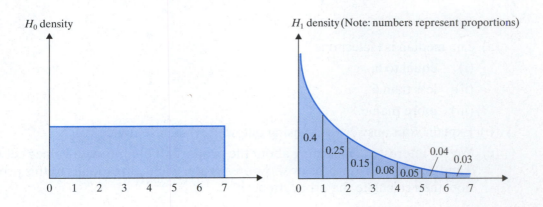

(a) What is the height of the density specified under the null hypothesis?

(b) Consider the following decision rule: Reject the null hypothesis if the lifetime of the one selected chip is 1 year or less.

   (i)   Compute the value of α.

   (ii) Compute the value of β.

(c) You select a chip at random from the population and the resulting lifetime is 0.5 years. Calculate the *p*-value for this test.

(d) Which of the following would lead to a reduction in the chance of committing a Type II error?

   (i)   decreasing α.

   (ii) increasing α.

(e) Is the following statement true or false? "The chance that the null hypothesis is true is equal to 1/7."

6.79 Consider two competing models for describing the population of values for the SAT scores in mathematics for students who received special training. Let X represent math SAT scores for the population under study.

H_0: The distribution of X is $U(200, 800)$.

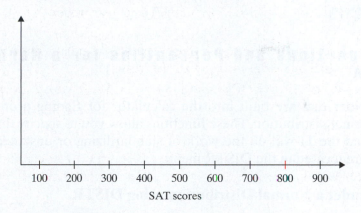

SAT scores

H_1: The distribution of X is given by

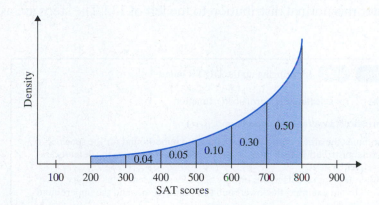

SAT scores

(a) Draw the distribution for X under the null hypothesis H_0. Be sure that you include labels for your axes and that you mark the height of the distribution on the y-axis. Also reproduce a sketch of the distribution for X under the alternative hypothesis H_1.

(b) What proportion of the students who receive special training will score less than 300 on the math SAT under the alternative distribution H_1?

(c) What is the direction of extreme for this testing problem? Select one:

(i) to the right.

(ii) to the left.

(iii) two sided.

(iv) can't tell.

(d) On your sketches given in (a), shade, clearly label, and compute α for the following decision rule: Reject H_0 if you observe a score of 700 or more extreme.

(e) On your sketches given in (a), shade, clearly label, and compute β for the decision rule in (d).

(f) Suppose that the observed value is a score of 400. What is the p-value for this observation?

(g) With the p-value from (f), were the data statistically significant? Explain.

TI Quick Steps

Finding Proportions and Percentiles for a Normal Distribution

The TI has functions that are built into the calculator for finding proportions and percentiles of any normal distribution. These functions allow you to specify the mean and standard deviation, and the TI will do the work of standardizing or unstandardizing for you. These functions are located in the DISTR menu.

Finding Areas under a Normal Distribution using DISTR

Our question is, "What proportion of 12-year-olds has IQ scores less than 132?" IQ scores follow a normal distribution with a mean of 100 and a standard deviation of 16. So we need to find the area under this normal distribution to the left of 132. The steps are as follows:

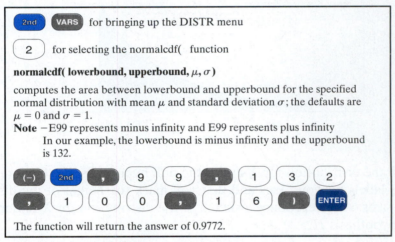

The function will return the answer of 0.9772.

Your screen should look like this:

```
normalcdf(-E99,
132,100,16)
          .977249938
```

Finding Percentiles for a Normal Distribution using DISTR

Our task is to find the IQ score necessary for a 12-year-old to place in the top 1% of the distribution of IQ scores. Since we want the area to the right of that score to be 0.01, the area to the left would be 0.99. That is, we want to find the 99th percentile of this IQ-score distribution. The TI steps are as follows:

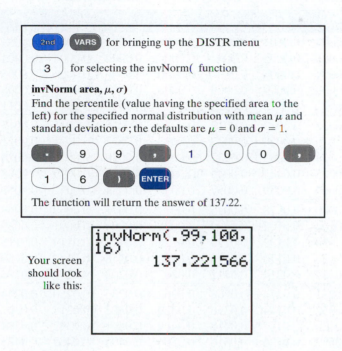

2nd VARS for bringing up the DISTR menu

3 for selecting the invNorm(function

invNorm(area, μ, σ)
Find the percentile (value having the specified area to the left) for the specified normal distribution with mean μ and standard deviation σ; the defaults are $\mu = 0$ and $\sigma = 1$.

. 9 9 , 1 0 0 ,

1 6) ENTER

The function will return the answer of 137.22.

Your screen should look like this:

```
invNorm(.99,100,
16)
           137.221566
```

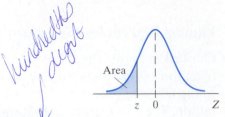

hundredths digit

Area

Table II *Areas under the Standard Normal Distribution*

z	.00	.01	.02	.03	.04	.05	.06	.07	.08	.09
−3.4	0.0003	0.0003	0.0003	0.0003	0.0003	0.0003	0.0003	0.0003	0.0003	0.0002
−3.3	0.0005	0.0005	0.0005	0.0004	0.0004	0.0004	0.0004	0.0004	0.0004	0.0003
−3.2	0.0007	0.0007	0.0006	0.0006	0.0006	0.0006	0.0006	0.0005	0.0005	0.0005
−3.1	0.0010	0.0009	0.0009	0.0009	0.0008	0.0008	0.0008	0.0008	0.0007	0.0007
−3.0	0.0013	0.0013	0.0013	0.0012	0.0012	0.0011	0.0011	0.0011	0.0010	0.0010
−2.9	0.0019	0.0018	0.0017	0.0017	0.0016	0.0016	0.0015	0.0015	0.0014	0.0014
−2.8	0.0026	0.0025	0.0024	0.0023	0.0023	0.0022	0.0021	0.0021	0.0020	0.0019
−2.7	0.0035	0.0034	0.0033	0.0032	0.0031	0.0030	0.0029	0.0028	0.0027	0.0026
−2.6	0.0047	0.0045	0.0044	0.0043	0.0041	0.0040	0.0039	0.0038	0.0037	0.0036
−2.5	0.0062	0.0060	0.0059	0.0057	0.0055	0.0054	0.0052	0.0051	0.0049	0.0048
−2.4	0.0082	0.0080	0.0078	0.0075	0.0073	0.0071	0.0069	0.0068	0.0066	0.0064
−2.3	0.0107	0.0104	0.0102	0.0099	0.0096	0.0094	0.0091	0.0089	0.0087	0.0084
−2.2	0.0139	0.0136	0.0132	0.0129	0.0125	0.0122	0.0119	0.0116	0.0113	0.0110
−2.1	0.0179	0.0174	0.0170	0.0166	0.0162	0.0158	0.0154	0.0150	0.0146	0.0143
−2.0	0.0228	0.0222	0.0217	0.0212	0.0207	0.0202	0.0197	0.0192	0.0188	0.0183
−1.9	0.0287	0.0281	0.0274	0.0268	0.0262	0.0256	0.0250	0.0244	0.0239	0.0233
−1.8	0.0359	0.0352	0.0344	0.0336	0.0329	0.0322	0.0314	0.0307	0.0301	0.0294
−1.7	0.0446	0.0436	0.0427	0.0418	0.0409	0.0401	0.0392	0.0384	0.0375	0.0367
−1.6	0.0548	0.0537	0.0526	0.0516	0.0505	0.0495	0.0485	0.0475	0.0465	0.0455
−1.5	0.0668	0.0655	0.0643	0.0630	0.0618	0.0606	0.0594	0.0582	0.0571	0.0559
−1.4	0.0808	0.0793	0.0778	0.0764	0.0749	0.0735	0.0722	0.0708	0.0694	0.0681
−1.3	0.0968	0.0951	0.0934	0.0918	0.0901	0.0885	0.0869	0.0853	0.0838	0.0823
−1.2	0.1151	0.1131	0.1112	0.1093	0.1075	0.1056	0.1038	0.1020	0.1003	0.0985
−1.1	0.1357	0.1335	0.1314	0.1292	0.1271	0.1251	0.1230	0.1210	0.1190	0.1170
−1.0	0.1587	0.1562	0.1539	0.1515	0.1492	0.1469	0.1446	0.1423	0.1401	0.1379
−0.9	0.1841	0.1814	0.1788	0.1762	0.1736	0.1711	0.1685	0.1660	0.1635	0.1611
−0.8	0.2119	0.2090	0.2061	0.2033	0.2005	0.1977	0.1949	0.1922	0.1894	0.1867
−0.7	0.2420	0.2389	0.2358	0.2327	0.2296	0.2266	0.2236	0.2206	0.2177	0.2148
−0.6	0.2743	0.2709	0.2676	0.2643	0.2611	0.2578	0.2546	0.2514	0.2483	0.2451
−0.5	0.3085	0.3050	0.3015	0.2981	0.2946	0.2912	0.2877	0.2843	0.2810	0.2776
−0.4	0.3446	0.3409	0.3372	0.3336	0.3300	0.3264	0.3228	0.3192	0.3156	0.3121
−0.3	0.3821	0.3783	0.3745	0.3707	0.3669	0.3632	0.3594	0.3557	0.3520	0.3483
−0.2	0.4207	0.4168	0.4129	0.4090	0.4052	0.4013	0.3974	0.3936	0.3897	0.3859
−0.1	0.4602	0.4562	0.4522	0.4483	0.4443	0.4404	0.4364	0.4325	0.4286	0.4247
−0.0	0.5000	0.4960	0.4920	0.4880	0.4840	0.4801	0.4761	0.4721	0.4681	0.4641

SOURCE: *Walpole/Myers/Myers*, Sixth Edition.

tenths digit

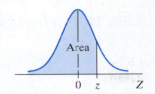

Table II   *Areas under the Standard Normal Distribution*

z	.00	.01	.02	.03	.04	.05	.06	.07	.08	.09
0.0	0.5000	0.5040	0.5080	0.5120	0.5160	0.5199	0.5239	0.5279	0.5319	0.5359
0.1	0.5398	0.5438	0.5478	0.5517	0.5557	0.5596	0.5636	0.5675	0.5714	0.5753
0.2	0.5793	0.5832	0.5871	0.5910	0.5948	0.5987	0.6026	0.6064	0.6103	0.6141
0.3	0.6179	0.6217	0.6255	0.6293	0.6331	0.6368	0.6406	0.6443	0.6480	0.6517
0.4	0.6554	0.6591	0.6628	0.6664	0.6700	0.6736	0.6772	0.6808	0.6844	0.6879
0.5	0.6915	0.6950	0.6985	0.7019	0.7054	0.7088	0.7123	0.7157	0.7190	0.7224
0.6	0.7257	0.7291	0.7324	0.7357	0.7389	0.7422	0.7454	0.7486	0.7517	0.7549
0.7	0.7580	0.7611	0.7642	0.7673	0.7704	0.7734	0.7764	0.7794	0.7823	0.7852
0.8	0.7881	0.7910	0.7939	0.7967	0.7995	0.8023	0.8051	0.8078	0.8106	0.8133
0.9	0.8159	0.8186	0.8212	0.8238	0.8264	0.8289	0.8315	0.8340	0.8365	0.8389
1.0	0.8413	0.8438	0.8461	0.8485	0.8508	0.8531	0.8554	0.8577	0.8599	0.8621
1.1	0.8643	0.8665	0.8686	0.8708	0.8729	0.8749	0.8770	0.8790	0.8810	0.8830
1.2	0.8849	0.8869	0.8888	0.8907	0.8925	0.8944	0.8962	0.8980	0.8997	0.9015
1.3	0.9032	0.9049	0.9066	0.9082	0.9099	0.9115	0.9131	0.9147	0.9162	0.9177
1.4	0.9192	0.9207	0.9222	0.9236	0.9251	0.9265	0.9278	0.9292	0.9306	0.9319
1.5	0.9332	0.9345	0.9357	0.9370	0.9382	0.9394	0.9406	0.9418	0.9429	0.9441
1.6	0.9452	0.9463	0.9474	0.9484	0.9495	0.9505	0.9515	0.9525	0.9535	0.9545
1.7	0.9554	0.9564	0.9573	0.9582	0.9591	0.9599	0.9608	0.9616	0.9625	0.9633
1.8	0.9641	0.9649	0.9656	0.9664	0.9671	0.9678	0.9686	0.9693	0.9699	0.9706
1.9	0.9713	0.9719	0.9726	0.9732	0.9738	0.9744	0.9750	0.9756	0.9761	0.9767
2.0	0.9772	0.9778	0.9783	0.9788	0.9793	0.9798	0.9803	0.9808	0.9812	0.9817
2.1	0.9821	0.9826	0.9830	0.9834	0.9838	0.9842	0.9846	0.9850	0.9854	0.9857
2.2	0.9861	0.9864	0.9868	0.9871	0.9875	0.9878	0.9881	0.9884	0.9887	0.9890
2.3	0.9893	0.9896	0.9898	0.9901	0.9904	0.9906	0.9909	0.9911	0.9913	0.9916
2.4	0.9918	0.9920	0.9922	0.9925	0.9927	0.9929	0.9931	0.9932	0.9934	0.9936
2.5	0.9938	0.9940	0.9941	0.9943	0.9945	0.9946	0.9948	0.9949	0.9951	0.9952
2.6	0.9953	0.9955	0.9956	0.9957	0.9959	0.9960	0.9961	0.9962	0.9963	0.9964
2.7	0.9965	0.9966	0.9967	0.9968	0.9969	0.9970	0.9971	0.9972	0.9973	0.9974
2.8	0.9974	0.9975	0.9976	0.9977	0.9977	0.9978	0.9979	0.9979	0.9980	0.9981
2.9	0.9981	0.9982	0.9982	0.9983	0.9984	0.9984	0.9985	0.9985	0.9986	0.9986
3.0	0.9987	0.9987	0.9987	0.9988	0.9988	0.9989	0.9989	0.9989	0.9990	0.9990
3.1	0.9990	0.9991	0.9991	0.9991	0.9992	0.9992	0.9992	0.9992	0.9993	0.9993
3.2	0.9993	0.9993	0.9994	0.9994	0.9994	0.9994	0.9994	0.9995	0.9995	0.9995
3.3	0.9995	0.9995	0.9995	0.9996	0.9996	0.9996	0.9996	0.9996	0.9996	0.9997
3.4	0.9997	0.9997	0.9997	0.9997	0.9997	0.9997	0.9997	0.9997	0.9997	0.9998

SOURCE: *Walpole/Myers/Myers*, Sixth Edition.

Table III *Selected Percentiles of the Standard Normal Distribution*

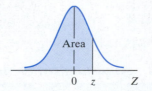

z	Area
−4.265	0.00001
−3.719	0.0001
−3.090	0.001
−2.576	0.005
−2.326	0.01
−2.054	0.02
−1.960	0.025
−1.881	0.03
−1.751	0.04
−1.645	0.05
−1.555	0.06
−1.476	0.07
−1.405	0.08
−1.341	0.09
−1.282	0.10
−1.036	0.15
−0.842	0.20
−0.674	0.25
−0.524	0.30
−0.385	0.35
−0.253	0.40
−0.126	0.45
0	0.50
0.126	0.55
0.253	0.60
0.385	0.65
0.524	0.70
0.674	0.75
0.842	0.80
1.036	0.85
1.282	0.90
1.341	0.91
1.405	0.92
1.476	0.93
1.555	0.94
1.645	0.95
1.751	0.96
1.881	0.97
1.960	0.975
2.054	0.98
2.326	0.99
2.576	0.995
3.090	0.999
3.719	0.9999
4.265	0.99999

Chapter 7

HOW TO MEASURE UNCERTAINTY WITH PROBABILITY

 7.1 Introduction

We sample from the population. Thus, our conclusions or inferences about the population will contain some amount of uncertainty. We call this measure of uncertainty **probability**. We are already familiar with some of the ideas of probability. In Chapter 1, we discussed the **chance** of a Type I error occurring in a decision-making process. We know that a *p*-value is a measure of the **likeliness** of the observed data, or data that show even more support for the alternative theory, computed under the null theory. In Chapters 2 and 3, we saw how **randomization** plays a role in the sampling of units and the allocation of units in studies. In Chapters 4 through 6, we learned that a model can provide a useful summary of the distribution of a variable that serves as a frame of reference for making decisions in the face of **uncertainty**.

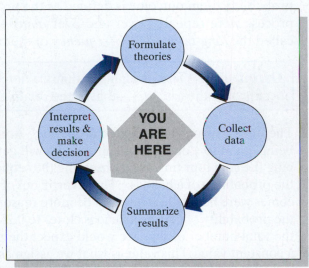

Probability statements are a part of our everyday lives. You have probably heard statements such as the following:

- If my parking meter expires, I will *probably* get a parking ticket.
- There is *no chance* that I will pass the quiz in tomorrow's class.
- The line judge flipped a fair coin to determine which player will serve the ball first, so each player has a *50-50 chance* of serving first.

Just what does it mean to have a 50-50 chance?

We begin our adventure into probability by first discussing just what probability means. Next, we will discover that probabilities may be estimated through simulation or found through more formal mathematical results. Simulation is a powerful technique especially when the problem at hand is difficult. Finally, in Section 7.5, we will merge the concept of probability with the ideas from Chapter 6 of a model for the distribution of a variable. The variables will be called *random variables*, their models will be called *probability distributions*, and these models will be used to find the probability that a randomly selected unit from the population will take on certain values. The material in this chapter on probability and the next on sampling distributions is preparing us for the final step in our cycle—drawing more formal statistical conclusions about a population based on the results from a sample.

7.2   What Is Probability?

You have a coin, on one side of which is a head and on the other a tail. The coin is assumed to be a *fair* coin; that is, the chance of getting a head is equal to the chance of getting a tail. You are going to flip the coin. Why do we say that the probability of getting a head is $\frac{1}{2}$? What does it mean? If we were to flip the fair coin two times, would we always get exactly one head? Of course not. If we were to flip the fair coin 10 times, we would not necessarily see exactly five heads. But if we were to flip the fair coin a large number of times, we *would expect* about half of the flips to result in a head. This use of the word *probability* is based on the **relative-frequency interpretation**, which applies to situations in which the conditions are exactly repeatable. The probability of an outcome is defined as the proportion of times the event would occur if the process were repeated over and over *many times* under the same conditions. This is also called the *long-term relative frequency* of the outcome.

> *Definition:*   The **probability** that an outcome will occur is the proportion of time it occurs over the *long run*—that is, the relative frequency with which that outcome occurs.

The emphasis on *long term* or *in the long run* is very important. The probability of a head being equal to $\frac{1}{2}$ does not mean that we will get one head in every two flips of the coin. Flipping the coin four times and observing the sequence THTT would not be strong evidence that the probability of a head is $\frac{1}{4}$. However, if, out of 1000 coin flips, approximately 25% of the outcomes were heads, then it would be more reasonable to conclude that the coin was biased and the probability of getting a head is closer to 0.25, rather than the fair value of $\frac{1}{2}$. As we increase the total number of flips, we would expect the proportion of heads to begin to settle down to a constant value. This value is what we assign as the probability of getting a head.

THINK ABOUT IT

People often flip a coin to make a random selection between two options, based on the assumption that the coin is fair (that is, the probability of getting a head is $\frac{1}{2}$). Suppose that you stand a penny on its edge and then make it fall over by using your hand with a downward stroke (palm side down) and striking the table. Would the probability of getting a head still be $\frac{1}{2}$?

How would you determine this probability of a head?

Would you use one penny or different pennies? How many repetitions would you do?

Try it and see what happens!

The relative-frequency approach to defining probabilities applies to situations that can be thought of as being repeatable under similar conditions. Some situations are not likely to be repeated under the same conditions. You are planning an outdoor party for the upcoming Saturday afternoon from 2 p.m. to 4 p.m. What is the probability that it will rain during the party? Two softball teams, the Jaguars and the Panthers, have made it to the final game of the tournament. What is the probability that the Jaguars will beat the Panthers? In such situations, a person would use his or her own experiences, background, and knowledge to assign a probability to the outcome. Such probabilities are called **personal** or **subjective probabilities**, which represent a person's degree of belief that the outcome will happen. Different people may arrive at different personal probabilities, all of which would be considered correct. Any probability, however, must be between 0 and 1 (or 0% and 100%). There are certain rules that should be met. We will learn about some of these rules in Section 7.4.

Probabilities help us make decisions. On the Friday night before the party, the weather forecast stated that on Saturday there would be periods of rain and a high temperature of 68 degrees. Even though it may not rain, based on this information, you decide to set up and hold the party indoors instead of outdoors. You need to fly to Chicago to attend a board meeting for Tuesday afternoon and wish to book a flight leaving Tuesday morning. There are two airlines that each offer a flight which, if on time, would allow you to make your meeting. One airline has a record that boasts that 88% of such flights to Chicago are on time. For the other airline, the probability of being on time is reported to be only 73%. These probabilities, along with other information, such as price or safety records, would help you decide which flight to reserve. However, no matter which airline is selected, your particular flight will either be on time, or it will not be on time. Probabilities cannot determine whether the outcome will occur for any individual case.

We will focus on the relative-frequency approach to defining probability. In the coin flipping example, there are two methods for determining the probability of getting a head that both fit the relative-frequency interpretation. We might assume that coins are made such that the two possible outcomes are equally likely, thus assigning the probability of $\frac{1}{2}$ to each outcome. We might actually observe the relative frequency of getting a head by repeatedly flipping the coin a large number of times and using the relative frequency as an estimate of the probability of getting a head. This process of estimating probabilities through simulation is our next topic.

7.3 Simulating Probabilities

One of the basic components in the study of probability is a **random process.** A random process is one that can be repeated under similar conditions. Although the set of possible outcomes is known, the exact outcome for an individual repetition cannot be predicted with certainty. However, there is a predictable long-term pattern such that the relative frequency for a given outcome to occur settles down to a constant value. Flipping a fair coin is an example of a random process. We have worked with other random process-selecting vouchers out of a bag, assigning subjects to receive one of two treatments, or selecting a registered voter at random from a population of registered voters.

> *Definition:*   A **random process** is a repeatable process whose set of possible outcomes is known, but the exact outcome cannot be predicted with certainty. However, there is a predictable long-term pattern of outcomes such that the relative frequency for a given outcome to occur settles down to a constant value.

Some probabilities can be very difficult or time consuming to calculate. We may be able to estimate the probability through **simulation**. To simulate means to imitate—to generate conditions that approximate the actual conditions. To simulate a random process, we could use any one of a number of different devices: a calculator, a computer program, or a table of random numbers. We would first need to specify the conditions of the underlying random circumstance (that is, provide a model that lists the possible individual outcomes and corresponding probabilities). Next, we need to outline how to simulate an individual outcome and how to represent a single repetition of the random process. Finally, you simulate many repetitions, say *n* repetitions are simulated, and determine the number of times that the event of interest occurred, say *x* times. The corresponding relative frequency, *x/n*, would be used to estimate the probability of that event.

Definition: A **simulation** is the imitation of random or chance behavior using random devices such as random number generators or a table of random numbers.

The basic steps for finding a probability by simulation are as follows:

Step 1: Specify a model for the individual outcomes of the underlying random phenomenon.

Step 2: Outline how to simulate an individual outcome and how to represent a single repetition of the random process.

Step 3: Simulate many repetitions, say, *n* times, determine the number of times *x* that the event occurred in the *n* repetitions, and estimate the probability of the event by its relative frequency, *x/n*.

Let's apply these basic steps to estimate some probabilities.

❖ EXAMPLE 7.1 How Many Heads? _____

Consider the random process of flipping a fair coin 10 times. One possible resulting sequence is HHTHHHTHHH. This sequence has a total of eight heads. Would a total of eight heads be considered unusual if the coin were actually fair? What is the probability of getting a total of eight heads in 10 flips of a fair coin? Let's simulate it.

Step 1: Specify a model for the individual outcomes.
The individual outcomes are a head and a tail, and for a fair coin, the probability of each is $\frac{1}{2}$.

Step 2: Simulate individual outcomes and a repetition.
A computer or calculator could be used to generate a random sequence of the integers 1 and 2 and define a 1 to represent a head and a 2 to represent a tail. We would need to simulate 10 flips of a fair coin to represent a single repetition.

For the TI graphing calculator, setting the seed value to 18 and using the randInt(1,2) function, the first 10 generated integers would be

$$1 \quad 1 \quad 1 \quad 1 \quad 1 \quad 2 \quad 1 \quad 1 \quad 1 \quad 2$$

This sequence would represent the coin flip sequence of HHHHHTHHHT, which does have a total of eight heads.

To represent the flipping of a fair coin using a random number table with digits 0, 1, 2, through 9, you might designate that the five odd digits will correspond to a head and the five even digits to a tail. Using Table I, starting at Row 10, Column 1, reading left to right, the first 10 digits are

$$8 \quad 5 \quad 4 \quad 7 \quad 5 \quad 3 \quad 6 \quad 8 \quad 5 \quad 7$$

This sequence would represent the coin flip sequence of THTHHHTTHH, which has a total of six heads, not eight.

Step 3:   Simulate many repetitions and estimate the probability.

The following table provides the results of 50 repetitions using the TI calculator with a seed value of 18 (the outcomes resulting in eight heads (or 1's) are highlighted in bold):

1111121112	1222222212	1122122111	2211112212	1222222122
1111122122	1121121211	1122121212	1221221221	1211111122
1212111111	1222111211	2111111222	1211122121	1212221121
2211222111	1121221211	2111212121	1221212221	1121221211
2122221222	1121121211	2212121121	2112111121	2222122221
2111211122	2112221222	1222122221	2221122222	2122211122
2212121111	2112211221	1222112212	1221112121	2211122121
1221122222	1112212112	2112221121	2212221221	1121221222
2211112221	2111211221	1212122211	1212121121	1211222221
1112212211	1121221212	1121121122	2121212122	2122121211

We have only 2 out of 50 repetitions resulting in a total of eight heads for an estimated probability of 0.04. In this simulation, we did pretty well. As it turns out, there are 45 ways to obtain a total of eight heads out of 10 flips of a fair coin and a total of 1024 equally likely possible sequences of 10 flips. So the actual probability of getting eight heads is 45/1024 = 0.043945. ❖

 ## **L**et's **do it!** **7.1** **A Family Plan**

A couple plans to have children. They would like to have a boy to be able to pass on the family name. After some discussion, they decide to continue to have children until they have a boy or until they have three children, whichever comes first. What is the probability that they will have a boy among their children? Let's simulate this couple's family plan and estimate this probability.

Step 1:   Specify a model for the individual outcomes.

The random process is to continue to have children until a boy is born or until there are a total of three children, whichever comes first. The individual random phenomenon is to "have a child" and the response of interest is its "gender." We need to start with some basic assumptions about these individual outcomes of "girl" and "boy." It seems fairly reasonable to assume that

- each child has probability $\frac{1}{2}$ of being a boy and $\frac{1}{2}$ of being a girl, and
- the gender of successive children is independent (that is, knowing the gender of a child does not influence the gender of any of the successive children).

Step 2:   Simulate individual outcomes and a repetition.

We will need to simulate the gender of a single child. We can use a calculator or computer with a built-in random number generator to simulate an individual outcome. There are only two possible outcomes, boy and girl, so we need to generate a random sequence of two values (for example, 1 and 2). We need to decide which value will represent a boy and which will represent a girl:

Let 1 = the child is a boy, then let _____ = the child is a girl.

To simulate one repetition of the family plan, we will use successive random values until either a boy (a "1") or three children (three girls, "222") are obtained.

Using the TI graphing calculator with a seed value of 102 and $N = 2$, we can write down the first few values and, below each, write either a "B" or a "G" to represent a boy or a girl outcome and then add a line to separate successive repetitions. The following is an example of a total of five repetitions of the family plan:

End of the
first repetition

End of the
fifth repetition

(a) In the first repetition, how many children did the couple have? ____One____

Did the couple have a boy? ____Yes____

(b) In the second repetition, how many children did the couple have? _____

Did the couple have a boy? _____

(c) In the third repetition, how many children did the couple have? _____

Did the couple have a boy? _____

Note: If you do not have a random number generator, you can use a table of random numbers. If we use a table of random numbers, we have 10 digits, 0 through 9. A single random digit can simulate the gender of a single child. We need to decide which five numbers will represent, for example, a boy:

let __0 2 4 6 8__ = the child is a boy,

then _____ = the child is a girl.

To simulate one repetition of the family plan, we will use successive random digits until either a boy or three children are obtained. Starting at Row 14, Column 1 of Table I, reading left to right, the first few digits are recorded with either a "B" or a "G" below each to represent a boy or a girl outcome and a line to separate successive repetitions.

1	0	3	6	5	6	1	1	2
G	B	G	B	G	B	G	G	B

End of the
first repetition

Step 3:   Simulate many repetitions and estimate the probability.

Working with a partner or in small groups, simulate many repetitions of the family plan and use the relative frequency of the event "the couple has a boy" to estimate its probability. Using your calculator with your group's choice of a seed or the random-number table with your group's choice of a starting point, simulate a total of 10 repetitions. Start by writing out a list of a number of generated values. Below each value, write either a "B" or a "G" to represent a boy or a girl outcome, then add a line to separate successive repetitions. You will generally need more than just 10 values. You need to generate enough values to be able to have 10 lines, representing 10 completed repetitions.

(a) Out of the first 10 repetitions, how many times did the couple have a boy? _____

(b) Your group's estimate of the probability that this strategy will produce a boy is _____.

Your group's relative frequency estimate in (b) is not a very precise estimate of the probability because only 10 repetitions were made. So let's combine the frequencies from various groups in the class and produce an estimate of the probability that this strategy will produce a boy.

Group	# Repetitions	# Times a Boy Was Born
1		
2		
3		
4		
5		
6		
7		
8		
9		
10		
TOTAL	$N =$	$\# B =$

So our *combined estimate of the probability* that this strategy will produce a boy is

$$\text{Estimated probability} = \frac{\# B}{N} = \underline{\hspace{5cm}}.$$

In Section 7.4, we will learn how to calculate the actual probability of having a boy, which is 0.875, using some of the basic rules of formal probability theory. How did your combined estimate compare to 0.875?

Our definition of probability as the proportion of time it occurs over the long run implies that, as more repetitions are used, the accuracy of using a simulation for estimating probabilities will increase. This, of course, is dependent on having stated the basic structure of the underlying model appropriately. In simulation, this underlying model is used as a basis for finding the probabilities of more complicated outcomes. In our next exercise, the underlying model for the individual outcomes is provided and you are asked to outline how to simulate an individual outcome.

 **Let's do it! 7.2** **Simulating Other Outcomes**

Select a random device, such as a random number generator on a calculator or a random number table, and state how you would assign values to simulate the following individual outcomes.

(a) How could you simulate an outcome that has probability 0.4 of occurring? Using (circle one)　　a calculator or a computer　　the random number table,

I would let _____ = the outcome occurs

and _____ = the outcome does not occur.

(b) How could you simulate a random process having four possible outcomes, represented by A, B, C, and D, with respective probabilities 0.1, 0.2, 0.3, and 0.4 of occurring?

Using (circle one)   a calculator or a computer   the random number table,

I would let_____ = Outcome A occurs,   _____ = Outcome B occurs,

let_____ = Outcome C occurs, and   _____ = Outcome D occurs.

(c) How could you simulate an outcome that has probability 0.45 of occurring? Using (circle one)   a calculator or a computer   the random number table,

I would let_____ = the outcome occurs

and _____ = the outcome does not occur.

Let's do it! 7.3 Maximum of 13

(a) Suppose that in a particular country couples are permitted to have only one child. We will refer to this as the **original ruling**. Assume that all couples can have children and each child has probability $\frac{1}{2}$ of being a boy and $\frac{1}{2}$ of being a girl.

Would you expect the number of boys to be more than, less than, or about the same as the number of girls under this original ruling? _____

What would the average number of children per family be under this original ruling? _____

(b) Consider a **new ruling** that allows couples to have children until they get a boy or until they have 13 children, whichever comes first.

Would you expect the number of boys to be more than, less than, or about the same as the number of girls under this new ruling? _____

What do you think the average number of children per family would be under this new ruling? _____

(c) Perform a simulation under the new ruling and present your results.

Let's do it! 7.4 The Three Doors

There are three doors. Behind one door is a car. Behind each of the other two doors is a goat. As a contestant, you are asked to select a door, with the idea that you will receive the prize that is behind that door. The game host knows what is behind each door. After you select a door, the host opens one of the remaining doors that has a goat behind it. Note that, no matter which door you select, at least one of the remaining doors for the host to open has a goat behind it. The host then gives you the following two options:

1. Stay with the door you originally selected and receive the prize behind it.
2. Switch to the other remaining closed door and receive the prize behind it.

What is the probability of winning the car if you stay with your original choice? What is the probability of winning the car if you switch? Will switching increase your chance of winning the car? Does your neighbor agree with you?

If the answer is not clear, you could carry out a simulation to estimate the probability of winning if you stay and the probability of winning if you switch.

Here is one way to simulate the game show: Working with a partner, designate one person to be the game host and the other the contestant (you can switch roles halfway through the simulation). The game host controls the three doors, represented by three index cards. These three cards are identical except that on the back of one of the cards there is a car and the back of the other two cards is a goat. (*Note*: Three cards from a standard deck will also work, a black-suited card as the car, and two red-suited cards as the goats.) The host will lay out the three cards blank side up, making sure he or she knows which has the car on the other side. You begin to take turns playing the game and, as you do, keep a record sheet, listing your strategy as either stay or switch, and the outcome as either win a car or win a goat. Once you have performed many repetitions, you can use the relative frequencies to estimate the corresponding probabilities.

Starting with the strategy of staying with the original door, simulate 20 outcomes of the game and tally the results in the accompanying table. Then simulate 20 outcomes of the game using the strategy of switching to the remaining door and tally the results in the second table shown.

Strategy = STAY		Strategy = SWITCH	
Win Car	Win Goat	Win Car	Win Goat

Summarize the Results

Of the 20 repetitions for which you stayed with the original door, what proportion of times did you win the car?

$$\frac{\text{number of wins using the stay strategy}}{20} =$$

Thus, your estimate of the probability of winning when you stay is _____.

Of the 20 repetitions for which you switched to the remaining door, what proportion of times did you win the car?

$$\frac{\text{number of wins using the switch strategy}}{20} =$$

Thus, your estimate of the probability of winning when you switch doors is _____.

Which strategy has the better chance of winning the car?

Combine the results for your class for better estimates of these probabilities. Which strategy appears to be the best?

Look at the Solution

Most people can readily understand that since you selected one of the three doors, if you stay, the probability of winning is $\frac{1}{3}$. What happens if you switch? Assuming that the host always opens a door that does not have the car (and this is a crucial assumption), you have a $\frac{2}{3}$ chance of winning if you switch. There are three equally likely possible orderings of the prizes behind the three doors, shown as A, B, or C.

		Original Door Selected		
		Door 1	*Door 2*	*Door 3*
Actual Situation	*Order A*	Car	Goat	Goat
	Order B	Goat	Car	Goat
	Order C	Goat	Goat	Car

Suppose that the player has selected Door 1. If the car is behind Door 1, as in Order A, the host will open either Door 2 or Door 3, and if the participant switches, he or she will get a goat. If the car is not behind Door 1, as in Order B or C, then the host will open the remaining door that has a goat, and if the player switches, he or she will get the car. Only Order A would result in a loss, that is, a goat. This same analogy also works if you start with the player selecting Door 2 or Door 3. The probability that player wins a car with a switch is $\frac{2}{3}$.

Exercises

7.1 For each of the following probabilities, state whether the relative-frequency approach or personal probability would be most appropriate for determining the probability:

(a) Tom, the manager of a small apartment complex, recently installed new doorbells for each apartment. According to Tom, about $\frac{1}{4}$ of the doorbells will not be working in 6 months and will need replacing.

(b) The manufacturer of the doorbells used by Tom in his apartment complex reports that the probability a doorbell will become defective within the 6-month warranty period is 0.03.

7.2 For each of the following probabilities, state whether the relative-frequency approach or personal probability would be most appropriate for determining the probability:

(a) The probability of getting 10 true or false questions correct on a quiz, if for each question you were simply guessing the answer.

(b) The probability that you will be living in a different state within the next two years.

7.3 In Section 7.2, two interpretations of probability were discussed, the long-run relative-frequency approach and personal probability. For some probabilities the long-run relative-frequency approach may not be appropriate. Provide an example and explain your answer.

7.4 Refer to Example 7.1 and use the results in the table of 50 repetitions to estimate the following probabilities:

(a) Estimate the probability of getting exactly five heads in 10 flips of a fair coin.

(b) Estimate the probability of getting fewer than three heads in 10 flips of a fair coin.

(c) Estimate the probability of getting more than three heads in 10 flips of a fair coin.

(d) Estimate the probability of getting a run of at least six consecutive heads in row in 10 flips of a fair coin.

7.5 Is Your Coin Fair?

(a) Take a coin, flip it 10 times, and record the number of times that resulted in a head.

(b) Repeat (a) an additional 9 times for a total of 100 flips, keeping track of the number of heads for each set of 10 flips and the cumulative proportion after each additional set of 10 flips.

(c) Make a series plot of the cumulative proportion of heads after each set of 10 flips.

(d) Did the proportion of heads start to settle down around a constant value? What is that approximate value? Do you think your coin is fair?

7.6 A Family Plan Revisited   Recall the couple who plans to continue to have children until they have a boy or until they have three children, whichever comes first. We estimated the probability that they will have a boy among their children. We compared our combined estimate to the actual probability of 0.875.

(a) Generate an additional 100 repetitions of this family plan (using a seed value of 102, or Row 25, Column 1 of the random number table). Report the total number of repetitions resulting in having a boy.

(b) Combine the results from "Let's do it! 7.1" with those in (a) and report an updated combined estimate of the probability that this strategy will produce a boy. How does this estimate compare to 0.875?

(c) Suppose that the family plan is to continue to have children until they have a boy or until they have four children, whichever comes first. Do you think the probability of having a boy with this strategy will be larger than, smaller than, or equal to 0.875? Perform a simulation and estimate this probability.

7.7 Planning a Family The Smiths are planning their family and both want an equal number of boys and girls. Mrs. Smith says that their chances are best if they plan on having two children. Mr. Smith says that they have a better chance of having an equal number of boys and girls if they plan on having four children.

(a) Assuming that boy and girl babies are equally likely, who do you think is correct? Mrs. Smith, Mr. Smith, or are they both correct?

(b) Check your answer to (a) by performing a simulation. To have comparable precision in your probability estimates, use the same number of repetitions for both Mrs. Smith's strategy and Mr. Smith's strategy. Provide all relevant details and a summary of your results.

7.8 ESP? A classic experiment to detect ESP uses a shuffled deck of five cards—one with a wave, one with a star, one with a circle, one with a square, and one with a cross. A total of 10 cards will be drawn, one by one, with replacement, from this deck. The subject is asked to guess the symbol on each card drawn.

(a) If a subject actually lacks ESP, what is the probability that he or she will correctly guess the symbol on a card?

(b) If a subject actually has ESP, should the probability that he or she will correctly guess the symbol on a card be smaller than, larger than, or the same as the probability in (a)?

(c) Julie thinks she has ESP. She wishes to test the following hypotheses:

H_0: Julie does not have ESP, so the probability of a correct answer is just 0.20.

H_1: Julie does have ESP, so the probability of a correct answer is greater than 0.20.

Julie participates in the experiment and is right in 6 of 10 tries. You are asked to test, at a 1% significance level, the hypothesis that Julie has ESP. Design and carry out a simulation to estimate the p-value, that is, the chance of getting 6 or more correct answers out of 10, if indeed Julie was just guessing and does not have ESP. Based on your estimated p-value, what is your conclusion?

7.9 Consider the process of playing a game in which the probability of winning is 0.20 and the probability of losing is 0.80.

(a) If you were to use your calculator or a random number table to simulate playing this game, what numbers would you generate and how would you assign values to simulate winning and losing?

(b) With your calculator (using a seed value of 72) or the random number table (Row 36, Column 1), simulate playing this game 50 times. Show the numbers generated and indicate which ones correspond to wins and which ones correspond to losses.

(c) From the simulated results, calculate an estimate of the probability of winning. How does it compare to the actual probability of winning of 0.20?

7.4 The Language of Probability

In this section, we turn to some of the basic ideas of probability and introduce some notation and rules. These rules will allow us to compute the probabilities of simple events and some more complex events. We begin by listing some of the key components in the study of probability.

 ### 7.4.1 Sample Spaces and Events

First, we have a random process. This could be tossing a fair coin three times, rolling a pair of fair dice, or picking a registered voter at random. Next, we have the **sample space** or **outcome set** for the random process. The sample space, denoted by S, is the set of all possible outcomes of the random process. If the random process were tossing a fair coin three times, then the outcomes that make up the sample space can be found in an orderly way using the "tree" method, as shown here.

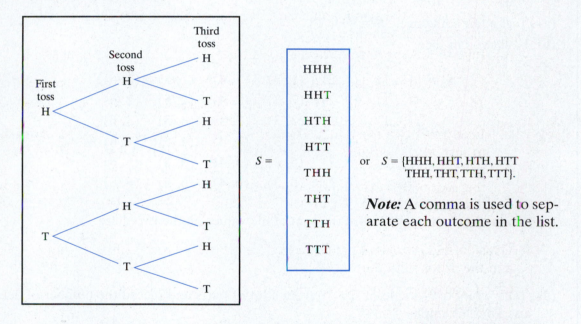

$$S = \{HHH, HHT, HTH, HTT, THH, THT, TTH, TTT\}$$

or $S = \{HHH, HHT, HTH, HTT, THH, THT, TTH, TTT\}$.

Note: A comma is used to separate each outcome in the list.

There are eight possible individual outcomes in this sample space. Since the coin is assumed to be fair, the eight outcomes can be assumed to be equally likely (that is, the probability assigned to each individual outcome is $\frac{1}{8}$).

If the random process were tossing a fair coin three times and the outcome is defined as the number of heads, the sample space is given by $S = \{0, 1, 2, 3\}$. There are four possible outcomes in this latest sample space. However, these four outcomes are not equally likely. Getting exactly one head is more likely to occur than getting zero heads, since three of the individual outcomes {HTT, THT, TTH} correspond to the outcome of exactly one head and only one individual outcome corresponds to zero heads {TTT}.

From these two examples we can see that

- the sample space does not necessarily need to be a set of numbers, although a coding scheme could be established if the outcome is not numeric.

- the definition of what constitutes an individual outcome is key in representing the sample space correctly.
- the individual outcomes in a sample space are not necessarily equally likely.

> *Definition:*   A **sample space** or **outcome set** is the set of all possible individual outcomes of a random process. The sample space is typically denoted by S and may be represented as a list, a tree diagram, an interval of values, a grid of possible values, and so on.

 **7.5** **Sample Spaces**

Give the sample space S for each of the following descriptions. Some are provided for you as examples.

(a) Toss a fair coin once: $S = \{H, T\}$.

(b) Roll two fair dice:

$$S = \{\ (1, 1) \quad (1, 2) \quad (1, 3) \quad (1, 4) \quad (1, 5) \quad (1, 6)$$
$$(2, 1) \quad (2, 2) \quad (2, 3) \quad (2, 4) \quad (2, 5) \quad (2, 6)$$
$$(3, 1) \quad (3, 2) \quad (3, 3) \quad (3, 4) \quad (3, 5) \quad (3, 6)$$
$$(4, 1) \quad (4, 2) \quad (4, 3) \quad (4, 4) \quad (4, 5) \quad (4, 6)$$
$$(5, 1) \quad (5, 2) \quad (5, 3) \quad (5, 4) \quad (5, 5) \quad (5, 6)$$
$$(6, 1) \quad (6, 2) \quad (6, 3) \quad (6, 4) \quad (6, 5) \quad (6, 6)\ \}.$$

(c) Roll two fair dice and record the sum of the values on the two dice:

$S = \{$

(d) Take a random sample of size 10 from a lot of parts and record the number of defectives in the sample:

$S = \{$

(e) Select a student at random and record the time spent studying statistics in the last 24-hour period:

$S = \{$ any time t between 0 hours and 24 hours (inclusive)$\}$ or $S = [0, 24]$.

(f) Select a bus commuter at random and record the waiting time between his or her arrival at a bus stop and the arrival of the next bus to that stop:

$S = \{$

Let's do it! 7.6 Voting Preference

Consider the process of randomly selecting two adults from Washtenaw County and recording the voting preference for each adult as Republican, Democrat, Independent, or Other. The two adults randomly chosen (in the order selected) are Ryan and Caitlyn. Which of the following gives the correct sample space for the set of possible outcomes of this experiment? Circle your answer.

(a) $S = \{$Ryan, Caitlyn$\}$.

(b) $S = \{$Republican, Democrat, Independent, Other$\}$.

(c) $S = \{$Republican, Independent$\}$.

(d) None of the above.

Did you select (b) in the preceding "Let's do it!" exercise? If so, you selected the correct sample space if the random process had been to randomly select exactly one adult from Washtenaw County and record his or her voting preference.

Did you select (c)? If so, you selected a set that represents just one of the possible individual outcomes, (R,I), which represents "Ryan is Republican and Caitlyn is Independent." The correct answer is (d), since the actual sample space contains a total of 16 possible individual outcomes. We should also note that the outcome (R,I) is different from the outcome (I,R), which represents "Ryan is Independent and Caitlyn is Republican." In other words, the *order* of the responses does matter. If we actually surveyed a larger number of adults and we were interested in learning about the proportion of adults for each of the political preference categories, we might not be concerned about the order of the responses.

Subsets of the sample space are called **events** and are typically denoted by capital letters at the beginning of the alphabet (A, B, C, and so on). In some cases, the sample space and events may be represented using a *Venn diagram*. The sample space is represented by the box and the events are a subset of the box.

Suppose that the outcome of the random process is a. Since outcome a is in the event A, we say that the event A **has occurred**. If the outcome is b, since b is not in the event A, we say that the event A has **not occurred**. If the random experiment were rolling a fair die, then the sample space is given by $S = \{1, 2, 3, 4, 5, 6\}$. Let the event A be defined as an odd outcome. Then, the event $A = \{1, 3, 5\}$ is a subset of S. If the die is rolled and a 1 is obtained, the event A has occurred. If the die is rolled and a 2 is obtained, the event A has not occurred.

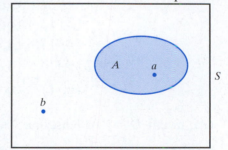

> **Definition:** An **event** is any subset of the sample space S. An event A is said to occur if any one of the outcomes in A occurs when the random process is performed once.

 **7.7** **Expressing Events**

Consider the experiment of rolling two fair dice. **Circle** the outcomes that correspond to the following events:

(a) Event A = "No sixes."

$$S = \{ \boxed{\begin{matrix} (1, 1) & (1, 2) & (1, 3) & (1, 4) & (1, 5) \\ (2, 1) & (2, 2) & (2, 3) & (2, 4) & (2, 5) \\ (3, 1) & (3, 2) & (3, 3) & (3, 4) & (3, 5) \\ (4, 1) & (4, 2) & (4, 3) & (4, 4) & (4, 5) \\ (5, 1) & (5, 2) & (5, 3) & (5, 4) & (5, 5) \end{matrix}} \begin{matrix} (1, 6) \\ (2, 6) \\ (3, 6) \\ (4, 6) \\ (5, 6) \end{matrix}$$
$$(6, 1) \quad (6, 2) \quad (6, 3) \quad (6, 4) \quad (6, 5) \quad (6, 6) \}$$

(b) Event B = "Exactly one six."

$$S = \{ (1, 1) \quad (1, 2) \quad (1, 3) \quad (1, 4) \quad (1, 5) \quad (1, 6)$$
$$(2, 1) \quad (2, 2) \quad (2, 3) \quad (2, 4) \quad (2, 5) \quad (2, 6)$$
$$(3, 1) \quad (3, 2) \quad (3, 3) \quad (3, 4) \quad (3, 5) \quad (3, 6)$$
$$(4, 1) \quad (4, 2) \quad (4, 3) \quad (4, 4) \quad (4, 5) \quad (4, 6)$$
$$(5, 1) \quad (5, 2) \quad (5, 3) \quad (5, 4) \quad (5, 5) \quad (5, 6)$$
$$(6, 1) \quad (6, 2) \quad (6, 3) \quad (6, 4) \quad (6, 5) \quad (6, 6) \}$$

(c) Event C = "Exactly two sixes."

$$S = \{ (1, 1) \quad (1, 2) \quad (1, 3) \quad (1, 4) \quad (1, 5) \quad (1, 6)$$
$$(2, 1) \quad (2, 2) \quad (2, 3) \quad (2, 4) \quad (2, 5) \quad (2, 6)$$
$$(3, 1) \quad (3, 2) \quad (3, 3) \quad (3, 4) \quad (3, 5) \quad (3, 6)$$
$$(4, 1) \quad (4, 2) \quad (4, 3) \quad (4, 4) \quad (4, 5) \quad (4, 6)$$
$$(5, 1) \quad (5, 2) \quad (5, 3) \quad (5, 4) \quad (5, 5) \quad (5, 6)$$
$$(6, 1) \quad (6, 2) \quad (6, 3) \quad (6, 4) \quad (6, 5) \quad (6, 6) \}$$

(d) Event D = "At least one six."

$$S = \{ (1, 1) \quad (1, 2) \quad (1, 3) \quad (1, 4) \quad (1, 5) \quad (1, 6)$$
$$(2, 1) \quad (2, 2) \quad (2, 3) \quad (2, 4) \quad (2, 5) \quad (2, 6)$$
$$(3, 1) \quad (3, 2) \quad (3, 3) \quad (3, 4) \quad (3, 5) \quad (3, 6)$$
$$(4, 1) \quad (4, 2) \quad (4, 3) \quad (4, 4) \quad (4, 5) \quad (4, 6)$$
$$(5, 1) \quad (5, 2) \quad (5, 3) \quad (5, 4) \quad (5, 5) \quad (5, 6)$$
$$(6, 1) \quad (6, 2) \quad (6, 3) \quad (6, 4) \quad (6, 5) \quad (6, 6) \}$$

Let's **d**o **i**t! **7.8** **Favor or Oppose**

In a group of people, some favor abortion (*F*) and others oppose abortion (*O*). Three people are selected at random from this group, and their opinions in favor or against abortion are noted. Assume that it is important to know which opinion came from each individual (that is, that order does matter).

(a) Write down the sample space for this situation.

$S =$

(b) Write out the outcomes that make up the event *A* = "at most one person is against abortion."

$A =$

(c) Write out the outcomes that make up the event *B* = "exactly two people are in favor of abortion."

$B =$

Sometimes we are interested in events that are not so simple. The event may be a combination of various events. The **union** of two events is represented by *A or B*, written mathematically as $A \cup B$ and shown by the shaded region in Figure 7.1. The union *A or B* contains the outcomes that are in the event *A* or in the event *B* or in both *A* and *B*. Sometimes the event of *A* or *B* is stated as **at least one of the two** events has occurred.

Figure 7.1 Union

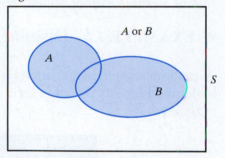

The **intersection** of two events is represented by *A and B*, written mathematically as $A \cap B$ and shown by the shaded region in Figure 7.2. The intersection *A and B* is comprised of only those outcomes that are in both the event *A* and the event *B*. Often the word **both** is used when describing the intersection of two events.

Figure 7.2 Intersection

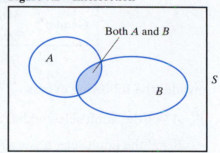

The **complement** of an event is represented by **not A**, written mathematically as A^C and shown by the shaded region in Figure 7.3. The complement of the event *A* is comprised of all outcomes that are not in the event *A*. If listing the outcomes that make up an event *A* seems a bit overwhelming, it may be easier to summarize the outcomes that make up the complement of the event *A*. Given the sample space *S*, every event *A* has a unique complementary event A^C in *S*.

Figure 7.3 Complement

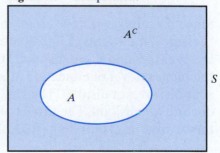

Two events *A* and *B* are said to be **disjoint** if they have no outcomes in common. Sometimes, instead of disjoint, the events are said to be **mutually exclusive**. In terms of mathematical notation, we would write $A \cap B = \varnothing$, where $\cap$ represents intersection, and $\varnothing$ represents the empty set (the set that contains no outcomes). Two events are disjoint if they *cannot* occur at the same time. Disjoint events can be shown using a Venn diagram. Figure 7.4 shows two events, *A* and *B*, that are disjoint.

Figure 7.4 Disjoint

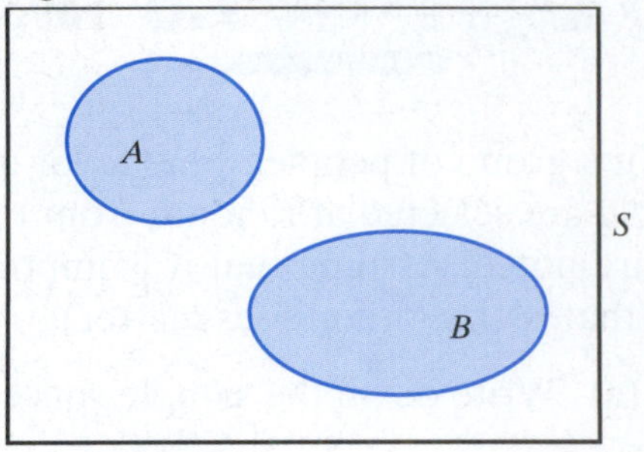

Definition: Two events *A* and *B* are **disjoint** or **mutually exclusive** if they have no outcomes in common. Thus, if one of the events occurs, the other cannot occur.

The notion of mutually exclusive events can be extended to more than two events. For example, we say that the events *A*, *B*, and *C* are mutually exclusive if the events *A* and *B* have no outcomes in common, the events *A* and *C* have no outcomes in common, *and* the events *B* and *C* have no outcomes in common. Note that these conditions imply that the intersection of all three events, "*A* and *B* and *C*," will also be empty.

❖ **EXAMPLE 7.2** **Disjoint Events** _____

A random sample of 200 adults is classified according to their gender, 88 male and 112 female, and highest education level attained, 83 elementary, 78 secondary, 39 college.

		Education		
		Elementary	*Secondary*	*College*
Gender	*Male*	38	28	22
	Female	45	50	17

Consider the following events:

   A = "the adult selected has a college level education."

   B = "the adult selected is a male with the highest level of education being secondary."

   C = "the adult selected is a female."

Since no adult can have as the highest level of education attained both "secondary" and "college," the events *A* and *B* are disjoint. Since an adult cannot be both male and female, the events *B* and *C* are disjoint. However, an adult can be both a female and have college as the highest level of education attained. In fact, there were 17 such adults. Hence, the events *A* and *C* are not disjoint. The row categories, male and female, are the various levels of gender and are disjoint. The three column categories—elementary, secondary, and college—are the various levels for highest level of education attained and are disjoint.   ❖

Let's do it! 7.9 Mutually Exclusive?

For each scenario and list of events, determine whether the events are mutually exclusive:

(a) A retail sales agent makes a sale:

$$A = \text{"the sale exceeds \$50."}$$
$$B = \text{"the sale exceeds \$500."}$$

(b) A retail sales agent makes a sale:

$$A = \text{"the sale is less than \$50."}$$
$$B = \text{"the sale is between \$100 and \$500."}$$
$$C = \text{"the sale exceeds \$1000."}$$

(c) Ten students are selected at random:

$$A = \text{"no more than three are female."}$$
$$B = \text{"at least seven are female."}$$
$$C = \text{"at most five are female."}$$

✳ Exercises

7.10 Packages delivered by a local mail service are delivered early (E), on time (O), or late (L). An evaluation of the service will be conducted by selecting two packages at random and noting the delivery status for each package. Which of the following is the correct sample space for this random experiment?

(a) $S = \{\text{package 1, package 2}\}$.

(b) $S = \{E, O, L\}$.

(c) $S = \{\text{package 1} = E, \text{package 2} = O\}$.

(d) None of the above sample spaces is correct.

7.11 Two chess players, Gabe and Ellie, decide to play several games of chess. They will stop playing if Gabe wins two matches or if they have played a total of three games in all. Each game can be won by either Gabe or Ellie, and there can be no ties.

(a) Give the sample space (all possible outcomes) for the series of games played by Gabe and Ellie.

(b) Let A be the event that no player wins two consecutive games. List the outcomes from the sample space in (a) that make up the event A.

7.12 Each day a bus travels from City A to City D by way of Cities B and C (as shown). André is a traveler who can get on the bus at any one of the cities and can get off at any other city along the route (except not the same city in which he got on the bus).

City A   City B   City C   City D

(a) Give the sample space (all possible outcome pairs) for representing the starting and ending points of André's journey.

(b) Let E be the event that André gets off at a city that comes after City B on the route of the bus. List the outcomes from the sample space in (a) that make up the event E.

7.13 A simple random sample of students will be selected from a student body, and college status, classified as either full time or part time, will be recorded.

(a) Give the sample space if just one student will be selected at random.

(b) Give the sample space if four students will be selected at random.

(c) Give the sample space if 20 students will be selected at random and the outcome of interest is the number of full-time students.

7.14 Replacement and Order A basket contains three balls, one green, one yellow, and one white. Two balls will be selected from the basket. For example, the outcome "G,Y" represents that the green ball was selected, followed by the yellow ball.

Write out the corresponding sample space if

(a) the sampling procedure was with replacement and order matters.

(b) the sampling procedure was with replacement and order doesn't matter.

(c) the sampling procedure was without replacement and order matters.

(d) the sampling procedure was without replacement and order doesn't matter.

7.15 Disjoint? Consider the experiment of drawing a card from a standard deck. Let A = "heart," B = "king," and C = "spade."

(a) Are the events A and B disjoint? Explain.

(b) Are the events A and C disjoint? Explain.

(c) Are the events B and C disjoint? Explain.

7.16 A computer-software retailer has made an inventory of last year's software to be placed on clearance. Of the 380 software packages, 270 were provided only on CD-ROM and the rest were stored only on disk. Of the CD-ROM software, 200 could be used on a Macintosh system, 180 could be used on a PC system, and 110 were dual platform (that is, they could be used both a Macintosh and a PC system).

(a) Use a Venn diagram to display the information regarding the clearance software. Be sure to label the events.

(b) Are the events "CD-ROM" and "Disk" disjoint?

(c) Are the events "Mac system" and "PC system" disjoint?

 7.4.2 Rules of Probabilities

We return to the idea of probability and relate it to events and the outcomes of a sample space. To any event A, we assign a number $P(A)$ called the **probability of the event A**. Recall that the probability of an event was defined as the relative frequency with which that event would occur in the long run. When the sample space contains a finite number of possible outcomes, we have another technique for assigning the probability of an event:

- Assign a probability to each individual outcome, each being a number between 0 and 1, such that the sum of these individual probabilities is equal to 1.
- The probability of any event is the sum of the probabilities of the outcomes that make up that event.

If the outcomes in the sample space are *equally likely* to occur, the probability of an event A is simply the proportion of outcomes in the sample space that make up the event A. Do not automatically assume that the outcomes in the sample space are equally likely—it will depend on the random process and the definition of the outcome that is being recorded. Our first exercise does involve equally likely outcomes. Soon we will learn some probability rules that will help us determine the probabilities of outcomes that are not equally likely.

 **Let's do it! 7.10 Assigning Probabilities to Events**

Consider the experiment of rolling two fair dice. Assume that the 36 points in the sample space are equally likely. What are the probabilities of the following events?

(a) Event A = "No sixes."

$$S = \{ \begin{array}{cccccc} (1, 1) & (1, 2) & (1, 3) & (1, 4) & (1, 5) & (1, 6) \\ (2, 1) & (2, 2) & (2, 3) & (2, 4) & (2, 5) & (2, 6) \\ (3, 1) & (3, 2) & (3, 3) & (3, 4) & (3, 5) & (3, 6) \\ (4, 1) & (4, 2) & (4, 3) & (4, 4) & (4, 5) & (4, 6) \\ (5, 1) & (5, 2) & (5, 3) & (5, 4) & (5, 5) & (5, 6) \\ (6, 1) & (6, 2) & (6, 3) & (6, 4) & (6, 5) & (6, 6) \} \end{array}$$

Since there are 25 of the 36 equally likely outcomes that comprise the event A, we have $P(A) = \frac{25}{36}$.

(b) Event B = "Exactly one six."

$P(B) =$

(c) Event C = "Exactly two sixes."

$P(C) =$

(d) Event D = "At least one six."

$P(D) =$

(e) Compare $1 - P(A)$ with $P(D)$. The events A and D are complementary events.

Basic Rules that Any Assignment of Probabilities Must Satisfy

1. Any probability is always a numerical value between 0 and 1. The probability is 0 if the event cannot occur. The probability is 1 if the event is a sure thing—it occurs every time; $0 \leq P(A) \leq 1$.

2. If we add up the probabilities of each of the individual outcomes in the sample space, the total probability must equal one; $P(S) = 1$.

3. The probability that an event occurs is 1 minus the probability that the event does not occur; $P(A) = 1 - P(A^C)$.

The third rule is called the *complement rule*. Any event and its corresponding complement are disjoint sets, which when brought back together give us the whole sample space S. The probability of the sample space S is 1, so the probabilities of the event and its complement must add up to 1. This rule can be very useful. If finding the probability of an event seems too difficult, see if finding the probability of the complement of the event is easier.

THINK ABOUT IT

Consider the experiment of tossing a fair coin 10 times. Think about what is the sample space S. Let A be the event of "at least 1 head." At least 1 head means exactly 1 head or exactly 2 heads or exactly 3 heads or exactly 4 heads or exactly 5 heads or exactly 6 heads or exactly 7 heads or exactly 8 heads or exactly 9 heads or all 10 heads. That is a lot of outcomes to try to count up. Think about what is the complement of A, and then find the probability of the event A using the complement rule.

 ## Let's do it! 7.11 A Fair Die?

A die, with faces 1, 2, 3, 4, 5, 6, is suspected to be unfair in the sense of having a tendency toward showing the larger faces. We wish to test the following hypotheses:

H_0: The die is fair (that is, it has an equal chance for all six faces).
H_1: The die has a tendency toward showing larger faces.

For the data, you will roll the die two times. Recall the 36 possible pairs of faces if you roll a die two times:

$$S = \{ (1, 1) \quad (1, 2) \quad (1, 3) \quad (1, 4) \quad (1, 5) \quad (1, 6)$$
$$(2, 1) \quad (2, 2) \quad (2, 3) \quad (2, 4) \quad (2, 5) \quad (2, 6)$$
$$(3, 1) \quad (3, 2) \quad (3, 3) \quad (3, 4) \quad (3, 5) \quad (3, 6)$$
$$(4, 1) \quad (4, 2) \quad (4, 3) \quad (4, 4) \quad (4, 5) \quad (4, 6)$$
$$(5, 1) \quad (5, 2) \quad (5, 3) \quad (5, 4) \quad (5, 5) \quad (5, 6)$$
$$(6, 1) \quad (6, 2) \quad (6, 3) \quad (6, 4) \quad (6, 5) \quad (6, 6) \}.$$

The sum of the two rolls will be the response used to make the decision between the two hypotheses.

(a) Consider the possible sum of 11. Circle the outcomes in the accompanying sample space that correspond to having a sum of 11. What is the probability of getting a sum of 11 on the next two rolls?

The suggested format for the decision rule is to reject H_0 if the **sum** of the two rolls is too large.

(b) The direction of extreme is (circle one)

one-sided to the right.   one-sided to the left.   two-sided.

(c) What is the *p*-value if the observed sum actually equals 11? (*Hint:* Think about the definition of a *p*-value.)

(d) Would an observed sum of 11 be statistically significant at the 5% significance level? at the 10% level?   Explain.

Our next basic rule tells us how to find the probability of the union of two events—that is, the probability that one or the other event occurs. The basis of this rule is easy to see by looking at the corresponding diagram. We start by taking those outcomes in the event A, and then add all of those outcomes that form the event B. The outcomes that occur in both A and B have been included twice, so we need to subtract them once.

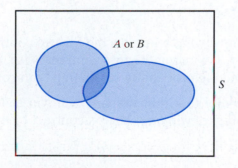

The Addition Rule

4. The probability that either the event A or the event B occurs is the sum of their individual probabilities minus the probability of their intersection.

$$P(A \text{ or } B) = P(A) + P(B) - P(A \text{ and } B)$$

If the two events A and B do not have any outcomes in common (that is, they are disjoint), then the probability that one or the other occurs is simply the sum of their individual probabilities.

If A and B are *disjoint* events, then $P(A \text{ or } B) = P(A) + P(B)$.

Note: This special case can be extended to more than two disjoint events. If the events A, B, and C are disjoint, then $P(A \text{ or } B \text{ or } C) = P(A) + P(B) + P(C)$.

❖ **EXAMPLE 7.3** **Gender versus Education** _____

Recall the data from a random sample of 200 adults classified by gender and by highest education level attained.

		Education			
		Elementary	*Secondary*	*College*	
Gender	*Male*	38	28	22	88
	Female	45	50	17	112
		83	78	39	200

Consider the following two events:

A = "adult selected has a college-level education."
C = "adult selected is a female."

What is the probability that an adult selected at random either has a college level of education or is a female?

$$P(A \text{ or } C) = P(A) + P(C) - P(A \text{ and } C) = \tfrac{39}{200} + \tfrac{112}{200} - \tfrac{17}{200} = \tfrac{134}{200} = 0.67 \qquad ❖$$

Let's do it! 7.12 Winning Contracts

A local construction company has entered a bid for two contracts for the city. The company feels that the probability of winning the first contract is 0.5, the probability of winning the second contract is 0.4, and the probability of winning both contracts is 0.2.

(a) What is the probability that the company will win at least one of the two contracts (that is, the probability of winning the first contract or the second contract)?

(b) The corresponding Venn diagram displays the events A = "win first contract" and B = "win second contract." Two of the probabilities have been entered. Note that the 0.2 and the 0.3 sum to the probability for the event A of 0.5. Add the remaining probabilities such that the total of all of the probabilities is equal to 1.

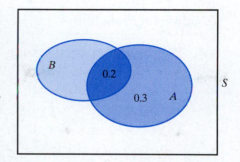

(c) What is the probability of winning the first contract but not the second contract? (*Hint*: Look at the portion of the diagram that represents the event of interest.)

(d) What is the probability of winning the second contract but not the first contract? (*Hint*: Look at the portion of the diagram that represents the event of interest.)

(e) What is the probability of winning neither contract? (*Hint*: Look at the portion of the diagram that represents the event of interest.)

Sometimes, we will have some given information about the outcome of the random process. We may wish to update the probability of a certain event occurring taking into account this given information. Consider rolling a single fair die one time. The sample space is $S = \{1, 2, 3, 4, 5, 6\}$, and each of the six outcomes is equally likely. The probability of getting the value of 1 is then $\frac{1}{6}$. But suppose that we know the outcome was an odd value: Now, what is the probability that the value is a 1? Since we know the outcome was an odd value, we no longer consider the original sample space as the set of possible outcomes. There are only three possible outcomes in the updated sample space—namely, $\{1, 3, 5\}$. Each of these three outcomes is now equally likely. Thus, the updated probability is $\frac{1}{3}$. What we have just computed is a **conditional probability**, the probability of the event $A = \{1\}$, given the event $B = \{ODD\}$ has occurred, represented by the general expression $P(A|B)$. In other words, conditioning on the fact that the event B has occurred, we wish to find the updated probability that the event A will occur.

Our next rule tells us how to find such conditional probabilities. The basis of this rule is easy to see by looking at the corresponding diagram. Since we know that the event B has occurred, we start by taking only those outcomes in the event B. This set of outcomes is our updated sample space and will form the base of our probability expression. We wish to find the probability of the event A occurring on this updated sample space. The only outcomes in the event A included in this updated sample space are those belonging to both the event A and the event B (that is, the outcomes that comprise the intersection between A and B).

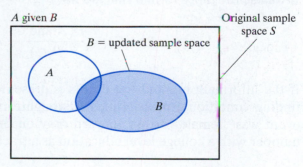

Conditional Probability

---

5. The conditional probability of the event *A* occurring, given that event *B* has occurred, is given by

$$P(A|B) = \frac{P(A \text{ and } B)}{P(B)}, \text{ if } P(B) > 0.$$

Note: We could rewrite this rule and have an expression for calculating an intersection, called the **multiplication rule**.

$$P(\text{both events will occur at the same time})$$
$$= P(A \text{ and } B) = P(B)P(A|B) = P(A)P(B|A)$$

The basis of this rule is as follows: For both events to occur, first we must have one occur (for example, the event *B*), and then given that *B* has occurred, the event *A* must also occur. Of course, the events *A* and *B* could be switched around, which gives us the last part of the preceding result.

❖ **EXAMPLE 7.4**   **Gender versus Education** _____

Recall the data from a random sample of 200 adults classified by gender and by highest education level attained.

		Education			
		Elementary	*Secondary*	*College*	
Gender	*Male*	38	28	22	88
	Female	45	50	17	112
		83	78	39	200

Consider the following two events:

   A = "adult selected has a college level education."

   C = "adult selected is a female."

What is the probability that an adult selected at random has a college level of education given that the adult is a female? Since we are given that the selected adult is a female, we only need to consider the 112 females as our updated sample space. Among the 112 females, there were 17 females who had a college level education.

$$P(A|C) = P(\text{college level education}|\text{female}) = \frac{17}{112} = 0.152$$

If we use the more formal rule, we have

$$P(A|C) = \frac{P(A \text{ and } C)}{P(C)} = \frac{\frac{17}{200}}{\frac{112}{200}} = \frac{17}{112} = 0.152.$$

If the information about the events is presented in a two-way frequency table of counts, finding conditional probabilities is straightforward. In the preceding example, the given event was "female," so we focused only on the female row of counts and expressed the number with a college level education as a fraction of the total number of females.   ❖

Let's do it! 7.13 Union Proposal

Before contract discussions, the state-wide union made a proposal that emphasizes fringe benefits rather than wage increases. The opinions of a random sample of 2500 union members are summarized in the following table:

		Opinion		
		Favor	*Neutral*	*Opposed*
Gender	*Female*	800	200	500
	Male	400	100	500

(a) Complete the table by computing the row and column totals.

(b) What is the probability that a *randomly selected union member* will be opposed?

$P(\text{opposed}) =$

(c) What is the probability that a *randomly selected **female** union member* will be opposed? Note that you are given that the selected union member is a female so you can focus on only the female union members and determine what proportion of them are opposed.

$P(\text{opposed}|\text{female}) =$

(d) Give an example of two events represented in the above table that are mutually exclusive (i.e., they are disjoint).

Let's do it! 7.14 Conditional Probabilities

Scenario I   The random process is rolling a fair die one time.
The sample space is $S = \{1, 2, 3, 4, 5, 6\}$.

(a) What is the probability of getting a 2? $P(2) =$

(b) Suppose that we know that the outcome was an even $P(2|\text{Even}) =$
value; now, what is the probability of getting a 2?

Scenario II   The random process is tossing a fair coin two times.
The sample space is $S = \{HH, HT, TH, TT\}$.

(a) What is the probability of getting a head on the $P(H \text{ on 2nd}) =$
second toss?

(b) What is the probability of getting a head on the $P(H \text{ on 2nd}|H \text{ on 1st}) =$
second toss, given it was a head on the first toss?

THINK ABOUT IT

In Scenario I of the preceding exercise, how do your answers to (a) and (b) compare?

In Scenario II, how do your answers to (a) and (b) compare?

What makes these two scenarios different?

Suppose that for events A and B we have $P(A|B) = 0.3$ and $P(A) = 0.3$. What does this tell us about the two events A and B? This is what happened in Scenario II of the previous exercise. If knowing that event B occurred does not change the probability of the event A occurring—that is, $P(A|B) = P(A)$—we say the two events are **independent**.

> **Definition:** Two events A and B are **independent** if $P(A|B) = P(A)$ or, equivalently, if $P(B|A) = P(B)$.
>
> If two events do not influence each other (that is, if knowing one has occurred does not change the probability of the other occurring), the events are independent. If two events are independent, the multiplication rule tells us that the probability of them both occurring together is found by multiplying their individual probabilities:
>
> If two events A and B are **independent** (only in this case), then $P(A \text{ and } B) = P(A)P(B)$.

❖ EXAMPLE 7.5 Gender versus Education

Recall the data from a random sample of 200 adults classified by gender and by highest education level attained.

		Education			
		Elementary	*Secondary*	*College*	
Gender	*Male*	38	28	22	88
	Female	45	50	17	112
		83	78	39	200

Consider the following two events:

A = "adult selected has a college level education."

C = "adult selected is a female."

Are these two events A and C independent?

From Examples 7.3 and 7.4, we have the following probabilities:

$P(A)$ = P(adult selected has a college level education) = $\frac{39}{200}$ = 0.195.

$P(A|C)$ = P(adult selected has a college level education *given* the adult is a female)

$= \frac{17}{112} = 0.152.$

Since knowing the adult is a female changes the probability that the adult will have a college level education (from 0.195 to 0.152), these two events A and C are *not* independent.

Alternatively, we could have checked for the independence of the two events using the multiplication rule for independent events. Is $P(A$ and $C)$ equal to $P(A)\,P(C)$? We have

$P(A$ and $C)$ = P(adult selected has a college level education and is a female)

$= \frac{17}{200} = 0.085$ and

$P(A)P(C)$ = $\left(\frac{39}{200}\right)\left(\frac{112}{200}\right)$ = 0.109.

Since $P(A$ and $C) \neq P(A)\,P(C)$, we can conclude that these two events are not independent.

We do not have to show both rules for assessing independence; just one is sufficient and will imply the same conclusion as the other rule. To determine which rule you might use for a given problem, take a look at the probabilities that you may have already computed. ❖

❖ EXAMPLE 7.6   A String of Lights

A string of lights contains 30 bulbs. If one bulb fails, the whole string fails. The probability that a single bulb operates for at least two years is 0.98. If the bulbs operate *independently*, what is the probability that the string of 30 bulbs will operate for at least two years?

$P(\text{string works})$ = $P(\text{all 30 bulbs work})$
$= P(\text{1st works and 2nd works and} \cdots \text{and 30th works})$
$= P(\text{1st works})P(\text{2nd works}) \cdots P(\text{30th works})$
$= (0.98)^{30} = 0.545.$

The bulbs operate independently, but there is a dependency here: that all 30 bulbs must be working in order for the string to operate. Here, we used the multiplication rule for independent events. ❖

Let's do it! 7.15 A Family Plan Revisited

Recall from "Let's do it! 7.1" that we discussed a couple that planned to have children until they have a boy or until they have three children, whichever comes first. Through simulation, we estimated the probability that they will have a boy among their children. What was your estimate?

We now have the probability background to be able to find the actual probability of 0.875. We assumed that

1. each child has probability 1/2 of being a boy and 1/2 of being a girl;

2. the gender of successive children is *independent*, that is, knowing the gender of a child does not influence the gender of any of the successive children.

First, we need to generate the sample space. Three of the four possible outcomes are listed in the accompanying diagram. These three outcomes result in the couple having a boy. *Fill in the one remaining possible outcome.* Next, we compute the probability for each of the outcomes, using the preceding assumptions. Some of the probabilities have been computed for you. *Find the remaining probabilities and verify that they add up to 1.*

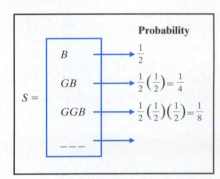

To find the probability that they will have a boy among their children, we need to add up the probabilities of the events that correspond to having a boy—namely, the probabilities for the first three outcomes in the sample space.

$$P(\text{having a boy among their children}) = \underline{\hspace{2in}}.$$

Convert your answer to decimal form, and you should have 0.875 as the probability.

Let's do it! 7.16 Getting Good Business?

The GOOD (not Better) Business Bureau conducts a survey of the quality of service offered by the 86 auto repair shops in a certain city. The results on **Service** and **Shop Type** are summarized in the following table:

		Service	
		Good	*Questionable*
Shop Type	*New-Car Dealership*	18	6
	Individual Shop	34	28

(a) What is the probability that a randomly selected shop provides good service?

(b) What is the probability that a randomly selected shop is an individual shop?

(c) What is the probability that a randomly selected shop is an individual shop *and* provides good service?

(d) What is the probability that a randomly selected *individual* shop provides good service?

(e) Are the events "individual shop" and "provides good service" mutually exclusive? Explain.

(f) Are the events "individual shop" and "provides good service" independent? Explain.

THINK ABOUT IT

Consider the random process of tossing a fair coin six times.

1. Which of the following sequences of tosses is more likely to occur?
   (a) THTHHT   (b) HHHTTT   (c) HHHHHH
   Check your answer by finding the probability of each of the sequences.
   Hint: Remember that successive tosses of a coin are assumed to be independent.

2. If we observe HHHHHH, is the next toss more likely to give a tail than a head?
   Hint: Think about whether or not the coin has any memory about the first six tosses.

Mutually Exclusive and Independent Events

Initially, it can be easy to confuse the two definitions of mutually exclusive events and independent events. It is important to keep the two definitions separate.

- The definition of **mutually exclusive** events is based on a *set* property. You can easily visualize mutually exclusive events by using a Venn diagram (the two sets do not overlap). So if two events are mutually exclusive and you are given that one of the events has occurred, the probability that the other event will occur is 0.

- The definition of **independent** events is based on a *probability* property. You cannot easily visualize independent events by using a Venn diagram, which must include probabilities that relate in a certain way. If two events are independent and you are given that one of the events has occurred, the probability that the other event will occur is not changed.

For mutually exclusive events	For independent events
$P(A \text{ and } B) = 0$	$P(A \text{ and } B) = P(A)P(B)$
$P(A \text{ or } B) = P(A) + P(B)$	$P(A \text{ or } B) = P(A) + P(B) - P(A)P(B)$
$P(A\|B) = 0$	$P(A\|B) = P(A)$

 Exercises

7.17 What Is the Error? Identify the error, if any, in each statement. If no error, write "no error."

(a) The probabilities that an automobile salesperson will sell 0, 1, 2, or at least 3 cars on a given day in February are 0.19, 0.38, 0.29, and 0.15, respectively.

(b) The probability that it will rain tomorrow is 0.40 and the probability that it will not rain tomorrow is 0.52.

(c) The probabilities that the printer will make 0, 1, 2, 3, or at least 4 mistakes in printing a document are 0.19, 0.34, −0.25, 0.43, and 0.29, respectively.

(d) On a single draw from a deck of playing cards, the probability of selecting a heart is $\frac{1}{4}$, the probability of selecting a black card (i.e., spade or club) is $\frac{1}{2}$, and the probability of selecting both a heart and a black card is $\frac{1}{8}$.

7.18 Students in a certain class were asked whether they had eaten at various local restaurants. Two of the Mexican restaurants on the questionnaire were Arriba and Bandido. The results were used to obtain the following probabilities:

The probability that a randomly selected student has eaten at Arriba is 0.30.

The probability that a randomly selected student has eaten at Bandido is 0.40.

The probability that a randomly selected student has eaten at Arriba, but not Bandido, is 0.20.

(a) Complete the following Venn diagram by filling in each empty box with the probability for the section that the box is in:

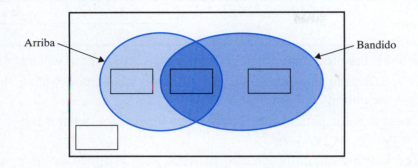

(b) What is the probability that a student selected at random will have eaten at both Arriba *and* at Bandido?

(c) What is the probability that a student selected at random will have *not* eaten at Arriba *nor* at Bandido?

7.19 Three teams, the Lions, the Tigers, and the Panthers, will play each other in a soccer tournament. The following probabilities are given about the upcoming games:

> The probability that the Lions will beat the Tigers is 0.40.
>
> The probability that the Lions will beat the Tigers and the Panthers is 0.20.
>
> The probability that the Panthers will beat the Lions is 0.30.

(a) What is the probability that the Lions will beat the Panthers?

(b) Complete the following Venn diagram by filling in each empty box with the probability for the section that the box is in:

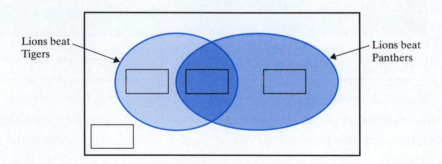

(c) What is the probability that the Lions will win at least one of the matches against the Tigers and the Panthers?

(d) Given that the Lions beat the Tigers, what is the probability that the Lions will beat the Panthers?

7.20 A study was conducted to assess the effectiveness of a new medication for seasickness as compared to a standard medication. A group of 160 seasick patients were randomly divided into two groups, with the first 80 selected at random assigned to receive the new medication and the other 80 assigned to the receive the standard medication. The results are presented in the following table:

	Medication	
	New	*Standard*
Response *Improved*	60	48
Did Not Improve	20	32

(a) What is the probability that a randomly selected patient will improve?

(b) Given that the patient received the standard medication, what is the probability that the patient will improve?

(c) Given that the patient received the new medication, what is the probability that the patient will improve?

(d) The first 80 randomly selected patients were assigned to the new medication. Using your calculator (with a seed value of 58) or the random number table (Row 24, Column 1), give the labels for the first five patients to receive the new medication.

7.21 There are a total of 50 people in a room. Each person can be classified as either married or single and as either unattractive or attractive. A total of 30 people are married, 20 are ugly, 15 are both married and attractive, and 5 are both single and unattractive.

(a) What is the probability of being single?

(b) What is the probability of being unattractive and married?

(c) Given that the person is unattractive, what is the probability that they are single?

7.22 The results of a study relating physical attractiveness to helping behavior are shown here:

		Physical Attractiveness	
		Attractive (A)	*Unattractive (NA)*
Helping Behavior	*Helped (H)*	34	26
	Did Not Help (NH)	46	54

(a) What is the probability of randomly selecting a subject who helped?

(b) Given that the person was attractive, what is the probability the subject helped?

(c) Given that the person was not attractive, what is the probability the subject helped?

7.23 The question of whether or not birth order is related to juvenile delinquency was examined in a large-scale study using a high school population. A total of 1160 girls attending the public high school were given a questionnaire that measured the degree to which each had exhibited delinquent behavior. Based on the answers to the questions, each girl was classified as either delinquent or not delinquent. The results are summarized in the following table:

		Birth Order				
		Oldest	*In Between*	*Youngest*	*Only Child*	**TOTAL**
Delinquency Status	*Yes*	20	30	35	25	110
	No	450	310	215	75	1050
	TOTAL	**470**	**340**	**250**	**100**	**1160**

(a) Calculate the probability that a high school girl selected at random from those surveyed will be delinquent.

(b) Calculate the probability that a high school girl selected at random from those surveyed will be delinquent given that they are the oldest child.

(c) Are the events "oldest child" and "delinquent" independent? Give support for your answer.

7.24 A survey of automobile ownership was conducted for 200 families in Houston, Texas. The results of the study showing ownership of automobiles of United States and foreign manufacturers are summarized as follows:

		Do you own a U.S. car?	
		Yes	*No*
Do you own a foreign car?	*Yes*	30	10
	No	150	10

(a) What is the probability that a randomly selected family will own a U.S. car and foreign car?

(b) What is the probability that a randomly selected family will not own a U.S. car?

(c) Given that a randomly selected family owns a U.S. car, what is the probability that they will also own a foreign car?

(d) Are the events "own a U.S. car" and "own a foreign car" independent events? Give support for your answer.

7.25 In an experiment to study the dependence of hypertension on smoking habits, the following data were collected on 180 individuals:

		Smoking Status		
		Nonsmoker	*Moderate smoker*	*Heavy smoker*
Hypertension Status	*Hypertension*	21	36	30
	No Hypertension	48	26	19

(a) What is the probability that a randomly selected individual is experiencing hypertension?

(b) Given that a heavy smoker is selected at random from this group, what is the probability that the person is experiencing hypertension?

(c) Are the events "hypertension" and "heavy smoker" independent? Give supporting calculations.

7.26 A random sample of 200 adults is classified according to gender and attained education level.

		Education		
		Elementary	*Secondary*	*College*
Gender	*Male*	38	28	22
	Female	45	50	17

If a person is picked at random from this group, what is the probability that

(a) the person is male, given that the person has a secondary education?

(b) the person does not have a college degree, given that the person is female?

(c) Are the events "male" and "secondary education" independent? Give supporting calculations.

7.27 Based on the orders of many of its customers, a bagel-store manager has found that 30% of the customers are male and 40% of the customers order coffee. We also have that the events "male" and "order coffee" are independent.

(a) Complete the following Venn diagram by filling in each empty box with the probability for the section that the box is in:

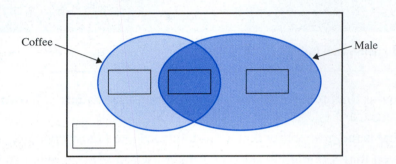

(b) What is the probability that a customer selected at random will be male and will not order coffee?

REAL DATA

7.28 An article entitled "Test Your Own Cholesterol" (SOURCE: *Ann Arbor News*, March 8, 1995) describes a home cholesterol test manufactured by Chem Trak. The article states that "Americans are becoming more savvy about cholesterol, ... 65 percent of adults have had their total cholesterol measured." Suppose that two American adults will be selected at random. Assume that the responses from these two adults are independent. What is the probability that both will have had their total cholesterol measured?

7.29 The utility company in a large metropolitan area finds that 82% of its customers pay a given monthly bill in full. Suppose that two customers are chosen (at random) from the list of all customers. Assume that the responses from these two customers are independent. What is the probability that at least one of the two customers will pay their next monthly phone bill in full?

7.30 Eighty percent of the residents of Washtenaw County are registered to vote. If four Washtenaw County residents who are old enough to vote are selected at random, what is the probability of obtaining no registered voters? (Assume that the responses of the four selected residents are independent.)

7.31 In a community college, 40% of the students oppose funding of special-interest groups. If a random sample of three students is selected and the opinions of these three students can be assumed to be independent, what is the probability that none of the three oppose funding of special-interest groups?

7.32 Suppose that you have invested in two different mutual funds. If the first mutual fund makes money, the probability that the second one will also make money is 0.7. If the first mutual fund loses money, the probability that the second one will make money is still 0.7. Given this information, we would say that the two mutual funds would be considered as _____. Select one of the following:

(a) mutually exclusive events.

(b) dependent investments.

(c) independent investments.

(d) complementary events.

7.33 Let A and B represent two events for which $P(A) = 0.40$ and $P(B) = 0.20$.

(a) If the events A and B are independent, find the probability that both events occur together—that is, $P(A \text{ and } B)$.

(b) If the events A and B are mutually exclusive, find the probability that at least one of the two events occurs—that is, $P(A \text{ or } B)$.

7.34 Consider two events, A and B, of a sample space such that $P(A) = P(B) = 0.6$.

(a) Is it possible that the two events A and B are mutually exclusive? Explain.

(b) If the events A and B are independent, what would be the probability that the two events occur together—that is, the probability of $P(A \text{ and } B)$?

(c) If the events A and B are independent, what would be the probability that at least one of the two events will occur—that is, the probability of $P(A \text{ or } B)$?

(d) Suppose that the probability of the event B occurring, given that event A has occurred, is 0.5—that is, $P(B|A) = 0.5$. What is the probability that at least one of the two events will occur—that is, the probability of $P(A \text{ or } B)$?

7.35 Determine whether each of the following statements is true or false:

(a) Two events that are disjoint (mutually exclusive) are also always independent.

(b) Suppose that among all students taking a certain test 10% cheat. If two students are selected at random (independently), the probability that both cheat is 0.01.

7.36 **A Coin and a Die**   A random process consists of flipping a fair coin, and then flipping it a second time if a head occurs on the first flip. If a tail occurs on the first flip, a fair die is tossed once.

(a) Give the sample space for this random process.

(b) Let the events A and B be defined as follows:

$$A = \text{``tail on first flip and 1 on die} = \{T1\}, \text{ and}$$
$$B = \text{``head on first flip and head of second flip''} = \{HH\}.$$

Find the probability of each of these events. (*Hint:* Use the multiplication rule.)

7.37 (a) If two events are mutually exclusive, are their complements also mutually exclusive? Explain.

(b) If two events are independent, are their complements also independent? Explain.

7.4.3 Partitioning and Bayes's Rule*

We wish to find the probability of the event A, but finding the probability is not always so straightforward. Suppose that we have information about the likelihood of the event A, not overall, but for two subgroups of the sample space, say $B1$ and $B2$, which are shown in the accompanying diagram. So we know $P(A|B1)$ and $P(A|B2)$. We would like to combine the probabilities of A for each piece to get the $P(A)$ overall. The two events, $B1$ and $B2$, form a **partition** of the sample space. In this case, the event $B2$ is actually the complement of the event $B1$.

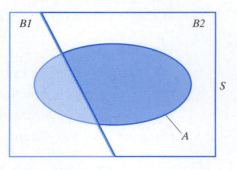

> *Definition:* The events $B1, B2, \ldots \, Bl$ form a **partition** of the sample space S if
>
>   1. the events $B1, B2, \ldots, Bl$ are mutually exclusive and
>   2. the union of $B1, B2, \ldots, Bl$ is the sample space S.
>
> *Note:* The events form a partition if each individual outcome of the sample space lies in exactly one of the events.

How can we find the overall probability of the event A? First, we consider the two disjoint parts of event A: $P(A) = P(A \text{ and } B1) + P(A \text{ and } B2)$. Next, we apply the multiplication rule for two events to each of the two probabilities on the righthand side.

$$P(A) = P(A|B1)P(B1) + P(A|B2)P(B2)$$

The two conditional probabilities of the event A, given that you are on a particular part of the sample space, are pooled together by weighing each by the chance of being on the particular part of the sample. This last line is called the **partition rule** or **law of total probability**.

Partition Rule

---

  6. If the events $B1$ and $B2$ form a partition of S, then

$$P(A) = P(A|B1)P(B1) + P(A|B2)P(B2).$$

This result can be extended to a partition of more than two events: If the events $B1, B2, \ldots, Bl$ form a partition of S, then

$$P(A) = P(A|B1)P(B1) + P(A|B2)P(B2) + \cdots + P(A|Bl)P(Bl).$$

---

*The probability results in this section are optional.

❖ EXAMPLE 7.7 Two-Stage Experiment

Suppose that there are two boxes, as shown. Consider the following two-stage experiment:

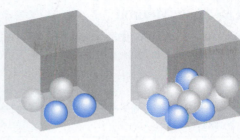

Stage 1: Pick a box at random (that is, with equal probability).

Stage 2: From that box, pick one ball at random.

Box I   Box II

What would be the probability of getting a teal ball? $P(\text{teal}) = ?$ It obviously depends on which box is selected—the two boxes form the partition.

(a) What is the chance of getting a teal ball from Box I? $P(\text{teal}|BI) = \frac{2}{4}$.

(b) What is the chance of getting a teal ball from Box II? $P(\text{teal}|BII) = \frac{4}{10}$.

Since in Stage 1 a box is selected at random, the probability that we select any given box is $\frac{1}{2}$.

$$P(BI) = P(BII) = \frac{1}{2}.$$

Finally, we use the partition rule to combine the two probabilities in (a) and (b).

$$P(\text{teal}) = P(\text{teal}|BI)P(BI) + P(\text{teal}|BII)P(BII)$$
$$= \left(\tfrac{2}{4}\right)\left(\tfrac{1}{2}\right) + \left(\tfrac{4}{10}\right)\left(\tfrac{1}{2}\right)$$
$$= 0.45. \hspace{4cm} ❖$$

Let's do it! 7.17 The Wrong Shoes

A company that manufactures athletic shoes has two factories. Factory A produces 75% of the company's athletic shoes and Factory B produces the remaining 25% of the company's athletic shoes. Approximately 1% of the shoes from Factory A are incorrectly labeled, while 2% of the shoes received from Factory B are incorrectly labeled. If you purchase a pair of athletic shoes from this company, what is the probability that your shoes are incorrectly labeled?

One interesting application of the partition rule is called the **Warner's randomized response model**, which is discussed in the next example. Suppose that we want to estimate the proportion of high school students in a school district who have engaged in illicit drug use. Asking this question directly would probably yield little useful information. How can we get accurate answers to a sensitive question that respondents might be reluctant to answer truthfully?

❖ **EXAMPLE 7.8**   **Getting Answers to Sensitive Questions** _____

The Basic Method

Set up a survey with two questions. Let Q1 be the sensitive question. Then, the second question, Q2, is any other question for which you know what proportion of "yes" responses you should expect to get. For example,

Survey		
Q1: "Have you ever shoplifted?"	YES	NO
Q2: "Is the second flip of your fair coin a head?"	YES	NO

Key Point

The respondent determines which question he or she answers using some probability device under his or her control. For example, we could have the respondent flip a fair coin two times. If the *first* flip results in a $\{H\}$, the respondent is to answer question Q1 truthfully, and if the flip results in a $\{T\}$, the respondent is to answer question Q2 truthfully. Since only the respondent knows which question he or she is answering, there should be no stigma attached to a "yes" or "no" response. Although everyone flips the fair coin two times, the result of the second flip would only be needed if the first flip was a $\{T\}$.

But can we still find out the proportion who respond "yes" to the sensitive question? We wish to learn about the proportion in the population for which the true response to question Q1 is "yes"—that is, the probability of a yes response, given question Q1. Applying the partition rule to the event "yes," with the partition being Q1 or Q2, we have

$$P(\text{YES}) = P(\text{YES}|\text{Q1})P(\text{Q1}) + P(\text{YES}|\text{Q2})P(\text{Q2}).$$

We would know, by the probability device used to determine which question is answered, the values of $P(\text{Q1})$ and $P(\text{Q2})$. We would also know the value of $P(\text{YES}|\text{Q2})$. For the second flip of a fair coin question, it would be 0.50. We would use the proportion of "yes" responses to estimate the $P(\text{YES})$. All that remains is the probability of interest, $P(\text{YES}|\text{Q1})$, which we can solve for.

For instance, if 41% of students responded "yes," then we would solve the preceding expression for the probability of "yes" for individuals who responded to question Q1.

$$P(\text{YES}) = P(\text{YES}|\text{Q1})P(\text{Q1}) + P(\text{YES}|\text{Q2})P(\text{Q2})$$

$$0.41 = P(\text{YES}|\text{Q1})(0.5) + (0.5)(0.5)$$

$$\rightarrow P(\text{YES}|\text{Q1}) = \big(0.41 - (0.5)(0.5)\big)/(0.5)$$

$$\rightarrow P(\text{YES}|\text{Q1}) = 0.32.$$

So we would estimate the $P(\text{YES}|\text{Q1})$—the proportion of students who have shoplifted—to be 32%. ❖

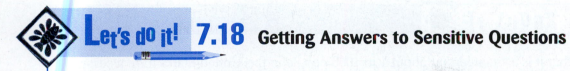

Let's do it! 7.18 Getting Answers to Sensitive Questions

Perform the Warner's randomized response model with the students in your class.

- Determine the sensitive question of interest and enter it in the survey below as Q1.
- Select two random mechanisms for the model—one for determining which of the two questions will be answered, and the other for developing a second question, Q2 in the survey, for which you know the proportion of time you expect someone to answer "yes" to it.

Since most students will have some type of coin, you could have everyone flip their coin two times, keeping the results of the two flips to themselves. The first flip of the coin can be used to determine which of the two questions is to be answered. For example, if the first flip is a head, answer Q1; otherwise, answer Q2. The second question will involve the result of the second flip of the coin, for example, "Was the second flip of the coin a head?"

Survey		
Q1: Sensitive question _____	YES	NO
Q2: On the second flip of the coin, did you get a head?	YES	NO

Procedure

Distribute slips of paper (small squares, all the same size work well), one to each student, for recording their response. Each student will answer one and only one question, either Q1 or Q2, based on the outcome of their first flip of the coin. So *only one word* should be written on each slip of paper. No names should be written on the paper, and certainly no indication as to which question was answered. For example, *do not* write "Q1 = No."

Collect the results: Number of responses = _____ .

Number of yes responses = _____ .

Proportion of yes responses = _____ = estimate of $P(\text{YES})$.

Plug all the values into the following expression and compute your estimate of $P(\text{YES}|\text{Q1})$.

$$P(\text{YES}) = P(\text{YES}|\text{Q1})P(\text{Q1}) + P(\text{YES}|\text{Q2})P(\text{Q2})$$

We would estimate that

THINK ABOUT IT

Suppose that it is estimated that about 1 in 1000 people in the United States have a certain rare disease. A diagnostic test is available for this disease that is 98% accurate (i.e., returns a positive result) for people who have the disease and 95% accurate (i.e., returns a negative result) for people who do not have the disease. Given that a person has tested positively for the disease, what do you think is the probability that the person actually has the disease? Do you think the probability is higher or lower than 50%?   *Higher*   *Lower*

As we shall see in our next example, the partition rule also plays an important part in determining the false positive rate for a diagnostic test, that is, the proportion of people who do not have the disease even though the test came back positive.

❖ EXAMPLE 7.9 Rare-Disease Testing

Suppose that a very reliable test has been developed for a particular rare disease such as AIDS or hepatitis. In particular, suppose that when the disease is present, the test is positive 98% of the time. When the disease is absent, the test is negative 95% of the time. The medical terminology for these two percentages is specificity and sensitivity, respectively. Also suppose that approximately 0.1% of the general population has the disease.

If the actual test is inexpensive to perform, should a large-scale public screening be offered? With the preceding numbers, should you be concerned if you test positive in a public screening?

What may be initially surprising is that without assuming any further information, the probability of actually having the disease given that you test positive is only about 2%. This is also the answer to the previous Think about it question.

Consider the following notation and diagram, which show the results for a population of 100,000 people:

$$D = \text{Have disease} \qquad + = \text{Test is positive}$$
$$ND = \text{Do not have disease} \qquad - = \text{Test is negative}$$

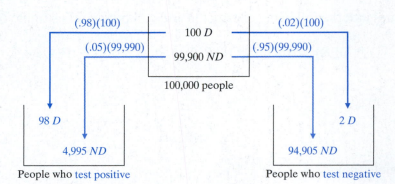

- Out of 100,000 people randomly selected from the population, we would expect about 0.1% or 100 to have the disease and thus the remaining 99,900 to not have the disease.

- Out of the 100 who have the disease, we would expect 98% or 98 to test positive and be placed in the population of people who test positive, while the remaining 2 are expected to test negative.

- Out of the 99,900 who do not have the disease, we would expect 95% or 94,905 (= 99,900 × 0.95) to test negative and the remaining 4995 to test positive and be placed in the population of people who test positive.

- Thus, out of the 5093 (= 4995 + 98) who test positive and were placed in the population of people who test positive, only 98 or about 2% actually have the disease.

Further testing would always be required for those who do test positive. Also consider the social and psychological impact of receiving a positive test result if you were unaware of the statistics. Thus, large-scale public screening for a serious rare disease is generally a bad idea. However, the key is that the disease is indeed rare. If the incidence of the disease were 10% instead of 0.1%, then the probability of having the disease given you tested positively would be approximately 69% instead of only 2%. Check out this number by making a corresponding diagram.   ❖

As you saw in Example 7.9, sometimes it is easier to determine probabilities by starting with a large population, using counts instead of probabilities. We then partitioned up the items in the large population and focused on the sequence that led to the outcome of interest. The probability information in Example 7.9 could have also been represented with tree diagram. The basic steps for making a **tree diagram** are given next and then demonstrated in Example 7.10.

Making a Tree Diagram

Step 1: Determine the first partition of the items. Create the first set of branches corresponding to the events that make up this first partition. For each branch, write the associated probability on the branch.

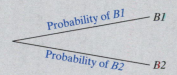

Step 2: Determine the second partition of the items. Starting from each branch in Step 1, create branches corresponding to the events that make up this next partition. On each of these appended branches, write the associated (conditional) probability on the branch. Continue this Step 2 process as necessary.

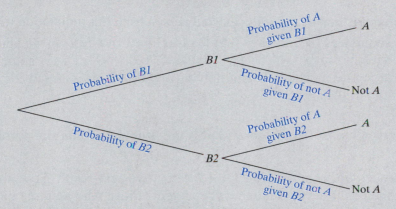

Step 3: Determine the probability of following any particular sequence of branches by multiplying the probabilities along those branches.

Step 4: The complete tree diagram can be used for finding various probabilities of interest.

❖ **EXAMPLE 7.10** **Tree Diagram for Rare-Disease Testing** _____

Step 1: The first partition of the general population is placed according to whether or not they have the disease. This is based on the 0.1% figure reported. The first set of branches correspond to these two events and the corresponding probability is written on each branch.

Step 2: The second partition is placed according to the test result, whether it is positive or negative. Starting from each branch in Step 1, branches corresponding to these events are appended and the conditional probabilities (the specificity and sensitivity) are written on the branches.

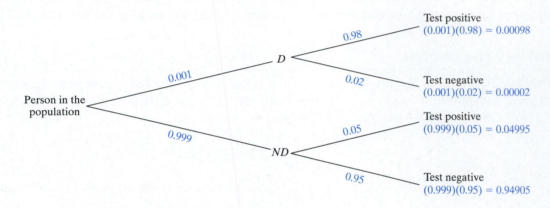

Step 3: Determine the probability of following any particular sequence of branches by multiplying the probabilities along those branches.

Step 4: The complete tree diagram can be used for finding various probabilities of interest.

What is the probability that a randomly selected person from this population will test positive?

There are two sequences of branches that lead to a positive test, so we add up these two probabilities: $P(\text{test positive}) = 0.00098 + 0.04995 = 0.05093$. This is actually an application of the partition rule, in which the multiplying of the weights is already done through the construction of the tree diagram.

Given that a person has tested positive, what is the probability that the person actually has the disease?

The two sequences of branches that led to a positive test result gave a probability of 0.05093. Only one of these two sequences represents those who actually have the disease (which had a probability of 0.00098). So our probability of interest is $P(D|\text{test positive}) = \frac{0.00098}{0.05093} = 0.01924$, for about 2%. ❖

Let's **do** it! **7.19** **Diagnostic Test for AIDS**

Suppose that it is estimated that about 0.1% of the population under study carries the AIDS antibody. A diagnostic test is available that screens blood for the presence of the AIDS antibody. The following are properties of this test:

- When the antibodies are present in the blood, the test gives a positive result with probability 0.99.
- When the antibodies are not present in the blood, the test gives a negative result with probability 0.96.

(a) Construct a tree diagram (similar to the one in Example 7.10) to portray the events and probability information for this AIDS-diagnostic test.

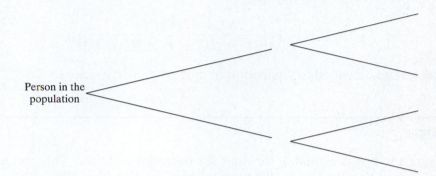

Person in the
population

(b) Using the provided information and the above tree diagram, answer the following probability questions:

(i) What is the **prior probability** that a person selected at random from this population actually has the disease?

$P(D) =$

(ii) What is the probability that a person selected at random from this population will test positive?

$P(test\ positive) =$

(iii) Given that a person selected at random from this population tests positive, what is the conditional probability that they actually have the disease?

$P(D|test\ positive) =$

This probability is also called the **posterior probability**.

(iv) Given that a person selected at random from this population tests positive, what is the probability that he or she actually *does not* have the disease?

$P(ND|test\ positive) =$

This probability is the complement of (iii), and is called the **false-positive** rate.

Note: The conditional probability found in (iii) is actually an application of the probability result called ***Bayes's rule***. The formula underlying this rule using the diagnostic-test notation is shown. In general, it is easier to work from the tree diagram directly than to apply this formula:

$$P(|test\ positive) = \frac{P(test\ positive|D)P(D)}{P(test\ positive|D)P(D) + P(test\ positive|ND)P(ND)}.$$

Bayes's Rule

7. Suppose that the events *B1* and *B2* form a partition (that is, they are complementary events whose probabilities sum to 1). Suppose that *A* is another event and we know the conditional probabilities of $P(A|B1)$ and $P(A|B2)$. Then **Bayes's rule** provides us with another conditional probability:

$$P(B1|A) = \frac{P(A|B1)P(B1)}{P(A|B1)P(B1) + P(A|B2)P(B2)}.$$

This result can be extended to a partition of more than two events.

 ## Exercises

7.38 A diagnostic test is available for dogs for bacterial infection. This test has the following properties: When applied to dogs that were infected, 98% of the tests indicated infection; when applied to dogs that were not infected, 95% of the tests indicated no infection. Suppose that 2% of the dogs are infected.

    (a) What proportion of the dogs that are not infected have tests indicating infection?

    (b) Of the dogs with tests that indicate an infection, what percentage actually is infected?

    (c) Of the tests indicating infection, what percentage is not infected?

    (d) If the population under study were a higher risk group with 20% instead of 2% actually infected, would the answer to (b) increase, decrease, or stay the same? Explain.

7.39 A box contains 5 red balls and 10 white balls. Two balls are drawn at random from this box, without replacement.

    (a) What is the probability that the second ball drawn is red?

    (b) What is the probability that the first ball drawn is white, given that the second ball drawn is red?

7.40 There are three classes of statistics students. Class A1 has 38% of the students, Class A2 has 30% of the students, and Class A3 has 32% of the students. Of the students in Class A1, 95% understand Bayes's rule. Of the students in Class A2, 84% understand Bayes's rule. Of the students in Class A3, only 8% understand Bayes's rule. Given that a randomly selected statistic student understands Bayes's rule, what is the probability that he or she is in Class A1?

7.41 Two different software corporations, say Corporation I and Corporation II, are contending to take over control of a third software corporation. The probability that Corporation I will succeed in taking control is estimated to be 0.7, and the probability that Corporation II will succeed in taking control is thus estimated to be 0.3. If Corporation I is successful, then the probability that a new software product will be introduced in the upcoming year is estimated at 0.8. If Corporation II is successful, then the probability that a new software product will be introduced in the upcoming year is estimated at only 0.4.

(a) What is the probability that a new software product will be introduced in the upcoming year?

(b) Given that a new software product is introduced in the upcoming year, what is the probability that Corporation I was successful in taking over control?

7.42 The membership for a small college is composed of 80% students and 20% faculty. The president of the college has introduced a proposal requiring all students to take an introductory statistical-reasoning course. Among faculty, 60% favor the proposal, 15% oppose the proposal, and 25% are neutral on the proposal. Among students, 40% are in favor, 50% oppose, and the rest are neutral. A letter (from a member of the college) was sent to the president regarding this proposal. Show all work for the following questions:

(a) What is the probability the writer of this letter favors the new proposal?

(b) Given that the writer of the letter states he/she is in favor of the new proposal, what is the probability the letter was sent by a faculty member?

7.43 A paint store sells three types of high-volume paint sprayers used by professional painters: Type I, Type II, and Type III. Based on records, 70% of the sprayer sales are of Type I, and 20% are of Type II. About 3% of the Type I sprayers need repair during warranty period, while the figure for Type II is 4% and the figure for Type III is 5%.

(a) Find the probability that a randomly selected sprayer purchased from this store will need repair under warranty.

(b) Find the probability that two randomly selected sprayers purchased from this store will both need repair under warranty.

(c) Find the probability that a sprayer is a Type III sprayer, given that it needed repair during warranty period.

7.44 A diagnostic test for drug use among Olympic athletes based on a blood sample was thought to be excellent. Among drug users, 98% had a positive test. Among nondrug users, 95% had a negative test. Suppose that 1% of the athletes are, indeed, drug users.

(a) What proportion of the athletes tested would have a positive test?

(b) Of those athletes whose test is positive, what proportion are really drug users?

7.45 Suppose that only two factories make Playstation machines. Factory 1 produces 70% of the machines and Factory 2 produces the remaining 30%. Of the machines produced at Factory 1, 2% are defective. Of the machines produced at Factory 2, 5% are defective.

(a) What proportion of Playstation machines produced by these two factories are defective?

(b) Suppose that you purchase a Playstation machine and it is defective. What is the probability that it was produced by Factory 1?

7.46 Suppose that it is known that, among students taking a certain entrance exam, 10% cheat on the exam and 90% do not cheat on the exam. Also, about 70% of the cheaters get a perfect score, while only 20% of the noncheaters get a perfect score.

(a) Create a tree diagram for this situation.

(b) What is the probability that a randomly selected student taking the exam will get a perfect score?

(c) Given that a student has a perfect score on the exam, what is the conditional probability that it was obtained by cheating?

7.5 Random Variables

Many gambling games use dice in which the number of dots rolled for a single die is equally likely to be 1, 2, 3, 4, 5, or 6. Consider the random experiment of rolling two fair dice. The corresponding sample space is shown. Assuming that all of these 36 points are equally likely, the probability of each point is equal to $\frac{1}{36}$. Now, let's look at the variable "the number of sixes." We actually discussed this variable, in terms of events, in "Let's do it! 7.7" and "Let's do it! 7.10." The number of sixes varies among the 36 points in the sample space, and the experiment may result in one of the possible values according to the random outcome of the experiment. That is why we refer to this variable as a **random variable**. Random variables are usually denoted by capital letters near the end of the alphabet ($\ldots, X, Y, Z$). Let X be the random variable denoting the number of sixes that occur when rolling two fair dice. The correspondence of the points in the sample space with the values of the random variable is as follows:

Sample Space					Random Variable X
(1,1)	(1,2)	(1,3)	(1,4)	(1,5)	
(2,1)	(2,2)	(2,3)	(2,4)	(2,5)	
(3,1)	(3,2)	(3,3)	(3,4)	(3,5)	$X = 0$
(4,1)	(4,2)	(4,3)	(4,4)	(4,5)	
(5,1)	(5,2)	(5,3)	(5,4)	(5,5)	
(1,6)	(2,6)	(3,6)	(4,6)	(5,6)	$X = 1$
(6,1)	(6,2)	(6,3)	(6,4)	(6,5)	
			(6,6)		$X = 2$

When we wrote out some of the possible sample spaces for various experiments, we observed that the sample space does not necessarily need to be a set of numbers. However, a coding could be established if the outcome is not numeric. As statisticians, we like to deal with numerical outcomes, and this leads us to our next definition.

> *Definition:*   A **random variable** is an uncertain numerical quantity whose value depends on the outcome of a random experiment. We can think of a random variable as a rule that assigns one (and only one) numerical value to each point of the sample space for a random experiment.

As we mentioned earlier, we use capital letters, such as X, to denote a random variable. We will use a lowercase letter to represent a particular value that the random variable can take on. For example, $x = 1$ tells us that a particular roll of the two fair dice resulted in exactly one six. Think of the capital letter X as *random*, the value of the variable before it is observed. Think of the lowercase letter x as *known*, a particular value of X that has been observed.

❖ **EXAMPLE 7.11** **Random Variables** _____

(a) If the experiment is tossing a fair coin three times, then the random variable X could correspond to the number of heads in the three tosses. The possible values of X would be $x = 0$, $x = 1$, $x = 2$, or $x = 3$.

$$X = \text{\# heads in three tosses of a fair coin.}$$

(b) If the experiment is picking a student at random from the population of students at a university, the random variable Y could correspond to the number of credit hours the student is registered for in the current term. The possible values of Y might be represented as $0, 1, 2, 3, \ldots, 20$. The maximum of 20 hours may be an upper limit set by the university.

$$Y = \text{\# credit hours registered for in the current term.}$$

(c) If the experiment is selecting a light bulb at random from the production line, the random variable Z could correspond to the lifetime of the light bulb measured in hours. The possible values of Z might be represented as the interval $[0, \infty)$; that is, any value greater than or equal to 0. Certainly, there may be some reasonable maximum, but without further information we would not know what value to use.

$$Z = \text{lifetime of light bulb, in hours.} \qquad ❖$$

Just as we distinguished in Chapter 4 between discrete and continuous variables, we also make the distinction between discrete and continuous random variables. A random variable having a finite or countable number of possible values is said to be *discrete*. A random variable is said to be *continuous* if its set of possible values is an interval or a collection of intervals of real numbers. In Example 7.11, the random variables X (number of heads) and Y (number of credit hours) are discrete, while the random variable Z representing lifetime is continuous.

> **Definitions:** A **discrete random variable** can assume at most a finite or infinite but countable number of distinct values. A **continuous random variable** can assume any value in an interval or collection of intervals.

In Chapter 6, we used mathematical models to describe the relationship between the possible values of a variable of interest and the proportion of items in the population having the various values. Such models for discrete variables were called mass functions, while density functions were used to model continuous variables. These ideas extend to modeling random variables as well.

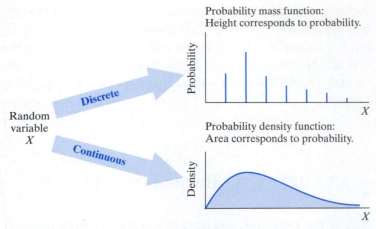

Associated with each random variable is a model—either its **probability mass function**, if discrete, or its **probability density** function, if continuous.

The primary difference between the models presented in this chapter and those presented in Chapter 6 is that earlier we spoke in terms of *proportion*, and now we can use the term *probability*. Next, we focus on how to construct the probability models for both discrete and continuous random variables.

 ### 7.5.1 Discrete Random Variables

A random variable X having a finite or countable number, k, of possible values is said to be *discrete*. A probability is assigned to each of the k possible values, where each probability must be between 0 and 1 and the sum of all of the probabilities must equal 1. The set of values along with their probabilities is called the **probability distribution of X or probability mass function for X**. If X is a discrete random variable taking on the values x_1, x_2, ... x_k, with probabilities p_1, p_2, ... p_k, the probability distribution of X is given by

Value of X $X = x$	x_1	x_2	...	x_k
Probability $P(X = x)$	p_1	p_2	...	p_k

We can also represent this function as

$$p(x) = \begin{cases} p_1 & if \quad x = x_1 \\ p_2 & if \quad x = x_2 \\ \vdots & \quad \vdots \\ p_k & if \quad x = x_k \end{cases}$$

> *Definition:* The probability distribution of a discrete random variable X is a table or rule that assigns a probability to each of the possible values of the random variable X. The values of a discrete probability distribution (the assigned probabilities) must be between 0 and 1 and must add up to 1, that is, $\sum p_i = 1$.

Recall the random experiment of rolling two fair dice and the random variable X defined to be "the number of sixes." The possible values for the random variable X are 0, 1, and 2. Since only one of the 36 equally likely outcomes corresponds to having exactly two sixes, the probability that the random variable takes on the value of 2 is equal to $\frac{1}{36}$. The probability that $X = 0$ is equal to $\frac{25}{36}$, since 25 of the outcomes correspond to obtaining no sixes. Likewise, the probability that $X = 1$ is equal to $\frac{10}{36}$. The probability distribution for X is given in the following table. Notice that the probabilities do sum up to one.

$X = x$	0	1	2
$P(X = x)$	$\frac{25}{36}$	$\frac{10}{36}$	$\frac{1}{36}$

/ Probability

A picture of the probability distribution of a discrete random variable X can be displayed using a stick graph or a probability bar graph, as shown in the next example.

❖ **EXAMPLE 7.12 People in a Household** _____

Let X be the number of people in a household for a certain community. Consider the following probability distribution for X, which assumes that there are no more than seven people in a household.

Value of X $X = x$	1	2	3	4	5	6	7
Probability $P(X = x)$	0.20	0.32	0.18	0.15	0.07	0.03	

(a) What must be the probability of seven people in a household for this to be a legitimate discrete distribution? Since the probabilities must sum to 1, the probability of seven people must be $1 - (0.20 + 0.32 + 0.18 + 0.15 + 0.07 + 0.03) = 0.05$.

(b) Display this probability distribution graphically. We could use a **stick graph** (Figure 7.5(a)) or a **probability bar graph** (Figure 7.5(b)).

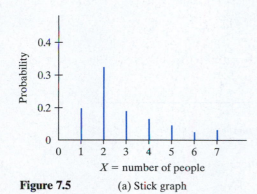

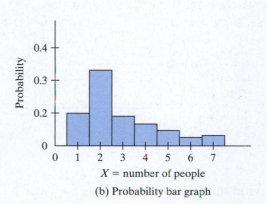

Figure 7.5 (a) Stick graph   (b) Probability bar graph

(c) What is the probability that a randomly chosen household contains more than five people? $P(X > 5) = 0.03 + 0.05 = 0.08$.

(d) What is the probability that a randomly chosen household contains no more than two people? $P(X \le 2) = 0.32 + 0.20 = 0.52$.

(e) What is $P(2 < X \le 4)$? It is the probability that a randomly selected household has more than two but at most four people. $P(2 < X \le 4) = 0.18 + 0.15 = 0.33$. ❖

 **7.20 Sum of Pips**

In a game called craps, two dice are rolled and the sum of the faces on the two dice determines whether the player wins, loses, or continues rolling the dice. Let X be the sum of the values on the two dice. Recall the 36 possible pairs of faces of the two dice.

$$S = \{ (1, 1) \quad (1, 2) \quad (1, 3) \quad (1, 4) \quad (1, 5) \quad (1, 6)$$
$$(2, 1) \quad (2, 2) \quad (2, 3) \quad (2, 4) \quad (2, 5) \quad (2, 6)$$
$$(3, 1) \quad (3, 2) \quad (3, 3) \quad (3, 4) \quad (3, 5) \quad (3, 6)$$
$$(4, 1) \quad (4, 2) \quad (4, 3) \quad (4, 4) \quad (4, 5) \quad (4, 6)$$
$$(5, 1) \quad (5, 2) \quad (5, 3) \quad (5, 4) \quad (5, 5) \quad (5, 6)$$
$$(6, 1) \quad (6, 2) \quad (6, 3) \quad (6, 4) \quad (6, 5) \quad (6, 6) \}.$$

(a) Give the probability distribution function of X—that is, list the possible values of X and their corresponding probabilities. Then, present the probability distribution function graphically by drawing a stick graph or a probability bar graph.

Value of X $X = x$	
Probability $P(X = x)$	

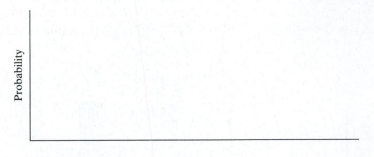

(b) Find the $P(X > 7)$.

(c) What is the probability of rolling a seven or an eleven on the next roll of the two dice?

(d) What is the probability of rolling at least a three on the next roll of the two dice? (Use the complement rule.)

Let's do it! 7.21 Couple 1 and Couple 2

Each of two couples, say Couple 1 and Couple 2, have decided that they will have a total of exactly three children. One possible outcome for either of the couples is the event GGB, which means the first two children are girls and the third child is a boy.

(a) Write down the sample space corresponding to the possible outcomes for Couple 1.

$S = \{$ $\}$.

(b) Let the random variable X be "the number of girls" for Couple 1. What are the possible values for X?

Possible values = _____.

(c) Assume that each child has probability $\frac{1}{2}$ of being a girl and $\frac{1}{2}$ of being a boy and that the gender of successive children is independent. Give the probability distribution for the random variable X = "number of girls" for Couple 1.

Value of X $X = x$	
Probability $P(X = x)$	

(d) Assuming that the outcome for Couple 1 is independent of the outcome for Couple 2, what is the probability that the two couples will have the same number of girls?

In Chapter 5, we discussed how to summarize a set of data using numerical summaries of the center and spread of the data. So before we turn to continuous random variables, we will learn how to summarize a discrete probability distribution by using the mean and the standard deviation. The **mean of a discrete random variable** X is the point at which the probability stick graph or bar graph would balance.

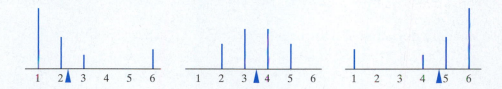

The mean of X is also referred to as the **expected value of X** and is calculated as follows:

> **Definition:** If X is a discrete random variable taking on the values x_1, x_2, $\ldots x_k$, with probabilities p_1, p_2, $\ldots p_k$, then the **mean or expected value of X** is given by
>
> $$E(X) = \mu = x_1 p_1 + x_2 p_2 + \cdots + x_k p_k = \sum x_i p_i.$$

A *note of caution* regarding the term *expected*: The expected value of a random variable does not necessarily have to equal one of the values that are possible for the random variable. The mean is the value that you would *expect*, on *average*—that is, the mean of observed values X in a very large number of repetitions of the random experiment.

The **standard deviation of a discrete random variable** X is a measure of the spread of the possible values from the mean. Examining the two distributions shown, we see that, although the mean of X is the same as the mean of Y (namely, 60), the standard deviation of X is smaller than the standard deviation of Y.

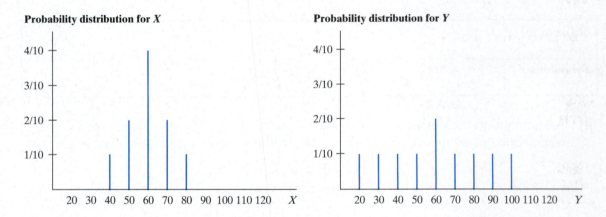

The expression for the variance of X is shown next. The equivalent right-hand-side expression is primarily used for calculation purposes, as shown in Example 7.13. If we take the square root of the variance, we have the standard deviation of X, interpreted as *roughly the average distance of the possible x-values from their mean.*

> **Definitions:** If X is a discrete random variable taking on the values x_1, x_2, $\ldots x_k$, with corresponding probabilities p_1, p_2, $\ldots p_k$, then the **variance of X** is given by
>
> $$\mathrm{Var}(X) = \sigma^2$$
>
> $$= E\big((X - \mu)^2\big) = \sum (x_i - \mu)^2 p_i = E(X^2) - \big(E(X)\big)^2 = \sum x_i^2 p_i - \mu^2$$
>
> and the **standard deviation of X** is given by
>
> $$SD(X) = \sigma = \sqrt{\sigma^2}.$$

❖ **EXAMPLE 7.13**   **People in a Household Revisited** _____

Based on this model, what is the expected number of people in a household?

Value of X $X = x$	1	2	3	4	5	6	7
Probability $P(X = x)$	0.20	0.32	0.18	0.15	0.07	0.03	0.05

$$E(X) = \mu = (1)(0.20) + (2)(0.32) + (3)(0.18) + (4)(0.15)$$
$$+ (5)(0.07) + (6)(0.03) + (7)(0.05) = 2.86.$$

Thus, we expect, on average, about 2.86 people per household.

To find the variance for the number of people in a household, we first need to find $E(X^2)$, following the basic form for computing expectations.

$$E(X^2) = \sum x_i^2 p_i = (1)^2(0.20) + (2)^2(0.32) + (3)^2(0.18) + (4)^2(0.15)$$
$$+ (5)^2(0.07) + (6)^2(0.03) + (7)^2(0.05) = 10.78.$$

So we have $\text{Var}(X) = \sigma^2 = E(X^2) - [E(X)]^2 = 10.78 - (2.86)^2 \approx 2.6$.

The standard deviation would be

$$SD(X) = \sigma = \sqrt{2.6} \approx 1.61.$$

Interpretation

If a household were selected at random from this community, we would expect the number of people to be about 2.86, give or take about 1.61.   ❖

Let's do it!   **7.22**   **Sum of Pips Revisited**

Consider the game called craps, in which two fair dice are rolled. Let X be the random variable corresponding to the sum of the two dice. Its probability distribution is given:

Value of X $X = x$	2	3	4	5	6	7	8	9	10	11	12
Probability $P(X = x)$	$\frac{1}{36}$	$\frac{2}{36}$	$\frac{3}{36}$	$\frac{4}{36}$	$\frac{5}{36}$	$\frac{6}{36}$	$\frac{5}{36}$	$\frac{4}{36}$	$\frac{3}{36}$	$\frac{2}{36}$	$\frac{1}{36}$

Calculate the mean of X, the expected sum of the values on the two dice.

You provided a graph of this distribution in "Let's do it! 7.20." Is your expected value consistent with the idea of being the balancing point of the probability stick graph or bar graph?

Let's do it! 7.23 Profits and Weather

A concert promoter has a decision to make about location for a concert for the next fall. The concert can be held in an indoor auditorium or a larger facility outdoors. The following are profit forecasts, recorded in thousands of dollars, for each facility given different weather conditions:

Weather	Profit Indoors	Profit Outdoors	Probability
Sunny and Warm	50	80	0.5
Sunny and Cold	50	60	0.2
Rainy and Warm	60	40	0.2
Rainy and Cold	40	6	0.1

(a) The expected profit for the indoor facility is $51,000. Compute the expected profit associated with the outdoor facility.

Outdoor facility expected profit = _____ (include your units).

(b) Which location is preferred if the objective is maximizing the expected profit?

(c) Which location is preferred if the promoter must clear a profit of at least $40,000?

(d) Which location has the least variability in the profits? Explain. No calculations are necessary—just picture the two distributions graphically.

7.5.2 Binomial Random Variables

At the start of Section 7.5, we learned that there are two broad types, of random variables, namely, discrete and continuous. Within each of these broad types, there are families of random variables that are common and have a special name. In this section, we will focus on a particular family of discrete random variables called **binomial random variables**. To understand the probability mass function for a binomial random variable, we must first review some basic counting problems that involve computing **combinations**. Then, we discuss an experiment in which there are only two possible outcomes, one referred to as a success and the other as a failure, with the probability of a success represented by p. This experiment is called a **Bernoulli** trial and forms the basis of the distribution of a binomial random variable.

Combinations

Suppose that you have exactly two friends, Emily and Caitlyn. You are going to consider inviting them to dinner. If you decide to invite two friends to dinner you have just **one choice**—invite them both. If you decide to invite exactly one friend to dinner, you have **two choices**—invite just Emily or invite just Caitlyn. If you decide to invite no friends to dinner you again have just **one choice**—invite no one. With this dinner invitation example, you have performed some basic counting problems called combinations. Let's look at this counting problem a little more formally.

Suppose that you have a set that contains just two distinguishable values, $S = \{1, 2\}$. How many subsets are there that would contain exactly two values? The answer is just one subset—namely, the set $\{1, 2\}$. How many subsets are there that would contain exactly one value? There are two possible subsets—namely, $\{1\}$ and $\{2\}$. How many subsets are there that contain exactly zero values (no values)? There is just one subset, called the **empty set**, represented by $\varnothing$. The following table summarizes these counting questions:

How many subsets of $S = \{1, 2\}$ are there that contain ...	Answer	The subsets are	Combination
... exactly zero values (no values)?	1	$\varnothing$	$\binom{2}{0} = 1$
... exactly one value?	2	$\{1\}, \{2\}$	$\binom{2}{1} = 2$
... exactly two values?	1	$\{1, 2\}$	$\binom{2}{2} = 1$

If we add up the "answer" column, we have a total of four possible subsets of the set $S = \{1, 2\}$. In general, the total number of possible subsets is computed as 2^n, where n is the number of elements in the original set. The last column presents the mathematical expression for determining the number of ways of selecting a certain number of items (without replacement) from a set of distinguishable items when the order of the selection is not important. For example, $\binom{2}{1}$ is the number of combinations of two items taking one at a time (that is, the number of ways you can select one item from two distinguishable items, which is equal to two).

In general, $\binom{n}{x}$ is the number of combinations of n distinguishable items taking x items at a time and is read "n choose x." The general formula for finding the number of combinations is as follows:

Combinations

"**n choose x**" represents the number of ways of selecting x items (without replacement) from a set of n distinguishable items when the order of the selection is not important and is given by

$$\binom{n}{x} = \frac{n!}{x!(n-x)!}, \text{ where } n! = n(n-1)(n-2)\cdots(2)(1).$$

Note: By definition, we have $0! = 1$ and $\binom{n}{x} = 0$ if $x < 0$ or $n < x$.

The TI graphing calculators have an operation called nCr for finding the number of combinations. This operation is found under the MATH PRB menu. The steps for finding the number of combinations of 20 items taking two at a time are as follows:

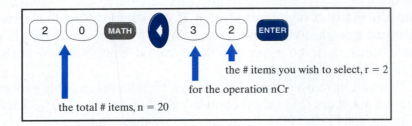

The answer is

$$\binom{20}{2} = \frac{20!}{2!18!} = 190.$$

Next, suppose that you have a set containing three distinguishable values, $S = \{1, 2, 3\}$. There is just one subset that contains exactly zero values—the empty set, $\emptyset$. There are three possible subsets that contain exactly one value—namely, $\{1\}$, $\{2\}$, and $\{3\}$. There are also three possible subsets that contain exactly two values—namely, $\{1, 2\}$, $\{1, 3\}$, and $\{2, 3\}$. Finally, there is just one subset, the set $\{1, 2, 3\}$, that contains exactly three values. The following table summarizes these counting questions:

How many subsets of $S = \{1, 2, 3\}$ are there that contain ...	Answer	The subsets are	Combination
... exactly zero values (no values)?	1	$\emptyset$	$\binom{3}{0} = 1$
... exactly one value?	3	$\{1\}, \{2\}, \{3\}$	$\binom{3}{1} = 3$
... exactly two values?	3	$\{1, 2\}, \{1, 3\}, \{2, 3\}$	$\binom{3}{2} = 3$
... exactly three values?	1	$\{1, 2, 3\}$	$\binom{3}{3} = 1$

The set $S = \{1, 2, 3\}$ has three values, so the total number of possible subsets is $2^3 = 8$, which is the sum of the "answer" column. Using the combinations formula, we can verify that there are three possible subsets of exactly two values:

$$\binom{3}{2} = \frac{3!}{2!(3-2)!} = \frac{(3)(2)(1)}{(2)(1)(1)} = 3.$$

Let's do it! **7.24** **Combinations of** $n = 4$

Suppose that you have a set that contains four distinguishable values, $S = \{1, 2, 3, 4\}$. The following table partially summarizes the counting questions:

How many subsets of $S = \{1, 2, 3, 4\}$ are there that contain ...	Answer	The subsets are	Combination
... exactly zero values (no values)?	1	Ø	
... exactly one value?		$\{1\}, \{2\}, \{3\}, \{4\}$	
... exactly two values?			$\binom{4}{2} = 6$
... exactly three values?	4		$\binom{4}{3} = 4$
... exactly four values?		$\{1, 2, 3, 4\}$	

(a) Complete the table by filling in the missing entries.

(b) The set $S = \{1, 2, 3, 4\}$ has four values, so the total number of possible subsets is $2^4 =$ _____. Confirm that this equals the total of the "answer" column.

 Let's do it! **7.25** **Office Arrangements**

The Statistics Department has a hallway with a total of 10 offices for faculty. There are three female faculty members and seven male faculty members that need to be assigned to these 10 offices.

Office 1	Office 2	Office 3	Office 4	Office 5	Office 6	Office 7	Office 8	Office 9	Office 10

(a) How many possible ways are there to select the three offices for the female faculty from these 10 offices?

(b) Give four possible selections by shading in the offices that would be assigned to the female faculty.

Office 1	Office 2	Office 3	Office 4	Office 5	Office 6	Office 7	Office 8	Office 9	Office 10

Office 1	Office 2	Office 3	Office 4	Office 5	Office 6	Office 7	Office 8	Office 9	Office 10

Office 1	Office 2	Office 3	Office 4	Office 5	Office 6	Office 7	Office 8	Office 9	Office 10

Office 1	Office 2	Office 3	Office 4	Office 5	Office 6	Office 7	Office 8	Office 9	Office 10

Bernoulli Variable

We shall consider some problems that involve a very simple type of discrete random variable, one that is dichotomous having exactly two possible outcomes. A variable with exactly two possible outcomes is also known as a Bernoulli variable. The two outcomes are often referred to as "success" and "failure." In many cases, it makes sense to assign probabilities to the two outcomes. The probability of a success is denoted by p, while the probability of a failure is denoted by $q = 1 - p$. Some examples of Bernoulli variables are as follows:

- A carnival game involves rolling a fair die one time. If you get a six you win a small prize. So a success is defined as getting the value six and a failure is not getting a six. If the die is fair, the probability of a success is $p = \frac{1}{6}$ and thus the probability of a failure is $q = \frac{5}{6}$.

- A medical treatment for lung-cancer patients involves radiation therapy. For any patient getting this treatment, there is a 70% chance of surviving at least 10 years. A success is defined to be that a patient survives at least 10 years after treatment. The probability of surviving at least 10 years for any one patient is given as $p = 0.70$.

- A true–false question was given to students on the first day of class. For any student with no idea of the correct answer who randomly guesses, there is a 50% chance of getting the correct answer. A success is defined to be that a student guesses the correct answer to the question. The probability of a success is $p = 0.50$.

> *Definition:* A **dichotomous** or **Bernoulli** random variable is one that has exactly two possible outcomes, often referred to as "success" and "failure." In this text, we will only consider such variables in which the success probability p remains the same if the random experiment were repeated under identical conditions.

(a) At a local community college, there are 500 freshmen enrolled, 274 sophomores enrolled, 191 juniors enrolled, and 154 seniors enrolled. An enrolled student is to be selected at random. If success is defined to be "senior," what is the probability of success?

$$p = \underline{\hspace{3cm}} .$$

(b) A standard deck of cards contains 52 cards, 13 cards of each of four suits. Each suit consists of four face cards (jack, queen, king, ace) and nine numbered cards (2 through 10). A card is drawn from a well-shuffled standard deck of cards. If success is defined to be getting a face card, what is the probability of success?

$$p = \underline{\hspace{3cm}} .$$

(c) A game consists of rolling two fair dice. If success is defined to be getting doubles, what is the probability of success?

$$p = \underline{\hspace{3cm}} .$$

Binomial Distribution

In many situations involving Bernoulli variables, we are really interested in learning about the total number of successful outcomes, x, in a set of n independent repetitions of the experiment. A repetition of an experiment is often referred to as a trial.

- A carnival game involves rolling a fair die one time. If you get a six, you win a small prize. The game attendant is interested in the total number of small prizes to be given away if the game is played $n = 1000$ times.

- A medical researcher is interested in the total number of lung-cancer patients out of a group of $n = 2000$ who survive at least 10 years after receiving a radiation treatment. For any lung-cancer patient getting this treatment, there is a 70% chance of surviving at least 10 years.

- A professor is interested in the total number of students in a class of $n = 200$ who answer a true–false question correctly, assuming that each student had no idea of the correct answer and randomly guessed.

> *Definition:* A **binomial** random variable X is the total number of successes in n independent Bernoulli trials, on which each trial, the probability of a success is p.
>
> **Basic Properties of a Binomial Experiment**
> - The experiment consists of n identical trials.
> - Each trial has two possible outcomes (success or failure).
> - The trials are independent.
> - The probability of a success, p, remains the same for each trial. The probability of a failure is $q = 1 - p$.
> - The binomial random variable X is the number of successes in the n trials; X is said to have a binomial distribution denoted by $Bin(n, p)$; X can take on the values $0, 1, 2, \ldots, n$.

Suppose that your next statistics quiz consists of four multiple-choice questions. Each of the four questions will have three possible answers to select from.

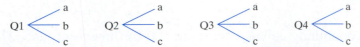

Each question is worth one point for a total possible score of four points. Suppose that a student will just be guessing the answer for each question independently. The probability of getting Question 1 correct would be just $\frac{1}{3}$. The probability of getting Question 2 correct is also $\frac{1}{3}$. In fact, the probability of getting any one question correct would be $\frac{1}{3}$. If we let the random variable X be the number of correct answers, then a model for X would be a binomial distribution with $n = 4$ and $p = \frac{1}{3}$—that is, X has a $Bin(4, \frac{1}{3})$ distribution. Let's work out this binomial probability distribution.

First, we consider the outcome $X = 0$ (that is, that the student gets all four questions incorrect for a score of 0). Since the student is assumed to be guessing the answer for each question independently, we can find the probability of getting all four questions incorrect by multiplying the individual probabilities for each question.

$$
\begin{aligned}
P(X = 0) &= P(\text{all 4 questions incorrect}) \\
&= P(\text{Q1 incorrect})P(\text{Q2 incorrect})P(\text{Q3 incorrect})P(\text{Q4 incorrect}) \\
&= \left(\tfrac{2}{3}\right)\left(\tfrac{2}{3}\right)\left(\tfrac{2}{3}\right)\left(\tfrac{2}{3}\right) = \left(\tfrac{2}{3}\right)^4 = 0.1975.
\end{aligned}
$$

Similarly, we would find the probability of getting all four questions correct by multiplying the individual probabilities for each question.

$$
\begin{aligned}
P(X = 4) &= P(\text{all 4 questions correct}) \\
&= P(\text{Q1 correct})P(\text{Q2 correct})P(\text{Q3 correct})P(\text{Q4 correct}) \\
&= \left(\tfrac{1}{3}\right)\left(\tfrac{1}{3}\right)\left(\tfrac{1}{3}\right)\left(\tfrac{1}{3}\right) = \left(\tfrac{1}{3}\right)^4 = 0.0123.
\end{aligned}
$$

How would we find the probability that the student will get a score of 1? The outcome $X = 1$ would occur for each of the following four possible situations:

Q1 correct, Q2 incorrect, Q3 incorrect, Q4 incorrect;

Q1 incorrect, Q2 correct, Q3 incorrect, Q4 incorrect;

Q1 incorrect, Q2 incorrect, Q3 correct, Q4 incorrect;

Q1 incorrect, Q2 incorrect, Q3 incorrect, Q4 correct.

Using our combinations formula, $\binom{4}{1}$ is the number of ways to select exactly one question to be answered correctly out of four possible questions, which is equal to four. Each of these four possible situations has probability $\left(\frac{1}{3}\right)^1\left(\frac{2}{3}\right)^3$, where $\left(\frac{1}{3}\right)^1$ corresponds to the one question answered correctly and $\left(\frac{2}{3}\right)^3$ corresponds to the three questions answered incorrectly. Summing up the probabilities for these four possible situations, we get

$$
P(X = 1) = \binom{4}{1}\left(\tfrac{1}{3}\right)^1\left(\tfrac{2}{3}\right)^3 = 4\left(\tfrac{1}{3}\right)^1\left(\tfrac{2}{3}\right)^3 = 0.3951.
$$

Using our combinations formula and the independence of the questions, we also have

$$P(X = 2) = \binom{4}{2}(\tfrac{1}{3})^2(\tfrac{2}{3})^2 = 6(\tfrac{1}{3})^2(\tfrac{2}{3})^2 = 0.2963;$$

$$P(X = 3) = \binom{4}{3}(\tfrac{1}{3})^3(\tfrac{2}{3})^1 = 4(\tfrac{1}{3})^3(\tfrac{2}{3})^1 = 0.0988.$$

The complete binomial distribution for X is provided in the following table (note that the probabilities do sum to 1, as they should for any discrete distribution):

Bin(4, $\tfrac{1}{3}$) *Probability Distribution*

Value of X X = x	0	1	2	3	4
Probability P(X = x)	0.1975	0.3951	0.2963	0.0988	0.0123

How would we find the probability that the student will get a score of 3 or more?

$$P(X \geq 3) = P(X = 3) + P(X = 4)$$

$$= \binom{4}{3}(\tfrac{1}{3})^3(\tfrac{2}{3})^1 + \binom{4}{4}(\tfrac{1}{3})^4(\tfrac{2}{3})^0 = 4(\tfrac{1}{3})^3(\tfrac{2}{3}) + 1(\tfrac{1}{3})^4 = 0.0988 + 0.0123 = 0.1111$$

The Binomial Probability Distribution

$$P(X = x) = \binom{n}{x}(p)^x(q)^{n-x}, \quad x = 0, 1, 2, \ldots, n$$

where p = probability of a success on each single trial;
$q = 1 - p$;
n = number of independent trials;
x = number of successes in the n trials.

In Section 7.5.1, we learned how to compute the mean and variance for any discrete random variable. We could apply these general formulas to any binomial distribution; however, a general relationship holds in the binomial case.

Mean, Variance, and Standard Deviation for a Binomial Random Variable

Mean: $\qquad\qquad E(X) = \mu = np$

Variance: $\qquad\qquad \text{Var}(X) = \sigma^2 = npq$

Standard Deviation: $\quad SD(X) = \sigma = \sqrt{npq}$

In our four-question-quiz example, if a student were just to guess the answer for each question independently, we would expect a total score of $\mu = (4)\left(\frac{1}{3}\right) = 1.33$ points, give or take about $\sigma = \sqrt{4\left(\frac{1}{3}\right)\left(\frac{2}{3}\right)} = 0.94$ points.

Many calculators can compute the probabilities for a specified binomial distribution. The TI graphing calculator has two built-in binomial functions that are located under the DISTR menu. The binompdf(function computes the probability of getting exactly x successes—that is, $P(X = x)$. We have that X has a $Bin\left(n = 4, p = \frac{1}{3}\right)$ distribution. The steps for finding the probability of getting exactly three successes, that is $P(X = 3)$, are shown here.

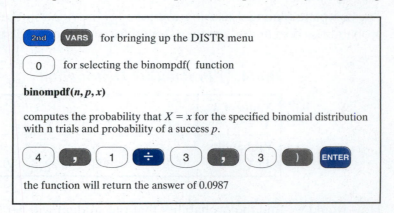

You could repeat the preceding steps to find $P(X = 4)$ and add up the two probabilities to find $P(X \geq 3)$.

The other built-in binomial function, binomcdf(, provides the probability of getting at most x successes—that is, it computes $P(X \leq x)$. We have that X has a $Bin\left(n = 4, p = \frac{1}{3}\right)$ distribution. The steps for finding the probability of at most two successes are shown here.

To find $P(X \geq 3)$, you could subtract your answer from 1:

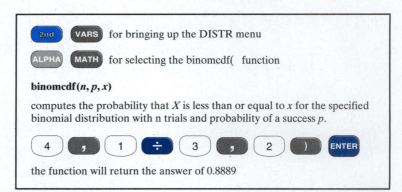

$$P(X \geq 3) = 1 - P(X < 3) = 1 - P(X \leq 2) = 1 - 0.8889 = 0.1111.$$

Let's do it! 7.27 Jury Decision

In a jury trial, there are 12 jurors. In order for a defendant to be convicted, at least 8 of the 12 jurors must vote "guilty." Assume that the 12 jurors act independently (i.e., how one juror votes will not influence how any other juror votes). Also assume that, for each juror, the probability that he or she will vote correctly is 0.85. If the defendant is actually guilty, what is the probability that the jury will render a correct decision?

Identify the following: A trial =

n = number of independent trials = _____.

p = probability of a success on each single trial = _____.

x = number of successes in the n trials.

$P(\text{correct decision by jury}) = P(X \qquad) =$

7.5.3   Continuous Random Variables

A random variable X that takes its values in some interval, or union of intervals, is said to be **continuous**. A probability cannot be assigned to each of the possible values since the number of possible values in an interval cannot be counted. So we must assign a probability of 0 to each individual value; otherwise, the sum of the probabilities would eventually be greater than 1. The idea is to **assign probabilities to intervals** of outcomes, not single values, and represent these probabilities as areas under a curve, called a probability density curve.

> *Definition:*   The **probability distribution of a continuous random variable X** is a curve such that the area under the curve over an interval is equal to the probability that the random variable X is in the interval. The values of a continuous probability distribution must be at least 0 and the total area under the curve must be 1.

That is, the probability distribution of a continuous random variable must be a **density curve**. We have looked at density curves in the past. Now, instead of referring to our area calculations as finding a proportion, we may state that we are finding a probability.

Suppose that the probability density function for the income of a randomly selected American adult is as given in the graph. How would you find the probability of selecting a person at random from this population with an income between \$30,000 and \$40,000?

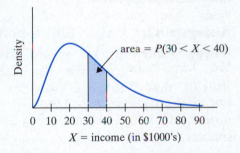

$$P[30 < X < 40] = ?$$

We would use the probability density function to find the area under it between 30 and 40.

The calculation of the mean and standard deviation when a probability density function is available is similar to the expressions for the discrete case, except that we need to use a branch of calculus called integration in place of summation. However, the location of the center of gravity or mean of the density may be evident when looking at the function.

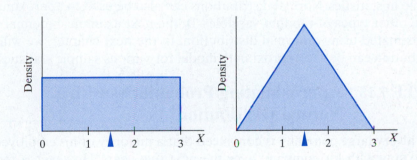

> *Definition:*   The **mean** or **expected value of a continuous random variable X** is the point at which the probability density function would balance.

❖ **EXAMPLE 7.14** **A Long Pregnancy** _____

Let X be the length of a pregnancy in days. Thus, X is a continuous random variable. Suppose that it has approximately a normal distribution with a mean of 266 days and a standard deviation of 16 days—that is, X is $N(266, 16)$. What is the probability that a pregnancy lasts at least 310 days?

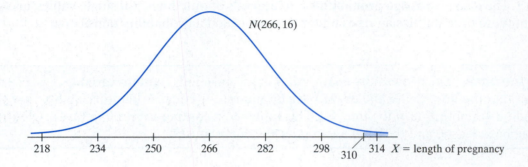

We want to compute $P(X > 310)$. We will do roughly the same calculations as we have done in the previous chapter for normal distributions, but using the language of probability instead of proportion.

Using the TI, $P(X > 310) = \text{normalcdf}(310, \text{E99}, 266, 16) = 0.00298$.

Using Table II, we would first compute the standard score for 310, namely, $z = \frac{310 - 266}{16} = 2.75$. We use Table II to find the area to the right of 2.75. Recall that Table II provides areas to the left of various values of Z. So $P(X > 310) = P(Z > 2.75) = 1 - 0.9970 = 0.0030$.   ❖

If a population of responses follows a normal distribution, and if X represents the response for a randomly selected unit from the population, then X is said to be a normal random variable—that is, the random variable X is said to have a normal distribution. This was the case in Example 7.14 for length of pregnancy. Normal random variables are the most common class of continuous random variables. There are many characteristics whose distribution follow a bell-shaped normal curve. The family of normal distributions plays a prominent role in statistics. Normal distributions can also be used to approximate probabilities for some other types of random variables. In the next example, binomial probabilities will be approximated using a normal distribution. In the next chapter, we will learn that a normal distribution can be the approximate model for various sample statistics.

❖ **EXAMPLE 7.15** **Approximating Probabilities with a Normal Distribution** _____

The CEO of a very large company is convinced that a majority of his employees are happy and expect to stay with his company over the next five years. However, a recent random sample of 400 employees resulted in only 192 saying they were happy and planned to stay. If indeed 50% of the employees are happy and plan to stay, how likely would it be to find 192 or fewer saying so in a random sample of 400 employees?

If we let $X =$ the number in the sample who are happy and plan to stay, and if 50% of all employees are happy and plan to stay, then X is a **binomial random variable** with $n = 400$ and $p = 0.50$. We would like to find $P(X \leq 192)$, but tying to compute this probability using the binomial probability distribution, as in Section 7.5.2, would take a long time. Instead, we will use the following result:

Normal approximation to the binomial If X is a binomial random variable based on n trials and success probability p, and the number of trials n is *large*, then the distribution of X can be approximated by a normal distribution with mean $= np$ and standard deviation $= \sqrt{np(1 - p)}$. ***Note***: This approximation works best when both np and $n(1 - p)$ are at least 5.

In our example, we have a mean $np = (400)(0.50) = 200$ and a standard deviation $\sqrt{np(1 - p)} = \sqrt{400(0.50)(1 - 0.50)} = \sqrt{100} = 10$. So we will do roughly the same calculations as we have done in the previous chapter for normal distributions to find our approximate probability.

Using the TI, $P(X \le 192) \approx \text{normalcdf}(-\text{E99}, 192, 200, 10) = 0.2119$.

Using Table II, we would first compute the standard score for 192, namely, $z = \frac{192 - 200}{10} = -0.8$. We use Table II to find the area to the left of -0.80. Recall that Table II provides areas to the left of various values of Z. So $P(X \le 192) \approx P(Z \le -0.8) = 0.2119$.

Thus, it is not very unusual to observe only 192 out of 400 employees who are happy and plan to stay, even if 50% of the employees are happy and plan to stay.   ❖

The family of normal distributions is not the only class of models for continuous random variables. Our final "Let's do it!" exercise returns us to the uniform density curve as the probability distribution for the continuous random variable $X =$ time to process a loan application.

Let's do it! 7.28 Applying for a Loan

Suppose that the time to process a loan application follows a uniform distribution over the range of 10 to 20 days.

(a) Sketch the probability distribution for $X =$ time to process a loan application, where X is $U(10, 20)$.

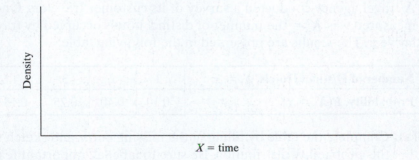

(b) What is the mean or expected processing time?

(c) Based on the distribution, what is the probability that a randomly selected loan application takes longer than two weeks to process?

(d) *Given* that the processing time for a randomly selected loan application is at least 12 days, what is the probability that it will actually take longer than two weeks to process?

 Exercises

7.47 The following table gives information regarding the distribution of the random variable X = number of visits to the hospital by adults aged 60–65 years in Flint, Michigan during the year 2000:

Number of Visits, $X = x$	0	1	2	3	4
Probability, $P(X = x)$	0.1	0.1	0.2	0.4	

(a) Complete the table by filling in the missing entry.

(b) What is the probability that a randomly selected adult will make exactly one hospital visits?

(c) What is the probability that a randomly selected adult will make fewer than three hospital visits?

(d) Given that an adult has fewer than three hospital visits, what is the probability that he or she will have only one hospital visit?

7.48 A travel agency conducted a survey of its customer travelers. One of the responses measured was X = the number of distinct hotels occupied by travelers in a particular year. The results are presented in the following table:

Number of Distinct Hotels, $X = x$	1	2	3	4	5
Probability, $P(X = x)$	0.10	0.40	0.25	0.15	

(a) Complete the table by filling in the missing entry, and sketch the stick graph for this probability distribution. Be sure to label all important features.

(b) What is the probability that a randomly selected traveler will use exactly five distinct hotels?

(c) What is the probability that a randomly selected traveler will use fewer than three distinct hotels?

(d) Given that a traveler uses fewer than three distinct hotels, what is the probability that he or she will use only one hotel?

7.49 *With* versus *without* **Replacement**

An urn contains three balls, of which one is teal and two are white. The experiment consists of selecting one ball at a time *without* replacement until the teal ball is chosen, at which point no further selections are made.

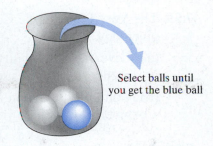

Select balls until you get the blue ball

(a) Write down the sample space for this experiment.

(b) Let the random variable X be the number of the draw on which the teal ball is selected—that is, X = the total number of balls selected. What are the possible values for this random variable X?

(c) Repeat (a) and (b) assuming that the ball is selected *with* replacement.

7.50 A fair coin is flipped five times and the number of heads is recorded. Let X be the number of heads. The probabilities of the possible outcomes are given.

Value of X						
$X = x$	0	1	2	3	4	5
Probability $P(X = x)$	$\frac{1}{32}$	$\frac{5}{32}$		$\frac{10}{32}$	$\frac{5}{32}$	$\frac{1}{32}$

(a) What probability value is missing in the table?

(b) What is the probability of getting exactly four heads, $P(X = 4)$?

(c) Suppose that you are told that the outcome resulted in an even number of heads. What is the probability now that $X = 4$?

(d) Suppose that the experiment is repeated twice (that is, the coin is flipped five times, and then another five times). What is the probability that $X = 4$ for both experiments?

7.51 Let X be the number of televisions in a randomly selected Zone City household. The mass function of X is given in the corresponding table.

Value of X					
$X = x$	0	1	2	3	4
Probability $P(X = x)$		0.05	0.30	0.40	0.20

(a) Complete the mass function of X and sketch the stick graph of the mass function of X. Be sure to label all important features.

(b) What is the probability that a randomly selected household has at least one TV?

(c) What is the expected number of televisions per household?

7.52 An automobile agency located in Beverly Hills, California, specializes in the rental of luxury automobiles. The distribution of daily demand (number of automobiles rented) at this agency is summarized. What is the expected value of daily demand (i.e., the expected number of automobiles rented)?

Value of X					
$X = x$	0	1	2	3	4
Probability $P(X = x)$	0.15	0.30	0.40	0.10	0.05

7.53 The Very Lite Bicycle Company is equally likely to sell 0, 1, 2, or 3 bicycles on any given day.

   (a) Give the probability distribution for the discrete random variable X = the number of bicycles sold in a given day.

   (b) Compute the mean and standard deviation of the number of bicycles sold in a given day. Include your units and a brief interpretation of these values.

7.54 Alfredo, a visitor to a casino, wishes to participate at a game of chance. He has two options.

   Option 1 A fair coin is tossed once. The player gets $4 if it falls heads and nothing otherwise. The entry fee for this option is $2.

Outcome	H	T
Net Return	$2	−$2
Probability	0.5	0.5

Expected Net Return = 0.

   Option 2 A fair coin is tossed twice. The player gets $8 if it falls heads on both tosses and nothing otherwise. The entry fee for this option is $3.

Using the expected net return as a criterion, which of the two options will be more attractive to Alfredo? Show all supporting details.

7.55 **An Interesting Case**

A random variable X takes on the values 1 and −1 with probability $\frac{2}{3}$ and $\frac{1}{3}$, respectively. Let $Y = X^2$.

   (a) What is the expected value for Y?

   (b) What is the standard deviation for Y?

7.56 The accompanying table shows the probability distribution for the random variable X = the number of pairs of shoes owned for a particular population.

Value of X					
$X = x$	2	3	4	5	6
Probability					
$P(X = x)$	0.1	0.1	0.4	0.3	0.1

   (a) The expected value of X (that is, the expected number of pairs of shoes owned) is (select one)

     (i) 4.

     (ii) 4.2.

     (iii) 5.

     (iv) cannot be determined.

   (b) Find $E(X^2)$.

   (c) Find the standard deviation of X.

7.57 Suppose that the number of cars, X, that pass through a car wash between 4:00 P.M. and 5:00 P.M. on any sunny Friday has the following probability distribution:

Value of X					
$X = x$	4	5	6	7	8	9
Probability					
$P(X = x)$	$\frac{1}{12}$	$\frac{1}{12}$	$\frac{1}{4}$	$\frac{1}{4}$	$\frac{1}{6}$	

(a) Complete the distribution.

(b) What is the probability that at least six cars will pass through the car wash between 4:00 P.M. and 5:00 P.M. on any sunny Friday?

(c) What is $E(X)$, the expected number of cars that pass through a car wash between 4:00 P.M. and 5:00 P.M. on any sunny Friday?

(d) In Chapter 5, we discussed linear transformations. We saw that the mean of a new variable Y, which is a linear transformation of the variable X, is found by plugging the mean of X directly into the linear transformation. So, if $Y = 2X - 1$ represents the amount of money, in dollars, paid to the attendant by the manager for each car, what is the attendant's expected earnings for this particular period?

7.58 Compute each of the following combinations and explain what the answer represents in a well-written sentence. (a), (b), and (c) should require no calculations; just think about what is being asked.

(a) $\binom{8}{0}$ (b) $\binom{8}{1}$ (c) $\binom{8}{8}$ (d) $\binom{20}{3}$ (e) $\binom{12}{2}$

7.59 **Combinations of $n = 5$.** Suppose that you have a set that contains five distinguishable values, $S = \{1, 2, 3, 4, 5\}$. The following table partially summarizes the counting questions:

How many subsets of $S = \{1, 2, 3, 4, 5\}$ are there that contain …	Answer	The subsets are	Combination
… exactly zero values (no values)?	1		
… exactly one value?		$\{1\}, \{2\}, \{3\}, \{4\}, \{5\}$	
… exactly two values?			$\binom{5}{2} = 10$
… exactly three values?	10		
… exactly four values?			
… exactly five values?		$\{1, 2, 3, 4, 5\}$	

(a) Complete the table by filling in the missing entries.

(b) The set $S = \{1, 2, 3, 4, 5\}$ has five values, so the total number of possible subsets is $2^5 = $ _____. Confirm that this value equals the sum in the "answer" column.

7.60 Suppose that the random variable X has a binomial distribution with $n = 10$ and $p = 0.4$. Find each of the following quantities:

(a) $P(X = 2)$.

(b) $P(X > 8)$.

(c) $P(X \leq 7)$.

(d) $P(X = 1.5)$.

(e) $E(X) = \mu$.

(f) $Var(X) = \sigma^2$.

7.61 In each given case, a random variable, X, is defined. You are to decide whether or not X has a binomial distribution. Justify your answer by referring to specific assumptions that are or are not satisfied.

(a) Six days a week for a year, you play the three-digit ("Pick Three") Michigan lottery. You win on a given day if your three-digit number matches the one selected at random by the State of Michigan. For the first six months, you select the number 123 each day; for the last six months, you select the number 234 each day. Let X be the number of days in the year that your choice is a winner.

(b) Studies have found that 80% of dorm residents get along with their roommates. Each room in a certain dorm has two occupants. A survey was conducted by selecting 10 rooms at random and asking each occupant "Do you get along with your roommate?" Let X be the number of "yes's" among the 20 responses obtained.

7.62 An experiment is designed to test whether a subject has ESP, extrasensory perception. A total of 96 cards is drawn, one by one, with replacement, from a well-shuffled ordinary deck of cards. The subject is asked to guess the suit of each card drawn. There are four possible choices for each card—namely, spades, hearts, diamonds, and clubs. We wish to test the null hypothesis that the subject is only guessing and does not have ESP against the one-sided alternative hypothesis that the subject has ESP.

(a) State the appropriate hypotheses in terms of $p =$ the probability of getting a correct response.

(b) The subject gets 35 correct out of the 96 cards. Give the p-value for this test. Think about the direction of extreme, based on your alternative hypothesis. Use your TI and the binomcdf(function to find the p-value.

(c) Give your decision using a significance level of $\alpha = 0.05$.

(d) State your conclusion using a well-written sentence.

Note: In Chapter 9, you will revisit this ESP scenario in a "Let's do it" exercise. You will apply another test to assess the significance of the results.

7.63 A tax attorney claims that *more than* 1% of all tax returns filed in the previous year were audited. To test this claim, a researcher took a random sample of 18 taxpayers and found that only one return was audited.

(a) Give the appropriate null and alternative hypotheses about the value of p, the proportion of audited tax returns.

(b) Find the p-value of the test, based on the researcher's sample results. Think about the direction of extreme, based on your alternative hypothesis.

(c) Carefully explain, in words, what the *p*-value means in this case, in terms of repeated samples.

(d) If you were to set a significance level 1%, would you reject or accept the null hypothesis? Explain.

7.64 In a computer lab, students must wait to get a computer terminal. It is assumed that the waiting time for the next available terminal is uniformly distributed between 0 and 15 minutes.

(a) Sketch the distribution of *X* (the waiting time, in minutes). Be sure to provide all important features.

(b) On average a student will have to wait _____ minutes for the next available terminal.

(c) What is the probability that a student will have to wait more than 10 minutes for the next available terminal?

(d) Students have been complaining that the waiting time is too long. So a test will be conducted to assess if the mean waiting time is greater than the average computed in (b).

    (i) What is the direction of extreme?

    (ii) Suppose that a randomly selected student was observed to wait 13 minutes for the next available terminal. Give the corresponding *p*-value for this observation.

    (iii) Using a 10% significance level, what is your decision? Explain.

7.65 No one likes to wait at a bus stop for very long. A city bus line states that the length of time a bus is late is uniformly distributed between 0 and 10 minutes.

(a) Sketch the distribution for the variable *X* = time late (in minutes). Be sure to provide all important features.

(b) For the distribution in (a), what is the mean or expected time a bus will be late—that is, what is $E(X)$?

(c) For the distribution in (a), what is the probability that a bus will arrive no more than three minutes late?

(d) The director has been receiving a number of complaints from its riders regarding how late buses have been running. He obtains the help of a statistician to test the alternative theory that the mean time a bus will be late is greater than that given in (b).

    (i) What is the direction of extreme?

    (ii) Suppose that, at a randomly selected bus stop, the bus arrives nine minutes late. What is the corresponding *p*-value for this observation?

    (iii) Using a 15% significance level, what is your decision?

REAL DATA

7.66 Even drugstore managers have their own headaches, due to chronic late and incomplete payments from insurers and other medical personnel. As reported in the March 25, 1995 *New York Times*, "Pharmacy Fund, a new company based in New York, thinks it has a cure. It buys bills from the previous day's prescription sales at a slight discount, paying the pharmacist immediately and eliminating weeks of cash flow delays for drugstores. The Pharmacy Fund then collects from the health plans." The article also reports that drugstores typically had to wait between 30 to 45 days to collect payments from insurers. Suppose that the waiting time, in days, to collect payment for non–Pharmacy Fund drugstores is uniformly distributed between 30 and 45 days.

(a) Sketch the distribution for the random variable X = waiting time in days to collect payment for non–Pharmacy Fund drugstores. Be sure to label and give appropriate values on the axes.

(b) What is the expected waiting time to collect payment for non–Pharmacy Fund drugstores?

(c) What is the probability that a non–Pharmacy Fund drugstore will have to wait at least six weeks?

(d) What is the probability that a non–Pharmacy Fund drugstore will have to wait at least six weeks, given that it has had to wait at least five weeks for payment?

(e) Based on your results to (c) and (d), are the events "wait at least six weeks" and "wait at least five weeks" independent events? Explain.

7.67 Based on the results of past races, the time to complete a 5000-meter track race is approximately normally distributed with a mean of 24 minutes and a standard deviation of 5 minutes.

(a) Participants who complete the race in under 18 minutes will receive a medal. What is the probability that a participant will receive a medal?

(b) If there are 200 participants, how many medals would you recommend the race committee have available? Explain.

7.68 Suppose that the time for an operator to place a long-distance call is normally distributed with a mean of 20 seconds and a standard deviation of 5 seconds.

(a) What is the probability that a long-distance call will go through in less than 10 seconds?

(b) What is the probability that it will take longer than 32 seconds for a long-distance call to go through?

7.69 The SAT scores of applicants to a certain university are normally distributed with a mean of 1170 and a standard deviation of 80. Let X represent the score of a randomly selected applicant.

(a) Explain in words what $P[1050 < X < 1250]$ means and compute this probability.

(b) Applicants whose SAT scores are in the upper 2.5% qualify for a scholarship.

(i) What percentile must an applicant's SAT score attain in order to qualify for a scholarship?

(ii) What SAT score must an applicant attain in order to qualify for a scholarship?

Chapter Summary

We sample from the population. Thus, our conclusions or inferences about the population will contain some amount of uncertainty and the evaluation of probabilities. In this chapter, we focused on the relative frequency interpretation of probability. We defined the probability of an outcome as the proportion of times it would occur over the long run. We discovered that probabilities can be found through simulation or through more formal mathematical results. The following is summary of basic rules for probability:

Basic Probability Rules

---

1. The probability that the event A occurs is denoted by $P(A)$. For any event A, $0 \leq P(A) \leq 1$.

2. The sum of the probabilities of each of the individual outcomes (for finite or infinitely countable S) in the sample space must equal one. $P(S) = 1$.

3. **Complement rule**. The probability that an event occurs is 1 minus the probability that the event does not occur. $P(A) = 1 - P(A^C)$.

4. **Addition rule**. The probability that either the event A or the event B occurs is the sum of their individual probabilities minus the probability of their intersection. $P(A \text{ or } B) = P(A) + P(B) - P(A \text{ and } B)$. If A and B are mutually exclusive (or disjoint) events, then $P(A \text{ or } B) = P(A) + P(B)$.

5. The **conditional probability** of the event A occurring, given the event B has occurred, is given by

$$P(A|B) = \frac{P(A \text{ and } B)}{P(B)}, \text{ if } P(B) > 0.$$

This leads to the following **multiplication rule**:

$P(\text{both events will occur at the same time}) = P(A \text{ and } B)$

$= P(B)P(A|B) = P(A)P(B|A).$

If two events do not influence each other—that is, if knowing that one has occurred does not change the probability of the other occurring—the events are independent. Two events A and B are **independent** if $P(A|B) = P(A)$ or, equivalently, $P(B|A) = P(B)$ or, equivalently, $P(A \text{ and } B) = P(A)P(B)$.

6. **Partition rule**. If the events *B1*, *B2*, ..., *BI* form a partition, then

$$P(A) = P(A|B1)P(B1) + P(A|B2)P(B2) + \cdots + P(A|BI)P(BI).$$

7. **Bayes's rule**. Let the events *B1*, *B2*, ..., *BI* form a partition. Suppose that *A* is another event and we know the conditional probabilities $P(A|B1)$, $P(A|B2)$, ...$P(A|BI)$. Then the conditional probability of say *B1* given that event *A* has occurred is given by

$$P(B1|A) = \frac{P(A|B1)P(B1)}{P(A)}$$

$$= \frac{P(A|B1)P(B1)}{P(A|B1)P(B1) + P(A|B2)P(B2) + \cdots + P(A|BI)P(BI)}.$$

Note: A tree diagram is often useful for computing partition rule or Bayes's rule probabilities.

Probability is an important tool in the decision-making process. Probability helps us quantitatively evaluate competing theories, to be able to assess whether the observed data are unlikely or probable under a given hypothesis (that is, to compute the *p*-value). To make a decision, we will use the sample data to compute sample statistics, such as the sample mean or sample proportion. The value for a sample statistic will vary in a random manner from sample to sample. A sample statistic is a *random variable* that has a distribution.

A sample statistic is a **random variable** that has a probability distribution. In this chapter, we studied two broad types of random variables—**discrete** and **continuous**. We learned how to characterize the pattern of the distribution of the values of a random variable via a **probability distribution**. We used the probability distributions for finding various probabilities and expectations. Within the two broad types of random variables, there are families of random variables that are common and have a special name. We looked at an important family of discrete random variables called the binomial random variables. A **binomial random variable** is basically a count of how many times an outcome occurs in a certain number of trials of a random experiment. In Section 7.5.3, we revisited two families of continuous probability distributions—the **normal** distributions and the **uniform** distributions.

In the next chapter, we will see the important role of the normal distributions in our study of the probability distributions of various sample statistics. Chapter 8, on sampling distributions, is our final topic to prepare us for the more formal statistical decision-making procedures presented in Chapters 9 through 15.

Key Terms

Be sure you can describe, in your own words, and give an example of each of the following key terms from this chapter:

probability	disjoint (or mutually	random variable (r.v.)
relative frequency	exclusive)	probability mass function
random process	equally likely	probability density function
simulation	complement rule	expected value (or mean)
sample space	addition rule	of a r.v.
(or outcome set)	conditional probability	standard deviation of a r.v.
event	multiplication rule	stick graph
Venn diagram	independence	probability bar graph
union	partition	binomial random variable
intersection	partition rule	Bernoulli trial combinations
complement	Bayes's rule	

Exercises

7.70 Consider the process of playing a game in which the probability of winning is 0.15 and the probability of losing is 0.85.

   (a) If you were to use your calculator or a random number table to simulate this game, what numbers would you generate and how would you assign values to simulate winning and losing?

   (b) With your calculator (using a seed value of 32) or the random number table (Row 4, Column 1), simulate playing this game 10 times. Show the numbers generated, and indicate which ones correspond to wins and which ones correspond to losses.

   (c) From the simulation, calculate an estimate of the probability of winning. How does it compare to the true probability?

7.71 In the March 19, 1995, edition of *The New York Times*, an article entitled "Struggling to Find Stability in Families Where Divorce Has Become a Way of Life" contains findings of a nationwide study. People were classified according to family setting—namely, either "No Divorce" or "Divorce" (if they had one or more divorces). An experiment consists of randomly selecting three people and recording their family setting. Give the sample space for the set of possible outcomes of this experiment.

7.72 The ELISA test is used to screen blood samples for antibodies to the HIV virus. It measures an "absorbency ratio." Blood samples from HIV-infected donors usually, but not always, give high ratios, while those from uninfected donors usually, but not always, give low ratios. Consider using this test on a sample of blood. The null and alternative hypotheses are defined as

   H_0: The blood sample is not infected with HIV.

   H_1: The blood sample is infected with HIV.

Suppose that the decision rule is as follows: If the observed absorbency ratio is greater than three, the blood sample is judged to be infected with the HIV virus (that is, the test is positive); otherwise, it is judged to be free of the HIV virus (that is, the test is negative). For this rule, it is known that 98% of HIV patients will test positive and 2% will test negative, while 7% of uninfected donors will test positive and 93% will test negative.

(a) What do Type I and Type II errors represent in this situation?

(b) What are the values of α and β for this procedure?

(c) Suppose that the decision rule was changed so that a higher absorbency ratio is required to judge a blood sample as testing positive.

   (i) The value of α would   increase.   decrease.   remain the same.

   (ii) The value of β would   increase.   decrease.   remain the same.

7.73 In a certain community, 12% of the families own an Alsatian dog, a type of German shepherd, 10% of the families own a Siamese cat, and 17% of the families own either an Alsatian dog or a Siamese cat. A family is to be randomly chosen. A Venn diagram may help in answering the following questions:

(a) What is the probability that the family owns both an Alsatian dog and a Siamese cat?

(b) Given that the family does not own a Siamese cat, what is the probability that it owns an Alsatian dog?

7.74 Consider the following game: Players select two digits from 0 through 9, inclusive and then sum the values of the two digits. A winning sum is determined by randomly selecting two digits and computing the sum. For example, if the two randomly selected digits are 3 and 5, then a winning ticket is any ticket in which the two digits sum to 8. Note that there are 100 choices for selecting two digits $(00, 01, \ldots, 98, 99)$. Assume that each of the 100 choices is equally likely to be selected for determining the winning sum.

(a) How many possible sums are there? Give the values for all possible sums.

(b) Find the probability of winning if the winning sum is 2.

(c) Find the probability of winning if the winning sum is 5.

(d) Which possible sum would result in the largest probability of winning?

7.75 **Matching Numbers.**

(a) Consider Game 1. The game host selects a number from 1 to 10 $(1, 2, 3, 4, 5, 6, 7, 8, 9, 10)$. You must also select a number from this same list of 1 to 10.

   You win if and only if you select the same number as the game host. What is the probability that you win?

(b) Consider Game 2. The game host selects a number from 1 to 10 $(1, 2, 3, 4, 5, 6, 7, 8, 9, 10)$. You and four of your friends must each select a number from this same list (with replacement).

   You win if and only if at least one among the five selects the same number as the game host. What is the probability that you win?

7.76 A gambler plays a sequence of games in which, for each game, he either wins (W) or loses (L). He stops playing as soon as he either wins two games or loses two games.

The outcomes of the games are independent and the probability that he wins any game is $\frac{1}{2}$.

(a) Give the sample space of all possible sequences of wins or losses until play ends.

(b) Assign each outcome in (a) its correct probability.

7.77 A group of 100 voters were classified in two ways: party affiliation and attitude toward a certain environmental proposal. The results are shown in the following table.

		Attitude Favor	Indifferent	Opposed
Affiliation	Democrat	27	15	18
	Republican	13	10	17

Suppose that one member of the group is to be selected at random.

    A = {Chosen member is a Democrat}.
    B = {Chosen member is Indifferent}.
    C = {Chosen member is Opposed}.

(a) What is the probability that the selected member will be a Democrat?

(b) What is the probability that the selected member will be oppose the proposal?

(c) What is the probability that the selected member will be a Democrat who opposes the proposal?

(d) Given that the selected member is a Democrat, what is the probability the member will be opposed to the proposal?

(e) What is the probability that the selected member will not favor the proposal?

(f) Are the events "member is a Democrat" and "member is opposed" independent? Explain.

7.78 Two draws are made at random without replacement from a box containing the following five tickets: What is the probability that the sum of the two tickets is 5?

7.79 The following table provides information regarding health status and smoking status of residents of a small community (health status for individuals was measured by the number of visits to the hospital during the year):

	Smoking Status = Smoker	Smoking Status = Nonsmoker	Total
Health Status = 0 visits	20	100	120
Health Status = 1–4 visits	70	90	160
Health Status = 5 or more visits	90	30	120
Total	180	220	400

(a) What is the probability that a randomly selected resident made zero visits to the hospital?

(b) What is the probability that a randomly selected nonsmoker made zero visits to the hospital?

(c) Are "nonsmoker" and "zero visits to the hospital" mutually exclusive (i.e., disjoint)? Show support for your answer.

(d) Are "nonsmoker" and "zero visits to the hospital" independent? Show support for your answer.

7.80 Do men and women treat male and female children differently? A study was conducted near the primate exhibit at the Sacramento Zoo during a particular weekend. The data are from 49 groups of three: one adult female, one adult male, and a toddler, in which the toddler was being carried (couples with more than one toddler were excluded from the study). Recorded in the following table is information about which adult (male or female) was carrying the toddler and the gender of the toddler:

		Gender of toddler	
		Boy (B)	*Girl (G)*
Gender of adult	*Male (M)*	15	20
carrying the toddler	*Female (F)*	6	8

The researcher will use these data to assess whether there is a relationship between the gender of adult and the gender of the toddler.

(a) Is this an observational study or an experiment?

(b) Based on these data, are the events "male adult carrying the toddler" and "girl toddler" mutually exclusive? Give support for your answer.

(c) Based on these data, are the events "male adult carrying the toddler" and "girl toddler" independent? Give support for your answer.

7.81 Attending College

Shown are data on 80 families in a midwestern town.

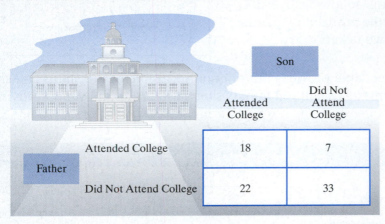

The data give the record of college attendance by fathers and their oldest sons.

(a) Are the events "father attended college" and "son attended college" independent events? Give supporting evidence.

(b) Are the events "father attended college" and "son attended college" disjoint? Give supporting evidence.

7.82 A group of 60 elderly patients visiting a certain clinic on an outpatient basis were categorized as being either "compliers" or "noncompliers" according to whether or not they followed their doctor's orders in taking medication. Each of them was also classified by religious affiliation. The following table summarizes the responses:

		Religious Affiliation Catholic	Religious Affiliation Protestant
Compliance	*Complier*	12	15
Status	*Noncomplier*	8	25

(a) What is the probability that a randomly selected elderly patient is Catholic?

(b) Are the events "Catholic" and "compliers" independent? Explain your answer.

(c) Give an example of two events that are disjoint (or mutually exclusive).

7.83 Suppose that we toss a fair coin twice. Which of the following pairs of events are independent?

(a) "Head on the first toss" and "Head on the second toss."

(b) "Head on the first toss" and "Tail on the first toss."

(c) "Head on the first toss" and "At least one head."

(d) "At least one head" and "At least one tail."

7.84 As part of pharmaceutical testing for headaches as a side effect of a drug, 300 patients were randomly assigned to one of two groups—200 patients to the group that received the drug and the other 100 patients to the group that received a placebo.

(a) Use your TI calculator (with a value of seed = 23), or Row 12, Column 1 of the random number table, and list (in the order selected) the first 10 patients selected to be in the group taking the drug.

The number of patients in each group who had a headache within the first hour after receiving the treatment was recorded. Results are recorded here:

		Treatment Group Drug	Treatment Group Placebo
Headache	*Headache*	44	12
Status	*No Headache*	156	88

(b) Suppose that a patient is selected at random. What is the probability that the patient did have a headache in the first hour after treatment?

(c) Suppose that a patient is selected at random. Given that the patient was given the drug, what is the probability that the patient did have a headache in the first hour after treatment?

(d) Are the events "drug" and "headache" mutually exclusive? Explain your answer.

(e) Are the events "drug" and "headache" independent events? Explain your answer.

7.85 Viviana applies to graduate school at two universities. Assuming that the probability of acceptance at University A is 0.4, the probability of acceptance at University B is 0.3, and the probability of acceptance at both universities is 0.1, answer the following questions:

(a) Is the event of acceptance at University A independent of the event of acceptance at University B? Explain.

(b) Is the event of acceptance at University A mutually exclusive of the event of acceptance at University B? Explain.

(c) What is the probability that Viviana is not accepted to either university?

7.86 Varivax, a vaccine for chicken pox, was approved by the FDA. One injection of the vaccine is recommended for children 12 months to 12 years old. Two injections, given four to eight weeks apart, are recommended for those 13 or older who have not had chicken pox. It was reported that 2.8% of all vaccinated children still develop chicken pox. However, such cases are generally mild, with an average of 50 lesions compared to an average of 300 lesions that develop from a natural chicken-pox infection.

(a) Suppose that we take a simple random sample of two children who have not yet had chicken pox, but have been vaccinated. What is the probability that both children still develop chicken pox?

(b) The McCoy family has two children, 2 and 4 years of age, both of whom received the vaccine since they had not yet had chicken pox. Does the probability calculated in (a) apply to these two children? Explain why or why not.

7.87 Suppose that 40% of students at a large university don't like statistics. If two students at this university are selected independently, what is the probability that at least one likes statistics?

7.88 Suppose that, in a certain country, 10% of the elderly people have diabetes. It is also known that 30% of the elderly people are living below poverty level, 35% of the elderly population fall into *at least one* of these categories, and 5% of the elderly population fall into *both* of these categories.

(a) Given that a randomly selected elderly person is living below the poverty level, what is the probability that she or he has diabetes?

(b) Are the events "has diabetes" and "living below the poverty level" mutually exclusive events in this elderly population? Explain.

(c) Are the events "has diabetes" and "living below the poverty level" independent events in this elderly population? Explain.

7.89 In a factory, there are three machines producing 50%, 30%, and 20%, respectively, of the total output. The following summarizes the percentages of defective items produced by the three machines:

- First machine 4%.
- Second machine 6%.
- Third machine 2%.

An item is drawn at random from the production line. What is the probability that it is defective?

7.90 A large shipment of batteries is accepted upon delivery if an inspection of three randomly selected batteries results in zero defectives.

(a) What is the probability that this shipment is accepted if 2% of the batteries in the shipment are actually defective?

(b) What is the probability that this shipment is accepted if 20% of the batteries in the shipment are actually defective?

7.91 Determine whether the following statements are true or false (a true statement is always true):

(a) Two events that are disjoint are also always independent.

(b) On two tosses of a fair coin, the probability of getting two heads in a row is the same as the probability of getting first a head and then a tail.

(c) The expected value of a random variable is always one of the possible values for the random variable.

(d) The expected value for the random variable X with the distribution shown is equal to 0.

Value of X $X = x$	-1	0	1
Probability $P(X = x)$	$\frac{1}{4}$	$\frac{1}{4}$	$\frac{1}{2}$

7.92 An investigator developed a screening test for cancer. He used this screening test on known cancer patients and known noncancer patients. He found that 5% of the noncancer patients had positive test results and that 20% of the cancer patients had negative test results. He is now going to apply this test to a population in which about 2% have undetected cancer.

(a) The sensitivity of a test is the probability of a positive test result given that the individual tested actually has the disease. What is the sensitivity of this screening test in the study?

(b) The specificity of a test is the probability of a negative test result given that the individual tested does not have the disease. What is the specificity of this screening test in the study?

(c) Given that a person selected at random from this population has tested positive, what is the probability that he or she will actually have cancer?

(d) Given that a person selected at random from this population has tested negative, what is the probability that he or she will not have cancer?

7.93 Customers at Pizza Palace order pizzas with different numbers of toppings. Let the random variable X be the number of toppings on the pizza ordered for a randomly selected customer.

Value of X $X = x$	0	1	2	3	4
Probability $P(X = x)$	0.15	0.30	0.35	0.15	

(a) Fill in the missing value in the preceding probability mass function.

(b) What is the probability that a customer orders at least one topping?

(c) Given that a customer ordered more than two toppings, what is the probability that a customer ordered four toppings?

7.94 The National Weather Service has the following model for the random variable X = the number of hurricanes that hit North Carolina in a year:

Value of X $X = x$	0	1	2	3	4	5
Probability $P(X = x)$	0.30	0.35	0.20	0.10	0.04	

(a) What is the probability that there will be more than three hurricanes in a year?

(b) What is the mean or expected number of hurricanes per year?

7.95 At a large university, the number of student problems handled by the dean for student affairs varies from semester to semester. The following table gives the probability distribution for the number of student problems handled by the dean for one semester:

(a) What is the probability that the dean handles no more than three student problems in a given semester?

Value of X $X = x$	0	1	2	3	4	5
Probability $P(X = x)$	0.10	0.15	0.30	0.25	0.10	0.10

(b) Find the mean of the distribution of the number of student problems.

7.96 A box contains two blue chips and six red chips. The experiment consists of selecting three chips at random from the box.

(a) Find the probability that all three chips will be the same color if the chips are selected randomly with replacement.

(b) Find the probability that all three chips will be the same color if the chips are selected randomly without replacement.

(c) Let the random variable X denote the number of blue chips selected when the sampling is conducted with replacement. Give the probability distribution for the random variable X by completing the following table:

Number of blue chips, $X = x$	0	1	2	3
Probability, $P(X = x)$				

(d) Let the random variable Y denote the number of blue chips selected when the sampling is conducted without replacement. Give the probability distribution for the random variable Y by completing the following table:

Number of blue chips, $Y = y$	0	1	2	3
Probability, $P(Y = y)$				

7.97 Suppose that the time a student spends studying for an exam (rounded to the nearest hour) is distributed as follows:

Value of X $X = x$	3	4	5	6	7	8	9	10
Probability $P(X = x)$	0.2	0.4	0.2	0.1	0.05	0.03	0.01	0.01

(a) Represent this distribution graphically (i.e., sketch the mass function).

(b) What is the mode of this distribution?

(c) What is the expected studying time?

(d) Is the median of this distribution (select one) equal to, greater than, or less than the mean? Explain.

(e) What is the probability that a student has spent at least four hours, but no more than six hours, studying?

(f) Given that a student has spent five hours or less studying, what is the probability that he or she spent more than three hours studying?

(g) The preceding distribution was based on data collected by the instructors. Mary, one of the students, thinks that students have a tendency to claim they studied more than they actually did in order to please the instructors. What possible type of bias is Mary thinking about?

7.98 One-fifth of a particular breed of rabbits carries a gene for long hair. Suppose that 10 of these rabbits are born.

(a) What is the expected number of long-haired rabbits?

(b) What is the probability that the number of long-haired rabbits will be more than three?

7.99 The number of typing errors per page for a certain typing pool has the distribution shown.

Value of X $X = x$	0	1	2	3	4	5	6
Probability $P(X = x)$	0.01	0.09	0.30	0.20	0.20	0.10	0.10

(a) If a page is selected at random, what is the probability that it has at least three typing errors?

(b) Find the expected number of typing errors on a page.

7.100 There are 10 washing machines, 3 of which are defective. Let the machines be represented as M_1, M_2, M_3, M_4, M_5, M_6, M_7, M_8, M_9, M_{10}, with M_1, M_2, M_3 denoting the defective machines.

(a) How many ways could you select 4 machines from the set of 10 machines?

(b) Among the selections of 4 machines out of the 10, how many would correspond to including all three defective machines? Explain your answer and list the possible subsets.

7.101 In the Midwest population of blood donors, 40% have blood type O. If four people are selected at random to be donors, what is the probability that at least one has blood type O?

7.102 A manufacturer of panel displays claims that a new manufacturing process has a *higher* success rate as compared to the old process. The success rate for the old process was 30%.

(a) State the appropriate null and alternative hypotheses about the value of p, that is, the proportion of successes with the new process.

(b) A recent sample of 10 panels made with the new process yielded 8 that worked (success) and 2 that did not. What is the corresponding *p*-value for the test in (a)? Think about the direction of extreme, based on your alternative hypothesis.

(c) Does it appear that the new process is better? Use a 1% significance level. Explain your answer.

7.103 The labeling on the package for a new treatment for warts states that 80% of all warts are successfully removed.

(a) A total of $n = 10$ subjects will use the new treatment on one wart. What is the probability that the treatment will successfully remove at least 90% of the treated warts? (*Hint*: 90% of 10 = 9 successes.)

(b) A total of $n = 100$ subjects will use the new treatment on one wart. What is the probability that the treatment will successfully remove at least 90% of the treated warts?

7.104 Assume that cholesterol levels for adults in the United States are normally distributed with a mean of 190 and a standard deviation of 20.

(a) What is the probability that a randomly selected adult will have a cholesterol level exceeding 240?

(b) What is the probability that a randomly selected adult will have a cholesterol level below 200?

(c) Let Y be the random variable representing the cholesterol level for a randomly selected adult. Consider the probability $P(Y > 240|Y > 200)$. Explain what this probability is asking for in your own words and compute the probability.

7.105 Each box of wine bottles shipped by the Grape Valley Wine Company contains two bottles of wine. The amount of wine dispensed into each bottle varies normally with a mean of 600 ml and a standard deviation of 16 ml. Assume that wine is dispensed into the bottles independently of one another.

(a) What is the probability that a randomly selected bottle will contain at least 592 ml of wine?

(b) For a box of wine bottles, what is the probability that both wine bottles will contain at least 592 ml of wine?

(c) Wine bottles are filled using a production process that is monitored over time. Bottles within a box were often filled consecutively (one right after the other) in the production line. If the production process happens to be running off its mark such that the bottles are slightly underfilled, then, if one bottle is underfilled, it is quite likely that the next bottle will also be underfilled. If this is the case, what does it imply about the assumption stated above and your answer to (b)?

7.106 The 8:00 A.M. bus that stops at the Union corner is always late. The amount of time, X, in minutes by which it's late is a continuous random variable that is uniformly distributed over the interval $[0, 5]$ minutes.

(a) Sketch the density of X. Be sure to label the axes and give some values on the axes.

(b) If a person arrives at 8:02 A.M. (that is, he or she is two minutes late), what is the probability that the person will have missed the bus?

TI Quick Steps

More on Simulation

In Chapter 2, we learned how to generate random integers between 1 and *N*. We first specified a starting point by setting the seed value. A **TI graphing calculator** can generate a list of random numbers between 1 and 50, starting with a **seed value of 33**, using the following keystroke steps:

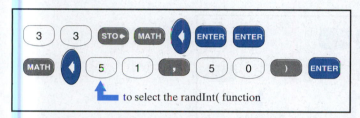

to select the randInt(function

Continue to push the `ENTER` button and your screen should look like

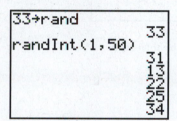

If you need to simulate your random process a large number of times, you need to generate a lot of random integers (that is, push the ENTER button many times). The randInt(function has a third entry that allows you to specify the number of values to be generated. To generate a list of five random integers between 1 and 2, starting with a **seed value of 33**, the keystroke steps are as follows:

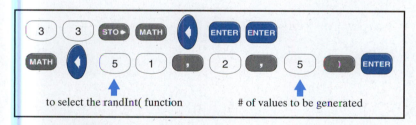

to select the randInt(function   # of values to be generated

Pushing the `ENTER` button a total of four times provides you with a total of 20 repetitions of a random process with two equally likely outcomes.

Your screen should
look like

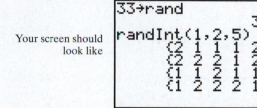

Suppose that you wish to simulate a random process for which the outcome of interest has a probability 0.30 of occurring. You could simulate random integers between 1 and 10 and assign three of the possible integers to represent the outcome occurring. A quicker way to perform this simulation with a TI is to use the randBin(function, which generates random numbers from specified binomial distributions. Binomial distributions are a family of discrete probability distributions used to model the number of successes out of *n* trials with a specified probability for a success *p*. Let 0 denote that the outcome did not occur and 1 indicate that the outcome did occur. Suppose that the probability of the outcome occurring is 0.30. To generate four repetitions of this random process, starting with a **seed value of 33**, the keystroke steps are as follows:

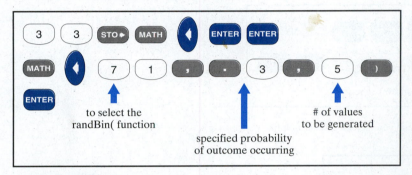

Pushing the **ENTER** button a total of four times provides you with a total of 20 repetitions of the random process.

Your screen should look like

```
33→rand
                33
randBin(1,0.3,5)
       {0 1 0 0 0}
       {0 0 0 1 0}
       {1 0 0 0 1}
       {0 0 0 0 1}
```

Chapter 8

SAMPLING DISTRIBUTIONS: MEASURING THE ACCURACY OF SAMPLE RESULTS

8.1   Introduction

In Chapters 1 through 3, we discussed various aspects of the decision-making process—formulating theories, taking samples and producing data, measuring the "likeness" of the information in the sample, and assessing the possibility of errors. In Chapters 4 through 6, we learned how to describe a sample through various numerical and graphical summaries and through the use of a statistical model that summarizes the distribution of a response of interest. In Chapter 7, we addressed the concept of chance, randomness, and likeliness, more formally called probability.

> Formulate theories → Collect data → Summarize results → Interpret results & make decision
>
> YOU ARE HERE

In this chapter, we enter the final step in our cycle—using the information from our sample to generalize and make decisions about the whole population. As we saw in Chapter 2, the value of a parameter is fixed and generally unknown, while the value of a statistic, although known once it is computed for one given sample, may vary from sample to sample. It would be nice if we could somehow measure the accuracy of this process, if we could say that the value of the statistic we will compute is expected to be very close to the corresponding parameter value. This is where the sampling technique used to select the sample, together with the sample size, play a role. We will see that a sample statistic from a simple random sample has a predictable pattern of variation that allows us to make statements about how close a sample statistic is expected to be to the true parameter, on average. To put these ideas at a more practical level, consider the example that follows.

❖ EXAMPLE 8.1 **Preparing for Employment** _____

REAL DATA

At some point in our lives, most of us face the task of finding employment—finding a job. Many pursue an education in order to aid in finding a job. Have the skill levels that are demanded in the workplace been on the increase over the past few years?

According to a nationwide Census Bureau survey of 3000 employers with more than 20 workers, including offices, factories, and construction sites, the answer is "Yes." Fifty-seven percent of the employers said the skill requirements of their workplaces had increased in the last three years. (SOURCE: "National Survey Shows a Rift Between Schools and Business," *New York Times*, February 20, 1995.)

The value of 57%, or the proportion of 0.57, is a sample statistic, since it was computed from a sample of 3000 employers. Imagine this process being repeated many times—the process being to take a sample of 3000 such employers and record the proportion of employers who say the skill requirements of their workplaces had increased in the last three years. Repeating this process many times would give us many proportions, not all equal to 0.57, with some values perhaps occurring more often than other values. What values for a sample proportion would be possible? Would a proportion of zero be possible? How about a proportion of one? Do you think any of the resulting proportions could be more than 0.75?

Based on a sample of 3000 employers, how good is this proportion at estimating the true proportion of employers with increasing skill levels? Do you know if this proportion of 0.57 is close to the true proportion? Could it be very different from the true proportion? If this value of 0.57 may not be the true proportion, why do you think these results were reported and conclusions drawn from them? ❖

To try to address some of the questions raised in the previous example, we might consider actually repeating the estimation process again and again and observing the variability of the values of the sample statistic computed from the many samples. We might try to describe this set of values of the statistic through various numerical and graphical summaries and through the use of a statistical model that summarizes the *distribution* of the possible values of the statistic. This distribution is called the **sampling distribution of the statistic**. With such a model in hand, we would have some idea of the accuracy of this estimation process and how well it predicts the value of a population parameter.

> *Definition:* The **sampling distribution of a statistic** is the distribution of the values of the statistic in all possible samples of the same size *n* taken from the same population.

The objective of this chapter is to study the sampling distribution of two statistics, the sample proportion and the sample mean. We will examine how these statistics vary in repeated *simple random sampling*. The randomness in the sampling procedure produces the sampling distributions presented in this chapter. We will examine the sampling distribution of a statistic for various sample sizes and for various values of the parameter. Once we have the information regarding the distribution of a sample statistic, we can use the results from one sample to make a good guess about the population parameter. We begin with the sampling distribution of a sample proportion.

8.2 Sampling Distribution of a Sample Proportion

We are interested in the true proportion of women in our population. So our population parameter of interest is the proportion of women in our population, denoted by

$$p = \frac{\text{number of women in the population}}{\text{population size}}.$$

Since the population of interest is too large and it is too costly to observe all of the elements in the population, we elect to estimate the true population value by taking a sample. The sample proportion is defined as

$$\hat{p} = \frac{\text{number of women in the sample}}{\text{sample size}}, \text{ and is read as "} p \text{ hat."}$$

Suppose for a moment that we know 50% of our population are women, although in general, we would not know this. What will happen if we take a simple random sample of 20 people from this population? Would we always have 10 women and 10 men? From our sampling exercises in Chapter 2 and our probability discussions in Chapter 7, we know that we will not.

The following are three different simple random samples of 20 people taken from a population with 50% women:

Sample 1 M W M M M M W W M W M W M M M W W M W W, proportion of women = $\hat{p} = \frac{9}{20} = 0.45$.
Sample 2 W W M W M W W M M M M W M M W W W M W W, proportion of women = $\hat{p} = \frac{11}{20} = 0.55$.
Sample 3 M M W W W M M W M W M W M W W M M M W W, proportion of women = $\hat{p} = \frac{10}{20} = 0.50$.

In the first sample, we had nine women and the sample proportion of women was $\hat{p} = \frac{9}{20} = 0.45$. This is a sample statistic. Doing this again, we obtained a different sample and a different value of $\hat{p}$. In fact, each of the preceding samples gave a different answer and the sample proportion did not always match the truth about the population. In the long run, we would expect 50% of our sample to be women, but a sample of 20 people does not constitute a long run.

In practice, researchers only take one sample. Although we cannot determine whether or not the results of that one sample are close to the truth about the population, sampling distributions help us quantify what kind of accuracy to expect for such samples.

The following activities will teach us about the **sampling distribution of a sample proportion** and **its usefulness**. First, we need to clarify that most of the activities in this chapter will involve simulating the estimation process and examining *many* possible samples, *not actually examining all* possible samples of the same size from the same population. The resulting distribution from these simulations is referred to as the **empirical sampling distribution**. These simulations will help us to understand better the theoretical results that are presented later.

❖ EXAMPLE 8.2   Proportion of Women _____

We will assume that 50% of the population are women (that is, the value of p is 0.50). We are going to take a simple random sample of size $n = 4$ people from this population and observe the proportion of women in the sample.

As we discussed in Chapter 7, there are many ways to simulate the outcome of a random experiment. We could flip four fair coins to simulate this sampling procedure. The outcome of the flip of a coin would correspond to the response of gender, with perhaps a head representing a woman and a tail representing a man. We could use a table of random digits, with the even digits of 0, 2, 4, 6, 8 representing a woman and the odd digits representing a man. We would select a starting place in the table and read off four successive digits to represent a simple random sample of size $n = 4$.

We could use the random number generating capabilities of a computer or calculator to simulate the gender of selected individuals from this population. All we need is two distinct values to represent the two possible gender outcomes. A common coding scheme for two outcomes is to use the values 0 and 1.

Let the value 0 represent a man and the value 1 represent a woman.

We would first select a starting point and then use our computer or calculator to generate a "random" list of 0's and 1's, where each value is equally likely to occur. Details on how to generate such a sequence with TI graphing calculators are presented in the ■ TI Quick Steps that follow the exercises at the end of this chapter.

Using a TI graphing calculator with a seed value of 91, the first four values generated by pressing the ENTER key four times are

$$1 \quad 1 \quad 1 \quad 1.$$

How many women did you get? **4.** What is the sample proportion of women? $\hat{p} = \frac{4}{4} = \mathbf{\mathit{1.00}}$.

We continue pressing the ENTER button, in sets of four times, to simulate a total of 50 trials or repetitions of this sampling procedure. For example, the next five trials result in the following:

0 1 0 1   for a total of 2 women, and a sample proportion of $\hat{p} = 0.50$.

1 0 0 0   for a total of 1 woman, and a sample proportion of $\hat{p} = 0.25$.

1 0 0 0   for a total of 1 woman, and a sample proportion of $\hat{p} = 0.25$.

1 0 0 1   for a total of 2 women, and a sample proportion of $\hat{p} = 0.50$.

0 1 0 0   for a total of 1 woman, and a sample proportion of $\hat{p} = 0.25$.

The results of the 50 trials have been tallied and summarized in the following table:

Number of Women	Sample Proportion	Tally	Frequency	Proportion of All Trials
0	0.00	||||	4	4/50 = 0.08
1	0.25	⟍⟍⟍ ⟍⟍⟍ ⟍⟍⟍ |	16	16/50 = 0.32
2	0.50	⟍⟍⟍ ⟍⟍⟍ ⟍⟍⟍ |	16	16/50 = 0.32
3	0.75	⟍⟍⟍ ⟍⟍⟍	10	10/50 = 0.20
4	1.00	||||	4	4/50 = 0.08
Total	———	—————	50	50/50 = 1.00

Based on this simulation,

(a) What was the most likely proportion of women in a sample? *either 0.25 or 0.50*

(b) What percentage of the time did we get

0 women, for a sample proportion of $\hat{p} = 0.00$? *8%*

1 woman, for a sample proportion of $\hat{p} = 0.25$? *32%*

2 women, for a sample proportion of $\hat{p} = 0.50$? *32%*

3 women, for a sample proportion of $\hat{p} = 0.75$? *20%*

4 women, for a sample proportion of $\hat{p} = 1.00$? *8%*

(c) Comment on the overall appearance of this empirical sampling distribution. *The tallies in the preceding table give us a quick picture of this empirical sampling distribution. We have that the sample proportions did vary from 0 to 1, while most are clustered and centered at about 0.50. In fact, 84% of the estimates were 0.25, 0.50, or 0.75, while only 16% of the estimates were as bad as the extremes of 0.00 and 1.00. The shape of the distribution is unimodal, centered at about 0.50, and mound shaped.* ❖

 Let's do it! 8.1 Proportion of Women

Now it is your turn. Work with a partner. Have one person select a starting point or seed value and generate the random list of 0's and 1's, reading off a set of four at a time. The other person can tally the results in the accompanying table. Continue until you have simulated a total of 50 trials of this estimation process.

Seed value or starting point = _____

Number of Women	Sample Proportion $\hat{p}$	Tally	Frequency	Proportion of All Trials
0	0.00			
1	0.25			
2	0.50			
3	0.75			
4	1.00			
Total	———	—————	50	50/50 = 1.00

(a) What was the most likely proportion of women in a sample? _____

(b) What percentage of the time did you get

        0 women, for a sample proportion of $\hat{p}$ = 0.00? _____

        1 woman, for a sample proportion of $\hat{p}$ = 0.25? _____

        2 women, for a sample proportion of $\hat{p}$ = 0.50? _____

        3 women, for a sample proportion of $\hat{p}$ = 0.75? _____

        4 women, for a sample proportion of $\hat{p}$ = 1.00? _____

(c) Comment on the overall appearance of your empirical sampling distribution.

(d) Were your results exactly equal to the previous 50 simulated trials? In what ways were they similar?

Combine your frequencies with those of one or more groups in your class and graph the frequency histogram on the axes given.

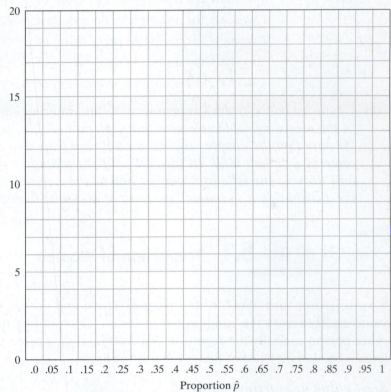

Proportion $\hat{p}$

 THINK ABOUT IT   **A Larger Sample Size**

If we randomly select a sample of four people from a population with 50% women, it is quite likely to have one woman (25%) in the sample, and it is possible to get all women in the sample. Suppose that you increase the sample size to 20 people.
Would you be surprised if only 5 (25%) of the 20 selected individuals were women?

Would you be surprised of all of the 20 selected individuals were women?

Let's do it! 8.2   Proportion of Women—Larger Sample Size

What do you think would happen if, instead of taking a simple random sample of size $n = 4$, we selected $n = 20$ people at random? We will still assume that 50% of the population are women. Repeat the steps in "Let's do it! 8.1," but take a simple random sample of size $n = 20$ people and observe the number of women in the sample. As before, we will let $1 = $ woman and $0 = $ man.

Select a starting point or seed value for generating a very long list of 0's and 1's. One person will generate a set of 20 values. Notice that with the 0 – 1 coding scheme, the sum of the 20 generated values will be the number of women in the sample of size 20. Tally this response in the following table and repeat for a total of 50 trials. You do not need to select a new seed each time. Just continue to read off or generate sequences of 0's and 1's.

A quicker way to perform this simulation with the TI is to use the randBin(function, which generates random numbers from specified binomial distributions. Binomial distributions (discussed in Chapter 7) are a family of discrete probability distributions used to model the number of successes out of n trials with a specified probability for a success p. To simulate five random samples each of size $n = 20$ from a population with proportion $p = 0.50$ of successes, starting with the seed value of 33, and report the total number of successes in each sample, the keystroke steps are as follows:

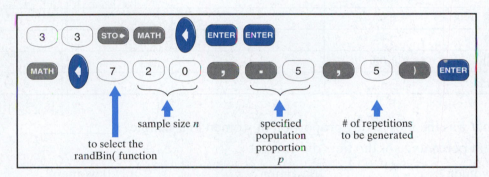

Your screen should look like this:

```
33→rand
              33
randBin(20,.5,5)
        {10 9 10 10 5}
```

So the first five random samples are quickly generated and the resulting sample proportions are

$$\hat{p} = \frac{10}{20} = 0.50, \; \hat{p} = \frac{9}{20} = 0.45, \; \hat{p} = \frac{10}{20} = 0.50, \; \hat{p} = \frac{10}{20} = 0.50, \text{ and } \hat{p} = \frac{5}{20} = 0.25.$$

Seed value or starting point = _____

Number of Women	Sample Proportion $\hat{p}$	Tally	Frequency	Proportion of All Trials
0	0.00			
1	0.05			
2	0.10			
3	0.15			
4	0.20			
5	0.25			
6	0.30			
7	0.35			
8	0.40			
9	0.45			
10	0.50			
11	0.55			
12	0.60			
13	0.65			
14	0.70			
15	0.75			
16	0.80			
17	0.85			
18	0.90			
19	0.95			
20	1.00			
Total	_____	_____	50	1.00

(a) What was the most likely proportion of women in a sample? _____

(b) What percentage of the time did you get

0 women? _____   10 women? _____   all 20 women? _____

(c) Comment on the overall appearance of this empirical sampling distribution. How does it compare to the empirical sampling distribution when $n = 4$, in terms of center, variability, and shape?

(d) Combine your frequencies with those of one or more groups in your class and graph the frequency histogram on the axes given. Compare this picture with the $n = 4$ picture (from "Let's do it! 8.1") and comment.

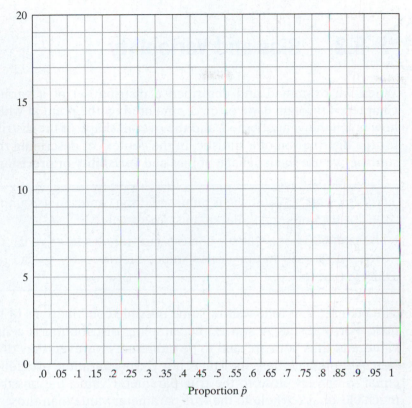

Proportion $\hat{p}$

Let's stop a moment and summarize the key ideas that we have seen in previous "Let's do it!" exercises concerning the sample proportion $\hat{p}$.

> ### What Do We Expect of Sample Proportions?
>
> - The **values** of the sample proportion $\hat{p}$ **vary** from random sample to random sample in a predictable way.
> - The **shape** of the distribution of $\hat{p}$ values is approximately symmetric and **bell-shaped**.
> - The **center** of the distribution of $\hat{p}$ values is at the **true proportion p**.
> - With a **larger sample size n**, the $\hat{p}$ values tend to be closer to the true proportion p. That is, the $\hat{p}$ **values vary less** around the true proportion p.

In these exercises, we knew something that in general we would not know—we knew the value of the true population proportion p to be 0.50. But even if we did not know the value of p, these properties of the sampling distribution for $\hat{p}$ would still hold. Although we can never guarantee that a particular sample proportion $\hat{p}$ value computed from a given simple random sample is close to the population proportion, we can be confident in using the sample proportion $\hat{p}$ as an estimate for p, because, most of the time, a simple random sample will give a value of $\hat{p}$ that is close to p, especially if the sample size is large. In the next section, we learn how to describe the accuracy of a statistic for estimating a parameter in terms of its bias and variability. We will discover and present the approximate model that describes the sampling distribution of a sample proportion. This model provides the frame of reference for making decisions about a population proportion based on a sample, as we shall see in Chapter 9.

8.3   Bias and Variability

When we examined and described a distribution of a variable, such as height back in Chapters 4 through 6, we often commented on the point at which the values were centered, how much the values varied, and the overall shape of the distribution. We do the same here for the distribution of a statistic. In the context of describing the sampling distribution of a statistic, we use the language of **bias** and **variability** or **precision**.

> *Definitions:*   A statistic is **unbiased** if the center of its sampling distribution is equal to the corresponding population parameter value.
>
> The **variability** of a statistic corresponds to the spread of its sampling distribution. A statistic whose distribution shows values that are very spread out and dispersed is said to **lack precision**.

The pictures that follow show some possible combinations of bias and variability for a statistic. For Statistics A and B, the sampling distribution is centered at the true parameter value, and thus both are said to be **unbiased**. However, they do differ with respect to precision. Although neither Statistic A nor B can guarantee providing an estimate that is exactly equal to or very close to the true parameter value, the possible values for Statistic A are much closer as a whole to the true parameter value than those for Statistic B.

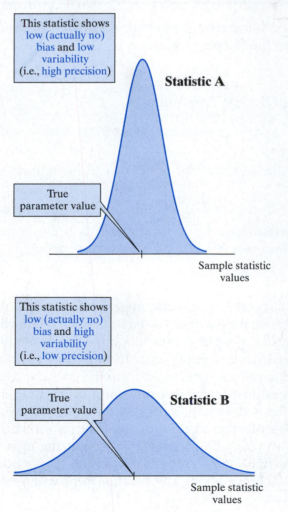

For Statistics C and D, the sampling distribution is not centered at the true parameter value. Statistic C consistently underestimates the true parameter value. On the other hand, Statistic D has a tendency to overestimate the true parameter value. Both Statistics C and D are **biased** estimators of the parameter. Although Statistic C is biased, it does have high precision; Statistic D suffers from both bias and low precision in terms of estimating the parameter.

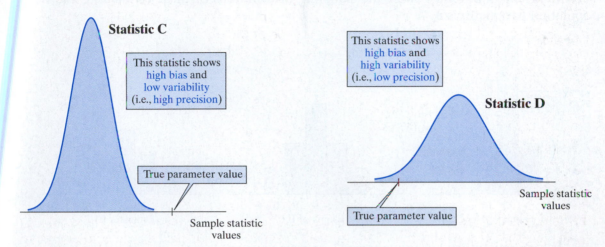

Statistic C

This statistic shows high bias and low variability (i.e., high precision)

True parameter value

Sample statistic values

This statistic shows high bias and high variability (i.e., low precision)

Statistic D

Sample statistic values

True parameter value

Ideally, we would like an estimator with low bias and low variability. One method we have seen for reducing the variability is to take a larger sample size. There is still one more aspect of variability of a sample statistic to be addressed.

The **variability of the sampling distribution does not depend on the size of the population**, as long as the size of the population is much larger than the sample size (say 100 times as large). This feature is illustrated in the next "Think About it" example.

THINK ABOUT IT 🐢 Taking Fabric Samples

Suppose that you have found a fabric with some print in a repeating pattern and you wish to get a swatch (a piece) of it to bring home to show to your roommate. How large of a piece do you need? If the size of the piece—the sample size—is too small, your roommate will not be able to "see" the pattern. You need to select a piece—a sample— just large enough to contain the pattern in the fabric. Once you have decided on how large the sample piece needs to be, it does not matter if you take that piece from a large bolt of the fabric, a large population, or a smaller bolt of the fabric, a small population. This is shown in the following picture:

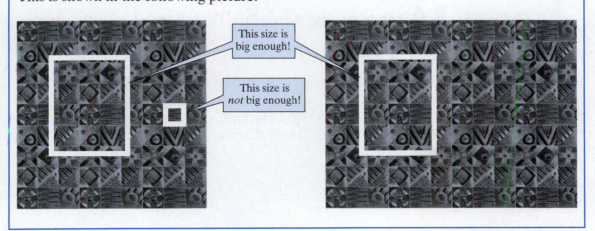

This size is big enough!

This size is *not* big enough!

Let's do it! 8.3 Three Estimators

The histograms that follow show the sampling distributions of three estimators. The true population parameter is 8.

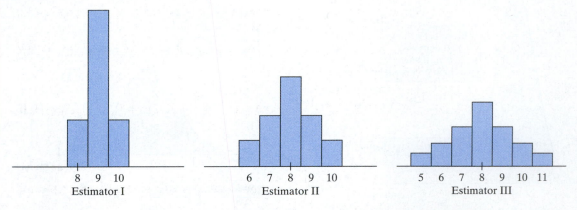

(a) Which estimator(s) is/are unbiased?   Circle your answer(s).   I   II   III
Explain.

(b) Is Estimator III more precise than Estimator I?   Circle one.   YES   NO
Explain.

We have been studying the process of estimating p, the true proportion of items in a population that have a certain characteristic, by taking a simple random sample from the population and computing the proportion in the sample with the characteristic of interest, the sample statistic $\hat{p}$. The value of $\hat{p}$ depends on the sample selected. We have discussed that the sample size plays a role in the precision of the statistic. Does the value of the true proportion p also influence the sampling distribution? In the next exercise, we will generate the empirical sampling distribution of $\hat{p}$ under various scenarios, compare the results, and make some generalizations. Generating sampling distributions can be a tedious task. However, if each person in the class participates and does a small portion, we can get a fairly accurate picture of the results we could expect in general.

Let's do it! 8.4 More on the Sampling Distribution of a Proportion

Step 1:   Divide the class up into five groups of approximately equal size.

Step 2:   Assign one of the following scenarios to each group (each group receives one transparency grid and one transparency marker, a different color for each group, for recording the results):

Scenario I   Assume the population proportion $p = 0.1$, the sample size $n = 50$, and the marker color = black.

Scenario II   Assume the population proportion $p = 0.3$, the sample size $n = 50$, and the marker color = purple.

Scenario III   Assume the population proportion $p = 0.7$, the sample size $n = 50$, and the marker color = blue.

Scenario IV   Assume the population proportion $p = 0.5$, the sample size $n = 50$, and the marker color = green.

Scenario V   Assume the population proportion $p = 0.5$, the sample size $n = 100$, and the marker color = red.

Step 3:   Discuss the following questions within each group:

(i)   What do you think will be the shape of the sampling distribution of $\hat{p}$? Sketch it.

(ii) What do you think will be the center—that is, the average or expected value—of the sampling distribution of $\hat{p}$?

(iii) What do you think will be the spread of the sampling distribution of $\hat{p}$? Give a possible range of values.

Step 4:   Each person in each group is to simulate 10 simple random samples of size n from a population with proportion p, based on the n and p corresponding to the scenario assigned to their group.

How Will You Do Step 4?

With the Random Number Table, the previous "Let's do it!" exercises have shown you how to simulate when the population proportion $p = 0.5$. When p is some other value, you need to modify which of the digits will correspond to a *yes* response. For example, if your $p = 0.1$ and your $n = 50$, then you select one digit from 0 to 9 to represent a *yes* response, say the digit 0, and the remaining digits would represent a *no* response. Using the random number table starting at Row 5, Column 1, the first 50 digits are as follows:

3 7 5 7 **0** 3 9 9 7 5 8 1 8 3 7 1 6 6 5 6 **0** 6 1 2 1
9 1 7 8 2 6 **0** 4 6 8 8 1 3 **0** 5 4 9 6 8 4 6 **0** 6 7 2.

Since the value of 0 occurred five times, the generated sample proportion is $\hat{p} = \frac{5}{50} = 0.1$. (A pretty good result, but you will not always get 5 out of 50.)

With the TI, the steps will be demonstrated for Scenario I but can be modified for the other scenarios. A quick way to perform this simulation is to use the randBin(function, which generates the random numbers from specified binomial distributions. Binomial distributions (discussed in Chapter 7) are a family of discrete probability distributions used to model the number of successes out of *n* trials with a specified probability for a success *p*. To simulate ten random samples each of size $n = 50$ from a population with proportion $p = 0.10$ of successes, starting with the seed value of 33, and report the proportion of successes in each sample, the keystroke steps are as follows:

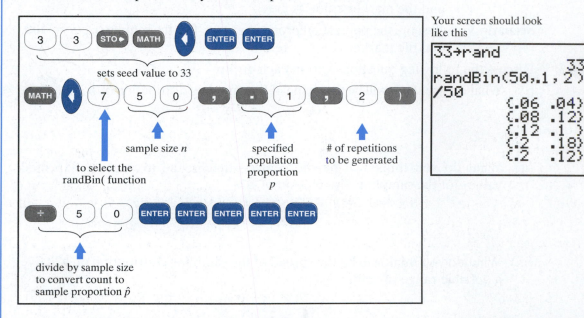

So the first ten random samples are quickly generated and the resulting sample proportions are: $\hat{p} = 0.06$, $\hat{p} = 0.04$, $\hat{p} = 0.08$, $\hat{p} = 0.12$, $\hat{p} = 0.12$, $\hat{p} = 0.10$, $\hat{p} = 0.20$, $\hat{p} = 0.18$, $\hat{p} = 0.20$, and $\hat{p} = 0.12$.

Note: Step 4 could be assigned at the end of class by asking students to bring their results back to their group for the beginning of the next class. The number of repetitions could also be adjusted depending on the size of the class.

Your 10 Results:

Result #	1	2	3	4	5	6	7	8	9	10
The sample proportion of yes responses $\hat{p}$:										

Step 5: Combine the results within each group. Each person in the group records the 10 responses by shading in a square for each response above the appropriate value on the transparency grid. Copies of all five grids follow. Be sure to use the color pen that is assigned to your group scenario. (A TI shortcut for performing Steps 4 and 5 all at once is given next.)

How did the results compare to the expected results stated in Step 3? Summarize the results for your group. Comment on the

(i) shape of the sampling distribution of $\hat{p}$.

(ii) center, average, or expected value of the sampling distribution of $\hat{p}$.

(iii) spread of the sampling distribution of $\hat{p}$.

TI Shortcut

Just as the TI steps shown in Step 4 were used to generate ten simple random samples of size 50 to find the corresponding values of $\hat{p}$, you can have the calculator generate *many* simple random samples, compute the corresponding values of $\hat{p}$, and store them all in a list for the making of a histogram. The steps for generating 200 simple random samples each of size $n = 50$ from a population with proportion $p = 0.10$ of successes, converting each to the value of the sample proportion $\hat{p}$ by dividing by $n = 50$, and then storing all of the results into the list L_1 are shown next. An arbitrary seed value of 66 was used. Note that it will take even the TI a few minutes to generate all of these values, so be patient.

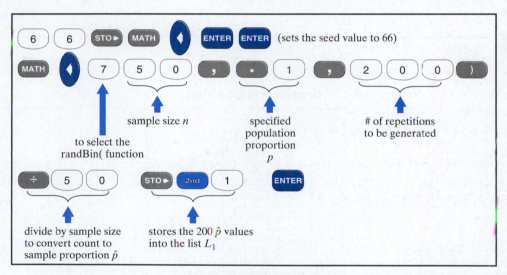

Your screen should look like this

```
66→rand
                 66
randBin(50,.1,20
0)/50→L₁
{.1 .06 .16 .06 ...
```

You can then repeat these steps with a new seed value for one of the other scenarios (keep the number of repetitions of 200 the same) and store them into a different list, say L_2. You can then set up the STAT PLOTS options so PLOT 1 is a histogram of the data in L_1 and PLOT 2 is a histogram of the data in L_2, and so on. Remember from Chapter 4 that, for histograms, you need to first specify the WINDOW settings and then push GRAPH (not ZOOM 9). The WINDOW settings that will work best for these histograms are shown here.

```
Xmin = −.05
Xmax = 1.05
Xscl = .05
Ymin = −11
Ymax = 200
Yscl = 0
Xres = 1
```

Step 6:   Each group presents the combined results to the entire class. Since each group used a different color pen to make their histogram, overlaying the transparencies is helpful to compare the distributions. Compare the results of the sampling distribution of $\hat{p}$ for the five scenarios. Make overall comments about the

(i)   shape of the sampling distribution of $\hat{p}$.

(ii)   center, average, or expected value of the sampling distribution of $\hat{p}$.

(iii)   spread of the sampling distribution of $\hat{p}$.

   (a)   Look at scenarios I, II, III, and IV. Does the value of p appear to influence the spread?

   (b)   Look at scenarios IV and V. Does the sample size n appear to influence the spread?

   (c)   Look at scenarios I and II. How does the distribution for $p = 0.3$ compare to that for $p = 0.7$?

Group I: $n = 50$ and $p = 0.1$

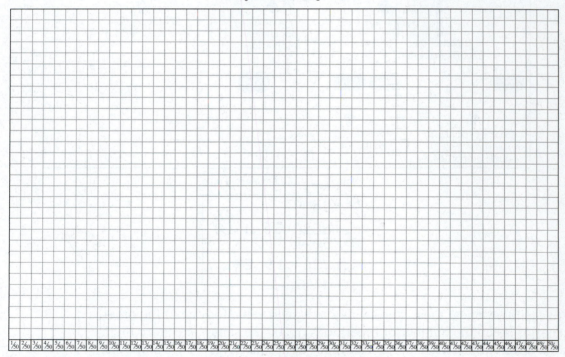

$\hat{p}$ = Sample proportion = count/50

Group II: *n* = 50 and *p* = 0.3

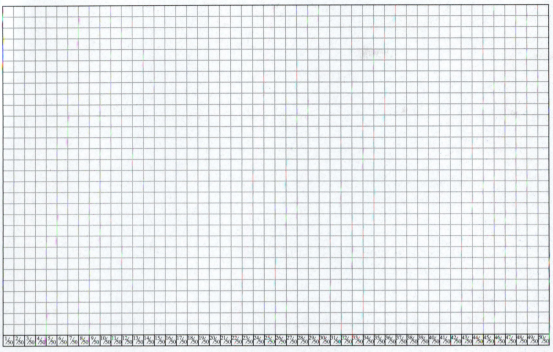

$\hat{p}$ = Sample proportion = count/50

Group III: *n* = 50 and *p* = 0.7

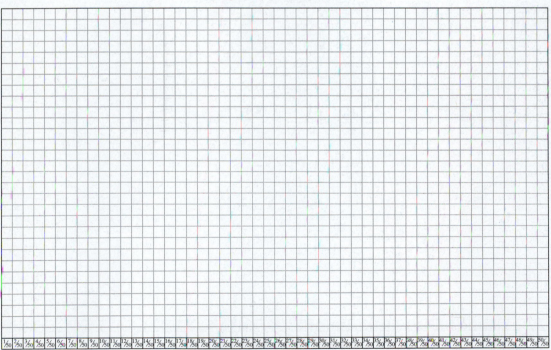

$\hat{p}$ = Sample proportion = count/50

Group IV: $n = 50$ and $p = 0.5$

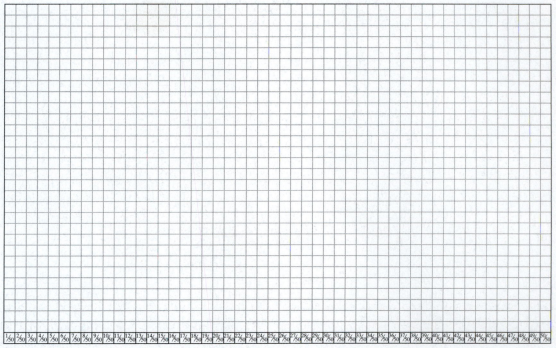

$\hat{p}$ = Sample proportion = count/50

Group V: $n = 100$ and $p = 0.5$

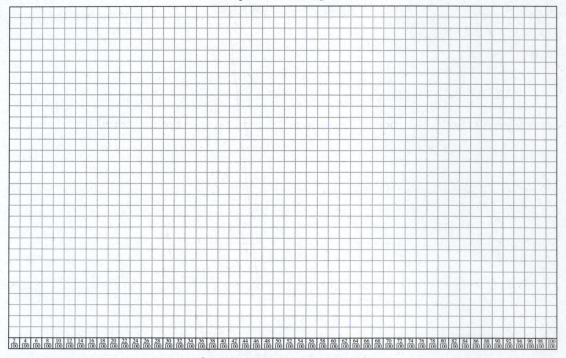

$\hat{p}$ = Sample proportion = count/100

Let's stop again and update our summary about the key ideas that we have seen concerning the sample proportion $\hat{p}$.

What Do We Expect of Sample Proportions?

- The **values** of the sample proportion $\hat{p}$ **vary** from random sample to random sample in a predictable way.

- **When the sample size n is large**, the sample proportion can take on many possible values in the range of 0 to 1, so the random variable $\hat{p}$ can be viewed as a **continuous** random variable with a **density curve** as its model.

- **When the sample size n is large**, the distribution of $\hat{p}$ can be modeled *approximately* with a **normal distribution**.

- The **center** of the distribution of the $\hat{p}$ values is at the **true proportion p** (for *any* sample size n and *any* value of p).

- With a **larger sample size n**, the $\hat{p}$ values tend to be closer to the true proportion p. That is, the $\hat{p}$ **values vary less** around the true proportion p. The variation also depends somewhat on the value of the true proportion p.

So far we have discussed the concept of variability for a sampling distribution somewhat loosely. We have visually compared two sampling distributions and commented on which statistic has less variability (that is, higher precision). We could, of course, actually summarize the spread of a sampling distribution with a number, one such number being the standard deviation of the sampling distribution, $\sigma_{\hat{p}}$. How do we find the standard deviation? We could compute the standard deviation for the generated $\hat{p}$ values for each of the scenarios in "Let's do it! 8.4" (which is left to an exercise). However, these would be approximations for only those particular scenarios. Using mathematics, it can be shown that, if a simple random sample of size n is selected from a large population with true proportion p, then the standard deviation for $\hat{p}$ is equal to

$$\sqrt{\frac{p(1 - p)}{n}},$$

which depends on the true population proportion p and the sample size n.

- If the sample size $n = 50$ and the population proportion $p = 0.50$, then the standard deviation of $\hat{p}$ is

$$\sqrt{\frac{p(1 - p)}{n}} = \sqrt{\frac{0.5(1 - 0.5)}{50}} \approx 0.07.$$

- If the sample size $n = 100$ and the population proportion $p = 0.50$, then the standard deviation of $\hat{p}$ is somewhat smaller:

$$\sqrt{\frac{p(1 - p)}{n}} = \sqrt{\frac{0.5(1 - 0.5)}{100}} = 0.05.$$

THINK ABOUT IT **The Standard Deviation for $\hat{p}$**

If the sample size is increased, how does this affect the standard deviation for $\hat{p}$?

If the true proportion moves closer to 0 or 1, how does this affect the standard deviation for $\hat{p}$?

For a fixed sample size, how does the standard deviation when $p = 0.3$ compare to the standard deviation when $p = 0.7$?

In general, for a fixed sample size, for what value of p will the standard deviation be maximized?

We have studied and simulated the sampling distribution of a sample proportion for a number of scenarios. We have seen some common results, such as the normal shape and centering at the population proportion p. Let us now state the main results more formally.

Sampling Distribution of $\hat{p}$, the Sample Proportion

Let p represent the proportion of elements in a large population having some characteristic—that is, the proportion of "successes," where "success" corresponds to having that characteristic. If simple random samples of size n are taken from a population in which the proportion of "successes" is p, then the sampling distribution of $\hat{p}$ has the following properties:

1. $\mu_{\hat{p}} = p$ The average of all the possible $\hat{p}$ values is equal to the parameter p. In other words, the center of the distribution of $\hat{p}$ is at the true proportion p, so $\hat{p}$ is an **unbiased** estimator of p.

2. $\sigma_{\hat{p}} = \sqrt{\dfrac{p(1 - p)}{n}}$ The **standard deviation** for $\hat{p}$ **decreases as the sample size n increases**.

                                   For a fixed sample size, the maximum standard deviation is attained at $p = 0.5$ (check this out).

3. If n is **"sufficiently" large**, the distribution of $\hat{p}$ eventually looks like a **normal distribution** with mean and standard deviation as given in properties 1 and 2.

$$\hat{p} \text{ is approximately } N\left(p, \sqrt{\dfrac{p(1 - p)}{n}}\right) \text{ for large sample size } n.$$

The necessary sample size n depends on the value of the population proportion. It must be large enough that $np \geq 5$ and $n(1 - p) \geq 5$. This guideline implies that we should expect to see at least five successes and at least five failures in a sample of size n.

So, now that we know the sampling distribution of $\hat{p}$, let's put it to use. We will use the approximate normal model for $\hat{p}$ to **compute probabilities** about the sample proportion $\hat{p}$ taking on certain values. In general, the value of the population proportion p is not known. We take a (large enough) random sample from the population and compute the sample proportion $\hat{p}$. We can use the sample proportion $\hat{p}$ to estimate the value of the population proportion p or to test a theory about the value of p. The foundation for the **estimation** and the **testing** about the population proportion p is the preceding sampling distribution for $\hat{p}$.

❖ **EXAMPLE 8.3** **Probabilities about the Proportion of Voters in Favor**

Suppose that, of all the voters in a state, 30% are in favor of Proposal A. As the voting day gets closer, many polls are taken. Suppose that a random sample of $n = 400$ voters will be obtained and the proportion $\hat{p}$ of sampled voters in favor of the proposal will be computed. This sample proportion $\hat{p}$ is a random variable and (with the large sample size) will have approximately a **normal** distribution with a **mean** of $p = \mathbf{0.30}$ and a **standard deviation** of $\sqrt{\frac{0.30(1 - 0.30)}{400}} = \sqrt{0.000525} = \mathbf{0.023}$. We can use the normal distribution to compute probabilities of seeing certain values for $\hat{p}$. It may be helpful to recall the steps for computing areas under a normal curve from Chapter 6.

(a) What is the probability that **less than 25%** of the voters sampled will be in favor of the proposal? In probability notation, we want to find $P(\hat{p} < 0.25)$.

Using the TI, $P(\hat{p} < 0.25) = \text{normalcdf}(-\text{E99}, 0.25, 0.30, 0.023) = 0.0149$.

Using Table II, we would first compute the standard score for 0.25, namely, $z = \frac{0.25 - 0.30}{0.023} = -2.17$. We use Table II to find the area to the left of -2.17. Recall that Table II provides areas to the left of various values of Z. So $P(\hat{p} < 0.25) = P(Z < -2.17) = 0.0150$.

(b) What is the probability that **between 25% and 35%** of the voters sampled will be in favor of the proposal—that is, the probability of getting a sample proportion within 5% of the true proportion? In probability notation, we want to find $P(0.25 < \hat{p} < 0.35)$.

Using the TI, $P(0.25 < \hat{p} < 0.35) = \text{normalcdf}(0.25, 0.35, 0.30, 0.023) = 0.9703$.

Using Table II, we already have the standard score for 0.25, namely -2.17. The standard score for 0.35 will be $z = \frac{0.35 - 0.30}{0.023} = 2.17$. We use Table II to find the area to the left of -2.17 and the area to the left of 2.17 and subtract the smaller area from the larger area to get the area left in between. So, $P(0.25 < \hat{p} < 0.35) = 0.9850 - 0.0150 = 0.9700$.

(c) Will the probability of getting a sample proportion $\hat{p}$ **between 27% and 33%** be larger or smaller than the probability calculated in part (b)? In probability notation, we have $P(0.27 < \hat{p} < 0.33)$.

Using the TI, $P(0.27 < \hat{p} < 0.33) = \text{normalcdf}(0.27, 0.33, 0.30, 0.023) = 0.807$.

Using Table II, the standard scores are -1.30 and 1.30. So, $P(0.27 < \hat{p} < 0.33) = 0.9032 - 0.0968 = 0.8064$. ❖

Probabilities about the Proportion of People with Type B Blood

Suppose that 9% of the United States population has type B blood. In a simple random sample of 400 people from the United States population, 12.5% were found to have type B blood.

(a) In this particular situation, what is the numerical value of the parameter? _____

(b) In this particular situation, what is the numerical value of the statistic? _____

(c) What is the probability that a new simple random sample of size 400 people from the United States population would contain at least 12.5% with type B blood? (*Hint:* First find the approximate distribution of $\hat{p}$.)

$\hat{p}$ is approximately $N($,   ) .

$P(\hat{p} \geq 0.125) =$

(d) Another simple random sample of eight people from the United States population is taken. Suppose that we wish to compute the probability that 12.5% or more of this sample have type B blood. Is it valid to use the same method of calculation as in (c)?

Circle one: Yes   No

Explain.

❖ EXAMPLE 8.4   Estimating the Proportion of Voters in Favor _____

In practice, we would not know that, of *all* the voters in a state, 30% are in favor of Proposal A. We would only have the results of various polls taken as the voting day gets closer. But the sampling distribution result does provide us with a useful measure of accuracy, the standard deviation; and from Chapter 5, we have a nice way to interpret the standard deviation.

> The **standard deviation** of $\hat{p} = \sqrt{\frac{p(1-p)}{n}}$ and is *roughly* the average distance of the possible $\hat{p}$ values from the population proportion p.

Recall from Chapter 6 on normal distributions that about 95% of observations will be within two standard deviations of the mean. In our case, about *95% of all random samples should result in a sample proportion $\hat{p}$ that is within two standard deviations of the population proportion p.* This works the other way too: Whenever the sample proportion $\hat{p}$ is within two standard deviations of the population proportion p, the population proportion p is within two standard deviations of the observed sample proportion $\hat{p}$. And this happens in about 95% of all random samples. So if we take our sample proportion $\hat{p}$ and compute the range covering two standard deviations each way, we would have an interval that we can be quite *confident* about it containing the (unknown) population proportion p.

This is quite a nice result, but we can't compute the standard deviation because the true population proportion p is not known. However, when we take our (large) random sample, we do get an estimate of the value of p, namely the sample proportion $\hat{p}$. So if we replace the value of p with $\hat{p}$ in the expression for the standard deviation, we have an *estimated standard deviation* for $\hat{p}$, called the **standard error** of $\hat{p}$.

The **standard error** of $\hat{p} = \sqrt{\frac{\hat{p}(1 - \hat{p})}{n}}$ and roughly *estimates* the average distance of the possible $\hat{p}$ values from the population proportion p.

Suppose that a random sample of $n = 400$ voters was obtained and the sample proportion $\hat{p}$ of voters in favor of Proposal A was 0.28. The standard error is $\sqrt{\frac{0.28(1 - 0.28)}{400}} = 0.022$. Since we know that the true population proportion p is quite often within two standard deviations of the observed proportion $\hat{p}$, we can be fairly confident that the population proportion p is within the range $\hat{p} \pm 2(\text{standard errors}) = 0.28 \pm 2(0.022) = 0.28 \pm 0.044$. *We are quite confident that the true proportion of voters who favor the proposal is between 0.236 and 0.324.* We have just computed, informally, our first *confidence interval estimate* for a population proportion. The ± 0.044 part of the interval is called the *margin of error*. You may have heard this term in the news when data from polls are presented. We will learn more about interval estimation for a population proportion in the next chapter.

❖

Estimating the Proportion of Patients with Side Effects

Let's do it! 8.6

A new medication is being studied for reducing high blood pressure. One of the issues in assessing the effectiveness of such drugs concerns side effects. In a recent study, out of the 250 patients who were taking the medication, 12% reported severe side effects. The researchers would like to use this sample information to provide an estimate of the proportion of all patients taking this medication that would experience severe side effects.

(a) The sample proportion $\hat{p}$ is an estimate of the population proportion p. What is the estimate of the proportion of all patients taking this medication that would experience severe side effects?

(b) The sample proportion $\hat{p}$ in (a) may not be equal to the population proportion p. But we can compute an estimate of the approximate average distance that we would expect a sample proportion $\hat{p}$ to be from the population proportion p. Compute this standard error.

The **standard error** of $\hat{p} = \sqrt{\frac{\hat{p}(1 - \hat{p})}{n}} =$

(c) Since we know that the true population proportion p is quite often within two standard deviations of the observed proportion $\hat{p}$, we can be fairly confident that the population proportion p is within the range $\hat{p} \pm 2(\text{standard errors})$. Compute this range and complete the following sentence:

We are quite confident that the true proportion of all patients who will experience severe side effects is between _____ *and* _____.

❖ **EXAMPLE 8.5** **Testing Hypotheses about the Proportion of Cracked Bottles** _____

A local market receives shipments of wine bottles from a supplier. In the past, about 10% of the bottles received in the shipment have been cracked and hence could not be sold to customers. The supplier is implementing a new shipping method that is expected to reduce the incidence of cracked bottles. The hypotheses to be tested are $H_0: p = 0.10$ versus $H_1: p < 0.10$, where p is the proportion of cracked bottles under the new shipping method. The alternative hypothesis tells us that the **direction of extreme** for this testing situation is *one sided to the left*. This will be important in finding the p-value for our test.

A random sample of 100 bottles is shipped with this new method, and the observed number of cracked bottles is only seven, for a sample proportion of $\hat{p} = 0.07$. Is this sufficient evidence at the 5% significance level to conclude the new method of shipping has reduced the incidence of cracked bottles?

(a) In testing hypotheses, we examine the data under the assumption that H_0 is true. If the true proportion of cracked bottles with the new method is still just 0.10, what is the distribution for the sample proportion $\hat{p}$?

This sample proportion $\hat{p}$ (with the large sample size) will have approximately a **normal** distribution with a **mean** of $p = $ **0.10** and a **standard deviation** of $\sqrt{\frac{0.10(1 - 0.10)}{100}} = $ **0.03**.

(b) The observed sample proportion $\hat{p} = 0.07$. This is indeed less than the 0.10 value, but is it significant?

Let's compute the **p-value**, the probability of getting a sample proportion $\hat{p}$ of 0.07 or less under the null hypothesis distribution stated in (a). In probability notation, we want to find $P(\hat{p} \leq 0.07)$.

Using the TI, $P(\hat{p} \leq 0.07) = \text{normalcdf}(-\text{E}99, 0.07, 0.10. 0.03) = 0.1586$.

Using Table II, we would first compute the standard score for 0.07, namely, $z = \frac{0.07 - 0.10}{0.03} = -1$. We use Table II to find the area to the left of -1. So, $P(\hat{p} \leq 0.07) = P(Z \leq -1) = 0.1586$.

(c) **Decision and Conclusion.**

The p-value of 0.1586 is larger than our significance level of 0.05, so we fail to reject the null hypothesis. Based on the results of this one sample, we cannot conclude that the new shipping method will significantly reduce the proportion of cracked wine bottles. ❖

Let's do it! **8.7** **Testing Hypotheses about Smoking Habits**

In a study on smoking habits in a large community, a scientist wishes to test the null hypothesis $H_0: p = 0.3$ against the alternative hypothesis $H_1: p > 0.3$, where p is the unknown proportion of smokers in the entire community. A simple random sample of size 50 is drawn from the community and the scientist decides to reject H_0 if and only if the sample proportion of smokers is 0.35 or more. Thus, large values for the sample proportion are considered extreme and lead to rejecting H_0.

(a) What is the approximate distribution for the sample proportion $\hat{p}$ if H_0 is true?

(b) What will be the approximate probability of a Type I error (the significance level)—that is, the probability of rejecting H_0 when H_0 is true?

(c) If the actual sample proportion of smokers equals 0.37, then what will be the approximate p-value of the test?

(d) From the given information, can you numerically compute the approximate probability of a Type II error associated with the decision rule? If not, what additional information is needed?

The Usefulness of Sampling Distributions

Did you discover the usefulness of sampling distributions?

As you see, in real life, you will *not know* the true population parameter, because if you did know it, you would not need to take a sample. You take just *one sample*, not many samples, as in constructing the empirical sampling distribution of its estimator. One sample may cost thousands of dollars or take a lot of time to select.

So, *what are we doing in this chapter*? We are discovering the behavior of a sampling distribution of an estimator in the case of *simple random sampling*. We have observed that the larger the sample size taken, the more "normal" the distribution looks, and also the more concentrated it is about the "true parameter value." This guarantees that the *larger the sample size* we take, the *greater the chance that the estimator* will be *close to the true parameter value.*

In general, we are taking a sample from a population because we do not know the true parameter value. If we have some theory regarding the value of the parameter, we can use the expected sampling distribution and the observed data to assess if the theory is supported or not.

We cannot say how close any one particular estimate is to the true parameter value. However, we can calculate the probability that the true parameter value will be contained in the range: $\hat{p} \pm$ "*something*." This "something" is called the *margin of error*, and we will learn more about it in the next chapter.

✳ Exercises

8.1 Explain briefly what is meant by the sampling distribution of a statistic.

8.2 To estimate the true parameter from a given population, three different estimators were proposed. The graphs of the sampling distributions of the three estimators are shown. Note that all three of the graphs are on the same scale.

(a) Which estimator(s) are unbiased? Explain.

(b) Of the two estimators I and II, which would you select? Why?

(c) Which estimator has the smallest variability? Explain.

(d) Based on the sampling distributions, which estimator is the best estimator of the true parameter? Explain.

8.3 A statistic is said to be an unbiased estimator of a parameter if (select one)

(a) a convenience sample is used to calculate the statistic.

(b) the mean or center of its sampling distribution is equal to the parameter being estimated.

(c) randomization is used to obtain the parameter.

(d) the variability of its sampling distribution is equal to the sample mean being estimated.

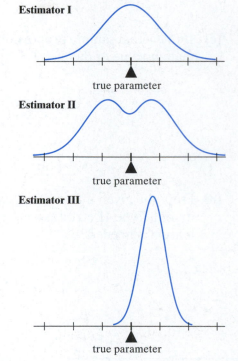

Estimator I

true parameter

Estimator II

true parameter

Estimator III

true parameter

8.4 Complete each of the following statements regarding the sampling distribution of the sample proportion of coffee drinkers in a simple random sample of size n adults from this population (select exactly one option in each part):

(a) As the sample size n increases, the standard deviation of the sampling distribution
  (i) decreases.
  (ii) increases.
  (iii) stays the same.
  (iv) not enough information to say for sure.

(b) As the sample size n increases, the mean of the sampling distribution
  (i) decreases.
  (ii) increases.
  (iii) stays the same.
  (iv) not enough information to say for sure.

(c) As the sample size n increases, the sampling distribution
  (i) looks more and more like the distribution from which the samples were drawn.
  (ii) looks more and more like a normal distribution.
  (iii) becomes more and more tightly clustered about its mean.
  (iv) both (2) and (3).
  (v) none of the above.

8.5 A population of 100,000 bearings is available, 20% of which were produced by Supplier I. For each question, refer to the four histograms given here.

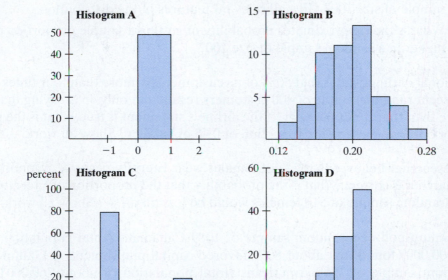

(a) Which histogram could represent the distribution of the variable X, where "$X = 1$" represents "the bearing is from Supplier I" and "$X = 0$" represents "the bearing is NOT from Supplier I"? Explain why you selected that histogram.

(b) Suppose that a statistical software package was used to simulate 50 repetitions of a simple random sample of 100 bearings from the population. Which histogram could represent the sampling distribution of $\hat{p}$, the sample proportion of bearings that are from Supplier I, for this situation? Explain why you selected that histogram.

8.6 A friend of yours sees you reading this chapter and asks, "When you take a random sample, you compute the sample proportion $\hat{p}$ and it is a number. A number doesn't vary nor have a distribution, so what do you mean by a *sampling distribution* for $\hat{p}$?" Give an answer to help your friend understand what a sampling distribution for $\hat{p}$ is.

8.7 For each of the following inference scenarios, determine whether or not you can use the approximate normal distribution for the sample proportion $\hat{p}$ to address the scenario.

(a) A librarian would like to estimate the proportion of the 8000 library-card holders that plan to use the new on-line reserve service being designed. A random sample of 200 card holders was obtained and 110 stated they plan to use the service.

(b) A new coin is minted to be used by the head referee at the start of the upcoming bowl games. You decide to flip the coin 20 times in order to assess whether it is a fair coin by testing $H_0: p = 0.50$ versus $H_1: p \neq 0.50$.

(c) In the past, it has been estimated that only 0.2% of the adult U.S. population are afflicted with a certain phobia. To assess whether the rate has changed, a random sample of 1000 adults was obtained in order to estimate the population proportion who are afflicted with this phobia.

8.8 Public opinion surveys can be used to generate time series that track the public's opinion on various topics. Suppose that the TRUE percentage of people in the population who support the death penalty is 60%. Suppose that a polling agency is commissioned to interview a simple random sample of 100 people and compute the sample proportion of people who support the death penalty.

(a) What is the (approximate) distribution of the sample proportion for a random sample of size 100? Give all relevant features of the distribution.

(b) What is the (approximate) probability of getting a sample proportion of 0.65 or higher in a random sample of size 100?

8.9 National Airlines states that 50% of its customers fly more than four times per year. A recent random sample of 400 customers resulted in only 45% saying that they fly more than four times per year. If the airline's statement is true, what is the probability of observing a sample proportion of 0.45 or smaller? Show all work.

8.10 A researcher believes that 2% of females wear Eternity perfume. Suppose that the researcher is correct. What is the probability that the proportion of Eternity wearers in a random sample of 510 females would be less than 1%? Show all work.

8.11 A Gallup poll of a random sample of 1089 Canadians (total population of about 26,000,000) found that about 80% favored capital punishment. A Gallup poll of a random sample of 1089 Americans (total population of about 260,000,000) also found that 80% favored capital punishment. Which one of the following statements is true? (Select one.)

(a) The Canadian poll is much more accurate, since a larger proportion of the total population was surveyed.

(b) The American poll is more accurate, since they have a larger total population.

(c) Both polls are almost equally precise, since they have the same sample size and the two populations are relatively large.

(d) You cannot compare the precision of the two polls, since we do not know the true population proportions.

(e) Both polls are equally precise, since in both polls 871 of respondents favored capital punishment and precision depends mainly upon the number in favor.

8.12 Suppose that 60% of all students at a large university access course information using the Internet.

(a) Sketch a picture of the distribution for the possible sample proportions you could get based on a simple random sample of 100 students.

(b) Use the 68–95–99.7 rule for normal distributions to complete the following statements:

(i) There is a 68% chance that the sample proportion is between _____ and _____.

(ii) There is a 95% chance that the sample proportion is between _____ and _____.

(iii) It is almost certain that the sample proportion is between _____ and _____.

(c) Would it be likely to observe a sample proportion of 0.50, based on a simple random sample of size 100, if the population proportion were 0.60? Explain.

(d) Sketch a picture of the distribution for the possible sample proportions you could get based on a simple random sample of 400 students.

(i) How does this picture differ from the one in (a)?

(ii) How will the increased sample size affect the range of values you gave in (i)–(iii) of (b)?

8.13 A method currently used by doctors to screen women for possible breast cancer detects cancer in 85% of the women who actually have the disease. A new method has been developed that researchers hope will be able to detect cancer *more* accurately. A simple random sample of 100 women known to have breast cancer were screened using the new method. Of these, the new method detected cancer in 90 women. Let p represent the proportion of women with cancer that is detected by the new method. Using a 5% significance level, test the hypotheses $H_0: p = 0.85$ versus $H_1: p > 0.85$.

(a) Suppose that the new method is just as good as the current method—that is, that the detection rate is $p = 0.85$. Draw a picture of the distribution of the possible sample proportions that would result from simple random samples of size 100 from a population with a proportion $p = 0.85$.

(b) What is the sample proportion of women with cancer detected by the new method?

(c) Would it be likely to see such a sample proportion, or one even more extreme, if indeed the detection rate is $p = 0.85$? That is, compute the p-value for testing the preceding hypotheses.

(d) Does the new method appear to be significantly more accurate at a 5% level? Explain.

8.14 A random sample of 300 holiday shoppers in a shopping mall is selected and 168 are in favor of having longer shopping hours. Is this sufficient evidence to conclude that a majority of all shoppers favor longer shopping hours? Using a 5% significance level, test $H_0: p = 0.50$ versus $H_1: p > 0.50$.

(a) If the true proportion of shoppers who favor longer hours is 0.50, what is the sampling distribution for the sample proportion $\hat{p}$?

(b) Compute the observed sample proportion $\hat{p}$. Using the sampling distribution in (a), report the corresponding p-value for this test.

(c) Give your decision and a written conclusion in the context of the problem.

8.15 In "Let's do it! 8.4," the sampling distribution of $\hat{p}$ was generated empirically for various values of the true population proportion $p = 0.1, 0.3, 0.5,$ or 0.7 and was based on a sample size of 50 or 100. Also, for large sample sizes, the sampling distribution of $\hat{p}$ is approximately normal, centered at the true population proportion p with a standard deviation of

$$\sqrt{\frac{p(1 - p)}{n}}.$$

Let's see if our empirically generated sampling distributions support this expression for the standard deviation.

(a) For each scenario, calculate the true standard deviation for $\hat{p}$, namely,

$$\sqrt{\frac{p(1 - p)}{n}},$$

and record these true standard deviations in the table that follows.

(b) For each scenario, compute the actual standard deviation for the generated $\hat{p}$ values. Enter all of the $\hat{p}$ values into your calculator or computer and find the standard deviation for this population of $\hat{p}$ values.

Note: We are treating the $\hat{p}$ values as a population of values, so use the population standard deviation σ, not the sample standard deviation s. Record these estimated standard deviations in the table.

Scenario	Sample Size n	Population Proportion p	Estimated Standard Deviation	True Standard Deviation
I	50	0.1		
II	50	0.3		
III	50	0.7		
IV	50	0.5		
V	100	0.5		

(c) Compare each estimated standard deviation to the corresponding true standard deviation. Do the empirically generated results support that the standard deviation for $\hat{p}$ is

$$\sqrt{\frac{p(1 - p)}{n}}?$$

8.16 Do you think it is important to limit access to information on the Internet? Suppose that 60% of American teenagers, aged 13 to 17, believe it is important to limit access to information on the Internet. That is, the true parameter $p = 0.60$.

(a) If we plan to choose a simple random sample of 20 American teenagers aged 13 to 17, how many would you expect to answer "yes?"

(b) Outline and describe a method for simulating this sampling procedure. You will need to state which random mechanism you will use and how you will use it. You will need to state which outcomes will correspond to a "yes" response, and which to a "no" response.

(c) Using your method described in (b), obtain your first simple random sample of size $n = 20$.

(i) How many "yes" responses did you get?

(ii) What is the sample proportion of "yes" responses?

(iii) Will you get this same sample proportion of "yes" responses in the next random sample of size 20?

(d) Enter the result for your first sample of size 20 in the given table. Continue your simulation until you have a total of 50 samples of size 20, and tally your results in the table. You may wish to work in a group and divide up this task.

Seed value or starting point = _____

Number of Women	Sample Proportion $\hat{p}$	Tally	Frequency	Proportion of All Trials
0	0.00			
1	0.05			
2	0.10			
3	0.15			
4	0.20			
5	0.25			
6	0.30			
7	0.35			
8	0.40			
9	0.45			
10	0.50			
11	0.55			
12	0.60			
13	0.65			
14	0.70			
15	0.75			
16	0.80			
17	0.85			
18	0.90			
19	0.95			
20	1.00			
Total	_____	_____	50	1.00

(e) In (d), you approximated another sampling distribution of $\hat{p}$ through simulation—the sampling distribution of the proportion of "yes" responses in a simple random sample of size $n = 20$ drawn from a population with 60% "yes" responses.

   (i)   What was the smallest number of "yes" responses in any one sample?

   (ii)   What was the largest number of "yes" responses in any one sample?

   (iii)   What is the most likely sample proportion of "yes" responses?

   (iv)   Comment on the overall appearance of this sampling distribution.

(f) Suppose that you have a simple random sample of 20 American adults. Based on the table of results in (d), make the following estimates:

   (i)   Estimate the probability that exactly 14 people think that it is important to control information on the Internet.

   (ii)   Estimate the probability that 10 or fewer think that it is important to control information on the Internet.

   (iii)   Estimate the probability that the sample proportion who think that it is important to control information on the Internet will be from 0.65 to 0.75, inclusive.

   (iv)   Estimate the probability that all 20 people think that it is important to control information on the Internet.

(g) How would taking a simple random sample of 40 American teenagers, instead of 20, affect the probability that the sample proportion will be from 0.65 and 0.75 inclusive?

The probability will (select one) increase. decrease. stay the same. Explain.

(h) Repeat (b) through (f) when the sample size is set at $n = 40$. Compare your results to those for when $n = 20$.

8.4 Sampling Distribution of a Sample Mean

We have been simulating what happens under repeated simple random sampling in order to examine the sampling distribution of a statistic. The sampling distribution summarizes the variability in the values of the statistic from sample to sample. Once the form of the sampling distribution is known for a statistic, it can be used to predict the accuracy of using the statistic as an estimator for the parameter. In the previous sections, we were interested in learning about the population proportion of items having a certain characteristic. We now turn to the sampling distribution of another common statistic—the sample mean.

❖ **EXAMPLE 8.6 Preparing for Employment** _____

REAL
DATA

As part of the Census Bureau's nationwide survey of 3000 employers with more than 20 workers, including offices, factories, and construction sites, employers were asked to rank how important various factors are with respect to making hiring decisions. The scale was from 1 to 5, with 1 being not important or not considered and 5 being very important. The factors that were ranked were, in alphabetical order, academic performance (grades), attitude, communication skills, experience or reputation of applicant's school, industry-based credentials certifying skills, previous work experience, recommendations from current employees, recommendations from previous employer, score on tests administered as part of interview, and teacher recommendations. (SOURCE: "National Survey Shows a Rift between Schools and Business," *New York Times*, February 20, 1995.)

The factor "attitude" received the highest mean ranking of 4.6, while "teacher recommendations" received the lowest mean ranking of 2.1. This mean rank of 4.6 is an observed sample mean because it is computed from the sample of employers surveyed. This observed sample mean of $\bar{x} = 4.6$ is an estimate of the population mean, μ.

Do you think that this sample mean of 4.6 is exactly equal to the true mean for all such employers? Your answer should be, "Probably not." We would not expect that to happen, but the Census Bureau would certainly be happy if these sample results were "close" to the population mean. Suppose that the Census Bureau were to repeat this survey with another 3000 such employers. Do you think the sample mean rank for "attitude" for these employers would have to be exactly 4.6? Again, your answer should be, "Probably not." Imagine this process being repeated many times—the process being to take a sample of 3000 such employers and compute the sample mean rank for "attitude." Repeating this process many times would give us many observed sample means, not all equal to 4.6, with some values perhaps occurring more often than other values. What values for a sample mean would be possible?

How good is this observed sample mean, based on a sample of 3000 employers, at estimating the population mean? Do you know if this sample mean of 4.6 is close to the population mean? Could it be very different from the population mean? If this value of 4.6 may not be equal to the population mean, why do you think these results were reported and conclusions drawn from them? ❖

Suppose that we have a large population of people, and we are interested in estimating the mean age. Our population parameter of interest is the population mean age, denoted by μ. If we were to take a simple random sample of size 20 from this population, we might get as our observed sample mean age $\bar{x} = 24.6$ years. This is the observed value of our statistic. If we were to do this again, we'd get a different sample, and most likely a different value for the sample mean. Each time we take a sample, we could get a different sample mean. Can we quantify in some way what we would expect those sample means to be? To do that, we need to learn about the sampling distribution of the sample mean. We will first simulate the sampling distribution of the sample mean, similar to what we did for the sample proportion. The patterns and results that we observe through simulation will then be extended to provide a general result that applies if the sample is random.

> *Definition:* The **sampling distribution of the sample mean** is the distribution of values of the sample mean in all possible random samples of the same size *n* taken from the same population.

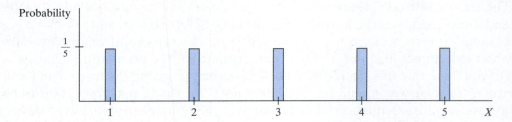

Let's do it! 8.8 Sampling Distribution of the Sample Mean

Consider a large population whose values of a variable X have the following distribution:

The variable X is discrete and uniformly distributed over the values of $1, 2, 3, 4,$ and 5. So the mean for this population, the expected value of X, is easily seen to be the value 3, $\mu = 3$. (It can also be shown that the population standard deviation is $\sigma = \sqrt{2}$.) Now suppose that you do not know the value of the population mean, but you can take a simple random sample of size $n = 2$ from this population. One possible simple random sample of size $n = 2$ is $\{1, 3\}$. For this particular sample, the sample mean is 2. This sample mean would be an estimate of the population mean, but it is not equal to the population mean.

What is another possible simple random sample of size $n = 2$ from this population? _____

What is the value of the sample mean for your sample of size $n = 2$? _____

Is the sample mean equal to the population mean? _____

Is the sample mean equal to the previous sample mean of 2? _____

What would we see if we were to take many simple random samples of the same size $n = 2$ from this same population and compute the sample mean for each sample? That is, what would the sampling distribution of the sample mean look like? Let's simulate this repeated sampling.

With the Random Number Table, we could read off consecutive digits from the table, skipping any that are 0, 6, 7, 8, or 9. The first two integers (that are between 1 to 5) would form the first simple random sample of size $n = 2$. Using the random number table starting at Row 5, Column 1, the first 24 digits are given below. The digits with the strike through them are those that are skipped, and parentheses mark a set of two that constitute a random sample.

$$\{3 \not{7} 5\} \not{7} \{0 3\} \not{9} \not{9} \not{7} \{5 \not{8} 1\} \not{8} \{3 \not{7} 1\} \not{6} \not{6} \{5 \not{6} 0\} \not{6} \{1 2\}$$

So the first six simple random samples of size $n = 2$ are generated and the resulting sample means are $\bar{x} = \frac{3 + 5}{2} = 4$, $\bar{x} = \frac{0 + 3}{2} = 1.5$, $\bar{x} = \frac{5 + 1}{2} = 3$, $\bar{x} = \frac{3 + 1}{2} = 2$, $\bar{x} = \frac{5 + 0}{2} = 2.5$, and $\bar{x} = \frac{1 + 2}{2} = 1.5$.

With the TI, we need to generate random integers between 1 and 5 in sets of $n = 2$. Once you have set your seed value, to say 52, you would use the command randInt(1,5,2) and successive enters would produce a set of two integers on each line. The TI screen would look like:

```
52→rand
             52
randInt(1,5,2)
          {4  3}
          {4  2}
```

These first two results have been entered in the table that follows. Complete the table by generating an additional 18 possible random samples of size $n = 2$ and compute the sample mean for each sample.

Sample #	Observations	Observed Sample Mean
1	4,3	3.5
2	4,2	3
3		
4		
5		
6		
7		
8		
9		
10		
11		
12		
13		
14		
15		
16		
17		
18		
19		
20		

The sample means listed in the third column are possible values that the sample mean could be if a simple random sample of $n = 2$ were selected from our population of interest.

Examine the empirical sampling distribution of the sample mean by making a frequency plot (using the axis given below) of these 20 observed sample mean values. Is the shape uniform, like the original population?

What is the approximate center of the empirical sampling distribution? Compute the average of the 20 observed sample mean values—that is, add up all 20 values and divide by 20. How does this average compare to the true population mean μ?

Take 20 more samples, calculate the sample mean for each, and add these results to your plot. Compute the average of the 40 observed sample mean values.

How does this average compare to the true population mean μ?

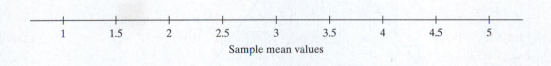

Sample mean values

❖ EXAMPLE 8.7 **Taking a Larger Sample Size**

Consider again a large population for which the response variable X is discrete and uniformly distributed over the values of 1, 2, 3, 4, and 5. Suppose that you do not know the value of the population mean, but you can take a simple random sample of size $n = 30$ from this population. What will the sampling distribution of the sample mean look like when the sample size is larger? A computer was used to generate a large number of random samples of size $n = 30$ from this population and a histogram of the resulting sample means is shown. A normal curve has been superimposed over this histogram.

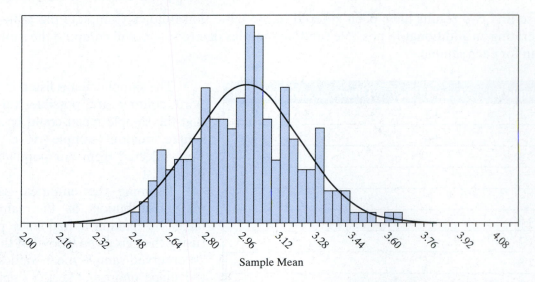

Sample Mean

With the larger sample size, the possible values for the sample mean are much more numerous (more continuous), and there is a prominent bell shape—even a normal-looking shape—to our sampling distribution. The center of this distribution again is approximately equal to the population mean value of 3. With samples of just $n = 2$, we saw sample means of 2 and 4 fairly often. With samples of size 30, sample means of 2 or 4 are not likely to occur at all. Most of the sample means are between 2.5 and 3.5. ❖

❖ **EXAMPLE 8.8   Sampling from a Normal Distribution** _____

Consider a large population for which the response variable X is now **normally distributed** with a mean of 3 and a standard deviation of $\sqrt{2}$. Suppose that you do not know the value of the population mean but you can take a simple random sample of size $n = 2$ from this population. *What will the sampling distribution of the sample mean look like when the original population is normal, but our sample size is small?* A computer was used to generate a large number of random samples of size $n = 2$ from this normal population and a histogram of the resulting sample means is shown. A normal curve has been superimposed over this histogram.

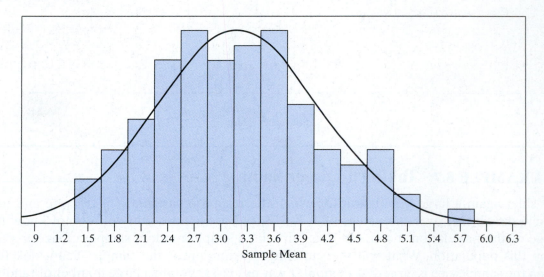

Sample Mean

Even with the small sample size, the distribution of the sample mean has a prominent bell shape—even a normal-looking shape. The center of this distribution is again approximately equal to the population mean value of 3. ❖❖

Let's stop a moment and summarize the key ideas that we have seen in the previous examples concerning the sample mean.

What Do We Expect of Sample Means?

- The **values** of the sample mean $\bar{x}$ **vary** from random sample to random sample in a predictable way.

- The **center** of the distribution of the $\bar{x}$ values is at the **true population mean** μ (for *any* sample size n).

- With a **larger sample size n**, the $\bar{x}$ values tend to be closer to the true **population mean** μ. That is, the $\bar{x}$ **values vary less** around the true mean μ. The variation will also depend on how much variation there was in the population, denoted by σ.

- **When the sample size n is large**, the sample mean can take on many possible values, so the sample mean can be viewed as a **continuous** random variable with a **density curve** as its model.

- **When the sample size n is large**, the distribution of the sample mean can be modeled *approximately* with a **normal distribution**.

- **For any sample size** (even small), if the distribution of the *original population is normal*, the distribution of the sample mean can be modeled with a **normal distribution**.

So far, we have discussed the concept of variability for a sampling distribution somewhat loosely. We have visually compared two sampling distributions and commented on which has less variability. We would like to quantify that variability; namely, we would like to know the standard deviation of the sample mean. Using mathematics, it can be shown that if a simple random sample of size n is selected from a large population with mean μ and standard deviation σ, then the **standard deviation for the sample mean** $\bar{x}$ is $\frac{\sigma}{\sqrt{n}}$, which depends on the original standard deviation in the population σ and the sample size n.

We have studied and simulated the sampling distribution of a sample mean. We have seen some common results, such as a normal shape (for large sample sizes or if the original population is normal) and centering at the population mean μ. Let us now state the main results more formally.

Sampling Distribution of $\overline{X}$, the Sample Mean

If simple random samples of size n are taken from a population having population mean μ and population standard deviation σ, then the sampling distribution of the sample mean has the following properties:

1. $\mu_{\overline{X}} = \mu$ The average of all the possible sample mean values is equal to the parameter μ. In other words, the center of the distribution of $\overline{X}$ is at the true mean μ, so the sample mean $\overline{X}$ is an **unbiased** estimator of μ.

2. $\sigma_{\overline{X}} = \dfrac{\sigma}{\sqrt{n}}$ The **standard deviation** of all of the possible sample mean values **decreases as the sample size n increases**.

   Note: Keep in mind the difference between σ (the standard deviation for the original population values) and $\frac{\sigma}{\sqrt{n}}$ (the standard deviation of the sample means).

3. **If the original population is normally distributed**, then for **any** sample size n the distribution of the sample mean is also **normal** with mean and standard deviation, as given.

$$\overline{X} \text{ is } N\left(\mu, \frac{\sigma}{\sqrt{n}}\right)$$

4. **If the original population is *not* necessarily normally distributed**, but n is "**sufficiently**" **large**, the distribution of the sample mean is *approximately* **normal** with mean and standard deviation, as given.

$$\overline{X} \text{ is approximately } N\left(\mu, \frac{\sigma}{\sqrt{n}}\right)$$

   Sufficiently large means that the normality does not hold exactly for any one sample size n, but as n increases, the distribution eventually starts to look more like a normal distribution. A sample size of 30 or more is often considered large enough. If there are extreme outliers in the observed data, it is better to have a larger sample size. This last result is very famous, known as the **central limit theorem**, and abbreviated **CLT**.

So, now that we know the sampling distribution of the sample mean, let's put it to use. We will sketch the (approximate) normal model for the sample mean and use it to *compute probabilities* about the sample mean taking on certain values. In general, the value of the population mean μ is not known. We take a random sample from the population and compute the observed sample mean $\overline{x}$. We can use the observed sample mean $\overline{x}$ to estimate the value of the population mean μ or to test a theory about the value of μ. The foundation for the *estimation* and the *testing* about a population mean μ is the preceding sampling distribution for the sample mean.

❖ EXAMPLE 8.9   Sketching the Distribution of the Sample Mean ____

Suppose that the variable X represents the filling weight of bins and has a normal distribution with a mean weight of 300 pounds and a standard deviation of five pounds. A picture of the distribution of X is provided. Suppose that a simple random sample of 25 bins will be obtained and the sample mean filling weight will be computed. The distribution of the sample mean weight is also sketched. Notice that both distributions are centered at the population mean weight of 300. The distribution of the sample mean is more tightly clustered around the population mean. Note that the standard deviation of the sample mean is

$$\frac{\sigma}{\sqrt{n}} = \frac{5}{\sqrt{25}} = 1.$$

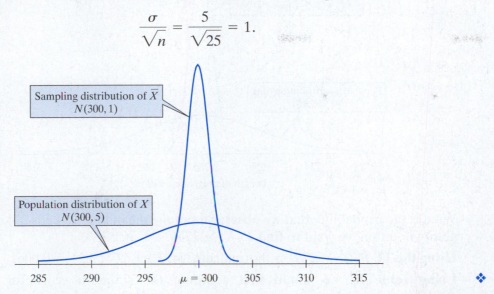

Sampling distribution of $\overline{X}$
$N(300, 1)$

Population distribution of X
$N(300, 5)$

285   290   295 $\mu = 300$ 305   310   315

❖ EXAMPLE 8.10   Probabilities about the Length of Pregnancy ———

Let X be the length of a pregnancy in days. Thus, X is a continuous random variable. Suppose that it has approximately a normal distribution with a mean of 266 days and a standard deviation of 16 days.

(a) Sketch the distribution for X.

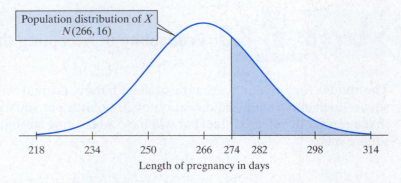

Population distribution of X
$N(266, 16)$

218   234   250   266   274   282   298   314
Length of pregnancy in days

(b) What is the probability that a randomly selected pregnancy lasts more than 274 days? The area corresponding to this probability is shaded on the sketch. Do you expect your answer to be greater than 0.5 or less than 0.5? In probability notation we want to find $P(X > 274)$.

Using the TI, $P(X > 274) = \text{normalcdf}(274, \text{E}99, 266, 16) = 0.3085$.

Using Table II, we would first compute the standard score for 274, namely, $z = \frac{274 - 266}{16} = 0.5$. We use Table II to find the area to the right of 0.5. Recall that Table II provides areas to the left of various values of Z. So $P(X > 274) = P(Z > 0.5) = 1 - 0.6915 = 0.3085$.

(c) Suppose that we have a random sample of $n = 25$ pregnant women. Is it more likely or less likely, as compared to (b), that we might observe a sample mean pregnancy length of more than 274 days? The answer is less likely. Draw a picture to support your answer. Sample means vary less than individuals, so observing an extreme value of 274 or larger is more likely for individuals — the picture shows that the probability for the sample mean will be smaller. Note that the standard deviation of the sample mean is $\frac{\sigma}{\sqrt{n}} = \frac{16}{\sqrt{25}} = 3.2$.

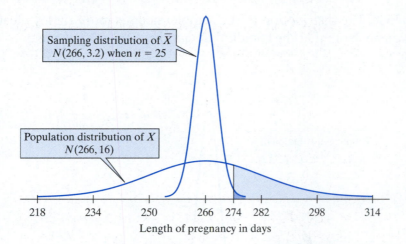

Length of pregnancy in days

What is the probability that we observe a sample mean that exceeds 274 days? In probability notation we want to find $P(\overline{X} > 274)$.

Using the TI, $P(\overline{X} > 274)$ = normalcdf(274, E99, 266, 3.2) = 0.0062.

Using Table II, we would first compute the standard score for 274, namely, $z = \frac{274 - 266}{3.2} = 2.5$. We use Table II to find the area to the right of 2.5. Recall that Table II provides areas to the left of various values of Z.

So $P(\overline{X} > 274) = P(Z > 2.5) = 1 - 0.9938 = 0.0062$.   ❖

 **8.9** **Probability of Accepting the Shipment**

The model for breaking strength of steel bars is normal with a mean of 260 pounds per square inch and a standard deviation of 20 pounds per square inch. What is the probability that a randomly selected steel bar will have a breaking strength greater than 250 pounds per square inch?

A shipment of steel bars will be accepted if the **sample mean** breaking strength of a random sample of 10 steel bars is greater than 250 pounds per square inch. What is the probability that a shipment will be accepted?

❖ **EXAMPLE 8.11 What Will the Sampling Distribution Look Like?** ___

The first histogram shows the distribution of some variable X for a population. This distribution is bimodal with a mean value of 6.

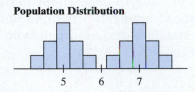

Population Distribution

There are three histograms given next, labeled Histogram A, B, and C. Which of these histograms best portrays the sampling distribution of the sample mean for a random sample of size $n = 50$ values from the original population?

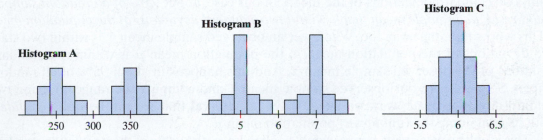

By the Central Limit Theorem, with a large sample size of 50, we expect the distribution of the sample mean to be approximately normal, centered around the value of 6. The histogram that best portrays this is Histogram C.

Note: The shape of the distribution of the sample mean is *not* expected to look like the shape of the distribution of the original population. For a large sample size, the shape of the distribution of the sample mean will be approximately normal, no matter what the original population looks like. ❖

Let's do it! 8.10 Mean Grocery Expenditures

Let X be the random variable representing the amount spent by shoppers at a grocery store. The distribution of the variable X is *skewed to the right* with a mean of $\mu = E(X) = \$60$ and population standard deviation $\sigma = \$35$. We plan to take a random sample of size $n = 100$ expenditures from the population and will consider the sample mean expenditure $\overline{X}$.

(a) What is the value of the mean, the value of the standard deviation, and the (approximate) shape of the distribution of $\overline{X}$, the sample mean expenditure?

mean = _____.

standard deviation = _____.

shape = _____.

(b) Consider the following probability statement:

$$P\left(60 - \frac{35}{\sqrt{100}} < \overline{X} < 60 + \frac{35}{\sqrt{100}}\right) \approx 0.68.$$

This statement is (circle one)   true.   false.   can't tell.

❖ **EXAMPLE 8.12** **Estimating the Population Mean**
Grocery Expenditure _____

In practice, we would not know the true population mean expenditure μ for all shoppers. We would only have the results of a survey on a sample of such shoppers and the observed value of a sample mean $\bar{x}$. But the sampling distribution result for $\overline{X}$ does provide us with a useful measure of accuracy, the standard deviation, and from Chapter 5, we have a nice way to interpret the standard deviation.

> **The standard deviation** of $\overline{X} = \frac{\sigma}{\sqrt{n}}$ and is *roughly* the average distance of the possible $\bar{x}$ values from the population mean μ.

Recall from Chapter 6 on normal distributions that about 95% of observations will be within two standard deviations of the mean. In our case, about *95% of all random samples should result in a sample mean that is within two standard deviations of the population mean* μ. This works the other way too: Whenever an observed sample mean $\bar{x}$ is within two standard deviations of the population mean μ, the population mean μ is within two standard deviations of the observed sample mean $\bar{x}$. And this happens in about 95% of all random samples. So if we take our observed sample mean $\bar{x}$ and compute the range covering two standard deviations each way, we would have an interval that we can be quite *confident* about its containing the (unknown) population mean μ.

This is quite a nice result, but we would need to know the value of the population standard deviation σ. In practice, the population standard deviation σ is also not known, so we would estimate it with the sample standard deviation s. Replacing σ with s would give us an *estimated standard deviation* for $\overline{X}$, called the standard error of $\overline{X}$. The **standard error of the mean**, abbreviated **SEM** $= \frac{s}{\sqrt{n}}$, would estimate how much the sample mean is in error as an estimate of the population mean, on average. You will see more on this standard error in Chapter 10.

Suppose that a random sample of $n = 100$ shopper expenditures was obtained and the sample mean expenditure was \$65 and the sample standard deviation in expenditures was $s = \$35$. The estimated standard deviation of $\overline{X}$ is $\frac{35}{\sqrt{100}} = 3.5$. Since we know that the true population mean μ is quite often within two standard deviations of the observed sample mean, we can be fairly confident that the population mean μ is within the range $\bar{x} \pm 2\,(\text{standard errors}) = 65 \pm 2(3.5) = 65 \pm 7$. *We are quite confident that the true mean expenditure of shoppers is between \$58 and \$72*. We have just computed, informally, our first "confidence interval estimate" for the mean population. The ± 7 in this interval is called the **margin of error**. We will learn more about interval estimation for a population mean in Chapter 10. ❖

Testing Hypotheses about the
8.11 Mean Weight of Nuts

One-pound cans of nuts are to contain a net weight of 16 ounces, but there is considerable variability from can to can. Past records indicate that net weights are normally distributed with a standard deviation of 0.5 pounds. The production manager would like to test the hypothesis of underfilling for the recently produced batch of cans—namely, $H_0: \mu = 16$ versus $H_1: \mu < 16$, where μ is the mean weight of all cans in the batch. The alternative hypothesis tells us that the **direction of extreme** for this testing situation is *one-sided to the left*. This will be important in finding the p-value for our test.

A random sample of six cans is obtained and the observed sample mean weight was 15.8 ounces. Is this sufficient evidence at the 5% significance level to conclude that the batch had a problem with underfilling?

(a) In testing hypotheses, we examine the data under the assumption that H_0 is true. If the true mean weight is 16 ounces, what is the distribution for the sample mean for a random sample of size $n = 6$?

(b) The observed sample mean was 15.8 ounces. This is indeed less than the 16-ounce value, but is it significant? Compute the *p*-**value**, the probability of getting a sample mean of 15.8 or less under the null hypothesis distribution stated in (a).

(c) State your decision and conclusion in terms of the problem.

Note: We performed this test with an unrealistic assumption that the population standard deviation σ is known to be 0.5. In practice, the population standard deviation σ is not known, so we would estimate it with the sample standard deviation s. Replacing the value of σ with s would change how we find the *p*-value (can't use a normal distribution anymore, but another distribution that is called a *t*-distribution is used). However, we will still use the *p*-value for making a decision in the same way. We will learn more about hypothesis testing for a population mean in Chapter 10.

 ## Exercises

8.17 Determine whether the following statement is true or false (a true statement must always be true): The central limit theorem ensures that the sampling distribution of $\overline{X}$ is a normal distribution for any sample size n.

8.18 The heights of adult males are modeled with a normal distribution with a mean of 69 inches and a standard deviation of 2.5 inches.

(a) Give the sampling distribution of the sample mean for a random sample of $n = 9$ adult males.

(b) Give the sampling distribution of the sample mean for a random sample of $n = 100$ adult males.

(c) What is the main difference between the sampling distributions in (a) and (b)?

REAL DATA

8.19 A histogram of the SAT scores of all first-year students at the University of Michigan for the academic year 1993–1994 is given. (SOURCE: *Michigan Daily*, December 10, 1993.)
Suppose that these data can be adequately modeled with a normal distribution with mean of 1250 and standard deviation of 150.

(a) Based on the model (not the histogram), what is the probability that a randomly selected first-year student has an SAT score between 1200 and 1400?

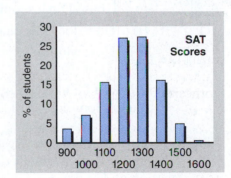

(b) A simple random sample of 36 first-year students is chosen. What is the probability that the sample mean SAT score for these students will be between 1200 and 1400?

(c) Explain, with the aid of a suitable sketch, why the answers to (a) and (b) are not the same.

8.20 Let X = "time to process a loan application," and suppose that X is normally distributed with a mean of 14 days and a standard deviation of 5 days. A simple random sample of 25 loan applications is to be obtained and their processing time to be recorded. Sketch and appropriately label the distribution for the sample mean processing time for a simple random sample of 25 loan applications.

8.21 Let X be the random variable representing the level of cholesterol in the blood of 14-year-old boys. Suppose that X follows a normal distribution with mean $\mu = 170$ mg/dl and standard deviation $\sigma = 30$ mg/dl; that is, $X \sim N(170, 30)$. Levels above 240 mg/dl may require medical attention.

(a) What is the probability that a randomly selected 14-year-old boy has more than 240 mg/dl? That is, find $P(X > 240)$.

(b) Suppose that a random sample of four 14-year-old boys is selected. State the distribution for the sample mean blood cholesterol level, $\overline{X}$. Include all important features.

(c) What is the probability that the mean blood cholesterol level of four randomly selected 14-year-old boys will exceed 190 mg/dl? That is, find $P(\overline{X} > 190)$.

8.22 The sticker on your new car states that the average number of miles per gallon (mpg) of gasoline is 28 for highway driving. Suppose that the distribution of mpg is approximately normal with a standard deviation of 2 mpg. You decide to calculate the sample mean mpg using data from the next four times you fill the gas tank.

(a) Sketch a picture of the distribution for the possible sample means you are likely to get, based on a simple random sample of four fill-ups.

(b) Use the 68–95–99.7 rule for normal distributions to complete the following statements:

   (i) There is a 68% probability that the sample mean mpg is between _____ and _____.

   (ii) There is a 95% probability that the sample mean mpg is between _____ and _____.

   (iii) It is almost certain that the sample mean mpg is between _____ and _____.

(c) Sketch a picture of the distribution for the possible sample means you are likely to get, based on a simple random sample of 16 fill-ups.

   (i) How does this picture differ from the one in (a)?

   (ii) How will the increased sample size affect the range of values you gave in (i) of (b)?

(d) Would it be likely to observe a sample mean mpg of 26 or less, based on a simple random sample of size 16, if the population mean mpg were 28? Explain.

8.23 A convention of sumo wrestlers is held at a hotel. The weights of the sumo wrestlers are known to be normally distributed with a mean of 540 pounds and a standard deviation of 45 pounds. The hotel elevator can accommodate nine wrestlers at a time. Suppose that a simple random sample of nine wrestlers enters the elevator. The elevator will fail to operate if the total weight of the occupants exceeds 5000 pounds. What is the chance the elevator will fail to operate? (*Hint*: Change the criterion for failing to operate from "total weight exceeding 5000 pounds" to "mean weight for nine people exceeding _____ pounds.")

8.24 Suppose that for a population, the response variable X has a $N(-1, 2)$ distribution.
   (a) Draw the distribution of X. Clearly indicate in your drawing the mean and the standard deviation.
   (b) Suppose that a simple random sample of size $n = 100$ is selected from this population. Let $\overline{X}$ represent the sample mean response.
       (i) What is the distribution of $\overline{X}$ for a simple random sample of size $n = 100$? Give all relevant features of the distribution.
       (ii) Draw the distribution of $\overline{X}$. Clearly indicate in your drawing the model, the mean, and the standard deviation.
       (iii) Calculate $P(\overline{X} > -0.9)$. Show all steps used to get your answer.

8.25 Assume that the random variable X has the following distribution with a mean of $\mu = E(X) = 12$ and a standard deviation of $\sigma = 6.93$:

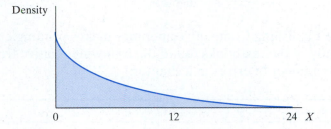

   (a) If we were to take simple random samples of size 48 from the above distribution and compute the sample mean, which of the following histograms best portrays what the sampling distribution of the sample mean $\overline{X}$ would look like?

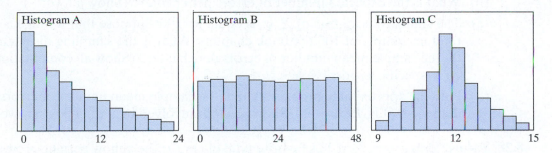

   (b) If we were to take simple random samples of size 480 from the above distribution and compute the sample mean, which of the following histograms best portrays what the sampling distribution of the sample mean would look like?

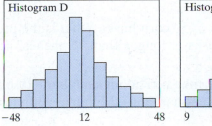
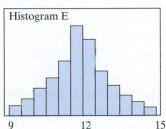
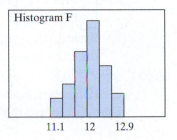

8.26 The distribution, called a triangular distribution, for a continuous random variable X is given.

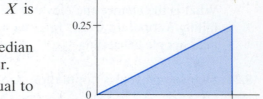

(a) Based on this distribution, is the median for X equal to 4? Explain your answer.

(b) Is the median of this distribution equal to its mean? If yes, explain why. If no, state which value is larger and explain.

(c) Suppose that we are going to take a simple random sample of size 500 from this distribution. Which of the following histograms best portrays the sampling distribution of $\overline{X}$? Explain.

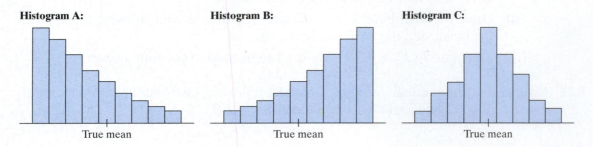

Histogram A: **Histogram B:**   **Histogram C:**

True mean   True mean   True mean

8.27 The College Publishing Company sometimes makes printing errors in college textbooks. A study of the textbooks found the following relative frequency distribution for X = the number of errors in a chapter:

Value $X = x$	*0*	*1*	*2*	*3*	*4*
Probability $P(X = x)$	0.16		0.13	0.07	0.03

(a) Complete the probability distribution.

(b) What is the expected number of errors per chapter? (Show all work.)

(c) The standard deviation of X, σ, is equal to 0.90. Suppose that we take a simple random sample of 100 textbook chapters. What is the sampling distribution of $\overline{X}$, the sample mean number of errors per chapter? (Show all computations and be as specific as possible.)

(d) Find the approximate probability that the sample mean number of errors is less than 1—that is, $P(\overline{X} < 1)$. Show all steps involved in getting your answer.

8.28 Farmer Bob grows corn. His favorite type of corn is the yellow corn he grows in the field right beside his farm. From previous years, he knows that the length of the husks in this field are *uniformly distributed* between 8 and 16 inches. Note that the mean length is 12 inches and the standard deviation of lengths is 2.3 inches.

(a) Farmer Bob randomly selects a corn husk from the field. What is the probability that it is longer than 13 inches?

(b) Farmer Bob selects a random sample of 50 corn husks. What is the probability that the *mean* length of these 50 husks is longer than 13 inches? (Show all work.)

(c) If Farmer Bob will only have time to take a random sample of 10 husks, is it valid to use the same method of calculation as in (b)? Explain your answer.

8.29 Suppose that you are in charge of student ticket sales for a major college football team. From past experience, you know that the number of tickets purchased by a student standing in line at the ticket window has a distribution with a mean of 2.1 and a standard deviation of 2.0. For today's game, there are 100 students standing in line to purchase tickets. If a total of 220 tickets remain, what is the approximate probability that all 100 students will be able to purchase the tickets they desire?

8.30 A random variable X has a mean of $E(X) = 10$ and a standard deviation, σ, equal to 3.

(a) Consider the following probability statement:

$$P(10 - 2(3) < X < 10 + 2(3)) \approx 0.95.$$

This probability statement is (select one)

(i) true.

(ii) false.

(iii) can't tell.

(b) Suppose that we take a simple random sample of size 400 from this population. What is the sampling distribution of $\overline{X}$, the sample mean?

(c) Find $P(\overline{X} > 10.3)$. Show all steps involved in getting your answer.

Chapter Summary

In this chapter we have studied how a statistic varies in repeated random sampling—that is, we have studied the sampling distribution of a statistic. It is the randomness in the sampling procedure that produces these sampling distributions. The idea is that, if nearly all of the samples give a result that is fairly close to the true parameter value, then we have *faith* in the results provided by our one sample even though we don't know whether or not our particular results are close to the truth. We have focused on the sampling distribution of the sample proportion and of the sample mean. A summary of these sampling distributions is provided.

Sampling Distribution of $\hat{p}$, the Sample Proportion

Let p represent the proportion of times some event occurs (that is, the proportion of "successes", where "success" corresponds to the event occurring). If a simple random sample of size n is taken from a population where the proportion of "successes" is p, and if n is large, then

$$\hat{p} \text{ is approximately } N\left(p, \sqrt{\frac{p(1-p)}{n}}\right).$$

The necessary size depends upon the value of the population proportion p. It must be large enough such that $np \geq 5$ and $n(1 - p) \geq 5$.

Sampling Distribution of $\overline{X}$, the Sample Mean

If a simple random sample of size n is taken from a population having population mean μ and population standard deviation σ, then

- if the original population is normally distributed, then the distribution of $\overline{X}$ is also normal:

$$\overline{X} \text{ is } N\left(\mu, \frac{\sigma}{\sqrt{n}}\right).$$

- if the original population is not necessarily normally distributed, but the sample size n is large, $(n \geq 30)$, then the distribution of $\overline{X}$ is approximately normal:

$$\overline{X} \text{ is approximately } N\left(\mu, \frac{\sigma}{\sqrt{n}}\right).$$

This is the **central limit theorem**.

Note: Using the normal model for the sample mean is *not appropriate* for small random samples unless the original population has a normal model. Using the (approximate) normal model for the sample mean is *not appropriate* if the sample is not a random sample (although, in practice, it is used if there is no apparent source of bias and the sample is representative).

These sampling distributions will be used in the next chapters when we turn to more formal statistical inference procedures. We will explore two basic inferencial techniques: formal **hypothesis testing** and **confidence interval estimation**. Knowing the distribution of a statistic will help us understand what kind of values to expect under a particular H_0. It will allow us to assess whether our observed results are consistent with H_0 or are unusual under H_0, using the *p*-value. The sampling distribution of a statistic will also allow us to create a confidence interval (that is, a range of values for which the researcher is confident it contains the population parameter).

 Key Terms

Be sure you can describe, in your own words, and give an example of each of the following key terms from this chapter:

sample statistic	empirical sampling	variability
population parameter	distribution	precision
sampling distribution	unbiased	bias
standard error	margin of error	central limit theorem

Exercises

8.31 Determine whether the following statements are true or false (a true statement is always true):

(a) A statistic that lacks precision is always unbiased.

(b) The sample mean is an unbiased estimator of the true parameter μ.

(c) The variance of the sampling distribution of the sample mean decreases if the size of the population from which you sample increases from 10,000 to 100,000.

(d) Suppose that the true proportion of households with a TV in your city is 0.99. With a sample of size $n = 20$, the distribution of $\hat{p}$ is approximately normal.

8.32 Consider another statistic, such as the sample median. Do you think the sample median will be an unbiased estimator for the population median? What do you think the sampling distribution of the median looks like? Explain briefly what is meant by the sampling distribution of the median, and explain how you would empirically generate this sampling distribution.

8.33 A statistical software package was used to simulate 50 repetitions of a simple random sample of size 100 from a lot of 100,000 bearings, 20% of which were produced by Supplier A. For each random sample, the sample proportion of bearings produced by Supplier A, $\hat{p}$, was computed.

(a) Which of the following histograms best represents the sampling distribution of $\hat{p}$ for this situation?

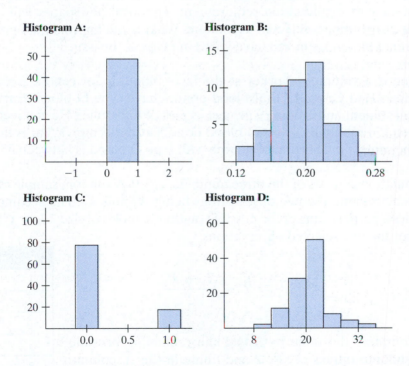

(b) Will the sampling variability of $\hat{p}$ change if in the description the 100,000 were increased to 200,000? Explain.

8.34 A statistical software package was used to simulate 50 repetitions of a simple random sample of size 100 from a lot of 250,000 bearings, 30% of which were produced by Supplier A. For each random sample, the sample proportion of bearings produced by Supplier A, $\hat{p}$, was computed.

(a) Which of the following histograms best represents the sampling distribution of $\hat{p}$ for this situation?

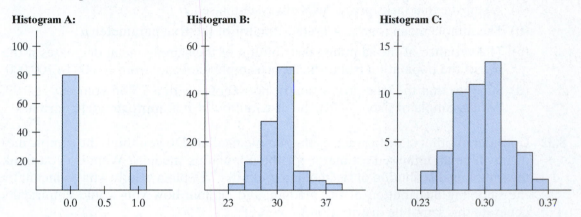

Histogram A: **Histogram B:**   **Histogram C:**

(b) How will the sampling variability of $\hat{p}$ change if, in the preceding description, the 250,000 were decreased to 100,000? Select one:

(i)   The sampling variability will increase.

(ii)   The sampling variability will decrease.

(iii)   The sampling variability will not change.

8.35 Suppose that 60% of the faculty voted in favor of a mandatory course in quantitative literacy as a graduation requirement. The local newspaper will be contacting 100 faculty members selected at random. What is the approximate probability that fewer than half of them will have voted in favor of the issue?

8.36 Enrique is a volunteer worker at the local blood-donor center. According to the American Red Cross, 42% of blood donors have type O blood. Enrique is also a medical student and is doing a project as part of his studies. He will record the blood type for 150 randomly selected blood donors at the center. What is the probability that the sample proportion of donors with type O blood is between 35% and 50%?

8.37 A population consists of the three numbers: 1, 5, 9. A random sample of two is taken *with* replacement, the two observations being X_1 and X_2. The accompanying table provides a partial listing of the possible random samples of size 2. Let $\overline{X}$ be the sample mean of the two selected observations,

$$\overline{X} = \frac{X_1 + X_2}{2}.$$

X_1	X_2	$\overline{X}$
1	1	
1	5	
1	9	
5	1	
5	5	
5	9	

(a) Complete the table by listing the remaining possible random samples of size 2 and filling in the $\overline{X}$ column for all possible random samples.

(b) Find the exact probability $P(\overline{X} = 5)$.

(c) What is the mean or expected value of $\overline{X}$? (Find the mean of all the sample mean values.)

8.38 Let the variable X represent the income (in dollars) of adult residents in a large county of Michigan. Suppose that the mean income for this population is $60,000, with a standard deviation of $12,000. The figure is a sketch of the density function for the distribution of X.

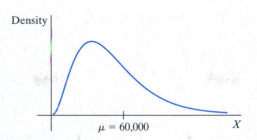

Suppose that we are going to take a random sample of size 300 from this population. Consider the sample mean income $\overline{X}$. Which of the following histograms best portrays the sampling distribution of $\overline{X}$? Explain.

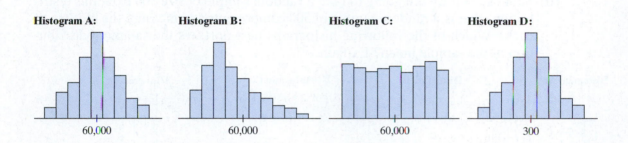

Histogram A: 60,000

Histogram B: 60,000

Histogram C: 60,000

Histogram D: 300

8.39 Consider the following set of even single-digit numbers: 0, 2, 4, 6, 8. These five values form a population.

(a) What is the mean of this population? What is the variance?

(b) Take 25 random samples of size $n = 2$ from this population with replacement, using a seed value of 83, or Row 10, Column 1 of the random number table. For each sample, calculate the sample mean.

(c) Make a histogram displaying the distribution of the observed sample means from (b).

(d) Take 25 random samples of size $n = 5$ from this population, using a seed value of 124, or Row 20, Column 1 of the random number table. For each sample, calculate the sample mean.

(e) Make a histogram displaying the distribution of the observed sample means from (d).

(f) Comment on and compare the two histograms for the sample means in (c) and (e).

8.40 The model for the weight of Big Mac sandwiches has a mean weight set at $\mu = 4.5$ ounces and a standard deviation for weight set at $\sigma = 0.6$ ounces. Suppose that a random sample of 36 Big Macs is obtained and the weight for each is recorded using a well-calibrated scale.

(a) What is the (approximate) probability that the sample mean weight will be at most 4.35 ounces?

(b) What is the name of the result that you used to answer the probability question in (a)?

8.41 Let the variable X represent the length of life, in years, for an electrical component. The density curve for the distribution of X is provided.

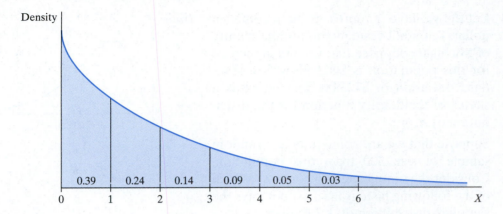

(a) What proportion of electrical components lasts longer than six years?

(b) Suppose that we are going to take a random sample of size 500 from this distribution (that is, a random sample of 500 components and measure the lifetime of each). Which of the following histograms best portrays the sampling distribution of the sample mean? Explain.

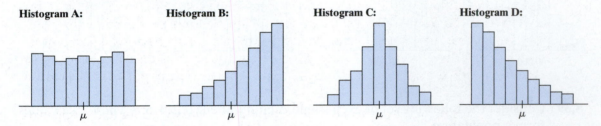

8.42 For adult females, the red-blood-cell count obtained in a medical blood test has a mean of 4,500,000 per mm³ and a standard deviation of 350,000 per mm³. Suppose that the cell counts are approximately normally distributed.

(a) What is the probability that a randomly chosen adult female has a red-blood-cell count of more than 5,000,000 per mm³?

(b) A simple random sample of 40 adult females will be chosen. What is the probability that the mean cell count $(\overline{X})$ for these 40 women will be 4,400,000 per mm³ or lower?

8.43 In human engineering and product design, it is important to consider the weights of people so that airplanes or elevators aren't overloaded. Based on data from the National Health Survey, the weight for adult males in the United States follows a normal distribution with a mean weight of 173 pounds and a standard deviation of 30 pounds.

(a) If one U.S. adult male is randomly selected, what is the probability that his weight will be greater than 180 pounds?

(b) If 36 different U.S. adult males are randomly selected, what is the probability that their sample mean weight will be greater than 180 pounds?

8.44 General Electric produces a "soft white" 100-watt light bulb, for which it states that the average life is 750 hours. Assume that the standard deviation is 120 hours.

(a) Suppose that a consumer agency randomly selects 100 of these bulbs and finds a sample mean life of 735 hours. Should the consumer agency doubt the manufacturer's claim? Provide supporting evidence.

(b) Approximately 95% of the possible sample means are expected to be $750 \pm$ *something*. What is the name for the "something?"

8.45 A random variable X has a normal distribution with a mean of $E(X) = -3$ and standard deviation, σ, equal to 4.

(a) Consider the following probability statement:

$$P(-3 - 2(4) < X < -3 + 2(4)) \approx 0.95.$$

This probability statement is (select one)

(i) true.

(ii) false.

(iii) can't tell.

(b) Suppose that we take a simple random sample of size 400 from this population. What is the sampling distribution of $\overline{X}$, the sample mean?

(c) Find $P(\overline{X} < -2.8)$. Show all steps involved in getting your answer.

8.46 A company packages paper clips in boxes labeled "100 pieces." We've counted the clips in some of these boxes and found that the number in a box varies from as low as 93 to over 100. Suppose that the number of clips in boxes produced by this company has a mean of 100 and a standard deviation of 8. A carton is made up of 64 boxes of clips. Suppose that a carton of 64 boxes can be considered to be a simple random sample of 64 boxes. Let $\overline{X}$ be the mean number of clips per box for a carton of 64 boxes.

(a) Use the approximate distribution of the sample mean to calculate the approximate probability that the sample mean number of clips per box will be between 98 and 100.

(b) Find the approximate probability that a carton of 64 boxes will contain fewer than 6300 clips in all, more than 100 shy of 6400.

8.47 Two independent public health research organizations take simple random samples of adult Michigan residents to estimate the mean cholesterol level of the population of Michigan adults. The first one takes a sample of 100 adults and the second one takes a sample of 1000 adults. Is the second organization likely to get a higher sample mean? Explain.

8.48 The following questions will be a review of the basic concepts underlying the idea of statistical inference (that is, saying something about the whole population based on the information contained in a sample from that population):

(a) What is a statistic?

(b) How does a statistic differ from a parameter?

(c) What is the sampling distribution of a statistic?

(d) How would you empirically generate the sampling distribution of a statistic?

8.49 Three students, Pedro, Eduardo, and Pablo, have been discussing the ideas presented in this chapter. Pedro thinks that it is the "number of samples used" that determines the spread (or variability) of an empirical sampling distribution. Eduardo thinks that it is the parameter value that determines the spread of an empirical sampling distribution. Pablo thinks that it is the "size of each sample used" that determines the spread of an empirical sampling distribution. They come to you for advice. Who is right? Explain your decision.

TI Quick Steps

Simulating Sampling Distributions

Suppose that the proportion of units in a large population that have a certain characteristic is $p = 0.50$. Let 1 denote a unit in the population that has the desired characteristic and 0 indicate a unit that does not have the desired characteristic. To simulate random samples of size $n = 4$ from this population, you need to generate lists of four random integers, either 0's or 1's. If you start with the **seed value of 33**, the **TI** keystroke steps are as follows:

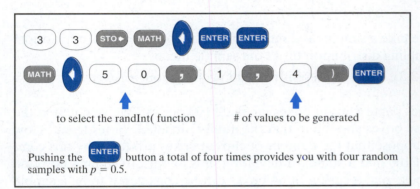

Your screen should look like:

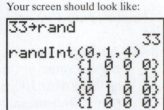

In the first scenario of "Let's do it! 8.4," you were to simulate random samples of size $n = 50$ from a population with proportion $p = 0.1$. You could simulate random integers between 1 and 10 and assign one of the possible integers to represent a success (that is, a unit that has the characteristic). For each random sample generated, you counted the number of successes and computed the sample proportion. A quicker way to perform this simulation is to use the randBin(function, which generates random numbers from specified binomial distributions. Binomial distributions (discussed in Chapter 7) are a family of discrete probability distributions used to model the number of successes out of n trials with a specified probability for a success p. To simulate the five random samples each of size $n = 50$ from a population with proportion $p = 0.10$ of successes, starting with the seed value of 33, and report the proportion of successes in each sample, the keystroke steps are as follows:

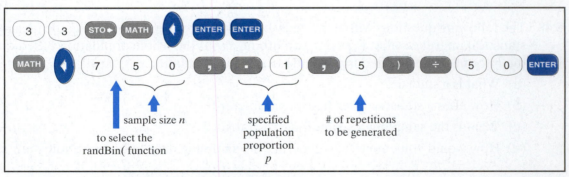

Your screen should look like this:

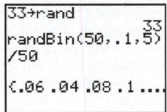

Chapter 9

MAKING DECISIONS ABOUT A POPULATION PROPORTION WITH CONFIDENCE

9.1 Introduction

Chapter 1 provided us with an overview of the statistical principles and procedures for obtaining and summarizing information in order to make decisions. We have competing *theories* about a population that we wish to test. We state these theories in terms of a null and an alternative hypothesis. We *collect data* to help us test the hypotheses. We look at the data and *summarize* the results. In particular, we compute a probability called a *p*-value, which is the chance of getting the observed results plus the chance of getting all of the more extreme results, assuming that the null hypothesis is true. We *interpret* these results and use the data to make a

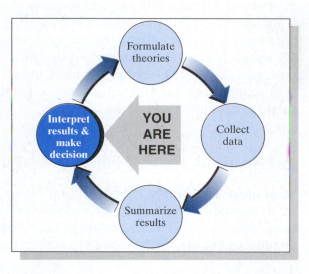

decision. The decision made could lead to an updated theory that may be tested at a later point in time.

In Chapters 2 through 6, we focused on various aspects that comprise the decision-making process. Chapters 7 and 8 provided the additional foundation of probability and sampling distributions needed to discuss formal statistical estimation and hypothesis testing. In the previous chapter, we saw that samples taken from the same population using the same basic method will not all yield the same results. We learned that the use of simple random sampling leads to a model for what we expect to see among the various samples that could be obtained from the same population. These models, formally called sampling distributions, are what allow us to use the statistical methods presented in the final chapters.

There are many situations for which researchers would like to use the results of a sample to decide something about a population. We learned in Chapter 3 that it is generally good practice to use a control group for making comparisons when conducting an experiment. The researcher would want to compare two populations (for example, patients receiving the actual treatment and patients receiving a placebo treatment). In this chapter, we will focus inference procedures for learning about a single population proportion p. We will also take the concept of estimation one step further and develop a confidence interval estimate for the population proportion p. We will be able to provide a statement, using probability, about how confident we are in our decision or interval estimate. Although we are studying the methods for drawing conclusions about a single population proportion p, the general concepts that we learn here apply to many other statistical inference scenarios that we encounter in later chapters. In the next chapter, we will illustrate the concepts of hypothesis testing and confidence interval estimation for a single population mean μ. In Chapter 11, we broaden the discussion of statistical inference to the situation of comparing two populations. Statistical methods for comparing more than two populations are presented in Chapters 12 and 14.

We begin in Section 9.2 with an overview of the two primary techniques of statistical inference—testing of hypotheses and confidence interval estimation—in the context of learning about a population proportion p.

9.2 Learning about a Population Proportion

Suppose that we are interested in learning about the value of the population proportion p. If we have a simple random sample from the population of interest, we would use the sample proportion $\hat{p}$ to **estimate** the population proportion p. Based on this information from the sample, we might like to know if there is evidence to say that the population proportion represents a majority—that is, p is more than 0.50—or how confident we are in our estimate $\hat{p}$ at representing the true proportion p. To put these ideas at a more practical level, consider the example that follows.

❖ EXAMPLE 9.1 Preparing for Employment _____

Most of us face the task, at some point in our lives, of finding employment—finding a job. Many pursue an education in order to aid in finding a job. Head-hunting firms, which match employers with potential employees, have noted that, recently, more skills are expected of employees. We may be interested in addressing the following questions:

1. ***Have the skill requirements increased for a majority of employers nationwide?*** Here, we are interested in making a **decision.** We can restate this question in terms of the following hypotheses about p, the population proportion of employers who say skill requirements have been on the increase:

$$H_0: \ p = 0.50 \text{ versus } H_1: \ p > 0.50.$$

We might take a simple random sample of employers, compute the sample proportion $\hat{p}$ as an estimate of p, and calculate the p-value for the test. If the p-value is "very small" (less than or equal to the significance level) we would reject the null hypothesis.

2. ***What proportion of employers say skill requirements have been on the increase?*** Here, we are interested in an **estimate** of *p*. We might take a simple random sample and compute the sample proportion $\hat{p}$ as an estimate of *p* or perhaps give a small range of values around $\hat{p}$ as an interval in which we think the population proportion *p* may lie.

REAL DATA

In Chapter 8, we were presented with some survey data regarding employers' views on skill levels. The Census Bureau surveyed 3000 employers nationwide. Fifty-seven percent of the employers said the skill requirements of their workplaces had increased in the last three years. The proportion of 0.57 is a sample proportion, a value of $\hat{p}$, based on a sample of size 3000. If the Census Bureau were to repeat this survey with another 3000 such employers, it is likely that the proportion of these employers that say the skill requirements of their workplaces have increased in the last three years will probably not be 0.57. The answers to questions such as (1) on *testing* and (2) on *estimation* will involve the use of the sampling distribution of the sample proportion $\hat{p}$ that we studied in Chapter 8. ❖

In Section 9.3, we focus on testing hypotheses about a population proportion *p*. We already have the foundation needed to do these tests; in fact, we have performed such tests in Example 8.5 and "Let's do it! 8.7." Our summary of the testing procedure here will be a little more formal, emphasizing the language that is used in statistical hypothesis testing. The other principal technique of statistical inference is presented in Section 9.4, using a confidence interval to estimate the population proportion *p*. Example 8.4 and "Let's do it! 8.6" provided an insight to this technique. Section 9.5 will discuss selecting the sample size for a survey, and we end with Section 9.6 on the connection between testing and confidence intervals.

9.3 Testing Hypotheses about a Population Proportion

The specific formulas involved in testing hypotheses depend on the type of response being measured and the type of sampling used. However, the underlying procedure for any statistical hypothesis testing can be outlined as follows:

- State the population(s) and corresponding parameter(s) of interest.
- State the competing theories—that is, the null and alternative hypotheses. Recall that these should be statements about the population(s), not the sample(s), often expressed in terms of the parameter(s).
- State the significance level for the test.
- Collect and examine the data and assess whether the assumptions seem valid.
- Compute a test statistic using the data and determine the corresponding *p*-value— that is, the chance of observing such a test statistic value or values more extreme if the null hypothesis were true. We will sketch a picture of the *p*-value as we did in Chapter 1.
- Make a decision and state a conclusion using a well-written sentence.

In Example 9.1, we wanted to assess whether the skill requirements increased for a majority of employers nationwide. The population under study was employers in the United States. The parameter of interest was the population proportion p of employers who say skill requirements have been on the increase. The hypotheses were expressed in terms of this population parameter p, not the sample proportion $\hat{p}$, namely, H_0: $p = 0.50$ versus H_1: $p > 0.50$. The value of 0.50 is called the *hypothesized value* for the population proportion under H_0 and is generally denoted by p_0. Before we perform the test in Example 9.2, let's practice setting up the hypotheses to be tested.

Let's do it! 9.1 Null and Alternative Hypotheses

State the null and alternative hypotheses that would be used to test the following statements—these statements are the researcher's claim, to be stated as the alternative hypothesis. All hypotheses should be expressed in terms of p, the population proportion of interest.

(a) More than half of the pregnancies in this country are not planned. H_0:   versus H_1:

(b) Less than 3% of children vaccinated for chicken pox still contract the disease. H_0:   versus H_1:

(c) The proportion of people in the United States who are lactose intolerant is different from 0.25. H_0:   versus H_1:

Once the hypotheses and the significance level have been established, the data are collected. The sample proportion $\hat{p}$ is then computed from the data. This sample proportion $\hat{p}$ is an estimate of the population proportion p and has a sampling distribution that is approximately normal for large sample sizes.

Sampling Distribution of $\hat{p}$, the Sample Proportion

Let p represent the proportion of elements in a large population having some characteristic. If a simple random sample of size n is taken from a population and if n is "sufficiently" large, then the distribution of $\hat{p}$ is approximately normal:

$$\hat{p} \text{ is approximately } N\left(p, \sqrt{\frac{p(1-p)}{n}}\right).$$

In testing hypotheses, we examine the evidence based on data—the sample proportion $\hat{p}$—under the assumption that the null hypothesis H_0 is true. Since the null hypothesis gives us a value for the population proportion p, we can determine the sampling distribution of $\hat{p}$ and use it to compute the p-value for our test, just as we did in Chapter 8. However, this approximate normal sampling distribution of $\hat{p}$ would need to be determined for each different value of p stated in a null hypothesis. In "Let's do it! 9.1," we had three different scenarios resulting in three different values for p in the null hypotheses, namely, (a) H_0: $p = 0.50$;

(b) H_0: $p = 0.03$; and (c) H_0: $p = 0.25$. Rather than reporting the value of $\hat{p}$, its sampling distribution under H_0, and then the p-value, a researcher will often report the value of $\hat{p}$ and its corresponding standard score. This standard score, or z-score, is the **test statistic**, and the standard normal distribution will be its approximate model.

Sampling Distribution of the Standard Score for $\hat{p}$

Let p represent the proportion of elements in a large population having some characteristic. If a simple random sample of size n is taken from a population and if n is "sufficiently" large, then the distribution of $\hat{p}$ is approximately normal:

$$\hat{p} \text{ is approximately } N\left(p, \sqrt{\frac{p(1-p)}{n}}\right),$$

and thus the standard score $Z = \dfrac{\hat{p} - p}{\sqrt{\dfrac{p(1-p)}{n}}}$ is approximately $N(0, 1)$.

Using the standard normal distribution provides readers of the results with a common frame of reference. A test statistic or standard score of $z = 2.8$ is somewhat unusually high and would tell us that the sample proportion $\hat{p}$ was nearly three standard deviations above the value hypothesized in H_0, while a test statistic or standard score of $z = -0.4$ implies that the sample proportion was only slightly below the hypothesized value. The p-value computed for the test statistic z using a standard normal distribution $N(0, 1)$ would be the same as the p-value computed for the sample proportion $\hat{p}$ using the approximate normal distribution for $\hat{p}$. So there is more than one way to find the p-value. In many of our examples, we will report the observed sample proportion and its corresponding test statistic value as a summary of the results. For the first example, we revisit the issue of whether skill requirements have increased for a majority of employers nationwide.

❖ EXAMPLE 9.2 **Preparing for Employment Revisited** _____

The hypotheses of interest regarding p, the population proportion of employers who say skill requirements have been on the increase, are H_0: $p = 0.50$ versus H_1: $p > 0.50$. We will perform this test using $\alpha = 0.05$. Based on a sample of 3000 employers, 57% said that skill requirements have been on the increase. The observed sample proportion is $\hat{p} = 0.57$, which does show some support for the alternative hypothesis, since it is more than 0.50. In general, the p-value is the probability of getting a value that is as extreme or more extreme than the observed value, assuming the null hypothesis H_0 is true. Because the alternative hypothesis includes values in only one direction, the test is called a **one-sided test**. The p-value will be computed using only the values in the direction specified in the alternative hypothesis. In this case, **the direction of extreme is to the right**, corresponding to values that are large.

How do we find the *p*-value?

Note that there are many ways to find the *p*-value. You do not need to do them all, but select one that works well for you.

With the TI

1. **Using the sampling distribution of the sample proportion $\hat{p}$ and the normalcdf(function**

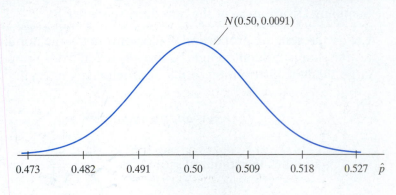

We consider the observed value to be the sample proportion $\hat{p} = 0.57$. If the null hypothesis is true, the sample proportion $\hat{p}$ will have an approximately normal distribution with a mean of $p = 0.50$ and a standard deviation of

$$\sqrt{\frac{0.50(1 - 0.50)}{3000}} = 0.0091.$$

A somewhat detailed sketch of this distribution for $\hat{p}$ is shown. The *p*-value will be the probability of getting a sample proportion $\hat{p}$ of 0.57 or more under this approximate normal distribution, which from the sketch will be quite small. We can find the area under this normal curve using the normalcdf(function.

$$p\text{-value} = P(\hat{p} \geq 0.57) = \text{normalcdf}(0.57, \text{E99}, 0.50, 0.0091)$$
$$= 7.3\text{E-}15 \text{ or about } 0.0000000000000073$$

2. **Using the distribution of the Z-test statistic and the normalcdf(function**

We standardize the observed sample proportion $\hat{p} = 0.57$ using the mean and standard deviation of its sampling distribution under H_0. We consider the observed value to be the corresponding test statistic $z = \frac{0.57 - 0.50}{0.0091} = 7.67$. If the null hypothesis is true, the Z-test statistic will have an approximate standard normal distribution $N(0, 1)$. A sketch of the standard normal distribution is shown. The *p*-value will be the probability of getting a Z-test statistic value of 7.67 or more under the $N(0, 1)$ distribution, which we know will be quite small. We can find the area under this normal curve using the normalcdf(function.

$$p\text{-value} = P(Z \geq 7.67) = \text{normalcdf}(7.67, \text{E99})$$
$$= 8.7\text{E-}15 \text{ or about } 0.0000000000000087.$$

Note: The difference between this *p*-value and the previous one is slight and is due to rounding of the standard deviation in the computations.

3. **Using the 1-PropZTest function under the STAT TESTS menu**

The TI has a TESTS menu located under the STAT menu that allows you to perform various hypothesis tests and generate various confidence intervals. To perform a test for a population proportion, called the *one-sample z-test for a population proportion* and abbreviated **1-PropZTest**, you need to specify the hypothesized proportion (the value in the null hypothesis), the number of "successes" in the sample (denoted by x), the sample size *n*, and the direction for the alternative hypothesis. Determining the value of x may require an extra step. For example, it was reported that 57% of the *n* = 3000 sampled employers responded "*Yes*" (what we were counting, what is deemed a success). So 57% of 3000 is $0.57 \times 3000 = 1710$ employers. This is the value for x. The steps are summarized and the corresponding input and output screens are shown here.

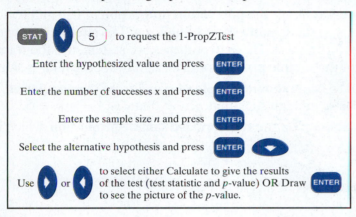

The input screen for Example 9.2 is given by:

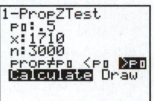

The corresponding test result screen is:

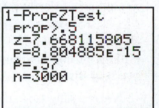

Note that the *Calculate* option produces an output screen that provides everything needed to report the results of the test and make a decision. You have which alternative hypothesis is being tested, the observed value of the standardized *z*-test statistic ($z = 7.668$), the *p*-value (8.8E-15 or 0.0000000000000088), the sample proportion $\hat{p} = 0.57$, and the sample size *n* = 3000. The *Draw* option would draw the normal distribution and shade in the area that corresponds to the *p*-value. The value of the *z*-test statistic and the *p*-value are also given. If you select the *Draw* option, just be careful that other graphing options have been turned off and cleared out.

With Table II

We need to compute the value of the *z*-test statistic—the standard score for $\hat{p} = 0.57$ using the mean and standard deviation of its sampling distribution under H_0. The mean is $p = 0.50$ and the standard deviation is $\sqrt{\frac{0.50(1 - 0.50)}{3000}} = 0.0091$, so the test statistic is $z = \frac{0.57 - 0.50}{0.0091} = 7.67$. The value of 7.67 tells us that the observed sample proportion of 0.57 is 7.67 standard deviations above the hypothesized proportion of 0.50. Then, we use Table II to find the area that corresponds to the *p*-value—in this case, the area to the right of 7.67. The largest entry in Table II is $z = 3.49$, which has an area to the left of 0.9998 and thus an area to the right of 0.0002. Our *p*-value is even smaller than this.

$$p\text{-value} = P(Z \geq 7.67) \text{ is less than } 0.0002.$$

What Is the Decision and Conclusion?

The *p*-value is very close to zero and is certainly less than $\alpha = 0.05$. If the null hypothesis were true and there really was no clear indication as to whether or not skill requirements have been on the increase, we would almost never expect to see such an extreme result. We would prefer to conclude that the proportion of employers who say that skill requirements have been on the increase is significantly more than a majority. ❖

We have just completed an example of a formal hypothesis test about the population proportion p, called a **one-sample z-test for a population proportion**. We use the capital Z when referring to the general test statistic as a random variable, and the small z represents the observed test statistic value. A summary of some of the key features of the test for a population proportion is given next.

Summary of the One-Sample z-test for a Population Proportion: The p-value Approach

- We want to **test hypotheses** about the **population proportion p**. The null hypothesis is H_0: $p = p_0$, where p_0 is the **hypothesized value** for p. The alternative hypothesis provides the direction for the test. These hypotheses are statements about the population proportion, not the sample proportion. The significance level α is selected.

- The data are assumed to be a **random sample** of size n from the population, where **n is large**. It must be large enough such that $np_0 \geq 5$ and $n(1 - p_0) \geq 5$.

- We base our decision about p on the **standardized sample proportion**, which is

$$z = \frac{\hat{p} - p_0}{\sqrt{\dfrac{p_0(1 - p_0)}{n}}}.$$

This z-score is the **test statistic**, and the distribution of the variable Z under H_0 is approximately $N(0, 1)$.

Note: The test statistic is the same no matter how the alternative hypothesis is expressed.

- We calculate the **p-value** for the test, which does depend on how the alternative hypothesis is expressed.

One-sided to the right	**One-sided to the left**	**Two-sided**
If H_1: $p > p_0$, then the p-value is the area to the right of the observed test statistic under the H_0 model.	If H_1: $p < p_0$, then the p-value is the area to the left of the observed test statistic under the H_0 model.	If H_1: $p \neq p_0$, then the p-value is the area in the two tails, outside of the observed test statistic under the H_0 model.

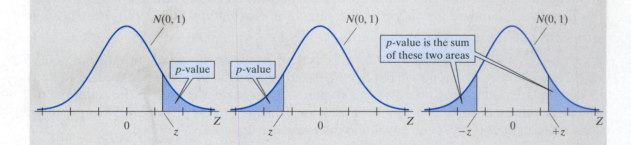

- **Decision.** A small p-value—namely, a p-value less than or equal to the significance level—leads to rejection of H_0.

Note: The TI graphing calculator has a STAT TESTS menu that allows you to perform hypothesis tests for a population proportion. See the ■ TI Quick Steps at the end of this chapter for details.

As you can see in the summary, it is important to set up the alternative hypothesis correctly. The direction of the alternative hypothesis corresponds to the direction of extreme for computing the *p*-value. In Example 9.2, we performed a one-sided, to-the-right (or upper-tailed) test. In our next example, we will also do a one-sided test, but the direction is to the left (or lower-tailed). And in Example 9.4, we will be assessing whether a coin is fair. Since a coin that either favors heads or favors tails would be deemed "not fair" or biased, we will conduct a two-sided test.

❖ EXAMPLE 9.3 Have More Residents Quit Smoking?

The proportion of Michigan adult smokers in 1992 was said to be 25.5%. The Michigan Department of Public Health conducted a survey in 1993 to assess whether the smoking level (that is, the proportion of Michigan adults who smoke) had decreased from the previous level of 25.5%. Let *p* be the proportion of Michigan adults who smoked in 1993. The department hypothesizes that the 1993 proportion *p* is lower than the previous year's proportion of 0.255. This is expressed in terms of the following hypotheses about *p*:

$$H_0: \; p = 0.255 \qquad \text{versus} \qquad H_1: \; p < 0.255.$$

REAL DATA

On March 28, 1995, the *Ann Arbor News* contained an article entitled "More Residents Quit Smoking," which described some of the results of the survey conducted in 1993 based on telephone interviews with 2400 adults. Some excerpts from this article are as follows:

- "The report from the Michigan Department of Public Health showed that 25 percent of Michigan adults smoked."

- "The 1993 smoking level (25 percent) compared to 25.5 percent the previous year (1992)."

- "A spokesman for the Tobacco Institute … said 'the change from 1992 was little to crow about. I think they're trying to make a story out of something that is really statistically indistinguishable from the previous year'."

Let's investigate the spokesman's claim that the 1993 smoking level of 25% is "statistically indistinguishable from the previous year."

THINK ABOUT IT

According to the spokesman's claim, was the null hypothesis, shown previously, rejected or accepted? Explain. (*Hint*: Think about which hypothesis supports the spokesman's remark that the change from 1992 to 1993 is "not a big change.")

If this test was performed using a 5% significance level, was the *p*-value > 0.05 or was the *p*-value ≤ 0.05? Explain.

The stated 1993 smoking level of 25% was based on a sample of size $n = 2400$. We will assume that it was a simple random sample of Michigan adults. The proportion of 0.25 is a sample statistic, denoted by $\hat{p}$. We know that the value of $\hat{p}$ will vary from sample to sample. We want to assess whether observing a sample proportion of $\hat{p} = 0.25$, or one even more extreme, showing even more support for H_1, is likely or is not likely had the proportion of adult smokers actually been $p = 0.255$ (that is, under the null hypothesis). This is another example of a *one-sided test*, since the alternative hypothesis includes values in only one direction. The p-value will be computed using only the values in the direction specified in the alternative hypothesis. In this case, **the direction of extreme is to the left**, corresponding to sample proportion values that are small.

How do we find the p-value?

With the TI

1. **Using the sampling distribution of the sample proportion $\hat{p}$ and the normalcdf(function**

We observed a sample proportion of $\hat{p} = 0.25$. Under H_0, $\hat{p}$ will be approximately normal with a mean of $p = 0.255$ and a standard deviation of $\sqrt{\frac{0.255(1-0.255)}{2400}} = 0.0089$. This distribution for $\hat{p}$ is shown in the accompanying diagrams. The p-value will be the probability of getting a sample proportion $\hat{p}$ of 0.25 or less under this approximate normal model,

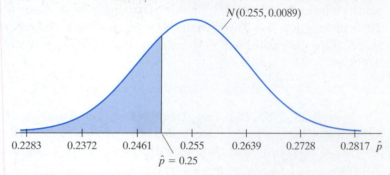

$$p\text{-value} = P(\hat{p} \le 0.25) = \text{normalcdf}(-\text{E99}, 0.25, 0.255, 0.0089) = 0.287.$$

2. **Using the distribution of the Z-test statistic and the normalcdf(function**

Standardizing the observed sample proportion $\hat{p} = 0.25$ using the mean and standard deviation of its sampling distribution under H_0, we get $z = \frac{0.25 - 0.255}{0.0089} = -0.56$. The p-value will be the probability of getting a Z-test statistic value of -0.56 or less under the $N(0,1)$ distribution.

$$p\text{-value} = P(Z \le -0.56) = \text{normalcdf}(-\text{E99}, -0.56) = 0.288.$$

3. **Using the 1-PropZTest function under the STAT TESTS menu**

In the TESTS menu, located under the STAT menu, you select option **5:1-PropZTest**. It was reported that 25% of the $n = 2400$ sampled adults smoked, so the value of x is $0.25 \times 2400 = 600$. The test input and test output screens are as follows:

Input screen:

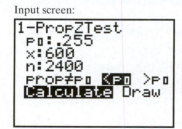

Output screen:

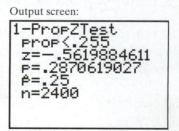

The observed value of the test statistic is $z = -0.562$ and the p-value is 0.2871.

With Table II

The observed test statistic is computed as $z = \frac{0.25 - 0.255}{0.0089} = -0.56$. The value of -0.56 tells us that the observed sample proportion of 0.25 is only 0.56 standard deviations below the hypothesized proportion of 0.255. Using Table II to find the area to the left of -0.56, we have

$$p\text{-value} = P(Z \le -0.56) = 0.2877.$$

What Is the Decision and Conclusion?

This p-value is somewhat large. The result that we observed, a sample proportion of 0.25, or something more extreme, would occur about 29% of the time in repeated sampling under the null hypothesis. The observed result is very likely under H_0. Recall that we compare the p-value to the significance level, α. If the p-value is less than or equal to α, the data are statistically significant and we would reject H_0. Since our p-value is quite large, larger than the common significance levels of 0.01, 0.05, and 0.10, we cannot reject H_0. Failing to reject H_0 supports the spokesman's claim that the 1993 level of 25% is "statistically indistinguishable from the previous year." ❖

❖ EXAMPLE 9.4 Evidence for a Fair Coin? _____

To assess whether a coin could be called a "fair" coin, the English mathematician John Kerrich once tossed a coin 10,000 times, and of those tosses, 5067 came up "heads." Is there evidence, at the 5% level, that the coin is not a fair coin?

In writing up the details about performing this test, we should state the hypotheses being tested and the significance level for the test, and summarize the data (the results of the sample). We should also report the observed test-statistic value, compute the corresponding p-value, report the decision, and state the conclusion.

If we let p be the true proportion of heads, for a coin to be fair, the proportion of heads should be 0.50.

The **hypotheses:** H_0: $p = 0.50$ versus H_1: $p \ne 0.50$.

The **significance level:** $\alpha = 0.05$.

The **results:**   The sample proportion of heads is $\hat{p} = \dfrac{5067}{10000} = 0.5067$.

Is 0.5067 close enough to 0.50 to say that the coin appears to be fair? How far away, in standard units, is 0.5067 from the hypothesized 0.50?

The **observed z-test statistic:** $z = \dfrac{0.5067 - 0.50}{\sqrt{\dfrac{0.50(1 - 0.50)}{10000}}} = 1.34.$

The 1.34 value means we observed a sample proportion, $\hat{p} = 0.5067$, that was 1.34 standard deviations above the hypothesized value of 0.50. Let's measure how likely such a standardized score is under the null hypothesis—that is, find the *p*-value. Here we have an example of a ***two-sided test***, since the alternative hypothesis includes values in both directions. The *p*-value will be computed using values in both directions. In this case, a test statistic of -1.34 would

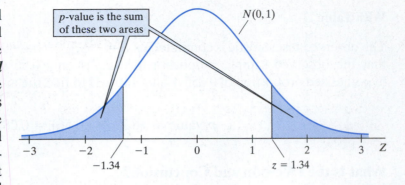

have been just as extreme as the observed test statistic of 1.34. The accompanying figure shows this two-sided *p*-value.

How do we find the *p*-value?

With the TI

Since the value of the *z*-test statistic has already been computed, we can either use the normalcdf(function for an $N(0, 1)$ distribution or enter the information into the TESTS option **5:1-PropZTest**.

$$p\text{-value} = P(Z \le -1.34) + P(Z \ge 1.34) = 2(\text{normalcdf}(1.34, \text{E99})) = 0.180245.$$

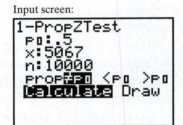

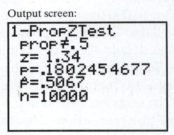

With Table II

We double the area to the left of -1.34:

$$p\text{-value} = P(Z \le -1.34) + P(Z \ge 1.34) = 2(0.0901) = 0.1802.$$

The *p*-value:	A *p*-value of 0.1802 tells us that a sample proportion of heads of 0.5067 (or something more extreme) in 10,000 tosses would occur about 18% of the time under the null hypothesis that the coin is fair. With a 5% significance level, we required that the result be so unusual under the null hypothesis that it would occur less than 5% of the time if H_0 were true.
The decision:	Since the *p*-value of 0.18 is larger than the significance level of 0.05, we fail to reject H_0.
The conclusion:	Based on the data, the coin appears to be fair at a 5% level. ❖

 **9.2** Improved Process?

A manufacturer of panel displays claims that the new process has a higher success rate as compared to the old process. The success rate for the old process was 30%. The appropriate null and alternative hypotheses regarding p, the population proportion of successes for the new process, are

$$H_0: p = 0.30 \quad \text{versus} \quad H_1: p > 0.30.$$

The significance level α is set at 1%. A recent sample of 80 panels made with the new process yielded 32 that worked (success), for a sample proportion of

$$\hat{p} = \frac{32}{80} = 0.40.$$

The **observed Z-test statistic** is given as

$$z = \frac{0.40 - 0.30}{\sqrt{\dfrac{0.30(1 - 0.30)}{80}}} = 1.95.$$

(a) Sketch the picture to show the p-value. Label the axis and give the name of the distribution.

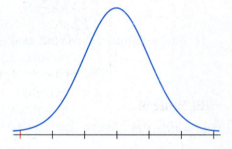

(b) Find the p-value for the test.

(c) Does it appear that the new process is better? Use the specified 1% significance level. Explain your answer.

Let's do it! **9.3** **ESP**

An experiment is designed to test whether a subject has ESP, extrasensory perception. A total of 96 cards is drawn, one by one, with replacement, from a well-shuffled ordinary deck of cards. The subject is asked to guess the suit of each card drawn. There are four possible choices for each card—namely, spades, hearts, diamonds, and clubs. We wish to test the null hypothesis that the subject is only guessing and does not have ESP against the one-sided alternative hypothesis that the subject has ESP. Let p represent the population proportion of correct responses if this experiment were repeated many times. Think about what the value of p should be if the subject is only guessing and does not have ESP and what the direction should be for values of p if the subject does have ESP. Use the following outline:

The **hypotheses**: H_0:   versus H_1:

The **significance level**: $\alpha = 0.05$.

The **results**:   The subject gets 35 correct out of 96 cards, for a sample proportion
                of correct responses $= \hat{p} =$ _____.

Using either the TI or Table II, conduct the test at the 5% significance level. Be sure to report the observed z-test statistic value, the corresponding p-value, the decision, and the conclusion in a well-written sentence. Show all work.

The **observed test statistic**: $z =$

The **p-value**:   You may wish to sketch the
                picture to show the p-value.
                Check for the direction of
                extreme in your alternative
                hypothesis.

p-value $=$

The **decision**:   At a significance level of 0.05, we would
                (circle one)   reject H_0 accept H_0
                because

The **conclusion**:

Let's do it! 9.4 Working Part Time

In the past, it has been stated that 60% of all students at our college work part time during the school year. A counselor, after speaking with many students over the past few weeks regarding schedules, feels that this percentage is actually too low. We are asked to assess whether the proportion of students working part time is indeed larger than 0.60. Use the following outline:

The **hypotheses:** H_0:   versus H_1:

The **significance level:** $\alpha = 0.05$.

The **results**:   Take a simple random sample of students—perhaps a sample of students from your class or poll your whole class. Do you think your class or a random sample from it could represent a random sample from the population of students from your college?
number of students sampled $= n =$ _____
number of students sampled who work part time $=$ _____
sample proportion of students $= \hat{p} =$ _____

The **observed test statistic:** $z =$

The **p-value:**   Sketch the picture to show the p-value. Check for the direction of extreme in your alternative hypothesis.

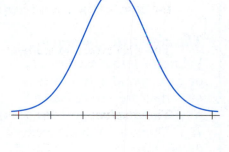

p-value =

The **decision:**   At a significance level of 0.05, we would (circle one)   reject H_0 accept H_0
because

The **conclusion:**

The approach that has been described in this section is sometimes referred to as the **p-value approach** or **probability value approach**. The p-value is reported and compared to the level of significance α in order to make the decision to reject H_0 or accept H_0. Knowing the p-value allows anyone to make the decision for whichever level of significance he or she chooses (selected before the results are observed). The p-value is very informative. A p-value of 0.049 is *just* statistically significant at the 5% level, while a p-value of 0.051 is *just not* statistically significant at the 5% level—sometimes stated as marginally significant—yet both p-values indicate nearly the same amount of evidence against H_0. There is a **classical approach** to hypothesis testing, sometimes referred to as the **fixed α-level approach**. We were introduced to this classical approach briefly in Chapter 1. Using the Bag A–Bag B scenario, we saw that, given a specified significance level for performing the test, we could determine an appropriate cutoff value and thus a corresponding decision rule for that level of significance. We will demonstrate this approach in the next example. We shall see that for a given α level, the two approaches will always reach the same decision. However, *reporting the p-value is more informative and will be the focus for the remainder of this chapter.*

❖ **EXAMPLE 9.5** **Brands at the Crossroads: The Classical Approach** *

Recently, more and more corporations are giving the primary responsibility for brand management to senior management. To assess if indeed a majority of corporations gives the primary responsibility for handling brands to senior management, the results of a new survey, called "Brands at the Crossroads," are examined. The primary responsibility for handling brands rests with senior management, according to 57% of the respondents. The respondents were 153 senior executives who answered 12-page questionnaires sent to corporations worldwide. Therefore, the 153 responses represent a sample from a larger population. Can we state, at a 5% significance level, that for a majority of corporations the primary responsibility for handling brands rests with senior management? If we let p be the proportion of all corporations with the primary responsibility resting with senior management, then to have a majority, the population proportion should be greater than 0.50.

The **hypotheses**: H_0: $p = 0.50$ versus H_1: $p > 0.50$.

The **significance level**: $\alpha = 0.05$.

We assume that the data can be treated as an observed random sample from the larger population of such corporations and that the sample size of 153 is sufficiently large.

The **results**: Observed sample proportion of $\hat{p} = 0.57$.

The **observed test statistic**: $z = \dfrac{0.57 - 0.50}{\sqrt{\dfrac{0.50(1 - 0.50)}{153}}} = 1.73$.

The classical approach

Values of $\hat{p}$ that are large, or correspondingly values of z that are large, begin to show some support for H_1. Thus, a preliminary decision rule is:

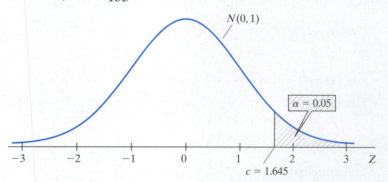

Reject H_0 if $z = \dfrac{\hat{p} - p_0}{\sqrt{\dfrac{p_0(1 - p_0)}{n}}}$ is too big—that is, if $z \geq c$.

The value of c is called the **cutoff or critical value**. How large is large enough? This is where the significance level α plays a role. Recall that we set $\alpha = 0.05$ and that α is the probability of a Type I error.

$\alpha = P(\text{Type I error}) = P(\text{reject } H_0 \text{ when } H_0 \text{ is true}) = P(Z \geq c \text{ when } Z \text{ is } N(0,1)) = 0.05$.

By the accompanying sketch, we see that c corresponds to the 95th percentile of a $N(0, 1)$ distribution—namely, $c = z_{0.95} = 1.645$. So our complete 5% level decision rule is:

Reject H_0 if $z = \dfrac{\hat{p} - p_0}{\sqrt{\dfrac{p_0(1 - p_0)}{n}}} \geq 1.645$.

The **decision**:   Since we observed a test statistic value of 1.73, which is larger than 1.645, we reject H_0 at the 5% significance level.

---

*This is an optional example.

The **conclusion**: Based on the data, it appears that a majority of corporations give the primary responsibility for handling brands to senior management, at a 5% level of significance.

The *p*-value approach

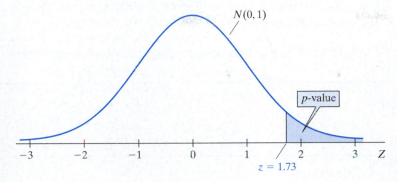

On the standardized scale, the observed sample proportion of $\hat{p} = 0.57$ corresponds to an observed test statistic value of $z = 1.73$. The *p*-value will be computed using only the values in the direction specified in the alternative hypothesis. In this case, the direction of extreme is to the right. The accompanying figure portrays the *p*-value for this test, which is calculated as follows:

$$p\text{-value} = P[\hat{p} \geq 0.57] = P[Z \geq 1.73] = 0.0419.$$

A sample proportion of 0.57 (or something larger) for a sample of size 153 would occur only about 4.2% of the time in repeated sampling if the null hypothesis were actually true. With a 5% significance level, we require that the result be so unusual under the null hypothesis that it would occur less than 5% of the time.

The **decision**: Since the *p*-value of 0.042 is smaller than the significance level of 0.05, we reject H_0.

The **conclusion**: Based on the data, it appears that a majority of corporations gives the primary responsibility for handling brands to senior management, at a 5% level of significance.

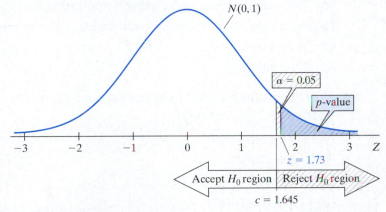

We came to the same decision whether we started with $\alpha = 0.05$ and compared the observed test statistic to the corresponding critical value or whether we reported the *p*-value corresponding to the observed test statistic and compared $\alpha = 0.05$ to the *p*-value. The figure shows why these two approaches will always reach the same decision for a given level α. The 5% critical value was 1.645. Any test statistic that falls to the right of 1.645 would lead us to reject H_0, and any such test statistic would have a corresponding *p*-value (tail area to the right) less than 0.05. Conversely, any test statistic that falls to the left of 1.645 would lead us to accept H_0, and any such test statistic would have a corresponding *p*-value of more than 0.05. In the classical approach, we are comparing values on the test-statistic axis—an observed test-statistic value z with a critical z-percentile. In the *p*-value approach, we compare the corresponding tail areas, the *p*-value, to the α level.

The decisions will be the same no matter which approach is taken. ❖

THINK ABOUT IT

The questionnaire was sent out to 1000 corporations nationwide, of which only 153 responded. What is the response rate for this study?

Do you have any concerns regarding the above conclusion?

Exercises

9.1 For each of the following statements, write out the appropriate null and alternative hypotheses:

(a) The proportion of Democrats in Wayne County is more than 0.50.

(b) The proportion of male births in Wayne County is not equal to the proportion of male births in Oakland County.

9.2 Determine whether each of the following statements is true or false (a true statement is always true):

(a) If the p-value is less than 0.01, then the results are statistically significant at the 0.05 level.

(b) The Type I error probability, α, is the probability that H_0 is true.

(c) To calculate the p-value, you need to know, among other things, the direction of extreme.

(d) The p-value can be determined without observing any data.

9.3 The probability of making a Type I error (select one)

(a) in general, is less when using a one-tailed than a two-tailed test.

(b) is increased by lowering your standard deviation.

(c) is equal to $1 - \alpha$.

(d) is equal to α.

9.4 If the decision is to fail to reject H_0, and H_0 is false, then what type of decision has been made? (Select one.)

(a) correct decision.

(b) Type I error.

(c) power.

(d) Type II error.

9.5 If $\alpha = 0.01$, which of the following p-values would lead to the rejection of the H_0? (Select all that apply.)

(a) p-value $= 0.02$.

(b) p-value $= 0.005$.

(c) p-value $= 0.15$.

(d) p-value $= 0.012$.

(e) None of the above.

9.6 For each of the hypotheses in "Let's do it! 9.1," state what would be a Type I error and a Type II error.

9.7 In recent years, there has been increasing concern about the health effects of computer terminals. It is known that the miscarriage rate under general conditions is about 20%. A random sample of 697 pregnant women working with a computer 1 to 20 hours per week was taken. For this sample, there were 155 miscarriages.

   (a) We wish to assess whether short-term exposure (1 to 20 hours per week) to computer terminals increases the rate of miscarriages. The hypotheses to be tested are H_0: $p = 0.20$ versus H_1: $p > 0.20$. These hypotheses are statements about a population proportion p. Clearly state what population this proportion is referring to.

   (b) Using a significance level of 0.01, test the hypotheses in (a). Show all of your steps. Give the formula for the test statistic, the test statistic value, and the corresponding p-value.

   (c) Are the results statistically significant at the 1% level? Explain.

9.8 Suppose that the unemployment rate in the United States is 0.05. A city mayor would like to determine if the unemployment rate for her city is higher than that for the United States. A random sample of 2000 adult residents of the city was collected, and 125 turned out to be unemployed.

   (a) State the appropriate null and alternative hypotheses.

   (b) Calculate the value of the test statistic for testing H_0 and H_1, and give the p-value.

   (c) State the results for a 1% level test. What is your decision? What does the decision suggest about the unemployment rate for this city?

9.9 In a survey of 439 working parents, with a total of 736 children younger than 12, 320 said that one parent would stay at home if money were not a factor.

   (a) What is the point estimate $\hat{p}$ of p, the population proportion of parents who would stay at home?

   (b) Do the survey results support the conjecture that p is greater than 0.70? Test the hypotheses H_0: $p = 0.70$ versus H_1: $p > 0.70$, using a significance level of 0.05. Show all steps clearly and state your conclusion carefully.

REAL DATA

9.10 What Makes a Good Teacher?   A survey of 735 students from nine business colleges randomly selected from the 1990–1991 American Association of Collegiate Schools of Business (AACSB) Handbook in the United States was taken to determine the instructor behaviors that students believe are more likely to contribute to their academic success. A total of 735 undergraduate students enrolled in business classes was asked to rate various instructor classroom behaviors with respect to the importance to their academic success. One of the instructor behaviors was "Encourage team or group work." A total of 423 students rated this behavior as important to their academic success. Could the researchers conclude that more than 50% of such students feel this behavior is important?

   (a) State the appropriate hypotheses.

   (b) Summarize the results in the form of a test statistic.

   (c) Find the p-value for the test.

   (d) State your decision and conclusion based on a 5% level of significance.

9.11 Alfredo thinks that he may have ESP. One experiment to detect ESP uses a shuffled deck of five cards in which each card has one of five different pictures on it (circle, cross, square, star, and wave). A card is selected from the deck. The subject is asked to guess the picture on the selected card. After the response is recorded the card is returned to the deck and the deck is shuffled. If a subject does not have ESP, then there is a $\frac{1}{5} = 0.20$ chance of being right by chance. If a subject does have ESP, there is a better chance of a correct answer. Alfredo wishes to test the hypotheses H_0: $p = 0.20$ versus H_1: $p > 0.20$. Alfredo participates in the experiment and is right in 36 of 50 tries. At $\alpha = 0.01$, test the hypothesis that Alfredo has ESP. Find the p-value and state your conclusion.

REAL DATA

9.12 It was believed that 20% of all workers were willing to work fewer hours for less pay to obtain more time for personal and leisure activities. A *USA Today*/CNN/Gallup poll with a sample of 600 respondents found that 14% of respondents were willing to work fewer hours for less pay to obtain more personal and leisure time (SOURCE: *USA Today*, April 10, 1995). Conduct the test of H_0: $p = 0.20$ versus H_1: $p < 0.20$, using a 5% significance level. Give all steps and include the observed test statistic value, p-value, decision, and how you made your decision.

9.13 Over the past few months, a supervisor has been attempting to decrease the number of status reports having discrepancies by implementing a new check system. These reports are being submitted to Headquarters Accounting for inventory reconciliation. Before the new system was in place, about 20% of the reports submitted had discrepancies. A recent random sample of 40 reports revealed only 4 having discrepancies. We wish to assess whether the population proportion of reports with discrepancies has significantly decreased from the past rate, using a 5% significance level. State the hypotheses, give the test statistic value, p-value, decision, and report whether the results are statistically significant. Show all steps.

9.14 A student decided to test the null hypothesis, H_0: $p = 0.50$, versus the one-sided alternative hypothesis, H_1: $p > 0.50$. A random sample of 500 observations from the population under study was obtained that gave an observed test-statistic value of $z = 0.44$. The corresponding p-value for the upper-tailed test was found to be 0.33.

(a) Based on the p-value, the student concludes that there is a 33% chance that the alternative hypothesis is true. Do you agree? Why or why not?

(b) What would be the p-value if the alternative hypothesis were two-tailed, namely, H_1: $p \neq 0.50$?

REAL DATA

9.15 The following information is from an article entitled "Finding Care Is Tough" (SOURCE: *Working Mother*, July 1995): Finding affordable child care is a problem for over half of mothers with young children, says the Women's Bureau of the U.S. Department of Labor. The bureau surveyed 1200 women, a representative sample of working women. Of those with children under six, 36% said finding care is their "most serious" or a "very serious" problem.

(a) What do you think "a representative sample" means?

(b) State the appropriate hypotheses for assessing whether the proportion of working mothers with young children who feel finding childcare is a problem, defined as "most serious" or "very serious," is different from 0.50.

(c) The survey states an observed sample proportion of 0.36. Is the information provided in this article sufficient for using this sample proportion to test the hypotheses in (b)? If so, perform the test. If not, explain what additional information you would need.

9.16 Many statistical computing packages automatically report the *p*-value for a two-sided test. Suppose that the reported two-sided *p*-value was 0.12.

(a) If the observed test statistic was $z = 1.555$ and the alternative hypothesis was a one-sided, upper-tailed test, what would the *p*-value be?

(b) If the observed test statistic was $z = -1.555$ and the alternative hypothesis was a one-sided, upper-tailed test, what would the *p*-value be?

9.17 Classical Approach (optional) In Example 9.5, the significance level was provided and a decision rule with a critical cutoff value was used to test the hypotheses. The test was a one-sided, upper-tailed test. The 0.05 level was placed in the upper tail of the normal distribution. The following is a summary of the classical approach to testing hypotheses:

One-sided, to the right
If H_1: $p > p_0$, then the α level decision rule is reject H_0 if $z \geq c$, where $c = z_{1-\alpha}$.

One-sided, to the left
If H_1: $p < p_0$, then the α level decision rule is reject H_0 if $z \leq c$, where $c = z_\alpha$.

Two-sided
If H_1: $p \neq p_0$, then the α level decision rule is reject H_0 if $z \geq c$ or $z \leq -c$, where $c = z_{1-\frac{\alpha}{2}}$.

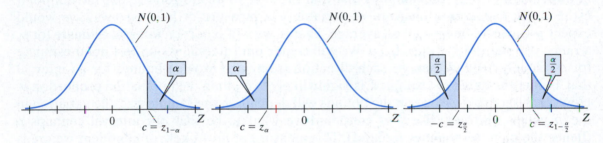

A Decision Rule for Assessing Internet Use   The gender balance of the Internet has long been believed to be nine men for every woman, corresponding to a proportion of women of 0.10. Has this balance changed? Is the proportion of women on the Internet different from 0.10? A yearly Internet survey will be conducted by Matrix Information & Directory Services to assess this question. State a decision rule for testing H_0: $p = 0.10$ versus H_1: $p \neq 0.10$ using the classical approach and a 1% significance level.

9.18 Classical Approach (optional)   For each of the hypotheses in "Let's do it! 9.1," write out the corresponding classical approach decision rule using a 5% significance level.

9.4 Confidence Interval Estimation for a Population Proportion

Up to this point, we have focused on the process of decision making using a procedure for testing hypotheses. However, we do not always have a hypothesized value for the parameter of interest in mind. Sometimes we just want to estimate the parameter.

- What proportion of adults actually suffers from chronic fatigue syndrome?
- What proportion of females expects to share household duties with her mate?
- What proportion of males expects to share household duties with his mate?
- What proportion of Northwest Airline flights to Chicago arrives on time?

Each of these questions is asking, "What is the value of p?" We have already discussed *point* estimation. We would estimate the value of p, the proportion in the population having some property, with the proportion in a sample having the property, $\hat{p}$. The question that remains is, "How good is the point estimate?" Is it equal to p? Probably not. Is it close to $\hat{p}$? How close?

We know that the value of $\hat{p}$ will vary from sample to sample. We have a model for that variation through the sampling distribution of $\hat{p}$. So we are going to expand our point estimate by including a measure of variability, a measure of how far away we would expect $\hat{p}$ values to be from p, on average. That is, we will report an **interval estimate** for p, a range of "plausible" values for p. We will center our interval at our best point estimate for p (namely, $\hat{p}$) and then go each direction from $\hat{p}$ to provide bounds for an interval that, before the data are examined, is highly likely to contain the population proportion p. How far we go from $\hat{p}$ to form the bounds will depend on the sample size, the variation in the $\hat{p}$ values, and how likely or **confident** we want to be that our interval contains p (hence the name **confidence interval**). This measure of how likely or confident we are is called the **confidence level**. We actually computed our first confidence interval back in Example 8.4 and "Let's do it! 8.6." Our discussion here will be more formal, emphasizing the language that is used in confidence interval estimation and including details on their correct interpretation.

> *Definitions:* The sample proportion $\hat{p}$ is a **point (single number) estimate** for the population proportion p.
>
> A **confidence interval estimate** for the population proportion p is an interval of values, computed from the sample data, for which we can be quite *confident* that it contains the population proportion p.
>
> The **confidence level** is the probability that the **estimation method** will give an interval that contains the parameter (p in this case). The confidence level is denoted by $1 - \alpha$, where common values of α are 0.10, 0.05, and 0.01, for 90%, 95%, and 99% confidence.

To develop the confidence interval for p, let's recall the sampling distribution of the sample proportion $\hat{p}$ for a large sample size n:

$$\hat{p} \text{ is approximately } N\left(p, \sqrt{\frac{p(1 - p)}{n}}\right).$$

By the 68–95–99.7 rule for normal distributions, approximately 95% of the $\hat{p}$ values are within two standard deviations of p. If we formed an interval around any of these $\hat{p}$ values, by going two standard deviations each direction from $\hat{p}$, we would have an interval that contains p. So *approximately* 95% of the confidence intervals formed by

$$\hat{p} \pm 2\sqrt{\frac{p(1-p)}{n}}$$

would contain p and 5% would not.

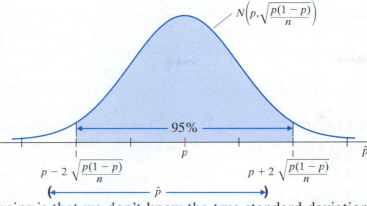

$$N\left(p, \sqrt{\frac{p(1-p)}{n}}\right)$$

95%

$$p - 2\sqrt{\frac{p(1-p)}{n}} \qquad p + 2\sqrt{\frac{p(1-p)}{n}}$$

The only problem that remains is that we don't know the true standard deviation for $\hat{p}$—namely, we don't know

$$\sqrt{\frac{p(1-p)}{n}}.$$

But we can estimate this standard deviation, to form the **standard error** of $\hat{p}$:

$$\text{SE}(\hat{p}) = \sqrt{\frac{\hat{p}(1-\hat{p})}{n}}.$$

So an approximate 95% confidence interval for p is given by

$$\hat{p} \pm 2\left(\sqrt{\frac{\hat{p}(1-\hat{p})}{n}}\right).$$

This confidence interval approximation for a proportion works best when both $np \geq 5$ and $n(1-p) \geq 5$ and can be checked by seeing if $n\hat{p} \geq 5$ and $n(1-\hat{p}) \geq 5$.

Definition: When data are used to estimate the standard deviation of a statistic, the result is called the **standard error** of the statistic.

Note: Sometimes the term "standard error" is used for the actual standard deviation of a statistic, that is, $\sqrt{\frac{p(1-p)}{n}}$, and the estimated value based on data, $\sqrt{\frac{\hat{p}(1-\hat{p})}{n}}$, is called the "estimated standard error." However, we will use the term "standard error" only when the standard deviation of a statistic is estimated from the data since many statistical computer packages use this meaning in the output.

❖ **EXAMPLE 9.6   Study: Chronic Fatigue Common** _____

REAL DATA

A study was performed to learn about the proportion of adults who suffer from chronic fatigue syndrome. For this study, 4000 randomly selected members of a health maintenance organization in Seattle were surveyed for the illness. Surveys were mailed asking respondents if they felt unusual fatigue that interfered with work or responsibilities at home for at least six months. Of the 3066 people who responded (possible nonresponse bias), 590 reported chronic fatigue.

We wish to estimate the proportion of adults who think they suffer from chronic fatigue syndrome. Out of the 3066 responses, 590 reported chronic fatigue, for a point estimate of $\hat{p} = \frac{590}{3066} = 0.1924$, or about 19.2%. An *approximate* 95% confidence interval estimate for the true proportion of adults who think they suffer from chronic fatigue syndrome is given by

$$0.192 \pm 2\sqrt{\frac{(0.192)(1 - 0.192)}{3066}} \rightarrow 0.192 \pm 0.014 \rightarrow (0.178, 0.206),$$

or a rate of 178 to 206 per 1000. This interval gives us *plausible values for the population proportion* of adults who think they are chronic fatigue sufferers, based on a sample of $n = 3066$. ❖

 **9.5** **When Do You Turn Off Your Cell Phone?**

REAL DATA

The following article appeared in the September 30, 2000 edition of the *Orlando Sentinel*:

Phones a turnoff at movies

BERLIN --- More Germans switch their cell phones off at the cinema than during sex, a poll published Friday showed. Only 45 percent of the 621 people questioned by opinion pollsters for the magazine Focus switched their cell phones off during sex, while 58 percent did so at the movies. About a quarter do not want to be disturbed by their "Handys" --- as cell phones are called in Germany --- while eating at home, while 46 percent switched them off visiting bars of restaurants.

Although the sample size of $n = 621$ is large enough to make a confidence-interval estimate, we are not given much detail as to how the respondents were selected. However, in practice, if the data can be considered representative of the responses in the population, inferences are generally made.

(a) *How many* of the Germans questioned stated that they switch off their cell phones when visiting bars or restaurants? *Note:* Percentages reported in the news are often rounded to the nearest percent, so you may have to round your number to the nearest integer.

(b) Based on the presented information, give an estimate of the proportion of Germans who switch off their cell phone at the movies.

(c) Compute the standard error for your estimate in (b). Fill in the following blanks to provide an interpretation of this standard error:

standard error of $\hat{p} = \sqrt{\frac{\hat{p}(1 - \hat{p})}{n}} =$ _____

We would *estimate* the *average distance* of the possible $\hat{p}$ values (obtained in repeated sampling) from the population proportion p to be *about* _____.

(d) Give an *approximate* 95% confidence interval estimate for the proportion of Germans who switch off their cell phones at the movies.

THINK ABOUT IT

Recall the approximate 95% confidence interval (0.178, 0.206) for the proportion of adults who think they suffer from chronic fatigue given in Example 9.6. Do you know if this 95% confidence interval contains the true proportion of adults who think they suffer from chronic fatigue?

Does the 95% confidence level mean there is a 95% chance the true proportion p is in the interval (0.178, 0.206)?

We should also note that only those members of a health maintenance organization in Seattle had the opportunity to respond. Do you think these results extend to the general U.S. adult population?

In Example 9.6, the approximate 95% confidence interval for the population proportion p went from 0.178 to 0.206. We can say that we are 95% *confident* that the interval contains the population proportion p. However, *we will never know whether the interval we have constructed contains the population proportion*. We can't tell unless we know the true proportion. Both the interval (0.178, 0.206) and the value of p are fixed, so either p is in the interval or it is not. So just what does a 95% confidence level mean?

Interpretation, Please!

Perhaps the most important part of confidence interval estimation is interpretation. The interval that we just constructed either contains p or it doesn't. It is *incorrect* to state that the probability that the interval (0.178, 0.206) contains p is 0.95. The value of the parameter p is fixed. It does not vary. Thus, p is either in the fixed interval of (0.178, 0.206) or it isn't. The interpretation of the 95% confidence level pertains to the proportion of times the procedure that we used to construct our interval would contain p if it were repeated over and over. A correct interpretation of the confidence level is given next and shown in the accompanying figure:

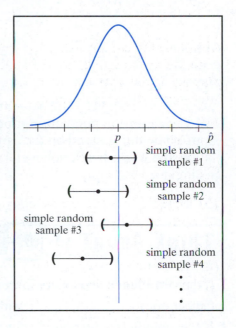

> If we repeated this procedure over and over, yielding many 95% confidence intervals for p, we would expect that approximately 95% of these intervals would contain p and approximately 5% would not.

Note the emphasis on more than one interval. The 95% accuracy does not apply to a particular interval, but to the many possible intervals that could be obtained with this method. Approximately 95% of the intervals generated with this method are expected to contain the true proportion p. The width of the interval will depend on the sample proportion obtained.

We should note that the value of 2 should more accurately be 1.96. The 1.96 is the actual value of the standard normal variable Z, which gives you the 97.5th percentile. (Check this with your calculator or the standard normal table.) This multiplier, generally denoted as z^*, would also change if you wished to form a confidence interval at a **different level of confidence**. The table that follows gives the appropriate z^* values for various confidence levels. Check some of these out with your calculator or the standard normal table. Notice that, as we increase the confidence level, the value of z^* increases, yielding a wider interval.

The corresponding graph shows how the confidence level of $(1 - \alpha)$ relates to the z^* percentile of a standard normal distribution. Note that the figure resembles those we drew when performing two-sided hypothesis tests.

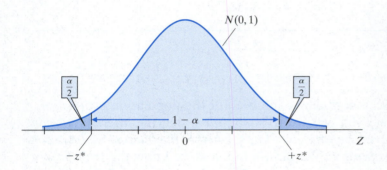

Confidence Level	
$1 - \alpha$	z^*
0.60	0.841
0.70	1.036
0.80	1.282
0.90	1.645
0.95	1.960
0.98	2.326
0.99	2.576
0.999	3.291

Confidence Interval for a Population Proportion p

$$\hat{p} \pm z^*\left(\sqrt{\frac{\hat{p}(1 - \hat{p})}{n}}\right),$$

where z^* is an appropriate percentile of the $N(0, 1)$ distribution. This interval gives potential values for the population proportion p based on just one sample proportion $\hat{p}$. This approximation is based on the assumption the data are a random sample from the population and works best when $np \geq 5$ and $n(1 - p) \geq 5$. This guideline can be checked by seeing if $n\hat{p} \geq 5$ and $n(1 - \hat{p}) \geq 5$.

THINK ABOUT IT

If the **confidence level were increased** from 95% to 99%, would the confidence interval be:

(select one)   wider   narrower   stay the same

If the **sample size increased** (but the sample proportion remained the same), would the confidence interval be:

(select one)   wider   narrower   stay the same

If the **population size increased** (but the population proportion remained the same), would the confidence interval be:

(select one)   wider   narrower   stay the same

Using the TI to Make a Confidence Interval

The TI has a TESTS menu located under the STAT menu that allows you to perform various hypothesis tests and generate various confidence intervals. To generate a (large sample) confidence interval for a population proportion, called the *one-sample z-interval for a population proportion* and abbreviated **1-PropZInt**, you need to specify the number of "successes" in the sample (denoted by x), the sample size n, and the desired confidence level (denoted by *C*-Level and is the proportion between 0 and 1, not the percent). Determining the value of x may require an extra step. For example, suppose that it was reported that 19.24% of the $n = 3066$ sampled members responded *yes*. So 19.24% of 3066 is $0.1924 \times 3066 = 589.8984$, which must be rounded to 590 members because the value for x must be an integer. The steps are summarized next along with the corresponding input and output screens for the Chronic Fatigue Example 9.6.

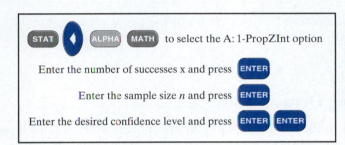

The input screen for
Example 9.6 is given by:

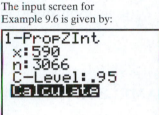

The corresponding
result screen is:

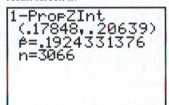

THINK ABOUT IT   Chronic Fatigue Revisited

In Example 9.6, a total of 590 of the 3066 respondents reported chronic fatigue. The researchers examined all survey questions and eliminated the chronic fatigue respondents who had other medical or psychiatric conditions that could explain the condition. The remaining 74 chronic fatigue respondents answered a questionnaire that would diagnose the syndrome. Only three fulfilled the most widely accepted definition of chronic fatigue syndrome, which is persistent fatigue for at least six months, impaired concentration and short-term memory, trouble sleeping, and muscle and joint pain.

Do you have any reservations about making a confidence interval estimate for the proportion of adults who suffer from chronic fatigue syndrome based on the most widely accepted definition? Explain.

Let's summarize the basic form of this confidence interval. The interval is centered at $\hat{p}$, our best point estimate for the true proportion p. The term

$$\sqrt{\frac{\hat{p}(1 - \hat{p})}{n}}$$

is called the **standard error** of $\hat{p}$. This quantity roughly estimates the average distance of the $\hat{p}$ values from their mean, which is p. So our confidence interval for the proportion has the form "$\hat{p}$ give or take a few, namely z^*, standard errors." We will use this basic form for developing other confidence interval estimates for other population parameters.

General Form for a Confidence Interval
_____

(point estimator) $\pm$ (a few)(standard errors of the point estimator)

❖ EXAMPLE 9.7 Drinking Occasionally _____

The following data are responses to the question "Do you drink beer, wine, or hard liquor at least occasionally?" for 40 adults residents selected at random from a large city listing:

Yes	No	Yes	Yes	Yes	Yes	No	Yes	No	Yes
Yes	Yes	Yes	Yes	No	Yes	No	Yes	Yes	Yes
Yes	Yes	Yes	Yes	Yes	Yes	Yes	No	No	Yes
Yes	No	Yes	No	No	No	No	No	Yes	No

- Counting the number of "*Yes*" responses gives a total of 26. So a **point estimate** for the proportion of all adult residents in the city who drink at least occasionally is $\hat{p} = \frac{26}{40} = 0.65$.

- Checking our **guidelines**, $n\hat{p} = 40(0.65) = 26 \geq 5$ and $n(1 - \hat{p}) = 40(0.35) = 14 \geq 5$.

- The **standard error** for our point estimate is
$$\sqrt{\frac{\hat{p}(1 - \hat{p})}{n}} = \sqrt{\frac{0.65(1 - 0.65)}{40}} = \sqrt{0.0056875} \approx 0.075.$$

- The **few** for a 95% confidence level is 1.96.

- The **95% confidence interval** is $0.65 \pm 1.96(0.075) \Rightarrow 0.65 \pm 0.148 \Rightarrow (0.502, 0.798)$.

TI Note: For the 1-PropZInt on the TI calculator, we would only need to specify the value of x = 26, the sample size *n* = 40, and the desired confidence level, 0.95. The resulting confidence interval would be (0.50219, 0.79781). ❖

 **Let's do it! 9.6** **Medical Care for Elderly**

REAL DATA

The *Ann Arbor News* article "U.S. Aged Fail to Get Best Care for Diabetes, Heart Attacks," May 18, 1995, states that America's elderly often fail to get optimal care for diabetes or the latest treatments for heart attacks, according to two studies of medical practices across the country. "Researchers in the diabetes study found that only a small proportion of elderly diabetics were getting recommended tests. The study focuses on 98,000 patients from Alabama, Iowa, and Maryland who were treated by primary care doctors. Eighty-four percent *weren't* getting special blood work, 54 percent weren't seeing an ophthalmologist, and nearly half received no cholesterol screening, the study found. The tests are considered essential by physicians who specialize in diabetes."

(a) What is the population under study?

(b) Based on the results of the study, provide a 90% confidence interval estimate for the proportion of patients who *were* getting special blood work.

(c) Interpret the interval. State the assumptions required for your interpretation to be valid.

(d) Construct a 95% confidence interval estimate for the proportion of patients who were getting special blood work. Is your interval wider or narrower than the interval in (b)? What role does the confidence level play in determining the width of your interval?

(e) What is the *half-width* (the amount that you add to and subtract from $\hat{p}$ in forming the interval)?

This half-width is called the **margin of error** for your interval.

In "Let's do it! 9.6," the term *margin of error* is introduced. Many survey and poll results report the point estimate and the corresponding margin of error, rather than the actual confidence interval. You might see results such as, "Fifty-two percent of the respondents said they have set aside money for retirement. The survey had a margin of error of three percentage points." The margin of error of plus or minus three percentage points describes the accuracy of the poll. It means that the researchers are very *confident* that between 49% and 55% of the population have saved for retirement.

Definition: The **margin of error** for a proportion is the half-width of the confidence interval, given by

$$E = z^* \left(\sqrt{\frac{\hat{p}(1 - \hat{p})}{n}} \right).$$

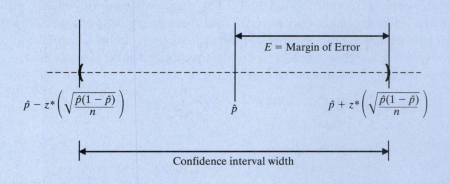

$$\hat{p} - z^* \left(\sqrt{\frac{\hat{p}(1 - \hat{p})}{n}} \right) \qquad \hat{p} \qquad \hat{p} + z^* \left(\sqrt{\frac{\hat{p}(1 - \hat{p})}{n}} \right)$$

E = Margin of Error

Confidence interval width

Let's do it! 9.7 ESP Revisited

Recall the experiment designed to test whether a subject has ESP (extrasensory perception). A total of 96 cards is drawn, one by one, with replacement, from a well-shuffled ordinary deck of cards, and the subject is asked to guess the suit of each card drawn (there are four possible choices for each card—spades, hearts, diamonds, and clubs). Let p be the proportion of correct responses if the subject has no ESP. The subject gets 35 correct out of the 96 cards, so the sample proportion of correct responses is $\hat{p}$ = _____ .

Give a 95% confidence interval for p based on the observed results.

Does your confidence interval contain the value of 0.25? If so, what does this imply? If not, what can you infer?

REAL
DATA

❖ EXAMPLE 9.8 Labor Unions Gain Sympathy_____

Americans' sympathy in labor disputes has tilted toward unions over companies in the past couple of years. A poll conducted for the Associated Press by ICR of Media, Pa. reported that 404 of the 1010 adults polled said that unions are now at about the right strength, whereas only 202 said they are too strong. (SOURCE: *Ann Arbor News*, August 30, 2001, A1.) We would estimate the population proportion of adults who think that unions are at about the right strength by the proportion in the sample, $\hat{p} = \frac{404}{1010} = 0.40$. Sometimes articles include the margin of error. What would be the margin of error for a 95% confidence interval for the population proportion of adults who think that unions are at about the right strength?

$$E = z^* \sqrt{\frac{\hat{p}(1 - \hat{p})}{n}} = 1.96 \sqrt{\frac{0.40(1 - 0.40)}{1010}} \approx 1.96(0.0154) = 0.0302$$

So the margin of error is about 0.03 or as was reported in the article "plus or minus three percentage points." If a 99% confidence level were used, the value of z^* would go from 1.96 to 2.576, so the margin of error would increase.   ❖

Let's do it! 9.8 What Is the Estimate?

The 99% confidence interval for p was calculated to be (0.27, 0.42).

(a) What is the value for the sample proportion $\hat{p}$?

(b) What is the value of the margin of error?

(c) Give two suggestions for how you might reduce the margin of error.

✳ Exercises

9.19 National Airlines wishes to estimate the proportion of its customers who fly more than four times yearly. In a random sample of 40 customers, 17 say that they fly more than four times per year.

(a) Give the estimate of the population proportion, p, of customers meeting this criterion.

(b) Give a 99% confidence interval for the population proportion p of customers meeting this criterion.

(c) What is the *margin of error* for your confidence interval in (b)?

9.20 **Voting "yes."** A simple random sample of 400 voters is taken after they have just voted "yes" or "no" on a certain proposal in a small city. Of the 400 polled, 36% voted "yes" and 64% voted "no." The total number of voters in the election is 40,000.

(a) Give a 95% confidence interval for the proportion of all voters in the election who voted "yes."

(b) Interpret your interval and explain what the 95% confidence level means.

(c) How would your interval change if the total number of voters in the election were 400,000 instead of 40,000?

(d) How would your interval change if the 4000 voters were polled (instead of 400) and 36% voted "yes?"

9.21 We wish to estimate the proportion of operating vehicles that are equipped with air bags. A random sample of size $n = 780$ was obtained and resulted in 70% of the sample operating vehicles having air bags.

(a) Give a 95% confidence interval for the proportion of all operating vehicles equipped with air bags.

(b) Your interval in (a) is a confidence interval for p. Clearly state what p represents in the context of this situation.

9.22 We wish to study various issues regarding legal counseling for students at a certain university. Let p denote the proportion of students who required such counseling last year. In a sample of $n = 280$ students, only 20 had legal counseling last year.

(a) Give a 90% confidence interval for p.

(b) Give a 95% confidence interval for p.

(c) Of the two intervals in (a) and (b), which interval is wider?

REAL DATA

9.23 The following excerpt is from an article in the March 27, 2001 edition of the *Ann Arbor News*.

Poll: Americans not getting enough sleep

The Associated Press

WASHINGTON --- Americans are sleep-deprived workaholics, with only about a third sleeping the recommended eight hours a night, and about 40 percent say they have trouble staying awake on the job, according to a poll released Tuesday.

The 2001 Sleep in America poll (by the National Sleep Foundation) of 1,004 adults found that 63 percent get less than eight hours a night, and about 31 percent get less than seven hours. Many, the poll found, try to catch up on sleep on the weekends, but even then the average slumber is 7.8 hours, still less than ideal. ...

... The survey was conducted for the foundation by WB&A Market Research. It is based on telephone polling of 1,004 Americans who were at least 18 years of age. The margin of error was plus or minus 3 percentage points.

(a) How many of the Americans polled stated that they get less than the recommended eight hours of sleep?

(b) Is the average slumber of 7.8 hours reporting the value of a parameter or a statistic?

(c) Based on the presented information, give an estimate of the proportion of Americans who get less than the recommended eight hours of sleep.

(d) Compute the standard error for your estimate in (c). Complete the following sentence to provide an interpretation of this standard error:

> *We would estimate that the average distance of the possible sample proportion values (obtained in repeated sampling) from the population proportion of adults who get less than the recommended eight hours of sleep to be about _____.*

(e) Give a 95% confidence interval estimate for the proportion of Americans who get less than the recommended eight hours of sleep.

(f) Compute the margin of error for your confidence interval in (e). Is it close to the stated margin of error of plus or minus three percentage points?

REAL DATA

9.24 In 1998, the Ann Arbor Transit Authority (AATA) completed a telephone survey of households in their service area. The survey was conducted by the University of Michigan Urban and Regional Planning Program. Its purpose was to measure the response of the general public to various transit issues in Washtenaw County. AATA uses this survey data to plan future service development and to gauge community support for their services.

Survey Says
Over **15%** of the sample had traveled by AATA bus in the past week.
Over **35%** had ridden an AATA bus in the past year.
81% of respondents think Ann Arbor has a traffic congestion problem.
52% of respondents think that traffic signals should give heavily loaded buses priority in getting through intersections.
73% of the respondents saw or heard an AATA advertisement in the past year.

(a) The following information was also provided regarding the source of the data:

> *The survey's methodology ensures that all telephone households in the geographic sampling area are given equal probability of selection, not just those listed in the local telephone directories. In total, 822 interviews were conducted.*

   (i) Does it appear that a random sample was selected? Why or why not?

   (ii) From what population was the sample selected?

   (iii) Some households may not be "telephone households." How do you think the proportion of users of the transit system compares between telephone households and households without a telephone? Could this bias the results?

(b) Based on the information presented, give an estimate of the proportion of Ann Arbor (telephone-household residents) who think Ann Arbor has a traffic congestion problem.

(c) Give a 95% confidence interval estimate for the proportion of Ann Arbor (telephone-household residents) who think Ann Arbor has a traffic congestion problem.

(d) Compute the margin of error for your confidence interval in (c). State what you could do to reduce the margin of error.

9.25 A new method for detecting a type of cancer has been developed. Among 80 adults who have this type of cancer, this method failed to detect the cancer in five of the adults. Provide a 92% confidence interval estimate for the failure rate for this method.

9.26 We wish to estimate the true proportion of women in a population. A random sample of size $n = 4$ produced a sample proportion of $\hat{p} = 0.25$. Consider the following statement:

A 95% confidence interval for p is

$$\left(0.25 - 1.96\sqrt{\frac{0.25(0.75)}{4}},\ 0.25 + 1.96\sqrt{\frac{0.25(0.75)}{4}} \right).$$

Is this a valid statement? Explain.

9.27 **What Is Wrong?** A box contains a large number of red and blue marbles, but the actual proportions are unknown. A student is assigned the task of constructing a 95% confidence interval for the true proportion of red marbles in the box. She takes a random sample of size $n = 100$ marbles, of which 53 turn out to be red. The student then calculates

$$\left(0.53 - 1.96\sqrt{\frac{0.53(0.47)}{100}},\ 0.53 + 1.96\sqrt{\frac{0.53(0.47)}{100}} \right).$$

Note that $1.96\sqrt{\frac{0.53(0.47)}{100}} \cong 0.10.$

The student wrote the following statement of her results in her final report: "The 95% confidence interval for the proportion of red marbles in the sample is (0.43, 0.63)." The instructor marked this statement, "WRONG!" Explain.

9.28 Suppose that a 95% confidence interval for the population proportion p is computed. The 95% confidence level implies that (select one)

(a) the probability that the population proportion is in this interval is 0.95.

(b) the probability that the population proportion will be outside this interval is 0.95.

(c) for 95 samples out of 100, my sample proportion will equal my population proportion.

(d) for 95 samples out of 100, the population proportion will be inside the confidence interval.

9.29 Just before an election, a national firm increases the size of its weekly random sample from the usual 1500 registered voters to 4000 registered voters.

(a) What is the sampling distribution of $\hat{p}$?

(b) Does the larger sample size lessen the bias of the sampling distribution of $\hat{p}$?

(c) Does the larger sample size lessen the variability of the sampling distribution of $\hat{p}$? Explain.

(d) Suppose that the proportion of people in the sample of 4000 registered voters who said they would vote for the Democratic candidate is 0.53, and that a 95% confidence interval for the true proportion of Democratic voters is the interval (0.51,0.55). Explain, in your own words, what this 95% confidence level means.

REAL DATA

9.30 Chronic Fatigue Syndrome Revisited. The study (described in Example 9.6) was stated as being a better design, since it went out into the community to try to determine how common the syndrome is. Previous studies relied on reports from doctors, thus counting only people who had sought medical help.

(a) The estimate for the proportion of chronic fatigue sufferers, based on this new study, was about 0.001. Would the bias in previous studies discussed earlier typically yield a proportion that was too large, overestimating p, or too small, underestimating p? Explain.

(b) It was reported that, based on previous studies, the syndrome affected some 4 to 10 adults out of every 100,000. Do these figures support your answer to (a)?

(c) Comment on the title of the article reporting these results—Study: Chronic fatigue *common*.

9.31 A psychologist gave a test measuring anxiety level, the State-Trait Anxiety Inventory (STAI), to a sample of 60 abused children, and found that 52 had above-average anxiety levels. An approximate 95% confidence interval for the true proportion of abused children with above-average anxiety levels is given by (0.781, 0.953).

(a) Explain, in your own words, what the 95% confidence interval of (0.781, 0.953) tells us.

(b) Explain, in your own words, what the 95% confidence level means.

9.5 Determining a Sample Size*

In general, a pollster establishes a desired confidence level and margin of error and determines the sample size required in order to estimate the population proportion with the desired accuracy. For example, we may wish to estimate the population proportion with 95% confidence and maintain a margin of error of three percentage points (that is, $E = 0.03$). What sample size should we take? If we start with the margin of error expression and solve for the sample size n, we have an expression for the minimum sample size required to meet the accuracy criteria.

$$\textbf{Sample size} = n = \frac{(z^*)^2 \hat{p}(1 - \hat{p})}{E^2}.$$

However, the righthand side of this expression involves $\hat{p}$. If we have not taken a sample, we will not have the value of $\hat{p}$. As it turns out, the quantity $\hat{p}(1 - \hat{p})$ will never be larger than $\frac{1}{4}$, and this maximum is attained at the value of $\hat{p} = \frac{1}{2}$. If we assign the value of $\frac{1}{2}$ to $\hat{p}$ in the sample size formula, we have a conservative expression for the sample size n that is required to meet the accuracy criteria.

---

*This section is optional.

THINK ABOUT IT Find the Maximum

Sketch the plot of $\hat{p}(1 - \hat{p})$ versus $\hat{p}$ on the accompanying graph. For example, plot $\hat{p}(1 - \hat{p}) = 0.25$ versus $\hat{p} = 0.50$.

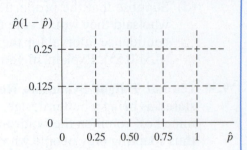

Verify that the maximum is attained at the value $\hat{p} = 0.5$.

Definition: The **required sample size** to estimate a population proportion with confidence level $1 - \alpha$ and margin of error of E is given by

$$n = \frac{(z^*)^2\left(\dfrac{1}{2}\right)\left(1 - \dfrac{1}{2}\right)}{E^2} = \frac{(z^*)^2\left(\dfrac{1}{2}\right)^2}{E^2} = \left(\frac{z^*}{2E}\right)^2,$$

where z^* is an appropriate percentile of the $N(0, 1)$ distribution.

❖ EXAMPLE 9.9 Blood Drive

Many colleges and universities hold a "blood drive" for such organizations as the American Red Cross. To help in the planning of an upcoming blood drive, we want to estimate the proportion of students and faculty who would be willing to donate a pint of blood. The Red Cross would like us to provide an interval estimate with a margin of error of 4%. A confidence level of 99% is desired. How large should the random sample be? We have $E = 0.04$ for the margin of error and $z^* = 2.576$ for the appropriate percentile from the table on p. 542.

$$n = \left(\frac{z^*}{2E}\right)^2 = \left(\frac{2.576}{2(0.04)}\right)^2 = 1036.84$$

So we would need to sample at least 1037 students and faculty. *If the calculation does not result in a whole number,* **round up to the nearest integer.** ❖

Let's do it! 9.9 What Is the Minimum Sample Size?

An admissions officer at a large university wants to estimate the proportion p of all undergraduates who are not happy with their decision to enroll there. How large a sample of undergraduates should she take in order to construct a 90% confidence interval for p with a margin of error of two percentage points (that is, $E = 0.02$)?

Let's do it! 9.10 What Proportion of Students ?

Group or Class Project What would you like to know about your fellow students?

Select an attribute for which you would like to investigate what proportion of students have that attribute.

Some possible examples are as follows: What proportion of students plans to go to graduate school? What proportion of students has declared a major? What proportion of students participates actively in a sport? What proportion of students has their own home pages on the Web?

Fill in the blank: What proportion of students _____?

Outline a procedure to produce a confidence interval estimate for the proportion of students who have the attribute.

Be sure to define your actual population of interest (all students? full time? part time? the students in your class? freshmen?)

What level of confidence will you use?

How will you gather your data?

Collect Data: A random sample of size $n = $ _____.

The number of those sampled with the attribute = _____.

The sample proportion with the attribute $p = $ _____.

Confidence interval estimate:

Underlying assumptions:

Interpretation:

✳ Exercises

REAL DATA

9.32 The EPIC-MRA-Mitchell Research poll found that 33% believed same-sex marriages should be allowed, while 63% disagreed. Four percent were undecided. These results were based on a telephone survey of 1000 registered voters from June 21 to 26, 1995 and have a margin of error of three percentage points either way.

(a) If the proportion of voters who agree is to be estimated with 90% confidence and a margin of error of three percentage points either way, what sample size would be required?

(b) Did the poll sample enough voters to meet these criteria?

(c) Give a 90% confidence interval for the proportion of voters who agree and explain what this interval tells us about the proportion of voters who agree.

(d) Would a 95% confidence interval, based on the same data, be wider or narrower? Explain.

9.33 A polling organization wants to obtain a 98% confidence interval for the proportion *p* of students who have taken out student loans. They will do this by taking a simple random sample of students. Suppose that they want their confidence interval to have a width of 0.06 (that is, a margin of error of 0.03 or three percentage points). How many students should be in their sample?

9.34 A study is to be conducted to estimate the proportion of primary-care physicians who have participated in some form of doctor-assisted suicide for terminally ill patients. A survey will be conducted. The researcher would like to use a 99% confidence interval with a margin of error of 0.03 to estimate this proportion. What is the minimum number of primary-care physicians that should be randomly selected and surveyed? Show all work.

9.35 A study was conducted to estimate *p*, the proportion of university professors who have high blood pressure. Based on the results of a random sample of size 267, a 95% confidence interval for *p* was found to be $(0.20, 0.32)$.

(a) For each statement, determine whether it is true or false.

   (i) About 95% of university professors have between a 20% and a 32% chance of having high blood pressure.

   (ii) There is a 95% probability that the sample proportion lies between 0.20 and 0.32.

   (iii) There is a 95% probability that the population proportion lies between 0.20 and 0.32.

   (iv) If repeated random samples of 267 university professors are obtained, we would expect 95% of the resulting samples to have a sample proportion between 0.20 and 0.32.

   (v) If repeated random samples of 267 university professors are obtained, we would expect 95% of the resulting intervals to contain the population proportion.

(b) What sample size would be necessary in order to have a 95% confidence interval for *p* with a margin of error of 0.04?

9.6   Using Confidence Intervals to Make Decisions

We have studied two forms of statistical inference in this chapter: hypothesis testing and confidence interval estimation. These two forms of inference are different, but they are related, and often both methods can be used to analyze the same situation. Suppose that a 95% confidence interval for a population proportion p goes from 0.82 to 0.90. You are "95% confident" that this interval contains the population proportion p because you used a method that produces a good interval (one that contains the population proportion) for 95% of all possible random samples. Suppose that someone claims that the population proportion p is 0.75. Your confidence interval can be used to assess this claim against the alternative that the population proportion is not 0.75. Your confidence interval provides a range of *plausible* values for the population proportion p, values for which you would *"accept"* as reasonable. Based on the 95% confidence interval for p of (0.82, 0.90), the claimed value of 0.75 is not plausible, since 0.75 falls outside the interval estimate. The general idea for evaluating or testing the plausibility of a parameter value is summarized below.

Making a Decision from a Confidence Interval

- If the claimed (or hypothesized) value for p *does not* fall in the confidence interval, it can be *rejected* as a plausible value for the population proportion p.

- If the claimed (or hypothesized) value for p *does* fall in the confidence interval, it can be *accepted* as a plausible value for the population proportion p.

The connection between testing and confidence intervals in the case of learning about a population proportion is not exactly one to one. The difference arises because the confidence interval uses the standard error in the margin of error, while the Z-test statistic has the true standard deviation under H_0 in its denominator (not the standard error). These guidelines do serve as an *informal* way to test (two-sided) hypotheses about a population proportion with a significance level of α using the corresponding $1 - \alpha$ level confidence interval. In Section 10.4, we will revisit this connection in the case of learning about the population mean, where the connection is exact and not just approximate.

❖ EXAMPLE 9.10   Tougher Gun Laws ____

REAL DATA

The following article appeared in the September 8, 1999 edition of the *Orlando Sentinel*. Although the sample size of $n = 1026$ is large enough to make a confidence interval estimate, we are not given much detail as to how the respondents were selected. However, if the data can be considered representative of the responses in the population, inferences can generally be made. Note that we are not provided with the number of women or the number of men polled, only the total number polled of 1026. Although not directly stated, most polls report the margin of error corresponding to a 95% confidence level.

Since 56% of American adults surveyed favored stricter gun laws and the margin of error was three percentage points, the approximate 95% confidence interval for the proportion of all American adults who favor stricter gun law could be found as

> **Women, men duel over gun control**
>
> The Associated Press
>
> **WASHINGTON** --- The fight over gun control reflects a battle of the sexes. American women say stricter weapons laws would curb violence, while men want better enforcement of existing laws, an Associated Press poll finds.
>
> ... The telephone survey by ICR of Media, Pa., found 56 percent of the American adults surveyed favored stricter gun laws and 39 percent opposed. Sixty six percent of the women favored the tougher laws, compared with 45 percent of the men. Thirty percent of the women and 49 percent of the men were opposed. ...
>
> ... The poll of 1,026 people taken Aug 27-31 had an error margin of plus or minus 3 percentage points.

Sample proportion $\pm$ margin of error $\rightarrow$ 0.56 $\pm$ 0.03 $\rightarrow$ (0.53, 0.59).

This survey was taken after the deadly school shootings in Colorado in April of 1999. The article also states that just before the Colorado shootings, 55% of adults favored tougher gun laws.

Does the confidence interval provide convincing evidence to say that the rate has *changed* from the previous proportion of 0.55? Since the value of 0.55 is in the above 95% confidence interval, the value of 0.55 cannot be rejected as a plausible value for the population proportion p. There does not appear to be convincing evidence to say that the rate has changed from the previous proportion of 0.55. ❖

✳ Exercises

9.36 Social workers claim that 15% of the beds in homeless shelters in San Francisco are assigned to families. In a simple random sample of 1395 shelter beds, 196 were assigned to families.

  (a) Give the estimate of the population proportion, p, of beds assigned to families among all homeless-shelter beds in San Francisco.

  (b) Give a 99% confidence interval for the population proportion p of beds assigned to families among all homeless-shelter beds in San Francisco. Show all steps involved in getting your answer.

  (c) What is the *width* of your confidence interval in (b)?

  (d) What is the *margin of error* for your confidence interval in (b)?

  (e) If you constructed a 90% confidence interval using the same data, the margin of error would be (select one) *the same, larger, smaller*.

  (f) Consider the hypotheses H_0: $p = 0.15$ versus H_1: $p \neq 0.15$, stated in terms of the population proportion p of beds reserved for families among all homeless-shelter beds in San Francisco. Use the confidence interval from (b) to test the hypotheses with a 1% significance level. Give your decision and explain how you made that decision.

9.37 As part of a quality improvement program, a mail-order company is studying the process of filling customers' orders. According to company standards, an order is shipped on time if it is sent within three working days of the time it is received. The company claims that 87% of the orders are shipped on time. In a random sample of 100 orders received in the past month, an audit reveals that 86 of these orders were shipped on time.

  (a) The preceding paragraph provides a proportion of $\frac{86}{100}$. Is this proportion a parameter or a statistic?

  (b) Give a 95% confidence interval for p, the true proportion of the month's orders that were shipped on time. Show all steps involved in getting your answer.

  (c) What is the margin of error of the 95% confidence interval compute in (a)?

  (d) Suppose that the company would like to report a 95% confidence interval for the true proportion p with a margin of error of 4% (i.e., $E = 0.04$). What sample size would be needed to produce such an interval estimate? Show all work.

  (e) A Better Business Bureau would like to test the theory that p, the population proportion of orders shipped on time, is less than the percentage given by the company. State the appropriate null and alternative hypothesis to conduct this test.

  (f) Conduct the test described in (e) using a 5% significance level. Show all work and give your decision and how you reached that decision.

9.38 What type of car, American or foreign, is driven more among college students? A study was conducted to estimate the proportion of college students who drive a foreign car. A random sample of 160 students were asked which type of car they drove and 96 students sampled responded that they drive a foreign car.

(a) Give a 95% confidence interval for the population proportion of college students who drive a foreign car.

(b) Based on the confidence interval, does it appear that foreign cars are driven *more* among college students? Explain.

(c) Is the following statement a correct interpretation of the meaning of the 95% level of confidence?

If repeated random samples of 160 college students are obtained, we would expect 95% of the resulting samples to have a sample proportion who drive a foreign car that falls in the interval computed in (a).

(d) What minimum sample size would be necessary to have a 95% confidence interval with a margin of error of 0.05?

Chapter Summary

In this chapter, we have studied two forms of statistical inference: hypothesis testing, which allows us to make a statistical decision about a population parameter, and confidence interval estimation of a population parameter.

Hypothesis Testing

We have two competing theories about a population parameter of interest—H_0 and H_1. We gather and summarize data to assess which theory is best supported. A test statistic is computed that measures the distance (in standardized units) between our estimate of the parameter and the hypothesized value for the parameter under H_0. The distribution of this test statistic shows us what to expect under H_0 and, the p-value allows us to assess whether our observed results are consistent with H_0 or are unusual under H_0. The p-value is the probability of getting a value as extreme or

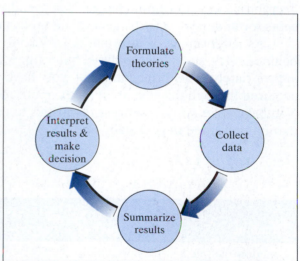

more extreme than the observed value if H_0 were true. The smaller the p-value, the more the evidence against H_0, in support of H_1. If the p-value is sufficiently small, less than or equal to the significance level, our decision is to reject the null hypothesis and the results are statistically significant. Once a decision has been made, it is possible that an error has occurred. Some users of statistics will claim that the statistical results prove that their product is better. But remember that the hypothesis test does not prove or disprove anything. Statistics allows us to make sound decisions based on information from data. The quality of the decision made will depend on the quality of the data obtained. We are able to quantify how much "faith" we have in our decision through the p-value.

Confidence Interval Estimation

Rather than estimating a population parameter with just a point (single number) estimate, we now have an interval estimate centered at the point estimate. The width of the interval depends on the variability in our data, the sample size, and how confident we want to be that our interval will contain the true parameter value. The general form of a confidence interval is

$$\text{(point estimator)} \pm \text{(a few)}\,\text{(standard errors of the point estimator)}.$$

It is important to understand just what that confidence level $1 - \alpha$ means. It is the probability that we will obtain a confidence interval that contains the parameter value. This probability applies only *before* we actually collect the data and compute our interval. The confidence level tells us what proportion of intervals constructed with the given method is expected to contain the parameter. Any one constructed interval either does or does not contain the parameter. A 95% confidence level tells us that we used a method that gives "correct" results for 95% of all possible samples.

In this chapter, we have focused our discussion on inferences for one population regarding a population proportion. These inference techniques are summarized in the table that follows. However, results reported in newspapers and journals are not always about a single proportion. There are statistical procedures available for means and for comparing the means or proportions for two or more populations. There are statistical methods available that do not require the assumption of normality. Which statistical method should be used depends on the number of populations under study, the type of responses measured, and the type of sampling used. Although the scenarios change, the way of thinking about decision making does not change. You are already equipped with the basic tools and concepts for understanding the general methods presented in the subsequent chapters.

As a consumer of information, your job is not to understand the detailed formulas behind every statistical analysis, but rather to understand how to interpret the results appropriately. The principles that you have learned and skills that you have already acquired through this text are directly applicable to understanding most statistical analyses. You have a sound foundation of statistics to understand and assess the value of the information presented in your daily life.

Situation	Parameter of Interest	Hypothesis Test Statistic	Confidence Interval
1 random sample from a population (large Sample)	population proportion p	$z = \dfrac{\hat{p} - p_0}{\sqrt{\dfrac{p_0(1 - p_0)}{n}}}$	$\hat{p} \pm z^* \left(\sqrt{\dfrac{\hat{p}(1 - \hat{p})}{n}} \right)$

Key Terms

Be sure you can describe, in your own words, and give an example of each of the following key terms from this chapter:

null hypothesis
alternative hypothesis
significance level
sampling distribution
z-test statistic for a
   proportion
upper-tailed test

lower-tailed test
two-tailed test
Type I error
Type II error
decision rule
p-value
point estimate

confidence interval estimate
level of confidence
width of the interval
margin of error
required sample size

Exercises

9.39 It has been believed that two-thirds of all women work outside of the home. A recent poll of 200 randomly selected women was conducted to assess if the proportion has changed from what was believed. State the appropriate hypotheses to be tested.

9.40 An official agency claims that the proportion of seniors (those aged 65 years or older) in a community is 0.20. A researcher doubts this claim and suspects that the proportion is actually lower. A simple random sample of 250 people is taken from this community, and among these 250 people, 40 are seniors.

(a) State the hypotheses to be tested.

(b) Report the observed test statistic value and the *p*-value for the test.

(c) Are the results statistically significant at the 5% level? Explain.

9.41 State the appropriate null and alternative hypotheses for each of the following situations (also, state whether the direction of extreme is one-sided to the right, one-sided to the left, or two-sided):

(a) The success rate for a standard treatment is 72%. A study is to be conducted to assess of the success rate for a new treatment is higher. Let *p* be the population proportion of successes with the new treatment.

(b) Approximately 90% of the population is right handed. A researcher would like to assess if people with a European background are less likely to be right handed. Let *p* be the population proportion of people with a European background that are right handed.

(c) Last term, approximately 60% of the students in a course attended office hours of the instructor. The instructor has changed the office hours and would like to assess if the proportion of students who will attend office hours this term will change. Let *p* be the population proportion of students this term who attend office hours of the instructor.

9.42 A random sample of $n = 12$ students from a large class of 1200 students are polled as to whether or not they have finished the current homework set due in one week. Explain why we cannot use the methods described in this chapter for learning about the population proportion of students in the class that have finished the current homework set.

9.43 An *Ann Arbor News* article entitled "Preventing Birth Defects," July 2, 1995, states that, in recent years, new information has been gathered about the relationship between living habits before pregnancy and problems such as miscarriage, birth defects, low birth weight, and premature infants. To assess the extent of prepregnancy problems among women of childbearing age, the March of Dimes Birth Defects Foundation sponsored a Gallup survey in early 1995. A national telephone sampling of 2010 women from 18 to 45 years of age was obtained. Some of the results are as follows:

- Among those who had been pregnant, 73% waited to see the doctor until after they thought they were pregnant.

- Nearly half the women questioned had never even heard of folic acid.

- Only 15% of the women were aware of the Public Health Service recommendation that all women of childbearing age should consume four-tenths of a milligram of folic acid a day.

- Of those who had heard of folic acid, 90% did not know why it was important. (Folic acid helps to prevent spina bifida and the fatal defect anencephaly.)

(a) Based on the results of the study, can you provide a 95% confidence interval estimate for the proportion of pregnant women who do not see the doctor until after they think they are pregnant? If so, do so, and interpret your interval. If not, why not?

(b) Based on the results of the study, can you provide a 95% confidence interval estimate for the proportion of women who are aware of the folic acid daily recommendation? If so, do so, and interpret your interval. If not, why not?

(c) Approximately how many of the women surveyed know why folic acid is important?

9.44 A city committee on crime wants to estimate the proportion of crimes related to firearms for the city. The committee randomly selects 60 files of recently committed crimes in the city and records the number in which a firearm was reportedly used. A 99% confidence interval for p, the true proportion of crimes in the city in which a firearm was reportedly used, is given by $(0.58, 0.68)$.

(a) The sample proportion of crimes in which a firearm was reportedly used is (select one)

    (i) 60.   (ii) 99.   (iii) 0.58.   (iv) 0.63.   (v) 0.68.

    (vi) cannot be determined from the information given.

(b) Consider the following statement: "The probability that the interval $(0.58, 0.68)$ contains p, the true proportion of crimes in the city in which a firearm was reportedly used, is 0.99." Is this a correct statement? (select one)   Yes   No   Explain.

9.45 Answer each of the following questions:

(a) You are reading an article in your field that reports several statistical analyses. The article says that the p-value for a significance test is 0.045. Is this result statistically significant at the 5% level?

(b) A result is statistically significant at the 5% level. Are such results always, sometimes, or never significant at the 1% level?

REAL DATA

9.46 A subheadline in the January 26, 1998 issue of the *Ann Arbor News* read as follows: "Polls show asking former intern to lie would be cause for leaving." An ABC News/*Washington Post* poll stated that 59% say President Clinton should resign if he lied. This result was based on a poll of 1537 adults surveyed and had a margin of error of plus or minus 3%.

(a) How many of the adults surveyed feel President Clinton should resign if he lied?

(b) What minimum sample size is required to estimate a proportion with 95% confidence and a 3% margin of error (i.e., within 0.03)? Did the ABC News/*Washington Post* poll sample enough adults to meet this requirement?
   Select one: Yes   No   Can't tell

(c) Using the stated margin of error of plus or minus 3%, give the corresponding confidence interval estimate for the *proportion* of adults who feel President Clinton should resign if he lied.

REAL DATA

9.47 The March 26, 1995, *New York Times* article entitled "Another Day Older, and the Clock Just Keeps On Ticking" stated that only half of all Americans have put aside money specifically for retirement. The actual percentage of those who have not put aside money reported in the article was 52%. This poll was based on telephone interviews conducted March 9 through 12 with 1156 adults around the United States, excluding Alaska and Hawaii. The sample of telephone exchanges called was selected by a computer from a complete list of exchanges in the country. For each exchange, the telephone numbers were formed by random digits, thus permitting access to both listed and unlisted numbers.

(a) Give a 95% confidence interval for the proportion of (nonretired) Americans who have *not* put aside money specifically for retirement.

(b) The article states, "in 19 cases out of 20 the results based on such samples will differ by no more than three percentage points in either direction." What does the 19 out of 20 correspond to? What does this 0.95 level mean? Did your 95% confidence interval in (a) have a margin of error of approximately 3%?

(c) Here is an example of good reporting of results. We are informed on how the poll was conducted. The possibilities of other errors are even acknowledged in the following excerpt: "In addition to sampling error, the practical difficulties of conducting any survey of public opinion may introduce other sources of error into the poll. Variations in question wording or other order of questions, for instance, can lead to somewhat different results." Can you describe another possible source of error?

(d) The confidence interval computed in (a) assumes that the data are a random sample. For this poll, the exchanges were chosen to assure that each region of the country was represented in proportion to its population. Within each household, one adult was designated by a random procedure to be the respondent for the survey. What type of sampling technique was used to select the 1156 adults?

9.48 A survey on the eating and food buying habits of American adults reported a 99% confidence interval for the proportion of adults who went on at least one diet in the past year, *p*, to be (0.275, 0.325).

(a) Does this mean $P(0.275 < p < 0.325) = 0.99$? Select one: Yes   No

(b) Explain briefly what the 99% confidence interval means.

(c) What is the value of the sample proportion $\hat{p}$?

(d) The results were based on a survey of 2210 American adults. Approximately *how many* adults in the sample went on at least one diet in the past year?

9.49 A newspaper article headline stated "Poll: Most workers see job switch in 5 years" (Associated Press, October 27, 1997). You are asked to use the data from the poll to assess if a majority of workers expect to leave their job within the next five years (that is, assess if the proportion is more than 0.50). A simple random sample of 714 workers was taken as part of a Harris Poll. Among these 714 workers, 380 stated they expect to leave their job within the next five years.

(a) State the hypotheses to be tested.

(b) Report the test statistic formula and its corresponding value and the *p*-value for the test.

(c) Are the results statistically significant at the 5% level?   Yes   No Explain your answer.

9.50 The "Ask Marilyn" column (*Parade* magazine, May 31, 1998) discusses the "margin of error" for estimating a population proportion reported in public opinion polls.

> *Reader's Question*: Can you explain how they arrive at the so-called "margin of error" in public opinion polls?
>
> *Marilyn's Response*: Good polling is a tricky business, but the guiding principle is simple: The larger the sample, the more accurate it is. After much data collection, pollsters have learned their numerical limits of accuracy and call them collectively the "margin of error." The individual numbers are considered standard. For this reason, the published margin of error on a particular poll merely tells us the size of the sample. It is based on past polls. For example, if a poll has a margin of error of plus or minus 3%, this usually tells us that about 1500 people were polled. That is, the margin-of-error percentage is assigned to the poll, not developed from it. Smaller samples have larger margins, and larger samples have smaller ones, but only slightly. For most purposes, a national sampling of 1500 is adequate. In fact, most public-opinion polls use samples ranging in size from only 1000 to 2000 people, but this is amazingly sufficient.

(a) Do you agree with the statement that the margin of error merely tells us the size of the sample? What factors determine the margin of error?

(b) What sample size is required to have a 95% confidence interval with a margin of error of plus or minus 3%? How does this sample size compare to the 1500 stated in Marilyn's response?

9.51 After once again losing a basketball game to the arch rival, a college's alumni association conducted a survey to see if alumni were in favor of firing the coach. A random sample of 200 alumni was taken and 65% were in favor of firing the coach.

(a) State the hypotheses for testing that more than 60% of all alumni are in favor of firing the coach.

(b) Using a significance level of 0.01, test the hypotheses in (a). Show all work including your test statistic formula, test statistic value, corresponding *p*-value, and report whether the results are statistically significant.

(c) Does the test you just performed require that the observations (data) come from a population having a normal distribution?

9.52 It is claimed that 75% of women wear shoes that are too small. The American Orthopedic Foot and Ankle Society in a study of 356 women found that 313 women were wearing shoes that were at least one size too small. Use the study results to test the hypotheses H_0: $p = 0.75$ versus H_1: $p \neq 0.75$ using a 1% significance level. Give your decision and explain how you made that decision.

REAL DATA

9.53 On January 22, 1997, the *Tampa Tribune* published a result from the National Center for Education Statistics. It stated that 48% of college students work to pay tuition and living expenses. The National Center had sampled 450 college students.

(a) Obtain a 90% confidence interval for the population proportion of college students who work to pay tuition and living expenses.

(b) What is the margin of error for your confidence interval in (a)?

(c) What would happen to the margin of error if the confidence level were increased to 99%? Verify your answer.

REAL DATA

9.54 *USA Today*, April 11, 1995, found that from a "random" sample of $n = 1993$ business travelers, 618 reported the airline's frequent flyer program the most important factor in choosing an airline carrier to fly.

(a) Find the point estimate of $p =$ the population proportion of business travelers who choose an airline to fly according to the airline's frequent flyer program.

(b) Use your point estimate in (a) to create a 95% confidence interval for p.

(c) How large of a sample would be needed for an error margin of 2% with 95% confidence?

9.55 Find an article that discusses the results of a poll in which a proportion (or percent) is reported and the error margin is included in the write-up.

(a) State the population under study.

(b) State the question of interest for which the margin of error applies. Include what the parameter p corresponds to.

(c) Give the corresponding confidence interval estimate for p using the stated margin of error.

(d) Compute the margin of error from the results and compare it to the stated margin of error.

(e) Give at least one question you have about the poll that would help you know if the confidence interval procedure is valid.

9.56 The head of an educational institution claims that his students are so intelligent that half of them have an intelligence quotient (IQ) as high as 140. On the basis of this claim, he approaches a funding agency and seeks additional grants for his institution. In an attempt to verify the claim, the funding agency selects a random sample of 50 students and administers an IQ test to the students in the sample. The frequency distribution of the IQ figures in the sample is as follows:

IQ Range	Frequency
110–119	4
120–129	12
130–139	19
140–149	11
150–159	4

(a) Write down the appropriate null and alternative hypotheses for the present problem.

(b) What will be your conclusion at the 5% level of significance?

(c) Suppose the sample size was 10 instead of 50. Would you have any difficulty in repeating what you did in (b)? Explain.

TI Quick Steps

Hypothesis Testing and Confidence Intervals with the TI

The TI has a TESTS menu located under the STAT menu that allows you to perform hypothesis tests and generate confidence intervals.

One-sample Z-test for a proportion

You need to specify the hypothesized proportion, the number of "successes" in the sample (denoted by x), the sample size n, and the direction for the alternative hypothesis. To perform a test, the steps are as follows:

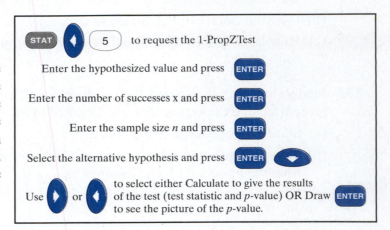

The input screen for Example 9.2 is given by:

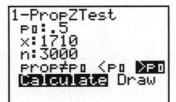

The corresponding test result screen is:

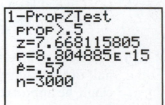

One-sample Z confidence interval for a proportion

You need to specify the number of "successes" in the sample (denoted by x), the sample size n, and the confidence level. To generate the confidence interval, the steps are as follows:

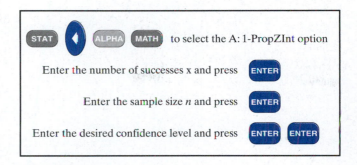

STAT ◀ ALPHA MATH to select the A: 1-PropZInt option

Enter the number of successes x and press ENTER

Enter the sample size n and press ENTER

Enter the desired confidence level and press ENTER ENTER

The input screen for Example 9.6 is given by:

```
1-PropZInt
 x:590
 n:3066
 C-Level:.95
 Calculate
```

The corresponding result screen is:

```
1-PropZInt
 (.17848,.20639)
 p=.1924331376
 n=3066
```

Chapter 10

MAKING DECISIONS ABOUT A POPULATION MEAN WITH CONFIDENCE

10.1 Introduction

In Chapter 9, we studied the methods for drawing conclusions about a single population proportion p. We learned the basic steps of hypothesis testing and performed a z-test for one proportion. The decision about whether to reject a null hypothesis was based on the sample through a test statistic, in this case, a standard score, or z-test statistic. This z-test statistic was a measure of how far away the sample proportion was from the hypothesized proportion in H_0, in which the distance was reported in standard deviation units. We assessed whether this distance was significant by computing a p-value, the probability of seeing such a distance or one even more extreme under H_0. If the p-value was small enough, the results

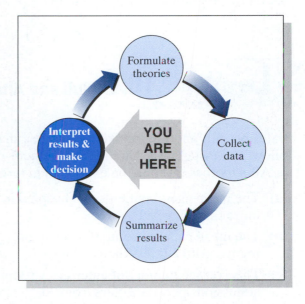

were declared to be *statistically significant*. We learned that a confidence interval for the proportion p is an interval of values computed from the sample that are reasonable values for the population proportion p. The confidence level reflects the amount of confidence we have in the *procedure* we used to make the interval. We can be quite confident that most of the intervals made with this procedure (using a new random sample of the same size n each time) will contain the population proportion p.

The general concepts that we learned in Chapter 9 apply to many other statistical inference scenarios. In this chapter, we will illustrate the concepts of hypothesis testing and confidence interval estimation for a single population mean μ. Suppose that we are interested in learning about the value of a population mean μ. If we have a random sample from the population of interest, we would use the observed sample mean $\bar{x}$ to estimate the population mean μ. Based on the information from the sample, we might like to know if there is evidence to say that the population mean has changed from some target level or how confident we are in our estimate at representing the true mean μ. In Section 10.2, we focus on testing hypotheses. We already have the foundation needed to do these tests. In fact, we performed one version of the tests in "Let's do it! 8.11." The summary of testing here will be a little more formal and will introduce a new version of the test for a mean that is primarily used in practice. Confidence interval estimation will be presented in Section 10.3. Example 8.12 provided an insight to the techniques in this section. Section 10.4 will revisit the connection between two-sided testing and confidence intervals.

10.2 Testing Hypotheses about a Population Mean

Many of the ideas from the previous chapter on testing for a population proportion carry over to testing for a population mean. We will start by stating the hypotheses about the population mean. The data will be collected and summarized in the form of a test statistic, which is again a type of standardized score. It will also be important to set up the alternative hypothesis correctly. The direction of the alternative hypothesis corresponds to the direction of extreme for computing the p-value. "Let's do it! 10.1" provides some practice at stating the hypotheses for testing about a population mean.

Let's do it! 10.1 Null and Alternative Hypotheses

State the null and alternative hypotheses that would be used to test the statements that follow. These statements are the researcher's claim, to be stated as the alternative hypothesis. All hypotheses should be expressed in terms of μ, the population mean of interest.

(a) The mean age of patients at a hospital is
more than 60 years. H_0:   versus H_1:

(b) The mean caffeine content in a cup of
regular coffee is less than 110 mg. H_0:   versus H_1:

(c) The average number of emergency room
admissions per day differs from 20. H_0:   versus H_1:

Once the hypotheses and the significance level have been established, the data are collected. The sample mean $\bar{x}$ is then computed from the data. This sample mean $\bar{x}$ is an estimate of the population mean μ and has a sampling distribution that is at least approximately normal in most cases. As we saw in the tests for a proportion, rather than reporting the value of $\bar{x}$, its sampling distribution under H_0, and then the p-value, a researcher might prefer to report the value of $\bar{x}$ and its corresponding standard score. This standard score, or z-score, would follow (approximately) a standard normal distribution.

Sampling Distribution of $\overline{X}$, the Sample Mean

If a simple random sample of size n is taken from a population having population mean μ and population standard deviation σ, and if the original population is normally distributed, then

$$\overline{X} \text{ is } N\left(\mu, \frac{\sigma}{\sqrt{n}}\right), \text{ and thus } Z = \frac{\overline{X} - \mu}{\frac{\sigma}{\sqrt{n}}} \text{ is } N(0, 1).$$

If the original population is not necessarily normally distributed, but the sample size n is large enough ($n \geq 30$), then

$$\overline{X} \text{ is } approximately \ N\left(\mu, \frac{\sigma}{\sqrt{n}}\right) \textbf{ (central limit theorem)},$$

and thus $Z = \dfrac{\overline{X} - \mu}{\frac{\sigma}{\sqrt{n}}}$ is *approximately* $N(0, 1)$.

In testing hypotheses, we examine the evidence based on data, the sample mean $\overline{x}$, under the assumption that the null hypothesis H_0 is true. The null hypothesis would give us a value for the population mean μ to use in the foregoing z-score expression, but computing the value of z also requires knowing the value of the population standard deviation σ, which in practice is rarely available. In the rare cases in which the population standard deviation is assumed to be known, a one-sample z-test for the mean can be performed using the z-score in the preceding sampling distribution summary as the test statistic. Just as with the one-sample z-test for the proportion in Chapter 9, we would use the standard normal distribution to find the p-value of the test. We present just one example of this z-test with a "Let's do it!" exercise to follow and then move directly to the more practical situation of estimating σ with the sample standard deviation s and using this modified standard score as our test statistic.

❖ EXAMPLE 10.1   Too Much Carbon Monoxide? (σ known)

REAL
DATA

The Federal Trade Commission annually rates varieties of domestic cigarettes according to their tar, nicotine, and carbon monoxide content. The U.S. Surgeon General considers each of these substances hazardous to a smoker's health. Suppose that in the past, the average carbon monoxide content has been 15 mg with a *standard deviation of 4.8 mg*. Have cigarettes improved on average with respect to carbon monoxide content (lower content means better)? If we let μ be the population mean carbon monoxide content for domestic cigarettes, and that in the past the mean was 15 mg, then we wish to know if the current mean μ is less than the past mean of 15 mg. This is expressed as the following hypotheses about μ, to be tested using $\alpha = 0.05$:

$$H_0: \mu = 15 \quad \text{versus} \quad H_1: \mu < 15.$$

Data on carbon monoxide content (in mg) for a sample of 25 domestic brand cigarettes have been obtained to perform this analysis. The histogram of the data support the assumption of a normal distribution for carbon monoxide content. We will assume that the population standard deviation has not changed and is still $\sigma = 4.8$ mg. The mean in the output provided was based on the sample of size $n = 25$ domestic brands of cigarettes. Hence, the 12.528 is an observed sample statistic, denoted by $\bar{x}$.

This sample mean of 12.528 does show some support for the alternative hypothesis, since it is less than 15. In general, the *p*-value is the probability of getting a value that is as extreme or more extreme than the observed value, assuming the null hypothesis H_0 is true. The direction of extreme in this case is **one-sided to the left** corresponding to values that are small.

Brand	Carbon Monoxide Content (mg)
Alpine	13.6
BensonHedges	16.6
BullDurham	23.5
CamelLights	10.2
Carlton	5.4
Chesterfield	15.0
GoldenLights	9.0
Kent	12.3
Kool	16.3
LandM	15.4
LarkLights	13.0
Marlboro	14.4
Merit	10.0
MultiFilter	10.2
NewportLights	9.5
Now	1.5
OldGold	18.5
PallMallLights	12.6
Raleigh	17.5
SalemUltra	4.9
Tareyton	15.9
True	8.5
ViceroyRichLights	10.6
VirginiaSlim	13.9
WinstonLights	14.9

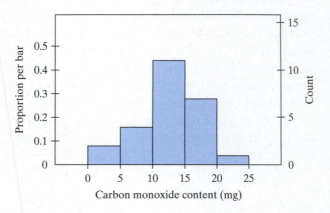

```
TOTAL OBSERVATIONS:        25

N OF CASES                 25
N OF CASES                 25
MINIMUM                 1.500
MAXIMUM                23.500
MEAN                   12.528
STANDARD DEV            4.740
STD. ERROR              0.948
MEDIAN                 13.000
```

How do we find the *p*-value?

With the TI

1. **Using the sampling distribution of the sample mean and the normalcdf(function**

We consider the observed value to be the sample mean $\bar{x} = 12.528$. If the null hypothesis is true, the sample mean will have (approximately) a normal distribution with a mean of $\mu = 15$ and a standard deviation of $\frac{\sigma}{\sqrt{n}} = \frac{4.8}{\sqrt{25}} = 0.96$. A somewhat detailed sketch of this distribution is shown. The *p*-value will be the probability of getting a sample mean of 12.528 or less under this normal distribution, which, from the sketch, will be somewhat small. We can find this area using the normalcdf(function.

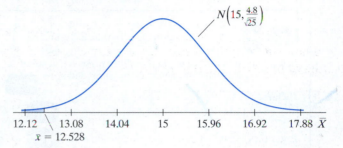

$$p\text{-value} = P(\overline{X} \le 12.528) = \text{normalcdf}(-\text{E99}, 12.528, 15, 0.96) = 0.00501.$$

2. **Using the distribution of the Z-test statistic and the normalcdf(function**

We standardize the observed sample mean $\bar{x} = 12.528$ using the mean and standard deviation of its sampling distribution under H_0. We consider the observed value to be the corresponding test statistic value $z = \frac{12.528 - 15}{0.96} = -2.58$. If the null hypothesis is true, the Z-test statistic will have approximately a standard normal distribution $N(0, 1)$.

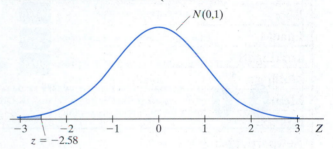

A sketch of the standard normal distribution is shown. The *p*-value will be the probability of getting a Z-test statistic value of -2.58 or less under the $N(0, 1)$ distribution, which we know will be somewhat small. We can find this area using the normalcdf(function.

$$p\text{-value} = P(Z \le -2.58) = \text{normalcdf}(-\text{E99}, -2.58) = 0.00494.$$

Note: The difference between this *p*-value and the previous one is slight and due to the rounding of the standard deviation in the computations.

3. **Using the Z-Test function under the STAT TESTS menu**

The TI has a TESTS menu located under the STAT menu that allows you to perform various hypothesis tests and generate various confidence intervals. The test for a population mean when the population standard deviation is known is Option 1 in this list and is called a *one-sample z-test for a population mean*, abbreviated **Z-Test**. You can either have the sample data entered into a list, say L1, or just enter the Stats (the sample mean, population standard deviation, and sample size). The steps are summarized and the corresponding input and output screens are shown.

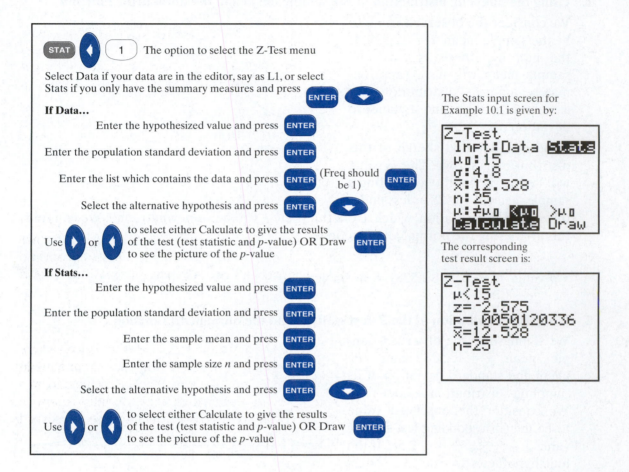

Note that the *Calculate* option produces an output screen that provides everything needed to report the results of the test and make a decision. It reports the alternative hypothesis being tested, the observed value of the standardized z-test statistic $z = -2.575$, the p-value = 0.0050120336, the sample mean $\bar{x} = 12.528$, and the sample size $n = 25$. The *Draw* option would draw the normal distribution and shade in the area that corresponds to the p-value. The value of z-test statistic and the p-value are also given. If you select the *Draw* option, just be careful that other graphing options have been turned off and cleared out, or you will receive an error message.

With Table II

We need to compute the value of the z-test statistic, the standard score for $\bar{x} = 12.528$ using the mean and standard deviation of its sampling distribution under H_0. The mean is $\mu = 15$ and the standard deviation is $\frac{\sigma}{\sqrt{n}} = \frac{4.8}{\sqrt{25}} = 0.96$, so the test statistic is $z = \frac{12.528 - 15}{0.96} = -2.58$. The value -2.58 tells us that the observed sample mean of 12.528 is 2.58 standard deviations below the hypothesized mean of 15. Then we use Table II to find the area that corresponds to the p-value—in this case, the area to the left of -2.58:

$$\text{p-value} = P(Z \le -2.58) = 0.0049.$$

What Is the Decision and Conclusion?

Thus, for the test of H_0: $\mu = 15$ versus H_1: $\mu < 15$, the *p*-value is about 0.005. This is quite a small *p*-value. The result that we observed, a sample mean of 12.528, or something more extreme, would occur only about 0.5% of the time in repeated sampling under the null hypothesis. The observed result is very unlikely under H_0. Recall that we compare the *p*-value to the significance level, α. If the *p*-value is less than or equal to α, the data are statistically significant and we would reject H_0. Since our *p*-value is smaller than the significance level of 0.05, we would reject H_0. Rejecting H_0 implies that there is evidence that domestic cigarettes have improved on average with respect to the carbon monoxide content. ❖

 **Let's do it! 10.2** **Completing a Maze**

Experiments on learning in animals sometimes measure how long it takes a mouse to find its way through a maze. The mean time is 18 seconds for one particular maze. A researcher thinks that a loud noise will cause the mice to complete the maze faster. The appropriate null and alternative hypotheses regarding μ, the population mean completion time **with a noise stimulus**, are

$$H_0: \mu = 18 \quad \text{versus} \quad H_1: \mu < 18.$$

The significance level for this test is set at $\alpha = 0.10$. The researcher decides to measure how long each of 10 mice takes to complete the maze with a noise as stimulus. The sample mean time to complete the race for these 10 mice was 17 seconds. Assume that the 10 mice form a random sample and that the population of completion times with a noise stimulus follows a normal distribution with a *known population standard deviation of 2 seconds*. The observed test statistic is given as

$$z = \frac{17 - 18}{\dfrac{2}{\sqrt{10}}} = -1.58.$$

(a) Sketch the picture to show the *p*-value. Label the axis and give the name of the distribution.

(b) Using either the TI or Table II, find the *p*-value for the test.

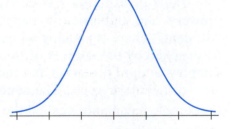

(c) Give your decision at a 10% level. Write up your conclusion in words that someone with no background in statistics would understand.

The testing procedure for the mean in the previous example and "Let's do it!" exercise requires that the population have a normal distribution with a ***known population standard deviation***. If we are interested in learning about the population mean (that is, the population mean is unknown), then it is not likely that the population standard deviation is known. Therefore, the remainder of this section focuses on the testing of hypotheses for the population mean in the case that the original population is normally distributed (or the sample size is large) the ***population standard deviation σ is unknown***, and a random sample from the original population is obtained.

If we do not know the true population standard deviation σ, we can estimate it by the sample standard deviation s. What happens if we replace σ in the standardized quantity

$$\frac{\overline{x} - \mu_0}{\frac{\sigma}{\sqrt{n}}}$$

by the sample standard deviation s? We need to first learn about the distribution of

$$\frac{\overline{x} - \mu_0}{\frac{s}{\sqrt{n}}}$$

under the null hypothesis. We know that the sample mean will vary from sample to sample and so will the value of the sample standard deviation. Thus, replacing σ by s results in a standardized quantity that will vary even more, since the sample standard deviation introduces another source of variation. The resulting **test statistic**

$$t = \frac{\overline{x} - \mu_0}{\frac{s}{\sqrt{n}}}$$

is called a **one-sample t-test statistic** and the distribution of the T variable under H_0 is called a **Student's t-distribution with $n - 1$ degrees of freedom**.

The t-distribution was derived by W. S. Gosset in 1908. Gosset worked in an Irish brewery and published a paper about the t-distribution under the pseudonym of "Student." There is actually a family of t-distributions, one for each sample size, which is determined by the degrees of freedom. Recall that "$n - 1$" is in the denominator of our sample standard deviation and can be thought of as an index for identifying the appropriate t-distribution to use. The denominator $\frac{s}{\sqrt{n}}$ is an estimate of the standard deviation of the sample mean and is called the **standard error of the mean**, often abbreviated **SEM**.

> **Definition:**   When data are used to estimate the standard deviation of a statistic, the result is called the **standard error** of the statistic.
>
> **Note:**   Sometimes the term "standard error" is used for the actual standard deviation of a statistic, that is, $\frac{\sigma}{\sqrt{n}}$, and the estimated value based on data, $\frac{s}{\sqrt{n}}$, is called the "estimated standard error." However, we will use the term "standard error" only when the standard deviation is estimated from the data since many statistical computer packages use this meaning in the output.

Properties of the *t*-distribution are summarized next. The figure shows how the standard normal distribution and *t*-distributions are related.

Properties of the *t*-distributions

- The *t*-distribution has a symmetric bell-shaped density centered at 0, similar to the $N(0, 1)$ distribution.
- The *t*-distribution is "flatter" and has "heavier tails" than the $N(0, 1)$ distribution.
- As the sample size increases, the *t*-distribution approaches the $N(0, 1)$ distribution.

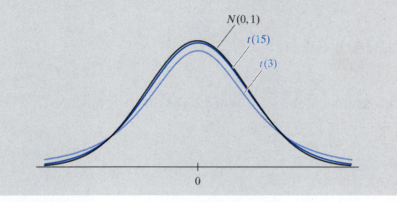

Many calculators have the ability to compute areas under a *t*-distribution and report percentiles of a *t*-distribution. The TI graphing calculator has a STAT TESTS menu that allows you to perform hypothesis tests for a population mean when the population standard deviation is unknown. For details on finding areas and percentiles for *t*-distributions with a TI graphing calculator, see the ■ TI Quick Steps at the end of this chapter. Table IV, presented at the end of this chapter, summarizes some percentiles for various *t*-distributions. Our next example shows how to perform a one-sample *t*-test for the population mean using the TI calculator, Table IV, and computer output.

❖ **EXAMPLE 10.2** **Waste Data**

REAL
DATA

A Levi Strauss & Co. clothing manufacturing plant in Albuquerque, NM, has a quality control department. Every week data are collected from its suppliers on the percentage waste (called runoff) relative to what can be achieved by computer layouts of patterns on cloth. A negative value for waste would indicate that the plant employees beat the computer in controlling waste for that week. The waste data for one of the supplier plants for 19 weeks are provided to

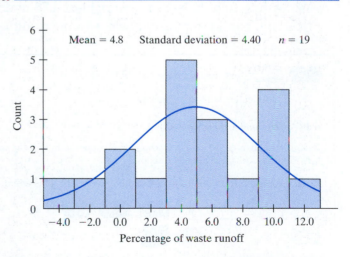

assess if, on average, this plant performs worse than the computer at controlling waste. If we let μ be the mean percentage of waste for this plant, then we wish to know if the mean μ is greater than 0%. This is expressed in terms of the following hypotheses about μ:

H_0: $\mu = 0$ versus H_1: $\mu > 0$. A significance level of 1% is to be used. From the alternative hypothesis, we see that in this case the direction of extreme is **one-sided to the right**. The data, histogram, and summary measures for the 19 weeks are provided.

Percentage of Waste Data

12.1, 9.7, 7.4, −2.1, 10.1, 4.7, 4.6, 3.9, 3.6, 9.6, 9.8, 6.5, 5.7, 5.1, 3.4, −0.8, −3.9, 0.9, 1.5

The mean of 4.8 is the observed sample mean $\overline{x}$. The standard deviation of 4.40 was also based on the sample results and is the observed sample standard deviation, s. Although descriptively the sample mean is more than 0, we need to address if it is *significantly more than* 0.

How do we find the *p*-value?

With the TI

1. **Using the distribution of the *t*-test statistic and the tcdf(function**

   We standardize the observed sample mean $\overline{x} = 4.8$ using the mean under H_0, namely, 0, and standard error of the mean, which is $\frac{s}{\sqrt{n}} = \frac{4.4}{\sqrt{19}} = 1.01$. We consider the *observed value* to be the corresponding test statistic $t = \frac{4.8 - 0}{1.01} = 4.76$. If the null hypothesis is true, the *t*-test statistic will have approximately a

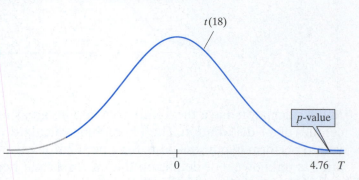

   t-distribution with $n - 1 = 19 - 1 = 18$ degrees of freedom. A sketch of the *t*-distribution is shown. The *p*-value will be the probability of getting a *t*-test statistic value of 4.76 or more under the $t(18)$ distribution. The TI has a tcdf(function that can compute various areas under a *t*-distribution with a specified degrees of freedom. This function is located under the DISTR menu (using the 2nd function and VARS button). We wish to find the area to the right of 4.76 under the *t*-distribution with 18 degrees of freedom. The TI steps are shown with the resulting output screen.

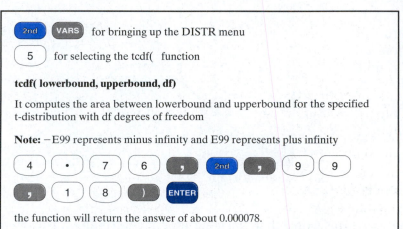

(2nd) (VARS) for bringing up the DISTR menu

(5) for selecting the tcdf(function

tcdf(lowerbound, upperbound, df)

It computes the area between lowerbound and upperbound for the specified t-distribution with df degrees of freedom

Note: −E99 represents minus infinity and E99 represents plus infinity

(4) (.) (7) (6) (,) (2nd) (,) (9) (9)
(,) (1) (8) ( )) (ENTER)

the function will return the answer of about 0.000078.

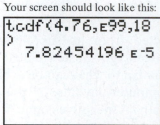

So the *p*-value $= P(T \geq 4.76) = \text{tcdf}(4.76, \text{E}99, 18) = 7.82454196\text{E-}5$ or 0.000078.

2. Using the T-Test function under the STAT TESTS menu

The TI has a TESTS menu located under the STAT menu that allows you to perform various hypothesis tests and generate various confidence intervals. The test for a population mean when the population standard deviation is **unknown** is called the *one-sample t-test for a population mean* and abbreviated **T-Test**. You can either have the sample data entered into a list, say, L1, or just enter the Stats (the sample mean, sample standard deviation, and sample size). The steps are summarized and the corresponding input and output screens are shown.

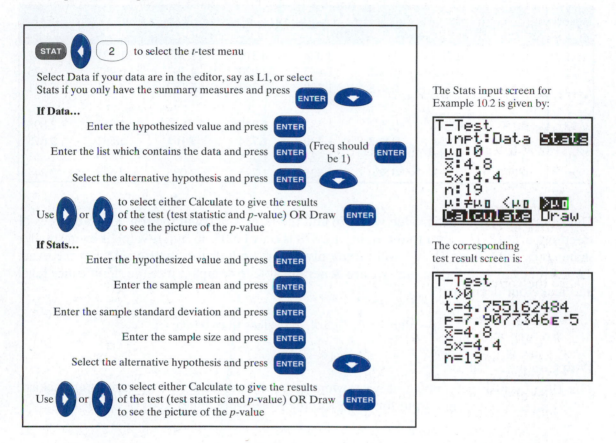

Note that the *Calculate* option produces an output screen that provides everything needed to report the results of the test and make a decision. You have which alternative hypothesis is being tested, the observed value of the standardized t-test statistic ($t = 4.76$), the p-value (7.9077346E-5 or 0.000079), the sample mean $\bar{x} = 4.8$, the sample standard deviation $s = 4.4$, and the sample size $n = 19$. The difference between this p-value and the previous one is slight and due to rounding of the standard deviation in the computations. The *Draw* option would draw the t-distribution and shade in the area that corresponds to the p-value. The value of that t-test statistic and the p-value are also given. If you select the *Draw* option, just be careful that other graphing options have been turned off and cleared out.

With Table IV

We standardize the observed sample mean $\bar{x} = 4.8$ using the mean under H_0, namely, 0, and the standard error of the mean, which is $\frac{s}{\sqrt{n}} = \frac{4.4}{\sqrt{19}} = 1.01$. The observed test statistic is $t = \frac{4.8 - 0}{1.01} = 4.76$. The value 4.76 tells us that the observed sample mean of 4.8 is 4.76 standard errors above the hypothesized mean of 0. This seems like a large difference. If the null hypothesis is true, the t-test statistic will have approximately a t-distribution with

$n - 1 = 19 - 1 = 18$ degrees of freedom. The *p*-value will be the probability of getting a *t*-test statistic value of 4.76 or more under the $t(18)$ distribution. In Table IV, we focus on the row for 18 degrees of freedom. Each value in this row has an area to the right that is specified by the column heading. If we found the exact value of our *t*-test statistic in this row, we would know the exact area to the right, but this does not often happen. In general, we find the value(s) in the row that are close to our observed test statistic value.

Table IV *Upper Percentiles of t-distributions*

df	\multicolumn{8}{c}{Area to the Right}							
	0.40	0.30	0.20	0.10	0.05	0.025	0.01	0.005
1	0.325	0.727	1.376	3.078	6.314	12.706	31.821	63.657
2	0.289	0.617	1.061	1.885	2.920	4.303	6.965	9.925
3	0.277	0.584	0.978	1.638	2.353	3.182	4.541	5.841
⋮	⋮	⋮	⋮	⋮	⋮	⋮	⋮	⋮
17	0.257	0.534	0.863	1.333	1.740	2.110	2.567	2.898
18	**0.257**	**0.534**	**0.862**	**1.330**	**1.734**	**2.101**	**2.552**	**2.878**
19	0.257	0.533	0.861	1.328	1.729	2.093	2.539	2.861
20	0.257	0.533	0.860	1.325	1.725	2.086	2.528	2.845

Our observed test statistic value was 4.76 and the largest entry in the row for 18 degrees of freedom is 2.878. The area to the right of 2.878 is only 0.005, so our *p*-value is even smaller than 0.005. With Table IV, we will often only report bounds for our *p*-value, not the exact value. But even with the bounds, we are generally able to compare it to the significance level and make our decision.

$$p\text{-value} = P(T \geq 4.76) \text{ is less than } 0.005.$$

With Computer Output

The 19 values for percentage of waste were entered into the statistical computer package SPSS and the following output for a one-sample *t*-test were generated:

One-Sample Test

	\multicolumn{6}{c}{Test Value = 0}					
					\multicolumn{2}{c}{95% Confidence Interval}	
	t	df	Sig. (2-tailed)	Mean Difference	Lower	Upper
WASTE	4.783	18	.00015	4.832	2.709	6.954

From the output, we have an observed *t*-test statistic of $t = 4.783$ and a two-sided *p*-value of 0.00015. Our one-sided area to the right of 4.783 will be half of the two-tailed value, *p*-value $= 0.00015/2 = 0.000075$. The output also provides a confidence interval estimate for the mean waste percentage. We will see more on confidence intervals for the mean in our next section.

What Is the Decision and Conclusion?

So at a 1% significance level we would reject H_0. We conclude that the data support that the plant performs worse than the computer at controlling waste, on average. ❖

We have just completed our first example of a **one-sample *t*-test** for the **population mean**. A summary of the key features of this *t*-test is given next.

Summary of the One-Sample *t*-test for a Population Mean: The *p*-value Approach

- We were interested in testing hypotheses about the **population mean** μ. The null hypothesis is $H_0: \mu = \mu_0$, where μ_0 is the **hypothesized value** for μ. The alternative hypothesis provides the direction for the test. These hypotheses are statements about the population mean, not the sample mean. The significance level α is selected.

- The data are assumed to be a **random sample** of size n from the population that has a **normal distribution** with **unknown** population **standard deviation** σ. The normality assumption is not crucial if the sample size is large.

- We base our decision about μ on the **standardized sample mean**, which is

$$t = \frac{\overline{x} - \mu_0}{\dfrac{s}{\sqrt{n}}}.$$

This is the **test statistic**, and the distribution of the variable T under H_0 is a *t*-distribution, with $n - 1$ degrees of freedom.

Note: The test statistic is the same no matter how the alternative hypothesis is expressed.

- We calculate the ***p*-value** for the test, which does depend on how the alternative hypothesis is expressed.

One-sided, to the right

If $H_1: \mu > \mu_0$, then the *p*-value is the area to the right of the observed test statistic under the H_0 model.

One-sided, to the left

If $H_1: \mu < \mu_0$, then the *p*-value is the area to the left of the observed test statistic under the H_0 model.

Two-sided

If $H_1: \mu \neq \mu_0$, then the *p*-value is the area in the two tails, outside of the observed test statistic under the H_0 model.

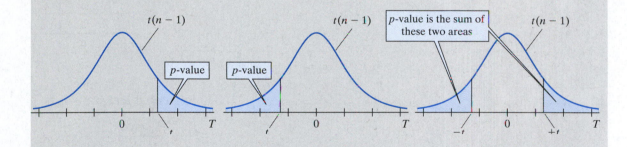

- **Decision:** A *p*-value less than or equal to the significance level leads to rejection of H_0.

Note: The TI graphing calculator has a STAT TESTS menu that allows you to perform hypothesis tests for a population mean. See the ■ TI Quick Steps appendix at the end of this chapter for details.

❖ **EXAMPLE 10.3** **Some Fishy Data**

Data were gathered on 159 fish from seven different species. The fish were caught in the Laengelmavesi Sea, near Tampere in Finland. Various measures of length and width and weight were obtained. * We wish to focus on the perch species and the width measurements. To adjust for differences in the size of a fish due to its age, the data on width were recorded as the percent of the length. We would like to assess if the average width percent of perch from this sea is less than 16 using a 10% significance level. If we let μ be the mean width percent for perch from this sea, then the hypotheses of interest are $H_0: \mu = 16$ versus $H_1: \mu < 16$. Of the 159 fish, 56 were perch. We will use a significance level of $\alpha = 0.05$. The data, histogram, and summary measures for the width percent of the 56 perch are provided.

The observed sample mean is $\bar{x} = 15.839$ and the observed sample standard deviation is $s = 1.362$. This sample mean of 15.839 does show some support for the alternative hypothesis, since it is less than 16. In general, the *p*-value is the probability of getting a value that is as extreme or more extreme than the observed value, assuming the null hypothesis H_0 is true. The direction of extreme is **one-sided to the left**, corresponding to values that are small.

Data:	
16.0	15.7
13.6	14.8
15.2	17.9
15.3	15.0
15.9	15.0
17.3	15.8
16.1	14.3
15.1	15.4
14.6	15.1
13.2	17.7
15.8	17.5
14.7	20.9
16.3	17.6
15.5	17.6
14.5	15.9
15.0	16.2
15.0	18.1
15.0	14.5
17.0	17.8
15.1	16.8
15.1	17.0
15.0	17.6
14.8	15.6
14.9	15.4
14.6	16.1
15.0	16.3
15.9	17.7
13.9	16.3

```
        WIDTH PERCENT

 N OF CASES           56
 MINIMUM          13.200
 MAXIMUM          20.900
 MEAN             15.839
 STANDARD DEV      1.362
 MEDIAN           15.550
```

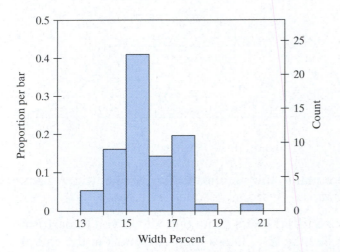

The **observed test statistic** is

$$t = \frac{15.839 - 16}{\dfrac{1.362}{\sqrt{56}}} = -0.885.$$

---

*SOURCE: Brofeldt, Pekka: Bidrag till kaennedom on fiskbestondet i vaera sjoear. Laengelmavesi. T. H. Jaervi: Finlands Fiskeriet Band 4, Meddelanden utgivna av fiskerifoereningen i Finland. Helsingfors, 1917.

This −0.885 tells us that the observed sample mean was 0.885 standard errors *below* the hypothesized mean of 16. We want to measure the chance of getting a test statistic of −0.885 or less under H_0. The model for our test statistic under H_0 is a *t*-distribution with 55 degrees of freedom. The accompanying picture portrays this chance as the area under the *t*-distribution to the left of the observed test statistic value.

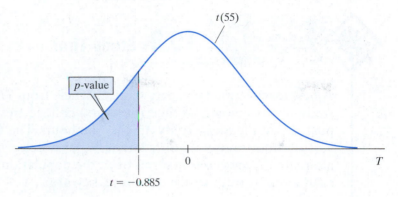

How do we find the *p*-value?

With the TI

1. **Using the distribution of the *t*-test statistic and the tcdf(function**

   We consider the *observed value* to be the corresponding test statistic $t = -0.885$. If the null hypothesis is true, the *t*-test statistic will have approximately a *t*-distribution with $n - 1 = 56 - 1 = 55$ degrees of freedom. The *p*-value will be the probability of getting a *t*-test statistic value of −0.885 or less under the $t(55)$ distribution. Using the tcdf(function on the TI, we have

   $$p\text{-value} = P(T \leq -0.885) = \text{tcdf}(-\text{E}99, -0.885, 55) = 0.1900047158.$$

2. **Using the T-Test function under the STAT TESTS menu**

   In the TESTS menu located under the STAT button, we select the **2:T-Test** option. With the sample mean of 15.839, the sample standard deviation of 1.362, and the sample size of $n = 56$, we can use the Stats option of this test. The corresponding input and output screens are shown.

   Input screen:
   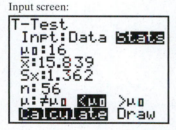

   Output screen:
   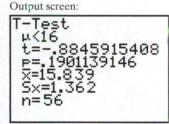

   $$p\text{-value} = P(T \leq -0.885) = 0.1901139146.$$

With Table IV

The *p*-value will be the probability of getting a *t*-test statistic value of −0.885 or less under the $t(55)$ distribution. In Table IV, we notice that the rows jump from 40 degrees of freedom to 60 degrees of freedom. We will be slightly conservative if we use the smaller degrees of freedom. Focusing on the df = 40 row, we imagine each value in this row to be negative instead of positive; then the column heading would represent the area to the left instead of to the right. Our observed test statistic value of −0.885 falls between the columns headed with 0.10 and 0.20. So we state that the *p*-value is between 0.10 and 0.20.

$$0.10 < p\text{-value} = P(T \leq -0.885) < 0.20$$

What Is the Decision and Conclusion?

The *p*-value is more than 0.05, so we cannot reject the null hypothesis. There is not sufficient evidence to support the average width percent for perch from this sea being less than 16. ❖

Let's do it! 10.3 Study Time

A new teacher at a university read an article from *The American Freshman* that discussed a study of the amount of time (in hours) college freshmen study each week. The study reported that the mean study time is 7.06 hours. The teacher feels that freshmen at her university study more than 7.06 hours per week on average. The appropriate null and alternative hypotheses in terms of μ, the population mean number of hours spent studying each week by freshmen at this university, are

$$H_0: \mu = 7.06 \quad \text{versus} \quad H_1: \mu > 7.06.$$

The teacher selected a simple random sample of 15 freshmen at her university and found the observed sample mean study time to be $\bar{x} = 8.43$ hours and the observed sample standard deviation to be $s = 4.32$. Assume that study time for freshmen at her university follow a normal distribution.

$$\text{The **observed test statistic** is } t = \frac{8.43 - 7.06}{\dfrac{4.32}{\sqrt{15}}} = 1.23.$$

(a) Sketch the picture to show the *p*-value. Label the axis and give the name of the distribution.

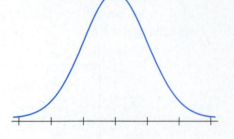

(b) Find the *p*-value for the test.

(c) Are the results statistically significant at the 10% level?   Yes   No
Explain.

(d) State your conclusion using a well-written sentence.

Let's do it! 10.4 pH Levels

A soil scientist is interested in studying the pH level in the soil for a certain field. In particular, she is going to examine a random sample of soil samples and measure their pH levels to assess if the mean field pH level is neutral (that is, equal to 7) versus the alternative hypothesis that the mean pH level is acidic (that is, less than 7). The significance level will be set at 5%.

(a) State the corresponding null and alternative hypotheses to be tested.

Suppose that it is reasonable to assume that the pH levels of all the possible samples that might be drawn are normally distributed. The scientist takes five randomly selected samples of soil from the field and measures the pH level in these samples. The pH levels in the sample were 5.8, 6.3, 6.9, 6.2, and 5.5.

(b) Find the observed sample mean pH level and the corresponding observed sample standard deviation.

(c) Using your TI or Table IV, find the observed test statistic and the corresponding *p*-value.

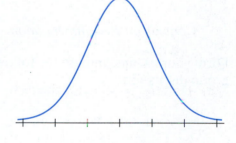

(d) The level of significance is 5%. Are the data statistically significant? Explain.

(e) State your conclusion using a well-written sentence.

Let's do it! 10.5 Apartment Rentals

A reporter from your college newspaper would like to write an article on the cost of attending college. As part of the article, she wants to discuss the cost of off-campus housing. A reference book about the cost of attending a college reports an average cost of $400 for monthly rental of a one-bedroom apartment within one mile of campus. You are asked to assess if this average cost is plausible for your college. That is, you are asked to test

$$H_0: \mu = 400 \qquad \text{versus} \qquad H_1: \mu \neq 400.$$

You may work in groups of two to five people. Your group must discuss how to gather data for performing this test and how many observations to take, test the assumptions with your data, and perform the test. Be sure to include your data and pertinent details in your brief summary. Your summary should include the following sections:

Introduction	A statement of the overall problem and the significance level of the test.
Data Collection	How the data were collected, why you believe it is a random sample, and from what population you did sample.
Data Summary and Checking Assumptions	Including various numerical and graphical summaries.
Test Results	The observed test statistic value and corresponding *p*-value.
Conclusion/Recommendation	

Form your groups, and use the following space to write down any notes and ideas from your group discussion.

❖ EXAMPLE 10.4 Factory Accidents (Paired Data)

It is claimed that an industrial safety program is effective in reducing the loss of working hours due to factory accidents. The following data are collected concerning the weekly loss of working hours due to accidents in six plants both before and after the safety program is instituted:

Plant Number	1	2	3	4	5	6
Before program	12	29	16	37	28	15
After program	10	28	17	35	25	16
Difference = Before − After	2	1	−1	2	3	−1

Initially, we might think that there are two populations to compare, the *before* program losses and the *after* program losses. However, the two sets of observations are related (that is, they are **paired**.) The first *before* observation from Plant 1 is meant to be compared directly to the first *after* observation from Plant 1. We are interested in analyzing the differences, which were computed as *before* minus *after*. If we let μ be the mean difference, we wish to test H_0: $\mu = 0$ versus H_1: $\mu > 0$ using a 10% significance level. The alternative hypothesis represents a reduction in the loss of hours on average.

The observed sample mean difference is 1.0 and the observed sample standard deviation of the differences is 1.673, giving an observed test statistic of

$$t = \frac{1 - 0}{\frac{1.673}{\sqrt{6}}} = 1.46.$$

Using a calculator and a t-distribution with $n - 1 = 6 - 1 = 5$ degrees of freedom, the corresponding p-value is 0.102. At a 10% significance level, the safety program was not effective in reducing the loss of hours due to factory accidents. This is an example of a **paired t-test**, which you will study further in the next chapter. ❖

The t-test for a population mean requires that the data are assumed to be a random sample from a population that has a normal distribution. The normality assumption is not crucial if the sample size is large. But what if the sample size is small and the distribution appears to be nonnormal? There are several techniques for handling such cases. You might be able to *transform* the data so that the distribution of the transformed data appears to be approximately normal. The t-test would then be conducted on the transformed data. A common transformation that tends to work for right-skewed data is the logarithm.

Another approach for handling nonnormality is to use a procedure which does not assume that the population distribution has any specific form, such as normal. Such procedures are called *nonparametric* or *distribution-free* procedures. The *sign test* is a nonparametric procedure that can be used for one-sample scenarios. The sign test is based on counts and the binomial distribution is used to compute the p-value. The last chapter addresses some of these techniques. We are primarily interested in understanding the basics of inference in order to assess the value of the information presented in our daily lives. In practice the t-test based on a normal distribution for the response is the most common.

 ## Exercises

10.1 For each of the following statements, write out the appropriate null and alternative hypotheses:

   (a) The mean temperature in Wayne County for the month of January is below 33 degrees.

   (b) The mean age of medical students at WSU is more than 26 years.

   (c) The mean score on an entrance exam differs from the target score of 200 set by the exam developers.

10.2 Experiments on concentration levels sometimes involve subjects completing a maze. For a particular maze, the mean time to completion is seven minutes for high school students. A researcher thinks that playing classical music will lead to increased concentration levels and thus the ability to complete the maze more quickly. State the appropriate null and alternative hypotheses to be tested. Include a clear statement about what the mean μ represents for this study.

10.3 *Z*-**test (σ assumed known).** A dairy refuses to accept raw milk having more than 5000 bacteria per milliliter (mL). The bacteria count varies from shipment to shipment. Assume that the count of bacteria per milliliter is normally distributed with a population standard deviation of 16. The dairy wants to test that the mean bacteria count is less than 5000 per mL for the next shipment. A simple random sample of 64 1-mL samples from the next shipment resulted in a mean bacteria count of 4995.

(a) State the null and alternative hypothesis for this problem.

(b) Calculate the observed test statistic value.

(c) Report the *p*-value for this test.

(d) Are the results statistically significant at the 1% level of significance? Explain.

10.4 *Z*-**test (σ assumed known).** A psychologist is studying the distribution of IQ scores of girls at an alternative high school. She wants to test the hypothesis that the average IQ score is 100 against the alternative that it is higher than 100. She assumes that the normal distribution with standard deviation 15 is a good model for these IQ scores. She takes a random sample of nine girls from this high school and measures their IQ. The average IQ score for these nine girls was 114. Assuming the appropriate assumptions are met, compute the test statistic and *p*-value for this test. State your decision at level $\alpha = 0.01$. and write up your conclusion in words.

10.5 *Z*-**test (σ assumed known).** Waiting times to be seated for dinner at a newly opened restaurant are thought to follow a normal distribution with a standard deviation of 20 minutes. We wish to test that the average waiting time for all dinner customers is *less* than one hour (that is, less than 60 minutes). A random sample of 25 customers yielded a sample mean waiting time of 50 minutes. Is there evidence to say that the average waiting time for all dinner customers is less than one hour? State the hypotheses and perform the test at a 10% significance level.

10.6 *Z*-**test (σ assumed known).** In a research report, it was claimed that mice, which generally have an average lifetime of 32 months, will live longer if 30% of the calories in their diet are replaced by vitamins and protein. We wish to assess whether there is evidence to support that the mean lifetime for mice fed this special diet is *more* than 32 months.

(a) State the corresponding hypotheses about the mean lifetime for mice fed this special diet.

A total of 64 randomly selected mice were fed this special diet and had an average lifetime of 38 months. Assume that the population standard deviation for lifetime is 5.8 months and that lifetime has a normal distribution.

(b) Perform the test using a 2.5% level of significance.

(c) Write up your conclusion in words.

10.7 For each of the following scenarios, sketch a picture of the area corresponding to the *p*-value, including the name of the distribution, and then find the *p*-value.

(a) The direction of extreme is one-sided to the right, $n = 10$, $t = 2.6$.

(b) The direction of extreme is one-sided to the left, $n = 50$, $t = -1.4$.

(c) The direction of extreme is two-sided, $n = 15$, $t = -2.0$.

10.8 For each of the following scenarios, sketch a picture of the area corresponding to the *p*-value, including the name of the distribution, and then find the *p*-value.

(a) The direction of extreme is one-sided to the right, $n = 25$, $t = 2.0$.

(b) The direction of extreme is one-sided to the left, $n = 8$, $t = -1.2$.

(c) The direction of extreme is two-sided, $n = 70$, $t = -2.5$.

10.9 A syrup company sells pure maple syrup in 16-ounce bottles. As part of the company's quality control program, random sampling is done weekly. If too little syrup is put in each bottle, customers will be dissatisfied, and if too much syrup is put in each bottle, the company will lose money. In either case, the equipment would be turned off to investigate the problem further. Testing is conducted using a 5% significance level.

(a) State the appropriate hypotheses.

(b) What assumptions are required to perform the test stated in (a)?

(c) How would a Type I error affect the company?

(d) How would a Type II error affect the company?

(e) The week's sample of 20 bottles had a mean of 15.8 ounces and a standard deviation of 0.7 ounces. Perform the test, and report the associated *p*-value.

(f) Based on the data and results of the test, what would be your recommendation?

10.10 The State Highway Patrol periodically samples vehicles at various locations on a particular highway to test the hypothesis that the average speed exceeds the set limit of 70 mph. These locations are the best locations for radar patrol cars to catch speeders. At a particular location, a sample of 16 vehicles shows a mean driving speed of 73.2 mph with a standard deviation of 5.1 mph. Is this location a candidate for a radar patrol? Test using a 5% significance level. Show all steps.

10.11 In the past, the average length of stay of tourists in a city's hotels has been 3.1 nights. A new marketing campaign to promote the attractiveness of the city has been in place for the last two months. An analyst is to test whether the new campaign has increased tourism. The analyst obtained a random sample of the number of nights spent by tourists in the city's hotels after the campaign started.

$$8 \quad 4 \quad 6 \quad 2 \quad 3 \quad 5 \quad 1 \quad 2 \quad 3 \quad 4 \quad 7 \quad 3$$

The number of nights spent by tourists at hotels is assumed to follow an approximately normal distribution. We wish to assess whether there is enough evidence to conclude that the mean number of nights spent at a hotel is higher than 3.1 (that is, we wish to test the hypotheses H_0: $\mu = 3.1$ versus H_1: $\mu > 3.1$).

(a) Give the value of the test statistic and the corresponding *p*-value for the test.

(b) What is your decision using a significance level of 10%?

(c) Could you have made a mistake? If so, what type of mistake could you have made?

REAL DATA

10.12 Some fishy data—a continuation of Example 10.3. Upon further investigation, it was discovered that the perch with the largest width percent of 20.9 had a very full stomach containing six roaches. Roach is a species of fish that happen to be smaller.

(a) Use the $1.5 \times$ IQR rule for assessing if this observation is an outlier.

(b) Remove this observation from the data set and repeat the test of H_0: $\mu = 16$ versus H_1: $\mu < 16$.

(c) How did removal of this observation affect the results? Comment. How did the sample mean change? the sample standard deviation? the p-value?

10.13 In a metropolitan area, the concentrations of cadmium in leaf lettuce were measured at six representative gardens at which sewage sludge was used for fertilizer. The measurements (in mg/kg) are 21, 38, 12, 15, 14, 8. The concentration of cadmium is assumed to follow a normal distribution. We wish to assess if there is enough evidence to conclude that the mean concentration of cadmium is higher than 12.

(a) State the appropriate hypotheses.

(b) Give the value of the test statistic and the corresponding p-value for the test.

(c) What assumptions are required for the test to be valid?

(d) Explain, in words, what the p-value found in (b) means.

(e) Would you reject the null hypothesis if the significance level is 0.05? Explain.

10.14 A Levi Strauss & Co. clothing manufacturing plant in Albuquerque, NM, has a quality control department. Every week, data are collected from its suppliers on the percentage waste (called runoff) relative to what can be achieved by computer layouts of patterns on cloth. A negative value for waste would indicate that the plant employees beat the computer in controlling waste for that week. The waste data for Supplier Plant 2 for 22 weeks are provided to assess whether, on average, this plant performed better than the computer at controlling waste.

REAL DATA

Plant 2 percentage of waste

1.2	10.1	−2.0	1.5	−3.0	−0.7	3.2	2.7	−3.2	−1.7	2.4
0.3	3.5	−0.8	19.4	2.8	13.0	42.7	1.4	3.0	2.4	1.3

(a) State the appropriate hypotheses.

(b) Report the test statistic and corresponding p-value for the test in (a).

(c) Are the results statistically significant at the 1% level?

(d) Make a histogram of these data and comment on the assumption of an underlying normal distribution for waste. What other assumption must we make about the data?

10.15 Consider the following hypotheses about μ = the population mean price for cellular phones (at the time of this study):

$$H_0: \mu = \$300 \quad \text{versus} \quad H_1: \mu > \$300.$$

In 1993, *Consumer Reports* obtained a simple random sample of 10 cellular phones. The prices (in dollars) for the 10 selected phones are as follows:

499	669	279	207	600	100	235	200	299	200

REAL DATA

Assume that the population of cellular-phone prices follows a normal distribution.

(a) Compute the p-value for this test. Show all steps involved in getting your answer.

(b) Based on your p-value in (a), would you accept or reject H_0 at a 5% significance level? Explain.

10.16 Authorities in Detroit, MI, suspect that the introduction of fluoride to the drinking water will significantly reduce the average number of cavities among children. Last year, the average number of cavities was $\mu = 2.6$. This year, fluoride was introduced, and, for a sample of 36 children, an average of 1.5 cavities was found with a standard deviation of 0.75.

(a) Using $\alpha = 0.05$, test the hypothesis of the authorities in Detroit.

(b) One of the assumptions made to perform the test in (a) is that the response is modeled by a normal distribution. Would the test still be valid if the distribution was not normal, but somewhat skewed to the right? Explain.

10.17 A statistics instructor has written a new placement exam. The department chair would like to use this new exam, but is concerned that it may be too long. Students will be given one hour to complete the exam. The exam will be used if there is sufficient evidence to conclude that the mean time to complete the exam is less than 50 minutes. The new exam is given to a random sample of 36 students and a sample mean time to complete the exam was 48 minutes with a sample standard deviation of 5. State the hypotheses to be tested and perform the test using a significance level of 0.05.

10.18 A real-estate appraiser wants to verify the market value for homes on the east side of the city that are very similar in size and style. The appraiser wants to test the popular belief that the average sales price is $37.80 per square foot for such homes. He will use a significance level of 0.01 and assumes a normal distribution is a good model for sales.

(a) State the appropriate null and alternative hypotheses about μ, the population mean sales price for such homes.

(b) Suppose that the random sample of six sales were selected. The sampled sales prices per square foot are $35.00, $38.10, $30.30, $37.20, $29.80, and $35.40. Does it appear that the popular perception of the market value is valid for this neighborhood? Give the value of the test statistic and the p-value.

(c) Based on your p-value in (b), are the results statistically significant at a 1% significance level?

10.3 Confidence Interval Estimation for a Population Mean

Each of the statements that follow is reporting an average. These averages were based on some data, a sample, and thus each average is an observed sample mean, $\bar{x}$.

- Studies have shown that those who have received the Varicella vaccine and still develop chicken pox experience an average of 50 lesions compared with about 300 that develop from natural chicken pox infection.

- The Centers for Disease Control and Prevention reported the average length of stay for hospital deliveries dropped to 2.6 days in 1992 from 4.1 days in 1970.

- The average annual return for an aggressive-growth mutual fund is 13% with a standard deviation of 10%.

We have already mentioned that a sample mean, based on a random sample, is a good point estimate for μ. The question that remains is, "How good is the point estimate?" Will the observed sample mean $\bar{x}$ be equal to μ? The answer is, probably not. Will it be close to μ? How close?

We know that the value of the sample mean $\bar{x}$ will vary from sample to sample. We have a model for that variation with the sampling distribution of the sample mean. So, as we did in estimating proportions, we are going to expand our point estimate by including a measure of variability, a measure of how far away we would expect sample mean values to be from μ (on average). That is, we will report an **interval estimate** for μ, a range of "plausible" values for μ. We will center our interval at our best point estimate for μ—namely, $\bar{x}$—and then go each direction from $\bar{x}$ to provide bounds for an interval that, before the data are examined, is highly likely to contain the population mean μ. How far we go from $\bar{x}$ to form the bounds will depend on the sample size, the variation, and the desired **confidence level**. We actually computed our first approximate confidence interval for a population mean in Example 8.12. Our discussion here will be a little more formal and will introduce a new version based on the *t*-distribution that is used in practice.

> **Definitions:** The sample mean $\bar{x}$ is a **point (single number) estimate** for the population mean μ.
>
> A **confidence interval estimate** for the population mean μ is an interval of values, computed from the sample data, for which we can be quite *confident* that it contains the population mean μ.
>
> The **confidence level** is the probability that the **estimation method** will give an interval that contains the parameter (μ in this case).

To develop the confidence interval for μ, we recall the sampling distribution of the sample mean:

$\overline{X}$ is (approximately) $N\left(\mu, \dfrac{\sigma}{\sqrt{n}}\right)$.

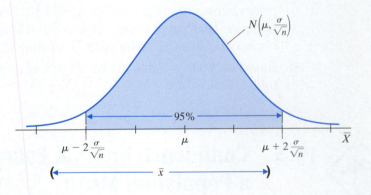

This result holds under random sampling and when either the original population is normal or the sample size n is large. By the 68–95–99.7 rule for normal distributions, approximately 95% of the $\bar{x}$ values are within two (actually 1.96) standard deviations of μ. If we formed an interval around any one of these $\bar{x}$ values by going about two standard deviations each direction from $\bar{x}$, we would have an interval that contains μ. So approximately 95% of the confidence intervals formed by

$$\bar{x} \pm 2\left(\frac{\sigma}{\sqrt{n}}\right)$$

would contain μ and 5% would not.

The general confidence interval, called a **one-sample z-interval**, is given by $\bar{x} \pm z^*\left(\frac{\sigma}{\sqrt{n}}\right)$, where z^* is an appropriate percentile of the $N(0, 1)$ distribution. The margin of error for the population mean is $E = z^*\left(\frac{\sigma}{\sqrt{n}}\right)$.

Once again, we have a dilemma. The formula for the confidence interval requires knowing the population standard deviation σ, which is rarely available. In practice, the population standard deviation σ is replaced by the sample standard deviation s, and the z^* percentile is replaced with a corresponding t^* percentile from an appropriate t-distribution with $n - 1$ degrees of freedom.

Confidence Interval for a Population Mean μ

$$\bar{x} \pm t^*\left(\frac{s}{\sqrt{n}}\right),$$

where t^* is an appropriate percentile of the $t(n - 1)$ distribution. This interval gives potential values for the population mean μ based on just one sample mean. This interval is based on the assumption that the data are a random sample from a normal population with *unknown* population standard deviation σ. If the sample size is large, the assumption of normality is not so crucial. However, outliers are always a concern.

The **margin of error** for the population mean when we use the data to estimate σ is given by

$$E = t^*\left(\frac{s}{\sqrt{n}}\right).$$

Example 10.5 and "Let's do it! 10.6" will be the only times you will see the one-sample z-interval. All remaining examples and exercises will involve the more practical confidence interval estimation for a mean that involves the t-distribution.

❖ EXAMPLE 10.5 Cereal Boxes

The Kellogg Corporation controls approximately a 43% share of the ready-to-eat cereal market worldwide. A popular cereal is Corn Flakes. Suppose the weights of full boxes of a certain kind of cereal are normally distributed with a population standard deviation of 0.29 ounces. A random sample of 25 boxes produced a mean weight of 9.82 ounces. A histogram of the data showed no strong departures from normality. We want to construct a 95% confidence interval for the true mean weight of such boxes.

With the TI

For the TI, a confidence interval for a population mean when the population standard deviation σ is known is called the *one-sample z-interval for a population mean* and abbreviated **ZInterval**. This is Option 7 under the TESTS menu obtained from the STAT button. You can have the sample data entered into a list, say, L1, or you can just enter the Stats (the sample mean, population standard deviation, and sample size). The steps are summarized next, and the corresponding input and output screens are shown. Note that the *Calculate* option produces an output screen that provides the confidence interval (9.7063, 9.9337), the sample mean $\bar{x} = 9.82$, and the sample size $n = 25$.

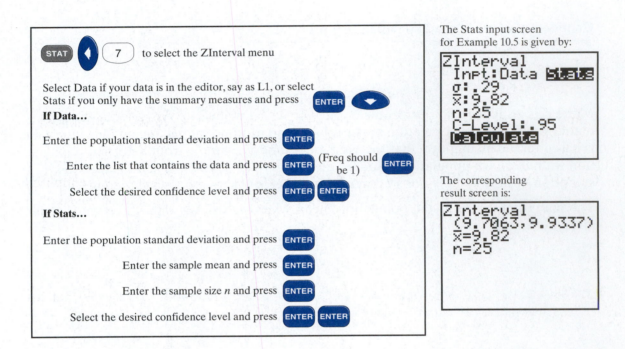

The Stats input screen for Example 10.5 is given by:

```
ZInterval
 Inpt:Data Stats
 σ:.29
 x̄:9.82
 n:25
 C-Level:.95
 Calculate
```

The corresponding result screen is:

```
ZInterval
 (9.7063,9.9337)
 x̄=9.82
 n=25
```

With the formula and Table III

We have the sample mean $\bar{x} = 9.82$, the assumed population standard deviation $\sigma = 0.29$, and the sample size $n = 25$. All that remains to complete the formula is the z^* value. For a 95% confidence interval, we may remember that it is 1.96 (as it is one of the most common confidence levels). If we do not remember it, we can look it up from the table provided on page 542 of Chapter 9 or from Table III, which provides selected percentiles for a standard normal distribution (we would need 95% in the middle, leaving 2.5% in each tail, so the 97.5th percentile gives us $z^* = 1.96$). Our 95% confidence interval for the true mean weight of such boxes is

$$\bar{x} \pm z^*\left(\frac{\sigma}{\sqrt{n}}\right) \to 9.82 \pm 1.96\left(\frac{0.29}{\sqrt{25}}\right) \to (9.706, 9.934).$$

What is the margin of error?

The width of the interval is $9.934 - 9.706 = 0.228$, so the half-width or margin of error is 0.114.

What does this 95% confidence interval tell us about the mean weight?

The interval of (9.706, 9.934) provides us with plausible values for the mean weight of full boxes of cereal, based on a sample of 25 boxes. Note that this interval does not contain the value of 10 ounces, the stated weight on the boxes. These data and results may lead to a decision to adjust the filling process.

How should we interpret the 95% confidence level?

Either our particular interval of (9.706, 9.934) contains μ, or it doesn't. However, we are 95% confident that μ is in this interval, because the interval was constructed with a method such that, if it is repeated, 95% of the confidence intervals produced are expected to contain μ. ❖

 **10.6** **How Much Beverage?**

A beverage-dispensing machine is calibrated so that the amount of beverage dispensed is approximately normally distributed with a population standard deviation of 0.15 deciliters (dL).

(a) Using your TI or the confidence interval formula, compute a 95% confidence interval for the population mean amount of beverage dispensed by this machine based on a random sample of 36 drinks dispensing an average of 2.25 dL.

(b) What is the margin of error for the 95% confidence interval in (a)? Recall that the margin of error is the half-width of the confidence interval.

(c) From the formula for the margin of error $E = z^* \left(\frac{\sigma}{\sqrt{n}} \right)$, we can solve for the sample size $n = \left(\frac{z^* \sigma}{E} \right)^2$ to have an expression for the required sample size needed for producing an interval with a desired confidence level and a desired margin of error. How large a sample would you need if you want the margin of error of the 95% confidence interval to be 0.02?

$$n = \left(\frac{z^* \sigma}{E} \right)^2 = \left(\frac{1.96(0.29)}{0.02} \right)^2 = $$

(remember to round up to the next integer)

The z-procedures used in Example 10.5 and "Let's do it! 10.6" required the unrealistic assumption of knowing the value of the population standard deviation and therefore are not used in practice. The interpretations are useful to see and they do extend to the more realistic t-procedures covered in the remainder of this chapter.

❖ EXAMPLE 10.6   Empty Seats Imply Dollars Lost

Unoccupied seats on flights cause airlines to lose revenue. Suppose that a large airline wants to estimate its average number of unoccupied seats per flight from Detroit to Minneapolis over the past month. To accomplish this, the records of 61 such flights were randomly selected, and the number of unoccupied seats was recorded for each of the sampled flights. The sample mean is 12.6 seats and sample standard deviation is 4.4 seats. Construct a 99% confidence interval estimate for the mean number of unoccupied seats.

With the TI

For the TI, a confidence interval for a population mean when the population standard deviation σ is **unknown** is called the *one-sample t-interval* and abbreviated **TInterval**. This is Option 8 under the TESTS menu obtained from the STAT button. You can have the sample data entered into a list, say, L1, or you can just enter the Stats (the sample mean, sample standard deviation, and sample size). The steps are summarized next, and the corresponding input and output screens are shown. Note that the *Calculate* option produces an output screen that provides the confidence interval (11.101, 14.099), the sample mean $\bar{x} = 12.6$, sample standard deviation $s = 4.4$, and the sample size $n = 61$.

The Stats input screen for Example 10.6 is given by:

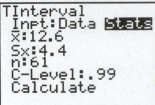

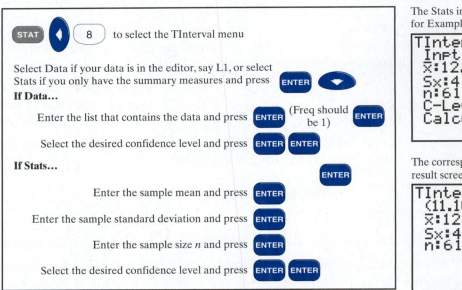

With the formula and Table IV

We have the sample mean $\bar{x} = 12.6$, the sample standard deviation $s = 0.29$, and the sample size $n = 61$. All that remains to complete the formula is the t^* value. For this, we look at the last row of Table IV (which labels the bottom of each column with the confidence levels) and find the 99% level of confidence. We also need to look in the row with the degrees of freedom of $n - 1 = 61 - 1 = 60$. So the t^* value for df = 60 and 99% confidence is 2.660. The 99% confidence interval for the population mean number of unoccupied seats is

$$12.6 \pm 2.660\left(\frac{4.4}{\sqrt{61}}\right) \rightarrow 12.6 \pm 1.5 \text{ or } (11.1, 14.1).$$

With 99% confidence, we estimate the mean number of unoccupied seats per flight to be between 11.1 and 14.1. ❖

THINK ABOUT IT **Reducing the Width**

A 95% confidence interval for the mean yield of a new variety of corn is (112, 148). Suppose that the researcher feels this interval is too wide. There are several ways the width of a confidence interval can be reduced. Give two ways.

❖ EXAMPLE 10.7   Leisure Time

A study is conducted to learn about the average number of hours a week adults in a community spend on leisure activities. A random sample of 20 adults from the community was selected and their time spent on leisure activities for a week was measured. The observed times were entered into SPSS, and part of the output is as follows:

One-Sample Statistics

	N	Mean	Std. Deviation	Std. Error Mean
Leisure Time	20	20.27	4.04	.90

One-Sample Test

	Test Value = 0	
	95% Confidence Interval	
	Lower	Upper
TIME	18.379	22.161

The **standard error of the mean** is given as 0.90. This value tells us that, if repeated samples of 20 adults are obtained, we would estimate the resulting sample means to differ from the true population mean μ by about 0.90 hours, on average. A 95% confidence interval is given as (18.379 hours, 22.161 hours). **With 95% confidence, we estimate the population mean** amount of leisure time spent by adults in this community to be between 18.4 hours and 22.2 hours. The **95% confidence level** pertains to the proportion of times the procedure produces an interval that actually contains μ when this procedure is repeated over and over using a new random sample of the same size n each time.   ❖

Let's do it! **10.7** **Customer Spending**

A restaurant owner believes that customer spending is below the usual spending level. The owner takes a simple random sample of 26 receipts from the previous week's receipts. The amount spent per customer served (in dollars) was recorded and some summary measures are provided.

$n = 26$, Sample mean $= 10.44$, Sample standard deviation $= 2.82$

Minimum $= 4$, Q1 $= 8.25$, Median $= 10$, Q3 $= 12.75$, Maximum $= 16$

(a) Assuming that customer spending is approximately normally distributed, compute a 90% confidence interval for the mean amount of money spent per customer served.

(b) Interpret what the 90% confidence level means.

(c) Use the preceding summary measures to sketch and label a corresponding boxplot.

Let's do it! **10.8** **Costs of an Education**

How much do full-time students pay for textbooks, on average, for a semester? We wish to produce a 90% confidence interval estimate for the mean cost of textbooks for all full-time students. Take a random sample of 25 full-time students from your class, and record how much each spent on textbooks for the semester.

Data

Summary Measures   observed sample mean $\bar{x}$ = _____.

observed sample standard deviation s = _____.

90% confidence interval

Interpretation

State the assumptions required and check the assumptions with an appropriate graph of the data.

❖ **EXAMPLE 10.8**   **Weight Change (Paired Data)**_____

A study was conducted to estimate the mean weight change of a female adult who quits smoking. The weights of eight female adults before they stopped smoking and five weeks after they stopped smoking were recorded. The differences were computed as "after – before." The difference in weight is assumed to follow a normal distribution.

Subject	1	2	3	4	5	6	7	8
After	154	181	151	120	131	130	121	128
Before	148	176	153	116	129	128	120	132
Difference	6	5	−2	4	2	2	1	−4

Here we have another example of a **paired design**. The response of interest is the difference in weight. The sample mean difference is 1.75 pounds and the sample standard deviation is 3.412 pounds. A 95% confidence interval estimate for the mean difference is

$$1.75 \pm 2.365\left(\frac{3.412}{\sqrt{8}}\right) \rightarrow 1.75 \pm 2.85 \text{ or } (-1.10, 4.60).$$

With 90% confidence, we estimate that the population mean weight change is between a loss of 1.1 pounds to a gain of 4.6 pounds.   ❖

Exercises

10.19 We want to learn about the average main-stem diameter of four-year-old pine trees in a given region. We took a random sample of 36 four-year-old pine trees and their main stem diameters (in inches) and generated the following output:

```
Descriptive Statistics
    DIAMETER
N of cases          36
Minimum             10.1
Maximum             16.2
Mean                13.5
Standard dev         1.2
```

(a) Give a point estimate of the average main-stem diameter for all four-year-old pine trees in this region.

(b) Give a 95% confidence interval estimate for the average main-stem diameter for all four-year-old pine trees.

(c) Interpret the confidence interval given in (b).

10.20 **Cuckoo eggs.** A study by E. B. Chance in 1940 called "The Truth About the Cuckoo" demonstrated that cuckoos return year after year to the same territory and lay their eggs in the nests of a particular host species. The eggs are then adopted and hatched by the host birds. The (ordered) data provided are the lengths (in millimeters) of cuckoo eggs found in the nests of meadow pipit birds.

19.65, 20.05, 20.65, 20.85, 21.65, 21.65, 21.65, 21.85, 21.85, 21.85, 22.05, 22.05, 22.05, 22.05, 22.05, 22.05, 22.05, 22.05, 22.05, 22.05, 22.05, 22.25, 22.25, 22.25, 22.25, 22.25, 22.25, 22.25, 22.25, 22.45, 22.45, 22.45, 22.65, 22.65, 22.85, 22.85, 22.85, 22.85, 23.05, 23.25, 23.25, 23.45, 23.65, 23.85, 24.25, 24.45

REAL
DATA

(a) Summarize these data numerically and graphically. In particular, comment on whether a normal model for egg length is plausible.

(b) Construct a 95% confidence interval estimate for the mean length of cuckoo eggs found in the nests of meadow pipit birds.

(c) What does your interval in (b) tell you about the mean length?

(d) Is there evidence that the mean length differs from 23 millimeters? Explain.

10.21 The value reported as lost for a random sample of $n = 20$ pickpocket offenses occurring in a city is given here.

883	447	207	627	214
313	844	253	587	217
768	1064	549	833	277
805	443	649	554	570

(a) Use the data to construct a 95% confidence interval for the mean value lost in all pickpocket offenses for this city.

(b) What is the margin of error for the interval estimate in (a)?

(c) Give an interpretation of the interval and of the confidence level.

10.22 On July 14, 1997, an article in *Time* magazine reported the results of a study by Arthur D. Little, Inc., which stated that 70% of mail received by a household in a week is advertisements. A sample of 20 households produced the following data:

Household	Number of Advertisements per Week	Total Number of Pieces of Mail per Week
1	12	16
2	18	24
3	15	24
4	15	25
5	21	33
6	21	24
7	17	22
8	20	27
9	16	28
10	13	19
11	20	25
12	20	23
13	17	20
14	13	20
15	23	33
16	15	28
17	9	12
18	18	30
19	9	14
20	24	35

(a) Find a point estimate for the mean number of advertisements received per week for all households.

(b) Give a 95% confidence interval estimate for the mean number of advertisements received per week for all households.

(c) Find a point estimate for the mean number of pieces of mail received per week for all households.

(d) Give a 95% confidence interval estimate for the mean number of pieces of mail received per week for all households.

(e) Are the results obtained in (a) through (d) in agreement with the statement reported in *Time*?

10.23 A 95% confidence interval for the mean reading achievement score for a population of third-grade students is (44.2, 54.2). Find the value of the sample mean and give the margin of error.

10.24 Super Sound music store sampled 49 customers to estimate the current mean age of their customers. Based on the data, a 90% confidence interval estimate for the mean age (in years) was found to be (24.8, 27.2).

(a) Does this mean $P(24.8 < \mu < 27.2) = 0.90$? Explain.

(b) What is the sample mean? Select one:

    (i)   24.8

    (ii) 27.2

    (iii) 49

    (iv) 26.0

    (v) cannot be determined based on the given information

10.25 A 95% confidence interval for the population mean test score is given by (68, 92). This interval was based on a random sample of 64 scores. Students in a statistics class were asked to explain what the 95% level of confidence means. The explanations given by some students are as follows:

■ Student 1: We are 95% confident that the population mean test score is in the interval (68, 92) because the probability that this interval contains the population mean test score is 0.95.

■ Student 2: A 95% confidence level means that 95% of the time the population mean test score will be in the interval (68, 92).

■ Student 3: A 95% confidence level means that 95% of the time the sample mean test score will be in the interval (68, 92).

■ Student 4: A 95% confidence level means that the confidence interval has been constructed by a procedure, which, if repeated, 95% of the intervals are expected to contain the population mean test score.

■ Student 5: A 95% confidence level means that the confidence interval has been constructed by a procedure, which, if repeated, 95% of the intervals are expected to contain the sample mean test score.

■ Student 6: A 95% confidence level means that 95% of the resulting sample means are expected to fall in the interval (68, 92).

Select all explanations that are correct.

10.26 A new type of fertilizer was used on a simple random sample of 40 plots of green beans. Assuming that yields are normally distributed, the 90% confidence interval for the mean yield μ (in bushels) is (118, 132). Which of the following statements gives a correct interpretation of the 90% confidence level?

(a) The chance that the mean yield μ falls in the interval (118, 132) is 0.90.

(b) If the process were repeated many times, the mean yield μ would lie in the interval (118, 132) 90% of the time.

(c) If the process were repeated many times, generating many 90% confidence intervals, 90% of these intervals would contain the mean yield μ.

(d) If the process were repeated many times, 90% of the sample mean yields would be between 118 and 132.

(e) 90% of the plots have a mean yield between 118 and 132.

10.27 A certain product is advertised as increasing the gasoline mileage of cars (mpg). To test the claim, 10 of these products are purchased and installed into 10 cars. The gasoline mileage for each car is recorded both before installation and after installation. The data are as follows:

Car	Before mpg	After mpg	Difference = After − Before
1	19.1	25.8	6.7
2	29.9	23.7	−6.2
3	17.6	28.7	11.1
4	20.2	25.4	5.2
5	23.5	32.8	9.3
6	26.8	19.2	−7.6
7	21.7	29.6	7.9
8	25.7	22.3	−3.4
9	19.5	25.7	6.2
10	28.2	20.1	−8.1

Give a 95% confidence interval for the mean difference in gas mileage where differences are computed as after less before (Diff = After − Before).

10.28 A statistician employed by a consumer-testing organization reports that a 95% confidence interval for the average content of 12-ounce bottles of root beer is (11.6 ounces, 12.2 ounces). The statistician obtained these results by taking a random sample of bottles of root beer and measuring the actual content of each. These data produced the following output:

(a) An estimate of the true average content of bottles of root beer is (select one)

    (i) 11.51 ounces.

    (ii) 1.38 ounces.

    (iii) 11.90 ounces.

    (iv) unknown, cannot be computed.

    (v) μ.

CONTENT	
Minimum	11.51
Maximum	12.43
Mean	11.90
Standard Dev	1.38
Std error	0.15

(b) The statistician, in his report, forgot to include the size of the sample he had selected. (Approximately) what sample size did he use?

10.4 Confidence Intervals and Hypothesis Testing

We have studied two forms of statistical inference in this chapter: hypothesis testing and interval estimation for a population mean. These two forms of inference are different, but they are related. For example, suppose that you have collected data to learn about the population mean response. A 95% confidence interval for the population mean based on these data went from 20.8 to 28.4. You are 95% confident that this interval captures the true population mean because you used a method that produces a good interval (one that contains the population mean) for 95% of all possible samples. This confidence statement does require that you have a random sample either from a normal population or of a large size n. Suppose that now someone claims that the population mean is 20. Your confidence interval can be used to test this claim against the alternative theory that the population mean is not 20. Your confidence interval provides a range of values that are regarded as plausible values for the population mean. Based on the 95% confidence interval (20.8, 28.4), the claimed value of 20 is *not plausible*, since it falls *outside* the interval estimate. We would reject the null hypothesis that the population mean is 20, using a 5% significance level in a two-tailed test.

Connection between Confidence Intervals and Two-sided Hypothesis Tests

You can test the hypothesis H_0: $\mu = \mu_0$ versus H_1: $\mu \neq \mu_0$ at the significance level α using the following decision rule:

Reject H_0 if the corresponding $(1 - \alpha)100\%$ confidence interval for the population mean μ *does not* contain the hypothesized value stated in H_0.

The test must be two-sided, and the confidence level and significance level must add up to a total of 100%.

❖ EXAMPLE 10.9 Qualified Applicants

An administrative assistant position has opened up at a large consulting company. Many applications have been submitted. Applicants must take a test, as part of the application process, as one measure of their qualifications. The preparer of the test maintains that qualified applicants should average 75.0 points. A random sample of scores of 41 applicants gave a mean score of 73.1 points and a standard deviation of 8.8 points. A 95% confidence interval for the mean score of all applicants is

$$73.1 \pm 2.021\left(\frac{8.8}{\sqrt{41}}\right) \rightarrow (70.32, 75.88).$$

Can the consulting company conclude that it is getting qualified applicants (as measured with this test)? Does it appear that a mean score of 75.0 is plausible? For the test of H_0: $\mu = 75$ versus H_1: $\mu \neq 75$, would you reject or accept H_0 at a 5% significance level? Since the 95% confidence interval for the population mean score does include the value of 75, the data do not refute the possibility of the population mean being equal to 75.   ❖

❈ Let's do it! 10.9 Fireflies Need a Rest Too!

The resting time between flashes of a random sample of 64 fireflies had a mean of 3.77 seconds and a standard deviation of 0.35 seconds. A 90% confidence interval for the mean resting time for this species of firefly is given by (3.70, 3.84). A friend plans to report this information in a paper he is writing. Consider the following two summaries:

I. "We can therefore conclude that if this procedure were repeated, the true mean will fall in the interval (3.70, 3.84) 90% of the time."

II. "We can therefore conclude that if this procedure were repeated, we'd expect 90% of the confidence intervals constructed to contain the true mean."

Your friend asks you to review his summaries. Your recommendation would be as follows:

(i) Use Summary I because this is the correct interpretation and Summary II is too vague.

(ii) Use Summary II because this is the correct interpretation.

(iii) Use either summary, since they both are correct interpretations of the results.

Your friend would like to assess whether or not the true mean resting time for this species differs from 3.90 seconds.

State the appropriate hypotheses: H_0: _____ H_1: _____

If your friend wants to use a significance level of 0.10, does it appear that the mean resting time is 3.90 seconds? Explain.

Let's do it! 10.10 Infant Sleep Patterns

A study of the sleep patterns of six-month-old infants in the United States reported a 95% confidence interval for the average amount of time infants sleep (out of every 24-hour period) to be (11.5 hours, 15.2 hours). Suppose that we want to test H_0: $\mu = 15$ versus H_1: $\mu \neq 15$.

At the $\alpha = 0.05$ level, we would (circle one)   reject H_0.   accept H_0.   can't tell. Explain.

At the $\alpha = 0.01$ level, we would (circle one)   reject H_0.   accept H_0.   can't tell. Explain.

✳ Exercises

10.29   In the testing of a new production method, 18 employees were selected at random and asked to try the new method. The sample mean production rate for the 18 employees was 80 parts per hour and the sample standard deviation was 10 parts per hour.

(a) Construct a 90% confidence interval for the true mean production rate. Give all steps and your resulting confidence interval.

(b) What is the margin of error of your confidence interval obtained in (a)?

(c) If you were to use the confidence interval calculated in (a) to test H_0: $\mu = 82$ versus H_1: $\mu \neq 82$, would you reject H_0 at $\alpha = 0.10$? Explain.

(d) If the sample size were to increase (but the sample mean and standard deviation were the same), the confidence interval would be (select one)

narrower.   wider.   can't tell.

10.30 The 99% confidence interval for the population mean μ was calculated to be $(54, 62)$.

(a) What is the probability that the sample mean used to calculate this confidence interval is in the confidence interval of $(54, 62)$?

(b) Suppose that you would like to test H_0: $\mu = 53$ versus H_1: $\mu \neq 53$ using the same data that gave the confidence interval of $(54, 62)$, but with $\alpha = 0.10$. Would you accept H_0? Explain.

(c) Suppose that you would like to test H_0: $\mu = 55$ versus H_1: $\mu \neq 55$ using the same data that gave the confidence interval of $(54, 62)$, but with $\alpha = 0.10$. Would you accept H_0? Explain.

(d) What is the margin of error?

10.31 A wine manufacturer sells a cabernet wine whose label asserts an alcohol content of 11%. Thirty bottles of this wine were selected at random and analyzed for alcohol content. A 95% confidence interval for the population mean alcohol content, μ, of such bottles of wine is $(9.92\%, 11.07\%)$. Answer the following questions:

(a) Does the sample mean lie in the interval $(9.92, 11.07)$?

Yes   No   Can't tell

(b) Does the population mean lie in the interval $(9.92, 11.07)$?

Yes   No   Can't tell

(c) For a future sample of 30 bottles, will the sample mean lie in the interval $(9.92, 11.07)$?

Yes   No   Can't tell

(d) If we use a 99% level of confidence, will the confidence interval calculation from the same data produce an interval narrower than $(9.92, 11.07)$?

Yes   No   Can't tell

(e) If we were to use the preceding data to test the hypotheses H_0: $\mu = 11$ versus H_1: $\mu \neq 11$ at a 5% significance level, would we accept the null hypothesis?

Yes   No   Can't tell

10.32 A market research company employs a large number of typists to enter data into a computer. A study is conducted to assess how quickly new typists learn the computer system. A random sample of 25 new typists was obtained and their times to learn the computer system were recorded. Based on these data, the 99% confidence interval for the population mean time to learn the computer system was calculated to be $(44, 92)$.

(a) What is the probability that the sample mean used to calculate this confidence interval is in the confidence interval of $(44, 92)$?

(b) Is it correct to say that the probability the population mean, μ, lies in the interval $(44, 92)$ is 0.99?

(c) What is the margin of error for this confidence interval?

(d) Suppose that you would like to test H_0: $\mu = 40$ versus H_1: $\mu \neq 40$ using the same data that gave the confidence interval of $(44, 92)$, with $\alpha = 0.01$. Would you accept H_0? Explain.

(e) Suppose you would like to test H_0: $\mu = 45$ versus H_1: $\mu \neq 45$ using the same data that gave the confidence interval of $(44, 92)$, but with $\alpha = 0.10$. Would you accept H_0? Explain.

Chapter Summary

In this chapter, we have studied two forms of statistical inference: hypothesis testing for making a decision about a population mean and confidence interval estimation of a population mean. These inference techniques are summarized in the table that follows. Note that the z-procedures are not often done in practice as the population standard deviation is not generally known.

Situation	Parameter of Interest	Hypothesis Test Statistic	Confidence Interval
One random sample from a normal population with σ known (not practical)	population mean μ	$z = \dfrac{\bar{x} - \mu_0}{\dfrac{\sigma}{\sqrt{n}}}$	$\bar{x} \pm z^* \left(\dfrac{\sigma}{\sqrt{n}} \right)$
One random sample from a normal population with σ unknown (practical)	population mean μ	$t = \dfrac{\bar{x} - \mu_0}{\dfrac{s}{\sqrt{n}}}$	$\bar{x} \pm t^* \left(\dfrac{s}{\sqrt{n}} \right)$

We must remember a few notes of caution as we consider applying one of these techniques to a set of data.

- An underlying assumption is that the data must be a random sample, although we are generally okay if the data can be viewed as representative of the population with respect to the response of interest.

- The techniques are not correct for other sampling designs that are more complex than random sampling. There are methods available for other designs. The general testing and interpretation ideas, however, do carry over.

- The results of the procedures extend only to the larger population represented by the sample.

- An underlying assumption is that the population has a normal model for the response, although we are generally okay if the sample size is moderate and there are no extreme outliers or very strong skewness. The larger the sample size, the more we will have the central limit theorem helping us out.

- If the population model is quite skewed and the sample size is not large, we might turn techniques that examine the more resistant measure of center, the median. These types of techniques are discussed in Chapter 15.

In Chapter 3, we discussed an important feature in studies, that of having a control group and a treatment group. Perhaps the control group receives a placebo or the standard treatment. If our response is quantitative, then we will be interested in comparing the mean response for the two groups. If our response is a categorical variable, we may be interested in comparing some proportion for the two groups. Our next chapter extends the ideas presented in Chapters 9 and 10 to comparing two means and comparing two proportions.

Key Terms

Be sure you can describe, in your own words, and give an example of each of the following key terms from this chapter:

null hypothesis	t-test statistic for a mean	p-value
alternative hypothesis	upper-tailed test	point estimate
significance level	lower-tailed test	confidence interval estimate
sampling distribution	two-tailed test	level of confidence
z-test statistic for a mean	Type I error	width of the interval
Student's t-distribution	Type II error	margin of error
degrees of freedom	decision rule	

Exercises

10.33 A significance test gives a p-value of 0.04. From this, we can (select one)

(a) reject H_0 with $\alpha = 0.01$.

(b) reject H_0 with $\alpha = 0.05$.

(c) say that the probability that H_0 is false is 0.04.

(d) say that the probability that H_0 is true is 0.04.

(e) none of the above.

10.34 Nutritionists stress that weight control generally requires significant reductions in the intake of fat. A simple random sample of 64 middle-aged men on weight control programs was selected to determine whether the mean intake of fat exceeds the recommended 30 grams per day. The sample mean and sample standard deviation are $\bar{x} = 37$, $s = 32$. Do the sample results indicate that the mean intake for middle-aged men on weight control programs exceeds 30 grams?

(a) Write the appropriate null and alternative hypothesis for this test.

(b) Report the p-value for this test.

(c) Are the data statistically significant at the 10% significance level ($\alpha = 0.10$)? Explain.

(d) If the significance level were changed to be $\alpha = 0.01$, would you reject or accept the null hypothesis? Explain.

10.35 For a particular species of snap beans, the mean germination time has been 17.1 days. A new species of snap beans has been created. An agricultural analyst is to test whether the new species has a faster germination time on average. The germination time for this new species is assumed to follow approximately a normal distribution. The analyst obtained the following random sample of the germination time (in days) for 10 seeds of the new species:

$$15, \quad 17, \quad 12, \quad 11, \quad 14, \quad 18, \quad 12, \quad 20, \quad 15, \quad 16.$$

(a) State the hypotheses to be tested.

(b) Report the test statistic formula, its corresponding value, and the *p*-value for the test.

(c) What is your decision if the significance level is $\alpha = 0.10$? Give your conclusion in a well-written sentence.

10.36 The producer of a certain brand of eggs claims that the eggs have reduced cholesterol content, with an average of only 4.6% cholesterol. A concerned health group wants to test whether the claim is true. The group believes that more cholesterol may be found, on average, in the eggs.

(a) State the null and alternative hypothesis that the health group wishes to test.

(b) A random sample of 100 eggs reveals a sample mean content of 5.2% cholesterol and a sample standard deviation of 2.8%. Calculate the value of the test statistic.

(c) Find the corresponding *p*-value for this test. Based on this value, does the health group's claim seem reasonable at the 5% significance level? Explain.

10.37 Suppose that, in the population of pregnant women who experience no complications, the length X of a pregnancy is normally distributed with a mean of 270 days and a standard deviation of 10 days.

(a) What proportion of pregnancies last 285 days or less?

(b) Let X_1, X_2, X_3, X_4 be a random sample of four pregnancy lengths.

(i) What is the probability that *at least one* of these lasts *more than* 285 days?

(ii) What is the probability that the sample mean pregnancy length is 285 days or less?

A researcher suspects that women who work outside the home during the entire time of their pregnancy tend to have pregnancies that are shorter, on average, than 270 days. A random sample of 15 such pregnancies gave a sample mean length of 265.8 days and a sample standard deviation of 12.2 days. We wish to use these data to assess the researcher's claim.

(c) State the appropriate null and alternative hypothesis in terms of μ, the population mean length of pregnancy for such women.

(d) Calculate the corresponding value of the test statistic and give the *p*-value.

(e) State the decision for a 5% level test.

(f) What does your decision suggest about the mean length of pregnancy for such women?

10.38 There are many ways to measure the reading ability of children. One frequently used test is the DRP, or Degree of Reading Power. The DRP scores for a simple random sample of 41 third graders in a suburban school district yielded a mean of 35.2 and a standard deviation of 11. Assume that DRP scores are normally distributed. We want to determine whether there is sufficient evidence at the 5% level to suggest that the mean score of all third graders in this district is higher than the national mean, which is 32.

(a) State the appropriate null and alternative hypothesis.

(b) Compute the value of the test statistic.

(c) Find the corresponding *p*-value for this test.

(d) Are the data statistically significant at the 5% level?

10.39 A new honors program was recently implemented at a college. A total of 250 students were accepted into the new honors program. One of the components of this program is a new interdisciplinary first-year honors course taught by a team of professors. Of the 80 students who took this course, 40 were males and 40 were females. The honors staff decided to take a random sample of 10 males and 20 females to be surveyed in detail about the quality of this first-year honors course.

(a) What type of sampling method is used to obtain the selected students to be surveyed?

(b) Use your calculator (with a seed value of 311) or the random number table (Row 18, Column 1) to give the labels for the selected males.

(c) The ratings by the 20 female students surveyed are provided (out of 10 points, with 10 being the best).

4.2	6.8	7.0	7.8	6.4	6.5	7.2	8.4	6.7	9.2
5.9	6.3	7.8	7.7	8.2	6.9	8.1	8.0	7.5	7.0

  (i) Give the mean and standard deviation of these data on course rating (up to two places after the decimal).

  (ii) Sketch a basic boxplot of these data. Be sure to label the axis and include the values and labels for the five-number summary.

(d) We wish to assess if there is sufficient evidence that the mean course rating by female students is less than 7.5. State the appropriate null and alternative hypothesis.

(e) Conduct the test in (d) using a 5% significance level. Show all steps and report the observed test statistic value, the *p*-value, your decision, and whether the results are statistically significant.

(f) The 10 male students surveyed gave the course a mean rating of 6.82. The female rating results were given in (c). Give the overall estimate of the average rating of the course for all students based on this sample information.

10.40 A supermarket advertises that the waiting time spent by customers waiting in line at the cash register is uniform between 0 and 6 minutes (with an average waiting time of 3 minutes and a standard deviation of 1.73 minutes). A number of customers complain that they wait in line too long for this to be true. Sparky, the manager, decides to test whether the complaints are justified by testing the hypotheses $H_0: \mu = 3$ versus $H_1: \mu > 3$.

(a) Sketch the distribution of waiting time when H_0 is true. Be sure to label and give appropriate values on the axes.

(b) One possible decision rule for Sparky is to reject H_0 if the next randomly chosen customer waits in line longer than five minutes. What is the chance of a Type I error using this decision rule?

(c) Another possible decision rule is to reject H_0 if the next two randomly chosen customers both wait in line longer than five minutes. What is the chance of a Type I error using this decision rule?

(d) Sparky's assistant, who has taken a statistics course, remembers that taking a large sample is generally a good thing to do, so she decides to take a random sample of 30 customers, and her decision rule is to reject H_0 if the average waiting time is longer than five minutes.

    (i) What is the approximate distribution for the average (sample mean) waiting time under the null hypothesis?

    (ii) What is the chance of a Type I error using this decision rule?

10.41 In February 1995, the mean cost for an airline round-trip ticket with a discount fare was \$258 (*USA Today*, March 30, 1995). A simple random sample of 15 round-trip discount fares during the month of March provided the following data:

    310, 260, 265, 255, 300, 310, 230, 250, 265, 280, 290, 240, 285, 250, 260.

REAL DATA

(a) Give the five-number summary of these data. Include the names and the values for each of the five-number summaries.

(b) Make a histogram of these data and comment on the underlying assumption of normality needed for performing a *t*-test.

(c) We wish to assess if there is sufficient evidence that the mean round-trip for discount fares has increased in March. Test at the 5% level of significance. State the appropriate null and alternative hypothesis.

(d) Conduct the test in (a) using a 5% significance level. Give all steps and provide the observed test statistic value, *p*-value, decision, and report whether the results are statistically significant.

10.42 The following statements are from the *New York Times* (May 21, 1995) article entitled "Travel Industry Anticipates Biggest Summer Season Ever":

> "Spurred by an improving economy, widespread discounts and growing consumer confidence, the travel industry is anticipating America's biggest ever summer vacation season."
>
> "In keeping with the recent trend toward shorter and more frequent getaways, 58 percent of the travelers surveyed by the association said they were planning more than one vacation this summer."
>
> "This summer a vacationing family of two adults and two children can expect to spend an average of \$221.80 per day for meals, lodging and vehicle operating costs. That compares to an inflation-adjusted \$245 in 1988."

REAL DATA

(a) Where do you think the \$221.80 figure came from? How about the \$245 figure?

(b) Part of the reasons cited for the increase in summer travel is lower costs. Suppose that we take the \$245 as the average spent (per day for such families, adjusted for inflation) in 1988. Write out the appropriate hypotheses for assessing whether there has been a significant decrease in the average amount spent.

(c) What additional information would you need to be able to perform the test in (b)? What assumptions would you need to assess and how might you assess them?

(d) If, in reporting the 58% of travelers estimated as planning more than one vacation this summer, the margin of error (or half-width) were stated as 6.8% with 95% confidence, what would be the corresponding 95% confidence interval for the true proportion of travelers planning more than one vacation this summer?

(e) Interpret what the 95% confidence level means.

10.43 In a study on the nutritional qualities of fast foods, the amount of sodium was measured for a random sample of 16 hamburgers for a certain fast-food restaurant. The mean and standard deviation for these 16 sampled hamburgers was found to be 642 milligrams and 38 milligrams, respectively.

(a) Compute a 90% confidence interval for the true mean sodium content for all hamburgers made by this restaurant.

(b) Determine whether the following statements are true or false (a true statement is always true):

(i) The true mean μ falls inside the confidence interval you calculated in (a) with probability 0.90.

(ii) A 90% confidence interval for μ based on a larger sample will be wider, since it contains more data.

10.44 The Four-minute Marriage. According to a recent study by the Center for Lifestyle Management, the average couple devotes four minutes per day to meaningful conversation. Could it really be that we spend so little time talking to each other? *Good Housekeeping* (August 1995) put the Four-minute Marriage to the test. They had wives keep daily and honest logs of their time spent in meaningful conversations with their spouse. The data are provided in minutes:

7, 6, 5, 6, 10, 8, 10, 19, 19.

REAL
DATA

(a) Provide a 90% confidence interval for the mean time spent in meaningful conversation per day.

(b) Interpret your interval from (a).

(c) What are the necessary assumptions to construct such an interval? Do you think these assumptions are valid? Explain.

(d) Based on your interval, does it appear that the average time spent in meaningful conversation per day could actually be only four minutes? Explain.

10.45 A beverage-dispensing machine is calibrated so that the amount of beverage dispensed is approximately normally distributed with a population standard deviation of 0.2 ounces. A random sample of 25 drinks dispensed an average of 11.5 ounces.

(a) Compute a 99% confidence interval for the mean amount of beverage dispensed by this machine.

(b) What is the margin of error for the 99% confidence interval based on the 25 observations?

10.46 The time to play the minute waltz is recorded for 25 different pianists. The sample mean playing time is $\bar{x} = 1.2$ minutes and the sample standard deviation is $s = 0.3$

minutes. Assuming that the distribution of playing time is approximately normal, construct a 99% confidence interval for the mean playing time.

10.47 A city planner randomly samples 100 apartments in a suburb to estimate the mean living area per apartment (in square feet). The sample yielded an average of 1325 square feet and a standard deviation of 42 square feet. Assume that the living area follows a normal distribution.

  (a) What is the population of interest?

  (b) What is the variable of interest?

  (c) Give a point estimate of the true mean living area per apartment.

  (d) Give a 95% confidence interval estimate of the true mean living area per apartment.

  (e) Is there a 95% chance that the true mean living area is within the interval you computed in (d)? Explain.

  (f) Give two ways to increase the width of the confidence interval.

10.48 Based on data from a random sample of $n = 36$ records of patients in a hospital (during a four-year period), the 95% confidence interval estimate for the mean length of stay (in days) of a patient in the non-intensive-care unit is given by $(3.5, 6.5)$.

  (a) Find the point *estimate* of μ, the mean length of stay (in days) of a patient in the non-intensive-care unit during the four-year period.

  (b) Find the point *estimate* of σ, the standard deviation for length of stay (in days) of a patient in the non-intensive-care unit during the four-year period.

10.49 Based on a random sample of size $n = 10$ salaries of recent college graduates, a 95% confidence interval for the population mean yearly salary of all recent college graduates was found to be ($28575, $32625). For each of the following statements, select your answer:

  (a) We can say that 95% of all recent college graduates earn between $28,575 and $32,625 per year.

            True   False   Can't tell

  (b) The probability that the true population mean yearly salary of all recent college graduates, μ, is inside this confidence interval ($28575, $32625) is 0.95.

            True   False   Can't tell

  (c) Based on the same random sample, a 90% confidence interval for μ would be wider than the 95% confidence interval.

            True   False   Can't tell

  (d) If we were to repeatedly take a random sample of the same size, from the same population, and for each sample construct a 95% confidence interval for μ, we would expect about 95% of these intervals to contain μ.

            True   False   Can't tell

  (e) If we want to test: H_0: $\mu = \$30,000$ versus H_1: $\mu \neq \$30,000$ using a 1% significance level, our decision would be

            Reject H_0 Accept H_0 Can't tell

10.50 A fast-food chain sells a hamburger that they claim has a sodium content of 650 milligrams. A simple random sample of 35 hamburgers was analyzed for sodium content. A 99% confidence interval for the population mean sodium content, μ, of such hamburgers is $(652, 672)$. Answer the following questions with "yes", "no", or "can't tell," and give an explanation for your answer:

(a) Does the population mean lie in the interval $(652, 672)$?

(b) Does the sample mean lie in the interval $(652, 672)$?

(c) For a future sample of 35 hamburgers, will the sample mean lie in the interval $(652, 672)$?

(d) If we use a 95% level of confidence, will the confidence interval calculation from the same data produce an interval narrower than $(652, 672)$?

(e) If we were to use the preceding data to test the hypotheses H_0: $\mu = 650$ versus H_1: $\mu \neq 650$ at a 1% significance level, would we reject the null hypothesis? Explain.

10.51 A study on the alcohol content of a certain type of wine reported a 95% confidence interval for the mean alcohol content, μ, of $(6.8, 8.0)$. This interval was based on data assumed to be a random sample from a normal distribution.

(a) What is the value of the sample mean alcohol content?

(b) At a 5% significance level, give the decision for testing
H_0: $\mu = 7$ versus H_1: $\mu \neq 7$.

  Select one: Reject H_0 Accept H_0 Can't tell

(c) At a 1% significance level, give the decision for testing H_0: $\mu = 7$
H_0: $\mu = 7$ versus H_1: $\mu \neq 7$.

  Select one: Reject H_0 Accept H_0 Can't tell

TI Quick Steps

Finding Areas Under a *t*-Distribution Using the TI

The TI has functions that are built into the calculator. One function, called the tcdf(function, can compute various areas under a *t*-distribution with a specified degrees of freedom. This function is located under the DISTR menu. Suppose that the observed test statistic value is $t = -0.885$ and we wish to find the area to the left of this value under a *t*-distribution with 55 degrees of freedom. The TI steps are as follows:

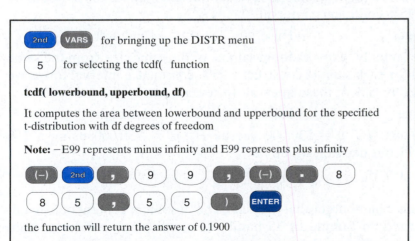

One-sample *z*-test for a mean

You can either have the one sample of *Data* entered, say as L1, or just enter the *Stats* (the sample mean, population standard deviation, and sample size). To perform a test, the steps are as follows:

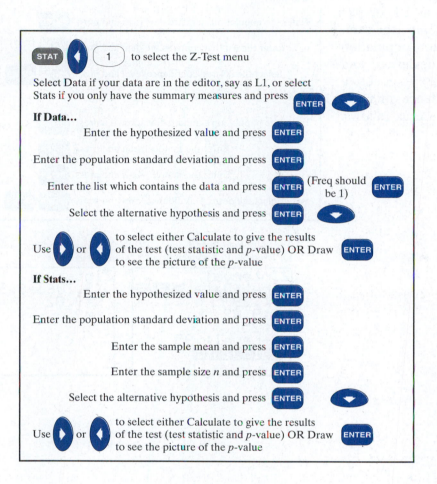

The Stats input screen for Example 10.1 is given by:

The corresponding test result screen is:

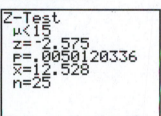

One-sample *z* confidence interval for a mean

You can either have the one sample of *Data* entered, say as L1, or just know the *Stats* (the sample mean, population standard deviation, and sample size). To generate the confidence interval, the steps are as follows:

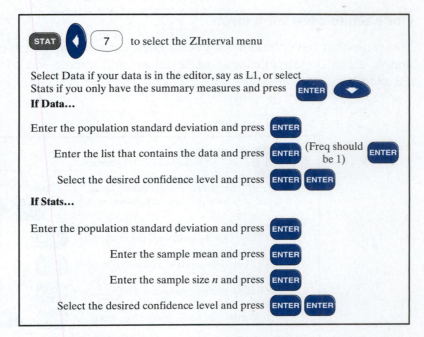

STAT ◀ 7   to select the ZInterval menu

Select Data if your data is in the editor, say as L1, or select Stats if you only have the summary measures and press ENTER ▼

If Data...

Enter the population standard deviation and press ENTER

Enter the list that contains the data and press ENTER   (Freq should be 1) ENTER

Select the desired confidence level and press ENTER ENTER

If Stats...

Enter the population standard deviation and press ENTER

Enter the sample mean and press ENTER

Enter the sample size *n* and press ENTER

Select the desired confidence level and press ENTER ENTER

The Stats input screen for Example 10.5 is given by:

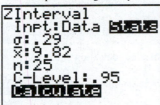

```
ZInterval
 Inpt:Data Stats
 σ:.29
 x̄:9.82
 n:25
 C-Level:.95
 Calculate
```

The corresponding result screen is:

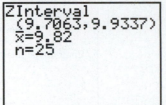

```
ZInterval
 (9.7063,9.9337)
 x̄=9.82
 n=25
```

One-sample *t*-test for a mean

You can either have the one sample of *Data* entered, say as L1, or just enter the *Stats* (the sample mean, sample standard deviation, and sample size). To perform a test, the steps are as follows:

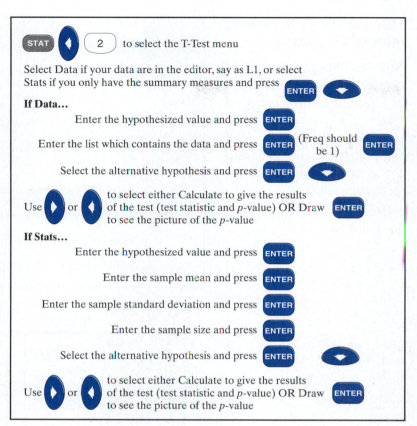

STAT ◄ 2 to select the T-Test menu

Select Data if your data are in the editor, say as L1, or select Stats if you only have the summary measures and press ENTER ▼

If Data...

Enter the hypothesized value and press ENTER

Enter the list which contains the data and press ENTER (Freq should be 1) ENTER

Select the alternative hypothesis and press ENTER ▼

Use ► or ◄ to select either Calculate to give the results of the test (test statistic and *p*-value) OR Draw to see the picture of the *p*-value ENTER

If Stats...

Enter the hypothesized value and press ENTER

Enter the sample mean and press ENTER

Enter the sample standard deviation and press ENTER

Enter the sample size and press ENTER

Select the alternative hypothesis and press ENTER ▼

Use ► or ◄ to select either Calculate to give the results of the test (test statistic and *p*-value) OR Draw to see the picture of the *p*-value ENTER

The Stats input screen for Example 10.2 is given by:

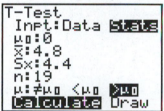

```
T-Test
 Inpt:Data Stats
 μ₀:0
 x̄:4.8
 Sx:4.4
 n:19
 μ:≠μ₀ <μ₀ >μ₀
 Calculate Draw
```

The corresponding test result screen is:

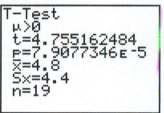

```
T-Test
 μ>0
 t=4.755162484
 p=7.9077346ᴇ-5
 x̄=4.8
 Sx=4.4
 n=19
```

One-sample *t* confidence interval for a mean

You can either have the one sample of *Data* entered, say as L1, or just know the *Stats* (the sample mean, sample standard deviation, and sample size). To generate the confidence interval, the steps are as follows:

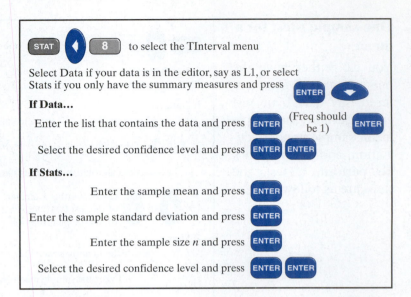

The Stats input screen for Example 10.6 is given by:

```
TInterval
 Inpt:Data Stats
 x̄:12.6
 Sx:4.4
 n:61
 C-Level:.99
 Calculate
```

The corresponding result screen is:

```
TInterval
 (11.101,14.099)
 x̄:12.6
 Sx:4.4
 n:61
```

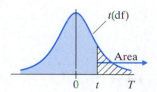

Table IV   *Upper Percentiles of t-distributions*

df	\multicolumn{8}{c}{Area to the Right}
	0.40	0.30	0.20	0.10	0.05	0.025	0.01	0.005
1	0.325	0.727	1.376	3.078	6.314	12.706	31.821	63.657
2	0.289	0.617	1.061	1.885	2.920	4.303	6.965	9.925
3	0.277	0.584	0.978	1.638	2.353	3.182	4.541	5.841
4	0.271	0.569	0.941	1.533	2.132	2.776	3.747	4.604
5	0.267	0.559	0.920	1.476	2.015	2.571	3.365	4.032
6	0.265	0.553	0.906	1.440	1.943	2.447	3.143	3.707
7	0.263	0.549	0.896	1.415	1.895	2.365	2.998	3.499
8	0.262	0.546	0.889	1.397	1.860	2.306	2.896	3.355
9	0.261	0.543	0.883	1.383	1.833	2.262	2.821	3.250
10	0.260	0.542	0.879	1.372	1.812	2.228	2.764	3.169
11	0.260	0.540	0.876	1.363	1.796	2.201	2.718	3.106
12	0.259	0.539	0.873	1.356	1.782	2.179	2.681	3.055
13	0.259	0.538	0.870	1.350	1.771	2.160	2.650	3.012
14	0.258	0.537	0.868	1.345	1.761	2.145	2.624	2.977
15	0.258	0.536	0.866	1.341	1.753	2.131	2.602	2.947
16	0.258	0.535	0.865	1.337	1.746	2.120	2.583	2.921
17	0.257	0.534	0.863	1.333	1.740	2.110	2.567	2.898
18	0.257	0.534	0.862	1.330	1.734	2.101	2.552	2.878
19	0.257	0.533	0.861	1.328	1.729	2.093	2.539	2.861
20	0.257	0.533	0.860	1.325	1.725	2.086	2.528	2.845
21	0.257	0.532	0.859	1.323	1.721	2.080	2.518	2.831
22	0.256	0.532	0.858	1.321	1.717	2.074	2.508	2.819
23	0.256	0.532	0.858	1.319	1.714	2.069	2.500	2.807
24	0.256	0.531	0.857	1.318	1.711	2.064	2.492	2.797
25	0.256	0.531	0.856	1.316	1.708	2.060	2.485	2.787
26	0.256	0.531	0.856	1.315	1.706	2.056	2.479	2.779
27	0.256	0.531	0.855	1.314	1.703	2.052	2.473	2.771
28	0.256	0.530	0.855	1.313	1.701	2.048	2.467	2.763
29	0.256	0.530	0.854	1.311	1.699	2.045	2.462	2.756
30	0.256	0.530	0.854	1.310	1.697	2.042	2.457	2.750
40	0.255	0.529	0.851	1.303	1.684	2.021	2.423	2.704
60	0.254	0.527	0.848	1.296	1.671	2.000	2.390	2.660
120	0.254	0.526	0.845	1.289	1.658	1.980	2.358	2.617
z^*	0.253	0.524	0.842	1.282	1.645	1.960	2.326	2.576
	20%	40%	60%	80%	90%	95%	98%	99%

Confidence Level

Chapter 11

COMPARING TWO TREATMENTS

◆ 11.1 Introduction

In Chapter 9 and 10, we studied inference procedures for learning about a single population. We tested hypotheses and developed confidence interval estimates for two parameters, the population proportion p and the population mean μ. We provided a statement, using probability, about how confident we were in our decision or interval estimate. The methods for drawing conclusions about one population can be extended to other scenarios. Although the scenarios change, the way of thinking about decision making does not change. The next page provides a review of the key components of testing hypotheses and confidence interval estimation, which form a foundation for this chapter.

In this chapter, we will learn how to use the results of two sets of data to decide something about two populations. We will first examine methods for comparing the means of two populations, μ_1 and μ_2. A primary issue will be to determine if the two sets of data collected are paired (dependent) or independent samples. In the last section we will discuss a procedure for comparing the proportions for two populations, p_1 and p_2. **From the summaries on the next page, we can see that for any given situation, we will need a point estimate for the parameter and a corresponding standard error (since the true standard deviation is not generally known).**

Suppose that we have a new teaching technique to be used to teach a group of students. We would like to assess whether the performance of such students improves using this new teaching technique. How will we know if there has been improvement? Before we can tell, we must have an **operational definition** of improvement. Perhaps improvement will

Basic Steps for Testing a Hypothesis about a Parameter

■ **State the population(s) and corresponding parameter(s) of interest**.

Remember that inference is valid only if the sample(s) is *representative* of the population(s) of interest.

■ **State the competing theories—that is, the null and alternative hypotheses**.

The null hypothesis gives a specific value for the parameter, called the *hypothesized value* or *null value*.

■ **State the significance level α for the test**.

This level should always be *set in advance* of examining the results.

■ **Collect and examine the data and assess if the assumptions seem valid**.

If assumptions do not seem reasonable, there may be *alternative procedures*, some of which are discussed in Chapter 15.

■ **Compute a test statistic using the data and determine the *p*-value**.

The *test statistic* is a measure of the distance between the *sample statistic* or *point estimate* of the parameter and the *hypothesized value* or *null value* for the parameter. The general form of a test statistic is that of a standard score:

$$\text{Test Statistic} = \frac{\text{Point Estimate} - \text{Null Value}}{\text{Null Standard Deviation or Error}}$$

If the denominator is the actual standard deviation under the null hypothesis, then the test statistic is a *z*-statistic.

If the denominator is a standard error (an estimated standard deviation) under the null hypothesis, then the test statistic is a *t*-statistic.

The *p*-value is found using the null distribution for the test statistic.

For a *z*-statistic we use a standard normal $N(0,1)$ distribution.

For a *t*-statistic we use a *t*-distribution with a certain degrees of freedom (recall in the one-sample *t*-test the degrees of freedom were $n - 1$).

The *p*-value is the probability of getting a test statistic value as extreme or more extreme than the observed value, assuming H_0 is true. The *direction of more extreme* is determined by the direction in H_1.

■ **Make a decision and state a conclusion using a well-written sentence**.

We compare the *p*-value to the significance level to make the decision.

Basic Form of a Confidence Interval for a Parameter

Point Estimate $\pm$ (*a few*) (Standard Error of the Point Estimate)

The "***a few***" is either a z^* or a t^* percentile, which depends on the confidence level desired and the sample size.

The *confidence level* describes how often the procedure (if repeated) provides an interval that actually contains the population parameter.

correspond to an increase in the mean test score at the end of the period of instruction. Perhaps the goal will be to reduce the variability in the test scores without necessarily increasing the mean score. Another measure of improvement might focus on the ability of students to retain what they have learned after a period of time.

In Chapter 3, we discussed that if we want to assess whether a result is better or worse, it is good practice to **compare** the results of the new "treatment" with the results of a **control group**, where the control group receives, if possible, a standard treatment or a **placebo** treatment. For assessing whether students' performance improves using the new teaching technique, we might design our study to include a control group of students who would receive instruction using the best standard teaching technique that is available. The two theories to be compared would be

H_0: The new teaching technique has no effect, with respect to the mean test score, when compared to the best standard teaching technique.

H_1: The new teaching technique is better, that is, has a higher mean test score, when compared to the best standard teaching technique.

Even when a decision is made, there may be new questions to address, new theories to be tested, and more data to be obtained and analyzed. We would continue to cycle through the steps of the scientific method. Suppose that the data are found to be statistically significant. Could we conclude that the new teaching technique would work for other teachers, with other students, in other countries, or for the same school two years from now?

The inference methods presented in this chapter provide useful statistics for making decisions. However, we must keep in mind which populations are actually under study. We should not blindly extend the results to other unknown, not studied, populations. When writing a report of such research, it is important to state pertinent details of the study and to discuss what may be the **valid extensions** of the results.

To set the stage for comparing two population means, we turn to an exercise that revisits some of the experimental design issues from Chapter 3.

Let's do it! 11.1 The Placebo Effect

REAL DATA

A *Wall Street Journal* article, "Baxter's Transplant Drug Doesn't Outperform Placebo," states that clinical trials showed that the medication Anti-CD45 had no statistically significant advantage over a placebo used as a control.

(a) What is the placebo effect?

(b) Why was a placebo used as a control in the experiments?

(c) Suppose the results of the experiments were used to compare the following hypotheses:

H_0: The new drug is equally effective as the placebo—success rates in warding off acute rejection are the same.

H_1: The new drug is more effective than the placebo—the new drug has a higher success rate than the placebo.

Baxter's Transplant Drug Doesn't Outperform Placebo

By a WALL STREET JOURNAL Staff Reporter

DEERFIELD, Ill. - **Baxter International Inc.** said its once-promising drug to ward off rejection episodes following kidney transplants showed no advantage over a placebo in preliminary results of a clinical trial.

Baxter said it obtained the rights to the medication, called Anti-CD45, from **Cantab Pharmaceuticals** PLC of Britain. Baxter contributed those rights to a partnership it owns with DNX Corp. While Baxter has several other major products in various stages of research, the antirejection medication clearly was one for which the company had high hopes, and the early results are a distinct disappointment.

Baxter said preliminary analysis of results from the Phase II — or moderately advanced — clinical trials showed that the medication had "no clinical advantage" over a placebo used as a control.

Baxter noted earlier that Anti-CD45 was believed to improve the success rate in warding off "acute rejection," meaning complete rejection of an organ within the first few months of transplant. One hope was that the drug could reduce costs of a transplant. Baxter said it hasn't decided whether to pursue any further development of the product.

Based on the conclusions stated in the article, if the preceding hypotheses were tested using a 5% significance level, would the *p*-value be larger than 0.05 or smaller than or equal to 0.05? Explain briefly.

(d) Suppose that Population 1 corresponds to the drug group that has true mean response μ_1 and Population 2 corresponds to the placebo group that has true mean response μ_2. The null hypothesis states that the new drug is *equally effective* as the placebo, that there is *no difference* between the two populations with respect to the mean response. How would you express this null hypothesis in terms of μ_1 and μ_2?

H_0:

11.2 Paired Samples versus Independent Samples

When comparing two populations, it would be ideal if the two populations were **similar in all respects** except for the two treatments that are imposed (that is, except for the factor under study).

Suppose that we are interested in comparing the mean salary of wives, the female head of household, to the mean salary of husbands, the male head of household. If we were to take a random sample of married couples, we would have **paired samples**, as shown in Paired Design 11.1.

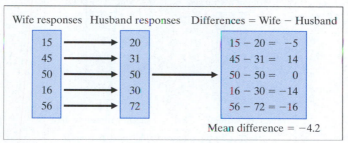

Paired Design 11.1

One main advantage of the paired design is to help reduce the effect of bias due to confounding factors. It helps us to compare apples to apples rather than apples to oranges. Subjects can be paired with respect to such factors as age, gender, initial health status, and so on. Such factors may influence the response. When we compare the paired observations, the effect of these factors on the response tend to cancel out.

In our wives-versus-husbands example, we should ask whether or not these two groups are similar with respect to other variables that could influence salary. For example, would they be similar with respect to the highest educational level attained or the number of years of job experience? If the conclusion was that the mean salary of husbands is higher than that for wives, we might not be able to determine which factors contributed and how the factors contributed to this difference. In such a case, the various factors are called **confounding factors**.

With paired data, we are interested in comparing the responses within each pair. We will analyze the *differences* in the responses that form each pair.

Paired Data: Response = Annual Salary (in $1000s)

Wife responses	Husband responses	Differences = Wife − Husband
15	20	15 − 20 = −5
45	31	45 − 31 = 14
50	50	50 − 50 = 0
16	30	16 − 30 = −14
56	72	56 − 72 = −16

Mean difference = −4.2

Another type of paired design is when each subject serves as its own control. If there are two treatments, each subject may receive both treatments with the order of the treatments randomized. If there is just one treatment, responses are measured for each subject before and after the treatment is administered, as depicted in Paired Design 11.2. In both cases, the two observations on each subject would be paired or dependent. For example, we would want to compare the blood pressure for the same individual before and after receiving some treatment. In this case, the treatment is the *explanatory* variable and the change in blood pressure is the *response* variable.

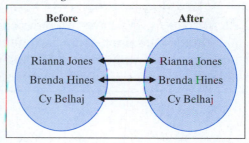

Paired Design 11.2

> *Definition:*   We have **paired** or **matched samples** when we know, in advance, that an observation in one data set is directly related to a specific observation in the other data set. It may be that the related sets of units are each measured once (Paired Design 11.1), or that the same unit is measured twice (Paired Design 11.2). In a paired design, the two sets of data must have the same number of observations.

Suppose that we are interested in comparing the mean salaries for female professors versus male professors employed at a particular university. Again, we would like the two groups to be as equal as possible with respect to variables that can influence salary. Such variables might include number of years since graduation, amount of teaching experience, or number of publications in prominent journals. But there may be other variables that can influence salaries, which we could not foresee and therefore could not "control" for. This is where random sampling can help to attempt to balance the two groups with respect to such variables. We would take a random sample of female professors from the female population of interest and take an independent random sample of male professors from the corresponding male population of interest. Thus we have **two independent samples**.

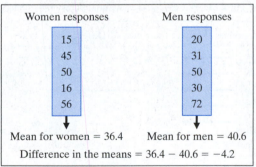

Independent Samples Design 11.3

Women	Men
Amy Lee	Pedro Hampton
Beth Thompson	
Cindy Merks	Mike Jennings
Sharida Banks	John Conner

In the two independent samples scenario, we will compare the responses of one treatment group as a whole to the responses of the other treatment group as a whole. We will calculate summary measures for the observations from one treatment group and compare them to similar summary measures calculated from the observations from the other treatment group.

Independent Samples Data:
Response = Annual Salary (in $1000s)

Women responses	Men responses
15	20
45	31
50	50
16	30
56	72

Mean for women = 36.4   Mean for men = 40.6

Difference in the means = $36.4 - 40.6 = -4.2$

> *Definition:*   We have **two independent samples** when two unrelated sets of units are measured, one sample from each population, as in Independent Samples Design 11.3. In a design with two independent samples, although the same sample size is often preferable, the sample sizes might be different.

In the next section, we will focus on inference techniques for the mean difference between *two dependent samples*. In Section 11.4, we will examine inference techniques for the difference between the means of *two independent samples*. We need to recognize when we have paired samples versus independent samples, so that we perform the appropriate analysis of our data. Let's turn to some exercises to practice distinguishing between these two cases.

Let's do it! 11.2 Paired Samples versus Independent Samples

(a) Three hundred registered voters were selected at random, 30 from each of 10 midwestern counties, to participate in a study on attitudes about how well the president is performing his job. They were each asked to answer a short multiple-choice questionnaire and then they watched a 20-minute video that presented information about the job description of the president. After watching the video, the same 300 selected voters were asked to answer a follow-up multiple-choice questionnaire. The investigator of this study will have two sets of data: the initial questionnaire scores and the follow-up questionnaire scores. Is this a paired or independent samples design?

Circle one: Paired   Independent

Explain.

(b) Thirty dogs were selected at random from those residing at the humane society last month. The 30 dogs were split at random into two groups. The first group of 15 dogs was trained to perform a certain task using a reward method. The second group of 15 dogs was trained to perform the same task using a reward–punishment method. The investigator of this study will have two sets of data: the learning times for the dogs trained with the reward method and the learning times for the dogs trained with the reward–punishment method. Is this a paired or independent samples design?

Circle one: Paired   Independent

Explain:

Exercises

11.1 Brown University is known for its advanced studies of the characteristics of twins. The investigators of a recent study selected 20 female students at random from among Brown University female students whose twin is a male. The other sample of size 20 is composed of the corresponding male twin of these female students. Are the data independent samples or paired samples? Explain. *Hint*: Think about how likely it would be to obtain such subjects if these samples were actually selected independently.

11.2 A study will be conducted at Decker University to compare the course loads of female versus male students. From among all female students registered full time for the current term, 20% were selected at random. From among all male students registered full time for the current term, 20% were selected at random.

(a) Are the data independent samples or paired samples? Explain. *Hint*: Think about what the sample sizes could be.

(b) Identify a possible confounding variable for this study (that is, a variable that is not of primary interest, and might not be measured, but is associated with the response of interest).

11.3 Dr. Jones was the investigator for a study that took a simple random sample of 10 women from the U.S. and then 10 men. The 10 men were the corresponding husbands of the 10 women. Are the data independent samples or paired samples? Explain. *Hint*: Think about how likely would it be to select 10 women at random, and then 10 men at random, independently from the sample of women, and actually get the 10 husbands of the 10 women selected first.

11.4 Ten dogs—Lilly, Libby, Jasmin, Maggie, Klaus, Peetey, Wrinkles, Spot, Johnny, and Sparky—are available for a study to compare two training methods. Design a random allocation scheme for splitting the 10 dogs into two groups of 5 and use your calculator (seed value = 23) or random number table (Row 18, Column 1) to carry out the randomization.

11.5 Eight students in a statistics course are available to serve as subjects in a study.

Name	Gender	Age
Ronald	M	20
Kerri	F	18
Emily	F	19
Lee	M	19
Kyle	M	20
Pablo	M	19
Monica	F	19
Sonya	F	18

(a) The researcher would like to conduct a paired design in which the subjects are paired by the same gender. List the possible ways to form pairs by gender.

(b) The researcher would like to conduct a paired design in which the subjects are paired by age. List the possible ways to form pairs by age.

(c) Use your calculator (seed value = 18) or random number table (Row 8, Column 1) to pair the subjects by gender. Explain your pairing scheme.

(d) Use your calculator (seed value = 28) or random number table (Row 10, Column 1) to pair the subjects by age. Explain your pairing scheme.

11.6 For each of the following research questions, briefly describe how you might design a study to address the question (discuss whether paired or independent samples would be obtained):

(a) Do sophomore students seek the advice of an academic advisor more often than freshmen students?

(b) Will taking a one-hour Kaplan SAT prep course improve test scores on average by 30 points?

(c) For twins, is the first born taller on average as compared to the second born when they reach the adult age of 18?

11.3 Paired Samples

In a paired design, the units in each pair are alike (in fact, they may be the same unit), whereas units in different pairs may be quite dissimilar. Pairing the units according to some factor helps to remove that source of variation from the study. By taking the difference between the two paired responses, we can focus on the effects of the treatments. Since we are interested in the differences for each pair, *the differences are what we analyze in paired designs*.

So we really have just one population under study, the population of differences. Thus, the paired data scenario is a special case of learning about a single population mean μ. To help us remember we are in a paired data scenario, we will refer to the **mean of the population of differences** as μ_D. If the population mean difference is zero, then the two treatments are said to be equivalent, that is, there is no treatment effect. If the population mean difference is not zero, then the responses to one treatment are generally higher than the responses to the other treatment and there is a treatment effect.

Population of Paired Observations
D = difference = treatment 1 − treatment 2

Observation for treatment 1

Observation for treatment 2

Population mean of the Differences = μ_D
Population standard deviation of the Differences = σ_D

By reducing the two sets of paired observations to one set of differences, we are back to the one-sample inference scenario for learning about a population mean from Chapter 10. Since we generally will not know the value of the population standard deviation σ_D, we will focus on the procedures based on the t-distribution. Our first summary box reminds us of the basic ideas behind estimating a population mean with a sample mean.

Learning about the Population Mean Difference μ_D

Parameter of interest: The population mean difference μ_D

Data: We have a *random sample of differences* from the population of all possible differences. The differences have a *normal distribution* with unknown mean μ_D and unknown standard deviation σ_D. Normality is not so crucial if the sample size n is large (≥ 30) due to the central limit theorem.

Point Estimate of μ_D: *Sample mean difference* $\bar{d}$, the average of the differences in the sample. This is just like $\bar{x}$, but we use the $\bar{d}$ notation to remember we have paired data.

Standard Deviation of a sample mean difference: $\dfrac{\sigma_D}{\sqrt{n}}$

Standard Error of a sample mean difference: $\dfrac{s_D}{\sqrt{n}}$ where s_D is the standard deviation computed on the differences in the sample.

❖ EXAMPLE 11.1 Weight Loss

The before-diet and after-diet weights (in pounds) of a random sample of 12 people in a special diet program are given, along with their differences, computed as "Before less After":

Before	155	168	172	180	174	142	135	196	200	149	165	139
After	152	170	166	180	160	143	132	180	189	147	168	138
diff = B − A	3	−2	6	0	14	−1	3	16	11	2	−3	1

Note that the differences computed this way represent the weight loss for an individual, with a positive value indicating weight loss and a negative value indicating weight gain. The sample mean difference or sample mean weight loss can be found by averaging the 12 differences.

$$\bar{d} = \frac{3 + (-2) + 6 + 0 + 14 + (-1) + 3 + 16 + 11 + 2 + (-3) + 1}{12} = 4.2.$$

Our (point) estimate of the population mean weight loss for all people on this diet program is 4.2 pounds. The sample standard deviation of the differences is given by

$$s_D = \sqrt{\frac{(3 - 4.2)^2 + (-2 - 4.2)^2 + (6 - 4.2)^2 + \cdots + (1 - 4.2)^2}{12 - 1}} = 6.3.$$

So the standard error of the mean is $\frac{s_D}{\sqrt{n}} = \frac{6.3}{\sqrt{12}} = 1.82$. We would interpret this standard error as follows: The average distance between the possible sample mean differences $\bar{d}$ and the population mean difference μ_D is roughly 1.82 pounds. ❖

With our point estimate and standard error in hand we can turn to a summary of the one-sample *t*-procedures performed on the differences, called the **paired *t*-test** and corresponding **confidence interval for the mean difference** μ_D. In the testing scenario, we are generally interested in assessing whether or not there is a differences between the two treatments, so the value in the null hypothesis for μ_D is generally set at zero. In the confidence interval scenario, it will often be of interest to see if the value of zero falls in the confidence interval as a plausible value for μ_D.

Paired *t*-Test

Assumptions: The sample of differences is a random sample from a larger population of differences. The model for the differences is normal, although this is less crucial if the sample size *n* is large.

Hypotheses: $H_0: \mu_D = 0$ versus $H_1: \mu_D \neq 0$ or $H_1: \mu_D > 0$ or $H_1: \mu_D < 0$.
The significance level α to be used is determined.

Data: The sample of *n* differences, generically written as $d_1, d_2, \ldots, d_n$, from which the sample mean difference $\overline{d}$ and the sample standard deviation of the differences s_D can be computed.

Observed Test Statistic: $t = \dfrac{\overline{d} - 0}{\dfrac{s_D}{\sqrt{n}}}$

and the *null distribution* for the *T* variable is the $t(n-1)$ *distribution*.

p-value: We find the *p*-value for the test using the $t(n-1)$ *distribution*. The *direction of extreme* will depend on how the alternative hypothesis is expressed.

Decision: A *p*-value less than or equal to α leads to rejection of H_0.

Notes:

■ If we are interested in assessing if μ_D is equal to some hypothesized value that is not 0, we would replace zero in the test statistic expression with this other null value.

■ The test statistic is the same no matter how the alternative hypothesis is expressed.

Confidence Interval for μ_D

Assumptions: The sample of differences is a random sample from a larger population of differences. The model for the differences is normal, although this is less crucial if the sample size n is large.

Data: The sample of n differences, generically written as $d_1, d_2, \ldots, d_n$, from which the sample mean difference $\bar{d}$ and the sample standard deviation of the differences s_D can be computed.

Confidence Interval: $\bar{d} \pm t^* \left(\dfrac{s_D}{\sqrt{n}} \right),$

where t^* is an appropriate percentile of the $t(n-1)$ distribution.

Interpretation of the Interval: The interval gives potential values for the population mean difference μ_D based on just one random sample of differences. The interval may be used to test hypotheses when the alternative hypothesis is two-sided.

Interpretation of the Confidence Level: The confidence level pertains to the proportion of times the procedure produces an interval that actually contains μ_D when this procedure is repeated over and over using a new random sample of the same size n each time.

Let's illustrate our paired t-test and confidence interval estimate through two examples. In each example, we will see the various ways these methods can be performed: using the TI graphing calculator, Table IV, or computer output.

❖ EXAMPLE 11.2 Comparing Test Scores

A group of 10 randomly selected children of elementary-school age among those in the Mankato County who were recently diagnosed with asthma was tested to see if a new children's educational video is effective in increasing the children's knowledge about asthma. A nurse gave the children an oral test containing questions about asthma before and after seeing the animated video. The test scores are as follows:

Child	1	2	3	4	5	6	7	8	9	10	
Before	61	60	52	74	64	75	42	63	53	56	Mean = 60
After	67	62	54	83	60	89	44	67	62	57	Mean = 64.5

Do the data provide sufficient evidence for us to conclude that the mean score after viewing the educational video is significantly higher than the mean score before the viewing, at the 1% level of significance? Since we have two observations from the same child, we have paired data. We are interested in examining the differences in the scores for each child. If we define our differences as "after − before," then positive differences would show some support that the video is effective in improving the mean test score. The corresponding hypotheses to be tested are $H_0: \mu_D = 0$ versus $H_1: \mu_D > 0$. The observed differences are as follows:

Child	1	2	3	4	5	6	7	8	9	10	
d = After − Before	6	2	2	9	−4	14	2	4	9	1	Mean difference = 4.5

The first observed difference is 6 and is represented by $d_1 = 6$, and the last difference is also positive and is represented by $d_{10} = 1$. Of the 10 differences, only 1 is negative and 9 are positive. The observed sample mean difference is $\bar{d} = 4.5$, which is our estimate of the unknown mean difference, μ_D. The observed sample standard deviation of the differences is $s_D = 5.126$, which is our estimate of the unknown population standard deviation σ_D.

THINK ABOUT IT **Scenario I versus Scenario II**

Why can't we simply reject the null hypothesis and support the alternative hypothesis, $H_1: \mu_D > 0$, since the observed sample mean difference $\bar{d} = 4.5$ is greater than 0?

Assuming that the differences follow a normal distribution, we proceed with our test. The **observed test statistic** is

$$t = \frac{4.5 - 0}{\dfrac{5.126}{\sqrt{10}}} = 2.78.$$

What does the observed test statistic value mean? It means that we observed a sample mean difference that was about 2.78 standard errors above the hypothesized mean of difference zero. Is this large enough (that is, far enough above zero) to reject the null hypothesis? Is there enough evidence to conclude that the results are statistically significant at the 5% significance level? To answer this question, we find the *p*-value and compare it to 0.05.

Finding the *p*-value

The *p*-value is the probability of getting a test statistic as large or larger than the observed test statistic of 2.78, computed under the null hypothesis model, which is a *t*-distribution with nine degrees of freedom.

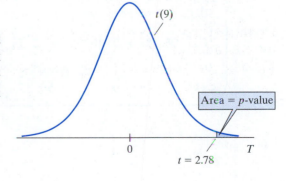

With the TI

1. Using the tcdf(function

Using the tcdf(function on the TI, we have

$$p\text{-value} = P(T \geq 2.78) = \text{tcdf}(2.78, \text{E99}, 9) = 0.0107.$$

2. Using the T-Test function under the STAT TESTS menu

In the TESTS menu located under the STAT button, we select the **2:T-Test** option. With the sample mean of 4.5, the sample standard deviation of 5.126, and the sample size of $n = 10$, we can use the Stats option of this test. The corresponding input and output screens are shown. Notice that the null or hypothesized value is zero.

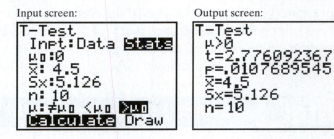

Input screen: Output screen:

$$p\text{-value} = P(T \geq 2.78) = 0.010761.$$

With Table IV

In Table IV, we focus on the df = 9 row and find that our observed test statistic value of 2.78 falls between the values in the columns headed with 0.025 and 0.01. So we state that the p-value is between 0.01 and 0.025; that is,

$$0.01 < p\text{-value} = P(T \geq 2.78) < 0.025.$$

With Computer Output

The Minitab computer output for performing the paired t-test on the video data is shown next. The data were entered and a variable called *diff* was created as the differences, defined as "after − before." The t-test was performed on the differences with a *greater than (G.T.)* alternative option selected. The observed mean difference of 4.5 and the observed standard deviation of 5.13 gave an observed test statistic of $t = 2.78$. The corresponding p-value for this one-sided to the right test is 0.011.

```
MTB > Let 'diff' = 'after' - 'before'
MTB > TTest 0.0 'diff';
SUBC>  Alternative 1.

TEST OF MU = 0.00 VS MU G.T. 0.00

            N      MEAN    STDEV    SE MEAN       T     P VALUE
diff        10     4.50    5.13       1.62     2.78       0.011
```

Decision and Conclusion

Since our *p*-value is less than 0.05, at the $\alpha = 0.05$ significance level we would reject H_0 and conclude that there is sufficient evidence to say that the mean score after viewing the educational video is significantly higher than the mean score before the viewing. ❖

❖ EXAMPLE 11.3 Comparing Two Burn Treatments

Two creams are available by prescription for treating moderate skin burns. A study to compare the effectiveness of the two creams is conducted using 15 randomly selected patients with moderate burns on their arms. Two spots of the same size and degree of burn are marked on each patient's arm. One of the two creams is selected at random and applied to the first spot, while the remaining spot is treated with the other cream. The number of days until the burn has healed is recorded for each spot. These data are provided along with the difference in healing time (in days).

Patient Number	1	2	3	4	5	6	7	8	9	10	11	12	13	14	15
Cream 1 = C1	16	2	10	7	6	10	5	4	19	7	12	9	10	20	12
Cream 2 = C2	14	4	10	4	5	12	5	6	23	10	12	7	11	24	10
Diff = C1 − C2	2	−2	0	3	1	−2	0	−2	−4	−3	0	2	−1	−4	2

Let's construct the 95% confidence interval for estimating the mean difference in healing time, μ_D. We have $n = 15$ paired observations with an observed sample mean of $\bar{d} = -0.53$ and sample standard deviation of $s_D = 2.26$.

With the TI

In the TESTS menu located under the STAT button, we select the **8:T-Interval**. You can either have the sample differences entered into a list, say L1, or just enter the Stats (the sample mean, sample standard deviation, and sample size). The input and output screens are shown. The output screen provides the confidence interval $(-1.782, 0.72155)$, the sample mean difference $\bar{d} = -0.53$, sample standard deviation $s_D = 2.26$, and the sample size $n = 15$.

```
Input screen:
TInterval
Inpt:Data Stats
x:-.53
Sx:2.26
n:15
C-Level:.95
Calculate
```

```
Output screen:
TInterval
(-1.782,.72155)
x=-.53
Sx=2.26
n=15
```

With the Formula and Table IV

We have the sample mean difference $\bar{d} = -0.53$, the sample standard deviation $s_D = 2.26$, and the sample size $n = 15$. All that remains to complete the formula is the t^* value. For this we look at the last row of Table IV, which labels the bottom of each column with the confidence levels, and find the 95% level of confidence. We also need the degrees of freedom of $n - 1 = 15 - 1 = 14$. So the t^* value for df = 14 and 95% confidence is 2.145. The 95% confidence interval for the mean number of unoccupied seats is

$$\bar{d} \pm t^* \left(\frac{s_D}{\sqrt{n}} \right) \Rightarrow -0.53 \pm (2.145) \left(\frac{2.26}{\sqrt{15}} \right) \Rightarrow -0.53 \pm 1.25 \Rightarrow (-1.78, 0.72).$$

The Minitab computer output for generating the paired samples confidence interval on the cream data is shown next. The data were entered and a variable called *diff* was created. The confidence interval was based on the differences, with a 95% confidence level selected. The observed mean difference of -0.533 and the observed standard deviation of 2.264 gave a 95% confidence interval of $(-1.787, 0.721)$.

```
MTB > Let 'diff' = 'cream 1'- 'cream 2'
MTB > TInterval 95.0 'diff'.

              N        MEAN      STDEV   SE MEAN     95.0 PERCENT C.I.
diff         15      -0.533      2.264     0.584   (-1.787,      0.721)
```

We are 95% confident that our interval contains μ_D because if this procedure were repeated many times, 95% of the confidence intervals produced are expected to contain μ_D. The margin of error associated with this confidence interval estimate is $2.145(0.58) = 1.244$. What could you do to reduce the margin of error? You might increase the sample size or perhaps use a lower confidence level.

Our 95% confidence interval does contain the value of zero. Thus we would accept the null hypothesis $H_0: \mu_D = 0$ when tested against the alternative hypothesis $H_1: \mu_D \neq 0$ at a 5% significance level. There appears to be no significant difference in the healing time for these two creams. The t-test was performed on the differences using MINITAB with a *not equal (N.E.)* alternative option selected. The observed test statistic of $t = -0.91$ gave a corresponding p-value of 0.38, which is more than 0.05.

```
MTB > TTest 0.0 'diff';
SUBC> Alternative 0.

TEST OF MU = 0.000 VS MU N.E.  0.000

              N        MEAN     STDEV    SE MEAN          T    P VALUE
diff         15      -0.533     2.264      0.584      -0.91       0.38
```

Let's do it! **11.3** **More on Confidence Intervals for Paired Data**

Consider the data and interval estimate for comparing the two burn cream treatments in Example 11.3.

(a) Would the 99% confidence interval estimate for μ_D, based on the same data, be wider or narrower than the 95% confidence interval? Explain.

(b) Would the 99% confidence interval contain the value of zero? Explain.

(c) Construct the 99% confidence interval to verify your answers.

(d) Repeat (a), (b), and (c) for a 90% confidence level.

Let's do it! 11.4 Pulse Rates and "Does Order Matter?"

Does performing an exercise, such as jumping in place, significantly increase a person's pulse rate? Let's gather some data to address this question. Take a random sample of 10 students (willing and able to participate in the experiment). Record just their names in the spaces provided.

Name	"Before" Pulse Rate	"After" Pulse Rate	Difference = After − Before
1.			
2.			
3.			
4.			
5.			
6.			
7.			
8.			
9.			
10.			

The students should take their pulse for one minute and record these "before" responses. Have these students jump in place for one minute and, immediately upon stopping, take their pulse for one minute. Record these "after" responses.

If differences are computed as "After − Before," complete the alternative hypothesis about μ_D for testing if the pulse rate is significantly higher after jumping. Circle the appropriate direction for H_1.

$$H_1: \mu_D(= \mu_{\text{After}} - \mu_{\text{Before}}) < 0$$

or

$$H_0: \mu_D(= \mu_{\text{After}} - \mu_{\text{Before}}) = 0 \quad \text{versus} \quad H_1: \mu_D(= \mu_{\text{After}} - \mu_{\text{Before}}) \neq 0$$

or

$$H_1: \mu_D(= \mu_{\text{After}} - \mu_{\text{Before}}) > 0$$

Compute the differences and perform the corresponding test. Provide the observed test statistic value, provide the *p*-value, give your decision using a 5% significance level, and state your conclusion in a well-written sentence.

Exercises

11.7 An educational psychologist has developed an instructional program for children that is designed to promote creative thinking. To test the effectiveness of this program, 14 fifth-grade children are selected at random from the fifth graders attending Lake Elementary and placed in matched pairs based on their IQ levels. One member of each pair is randomly assigned to the regular instructional program and the other member to the new program. The measure of creativity is based on the number of ways a child can think to use common, everyday objects. The scores are as follows (a higher score implies that the child is more creative):

Pair	1	2	3	4	5	6	7
New Program	21	11	16	5	20	7	16
Regular Program	15	8	19	4	16	9	11
Diff = New − Regular	6	3	−3	1	4	−2	5

The psychologist would like to use a paired *t*-test to assess if the new program has higher creativity scores on average as compared to the regular program.

(a) The null hypothesis is $H_0: \mu_D = 0$, where $\mu_D = \mu_{New} - \mu_{Regular}$. Which alternative hypothesis is appropriate to test the psychologist's theory?

(b) State the assumptions needed for the test to be valid.

(c) Test the hypotheses using $\alpha = 0.05$. Compute the observed test statistic value, the corresponding *p*-value, and state your conclusion using a well-written sentence.

(d) The paired design was chosen because (select one)

   (i)   the psychologist thought that there might be a great deal of variability in creativity scores for children with different IQ levels.

   (ii) the psychologist thought that there would hardly be any variability in creativity scores for children with different IQ levels.

   (iii) the psychologist thought that the normality assumption was violated.

   (iv) the population variance was known.

11.8 A company advertises that food preparation time can be significantly reduced with the Mega Slicer. A random sample of 12 people present at the Local Food Prep Conference prepared the ingredients for a meal with and without the slicer. For each person, the order in which the measurements were obtained was also randomized (i.e., flip a fair coin, if heads, then do "without" first; otherwise, do "with" first), to help avoid any bias that may arise due to a learning effect over time. The preparation times in minutes are given in the following table:

Person	1	2	3	4	5	6	7	8	9	10	11	12
Without Slicer	22	18	18	22	19	19	18	12	18	25	26	23
With Slicer	20	11	20	14	19	20	19	15	22	19	21	20
Differences	2	7	−2	8	0	−1	−1	−3	−4	6	5	3

Suppose that the difference between the preparation times (without − with) follows a normal density.

(a) Which test should be used?

(b) Give the appropriate null and alternative hypotheses for testing that the company's claim is true.

(c) Compute the value of the test statistic, give the corresponding p-value, and state your decision using a 5% significance level.

11.9 The data shown in the accompanying table were obtained in an experiment to check whether there is a significant difference in the weights with two different scales. The weight is measured in grams. The same six objects were weighed on the two scales.

(a) State the appropriate null and alternative hypotheses for assessing if there is a significant difference in the weights for these two scales.

(b) State the assumptions for performing the test in (a).

(c) Examine the data in the table and assess whether there are any inconsistencies.

Scale 1	45	47	62	39	56	64
Scale 2	40	45	63	34	56	63

Upon further examination, it was determined that the third pair of observations (62, 63) was not collected according to protocol—that is, a person who was not trained in proper use of the scales made the measurements. Answer (d) excluding this observation.

(d) Calculate a 95% confidence interval for the mean difference in weight for Scale 1 versus Scale 2 (excluding the third pair of observations).

(e) If the test in (a) were performed, the p-value would be (select one)

(i) less than or equal to 0.05.

(ii) greater than 0.05.

(iii) can't tell.

Explain your answer.

11.10 Some statistical computer packages do not provide an option to select the direction for the alternative hypothesis. Instead, only the p-value for a two-sided rejection region is reported. The output for the video data of Example 11.1 is presented. The observed **MEAN DIFFERENCE**, $\bar{d}$, is given as 4.5; the observed SD DIFFERENCE (standard deviation of the differences), s_D, is 5.126, for an observed test statistic $t = 2.776$. The PROB is the p-value for the two-sided test—that is, when $H_1: \mu_D \neq 0$. Since our observed test statistic was positive, the p-value for testing $H_1: \mu_D > 0$ would be

$$p\text{-value} = \frac{\text{PROB}}{2} = \frac{0.022}{2} = 0.011.$$

```
PAIRED SAMPLES T-TEST ON    AFTER   VS  BEFORE   WITH   10 CASES

  MEAN DIFFERENCE =      4.500
  SD DIFFERENCE =        5.126
  T =       2.776      DF =   9    PROB =      0.022
```

What would be the *p*-value if the alternative hypothesis was $H_1 : \mu_D < 0$? You may wish to sketch a picture of the *p*-value first.

11.11 The weights of eight people before they stopped smoking and five weeks after they stopped smoking were obtained and entered into SPSS. The researcher would like to assess if there is enough evidence to conclude that, on average, weight increases if one quits smoking—that is, we wish to test the hypotheses $H_0 : \mu_D = 0$ versus $H_1 : \mu_D > 0$, where the differences were computed as "weight after − weight before." The difference in weight is assumed to follow a normal distribution.

(a) What does the observed test statistic value of 1.45 mean? (Do not give a decision.)

(b) What is the corresponding *p*-value for testing $H_0 : \mu_D = 0$ versus $H_1 : \mu_D > 0$?

(*Note:* The output provides the *p*-value when the alternative hypothesis is two-sided. You may wish to review Exercise 11.10.)

(c) Would you reject the H_0 if the significance level is 0.05? Explain.

(d) Could you use the 95% confidence interval for μ_D of (−1.103, 4.603) to test the hypotheses in (b)? Explain.

```
t-tests for Paired Samples
                   Number of             2-tail
Variable             pairs     Corr      Sig          Mean        SD    SE of Mean
-----------------------------------------------------------------------------------
AFTER    weight after                            139.5000    20.983        7.419
                       8        .987     .000
BEFORE   weight before                           137.7500    19.977        7.063
-----------------------------------------------------------------------------------
          Paired Differences          |
    Mean         SD     SE of Mean    |      t-value         df      2-tail Sig
-------------------------------------|---------------------------------------------
   1.7500       3.412      1.206      |        1.45           7            .190
 95% CI (-1.103,  4.603)             |
```

11.12 In a study to investigate the effect of diet on reducing cholesterol levels, a random sample of 12 subjects known to have high cholesterol had their cholesterol measured at the beginning of the study and again after six months on a special diet. Suppose that you could assume that the differences in cholesterol (before minus after) are a random sample from some approximately normal population.

(a) What test would you use to assess if the special diet is effective at reducing cholesterol levels? State clearly the appropriate hypotheses.

(b) Suppose that the mean difference (before minus after) is 8 and the standard error of the mean difference is 4.88. Interpret this value of 4.88.

(c) Compute the test statistic and the corresponding *p*-value for the hypotheses in (b) and give your decision at a 5% level of significance.

(d) An important assumption in performing the above test was that the distribution of the differences is approximately normal. Explain how you should assess the validity of this assumption.

11.13 Do students improve their SAT mathematics score the second time they take the test? A random sample of four students who took the test twice received the following scores:

Student	1	2	3	4
First Score	450	520	720	600
Second Score	440	600	720	630

(a) We wish to test whether students improve their SAT mathematics score the second time using a paired *t*-test. State the appropriate null and alternative hypotheses regarding the mean difference.

(b) Clearly state the assumptions that must be made to perform the test in (a).

(c) The sample standard deviation of the differences (second score − first score) in the scores is 41.4 points. Compute the standard error of the mean difference and explain clearly what this standard error means in terms of an average distance.

(d) Test the hypothesis in (a). Compute the observed test statistic value. Report the corresponding *p*-value. Give your decision using a 5% level of significance and write a conclusion regarding the improvement.

(e) If H_0 is true and we repeatedly took a random sample of four students who took the SAT twice and conducted the hypothesis test at the 5% significance level, then we will reject H_0: (select one)

(i) about 95% of the time.

(ii) about 5% of the time.

(iii) about 2.5% of the time.

(iv) more than 5% of the time.

(v) can't tell—need more information.

11.14 An investigator believed that people who smoke tend to smoke more during periods of stress. He compared the number of cigarettes ordinarily smoked by a group of 18 randomly selected students with the number they smoked during the 24 hours prior to a final examination. Test the appropriate hypotheses using a significance level of 10%. Show all steps.

Person	Usual #	# Prior to Final	Person	Usual #	# Prior to Final
1	9	13	10	14	14
2	16	15	11	10	16
3	23	23	12	11	13
4	12	17	13	13	19
5	30	32	14	6	12
6	8	16	15	11	13
7	14	18	16	17	11
8	21	23	17	36	41
9	11	15	18	32	40

11.15 Although the Occupational Safety and Health Act (OSHA) is not very popular with management because it is expensive to implement, some sources claim that it has been effective in reducing industrial accidents. The following data were collected on lost-time accidents both before and after OSHA came into effect (the figures given are mean man-hours lost per month over a one-year period) at six different industrial plants:

Plant	*1*	*2*	*3*	*4*	*5*	*6*
Before OSHA	38	64	42	70	58	30
After OSHA	31	58	43	65	52	29
Difference (B − A)						

(a) A test is to be performed to see if OSHA has been effective in reducing the industrial accidents. Differences are calculated as "Before − After." Enter those differences in the table.

(b) What was the average reduction in man-hours lost per month in the sample?

(c) State the hypotheses to test if OSHA has been effective in reducing the average number of man-hours lost per month.

(d) One assumption you have to make in order to perform this test is that the population distribution is normal. What population, or what random variable, does this assumption refer to?

(e) The sample standard deviation equals 3.225 hours. Calculate the value of the observed test statistic for the test stated in (c).

(f) Give the *p*-value of the test.

(g) Are the data statistically significant at the 5% significance level?

(h) What type of mistake could you have made? Use the "official" statistical name to identify this mistake.

11.16 A restaurant manager wishes to compare mean daily sales between two of his restaurants located several miles apart in the same city. For a one-week period, the following sales data are recorded:

Day	*Sunday*	*Monday*	*Tuesday*	*Wednesday*	*Thursday*	*Friday*	*Saturday*
Restaurant 1	1916	991	1112	1212	1461	2017	2406
Restaurant 2	1771	904	996	1258	1385	1867	2192
Difference	145	87	116	−46	76	150	214

SPSS was used to provide the following output:

Paired Samples Test

		Paired Differences					
		Mean	Std. Deviation	Std. Error Mean	t	df	Sig. (2-tailed)
Pair 1	REST1 - REST2	106.00	81.23	30.70	3.45	6	.014

(a) Explain why these data can be considered matched pairs.

(b) Clearly state the hypotheses to test whether the mean daily sales for Restaurant 1 is higher. Make sure to identify all parameters.

(c) Finish the following sentence: The standard error of 30.70 from the previous output means that if we were to take repeated samples (of the same size, from the same population), we would estimate ...

(d) What assumption would we have to make about the distribution of the difference in sales between the two restaurants in order for the *t*-test to be valid?

11.4   Independent Samples: Comparing Means

In this section, we discuss the technique for comparing the means of two treatments or populations on the basis of independent samples. In a two independent samples design, we have two unrelated sets of units that are measured, one sample from each of the two populations. One example of this type of design occurs when the subjects are randomly divided into two groups. One group is assigned to Treatment 1 and the other group to Treatment 2. The responses actually measured for each treatment are from unrelated groups of subjects. We think of Population 1 as the set of possible responses of all potential subjects given Treatment 1. Likewise, Population 2 is the set of possible responses of all potential subjects given Treatment 2.

The independent samples design also arises in the case of comparing, for example, selling prices of homes for two counties. The set of all homes sold in each county (over a certain time period) would form the two populations. Although there is no random allocation of homes to the two counties, independent samples of homes sold from each county would be obtained. Most often, having equal sample sizes is preferred. However, as we discussed in stratified random sampling, we may decide to select in proportion to the size of the population.

Two Independent Random Samples

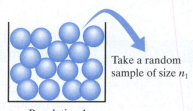

Population 1

Population 1 mean = μ_1
Population 1 standard deviation = σ_1

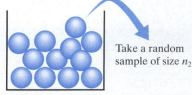

Population 2

Population 2 mean = μ_2
Population 2 standard deviation = σ_2

So we have two populations under study: Population 1, which has mean μ_1 and standard deviation σ_1, and Population 2, which has mean μ_2 and standard deviation σ_2. The data will consist of a random sample of size n_1 from Population 1 and a random sample of size n_2 from Population 2. The two samples are assumed to be independent.

We are interested in comparing the population means μ_1 and μ_2, so the parameter of interest is the difference $\mu_1 - \mu_2$. If the difference in the population means is zero, then the two treatments are said to be equivalent in terms of the mean response. If the difference in the population means is positive $\mu_1 - \mu_2 > 0$, then the responses to Treatment 1 are higher than those for Treatment 2, on average.

We need to make some additional assumptions to compare the two population means by performing a test of hypotheses or constructing a confidence interval estimate. We assume that the first set of responses is a random sample of size n_1 from a normal population with population mean μ_1 and population standard deviation σ, the second set of responses is a random sample of size n_2 from a normal population with population mean μ_2 and the same population standard deviation σ, and the two samples are independent. If the sample sizes are large enough ($n_1 \geq 30$ and $n_2 \geq 30$), the assumption of normal populations is less crucial. Also note that the population standard deviations are assumed to be equal. Additional comments regarding the assumptions are made later in this section.

Assumptions for a Two Independent Samples Design

We have a **random sample** of n_1 observations from a $N(\mu_1, \sigma)$ population.

We have a **random sample** of n_2 observations from a $N(\mu_2, \sigma)$ population.

The two random samples are **independent** of each other.

Before we continue with the background, let's look at a little notation that will be standard in the two independent samples scenario for representing the information about the observed data and corresponding sample statistics.

Notation in a Two Independent Samples Design

n_1 = sample size for the first sample (the number of observations from Population 1).

n_2 = sample size for the second sample (the number of observations from Population 2).

$\overline{x}_1$ = observed sample mean for the first sample.

$\overline{x}_2$ = observed sample mean for the second sample.

s_1 = observed sample standard deviation for the first sample.

s_2 = observed sample standard deviation for the second sample.

Since we are trying to compare two population means and we have a sample from each population, the logical idea would be to look at the difference in the sample means, $\overline{x}_1 - \overline{x}_2$. An observed difference in the sample means that is "close" to zero, incorporating in the measure of "close" some allowance for sampling variation, would show some support for no difference between the two treatments. The results and subsequent formulas needed to make inferences about $\mu_1 - \mu_2$ in the two independent samples scenario may look somewhat complicated. However, the basic ideas underlying how to make a confidence interval and how to test hypotheses are the same as those you studied in Chapters 9 and 10. Inferences about the mean μ of a single population were based on the sampling distribution of the sample mean $\overline{X}$. So we need to know the sampling distribution of $\overline{X}_1 - \overline{X}_2$ when we have independent random samples from normally distributed populations with the same population standard deviation. The distribution is based on three main statistical results that we will not prove here:

1. The mean of the difference of two random variables is the difference of the means.
2. The variance of the difference of two *independent* random variables is the sum of the variances.
3. The difference of two independent normally distributed random variables is also normally distributed.

With these statistical results we have the following results:

Distribution of $\overline{X}_1 - \overline{X}_2$ for the Two Independent Samples Scenario

The **mean** of $\overline{X}_1 - \overline{X}_2$ is $\mu_1 - \mu_2$.

The **standard deviation** of $\overline{X}_1 - \overline{X}_2$ is $\sigma\sqrt{\left(\dfrac{1}{n_1} + \dfrac{1}{n_2}\right)}$.

The **distribution** of $\overline{X}_1 - \overline{X}_2$ is normal.

Thus, $\overline{X}_1 - \overline{X}_2$ is $N\left(\mu_1 - \mu_2, \sigma\sqrt{\left(\dfrac{1}{n_1} + \dfrac{1}{n_2}\right)}\right)$.

The general formulas for a confidence interval estimate of μ and standardized test statistic require an estimate and its standard error. We have $\overline{x}_1 - \overline{x}_2$ as the point estimate of $\mu_1 - \mu_2$ and its standard deviation is $\sigma\sqrt{\dfrac{1}{n_1} + \dfrac{1}{n_2}}$. Replacing the unknown population standard deviation σ with an estimate will give us the standard error. From the data we have two sample standard deviations, s_1 and s_2, which are *both estimates of σ*. Thus, it makes sense to combine them in some way to form a better, overall estimate of σ, the common population standard deviation.

THINK ABOUT IT

Should we estimate the common population standard deviation σ by just averaging the two sample standard deviations? What if the first sample is of size 10 but the second sample is of size 100? Would you want to weight the corresponding sample standard deviations equally? Which sample standard deviation, s_1 or s_2, would you "trust" more?

We use a weighted average to form our **pooled** (s_p) estimate of σ. Note that each **sample variance** is weighted by its associated degrees of freedom, and the denominator is the total degrees of freedom, $(n_1 - 1) + (n_2 - 1) = n_1 + n_2 - 2$.

$$\text{Estimate of } \sigma: \quad s_p = \sqrt{\frac{(n_1 - 1)s_1^2 + (n_2 - 1)s_2^2}{(n_1 + n_2 - 2)}}$$

Therefore, our standard error of $\overline{X}_1 - \overline{X}_2$ is given by

$$s_p\sqrt{\frac{1}{n_1} + \frac{1}{n_2}}.$$

Recall that when we replace the standard deviation by an estimator, the resulting standardized quantity no longer follows a normal distribution, but rather had a *t*-distribution.

Distribution of the Standardized $\overline{X}_1 - \overline{X}_2$ for the Two Independent Samples Scenario

$$\text{If } \overline{X}_1 - \overline{X}_2 \text{ is } N\left(\mu_1 - \mu_2, \; \sigma\sqrt{\left(\frac{1}{n_1} + \frac{1}{n_2}\right)}\right),$$

then

$$Z = \frac{(\overline{X}_1 - \overline{X}_2) - (\mu_1 - \mu_2)}{\sigma\sqrt{\dfrac{1}{n_1} + \dfrac{1}{n_2}}} \text{ is } N(0, 1).$$

If the true standard deviation σ is unknown and replaced by an estimate s_p, where

$$s_p = \sqrt{\frac{(n_1 - 1)s_1^2 + (n_2 - 1)s_2^2}{(n_1 + n_2 - 2)}},$$

then the standardized quantity

$$T = \frac{(\overline{X}_1 - \overline{X}_2) - 0}{s_p\sqrt{\dfrac{1}{n_1} + \dfrac{1}{n_2}}}$$

has a t-distribution with $n_1 + n_2 - 2$ degrees of freedom.

Our test statistic to test hypotheses about $\mu_1 - \mu_2$ and our confidence interval for $\mu_1 - \mu_2$ are based on the following general forms:

General test statistic: $\dfrac{\text{Point Estimate} - \text{Null Value}}{\text{Null Standard Deviation or Error}}$

General confidence interval: Point Estimate $\pm$ (a few) (Standard Errors of the Point Estimate)

The following is a summary of the two independent samples test of $H_0: \mu_1 - \mu_2 = 0$, sometimes referred to as the Student's t-Test, and the corresponding confidence interval for $\mu_1 - \mu_2$:

Two Independent Samples t-Test

Assumptions: The first sample is a random sample from a normal population with mean μ_1 and standard deviation σ. The second sample is a random sample from a normal population with mean μ_2 and standard deviation σ. The two samples are independent. Normality is less crucial if the sample sizes n_1 and n_2 are large, and preferably $n_1 = n_2$.

Hypotheses: $H_0: \mu_1 - \mu_2 = 0$ versus

$H_1: \mu_1 - \mu_2 \neq 0$ or $H_1: \mu_1 - \mu_2 > 0$ or $H_1: \mu_1 - \mu_2 < 0$.

The significance level α to be used is determined.

Data: The two sets of data from which the two sample means $\bar{x}_1$ and $\bar{x}_2$ and the two sample standard deviations s_1 and s_2 can be computed.

Observed Test Statistic: $\quad t = \dfrac{\bar{x}_1 - \bar{x}_2}{s_p\sqrt{\dfrac{1}{n_1} + \dfrac{1}{n_2}}}$, where $s_p = \sqrt{\dfrac{(n_1 - 1)s_1^2 + (n_2 - 1)s_2^2}{n_1 + n_2 - 2}}$

and the *null distribution* for the T variable is the $t(n_1 + n_2 - 2)$ *distribution*.

p-value: We find the p-value for the test using the $t(n_1 + n_2 - 2)$ *distribution*. The *direction of extreme* will depend on how the alternative hypothesis is expressed.

Decision: A p-value less than or equal to α leads to rejection of H_0.

Notes:

■ If we are interested in assessing if $\mu_1 - \mu_2$ is equal to some hypothesized value that is not zero, we would replace zero in the test statistic expression with this other null value.

■ The test statistic is the same no matter how the alternative hypothesis is expressed.

Confidence Interval for $\mu_1 - \mu_2$

Assumptions: The first sample is a random sample from a normal population with mean μ_1 and standard deviation σ. The second sample is a random sample from a normal population with mean μ_2 and standard deviation σ. The two samples are independent. Normality is less crucial if the sample sizes n_1 and n_2 are large, and preferably $n_1 = n_2$.

Data: The two sets of data from which the two sample means $\bar{x}_1$ and $\bar{x}_2$ and the two sample standard deviations s_1 and s_2 can be computed.

Confidence Interval: $\quad (\bar{x}_1 - \bar{x}_2) \pm t^*\left(s_p\sqrt{\dfrac{1}{n_1} + \dfrac{1}{n_2}}\right),$

where $s_p = \sqrt{\dfrac{(n_1 - 1)s_1^2 + (n_2 - 1)s_2^2}{n_1 + n_2 - 2}}$ and

t^* is an appropriate percentile of the $t(n_1 + n_2 - 2)$ *distribution*.

Interpretation of the Interval: The interval gives potential values for the difference in the population means $\mu_1 - \mu_2$ based on independent random samples, one from each population. The interval may be used to test hypotheses when the alternative hypothesis is two sided.

Interpretation of the Confidence Level: The confidence level pertains to the proportion of times the procedure produces an interval that actually contains $\mu_1 - \mu_2$ when this procedure is repeated over and over using new independent random samples of the same sizes each time.

Let's illustrate our two independent samples *t*-test and confidence interval estimate through three examples. In each example, we will see various ways these techniques can be performed: using the TI graphing calculator, Table IV, or computer output.

❖ EXAMPLE 11.4 Comparing Two Headache Treatments

Medical researchers are comparing two treatments for migraine headaches. They wish to perform a **double-blind** experiment to assess if Treatment 2 (the new treatment) is significantly better than Treatment 1 (the standard treatment) using a 5% significance level. Twenty subjects were available for the study and were randomized to one of the two treatment groups. Treatment 1 was administered to 10 subjects, while Treatment 2 was administered to the remaining 10 subjects. Each subject was instructed to take the medication at the  onset of a migraine headache and to record the time that elapsed until relief, defined as a reduction in throbbing. The mean time to relief for the Treatment 1 subjects was 22.6 minutes, with a standard deviation of 5.2 minutes. The mean time to relief for the Treatment 2 group was 19.4 minutes, with a standard deviation of 4.9 minutes.

Suppose that the conditions for performing a two independent samples *t*-test are satisfied. (Can you recall the assumptions for this test?) First, we need to state the appropriate hypotheses to test. If smaller responses imply a better treatment, the hypotheses would be $H_0: \mu_1 - \mu_2 = 0$ versus $H_1: \mu_1 - \mu_2 > 0$. The data are summarized next.

$$n_1 = 10 \quad \overline{x}_1 = 22.6 \quad s_1 = 5.2$$
$$n_2 = 10 \quad \overline{x}_2 = 19.4 \quad s_2 = 4.9$$

Recall that one of the assumptions for performing this test is equal population standard deviations. Even though the *sample* standard deviations of 5.2 and 4.9 are not equal, this does not mean the equal *population* standard deviations assumption has been violated. Examining the relative magnitude of the two sample standard deviations is a quick check for this assumption.

An estimate of the equal population standard deviation is

$$s_p = \sqrt{\frac{(10-1)(5.2)^2 + (10-1)(4.9)^2}{10 + 10 - 2}} = 5.05.$$

Note: Our estimate should make sense. Since it is a weighted average, it should be between the two sample standard deviations of 5.2 and 4.9. When the sample sizes are equal, the pooled variance estimate is the average of the two sample variances.

Now we can compute our test statistic to assess if the observed difference in the sample means of $22.6 - 19.4 = 3.2$ is significantly greater than zero.

$$\text{The } \textbf{observed test statistic} \text{ is } t = \frac{22.6 - 19.4}{5.05\sqrt{\dfrac{1}{10} + \dfrac{1}{10}}} = 1.416.$$

What does the observed test statistic value mean? It means that we observed two sample means that are about 1.4 standard errors apart. Is this a large enough difference to reject the null hypothesis at a 5% significance level? Let's find the corresponding *p*-value and compare it to 0.05.

Finding the *p*-value

The *p*-value is the probability of observing a test statistic as large or larger than the observed value of 1.416, computed under the null distribution, which is the *t*-distribution with 18 degrees of freedom.

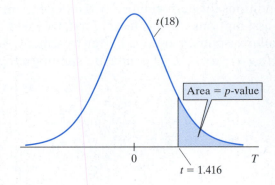

With the TI

1. Using the tcdf(function

Using the tcdf(function on the TI, we have,

$$\text{*p*-value} = P(T \geq 1.416) = \text{tcdf}(1.416, \ \text{E99}, \ 18) = 0.0869.$$

2. Using the 2-SampTTest function under the STAT TESTS menu

In the TESTS menu located under the STAT button, we select the **4:2-SampTTest** option. With the sample means of 22.6 and 19.4, the sample standard deviations of 5.2 and 4.9, and the sample sizes of 10 and 10, we can use the Stats option of this test. The steps and corresponding input and output screens are shown. Notice that you must specify **Yes** under the **Pooled** option. The **No Pooled** option is discussed at the end of this section as another version of our test.

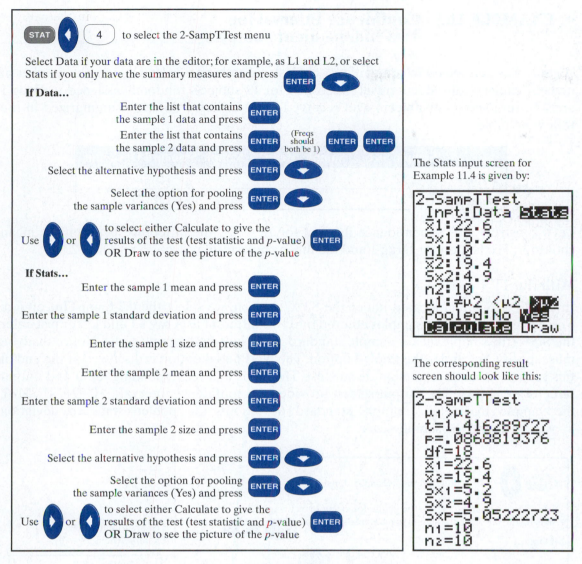

The Stats input screen for Example 11.4 is given by:

```
2-SampTTest
 Inpt:Data Stats
x1:22.6
Sx1:5.2
n1:10
x2:19.4
Sx2:4.9
n2:10
μ1:≠μ2 <μ2 >μ2
Pooled:No Yes
Calculate Draw
```

The corresponding result screen should look like this:

```
2-SampTTest
μ1>μ2
t=1.416289727
P=.0868819376
df=18
x1=22.6
x2=19.4
Sx1=5.2
Sx2=4.9
SxP=5.05222723
n1=10
n2=10
```

$$p\text{-value} = P(T \geq 1.416) = 0.08688.$$

With Table IV

In Table IV, we focus on the df = 18 row and find that our observed test statistic value of 1.416 falls between the values in the columns headed with 0.10 and 0.05. So we state that the p-value is between 0.05 and 0.10.

$$0.05 < p\text{-value} = P(T \geq 1.416) < 0.10.$$

Decision and Conclusion

Our decision at the 5% significance level is to accept H_0. Our conclusion is that, at the 5% significance level, the claim that Treatment 1 is as effective as Treatment 2, in terms of the mean response, cannot be rejected. Based on the data, it appears the two treatments are equally effective. This does not mean that we are not going to use the new treatment. It might be that the new treatment is less expensive or has fewer side effects for patients, in which case, since both treatments are equivalent in terms of time to relief, it may be reasonable to use the new treatment. ❖

❖ EXAMPLE 11.5 Confidence Interval for Two Independent Samples

A study was performed to compare two cholesterol-reducing drugs. Data on the number of units of cholesterol reduction were recorded for 12 subjects randomly assigned to Drug 1 and the remaining 14 subjects who received Drug 2. These data are summarized in the following table:

	Sample Size	Sample Mean	Sample Standard Deviation
Drug 1	12	5.6	1.3
Drug 2	14	5.0	1.8

Let's construct a 95% confidence interval for $\mu_1 - \mu_2$, the difference between the mean cholesterol reduction for Drug 1 and the mean cholesterol reduction for Drug 2.

With the TI

In the TESTS menu located under the STAT button, we select the **0:2-SampTInt** option. You can either have the samples entered into two different lists, say L1 and L2, or just enter the Stats (the sample means, sample standard deviations, and sample sizes). Notice that you must specify **Yes** under the **Pooled** option. The **No Pooled** option is discussed at the end of this section as another version of our test. The steps and corresponding input and output screens are shown. The output screen provides the confidence interval (−0.6914, 1.8914), the sample means, the sample standard deviations, the pooled standard deviation $s_p = 1.5905$, and the sample sizes.

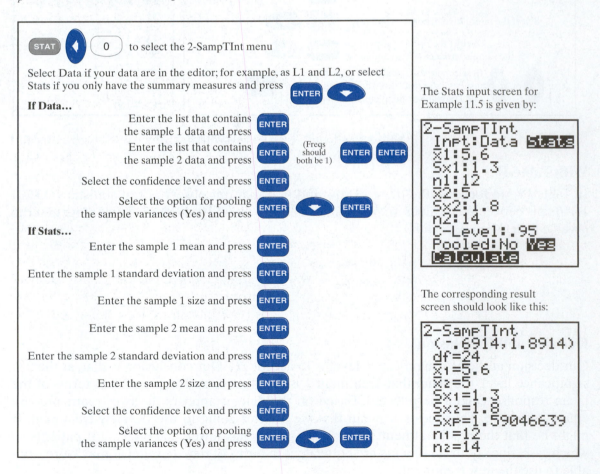

With the Formula and Table IV

We have the sample means, the sample standard deviations, and the sample sizes. We need to first find the pooled standard deviation

$$s_p = \sqrt{\frac{(n_1 - 1)s_1^2 + (n_2 - 1)s_2^2}{n_1 + n_2 - 2}} = \sqrt{\frac{(12 - 1)(1.3)^2 + (14 - 1)(1.8)^2}{12 + 14 - 2}} = 1.5905.$$

All that remains to complete the formula is the t^* value. For this we look at the last row of Table IV, which labels the bottom of each column with the confidence levels, and find the 95% level of confidence. We also need to find the row with the degrees of freedom, $12 + 14 - 2 = 24$. So the t^* value for df $= 24$ and 95% confidence is 2.064. The 95% confidence interval is

$$(\bar{x}_1 - \bar{x}_2) \pm t^* \left(s_p \sqrt{\frac{1}{n_1} + \frac{1}{n_2}} \right) \Rightarrow (5.6 - 5.0) \pm 2.064 \left(1.5905 \sqrt{\frac{1}{12} + \frac{1}{14}} \right)$$

$$\Rightarrow 0.6 \pm 1.2914 \Rightarrow (-0.6914, 1.8914).$$

We are 95% confident that this interval contains $\mu_1 - \mu_2$, because if we repeated this procedure many times, we would expect 95% of such confidence intervals produced to contain the difference in the population means $\mu_1 - \mu_2$. Suppose that the researcher would like to assess if the difference between the mean cholesterol reduction for Drug 1 and the mean cholesterol reduction for Drug 2 is statistically significant at the 5% level. Our 95% confidence interval for $\mu_1 - \mu_2$ does contain zero as a plausible value for $\mu_1 - \mu_2$. Thus, if the test of $H_0: \mu_1 - \mu_2 = 0$ versus $H_1: \mu_1 - \mu_2 \neq 0$ were performed, we would accept H_0 since the *p*-value would be greater than 0.05. ❖

❖ EXAMPLE 11.6   Computer Output for Two Independent Samples

Two different emission-control devices were being tested to determine the average amount of nitric oxide being emitted by a car over a one-hour period of time. Twenty cars of the same model and year were selected at random for the study. Ten cars were randomly selected and equipped with a Device 1 emission control, and the remaining cars were equipped with a Device 2 emission control. Each of the 20 cars was then monitored for a one-hour period to determine the amount of nitric oxide emitted. Assume that for each type of device, the amount of nitric oxide emitted follows a normal distribution and that the two population standard deviations are equal. We wish to test the research hypothesis that the mean level of emission for Device 1 is greater than the mean emission level for Device 2 using a two independent samples *t*-test. The appropriate hypotheses to be tested are $H_0: \mu_1 - \mu_2 = 0$ versus $H_1: \mu_1 - \mu_2 > 0$. The observed emissions were analyzed using MINITAB.

Side-by-side boxplots of the emission data are examined first. We can see graphically that there is some difference in the median emission levels, with Device 1 producing the higher levels overall. We also can informally check the equal population standard deviations assumption by noting that the two sample standard deviations as well as the sample IQRs (lengths of the boxes) are quite similar.

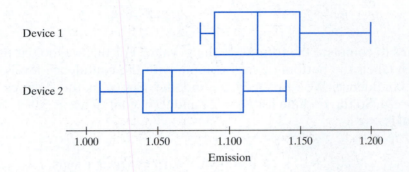

The two independent samples *t*-test was performed using MINITAB assuming equal population variances (*Pooled* option) with a *greater than (GT)* alternative option selected. The following output was produced:

```
MTB > TwoSample 95.0 'device 1' 'device 2';
SUBC>    Alternative 1;
SUBC>    Pooled.

TWOSAMPLE T FOR device 1 VS device 2
              N      MEAN     STDEV    SE MEAN
device 1     10     1.1280    0.0413     0.013
device 2     10     1.0710    0.0443     0.014

95 PCT CI FOR MU device 1 - MU device 2: (0.017, 0.097)

TTEST MU device 1 = MU device 2 (VS GT): T= 2.97   P=0.0041   DF=  18

POOLED STDEV =      0.0428
```

In the output, the sample size (N), the observed sample mean (MEAN), and the observed sample standard deviation (STDEV) for each group (that is, device) are provided. The mean amount of nitric oxide emitted by the 10 Device 1 controls was 1.128, with a standard deviation of $s_1 = 0.041$. The observed test statistic of 2.97 and the corresponding *p*-value of 0.0041 for the one-sided to the right alternative hypothesis are provided. Does it appear that the mean level of emission for Device 1 controls is greater than the mean emission level for Device 2 controls? For any $\alpha > 0.0041$ we would conclude the mean level of emission is greater for Device 1 controls.

Notice that the 95% confidence interval for $\mu_1 - \mu_2$ is also provided by default. The estimate of the common population standard deviation is given by s_p = POOLED STDEV = 0.0428. ❖

Let's do it! 11.5 Sheep Treatment

An experiment was conducted to compare the mean number of tapeworms in the stomachs of sheep that had been treated for worms against the mean number in those untreated. A sample of 14 worm-infected lambs was randomly divided into two groups. Seven were injected with the drug and the remainder were left untreated. After a six-month period, the lambs were slaughtered, and the following counts were recorded:

Group 1	Drug-treated sheep	18	43	28	50	16	32	13
Group 2	Untreated sheep	40	54	26	63	21	37	39

Assume that these data are the observed values of independent random samples, in which the two distributions are normal with equal population standard deviation.

(a) Side-by-side boxplots for the two sets of data are provided. The boxplot for the untreated sheep responses *establishes* that a normal distribution is the appropriate model for number of worms.

Circle one: **True** **False**

Explain:

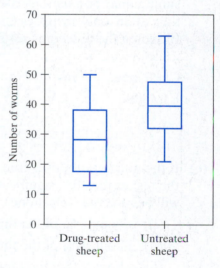

(b) Based on these boxplots, does the assumption of equal population standard deviation seem plausible?

Circle one: **Yes** **No**

Explain:

(c) State the hypotheses for testing that the mean number of worms for the treated lambs is *less than* the mean number of worms for the untreated lambs.

H_0:_____ H_1:_____

Computer output from the SPSS computer package based on the sheep data is provided.

```
t-tests for Independent Samples of SHEEP     Type of sheep
                                Number
Variable                        of Cases      Mean         SD     SE of Mean
-------------------------------------------------------------------------------
WORMS   number of worms
drug-treated                       7         28.5714     14.093      5.327
untreated                          7         40.0000     14.674      5.546
-------------------------------------------------------------------------------
        Mean Difference = -11.4286
        Levene's Test for Equality of Variances: F= .022    P= .885
   t-test for Equality of Means                                    95%
Variances   t-value        df     2-Tail Sig    SE of Diff      CI for Diff
-------------------------------------------------------------------------------
Equal        -1.49         12        .163         7.690      (-28.184, 5.326)
Unequal      -1.49       11.98       .163         7.690      (-28.187, 5.330)
-------------------------------------------------------------------------------
```

(d) Since we are assuming equal population variances, we should look at the results presented in the line headed by "*Equal*" under the variances column, shown in color. The information provided in the line headed "*Unequal*" is based on a test that does not assume equal population variances, which is briefly discussed on page 658. Also note that SPSS only provides the *p*-value when the alternative hypothesis is two sided. Based on the preceding results, would you reject the null hypothesis at the 10% significance level?

Circle one: **Yes**   **No**

Explain.

(e) If the null hypothesis is true, and the experiment of sampling 14 sheep and conducting the test at the 10% significance level is repeated many times, then the null hypothesis will be rejected (select one)

    (i)   about 90% of the time.

    (ii)   about 10% of the time.   **Answer**:_____

    (iii) about 5% of the time.

    (iv) more than 10% of the time.

    (v)   can't tell—need more information.

(f) Suppose that, instead of randomly assigning the 14 sheep to the two groups, the experimenter just went out to the pen and put the first seven lambs that he could *catch easily* into the placebo group. Explain to the experimenter why that might lead to biased results. Your answer should be specific to this example.

REAL
DATA

❖ EXAMPLE 11.7 Reducing Variability with Pairing—Boys' Shoes ___

One advantage of the paired design is the increased precision gained by controlling for confounding factors. When units are paired with respect to factors that could influence the response, the variability among units due to these factors can be reduced by taking differences. The boys' shoes experiment, presented in Box, Hunter, and Hunter's text *Statistics for Experimenters* (John Wiley & Sons, 1978, pp. 97–101), illustrates the reduced variability with a paired design.

A manufacturer was interested in comparing its standard synthetic material, Material 1, for soles of shoes to a cheaper substitute, referred to as Material 2. The manufacturer was concerned about the amount of wear and wished to assess if there was evidence of increased wear with Material 2. The data are measurements of the amount of wear of the soles of shoes worn by 10 boys, where a larger value indicates more wear.

Side-by-side boxplots of these data suggest that the two materials are comparable with respect to amount of wear. The similar IQRs, as evident by the length of the boxes, support the assumption of equal population standard deviation.

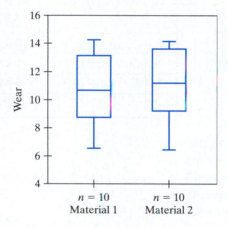

The SPSS output for a two independent samples *t*-test on these data follows:

```
t-tests for Independent Samples of MATERIAL
                              Number
Variable                      of Cases      Mean           SD     SE of Mean
-----------------------------------------------------------------------------
WEAR
MATERIAL 1                        10        10.6300        2.451       775
MATERIAL 2                        10        11.0400        2.518       796
-----------------------------------------------------------------------------
          Mean Difference = -.4100
          Levene's Test for Equality of Variances: F= .019    P= .892

t-test for Equality of Means                                          95%

Variances   t-value      df    2-Tail Sig    SE of Diff        CI for Diff
-----------------------------------------------------------------------------
Equal          -.37      18         .716         1.111     (-2.745, 1.925)
Unequal        -.37   17.99         .717         1.111     (-2.745, 1.925)
-----------------------------------------------------------------------------
```

The *p*-value of $\frac{0.716}{2} = 0.358$ for testing the hypotheses $H_0: \mu_1 - \mu_2 = 0$ versus $H_1: \mu_1 - \mu_2 < 0$ supports the conclusion of no significant difference in the mean wear for the two materials.

However, this experiment was actually conducted as a *paired design*. Boys do not run, jump, scuff the same surfaces the same amount of time. So the rates at which they wear out the soles of their shoes will vary from boy to boy. To reduce the variability between boys, each boy wore a pair of shoes with one sole made with Material 1 and the other sole made with Material 2. Randomization was used to determine which sole (left or right) was to be made with Material 1.

The wear data, the corresponding differences, computed as Material 1 − Material 2, and the randomization results of which shoe (left or right) was made with which material are shown in the table that follows. Only two of the differences were positive. The histogram of the differences shows evidence that Material 1 has less wear than Material 2.

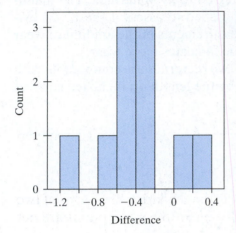

Boy	Material 1	Material 2	Difference = 1 − 2
1	13.2 (L)	14.0 (R)	−0.8
2	8.2 (L)	8.8 (R)	−0.6
3	10.9 (R)	11.2 (L)	−0.3
4	14.3 (L)	14.2 (R)	0.1
5	10.7 (R)	11.8 (L)	−1.1
6	6.6 (L)	6.4 (R)	0.2
7	9.5 (L)	9.8 (R)	−0.3
8	10.8 (L)	11.3 (R)	−0.5
9	8.8 (R)	9.3 (L)	−0.5
10	13.3 (L)	13.6 (R)	−0.3

A scatterplot of the paired observations shows a relationship between the Material 1 wear measurements and the Material 2 measurements. Boys with low wear measurements with Material 1, worn on one of their shoes, generally had low wear measurements with Material 2, worn on the other shoe. This plot supports that the rates at which boys wear out the soles of their shoes will vary from boy to boy and that pairing will help eliminate much of this boy-to-boy variation.

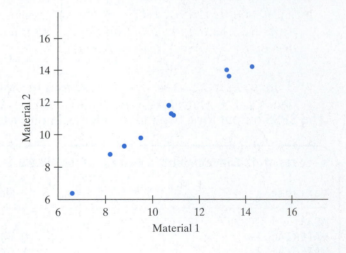

The SPSS output for a paired samples *t*-test on these data is as follows:

```
t-tests for Paired Samples
                Number of           2-tail
Variable          pairs      Corr   Sig          Mean          SD      SE of Mean
-----------------------------------------------------------------------------------
MAT_1                                           10.6300       2.451          .775
                   10       .988   .000
MAT_2                                           11.0400       2.518          .796
-----------------------------------------------------------------------------------
          Paired Differences        |
    Mean          SD    SE of Mean  |     t-value              df       2-tail Sig
------------------------------------|----------------------------------------------
  -.4100        .387       .122     |      -3.35                9             .009
95% CI (-.687, -.133)               |
```

The mean difference for the 10 boys was -0.41 and the sample standard deviation of the differences was 0.387. The corresponding observed test statistic was -3.35. The *p*-value for testing $H_0: \mu_D = 0$ versus $H_1: \mu_D < 0$ is very small; *p*-value $= \frac{0.009}{2} = 0.0045$. We conclude that a highly statistically significant increase in the amount of wear is associated with the use of the cheaper Material 2. ❖

THINK ABOUT IT

Sometimes the amount of wear for a right shoe versus a left shoe by the same person can differ. How might you modify the study to control for this potential confounding?

Some Notes Regarding the Assumptions

The two independent samples *t*-test is a powerful technique for comparing the means of two populations, provided that the assumptions hold. What if some of the assumptions are not satisfied?

- If the **sample sizes are large** enough to apply the central limit theorem, ($n_1 \geq 30$ and $n_2 \geq 30$), then the assumption of normal populations is less crucial, since the distribution of $\overline{X}_1 - \overline{X}_2$ will be approximately normal.

- If the **sample sizes are equal**, the assumption of common population standard deviation is not so crucial. The *t*-test statistic will still follow a *t*-distribution approximately.

- If the **population standard deviations** were assumed to be **known and not necessarily equal**, then we could base our test on the statistic

$$z = \frac{(\overline{x}_1 - \overline{x}_2) - (\mu_1 - \mu_2)}{\sqrt{\dfrac{\sigma_1^2}{n_1} + \dfrac{\sigma_2^2}{n_2}}},$$

where $\mu_1 - \mu_2$ is the hypothesized value under H_0, and the Z variable would follow an $N(0, 1)$ distribution. If the sample sizes are large, the normal approximation for this test statistic remains valid even if the population standard deviations are estimated by the sample standard deviations.

■ If the *population standard deviations* were *not known and not assumed to be equal*, then it seems natural to base our hypothesis test on the following statistic:

$$t = \frac{(\overline{X}_1 - \overline{X}_2) - (\mu_1 - \mu_2)}{\sqrt{\dfrac{s_1^2}{n_1} + \dfrac{s_2^2}{n_2}}}.$$

The sampling distribution of this T variable, under H_0, is not a t-distribution and is rather complicated. This situation is called the **Behrens–Fisher** problem. The results of this test are often reported in standard computer output. In Example 11.7, the results are provided in the line labeled "*Unequal.*" One approach to the complicated distribution is to use the t-distribution with an approximate number of degrees of freedom. The expression for the degrees of freedom, called **Welch's approximation**, is

$$df = \frac{\left(\dfrac{s_1^2}{n_1} + \dfrac{s_2^2}{n_2}\right)^2}{\dfrac{\left(\dfrac{s_1^2}{n_1}\right)^2}{n_1 - 1} + \dfrac{\left(\dfrac{s_2^2}{n_2}\right)^2}{n_2 - 1}}.$$

Luckily, many software packages and calculators will determine degrees of freedom and will find the appropriate t^* value for performing the approximate test or constructing the approximate confidence interval estimate. Exercises 11.27 and 11.47 comment on this version of the two independent samples t-test. A second, somewhat conservative, approach is to use the t-distribution, with degrees of freedom being the smaller of $(n_1 - 1)$ and $(n_2 - 1)$ as the approximate sampling distribution under H_0. Then the true significance level is always smaller than the level used with this t-distribution.

■ If there is concern that the assumptions are not satisfied, there is another test procedure that requires fewer or less stringent assumptions about the underlying populations. Such tests are called **nonparametric** or *distribution-free* procedures. The **Wilcoxon rank sum test** is a nonparametric procedure that can be used for two independent samples scenarios, which is based on ranks of the data. This nonparametric procedure can be effective and is discussed in Chapter 15. In practice, the presented t-tests based on normal distributions are the most common.

Which Version of a Two Independent Samples Test to Use?

Each scenario presents a picture of the distributions of the two populations being compared. Based on these distributions, determine which version of the two independent samples test to use.

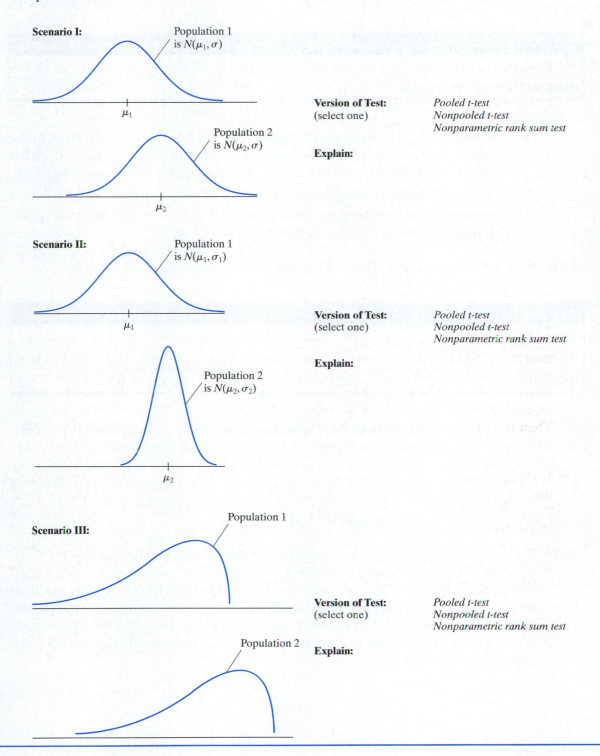

Scenario I:

Population 1 is $N(\mu_1, \sigma)$

μ_1

Population 2 is $N(\mu_2, \sigma)$

μ_2

Version of Test: (select one)

Pooled t-test
Nonpooled t-test
Nonparametric rank sum test

Explain:

Scenario II:

Population 1 is $N(\mu_1, \sigma_1)$

μ_1

Population 2 is $N(\mu_2, \sigma_2)$

μ_2

Version of Test: (select one)

Pooled t-test
Nonpooled t-test
Nonparametric rank sum test

Explain:

Scenario III:

Population 1

Population 2

Version of Test: (select one)

Pooled t-test
Nonpooled t-test
Nonparametric rank sum test

Explain:

Exercises

11.17 A teacher claims that use of *discovery* improves performance on the standard Iowa test given to third-grade students. She decides to compare the scores of 20 students in one of her sections (taught by her using *discovery*) with the scores of 20 students in a second section who have been taught by her without the *discovery* process. The data, summarized in the following table, are assumed to be independent random samples from two normal distributions:

	Sample Size	Sample Mean	Sample Standard Deviation
With Discovery (Group 1)	16	600	90
Without Discovery (Group 2)	16	550	100

(a) The teacher plans to compute a 90% confidence interval for the difference in the population mean scores (with *discovery* minus without *discovery*) $\mu_1 - \mu_2$. What additional assumption is needed for the 90% confidence interval to be valid?

(b) Does the additional assumption in (a) seem plausible? Explain.

(c) Assuming the assumptions are met, construct the 90% confidence interval for the difference in the population mean scores $\mu_1 - \mu_2$.

11.18 Recall the boys' shoes Example 11.7. The data are reproduced as follows:

Boy	*1*	*2*	*3*	*4*	*5*	*6*	*7*	*8*	*9*	*10*
Material 1	13.2	8.2	10.9	14.3	10.7	6.6	9.5	10.8	8.8	13.3
Material 2	14.0	8.8	11.2	14.2	11.8	6.4	9.8	11.3	9.3	13.6
Differ = 1 − 2	−0.8	−0.6	−0.3	0.1	−1.1	0.2	−0.3	−0.5	−0.5	−0.3

(a) The first analysis presented in Example 11.7 was based on the incorrect two independent samples design. Using the data for Material 1 and Material 2, verify the results of the test of $H_0: \mu_1 - \mu_2 = 0$ versus $H_1: \mu_1 - \mu_2 < 0$ using a 0.05 level.

(b) The second analysis presented in Example 11.7 was based on the correct paired samples design. Using the differences, verify the results of the test of $H_0: \mu_D = 0$ versus $H_1: \mu_D < 0$ using a 0.05 level.

11.19 The power of supercomputers derives from the idea of parallel processing. Engineers at a computer research facility are interested in comparing the mean computing time for their new parallel processing design (DESIGN = 1) to their standard design (DESIGN = 2). The output that follows summarizes the results, in seconds, of independent random computation times using the two designs. Assume that the distributions of computation times are normal with common population standard deviation.

```
Independent samples t-test on COMPTIME  grouped by  DESIGN
         Group        N         Mean          SD
         1.000        15        2.173        0.375
         2.000        13        2.523        0.365
Separate variances t = -2.495 df =   25.6 prob = .020
  Pooled variances t = -2.490 df =     26 prob = .019
```

(a) Produce a 95% confidence interval estimate for $\mu_1 - \mu_2$.

(b) Interpret your interval in (a). Explain what the 95% confidence level means.

(c) What hypotheses H_0 and H_1 can you test with the confidence interval in (a)?

11.20 An experiment was conducted to compare the mean absorption of two drugs in muscle tissue. Muscle tissue specimens were randomly divided into two groups of equal size. Group 1 was tested with Drug 1 and Group 2 was tested with Drug 2. Assuming that the absorption rate is approximately normally distributed for each drug, with equal population standard deviation, a 99% confidence interval estimate for $\mu_1 - \mu_2$ is $(-0.1067, 0.5407)$. Consider the potential interpretations that follow for the 99% confidence interval. Which of the interpretations are correct? If it is incorrect, explain why it is wrong.

(a) If we repeated the procedure many times, we'd expect 99% of the intervals produced to contain $\bar{x}_1 - \bar{x}_2$.

(b) If we redid the procedure many times, we would expect to see 99% of the confidence intervals to contain $0.217 \pm 0.3237 \rightarrow (-0.1067, 0.5407)$.

(c) We expect 99% of the time, $\bar{x}_1 - \bar{x}_2$ will fall between $(-0.1067, 0.5407)$.

(d) With repeating samples, we can be confident that $\bar{x}_1 - \bar{x}_2$ will be within the sample 99% of the time.

(e) This tells us that if the procedure were repeated, we would expect 99% of the confidence intervals produced to contain $\mu_1 - \mu_2$.

REAL DATA

11.21 Albert Shanker (President, American Federation of Teachers), in his *Where We Stand* commentary (SOURCE: *New York Times*, May 28, 1995), addresses the issue that teachers in other countries work on average many more days a year, but teach substantially fewer hours. In the United States, the average length of the school year is 180 days, while for Germany it is 225 days, in Italy, 215, in England, 195, and in Norway, 190. A survey by the international Organization for Economic Cooperation and Development (OECD) shows that United States teachers pack more than 1000 hours of classroom instruction into every school year—hundreds of hours more than teachers in other countries, even though they work fewer days. Here are the findings:

Teaching Hours Per Year

Nation	Primary	Lower Secondary	Upper Secondary
Austria	780	747	664
Belgium	840	720	660
Finland	874	798	760
France	944	632	—
Germany	790	761	673
Ireland	951	792	792
Italy	748	612	612
Netherlands	1000	954	954
New Zealand	790	897	813
Norway	749	666	627
Portugal	882	648	612
Spain	900	900	630
Sweden	624	576	528
Turkey	900	1080	1080
United Kingdom	—	669	—
United States	1093	1042	1019
Mean	**858**	**781**	**745**

(a) Give a point estimate for difference in the average number of primary teaching hours per year for the United States versus Germany.

(b) Does the difference [the answer in (a)] in average number of teaching hours per year seem to suggest that U.S. primary teachers spend significantly more hours teaching on average than Germany primary teachers? Explain. What other information is needed in order to determine if the difference is significant at say the 5% level?

11.22 A business manager of a local health clinic is interested in estimating the difference between the fees for extended office visits at her clinic (OFFICE = 1) and the fees for extended office visits at a newly opened group practice (OFFICE = 2). She obtained data on the fees for 15 visits from her clinic and for 15 visits from the group practice, recorded in dollars. Assume that fees at each office are values of independent random samples, where the two distributions are approximately normally distributed with equal population standard deviation. The following computer output was produced:

```
The following results are for:      The following results are for:
OFFICE = 1 (Health Clinic Fees)      OFFICE = 2 (Group Practice Fees)
  N of cases          15               N of cases          15
  Minimum         16.000               Minimum         14.000
  Maximum         27.000               Maximum         24.000
  Mean            20.867               Mean            19.067
  Standard dev     3.248               Standard dev     2.915
```

(a) No details are provided on how the manager obtained the data. Comment on what factors might be important to consider when gathering such data (which days, which patients, etc.).

(b) Give an estimate of the common population standard deviation, σ.

(c) Develop a confidence interval estimate for the difference between the average fees of the two offices. Use a confidence level of 0.90.

(d) Based on this data, is there a significant difference between the average fees of the two offices at a level $\alpha = 0.10$? Explain.

11.23 A researcher was interested in comparing body weights for two strains of laboratory mice. She observed the following weights (in grams) of adult mice:

Strain 1: 32 35 36 37 38 41 43.   **Strain 2:** 38 39 39 40 44 46 47 47.

The researcher was interested in seeing whether these data provide evidence of a difference in mean body weight of the two strains. Assume that the data are the observed values of independent random samples—that is, the two distributions are normal with common population standard deviations.

(a) State the appropriate null and alternative hypotheses H_0 and H_1.

(b) Calculate the value of a test statistic for testing H_0 and H_1, and give the p-value.

(c) State the results for a test with significance level 0.05. Give your decision and a conclusion in terms of the mean body weight for the two strains.

(d) A 95% confidence interval for the difference between the mean weight for Strain 1 and the mean weight for Strain 2 is given by $(-9.32, -0.82)$.

   (i) What does the 95% confidence level tell us about the quality of the confidence interval?

   (ii) Using the 95% confidence interval, would you accept or reject the null hypothesis at the 5% level that the mean weight of Strain 2 is 9 grams higher than the mean weight of Strain 1? That is, test the hypotheses $H_0: \mu_1 - \mu_2 = -9$ versus $H_1: \mu_1 - \mu_2 \neq -9$.

11.24 A toothpaste manufacturer claims that children brushing their teeth daily with his company's new toothpaste product will have fewer cavities than children using a competitor's brand. In a carefully supervised study in which children were randomly assigned to one of the two brands of toothpaste for a 2-year period, the number of cavities for children using the new brand was compared with the number of cavities for children using the competitor's brand. The results are as follows:

    New Brand: 2, 1, 1, 2, 3, 1, 2, 3, 4, 1, 2.

    Competitor Brand: 3, 1, 1, 4, 0, 7, 1, 1, 4, 6, 1.

Test the manufacturer's claim using a significance level of 0.01.

11.25 An instructor hypothesized that the students who earn a "C" or higher spend more hours per week outside of class on course work than those who receive a "D" or an "E." She collects the following data on two independent samples of students:

    Students with a "C" or higher: 9, 4, 7, 11, 1, 5, 3, 2, 1, 3.

    Students with a "D" or an "E": 3, 2, 0, 3, 2, 1, 4, 3.

(a) State the appropriate hypotheses.

(b) Using a significance level of 10%, perform the test.

(c) State the assumptions underlying the test and perform an assessment of the assumptions.

11.26 Before doing this exercise, you may wish to review Section 5.4 on linear transformations. Two properties for a random variable X are

$$\text{Mean}(aX + b) = a\text{Mean}(X) + b \text{ and } \text{Variance}(aX + b) = a^2\text{Variance}(X).$$

Suppose that the variable X takes on just four equally likely values: 1, 2, 3, 4. Verify these properties for this variable X when $a = 2$ and $b = 5$. If you have a TI, enter the four values in L1, define L2 = 2(L1) + 5, and obtain the 1-Var Stats for L1 and L2. The preceding result can be used to show that for two independent random variables X and Y we have

$$\text{Mean}(X - Y) = \text{Mean}(X) - \text{Mean}(Y);$$
$$\text{Variance}(X + Y) = \text{Variance}(X) + \text{Variance}(Y);$$
$$\text{Variance}(X - Y) = \text{Variance}(X) + (-1)^2\text{Variance}(Y)$$
$$= \text{Variance}(X) + \text{Variance}(Y).$$

These results underlie the statement regarding the sampling distribution of $\overline{X}_1 - \overline{X}_2$ on page 644.

11.27 An educator wishes to compare the efficiency of a Strat method for teaching reading to the standard Basal method. A class of 44 students is randomly divided into two groups of 22 each. One group is taught with the Strat method (Group 1), while the other group is taught with the Basal method (Group 2). Is the mean reading score with the Strat method higher than with the Basal method? The two independent samples t-test was conducted. The SPSS output is provided.

Group Statistics

	Teaching Method	N	Mean	Std. Deviation	Std. Error Mean
Reading Score	Strat	22	43.0691	5.5641	1.1863
	Basal	22	40.0768	5.2247	1.1139

Independent Samples Test

		Levene's Test for Equality of Variances		t-test for Equality of Means					95% Confidence Interval of the Difference	
		F	Sig.	t	df	Sig. (2-tailed)	Mean Difference	Std. Error Difference	Lower	Upper
Reading Score	Equal variances assumed	.006	.939	1.839	42	.073	2.9923	1.6273	−.2917	6.2762
	Equal variances not assumed			1.839	41.835	.073	2.9923	1.6273	−.2921	6.2766

We wish to determine whether the mean reading score for students taught with the Strat method is higher than the mean reading score for students taught with the Basal method. A 5% significance level will be used.

(a) State the appropriate hypotheses.

(b) Assume that the data are independent random samples from normal populations with equal population variances. Using the "Equal variances assumed" row in the output, report the test statistic value, the p-value, and your decision and conclusion in terms of the scenario.

(c) Suppose that we cannot assume that the population variances are equal. Using the "Equal variances not assumed" row in the output, report the test statistic value, the p-value, and your decision and conclusion in terms of the scenario.

(d) Are the results for (b) and (c) similar or different? Looking at the two sample standard deviations reported in the output for the two sets of data, does your answer make sense? Explain.

11.28 Suppose that a commuter, Donna, wants to determine whether a second possible route would get her to work more quickly. She decides to consider the next 80 working days and randomly divides them into two groups of 40 days each. She drives her standard route (Route 1) for one set of 40 days and drives the new route (Route 2) for the other set of 40 days, recording for each day the commuting times (in minutes). Assume that for each route, commuting times are normally distributed. Some SPSS output is also provided.

Group Statistics

	Route	N	Mean	Std. Deviation	Std. Error Mean
TIME	Route 1	40	31.945	5.87	.928
	Route 2	40	28.105	6.24	.986

Independent Samples Test

	Levene's Test for Equality of Variances		t-test for Equality of Means					95% Confidence Interval of the Difference	
	F	Sig.	t	df	Sig. (2-tailed)	Mean Difference	Std. Error Difference	Lower	Upper
Time Equal variances assumed	.144	.705	2.835	78	.006	3.840	1.354	1.143	6.537
Equal variances not assumed			2.835	77.713	.006	3.840	1.354	1.143	6.537

(a) What is the quickest commuting time when driving the new route (Route 2)?

(b) Compute the pooled estimate of the common population standard deviation. Show all work.

(c) Does the assumption of common population standard deviation seem plausible? Explain.

(d) State the hypotheses to test if the Route 2 is quicker than Route 1 on average.

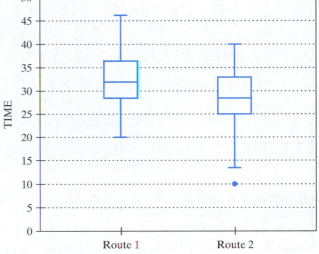

(e) Give the test statistic value, the p-value, and decision for the test in (d) using $\alpha = 0.05$.

(f) Complete the conclusion by selecting the correct phrase: Thus, Route 2 (DOES, DOES NOT) appear to be significantly quicker than Route 1 on average.

11.5 Independent Samples: Comparing Proportions

Much of the data reported in newspaper and journals focus on the proportion or percentage of individuals having some characteristic. Therefore, we may be interested in comparing the proportions associated with two populations from which we have selected independent samples from each population. In this section, we will learn how to test whether the two population proportions are the same or different and how to estimate the difference in the two population proportions with a confidence interval.

Since we are trying to compare two population proportions, p_1 and p_2, a point estimator of this parameter should be logical. We would estimate the difference between two population proportions $p_1 - p_2$ by using the difference between the two corresponding sample proportions, $\hat{p}_1 - \hat{p}_2$. The results and subsequent formulas needed to make inferences about $p_1 - p_2$ in the two independent samples scenario may look somewhat complicated. However, the basic ideas underlying how to make a confidence interval and how to test hypotheses are the same as you studied in Chapter 9. Inferences about the proportion p of a single population were based on the sampling distribution of the sample proportion $\hat{p}$. Recall that inferences about a proportion **required a large sample size**. So we need to know the sampling distribution of $\hat{p}_1 - \hat{p}_2$ when we have large independent random samples. The distribution is based on three statistical results regarding the difference of two independent approximately normally distributed random variables:

1. The mean of the difference of two random variables is the difference of the means.
2. The variance of the difference of two *independent* random variables is the sum of the variances.
3. The difference of two independent normally distributed random variables is also normally distributed.

With these statistical results, we can conclude the following:

Distribution of $\hat{p}_1 - \hat{p}_2$ for the Two Large Independent Samples Scenario

The **mean** of $\hat{p}_1 - \hat{p}_2$ is $p_1 - p_2$.

The **standard deviation** of $\hat{p}_1 - \hat{p}_2$ is $\sqrt{\dfrac{p_1(1 - p_1)}{n_1} + \dfrac{p_2(1 - p_2)}{n_2}}$.

The **distribution** of $\hat{p}_1 - \hat{p}_2$ is *approximately* normal.

Thus, $\hat{p}_1 - \hat{p}_2$ is *approximately* $N\left(p_1 - p_2, \sqrt{\dfrac{p_1(1 - p_1)}{n_1} + \dfrac{p_2(1 - p_2)}{n_2}}\right)$.

The unknown standard deviation of $(\hat{p}_1 - \hat{p}_2)$ can be estimated from the data to give the standard error of $(\hat{p}_1 - \hat{p}_2)$:

$$\text{Standard Error of } (\hat{p}_1 - \hat{p}_2) = \sqrt{\dfrac{\hat{p}_1(1 - \hat{p}_1)}{n_1} + \dfrac{\hat{p}_2(1 - \hat{p}_2)}{n_2}}.$$

We will use this standard error in constructing our confidence interval estimate for $\hat{p}_1 - \hat{p}_2$.

In hypothesis testing, we are generally interested in assessing whether or not there is a difference between the two population proportions, so the null hypothesis is $H_0: p_1 = p_2$ (or, equivalently, $p_1 - p_2 = 0$). The alternative hypothesis might be two-sided, $H_1: p_1 \neq p_2$, for assessing whether there is any difference; or one-sided upper or lower, $H_1: p_1 > p_2$ or $H_1: p_1 < p_2$, for assessing a particular direction for the difference in the proportions. For the testing problem, we will not use the same standard error of $\hat{p}_1 - \hat{p}_2$ that we do in our confidence interval estimator. The reason is that under the assumption that H_0 is true, the two population proportions are equal (for example, to some proportion $p_1 = p_2 = p$). So we can view the data as coming from one overall population, not two populations.

We estimate this overall population proportion p by combining or pooling the data from both samples. If we let x_1 be the number of items in Sample 1 that have the given characteristic and x_2 be the number of items in Sample 2 that have the given characteristic, then the combined or **pooled estimate of p** is

$$\hat{p} = \frac{x_1 + x_2}{n_1 + n_2} = \frac{n_1\hat{p}_1 + n_2\hat{p}_2}{n_1 + n_2}.$$

We would also estimate the standard deviation, under the null hypothesis, by substituting $\hat{p}$ for $\hat{p}_1$ and $\hat{p}_2$ in our previous standard error expression. We will call this the **null standard error** for $\hat{p}_1 - \hat{p}_2$, which is given by

$$\text{Null Standard Error of } \hat{p}_1 - \hat{p}_2 = \sqrt{\hat{p}(1 - \hat{p})\left(\frac{1}{n_1} + \frac{1}{n_2}\right)}.$$

Our test of equal population proportions would be based on the **z-test statistic**

$$z = \frac{(\hat{p}_1 - \hat{p}_2)}{\sqrt{\hat{p}(1 - \hat{p})\left(\frac{1}{n_1} + \frac{1}{n_2}\right)}},$$

and the z variable follows approximately an $N(0, 1)$ distribution under H_0.

The summary that follows is of the two independent (large) samples test of $H_0: p_1 - p_2 = 0$ and the confidence interval for $p_1 - p_2$. Note that there is no corresponding t-test if the sample sizes are too small.

Two Independent Samples z-Test for Proportions (Large Samples)

Assumptions: The first sample is a random sample of size n_1 from the first population. The second sample is a random sample of size n_2 from the second population. The two samples are independent. The sample sizes n_1 and n_2 are large. It is best to have at least five successes in each sample.

Hypotheses: $H_0: p_1 - p_2 = 0$ (that is, $p_1 = p_2$) versus
$H_1: p_1 - p_2 \neq 0$ or $H_1: p_1 - p_2 > 0$ or $H_1: p_1 - p_2 < 0$.
The significance level α to be used is determined.

Data: The two sets of data from which the two sample proportions $\hat{p}_1$ and $\hat{p}_2$ can be computed and the **pooled estimate** of the common **population proportion** p (under H_0) can be computed as

$$\hat{p} = \frac{x_1 + x_2}{n_1 + n_2} = \frac{n_1\hat{p}_1 + n_2\hat{p}_2}{n_1 + n_2}.$$

Observed Test Statistic: $z = \dfrac{\hat{p}_1 - \hat{p}_2}{\sqrt{\hat{p}(1 - \hat{p})\left(\dfrac{1}{n_1} + \dfrac{1}{n_2}\right)}}$

and the *null distribution* for this Z variable is a *standard normal $N(0, 1)$ distribution*.

***p*-value**:   We find the *p*-value for the test using the *N(0, 1) distribution*.

The *direction of extreme* will depend on how the alternative hypothesis is expressed.

Decision: A *p*-value less than or equal to α leads to rejection of H_0.

Confidence Interval for $p_1 - p_2$

Assumptions: The first sample is a random sample of size n_1 from the first population. The second sample is a random sample of size n_2 from the second population. The two samples are independent. The sample sizes n_1 and n_2 are large. It is best to have at least five successes in each sample.

Data: The two sets of data from which the two sample proportions $\hat{p}_1$ and $\hat{p}_2$ can be computed.

Confidence Interval: $(\hat{p}_1 - \hat{p}_2) \pm z^* \sqrt{\dfrac{\hat{p}_1(1 - \hat{p}_1)}{n_1} + \dfrac{\hat{p}_2(1 - \hat{p}_2)}{n_2}}$

where z^* is an appropriate percentile of the *standard normal N(0, 1) distribution*.

Interpretation of the Interval: The interval gives potential values for the difference in the population proportions $p_1 - p_2$ based on independent random samples, one from each population. The interval may be used to test hypotheses when the alternative hypothesis is two-sided.

Interpretation of the Confidence Level: The confidence level pertains to the proportion of times the procedure prod uces an interval that actually contains $p_1 - p_2$ when this procedure is repeated over and over using new independent random samples of the same sizes each time.

Margin of Error: The margin of error $E = z^* \sqrt{\dfrac{\hat{p}_1(1 - \hat{p}_1)}{n_1} + \dfrac{\hat{p}_2(1 - \hat{p}_2)}{n_2}}$.

Minimum (common) sample sizes: n_1 and n_2 needed for a confidence interval with desired margin of error E is $n_1 = n_2 = \dfrac{1}{2}\left(\dfrac{z^*}{E}\right)^2$.

Let's illustrate our two independent samples *z*-test for proportions through an example.

❖ EXAMPLE 11.8   Feeling Successful: Women versus Men _____

REAL
DATA

A survey by Louis Harris & Associates, Inc., has found that working women are providing about half of their families' income. A summary of the findings of the survey appeared in the *New York Times*, May 11, 1995 issue. The survey is based on telephone interviews with a representative national sample of 1502 women ages 18 to 55, conducted November and December (1994). Two-thirds (or 1001) of the women are employed outside of the home at least part time.

The survey was designed so that for some questions, the women's answers could be compared to those of an independent sample of 460 men. One research question to be addressed by the survey was if the proportion of women who mentioned the paycheck when asked what makes them feel successful is significantly lower than the proportion of men who mentioned the paycheck in addressing what makes them feel successful. We are interested in testing about the difference between two population proportions, $p_1 - p_2$. In this case, we can define 1 = women and 2 = men. The hypotheses of interest are $H_0: p_1 = p_2$ and $H_1: p_1 < p_2$.

The data showed that only 7% of the women (who were employed outside the home at least part time) mentioned the paycheck, while 26% of the men talked about their paycheck. Since we had 7% of 1001 women, or 70 women, who mentioned paycheck, and 26% of the 460 men, or 120 men, who mentioned paycheck, our combined estimate of the proportion who mentioned paycheck is

$$\hat{p} = \frac{70 + 120}{1001 + 460} = 0.13.$$

The observed z-test statistic is

$$z = \frac{0.07 - 0.26}{\sqrt{0.13(1 - 0.13)\left(\frac{1}{1001} + \frac{1}{460}\right)}} = \frac{-0.19}{0.019} = -10.$$

The observed difference in the proportions of -0.19 is about 10 standard errors below the hypothesized difference of zero. Based on our frame of reference about normal distributions, we know this result is very extreme.

The p-value measures how extreme the observed data are under H_0. In this case, we want to find the probability of getting a test statistic less than or equal to -10, under the $N(0, 1)$ model. We know that this p-value is essentially zero, but let's see the process of determining the p-value using a TI or the Standard Normal Table.

Finding the p-value
With the TI

1. **Using the distribution of the Z-test statistic and the normalcdf(function**
    We can find this area under this normal curve using the normalcdf(function.

    $$p\text{-value} = P(Z \leq -10) = \text{normalcdf}(-\text{E}99, -10) = 0.$$

2. **Using the 2-PropZTest function under the STAT TESTS menu**
    In the TESTS menu under the STAT button, we select the **6:2-PropZTest** option. You need to specify the number of "successes" in each sample (denoted by the x's), the sample sizes, and the direction for the alternative hypothesis. The steps are summarized next and the corresponding input and output screens are shown.

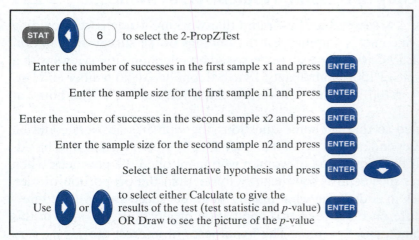

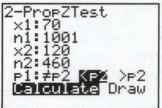

The corresponding result
screen should look like this:

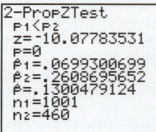

Note that the *Calculate* option produces an output screen which provides everything needed to report the results of the test and make a decision. You have which alternative hypothesis is being tested, the observed value of the standardized z-test statistic ($z = -10.0778$), the p-value (0), the sample proportions $\hat{p}_1 = 0.07$ and $\hat{p}_2 = 0.26$, and the sample sizes. The *Draw* option would draw the normal distribution and shade in the area that corresponds to the p-value, and the value of z-test statistic and the p-value are also given. If you select the *Draw* option, just be careful that other graphing options have been turned off and cleared out.

With Table II

Our test statistic is $z = -10$. The smallest entry in Table II is $z = -3.49$, which has an area to the left of 0.0002. Our p-value is even smaller than this.

$$p\text{-value} = P(Z \leq -10) \text{ is less than } 0.0002$$

Decision and Conclusion

The p-value is very small, nearly zero. We would reject H_0 at any reasonable level of significance and conclude that the proportion of women who mentioned the paycheck appears to be significantly lower than the proportion of men who mentioned the paycheck, in addressing what makes them feel successful. ❖

THINK ABOUT IT

Could we have constructed and used a 99% confidence interval for $p_1 - p_2$ to test our alternative hypothesis $H_1: p_1 < p_2$ that the proportion of women who mentioned the paycheck is significantly lower than the proportion of men who mentioned the paycheck in addressing what makes them feel successful? Explain.

Let's do it! 11.7 HMOs on the Rise

In the March 24, 1995, edition of the *New York Times*, it was reported that a quarter of the state's residents have joined health maintenance organizations (HMOs). From March 1993 to January 1995, the percentage of New York State residents who joined HMOs rose to 25% from 19%. The article "HMOs on Rise in New York State" also reported results from a Foster Higgins National Survey of Employer Sponsored Health Plans, not just for New York, but by regions of the United States. Survey results indicate that 37% of insured workers in the Northeast are enrolled in an HMO program, while 42% of the insured workers in the South are enrolled in an HMO program.

(a) How does the Northeast compare to the South with respect to the percentage of insured workers enrolled in an HMO program? What additional information would we need to assess if the proportion enrolled in an HMO program is statistically the same for these two regions?

(b) Suppose that the above rates of 37% and 42% were based on independent random samples of 2000 insured workers from each region. Provide a 90% confidence interval for the difference in the proportions of insured workers enrolled in an HMO program (South Region − Northeast Region). Interpret your confidence interval estimate.

TI Note: For creating this confidence interval with the TI calculator, see the Quick Steps at the end of this chapter.

THINK ABOUT IT

A survey asked respondents the following question regarding retirement plans: "Looking ahead to retirement, what do you expect to be your major source of income?" Of those sampled, 39% said *Savings*, while 22% reported *Social Security*.

Could we use the techniques described in this section to test whether the true proportion reporting *Savings*, p_1, is statistically different from the true proportion reporting *Social Security*, p_2? Explain.

Let's do it!   11.8 Favorite Pizza Topping

Pizza is often a familiar and favorite food for college students. Conduct a survey of the students in your class to assess whether the proportion of females whose favorite pizza topping is pepperoni is different from the proportion of males whose favorite pizza topping is pepperoni. You will need to assume that the female students in your class represent a simple random sample from the population of female students at your school. Also assume that the male students in your class represent an independent simple random sample from the population of male students at your school. Let p_1 represent the proportion of all female students whose favorite pizza topping is pepperoni and p_2 represent the proportion of all male students whose favorite pizza topping is pepperoni.

Hypotheses:

$$H_0: p_1 = p_2 \qquad H_1: p_1 \neq p_2$$

Data:

$$\hat{p}_1 = \frac{x_1}{n_1} = \frac{\text{number of female students interviewed whose favorite topping is pepperoni}}{\text{number of female students interviewed}} =$$

$$\hat{p}_2 = \frac{x_2}{n_2} = \frac{\text{number of male students interviewed whose favorite topping is pepperoni}}{\text{number of male students interviewed}} =$$

Test Results:

Combined estimate for the common population proportion: $\hat{p} = \dfrac{x_1 + x_2}{n_1 + n_2} =$

Observed test statistic: $z = \dfrac{(\hat{p}_1 - \hat{p}_2)}{\sqrt{\hat{p}(1 - \hat{p})\left(\dfrac{1}{n_1} + \dfrac{1}{n_2}\right)}} =$

p-value:

Are these data statistically significant at the 5% level?   Yes   No

State your conclusion in words:

✳ Exercises

REAL
DATA

11.29 Louis Pasteur had conducted a series of experiments that demonstrated the roles that yeast and bacteria play in the fermentation process. These results gave Joseph Lister, a British physician, the idea that human infections might have a similar organic origin. Lister developed a theory that using carbolic acid as a surgical room disinfectant would improve the postoperative survival rates for surgical patients. Out of the 40 patients amputated with the use of carbolic acid, 34 survived. Out of the 35 patients amputated without the use of carbolic acid, 19 survived. (SOURCE: *The Conquest of Epidemic Disease*, by Charles Winslow, Princeton University Press, 1943, p. 303.)

(a) State the appropriate hypotheses for assessing if the use of carbolic acid improves the postoperative survival rates.

(b) What proportion of patients amputated with the use of carbolic acid survived? What proportion of patients amputated without the use of carbolic acid survived?

(c) Test the hypotheses in (a) using a 5% significance level.

REAL
DATA

11.30 Back in 1968, George Wallace and his supporters were strongly committed to strict law enforcement. A group of sociologists in Nashville, Tennessee conducted a survey to assess if Wallace supporters actually practiced what they preached. A county law had been passed that required local vehicles to display a $15 "Metro sticker" on the windshield. The sociologists made spot checks at various parking lots over several days. They recorded the number of cars in support of each candidate (based on factors such as displaying a bumper sticker for a candidate) and how many of them had the required sticker. These data are provided in the following table (SOURCE: "Wallace Supporters and Adherence to 'Law and Order'," by Lawrence S. Wrightman, in *Human Social Behavior*, Dorsey Press, Homewood, Ill., 1971):

In Support of	Number of Cars	Number with Metro Sticker
Wallace	361	270
Humphrey	178	154

(a) The sociologists were interested in testing if p_1, the proportion of Wallace supporters in compliance, differed from p_2, the proportion of Humphrey supporters. State the hypotheses more formally in terms of p_1 and p_2.

(b) What proportion of Wallace supporters sampled were in compliance? What proportion of Humphrey supporters sampled were in compliance?

(c) Test the hypotheses in (a) using a 1% significance level. Provide your conclusion in words.

(d) Would a 99% confidence interval for $p_1 - p_2$ contain the value of zero? Explain. Then compute the interval to check your answer.

(e) Compute the margin of error for your interval in (d).

REAL
DATA

11.31 Louis Harris & Associates have conducted a survey over the many years regarding the health habits of Americans. (SOURCE: *Americans Round Retreating*, March 14, 1993.) One of the issues measured was the proportion of those who avoid excess salt in their diet. The 1983 survey results stated that 53% of those surveyed avoided excess salt (although not given in the article, assume that the 1983 survey was based on 1251 responses). Physicians are worried that Americans are not as concerned about their salt intake as they had been in the past. They would like to assess whether there has been a significant decrease in the proportion of Americans who avoid excess salt in their diet. The 1992 survey results stated that 46% of the 1250 people surveyed avoided excess salt.

(a) State the hypotheses more formally in terms of p_1, the proportion of Americans in 1993 who avoided excess salt, and p_2, the proportion of Americans in 1982 who avoided excess salt.

(b) Test the hypotheses in (a) using a 1% significance level. Provide your conclusion in words.

11.32 In testing hypotheses regarding two population proportions, the null hypothesis can be expressed as $H_0: p_1 = p_2$. In order to get the value of the test statistic, the following expression is computed: $\dfrac{x_1 + x_2}{n_1 + n_2} = \dfrac{n_1\hat{p}_1 + n_2\hat{p}_2}{n_1 + n_2}$. State clearly what parameter this expression estimates.

11.33 Determine the minimum (common) sample sizes $n_1 = n_2$ needed for a 95% confidence interval for $p_1 - p_2$ with a margin of error of ≤ 0.01.

11.34 Determine the minimum (common) sample sizes $n_1 = n_2$ needed for a 99% confidence interval for $p_1 - p_2$ with a margin of error of ≤ 0.05.

Chapter Summary

In any decision-making process, we are making comparisons. In this chapter, we have studied techniques that allow us to compare the results of two populations. We began with the objective of comparing the means of two populations. The primary issue was to distinguish between paired samples versus two independent samples. In our last section, we discussed a procedure for comparing the proportions for two populations. In all sections, we learned how to decide whether significant differences existed and how to estimate the differences. The results and subsequent formulas were based on the underlying ideas of how to test hypotheses and how to make a confidence interval that were developed in Chapters 9 and 10. The various components of the tests and confidence intervals for making inferences involving two populations are summarized in the following table:

Situation	Parameter of Interest	Null Hypothesis and Test Statistic	Confidence Interval Estimate
Two Dependent Means	μ_D	$H_0: \mu_D = 0$ $t = \dfrac{\bar{d} - 0}{\dfrac{s_D}{\sqrt{n}}}$	$\bar{d} \pm t^*\left(\dfrac{s_D}{\sqrt{n}}\right),$ where df $= n - 1$
Two Independent Means (equal population variance)	$\mu_1 - \mu_2$	$H_0: \mu_1 - \mu_2 = 0$ $t = \dfrac{(\bar{x}_1 - \bar{x}_2) - 0}{s_p\sqrt{\dfrac{1}{n_1} + \dfrac{1}{n_2}}}$ where $s_p = \sqrt{\dfrac{(n_1 - 1)s_1^2 + (n_2 - 1)s_2^2}{(n_1 + n_2 - 2)}}$ and df $= n_1 + n_2 - 2$	$(\bar{x}_1 - \bar{x}_2) \pm t^*\left(s_p\sqrt{\dfrac{1}{n_1} + \dfrac{1}{n_2}}\right),$
Two Independent Proportions (large sample sizes)	$p_1 - p_2$	$H_0: p_1 - p_2 = 0$ $z = \dfrac{(\hat{p}_1 - \hat{p}_2)}{\sqrt{\hat{p}(1 - \hat{p})\left(\dfrac{1}{n_1} + \dfrac{1}{n_2}\right)}}$ where $\hat{p} = \dfrac{x_1 + x_2}{n_1 + n_2}$	$(\hat{p}_1 - \hat{p}_2) \pm z^*\sqrt{\dfrac{\hat{p}_1(1 - \hat{p}_1)}{n_1} + \dfrac{\hat{p}_2(1 - \hat{p}_2)}{n_2}}$

Key Terms

Be sure you can describe, in your own words, and give an example of each of the following key terms from this chapter:

paired samples
control group
independent samples
paired *t*-test
standard error for mean
  difference

mean difference
difference in two means
difference in two
  proportions
z-test for difference in two
  proportions
standard error for difference
  in two means

pooled proportion estimate
pooled standard error
pooled two independent
  samples *t*-test
standard error for difference
  in two proportions

Exercises

**REAL
DATA**

11.35 SIDS is the sudden, unexpected death of an apparently healthy infant that remains unexplained even after investigation. A study was published in the *Journal of the American Medical Association* (March 1995), which concludes that the greater the number of smokers or the greater the number of cigarettes an infant is exposed to, the higher the SIDS risk. A summary of a newspaper's account of the study is as follows:

- The researchers interviewed parents of 200 infants who died of SIDS in Southern California between January 1989 and December 1992, and parents of 200 similar healthy infants.

- If more than one adult smoked, the risk of the infant dying was 3.5 times higher than for infants in smoke-free households.

(a) Was this an experiment or an observational study? Explain.

(b) Two primary populations are being compared. Name them.

(c) How were the data obtained? Were the samples paired or independent?

11.36 Three hundred students are enrolled in an introductory quantitative reasoning course for the next term. You wish to assess students' quantitative reasoning skills by comparing a pretest, to be given at the start of the course, to a posttest, to be given at the end of the course.

(a) Describe how you would obtain two independent samples of size 30 from these 300 students to perform this assessment.

(b) Describe how you would obtain your observations for a paired design.

11.37 The following data represent the number of computer units sold per week by a sample of eight salespersons before and after a bonus plan was introduced.

Salesperson	1	2	3	4	5	6	7	8
Before	54	25	80	76	63	82	94	72
After	53	30	79	79	70	90	92	81
Difference	−1	5	−1	3	7	8	−2	9

```
t-tests for Paired Samples
                 Number of            2-tail
 Variable          pairs    Corr      Sig        Mean           SD      SE of Mean
------------------------------------------------------------------------------------
 AFTER                                          71.7500       20.810      7.358
                     8       .978     .000
 BEFORE                                         68.2500       21.265      7.518
------------------------------------------------------------------------------------
          Paired Differences        |
   Mean         SD    SE of Mean    |     t-value              df      2-tail Sig
------------------------------------|-----------------------------------------------
  3.5000      4.408     1.558       |      2.25                 7          .060
 95% CI (-.185, 7.185)              |
```

(a) What assumptions are required in order to perform a paired *t*-test on these data?

(b) Suppose that the assumptions are valid. At a 5% level, did the bonus plan significantly increase sales, on average? Explain.

11.38 To evaluate the effectiveness of a tranquilizer drug, a doctor assigned "anxiety scores" to nine patients before and after administration of the drug. (High scores indicate greater anxiety.) The doctor was interested in seeing whether the results supported the manufacturer's claim that the drug lessens anxiety. The data, together with the difference in scores (Before − After), are given in the following table:

Patient	1	2	3	4	5	6	7	8	9
Before	23	18	18	16	22	26	15	21	20
After	20	12	14	18	21	20	20	24	13
Difference	3	6	4	−2	1	6	−5	−3	7

Assume the differences are values of a random sample, where the distribution is normal.

(a) State the appropriate hypotheses for a test aimed at determining if the drug lessens anxiety.

(b) Calculate the observed test statistic value and give the p-value for your test in (a).

(c) If you were to use a significance level of 0.10, what would your decision and conclusion be?

11.39 One factor in the process of purchasing a new car is the economy, or miles-per-gallon, rating. A car manufacturer has developed a new engine model for a particular car line, with the hope that it will be more economical than the old engine model. Consider the following two scenarios:

Scenario 1 Twenty cars from the same car line were available—10 were equipped with the new engine model and 10 with the old engine model. Twenty drivers (one per car) drove the cars over a designated stretch of highway, and the corresponding miles per gallon were obtained and recorded as follows:

Driver	1	2	3	4	5	6	7	8	9	10	
Old Engine	20	22	19	22	24	25	22	21	23	25	Mean = 22.3 miles per gallon
Driver	**11**	**12**	**13**	**14**	**15**	**16**	**17**	**18**	**19**	**20**	
New Engine	18	19	20	26	24	24	23	28	25	24	Mean = 23.1 miles per gallon

Scenario 2 Twenty cars from the same car line were available—10 were equipped with the new engine model and 10 with the old engine model. Ten drivers were available and randomly assigned to a pair of cars, one of each engine model. The drivers drove one of the two cars (selected at random) over a designated stretch of highway, and the corresponding miles per gallon were obtained. Then, the drivers repeated the process using the remaining assigned car. The data are given as follows:

Driver	1	2	3	4	5	6	7	8	9	10	
Old Engine	21	20	22	24	24	25	22	23	23	26	Mean = 23 miles per gallon
New Engine	19	21	20	26	25	27	24	28	25	25	Mean = 24 miles per gallon

(a) In Scenario 1, is the value 22.3 miles per gallon the sample mean, $\bar{x}_1$, or the population mean, μ_1?

(b) How are the data collected in Scenario 1 different from the data collected in Scenario 2?

(c) For which scenario might it be easier to assess if there is a significant difference between the two groups? Explain.

(d) For Scenario 1, test the following hypotheses: H_0: The two engines perform the same. H_1: The new engine performs better than the old engine. Test them at a 5% significance level and state the assumptions required to perform the test.

(e) For Scenario 2, provide a 99% confidence interval estimate for the difference in the mean miles per gallon for the new engine versus the old engine. Interpret your confidence interval.

Exercises 11.40 and 11.41 remind us that judging a study based on the information presented in a new article is not always easy.

11.40 Skin cancer, or melanoma, caught in the early stages is quite treatable and curable. But when the melanoma reaches Stage II (tumors thicker than 4 mm) or Stage III (which involves lymph nodes), it often recurs and metastasizes after surgery. Research scientists have been trying to develop and test postoperative therapies designed to prolong survival for such patients. A report in The Skin Cancer Foundation journal, the *Melanoma Letter*, states promising results from a protein–carbohydrate compound produced by white blood cells, called interferon-alpha (IFN-alpha). In a study at the University of Vienna (Austria), 52 Stage II melanoma patients and 15 Stage III melanoma patients received IFN-alpha for 20 months after surgery. Of the Stage II patients, 92% had a 20-month relapse-free survival rate. This is compared to 81.4% for patients who did not receive IFN-alpha.

(a) We have two percentages presented in these results. (SOURCE: *Ann Arbor News*, July 5, 1995.) The first percentage of 92% corresponds to the sample of 52 Stage II melanoma patients. To what group does the 81.4% correspond to? Do we know if this group contains only Stage II melanoma patients? How many patients are in this group?

(b) Consider the following two populations: Population 1 = Stage II melanoma patients who received IFN-alpha; Population 2 = Stage II melanoma patients who did not receive IFN-alpha. Suppose that the 81.4% corresponds to a sample of 52 Stage II melanoma patients who did not receive IFN-alpha. Is the survival rate for Population 1 significantly higher than the survival rate for Population 2? Test using a 5% significance level. State your hypotheses and assumptions, and report your test statistic, p-value, and your conclusion.

11.41 An article entitled "Does Sugar Make Kids Squirrelly? Study Says No"(SOURCE: *Associated Press*, February 3, 1994) states that several small studies have looked at sugar's effect on children. In general, they found no indication that sugar made the children hyperactive. A new study was conducted on 25 normal preschoolers, ages 3 to 5, and on 23 children ages 6 to 10. The older children were described by their parents as being sensitive to sugar. Dietitians went to the children's homes and took away all food. Then, they provided prepared meals for the children and their families.

(a) Do these data constitute paired or independent samples? Explain.

(b) Further details of the study reveal that the results (various behavioral measures) for these two groups of children were never directly compared. Rather, the families were fed different diets (sweetened with sugar, sweetened with saccharin) each lasting the same number of weeks. Behavioral measures were obtained on all of the children after each diet. Comparisons were made within each group (not across groups) on the behavioral measures after Diet 1 versus the behavioral measures after Diet 2. Do these data constitute paired or independent samples? Explain.

11.42 A new method of storing snap beans is believed to retain more ascorbic acid (vitamin C) than an old method. In an experiment, snap beans were harvested under uniform conditions and frozen in 32 equal-sized packages. Sixteen of these packages were randomly selected and stored according to the new method, and the other 16 packages were stored by the old method. Subsequently, ascorbic acid determinations (in mg/kg) were made, and the following summary statistics were calculated (assume that ascorbic acid retention is approximately normally distributed for both methods).

Ascorbic Acid	Group 1—New	Group 2—Old
Sample Mean	449	410
Sample Standard Deviation	25	30

(a) We wish to assess if these data substantiate the claim that more ascorbic acid is retained under the new method of storing. State the appropriate hypotheses to be tested.

(b) Perform the appropriate test at $\alpha = 0.05$. Report your observed test statistic value, your decision, and how you arrived at your decision, and state a conclusion in terms of the study.

REAL DATA

11.43 The Helsinki Heart Study reported a reduction in the incidence of cardiac events in middle-aged men with high cholesterol levels, but no known coronary artery disease. Of the 2051 subjects who spent five years in a treatment group receiving the cholesterol-reducing drug, 56 had experienced a cardiac event. Of the 2030 subjects who spent five years in a placebo group, 84 had experienced a cardiac event. (SOURCE: Article by K. G. Marshall in the *Canadian Medical Association Journal*, May 15, 1996.)

(a) What is the incidence of cardiac events for treatment group subjects—that is, what is the proportion of treatment group subjects who experienced a cardiac event? What is the incidence (proportion) of cardiac events for placebo group subjects?

(b) At the 1% level of significance, is there sufficient evidence of a difference in the incidence rates of cardiac events for the two groups? Give supporting details.

(c) In reporting on this study, it was stated that the cholesterol-reducing drug resulted in a 34% reduction in the incidence of cardiac events. Many studies report the results by taking the difference in the incidence rates for the two groups (treatment group rate − placebo group rate) and divide this difference by the placebo group rate as a base. Check that this 34% figure was computed this way from the study results.

REAL DATA

11.44 Caffeine levels in coffee are often checked very closely. Four freeze-dried brands of instant coffee and eight spray-dried brands of instant coffee were obtained. The caffeine content of each (in grams per 100 grams of dry matter) was recorded using a liquid chromatography method. (SOURCE: "Determination of Purine Alkaloids and Trigonelline in Instant Coffee and Other Beverages Using High Performance Liquid Chromatography," *Journal of the Science of Food and Agriculture*, 34 (1983), pp. 300–306.)

Freeze-Dried Instant	3.7	2.8	3.4	3.7				
Spray-Dried Instant	4.8	4.0	3.8	4.3	3.9	4.6	3.1	3.7

(a) What assumptions are required to construct a 95% confidence interval for the difference in the mean caffeine content $\mu_1 - \mu_2$? Assume these assumptions are valid for answering (b) through (d).

(b) Construct a 95% confidence interval for the difference in the mean caffeine content $\mu_1 - \mu_2$.

(c) What does the 95% confidence level mean?

(d) What does your 95% confidence interval in (b) tell you about the decision for the test of $H_0: \mu_1 - \mu_2 = 0$ versus $H_1: \mu_1 - \mu_2 \neq 0$ using a 5% significance level?

11.45 Title IX was created to ensure equal opportunity for male and female student athletes. All colleges and universities are supposed to abide by this rule. A random sample of private schools was taken and the number of sports offered for male student athletes and the number of sports offered for female student athletes were recorded. We wish to test whether for private schools the Title IX seems to be upheld against the alternative that the male sports still dominate. SPSS output follows.

Paired Samples Statistics

		Mean	N	Std. Deviation	Std. Error Mean
Pair 1	Intercollegiate Sports for Males	9.74	68	5.79	.70
	Intercollegiate Sports for Females	8.63	68	4.41	.53

Paired Samples Test

		Paired Differences							
					95% Confidence Interval of the Difference				Sig.
		Mean	Std. Deviation	Std. Error Mean	Lower	Upper	t	df	(2-tailed)
Pair 1	Intercollegiate Sports for Males Intercollegiate Sports for Females	1.10	4.10	.50	.11	2.10	2.216	67	.030

(a) What was the sample mean number of sports offered to male student athletes at the sampled private schools?

(b) Clearly state the appropriate null and alternative hypotheses.

(c) Perform the test at a 2% significance level. Clearly state the value of the test statistic, the *p*-value, your decision, and conclusion in words.

11.46 An investigator is interested in comparing two states with respect to the proportion of employees for which an employer-supported smoking cessation program is available. One of the health objectives for the United States was to have such a program available to at least 35% of all employees. Suppose that a random sample of 260 employees from State 1 resulted in 83 employees with access to such a program. An independent random sample of 190 employees from State 2 resulted in 74 employees with access to such a program.

(a) What proportion of State 1 employees sampled have an employer-supported smoking cessation program available to them? What proportion of State 2 employees sampled have an employer-supported smoking cessation program available to them?

(b) State the hypotheses for assessing if there is a difference in the proportion of employees for whom an employer-supported smoking cessation program is available for the two states.

(c) If the null hypothesis is true, the proportions for the two states are equal. What is the combined estimate for this overall proportion of employees for whom an employer-supported smoking cessation program is available? How does it compare to the 35% objective?

(d) Based on the data, is there a significant difference in the proportion of employees for whom an employer-supported smoking cessation program is available for the two states? Use a 5% level of significance. Give supporting details.

11.47 The starting salaries of 10 male and 10 female college graduates were randomly selected and entered into SPSS to produce the following summary and output (salaries are in dollars $):

Group Statistics

	Gender	N	Mean	Std. Deviation	Std. Error Mean
SALARY	male	10	39630.00	3247.24	1026.87
	female	10	37153.00	3087.71	976.42

Independent Samples Test

		Levene's Test for Equality of Variances		t-test for Equality of Means					95% Confidence Interval of the Difference	
		F	Sig.	t	df	Sig. (2-tailed)	Mean Difference	Std. Error Difference	Lower	Upper
SALARY	Equal variances assumed	.012	.913	1.748	18	.097	2477.00	1416.99	−499.98	5453.98
	Equal variances not assumed			1.748	17.955	.098	2477.00	1416.99	−500.52	5454.52

Assume that the salaries represent two independent random samples. We wish to assess whether the mean salary of men is higher than the mean salary of women using a 5% significance level.

(a) Clearly state the appropriate null and alternative hypotheses to be tested.

(b) To perform the pooled two independent samples *t*-test, we needed to assume the distribution of salaries is normal for both populations and the standard deviations for these two populations are equal. What is the value of the estimate for this common standard deviation?

(c) Perform the test using a significance level of 5%. Give the value of the test statistic, the *p*-value, and clearly state your decision.

(d) According to our data, do men make more money than women? It appears that (select one)   they do   they don't.

(e) The 95% confidence interval for the difference in the mean salary (males less females) is given by (−499.98, 5453.98). Explain what the 95% confidence level means.

(f) Suppose that we cannot assume equal population variances as stated in (b). Using the "Equal variances not assumed" row in the output, report the test statistic value, the *p*-value, and your decision using a 5% level of significance. Also report the corresponding 95% confidence interval. How do these results compare to those in (c), (d), and (e)?

11.48 The elasticity of plastic can vary depending on the process by which it is prepared. To compute the elasticity of plastic produced by two different processes, four pieces from each process were analyzed. The data are given here:

Process 1	6.9	8.1	7.8	7.6
Process 2	8.2	7.9	8.0	8.4

Suppose that we can assume that the measurements constitute independent random samples from normal populations with equal standard deviations.

(a) State the appropriate hypotheses to test that Process 1 produces plastic with lower elasticity on average compared to Process 2.

(b) If the mean elasticity of plastic for Process 1 is the same as that for Process 2, what is the distribution of the test statistic?

(c) The value of the test statistic for performing the test in (a) was found to be -1.89. Give the corresponding *p*-value and your decision using a 10% level of significance.

(d) The 90% confidence interval for the difference in the mean elasticity for the two processes $(1 - 2)$ is given by $(-1.065, 0.015)$. Interpret the meaning of the 90% confidence level.

TI Quick Steps

Finding Areas under a *t*-Distribution Using the TI

Recall from Chapter 10 that the TI has the tcdf(function built into the calculator, which can compute various areas under a *t*-distribution with a specified degrees of freedom. This function is located under the DISTR menu. Please see page 612 for a review on the use of this function for the TI.

Hypothesis Testing and Confidence Intervals with the TI

The TI has a TESTS menu located under the STATS menu that allows you to perform hypothesis tests and generate confidence intervals.

Paired *t*-test for the Mean Difference

Recall that the paired *t*-test for the mean difference is actually just a one-sample *t*-test performed on the differences. So the TI for a one-sample *t*-test in Chapter 10 will apply again. If you enter the paired observations into two columns (for example, L1 and L2), then you will first need to compute the differences using a linear transformation. The sequence of buttons for computing the differences as L3 = L2 − L1 are provided next.

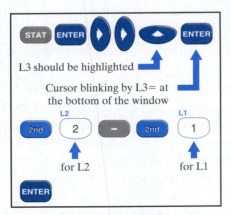

If the differences have already been computed, you can enter them directly, say as L1, and the test can be performed on the one sample of *Data* (differences) entered. If you only have the summary statistics for the differences, you can enter these *Stats* (the sample mean difference, sample standard deviation of the differences, and sample size). The steps for performing a paired *t*-test are as follows:

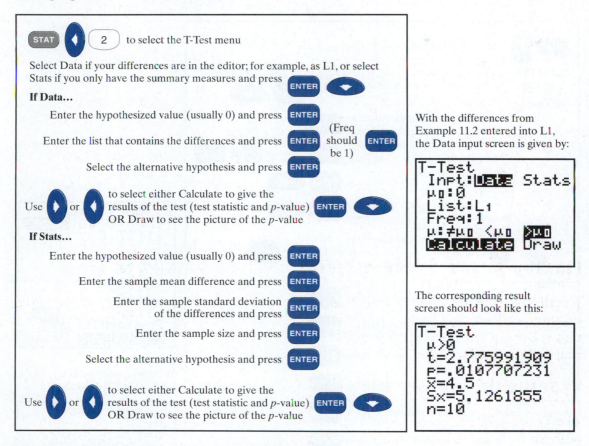

With the differences from Example 11.2 entered into L1, the Data input screen is given by:

```
T-Test
  Inpt:Data Stats
  μ₀:0
  List:L₁
  Freq:1
  μ:≠μ₀  <μ₀  >μ₀
  Calculate Draw
```

The corresponding result screen should look like this:

```
T-Test
  μ>0
  t=2.775991909
  p=.0107707231
  x̄=4.5
  Sx=5.1261855
  n=10
```

Paired Confidence Interval for the Mean Difference

You can either have the one sample of *Data* (differences) entered (for example, as L1) or just know the *Stats* (the sample mean difference, sample standard deviation of the differences, and sample size). To generate the confidence interval, the steps are as follows:

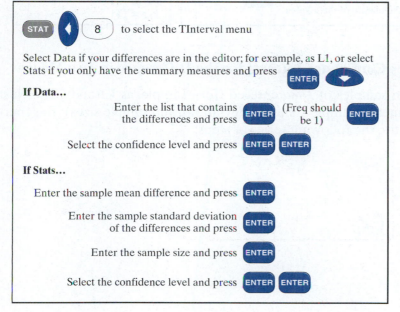

The Stats input screen for Example 11.3 is given by:

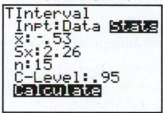
```
TInterval
  Inpt:Data Stats
  x̄: -.53
  Sx:2.26
  n:15
  C-Level:.95
  Calculate
```

The corresponding result screen should look like this:

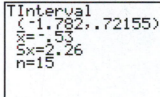
```
TInterval
  (-1.782,.72155)
  x̄=-.53
  Sx=2.26
  n=15
```

Two Independent Samples *T*-test for the Difference in Two Means

You can either have the two samples of *Data* entered (for example, as L1 and L2) or just enter the *Stats* (the sample means, sample standard deviations, and sample sizes). To perform a test, use the following steps:

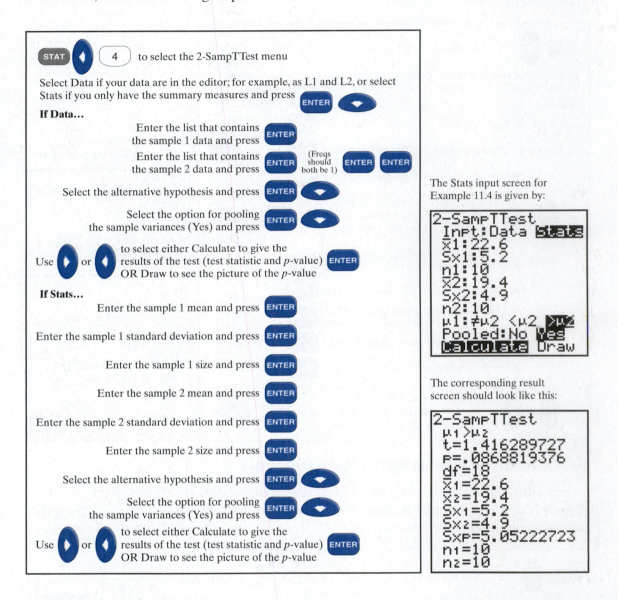

STAT ◀ 4 to select the 2-SampTTest menu

Select Data if your data are in the editor; for example, as L1 and L2, or select Stats if you only have the summary measures and press ENTER ▼

If Data...

Enter the list that contains the sample 1 data and press ENTER

Enter the list that contains the sample 2 data and press ENTER (Freqs should both be 1) ENTER ENTER

Select the alternative hypothesis and press ENTER ▼

Select the option for pooling the sample variances (Yes) and press ENTER ▼

Use ▶ or ◀ to select either Calculate to give the results of the test (test statistic and *p*-value) ENTER OR Draw to see the picture of the *p*-value

If Stats...

Enter the sample 1 mean and press ENTER

Enter the sample 1 standard deviation and press ENTER

Enter the sample 1 size and press ENTER

Enter the sample 2 mean and press ENTER

Enter the sample 2 standard deviation and press ENTER

Enter the sample 2 size and press ENTER

Select the alternative hypothesis and press ENTER ▼

Select the option for pooling the sample variances (Yes) and press ENTER ▼

Use ▶ or ◀ to select either Calculate to give the results of the test (test statistic and *p*-value) ENTER OR Draw to see the picture of the *p*-value

The Stats input screen for Example 11.4 is given by:

```
2-SampTTest
 Inpt:Data Stats
x̄1:22.6
Sx1:5.2
n1:10
x̄2:19.4
Sx2:4.9
n2:10
µ1:≠µ2 <µ2 >µ2
Pooled:No Yes
Calculate Draw
```

The corresponding result screen should look like this:

```
2-SampTTest
µ1>µ2
t=1.416289727
p=.0868819376
df=18
x̄1=22.6
x̄2=19.4
Sx1=5.2
Sx2=4.9
SxP=5.05222723
n1=10
n2=10
```

Two Independent Samples Confidence Interval for the Difference in Two Means

You can either have the two samples of *Data* entered (for example, as L1 and L2) or just enter the *Stats* (the sample means, sample standard deviations, and sample sizes). To generate the confidence interval, use the following series of steps:

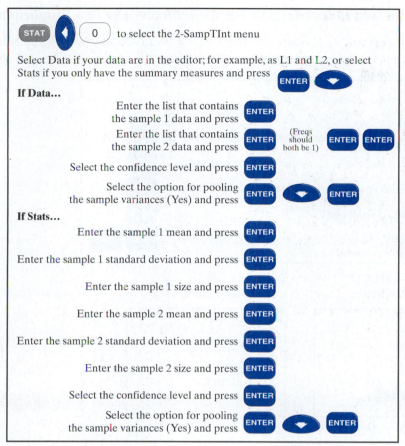

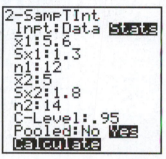

The Stats input screen for Example 11.5 is given by:

```
2-SampTInt
 Inpt:Data Stats
 x̄1:5.6
 Sx1:1.3
 n1:12
 x̄2:5
 Sx2:1.8
 n2:14
 C-Level:.95
 Pooled:No Yes
 Calculate
```

The corresponding result screen should look like this:

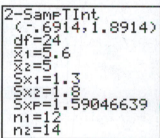

```
2-SampTInt
 (-.6914,1.8914)
 df=24
 x̄1=5.6
 x̄2=5
 Sx1=1.3
 Sx2=1.8
 SxP=1.59046639
 n1=12
 n2=14
```

Two Independent Samples Z-test for the Difference in Two Proportions

You need to specify the number of "successes" in each sample (denoted by the *x*'s), the sample sizes, and the direction for the alternative hypothesis. To perform a test follow these steps:

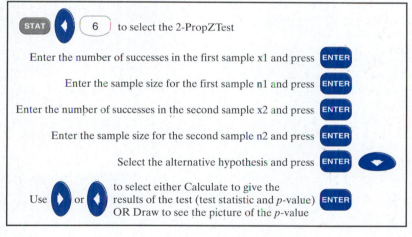

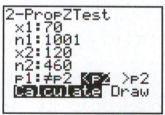

The Stats input screen for Example 11.8 is given by:

```
2-PropZTest
 x1:70
 n1:1001
 x2:120
 n2:460
 p1:≠p2 <p2 >p2
 Calculate Draw
```

The corresponding result screen should look like this:

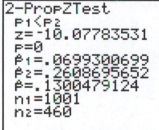

```
2-PropZTest
 p1<p2
 z=-10.07783531
 p=0
 p̂1=.0699300699
 p̂2=.2608695652
 p̂=.1300479124
 n1=1001
 n2=460
```

Two Independent Samples Confidence Interval for the Difference in Two Proportions

You need to specify the number of "successes" in each sample (denoted by the x's), the sample sizes, and the confidence level. To generate the confidence interval, use the following steps:

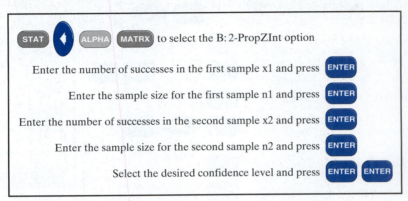

The Stats input screen for Let's do it! 11.7 is given by:

```
2-PropZInt
 x1:840
 n1:2000
 x2:740
 n2:2000
 C-Level:.9
 Calculate
```

The corresponding result screen should look like this:

```
2-PropZInt
 (.02461,.07539)
 p̂1=.42
 p̂2=.37
 n1=2000
 n2=2000
```

Chapter 12

COMPARING MANY TREATMENTS

 ## 12.1 Introduction

Many research questions in education, psychology, business, industry, and the natural sciences involve several groups or treatments. For example, we may want to decide on the basis of sample data if there is a significant difference in the effectiveness of three methods for teaching statistics, or we may want to determine if there really is a difference in the lifetime of four brands of batteries.

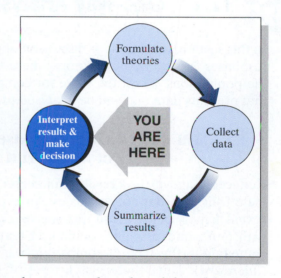

Suppose that the three teaching methods are as follows: Method (1), which uses a noninteractive standard text in a large lecture format classroom; Method (2), which uses an interactive text with group work in the classroom; and Method (3), which uses self-paced, computer-based modules with no lecture. The effectiveness of the three methods will be measured based on comparing mean test scores at the end of the term. The two theories to be tested would be

H_0: There is no difference between the three teaching methods with respect to the population mean test score—that is, all three population mean test scores are the same.

H_1: There is a difference between the three teaching methods with respect to the population mean test score—that is, at least one of the three population mean test scores is different from the others.

Even when a decision is made, there may be new questions to address, new theories to be tested, and more data to be obtained and analyzed. We would continue to cycle through the steps of the scientific method. Suppose that the data are found to be statistically significant.

We have evidence that at least one of the population means is different, but we would like to know more. Which mean or means are different and how different? Is there one method that appears to have a higher mean test score when compared to the other two methods?

In Chapter 11, we studied the technique for comparing the means of two treatments or populations on the basis of independent samples. In the two independent samples design, we have two unrelated sets of units that are measured, one sample from each of the two populations. We tested hypotheses and developed a confidence interval estimate for the difference between the two population means $\mu_1 - \mu_2$. In this chapter, we focus on a technique for assessing if there is a significant difference between two or more population means. We will learn how to use the results of sets of data to decide something about the population means. The procedure for determining whether there are differences between the population means is part of a technique called **analysis of variance**, often referred to by its acronym **ANOVA**. There are actually several different kinds of analysis of variance. In Section 12.2, we begin discussing the ANOVA procedure called **one-way ANOVA**. The *"one-way"* refers to having only one explanatory variable with at least two, but usually more, levels. While most of this chapter will focus on the methods involved in one-way ANOVA, Section 12.7 will present an overview of **two-way ANOVA**, which examines the effect of two explanatory variables on the mean response. An important feature in two-way analyses is the possibility of **interaction** between the two explanatory variables.

12.2   One-Way Analysis of Variance

In this section, we begin our discussion of one-way ANOVA. We wish to compare the mean responses for several populations where the levels of a single explanatory variable define the populations. To set the stage for comparing population means, we turn to an example that presents data that will be analyzed using a one-way ANOVA later in this section.

❖ **EXAMPLE 12.1**   **Does Increased Reproduction Reduce the Lifespan of Male Fruit Flies?**_____

REAL
DATA

Reduced longevity as a result of increased reproduction has been shown for female fruit flies, but not for males. An experiment was conducted in which the sexual activity of male fruit flies was manipulated by supplying individual males with none, one, or eight companion females per day. Of those male fruit flies that had companions, the type of companion was either newly pregnant (*nonreceptive*) or virgin (*interested*). Thus, there was a total of five groups: Group 1 = no companion, Group 2 = one newly pregnant companion, Group 3 = one virgin companion, Group 4 = eight newly pregnant companions, and Group 5 = eight virgin companions. There were 25 males randomized to each of the five groups. All 125 male fruit flies were treated identically with respect to the number of anesthetizations and the provision of fresh food medium. Responses measured for each fly were as follows: longevity in days, thorax length in millimeters, and percentage of day spent sleeping.

We are interested in comparing the distributions of longevity between the various groups. Do active males, those with eight virgin companions per day, have shorter lifespans on average as compared to nonactive males with no companion or compared to less active males who only have one companion per day? Is there a difference in longevity for males with newly pregnant companions versus virgin companions? How does thorax length and time sleeping relate to longevity? There are many hypotheses that could be examined with these data. We begin with the basic question: Is there a difference in the average lifetime for the five groups or populations? One of the first steps in any data analysis is to examine the data graphically. Figure 12.1 shows the side-by-side boxplots for longevity in days for the five groups.

In terms of the medians, those male fruit flies that were supplied eight interested companions per day had shorter lifespans than any other group. Also, interested companions, Groups 3 and 5, led to a lower median lifespan as compared to no companions, Group 1, and nonreceptive companions, Groups 2 and 4. But are these descriptive differences significant?

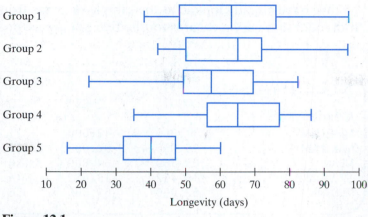

Figure 12.1

If we let μ_i represent the average lifetime for male fruit flies from population i, then we are interested in testing the following hypotheses:

H_0: $\mu_1 = \mu_2 = \mu_3 = \mu_4 = \mu_5$.

H_1: The population means are not all equal.

A natural first step in assessing the equality of the population means would be to examine the sample means. The data are shown in the following table:

Group 1	Group 2	Group 3	Group 4	Group 5
40	46	21	35	16
37	42	40	37	19
44	65	44	49	19
47	46	54	46	32
47	58	36	63	33
47	42	40	39	33
68	48	56	46	30
47	58	60	56	42
54	50	48	63	42
61	80	53	65	33
71	63	60	56	26
75	65	60	65	30
89	70	65	70	40
58	70	68	63	54
59	72	60	65	34
62	97	81	70	34
79	46	81	77	47
96	56	48	81	47
58	70	48	86	42
62	70	56	70	47
70	72	68	70	54
72	76	75	77	54
75	90	81	77	56
96	76	48	81	60
75	92	68	77	44

We have already looked at the sample medians from the boxplots for the five groups. Additional summary measures for the five groups are provided as follows:

```
TOTAL OBSERVATIONS:      125
                     LONG1        LONG2        LONG3        LONG4        LONG5
N OF CASES              25           25           25           25           25
MINIMUM             37.000       42.000       21.000       35.000       16.000
MAXIMUM             96.000       97.000       81.000       86.000       60.000
MEAN                63.560       64.800       56.760       63.360       38.720
STANDARD DEV        16.452       15.652       14.928       14.540       12.102
MEDIAN              62.000       65.000       56.000       65.000       40.000
```

The sample mean lifetime for Group 5 males who had eight interested companions per day was only 38.72 days, as compared to 63.56 days for Group 1 males that were forced to live alone. The sample means are not all equal, but there is variation in these sample measurements. How can we assess if there is enough evidence to say that the mean longevities for male fruit flies for these five populations are not all equal?   ❖

As seen in Example 12.1, the null hypothesis states that all populations have the same mean—that is, the population means are all equal. The alternative hypothesis is that the population means are not all equal. This alternative hypothesis does *not* imply that each population mean must be different from any other of the population means. Indeed, it could be that the first population mean is different and the other remaining population means are equal to each other. So, another way to state the alternative hypothesis is to say *at least one population mean is different*. If we let *I* represent the number of populations under study, then the generic one-way ANOVA hypotheses can be expressed as follows:

H_0: $\mu_1 = \mu_2 = \ldots = \mu_I$.
H_1:   The population means are not all equal.

So we use a technique called analysis of *variance* to test a null hypothesis that talks about equality of the *means*. Indeed while the aim of ANOVA is to assess if there are differences among several population means, the method involves comparing two measures of variation in the sampled data. Let's turn to an example with some pictures to help us see the logic behind this analysis of variance.

❖ EXAMPLE 12.2   One-Way ANOVA Logic

The two figures that follow provide frequency plots for data obtained by taking independent random samples of size 10 from three populations. The three populations each had a normal distribution (normality is one of the assumptions of ANOVA) and the population means were 60, 65, and 70, respectively. So the population means are indeed not all equal. In Scenario I (Figure 12.2), the population standard deviations were all equal to 1.5. In Scenario II (Figure 12.3), the population standard deviations were all equal to 3. Another assumption for ANOVA is that the populations have equal standard deviations. We will discuss the assumptions for performing a one-way ANOVA in the next section.

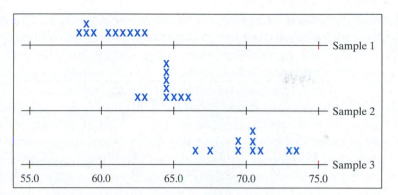

Figure 12.2
Scenario I

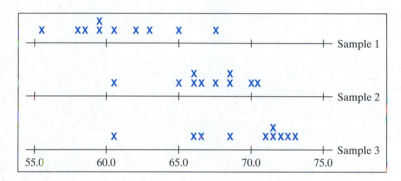

Figure 12.3
Scenario II

 THINK ABOUT IT | **Scenario I versus Scenario II**

There is an obvious difference between Scenario I and Scenario II.
What is that difference?

Just looking at the frequency plots, which of the two scenarios do you think would provide more evidence that at least one of the population means is different from the others?
Scenario I   or   Scenario II

Explain.

The summary measures, produced using MINITAB, for all six samples are shown as follows:

Scenario I	N	MEAN	MEDIAN	STDEV	MIN	MAX	Q1	Q3
Sample 1	10	60.6	60.7	1.6	58.6	62.8	59.1	62.2
Sample 2	10	64.5	64.7	1.1	62.3	66.1	64.0	65.1
Sample 3	10	70.2	70.3	2.2	66.3	73.3	69.0	71.6
Scenario II	N	MEAN	MEDIAN	STDEV	MIN	MAX	Q1	Q3
Sample 1	10	60.9	60.3	3.5	55.7	67.3	58.2	63.6
Sample 2	10	66.9	66.9	2.9	60.6	70.3	65.8	69.1
Sample 3	10	69.4	71.1	3.7	61.2	73.2	66.5	72.2

Notice first that each sample mean is not exactly equal to the mean of the population from which the sample was generated (recall that Population 1 mean was 60, Population 2 mean was 65, and Population 3 mean was 70). Also note that the two Sample 1 means of 60.6 and 60.9 were not equal to each other nor equal to the population mean of 60. Next, compare the sample standard deviations in Scenario I to those in Scenario II. The samples in Scenario I were generated from populations whose natural variation within each population was smaller compared to the natural variation within each population in Scenario II. For each scenario, even though the population standard deviations were equal, the sample standard deviations were not exactly equal, but they were comparable.

So the sample means do vary, from about 60 to about 70, for each scenario. There is a good deal of variation between the sample means. In Scenario I, the variation between the sample means is more noticeable, since the natural variation within the samples is smaller. Remember that in general we would not know the population parameters — that is, we would not know the true population means and the true population standard deviation. However, by examining the data in Scenario I, we do see strong evidence that the population means are different. In contrast, for Scenario II, the variation between the sample means is not as noticeable, since the natural variation within the samples is larger. Even if we did not know the values of the population parameters, the data in Scenario II do *not* provide as much evidence that the population means are different. It is not clear whether the observed differences in the sample means in Scenario II is due to the differences in the population means or to the larger natural variation within the populations.

Finally, in Figure 12.4, we have frequency plots for three independent random samples of size 10 each taken from a population with a normal model with a population mean of 65 and population standard deviation of 1.5. So in this Scenario III, the population means are indeed all equal—that is, the null hypothesis tested in one-way ANOVA is true.

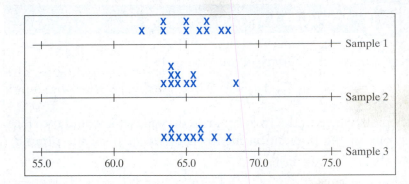

Figure 12.4
Scenario III

Notice that, although the population means were all equal, there is still a small amount of variation between the sample means. In Scenario III, there is still natural variation within the samples and the slight variation between the sample means is hardly noticeable. The data in Scenario III do *not* provide evidence that the population means are different. ❖

As seen in Example 12.2, the decision about equality of the population means will be based on examining the variability between the sample means. This *between variability* will be contrasted to how much *natural variation there is within the groups*. The test statistic actually used to make the decision is called an ***F*-statistic** and can be loosely viewed as follows:

$$F = \frac{\text{Variation BETWEEN the sample means}}{\text{Natural Variation WITHIN the samples}}.$$

THINK ABOUT IT 🐢

> **Case A:** If all the sample means were exactly the same, what would be the value of the numerator of the F-statistic?
>
> **Case B:** If all the sample means were very spread out and very different from each other, how would the variation between the sample means compare to the value in Case A?
>
> So what values could the F-statistic take on? Could you get an F-statistic that is negative? What type of values of the F-statistic would support the alternative hypothesis (would lead you to reject the null hypothesis)?

The F-statistic will be sensitive to differences between the sample means. The larger the variation between the sample means, the larger the value of the F-statistic. Since larger values of the F-statistic provide more support for rejecting the null hypothesis (more support toward the alternative hypothesis), we have that the ***direction of extreme*** for our one-way ANOVA F-test is ***one-sided to the right***. As we shall see in Section 12.4, the p-value for an F-test can be found using common tables, or with some statistical calculators and statistical computing packages. As with any statistical test, the smaller the p-value, the more evidence against the null hypothesis. If the p-value is smaller than the preset significance level α, we would conclude that the data support that the populations do not all have the same mean.

❖ EXAMPLE 12.3   One-Way ANOVA Logic Revisited

Recall the three scenarios that were presented in Example 12.2.

> ***Scenario I:*** Independent random samples from three normal populations with population means of 60, 65, and 70, respectively, and population standard deviations all equal to 1.5.
>
> ***Scenario II:*** Independent random samples from three normal populations with population means of 60, 65, and 70, respectively, and population standard deviations all equal to 3.
>
> ***Scenario III:*** Independent random samples from three normal populations with population means all equal to 65 and population standard deviations all equal to 1.5.

The variation between the sample means was greatest for Scenarios I and II compared to Scenario III. The natural variation within the samples was greatest for Scenario II compared to Scenarios I and III. So which of the three scenarios would you expect to result in the largest value of the F-statistic? The following table provides the values of the F-statistic for the one-way ANOVA test of equality of the population means (calculation of the F-statistic is discussed in more detail in Section 12.5):

Scenario	Value of the F-statistic for testing H_0: $\mu_1 = \mu_2 = \mu_3$	p-value
Scenario I H_1 is true	$F \cong 80.4$	**p-value** $\cong 0$
Scenario II H_1 is true	$F \cong 16.4$	**p-value** $\cong 0.01$
Scenario III H_0 is true	$F \cong 0.17$	**p-value** $\cong 0.84$

Note the value of the F-statistic is smallest and the p-value the largest when the null hypothesis is true (Scenario III). For Scenarios I and II, the population means are different, but the smaller population standard deviation in Scenario I accentuates the differences by producing a larger F-ratio and an extremely small p-value. The larger value of the F-statistic corresponds to more evidence (based on the data) that the population means are not all equal. ❖

Let's do it! 12.1 **Would We Reject the Null Hypothesis?**

Two sets of side-by-side boxplots are shown. Each set represents the boxplots based on independent random samples selected from three normal populations with possibly different population means, but equal population standard deviations. Assume that the response axis for each set of boxplots has the same scale, so that the two sets can be readily compared. Although the sample medians shown in the boxplot may not be exactly equal to the corresponding sample means, they do reflect the center of the distribution for each group. Similarly, the sample IQRs give a reflection of the variability within each group. In answering the accompanying questions, remember that the *F*-test in ANOVA is based on comparing the variation between the sample means to the natural variation within the samples.

Set I

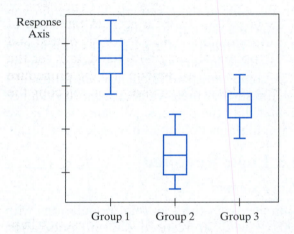

Set II

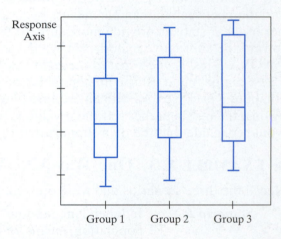

(a) Based on the boxplots in Set I, do you think the null hypothesis H_0: $\mu_1 = \mu_2 = \mu_3$ will be rejected using a one-way ANOVA *F*-test? Explain your answer.

(b) Based on the boxplots in Set II, do you think the null hypothesis H_0: $\mu_1 = \mu_2 = \mu_3$ will be rejected using a one-way ANOVA *F*-test? Explain your answer.

> *Definition:* **One-way Analysis of Variance (ANOVA)** is a statistical technique for comparing the means for several populations, where the levels of a single explanatory variable define the populations.
>
> **The basic concept underlying ANOVA**: To assess whether the population means are equal, the variation between the sample means is compared to the natural variation of the observations within the samples through their ratio, which is called the *F*-**statistic**. The larger the **variation between the sample means** compared to the **natural variation within the samples**—that is, the larger the value of the *F*-statistic—the more support there is of a difference in the population means.

In Section 12.3, we step back and discuss the assumptions that are needed in order for the *F*-test results to be valid. In Section 12.4, we will discuss the distribution of the *F*-statistic, called the *F*-distribution, and see how to find the *p*-value for testing the hypothesis of equality of the population means. We will let the computer or calculator do the work of computing the observed *F*-statistic value and focus on understanding the basic output and interpreting the results. Section 12.5 provides the (optional) computational details for the *F*-statistic. Section 12.6 will discuss why we need an ANOVA technique and the procedure involving ***multiple comparisons*** that is performed should the decision result in rejecting the null hypothesis.

12.3 Assumptions in One-Way ANOVA

One-way ANOVA is basically an extension of the two independent samples *t*-test of Chapter 11. Thus, the assumptions required for performing the *F*-test are simply the assumptions for the two independent samples *t*-test extended to the *I* populations. Before we continue with the assumptions, let's look at the notation that will be standard in ANOVA for representing information about the observed data and corresponding summary statistics.

Notation in ANOVA

I = number of groups or populations under study (the number of levels of the explanatory variable).

n_i = sample size for the *i*th sample (the number of observations in the *i*th group).

n = total sample size across all *i* samples ($n = n_1 + n_2 + \cdots + n_I$).

$\bar{x}_i$ = observed sample mean for the *i*th sample (the average of the observations in the *i*th group).

s_i = observed sample standard deviation for the *i*th sample (the standard deviation of the observations in the *i*th group).

We assume that each set of observations is a simple random sample from a normal population with possible different population means, but the same population standard deviation, and that the samples are independent. A summary of the assumptions in words and in picture form is provided next.

Assumptions for ANOVA

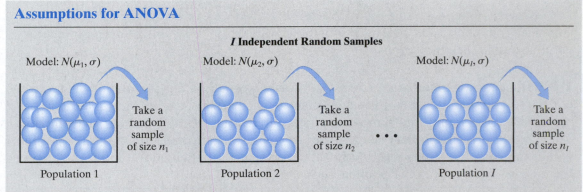

_I_ Independent Random Samples

Model: $N(\mu_1, \sigma)$ Model: $N(\mu_2, \sigma)$ Model: $N(\mu_I, \sigma)$

Take a random sample of size n_1

Take a random sample of size n_2

Take a random sample of size n_I

Population 1   Population 2   Population I

We have a simple **random sample** of n_1 observations from a $N(\mu_1, \sigma)$ population.

We have a simple **random sample** of n_2 observations from a $N(\mu_2, \sigma)$ population.

⋮

We have a simple **random sample** of n_I observations from a $N(\mu_I, \sigma)$ population. The I random samples are **independent** of each other.

Note: The **population standard deviations** are all assumed to be **equal** to σ, called the **common population standard deviation**.

Screening for errors in the data set and checking for departures from the ANOVA assumptions should be done *before* the analysis is performed. An unusually small or large sample mean or sample standard deviation could indicate a possible outlier. Examining the magnitude of the sample standard deviations or comparing the interquartile ranges (with side-by-side boxplots) can provide a simple check on the common population standard deviation assumption. Unequal population standard deviations are less serious if you have a balanced design (that is, an equal number of observations in each treatment group). A check for normality could be performed by examining a histogram for each sample of data. If the normality assumption is not valid, a transformation of the data might help. If the sample sizes are large enough, the assumption of normal populations is less crucial. The assumption of independent samples is often ensured by random allocation of the units to the treatment groups. Lack of independence can be the most serious violation and invalidate the level of significance, thus proper randomization is crucial to performing a valid analysis.

Recall that the hypotheses for comparing the I population means are given by

$$H_0: \quad \mu_1 = \mu_2 = \mu_3 = \cdots = \mu_I.$$
$$H_1: \quad \text{The population means are not all equal.}$$

In the null hypothesis, we are saying that all of the treatments have the same effect on the mean response. Pictures of the population distributions under the null and alternative hypotheses are shown next in the case of $I = 3$ populations. Notice that under the null hypothesis there is just one overall population with one common mean and standard deviation. Under the alternative hypothesis, there are many possible pictures. In this H_1 picture, the first two populations have the same mean and the third population mean is different.

H_0: **The population means are all equal.** H_1: **At least one population mean is different.**

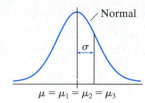

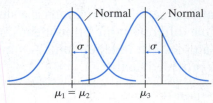

❖ **EXAMPLE 12.4   Assessing the Assumptions for the Lifetime of Male Fruit Flies Data**

REAL
DATA

Recall that Example 12.1 described a study to compare the longevity for male fruit flies for five different groups, in which the groups varied in number and type of female fruit-fly companions. Side-by-side boxplots for the five groups were presented in Figure 12.1 and the data with summary measures were also provided. Although boxplots can hide the actual shape of the distribution, and symmetric boxplots cannot "prove" normality, boxplots can be used to check for evidence of departures from normality, such as the presence of outliers and nonsymmetry. The equal population standard deviation assumption can be assessed by comparing the lengths of the boxes—that is, the interquartile ranges (IQRs).

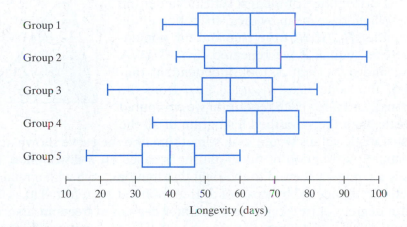

The corresponding boxplots do not indicate any observations as being outliers, and there is no evidence of strong skewness for any of the groups. The standard deviations are all comparable, ranging from 12.1 to 16.5, and the IQRs are also similar. Thus, the assumption of equal population standard deviations seems reasonable. We also have a randomized balanced design with a moderate sample size of 25 observations in each group. To better assess normality for each population we might examine a histogram for each of the five sets of data.   ❖

As you begin to work with real data, you will discover that the assumptions for performing an ANOVA *F*-test may not always be valid. In Chapter 15, we discuss methods that can be used when one or more of the assumptions about the population models in ANOVA are violated. The Chapter 15 methods are called ***nonparametric*** as there are no assumptions made about a specific distribution for the population of responses. Two nonparametric tests for comparing several treatments are the **Kruskal–Wallis** test and the **Mood's Median** test. Finally, the data can be transformed so that the ANOVA assumptions are reasonable for the transformed data. However, interpretation of the results is often difficult as they apply to the transformed data, not the original data.

12.4   The *F*-distribution

Recall that the test statistic in ANOVA is essentially the ratio of two measures of variation in the sample data. The variation *between* the sample means is compared to the natural variation of the observations *within* the samples through their ratio, called the *F*-statistic. Formally, these two measures of variation are called **mean squares**, with the numerator referred to as the **mean square between the groups (MSB)** and the denominator referred to as the **mean square within the groups (MSW)**.

$$F = \frac{\text{Variation BETWEEN the sample means}}{\text{Natural Variation WITHIN the samples}} = \frac{MSB}{MSW}$$

The larger the variation between the sample means (MSB), as compared to the natural variation within the samples (MSW), the more support there is for a difference in the population means. Since only larger values for our test statistic support the alternative hypothesis, all ANOVA *F*-tests are *one-sided tests* with the direction of extreme being *to the right*. The *p*-value will be the probability of observing a test statistic as large or larger than the observed *F*-test statistic value, assuming the null hypothesis is true. The *p*-value would then be compared to the significance level, α, to decide which hypothesis to support.

As in previous chapters, to determine the *p*-value we need to know the distribution of the test statistic under the null hypothesis. What do we mean by the distribution of the *F*-test statistic under H_0? If H_0 were true and the population means were all equal, if we repeated this ANOVA procedure over and over (that is, took independent random samples of the same sizes $n_1, n_2, \ldots, n_I$, from the same populations), and if for each repetition we computed the *F*-test statistic, then the smoothed histogram of all the resulting *F*-test statistic values would look something like the curve shown at the right.

The probability distribution of the *F*-statistic is called an **F-distribution**. The family of *F*-distributions is a family of skewed to the right distributions, each with a minimum value of zero. *F*-distributions are indexed by a pair of degrees of freedom, referred to as the numerator and denominator degrees of freedom. The **numerator degrees of freedom**, associated with the MSB, are computed as $I - 1$ (number of groups -1). The **denominator degrees of freedom**, associated with the MSW, are computed as $n - I$ (number of observations $-$ *number* of groups). So in one-way ANOVA, the *F*-distribution is written as $F(I - 1, n - I)$. Note that the order of the degrees of freedom is important—the $F(2, 15)$ distribution is not the same as the $F(15, 2)$ distribution. A summary of the properties of an *F*-distribution is given below along with a graph of two different *F*-distributions.

Properties of the *F*-distribution *F*(numerator DF, denominator DF)

- The total area under the curve is one (as it is a density curve).
- The distribution is skewed to the right.
- The values are nonnegative, start at zero and extend to the right—the curve approaches, but never touches, the horizontal axis.
- There is a different *F*-distribution for each different set of degrees of freedom.

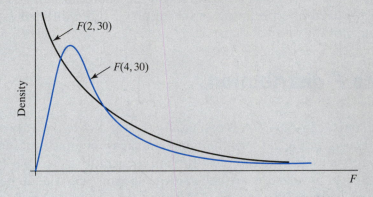

Distribution of the *F*-statistic for One-Way ANOVA

If H_0 is true, then the test statistic, $F = \dfrac{MSB}{MSW}$, follows an $F(I - 1, n - I)$ distribution.

Note: We will use $F(I - 1, n - I)$ to denote an *F*-distribution. We will use the English letter $F = \frac{MSB}{MSW}$ when referring to the general test statistic, and F_{OBS} will represent an observed test statistic value.

Many calculators have the ability to compute areas under an *F*-distribution. The TI graphing calculators can draw an *F*-distribution, can compute areas under an *F*-distribution, and have an option under the STAT TESTS menu that allows you to perform the one-way ANOVA test of equality of the population means. Details are provided in the ■ TI Quick Steps at the end of this chapter. Table V, also presented at the end of this chapter, provides us with the 90th, the 95th, and the 99th percentiles for various *F*-distributions. Our next example will demonstrate finding a *p*-value for an ANOVA *F*-test using both the TI graphing calculator and Table V.

❖ **EXAMPLE 12.5** **Working with the *F*-Distribution** _____

A marketing research company must select between $I = 3$ different ads, Ad A, Ad B, and Ad C, for promoting a new product. A study was conducted to assess the effectiveness of the three ads. Of the $n = 43$ subjects available for the study, 14 were shown Ad A, 14 were shown Ad B, and the remaining 15 were shown Ad C. A measure of effectiveness was obtained for each subject based on each subject's responses to questions on a questionnaire. A total score for each questionnaire was computed, where a larger score indicates the subject looked more favorably on the product. These data and corresponding summaries are provided in the following table:

		Mean	Standard Deviation
Ad A	16 17 26 33 11 26 28 5 52 33 36 8 28 20	24.2	12.5
Ad B	33 37 41 44 18 25 38 14 46 38 26 25 13 18	29.7	11.3
Ad C	8 16 11 28 22 41 23 31 22 9 15 19 26 14 24	20.6	8.9

Is there evidence at a 5% level to conclude that any one ad is more effective than the other two ads? The hypotheses for this example are

$$H_0: \quad \mu_1 = \mu_2 = \mu_3.$$
$$H_1: \text{ The three population means are not all equal.}$$

Suppose that the ANOVA assumptions hold and the MSB = 303.7 and the MSW = 120.4. Then the observed value of the *F*-test statistic is

$$F_{OBS} = \frac{MSB}{MSW} = \frac{303.7}{120.4} = 2.52.$$

We want to measure the chance of getting an *F*-statistic value of 2.52 or larger under the null hypothesis of equality of the population means. The distribution for the test statistic under H_0 is an F-distribution with $I - 1 = 2$ and $n - I = 43 - 3 = 40$ degrees of freedom. The following picture portrays this chance, the p-value, as the area under the F-distribution to the right of the observed value of 2.52.

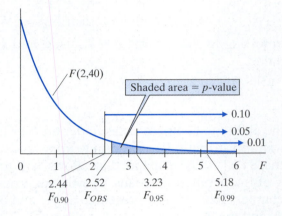

Using the TI

The TI has a function that is built into the calculator for finding areas under any F-distribution. This function is located under the DISTR menu and allows you to specify the set of degrees of freedom. The TI steps to compute the area to the right of 2.52 under the $F(2, 40)$ distribution are shown.

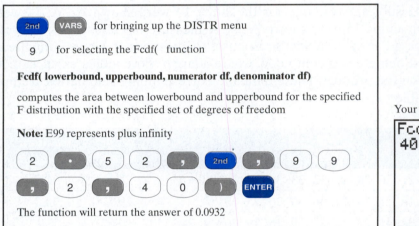

Using Table V

Table V covers three pages (739 to 741) to provide three percentiles (the 90th, the 95th, and the 99th) for various F-distributions. For each table, the columns designate the numerator degrees of freedom (in one-way ANOVA this is $I - 1$) and the rows designate the denominator degrees of freedom ($n - I$ in one-way ANOVA). Recall that the 90th percentile of a distribution would be the value (on the horizontal axis) such that the area to the left of that value is 0.90, or equivalently, the area to the right of that value is 0.10. From Table V for an $F(2, 40)$ distribution, we have that the 90th percentile is the value 2.44 (written as $F_{0.90} = 2.44$ or $F_{0.90}(2, 40) = 2.44$). Note that our observed F-statistic of 2.52 is a little larger than this percentile. The other two percentiles for this F-distribution are $F_{0.95} = 3.23$ and $F_{0.99} = 5.18$, and our test statistic value is smaller than both of these percentiles. Be sure you can find these percentiles from Table V yourself. These percentiles and their corresponding areas to the right are shown on the previous picture, along with the *p*-value area. Since the F_{OBS} value of 2.52 is larger than the 90th percentile, but smaller than the 95th (and then of course the 99th) percentile, using only Table V we would report that the *p*-value is between 0.05 and 0.10.

Decision and Conclusion

At a 5% significance level we would accept H_0. We conclude that it appears that the three ads are equally effective. ❖

Let's do it! 12.2 Working with the *F*-Distribution

An ANOVA was performed to test the equality of $I = 4$ population means. Independent random samples of size seven were obtained from each of the four populations. The *F*-test resulted in an observed test statistic value of 3.28. You are asked to find the *p*-value for this test and make a decision using a 5% significance level. Recall that all *F*-tests are one-sided tests with the direction of extreme being to the right. So the *p*-value will be the probability of observing a test statistic value as large or larger than the observed *F*-statistic value, assuming that the null hypothesis is true.

(a) Complete the picture with the proper labeling for the distribution under H_0 and shade the area corresponding to the *p*-value.

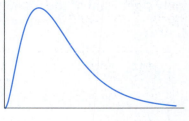

(b) Give the degrees of freedom.

Numerator degrees of freedom $= I - 1 =$ _____.

Denominator degrees of freedom $= n - I =$ _____.

(c) Find the *p*-value for this test.

(d) Are the results statistically significant at the 5% level? Yes   No
Explain.

As will be shown in Section 12.5, one-way analysis of variance is an involved technique, and the computations, if done by hand, are quite time consuming. In our next example, we will let the computer do the computations and provide the observed *F*-statistic value and the corresponding *p*-value.

❖ EXAMPLE 12.6 Comparing Three Advertisements

Example 12.5 described a market research study to compare the mean effectiveness of three different ads for promoting a new product. The response was the total score on a questionnaire as the measure of effectiveness, with a larger score implying "more effective." We want to assess whether there is a significant difference in the mean scores for the three different ads using a 5% significance level. From Example 12.5, we already know that the observed *F*-statistic was 2.52 and the resulting *p*-value is more than 0.05. The calculations leading up to the *F*-statistic are summarized in what is called an **ANOVA table**. The ANOVA table produced using the *Oneway ANOVA* command in MINITAB is shown.

```
MTB > Oneway 'score' 'ad'.

ANALYSIS OF VARIANCE ON score
SOURCE      DF        SS        MS         F         p
ad           2       607       304      2.52     0.093
ERROR       40      4815       120
TOTAL       42      5422
```

The two sources of variation are called **ad**, for *between the ad groups*, and **ERROR**, for the *within group variation*. The SS (sum of squares), DF (degrees of freedom), and MS (mean squares) are provided, which lead to the test statistic value of $F_{OBS} = \frac{MSad}{MSERROR} = \frac{304}{120} = 2.52$. The computational details of the sum of squares and mean squares are provided in Section 12.5. Is the value of 2.52 large enough to support the alternative hypothesis? From the computer output, the actual *p*-value is shown to be 0.093, which agrees with Example 12.5. Thus, at the 0.05 significance level, we would accept H_0 and conclude that it appears that the three ads are equally effective. The data are not statistically significant at the 5% level.

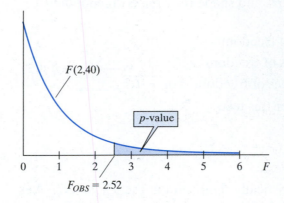

In order for the ANOVA *F*-test to be valid, certain assumptions must be met. We must *first* examine the data to check these assumptions. The side-by-side boxplots show one large outlier in Ad C scores. Apart from this value, there are no strong nonnormal patterns evident. The comparable IQRs (length of the boxplots) and similar sample standard deviations appear to support the equal population standard deviation assumption.

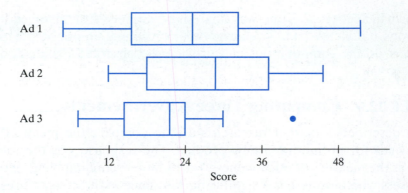

Let's do it! 12.3 Lifetime of Male Fruit Flies Revisited

Recall that Example 12.1 described a study to compare the longevity for male fruit files for five different groups, where the groups varied in number and type of female fruit fly companions. Side-by-side boxplots for the five groups were presented and the data with summary measures. The corresponding ANOVA computer output is as follows:

```
          - - - - -   O N E W A Y   - - -
Variable LONGEVIT  By Variable GROUP
                     Analysis of Variance
                       Sum of        Mean        F        F
        Source      D.F.   Squares     Squares     Ratio    Prob.
Between Groups        4    11939.280   2984.820   13.612    .000
Within Groups       120    26313.520    219.279
Total               125    38252.800
```

Based on these data, is there evidence of a difference in the mean longevity for the five populations of male fruit flies at the 1% significance level? Give support to your answer by providing relevant summaries and performing the actual test.

H_0:

H_1:

Observed test statistic: F_{OBS} = _____. p-value = _____.

Decision using $\alpha = 0.01$ and conclusion:

 ## Exercises

12.1 What is the main principle behind analysis of variance? That is, explain why we call a technique for testing equality of the population *means*, analysis of *variance* (ANOVA)?

12.2 A survey of college students was conducted to determine how much money they spend per year on textbooks and to compare students by year of study. From lists of students provided by the register, random samples of size 16 were chosen from each of the four classes (freshmen, sophomores, juniors, seniors). The students selected were asked how much they spent on textbooks during the previous year. A one-way ANOVA will be performed on the data.

(a) State the appropriate hypotheses.

(b) What are the assumptions for a one-way ANOVA?

(c) Suppose that the assumptions for a one-way ANOVA hold. What is the distribution of the F-test statistic under the assumption that H_0 is true?

12.3 A new computer software package has been developed to help systems analysts reduce the time required to design and implement an information system. To evaluate the benefits of the new software package, 33 systems analysts were randomly divided into three groups of 11 each. All 33 analysts were given specifications for a hypothetical information system. Group 1 of 11 analysts was instructed to produce the information system using current software package 1. Group 2 of 11 analysts was instructed to produce the information system using current software package 2. Group 3 of 11 analysts was first trained in the use of a new software package and then instructed to use it to produce the information system. The researcher wants to assess whether or not the three software packages provide equal mean completion times using an ANOVA—that is, to test H_0: $\mu_1 = \mu_2 = \mu_3$ versus H_1: The population means are not all equal.

(a) Sketch a picture of the distribution for time to completion for the three populations, first under the null hypothesis and then under the alternative hypothesis. Be sure to include appropriate labels for the distributions.

(b) Sketch a picture of the distribution of the test statistic under the null hypothesis. Be sure to include the appropriate label for the distribution.

12.4 Consider the following notation for a percentile for an F-distribution: $F_{0.95}(4, 10)$.

(a) In the notation $F_{0.95}(4, 10)$, the value 4 represents (select one)

(i) the degrees of freedom for the numerator.

(ii) the degrees of freedom for the denominator.

(iii) the p-value.

(b) In the notation $F_{0.95}(4, 10)$, the value 10 represents (select one)

(i) the degrees of freedom for the numerator.

(ii) the degrees of freedom for the denominator.

(iii) the p-value.

(c) The value for this percentile is $F_{0.95}(4, 10) = 3.48$. Sketch the $F(4, 10)$ distribution and mark the approximate location for the value of 3.48 on the horizontal axis. Shade in the area under this density to the right of 3.48. What is the value of this area to the right?

12.5 Find the numerical value for each of the following percentiles:

(a) $F_{0.95}(6, 16)$.

(b) $F_{0.95}(3, 25)$.

(c) $F_{0.99}(5, 10)$.

(d) $F_{0.99}(1, 22)$.

(e) $F_{0.90}(4, 16)$.

(f) $F_{0.90}(7, 30)$.

12.6 For each of the following scenarios, report the p-value for the one-way ANOVA F-test:

(a) The observed F-statistic value is 3.2 and the null distribution is the $F(6, 15)$ distribution.

(b) The observed F-statistic value is 1.7 and the null distribution is the $F(5, 10)$ distribution.

(c) The observed *F*-statistic value is 7.0 and $I = 3$ and $n = 24$.

(d) The observed *F*-statistic value is 7.0 and six subjects were randomly assigned to each of the four treatment groups.

12.7 An ANOVA to compare various methods for washing a particular type of fabric resulted in $F_{OBS} = 25.4$, based on 3 and 60 degrees of freedom.

(a) How many washing methods are being compared?

(b) How many observations (that is, pieces of fabric) were used in this study?

(c) Are the results statistically significant at a 1% significance level? Explain.

12.8 Consider the three ads study in Examples 12.5 and 12.6.

(a) What would you recommend that the researcher do with the data to check the assumptions for an ANOVA?

(b) Suppose that 15 subjects were used in each group, for a total of 45 subjects (instead of 43). What are the corresponding degrees of freedom now? Table V only has entries for 40 and then 60 degrees of freedom for the denominator. In general, if the degrees of freedom are not in the table, using the smaller degrees of freedom will result in conservative results.

12.9 An experimenter was interested in studying the effect of sleep deprivation on the ability to detect moving objects on a radar screen. A total of 20 subjects were available for the study. The first five randomly selected subjects were assigned to Treatment Group 1 (4 hours without sleep), the next five randomly selected subjects were assigned to Treatment Group 2 (12 hours without sleep), the next five randomly selected subjects were assigned to Treatment Group 3 (20 hours without sleep), and the remaining five subjects were assigned to Treatment Group 4 (28 hours without sleep). After being deprived of sleep, each subject was given a test and the number of times they failed to spot a moving object during the set test period was recorded. The results are given in the following table:

4 Hours	12 Hours	20 Hours	28 Hours
36	38	46	76
21	45	74	66
20	46	67	62
26	22	61	44
21	40	59	61

(a) Is this study an experiment or an observational study?

(b) Give a name for the explanatory variable in this study.

(c) Give a name for the response variable in this study.

(d) It stated that the 20 subjects were randomly assigned to one of four treatment groups. Using your calculator (with a seed value of 14) or the random number table (Row 10, Column 1), perform the randomization. Report the labels of the subjects assigned to each of the four groups.

(e) State the appropriate null and alternative hypotheses for assessing whether the mean number of failures is the same for the four sleep-deprivation conditions.

12.10 A realtor wants to assess whether a difference exists between home prices in three subdivisions. Independent samples of homes from each of the three subdivisions are obtained and their prices are recorded. The SPSS analysis of variance output for comparing these prices is provided.

```
- - - - -  O N E W A Y  - - - - -
Variable PRICE By Variable SUBDIV
                        Analysis of Variance
                        Sum of        Mean        F       F
Source             D.F. Squares       Squares     Ratio   Prob.
Between Groups      2   157.4444      78.7222     4.6581  .0267
Within Groups      15   253.5000      16.9000
Total              17   410.9444
```

(a) How many homes were sampled in total?

(b) Give the hypotheses for assessing whether there is a difference in the mean prices for the three subdivisions.

(c) If the null hypothesis is true, then the appropriate model for prices for the three populations is (select one)

   (i) $N(0, 1)$ distribution.   (ii) $N(\mu, \sigma)$ distribution.   (iii) $F(2, 15)$ distribution.

(d) If the null hypothesis is true, then the appropriate model for the test statistic is (select one)

   (i) $N(0, 1)$ distribution.   (ii) $N(\mu, \sigma)$ distribution.   (iii) $F(2, 15)$ distribution.

(e) Test the hypotheses in (b). Report the observed test-statistic value, the p-value, and your conclusion using a 1% significance level.

12.11 The number of units of production was recorded for a random sample of 10 hourly periods from the three bottling assembly lines of a plant. A one-way ANOVA was performed on these data to assess whether there are any differences between the mean production for the three assembly lines.

```
Dep var:   UNITS   N:  30   Multiple R:  .529   Squared multiple R:  .280
Analysis of Variance
   Source    Sum-of-squares    DF  Mean-square    F-ratio       P
   LINE          1431.267       2    715.633       5.241      0.012
   Error         3686.900      27    136.552
```

(a) Based on this output, is there evidence at the 5% significance level to indicate any differences among the mean productions of the three lines? State the appropriate hypotheses and explain your answer.

The side-by-side box plots for these data are shown.

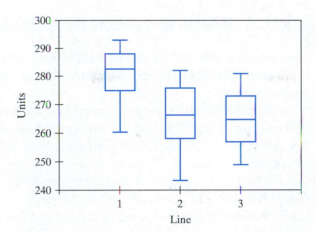

(b) Based on the sample results, fifty percent of the time, Assembly Line 1 produced more than or equal to _____ units.

(c) Briefly comment on the validity of the assumptions for performing this ANOVA.

12.12 A malfunctioning communication system within the brain might predispose some people to become alcoholics, a study says. The study also suggests that medications that help this system work better might help nonviolent alcoholics stop drinking but may not help violent ones. The study focused on a brain system in which brain cells signal each other by releasing bursts of a chemical called dopamine. Once each burst has done its job, it is pulled back into the signaling brain cell by a structure called a transporter. The study examined a part of the brain called the striatum to record the density of these dopamine transporters. Researchers scanned the brains of 19 habitually violent alcoholics whose alcoholism usually developed before age 25, and 10 nonviolent alcoholics, plus a comparison group of 19 nonalcoholics. The study found that the transporter density was markedly lower than normal in the nonviolent alcoholics. In the violent alcoholics, it was slightly higher than normal.

(a) What is the response variable in this study?

(b) How many populations are being compared in this study?

(c) What assumptions are required to perform an ANOVA on the responses for these three groups?

(d) If an ANOVA had been performed, what was the decision using a 5% significance level? Select one: reject H_0.   accept H_0.   don't know. Explain your answer.

(e) Comment on the following two theories with respect to the idea of cause and effect and whether the study results prove one theory over the other:

> **By researcher**: The reduced density in the nonviolent alcoholics was probably in place before the alcoholism began, and it may be a sign that the dopamine system is not working normally. The malfunctioning system may have produced a vulnerability to alcoholism.

> **Another alcohol expert**: The abnormality found in the study is a result of alcoholism rather than a sign of a predisposition to it.

 ## 12.5   Computational Details of One-Way ANOVA*

Nearly all statistical packages and enhanced statistical calculators have the ability to perform a one-way analysis of variance. In this section, we provide the *computational details for the F-statistic*. In general, it is not necessary to do the computations "by hand," but a look at the formulas will provide some deeper understanding about analysis of variance and the *F*-test.

* This section is optional.

12.5.1 How Do We Compute the *F*-statistic?

The one-way ANOVA *F*-statistic is the ratio of two measures of variability, the mean square between MSB and the mean square within MSW. Just how are these two mean squares computed?

Recall that when we have a single random sample from a population with an unknown variance σ^2, we estimate σ^2 by the sample variance s^2. The sample variance s^2 is computed by taking a sum of squared deviations about the mean (SS) and dividing by the degrees of freedom, $n - 1$.

The two measures of variability in ANOVA, the MSB and the MSW, also have this form. Mean squares are formed by taking a sum of squared deviations about a mean (SS) and dividing by a corresponding degrees of freedom (DF).

$$\text{Mean square (MS)} = \frac{\text{Sum of squares (SS)}}{\text{Degrees of freedom (DF)}}.$$

Each of the random samples, one from each of the *I* populations, gives rise to a sample mean and a sample variance, calculated in the usual way, and represented by $\bar{x}_1, \bar{x}_2, \ldots, \bar{x}_I$ and $s_1^2, s_2^2, \ldots, s_I^2$. These are just the observed sample mean and the observed sample variance for each sample. These statistics are used to compute the mean squares as shown in the next sections.

12.5.2 Measuring Variation Between the Groups (MSB)

Under the null hypothesis, the population means are all equal. If this hypothesis were true, it would be reasonable to average all of the observations to have an estimate of the overall common population mean. The overall sample mean, denoted by $\bar{x}$, can be found by averaging all of the *n* observations or by computing the weighted average of the sample means,

$$\bar{x} = \frac{n_1\bar{x}_1 + n_2\bar{x}_2 + \cdots + n_I\bar{x}_I}{n}.$$

In general, the sample means for the *I* groups will fluctuate around this overall sample mean. The variation of the sample means around the overall sample mean will also provide an estimate of σ^2. Since there are *I* sample means varying around the overall sample mean, the degrees of freedom associated with the MSB is $I - 1$. So the MSB is a measure of the variation **between** the sample means. The mean square between MSB is sometimes called the *mean square for groups* or the *treatment mean square*. The numerator of the *F*-statistic, the mean square between (MSB), is computed as $MSB = \frac{SSB}{I-1}$, where the sum of squares between the sample means (SSB) is computed as

$$SSB = n_1(\bar{x}_1 - \bar{x})^2 + n_2(\bar{x}_2 - \bar{x})^2 + \cdots + n_I(\bar{x}_I - \bar{x})^2 = \sum_{groups} n_i(\bar{x}_i - \bar{x})^2$$

$$= \sum_{groups} (\text{group sample size})(\text{group sample mean} - \text{overall sample mean})^2.$$

12.5.3 Measuring Variation Within the Groups (MSW)

One of the assumptions of ANOVA is that the I populations have the same variance σ^2. Each of the sample variances, s_i^2, $i = 1, 2, \ldots I$, is an estimate of this common population variance σ^2, regardless of whether the null hypothesis is true. The degrees of freedom for each sample variance is the sample size minus 1, $n_i - 1$. The MSW essentially combines or pools the sample variances to form an overall estimate of σ^2 that is unbiased. The total degrees of freedom associated with the MSW is given by $(n_1 - 1) + (n_2 - 1) + \cdots + (n_I - 1) = (n_1 + n_2 + \cdots + n_I) - I = n - I$. So the MSW is a measure of the variation of the responses **within** each group or treatment.

The mean square between MSW is sometimes called the *mean square for error* or the *error mean square*. The denominator of the F-statistic, the mean square between (MSW), is computed as $MSW = \frac{SSW}{n-I}$, where the sum of squares between the sample means (SSW) is computed as

$$SSW = (n_1 - 1)s_1^2 + (n_2 - 1)s_2^2 + \cdots + (n_I - 1)s_I^2 = \sum_{groups} (n_i - 1)s_i^2$$

$$= \sum_{groups} (\text{group sample size} - 1)(\text{group sample variance}).$$

Note that the MSW is just a weighted average of the sample variances for the I groups, with the weights being the corresponding degrees of freedom for each group. The square root of the MSW is denoted by $s = \sqrt{MSW}$ and is called the **pooled standard deviation**. This pooled standard deviation is the **estimate of the common population standard deviation** σ, that is, $s = \hat{\sigma}$. We saw a pooled standard deviation back in Chapter 11 for the two independent samples t-test. In fact, if you had just $I = 2$ groups, the pooled standard deviations for the t-test and the F-test are equal. When $I = 2$, the two sample t-test is preferred over the F-test, because the t-test is robust for the equal population variance assumption, while the F-test can perform poorly if the population variances are not equal.

 ### 12.5.4 Measuring Total Variation

In one-way ANOVA, the total variance of all of the observations is given by the total sum of squares, SST, which measures the variation of each observation from the overall sample mean for all observations. Let an individual observation be represented by x_{ij} and read as the jth observation from the ith group. Then, a formula for the total sum of squares (SST) is

$$SST = \sum_{all\ observations} (x_{ij} - \bar{x})^2 = \sum_{all\ observations} (\text{observation} - \text{overall sample mean})^2.$$

This total sum of squares looks like the numerator of the sample variance if we treat all n observations as one large sample. This total variation can be partitioned up into the two sources of variation, between and within. The relationship between the three sums of squares is $SST = SSB + SSW$. So if you are given two of the three sum of squares values, you can easily find the value for the third sum of squares.

The steps involved in calculating the mean squares, and thus the test statistic, are given as follows:

Summary of Steps for One-Way ANOVA

Step 1: Calculate the I sample means — $\bar{x}_1, \bar{x}_2, \ldots, \bar{x}_I$, and the I sample variances — $s_1^2, s_2^2, \ldots, s_I^2$.

Step 2: Calculate the overall sample mean $\bar{x}$.

Step 3: Calculate the *sum of squares between the sample means* (SSB).

Step 4: Calculate the *sum of squares within* (SSW).

Step 5: *Optional*—calculate the *sum of squares total* (SST).

As a check on your calculations, you can verify that $SST = SSB + SSW$. This is an optional step, and many computer packages do not even include a TOTAL line in the ANOVA table output.

Step 6: Summarize the calculations leading up to getting our two estimators of the variance and complete the F-test statistic by completing the following ANOVA table:

Source of Variation	Sum of Squares	Degrees of Freedom	Mean Square	F
Between (groups)	SSB	$I - 1$	$MSB = \frac{SSB}{I - 1}$	$F = \frac{MSB}{MSW}$
Within (error)	SSW	$n - I$	$MSW = \frac{SSW}{n - I}$	
Total	SST			

The F-test statistic has an F-distribution with $I - 1$ and $n - I$ degrees of freedom under the null hypothesis.

❖ EXAMPLE 12.7   Three Treatments

We wish to compare three drugs for the treatment of headache pain, with the response variable of interest being the time to relief in minutes. Nineteen patients were randomly assigned to the three drug or treatment groups. The data are provided in the table. We assume that the three columns represent the values of independent random samples drawn from normal populations, with means μ_1, μ_2, and μ_3, and common population variance σ^2. We are interested in testing these hypotheses:

Drug 1	Drug 2	Drug 3
7.3	7.1	5.8
8.2	10.6	6.5
10.1	11.2	8.8
6.0	9.0	4.9
9.5	8.5	7.9
	10.9	8.5
	7.8	5.2

H_0: $\mu_1 = \mu_2 = \mu_3$.

H_1:   The population means are not all equal.

Step 1: The sample means for the three groups are

$$\bar{x}_1 = 8.22, \ \bar{x}_2 = 9.30, \ \bar{x}_3 = 6.80,$$

and the sample variances for the three groups are

$$s_1^2 = 2.737, \ s_2^2 = 2.613, \ s_3^2 = 2.56.$$

Step 2: The overall sample mean is

$$\bar{x} = \frac{7.3 + 8.2 + \cdots + 8.5 + 5.2}{19} = 8.095.$$

Step 3: Calculate the *sum of squares between* (SSB):

$$SSB = \sum_{groups} n_i(\bar{x}_i - \bar{x})^2 = \sum_{groups} (\text{group sample size})(\text{group sample mean} - \text{overall mean})^2$$

$$= 5(8.22 - 8.095)^2 + 7(9.30 - 8.095)^2 + 7(6.80 - 8.095)^2 = 21.98.$$

Step 4: Calculate the *sum of squares within* (SSW):

$$SSW = \sum_{groups} (n_i - 1)s_i^2 = \sum_{groups} (\text{group degrees of freedom})(\text{group sample variance})$$

$$= (4)(2.737) + (6)(2.613) + (6)(2.56)$$

$$= 41.99$$

Step 5: *Optional*—calculate the *sum of squares total* (SST):

$$SST = \sum_{all\ observations} (x_{ij} - \bar{x})^2 = \sum_{all\ observations} (\text{observation} - \text{overall sample mean})^2$$

$$= (7.3 - 8.095)^2 + (8.2 - 8.095)^2 + \cdots + (5.2 - 8.095)^2$$

$$= 63.97$$

Does SST = SSB + SSW?
We have 21.98 + 41.99 = 63.97, so our calculations *do* check out.

Step 6: Complete the ANOVA table.

Source	SS	DF	MS	F
Between (groups)	21.98	$I - 1 = 2$	$MSB = \frac{21.98}{2} = 10.99$	$F = \frac{10.99}{2.62} = 4.19$
Within (error)	41.99	$n - I = 16$	$MSW = \frac{41.99}{16} = 2.62$	
Total	63.97			

The *p*-value is 0.034. This *p*-value can be found by using the TI, as shown in Example 12.5. Table V could be used to report bounds for the *p*-value of between 0.01 and 0.05. Either way, at the 5% level, we reject the hypothesis of equality of the means and conclude that there appears to be a difference in the mean time to relief for the three drugs. ❖

 12.5.5 Performing an ANOVA *F*-test Using the TI

The TI has an ANOVA option under the TESTS menu located under the STATS menu that allows you to perform an ANOVA *F*-test. The data must first be stored in the editor with each sample entered as a separate list. The input to the ANOVA option is to provide the lists, separated with a comma, that contain the data. The *F*-test statistic, the *p*-value, the sum of squares, the mean squares, the degrees of freedom, and the estimated common population standard deviation are reported. Suppose that we enter the data for the three headache drugs from Example 12.7 under the lists L_1, L_2, and L_3. The following steps can be used to perform the one-way ANOVA on these data:

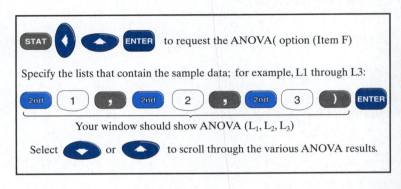 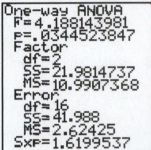

With the drug data from Example 12.7 entered into L1, L2, and L3, the resulting screen of results should look like

Note: The term *Factor* represents the *between* source of variation. The term *Error* represents the *within* source of variation. So we have $MSB = 10.9907368$ and $MSW = 2.62425$.

In the next example, we are not given the original data, but only the summary measures for each group. The overall sample mean must be computed as a weighted average, and the total sum of squares cannot be found. Programs are available for TI graphing calculators to perform such an analysis from only the summary measures. (See the ■ TI Quick Steps for more details.)

❖ **EXAMPLE 12.8   Quieting the Cough** _____

Suppose that five cough formulas are compared in an experiment, with the variable measured being hours of relief from coughing. Do these cough formulas have the same mean hours of relief?

Drug i	Sample Size n_i	Sample Mean $\bar{x}_i$	Sample Variance s_i^2
1	5	8.5	5.29
2	9	8.0	4.00
3	7	6.0	3.61
4	8	6.4	4.84
5	6	7.5	6.25

The total sample size is $n = 35$. Let's run through the steps involved in calculating the *F*-test statistic.

Step 1:   The five sample means and the five sample variances have already been computed.

Step 2:   The overall sample mean is

$$\bar{x} = \frac{5(8.5) + 9(8.0) + 7(6.0) + 8(6.4) + 6(7.5)}{35} = 7.22.$$

Step 3: Calculate the *sum of squares between* (SSB):

$$SSB = 5(8.5 - 7.22)^2 + 9(8.0 - 7.22)^2 + 7(6.0 - 7.22)^2$$
$$+ 8(6.4 - 7.22)^2 + 6(7.5 - 7.22)^2 = 29.936.$$

Step 4: Calculate the *sum of squares within* (SSW):

$$SSW = 4(5.29) + 8(4.00) + 6(3.61) + 7(4.84) + 5(6.25) = 139.95.$$

Step 5: Optional—calculate the *sum of squares total* (SST):
We cannot find SST from only the summary measures given. We would need to have the original observations. We can find SST by summing SSB and SSW, but we do not have an independent check on our calculations.

Completing the ANOVA table and making the decision are left for the next "Let's do it!" exercise.   ❖

Lₑₜ's do it! 12.4 Quieting the Cough ANOVA Table

(a) Use the computations provided in the previous example to complete the ANOVA table.

Source	SS	DF	MS	F
Between (groups)				
Within (error)				
Total				

(b) What is the *p*-value for testing the hypothesis of equality of the means?

(c) Are the results statistically significant at the 5% level? Explain.

✳ Exercises

12.13 In a study of the effects of exercise on physiological and psychological variables, four groups of male subjects were studied. A total sample of $n = 24$ was taken. One of the variables measured at the end of the program was a physical fitness score. Part of the ANOVA table used to analyze these data is given.

Source	SS	DF	MS	F
Between (groups)	104,855			
Within (error)	70,500			
Total				

(a) Fill in the missing entries in the ANOVA table.

(b) Give an estimate of the common population standard deviation.

(c) State the appropriate hypotheses.

(d) Find the *p*-value for this test.

(e) What do you conclude at a 1% level?

12.14 Data were collected on the weight in kilograms of men in three different age groups. The sample standard deviations for the groups were 12, 9, and 11. The sample sizes for the groups were 32, 38, and 34. Compute the within groups variance estimate (that is, the MSW).

12.15 General Motors used ANOVA to compare the mean number of cars sold in three regions by randomly selecting 11 dealerships in each region. The ANOVA table is given below.

```
Dep var:  SOLD    N: 33   Multiple R:  .365    Squared multiple R:   .133

                        Analysis of Variance
   Source   Sum-of-squares   DF  Mean-square    F-ratio     P
   DEALER           14.970    2       7.485      2.308    0.117
   Error            97.273   30       3.242
```

(a) Test the hypothesis that the mean number of cars sold in the three regions are equal at the 10% level and explain your decision.

(b) One of the assumptions in one-way ANOVA is that the random samples in the three regions are drawn from distributions with the same standard deviation. What is our estimate of this standard deviation?

(c) Explain why a large *F*-ratio is considered evidence against the null hypothesis that the means are all equal.

12.16 Recall Exercise 12.9 in which an experimenter was interested in studying the effect of sleep deprivation on the ability to detect moving objects on a radar screen. A total of 20 subjects was available for the study; five subjects were randomly assigned to each of the four treatment groups (Group 1 = 4 hours without sleep, Group 2 = 12 hours without sleep, Group 3 = 20 hours without sleep, and Group 4 = 28 hours without sleep). After being deprived of sleep, each subject was given a test and the number of times they failed to spot a moving object during the set test period was recorded. The results are given in the following table:

4 Hours	12 Hours	20 Hours	28 Hours
36	38	46	76
21	45	74	66
20	46	67	62
26	22	61	44
21	40	59	61

The hypothesis for assessing whether the mean number of failures is the same for the four sleep-deprivation conditions can be written as $H_0: \mu_1 = \mu_2 = \mu_3 = \mu_4$. Use a 1% significance level to perform the one-way ANOVA F-test of this hypothesis. Report the observed value of the F-statistic, its corresponding p-value, your decision, and conclusion using the language of this study.

12.17 A nutritional expert randomly assigns 18 bicyclers to three groups (with 6 to each group). Group B is given a vitamin supplement and Group C is given a diet of health foods. Group A is instructed to eat as they normally do. After a set time period on the prescribed diets, the expert subsequently records the number of minutes it takes each person to ride six miles. The data are provided in the following table:

Group A	Group B	Group C
19	16	12
18	12	15
16	14	12
18	15	14
14	14	10
17	13	13

State the appropriate null and alternative hypotheses and conduct a one-way analysis of variance using $\alpha = 0.01$. Show all work and provide a sound written conclusion using the language of this study.

12.18 A study is conducted on child stars (famous child performers) who have *gone bad*. A researcher compares child stars from the '60s, '70s and '80s and measures the number of maladaptive behaviors each engaged in.

'60s Decade	'70s Decade	'80s Decade
7	13	3
11	18	12
4	20	5
12	1	31
7	2	0
0	6	12
2	0	15
0	21	2
0	3	21
0	0	6

You are to assess whether there are any differences between the three decades in terms of the mean number of maladaptive behaviors. Use a one-way ANOVA F-test with a significance level of $\alpha = 0.05$. Include a summary (with graphs) regarding the validity of the assumptions.

12.6   Multiple Comparisons*

In one-way ANOVA, we are trying to compare many population means—that is, we are trying to do multiple comparisons. However, we would like to do the many comparisons at one time and still be able to attach to our conclusions some overall measure of confidence. The statistical approach to this problem of multiple comparisons is to first do an *overall test* to assess whether there are any differences between the population means, and if we accept that **there is** a difference, then do a *follow-up* analysis that helps determine which of the means differ and estimate by how much they differ. The *F*-test described in the preceding sections is this overall test in ANOVA. In this section, we discuss the need for this overall *F*-test in ANOVA and present some multiple comparison methods that are part of the follow-up analysis should our decision be to reject H_0.

12.6.1   Why Do We Need ANOVA?

Why do we need a new method to test for differences about the population means? Why can't we just use our two independent samples *t*-test to make many comparisons of two population means at a time? If there are $I = 4$ populations, then there would be six possible comparisons—that is, six *t*-tests to perform: 1 versus 2, 1 versus 3, 1 versus 4, 2 versus 3, 2 versus 4, and 3 versus 4. The problem with this approach concerns the significance level—as the number of *t*-tests increases, the overall significance level increases above the individual significance level.

THINK ABOUT IT

Let A be the event of rejecting H_0 when H_0 is true using Test 1. Let B be the event of rejecting H_0 when H_0 is true using Test 2. Assume that the events A and B are independent. Suppose that $P(A) = 0.05$ and $P(B) = 0.05$.

Then, the overall significance level =
$P(\text{At least one of the two tests results in a Type I error}) = P(A \text{ or } B)$
$$= 1 - P(\text{not } A \text{ and not } B)$$
$$= 1 - (1 - 0.05)(1 - 0.05)$$
$$= 0.0975 > 0.05.$$

If we extend this idea to three tests, so that we have three independent events, *A*, *B* and *C*, and each has probability 0.05 of occurring, $P(A) = P(B) = P(C) = 0.05$, what is the overall significance level—that is, the probability that at least one of the three tests results in a Type I error?

$P(\text{At least one of the three tests results in a Type I error}) = P(A \text{ or } B \text{ or } C)$
$$= 1 - P(\text{none of them occur})$$
$$= 1 - P(\text{not } A \text{ and not } B \text{ and not } C)$$
$$=$$

Is the resulting overall probability greater than 0.05?   Yes   No

---

*This section is optional.

Suppose that for each of the six t-tests, the significance level was to be set at $\alpha = 0.05$, that is, the probability of concluding that a difference exists between a specific pair of means when there is no difference is 0.05. Then the overall probability that *at least one of the six tests makes a Type 1 error* is $1 - (1 - 0.05)^6 \approx 0.265$, which is much larger than 0.05. Clearly this is not acceptable if we want to keep the significance level under control. We need a test for which we can choose the significance level in advance. Sir Ronald Fisher, an English statistician figured out how to avoid this difficulty when he invented ANOVA. The test statistic for ANOVA is called an F-test statistic in honor of Fisher's contributions to the field of statistics.

12.6.2 Multiple Comparisons

When we perform an ANOVA to test whether the population means are equal, we must remember that if we reject the null hypothesis we can only conclude that at least one of the population means appears to be different from the other population means. Thus, the ANOVA F-test is only the first step in our analysis. If significant differences exist among the treatment means, we would like to investigate which means are significantly different and perhaps measure how different they are.

We can determine where the differences among the means seem to occur by conducting **multiple comparisons**. The most common set of multiple comparisons performed is the set of all *pairwise comparisons*. In an $I = 3$ population ANOVA, if we reject the null hypothesis of equality of the three population means, we would like to examine the three pairwise comparisons (or contrasts) of the population means, $\mu_1 - \mu_2$, $\mu_1 - \mu_3$, and $\mu_2 - \mu_3$. We could construct a confidence interval for each pairwise difference in the means, in this case three confidence intervals. A confidence interval that *does not contain zero* would imply a significant difference between the corresponding means. Alternatively, for each pair we could form a t-test statistic and perform a test to determine if the two means differed significantly. As we saw in the previous "Think about it," if we were to conduct each test using a 5% significance level, the overall significance level would be inflated. There are many multiple comparison methods presented in the statistical literature designed to provide better control over the overall significance level, sometimes referred to as the family error rate. Two such methods are the Tukey method and the Bonferroni method. Likewise, these methods can be used to control for the overall confidence level, called the family confidence level, when multiple comparisons are made by examining the family of confidence intervals. The general formulas for the confidence interval and the t-test statistic are given below. The critical value t^* depends upon which multiple comparison method is used and is sometimes found in special tables. We will let the computer perform the multiple comparisons analysis and focus on interpreting the results.

Multiple Comparisons

To perform a multiple comparisons procedure, you could …

…compute the set of t-test statistics $t = \dfrac{\overline{x}_i - \overline{x}_j}{\sqrt{MSE\left(\frac{1}{n_i} + \frac{1}{n_j}\right)}}$

and if $t \geq$ critical t^*, then we conclude that the corresponding population means μ_i and μ_j are different; or

…compute the set of confidence intervals $(\overline{x}_i - \overline{x}_j) \pm t^*\sqrt{MSE\left(\frac{1}{n_i} + \frac{1}{n_j}\right)}$

and if an interval does not contain zero, then we conclude that the corresponding population means μ_i and μ_j are different.

❖ EXAMPLE 12.9 Lifetime of Male Fruit Flies: Which Group Means Were Different?

In our ANOVA of the fruit fly data (Example 12.1 and "Let's do it! 12.3"), the p-value was very small. We rejected the null hypothesis of equality of the population means at the 1% level of significance. What does reject H_0 mean? It simply tells us that not all the population means are equal. It does not tell us which means are different, which means are nearly the same, or the size of the differences.

The side-by-side boxplots gave us some idea of where the differences occur. Figure 12.5, called **a profile plot** of the means, can also provide some idea of where the differences occur.

As we noted earlier, the mean lifetime for Group 5 male fruit flies, which had eight interested females per day, is much lower as compared to any other groups. The ordering of the groups with respect to sample mean longevity, from lowest to highest, is

Figure 12.5 Profile Plot of the Sample Means

Group 5 (8 interested females per day).

Group 3 (1 interested female per day).

Group 4 (8 nonreceptive females per day).

Group 1 (no companions).

Group 2 (1 nonreceptive female per day).

The output that follows was generated with the SPSS statistical computing package by requesting pairwise comparisons using the Tukey method with a family error rate of 5% (or, equivalently, a family confidence level of 95%). The sample mean difference is reported along with the standard error, p-value for comparing the two means, and the confidence interval estimate for the difference in the two means. Note that there are two confidence intervals for comparing Group 1 and Group 2, the confidence interval for $\mu_1 - \mu_2$ is -12.88 to 10.32 and the confidence interval for $\mu_2 - \mu_1$ is -10.32 to 12.88. Both intervals provide the same basic information of no significant difference between these two groups. An asterisk (*) in the Mean Difference column indicates a significant difference

Multiple Comparisons

Dependent Variable: LONG
Tukey HSD

(I) GROUP	(J) GROUP	Mean Difference (I−J)	Std. Error	Sig.	95% Confidence Interval Lower Bound	95% Confidence Interval Upper Bound
1	2	−1.28	4.19	0.998	−12.88	10.32
	3	6.76	4.19	0.491	−4.84	18.36
	4	0.16	4.19	1.000	−11.44	11.76
	5	24.80*	4.19	0.000	13.20	36.40
2	1	1.28	4.19	0.998	−10.32	12.88
	3	8.04	4.19	0.312	−3.56	19.64
	4	1.44	4.19	0.997	−10.16	13.04
	5	26.08*	4.19	0.000	14.48	37.68
3	1	−6.76	4.19	0.491	−18.36	4.84
	2	−8.04	4.19	0.312	−19.64	3.56
	4	−6.60	4.19	0.515	−18.20	5.00
	5	18.04*	4.19	0.000	6.44	29.64
4	1	−0.16	4.19	1.000	−11.76	11.44
	2	−1.44	4.19	0.997	−13.04	10.16
	3	6.60	4.19	0.515	−5.00	18.20
	5	24.64*	4.19	0.000	13.04	36.24
5	1	−24.80*	4.19	0.000	−36.40	−13.20
	2	−26.08*	4.19	0.000	−37.68	−14.48
	3	−18.04*	4.19	0.000	−29.64	−6.44
	4	−24.64*	4.19	0.000	−36.24	−13.04

* The mean difference is significant at the 0.05 level.

between the corresponding two group means at the specified significance level. The only confidence intervals that did not contain zero involved Group 5. We conclude that there is a significant difference between the mean lifetime for Group 5 male fruit flies (eight interested femails per day) and each of the other groups. If the Bonferroni method had been requested, the confidence intervals would have been slightly different, but the same conclusion would have been reached (this would not always be the case). ❖

Let's do it! 12.5 Three Treatments Revisited

Drug 1	Drug 2	Drug 3
7.3	7.1	5.8
8.2	10.6	6.5
10.1	11.2	8.8
6.0	9.0	4.9
9.5	8.5	7.9
	10.9	8.5
	7.8	5.2

In Example 12.7, we compared three drugs for the treatment of headache pain, with the response variable of interest being the time to relief in minutes. Nineteen patients were randomly assigned to the three drug or treatment groups. The researcher was initially interested in examining all three possible pairwise comparisons: Drug 1 versus Drug 2, Drug 1 versus Drug 3, and Drug 2 versus Drug 3. The sample mean times for the three drug groups were $\bar{x}_1 = 8.22$, $\bar{x}_2 = 9.30$, $\bar{x}_3 = 6.80$. The ANOVA resulted in an observed test statistic of $F_{OBS} = 4.19$ with a p-value of 0.034. So at the 5% level, we reject the hypothesis of equality of the means and conclude that there appears to be a difference in the mean time to relief for the three drugs.

Source	SS	DF	MS	F
Between (groups)	21.98	$I - 1 = 2$	$MSB = \frac{21.98}{2} = 10.99$	$F = \frac{10.99}{2.62} = 4.19$
Within (error)	41.99	$n - I = 16$	$MSW = \frac{41.99}{16} = 2.62$	
Total	63.97			

The multiple comparisons output using Bonferroni's method is provided next.

Multiple Comparisons

Dependent Variable: TIME
Bonferroni

(I) DRUG	(J) DRUG	Mean Difference (I−J)	Std. Error	Sig.	95% Confidence Interval Lower Bound	Upper Bound
1	2	−1.08	0.95	0.815	−3.62	1.46
	3	1.42	0.95	0.462	−1.12	3.96
2	1	1.08	0.95	0.815	−1.46	3.62
	3	2.50*	0.87	0.032	0.19	4.81
3	1	−1.42	0.95	0.462	−3.96	1.12
	2	−2.50*	0.87	0.032	−4.81	−0.19

* The mean difference is significant at the 0.05 level.

(a) Using the output, report the confidence interval for each pairwise comparison:
Confidence interval for $\mu_1 - \mu_2$: _____ Does it contain 0?_____
Confidence interval for $\mu_1 - \mu_3$: _____ Does it contain 0?_____
Confidence interval for $\mu_2 - \mu_3$: _____ Does it contain 0?_____

(b) State your conclusions regarding the differences between the mean response for the three drug groups based on the Bonferroni multiple comparison method.

Exercises

12.19 The problem with computing multiple independent t-tests for comparing I sample means is that, as the number of t-tests increases, the overall Type I error rate (overall significance level or family error rate) increases above the individual significance level. The overall Type I error rate $= 1 - (1 - \alpha)^c$, where c is the number of independent t-tests. For a significance level of 0.05 and for $c = 5, 10, 25, 50, 100$, find the overall Type I error rate.

12.20 A recent study was conducted to assess the effectiveness of a new antibiotic treatment for strep throat. Children ages 6 to 14 who meet the entry criteria, were randomized to one of three treatment groups. The new antibiotic treatment was compared with two other standard antibiotic treatments. The following data, in days to cure, were reported:

New	4	5	6	5	9	4	8	7	5	8	9
Standard 1	6	5	5	8	10	8	8	5	8	9	11
Standard 2	9	10	14	8	7	5	9	8	10	12	7

The output for performing the one-way ANOVA and follow-up multiple comparisons are provided.

Descriptives

DAYS

	N	Mean	Std. Deviation	Std. Error	95% Confidence Interval for Mean Lower Bound	95% Confidence Interval for Mean Upper Bound	Minimum	Maximum
New	11	6.36	1.91	0.58	5.08	7.65	4.00	9.00
Standard 1	11	7.55	2.07	0.62	6.16	8.93	5.00	11.00
Standard 2	11	9.00	2.49	0.75	7.33	10.67	5.00	14.00
Total	33	7.64	2.37	0.41	6.80	8.48	4.00	14.00

ANOVA

DAYS

	Sum of Squares	df	Mean Square	F	Sig.
Between Groups	38.364	2	19.182	4.073	0.027
Within Groups	141.273	30	4.709		
Total	179.636	32			

Multiple Comparisons

Dependent Variable: DAYS
Tukey HSD

(I) treatment	(J) treatment	Mean Difference (I−J)	Std. Error	Sig.	95% Confidence Interval Lower Bound	95% Confidence Interval Upper Bound
New	Standard 1	−1.18	0.93	0.419	−3.46	1.10
	Standard 2	−2.64*	0.93	0.021	−4.92	−0.36
Standard 1	New	1.18	0.93	0.419	−1.10	3.46
	Standard 2	−1.45	0.93	0.273	−3.74	0.83
Standard 2	New	2.64*	0.93	0.021	0.36	4.92
	Standard 1	1.45	0.93	0.273	−0.83	3.74

* The mean difference is significant at the 0.05 level.

(a) Are the mean times to cure in days the same for the three treatment groups? Conduct the test using the 0.05 level of significance.

(b) Is the new antibiotic treatment significantly better than the Standard 1 antibiotic treatment? How does the new treatment compare to the Standard 2 antibiotic treatment? How does the Standard 1 treatment compare to the Standard 2 treatment? Use the output for Tukey's method to carry out the tests for the three pairwise comparison such that the overall significance level is 0.05.

12.21 Sixty rats were used in an experiment to determine the effect of ethanol on sleep. Twenty of the rats were given 1 g of ethanol per kilogram of body weight and called Group 1, another 20 rats were given 2 g/kg and called Group 2, and 20 more rats were given no ethanol and called Group 3, the control group. The following Tukey test results were obtained for the three pairwise comparisons:

Difference	$\mu_1 - \mu_2$	$\mu_1 - \mu_3$	$\mu_2 - \mu_3$
t-test statistic (p-value)	−2.64 (0.011)	−3.9 (0.0003)	−1.2 (0.24)

Using a 5% significance level, which of the following statements best describes the relationship between μ_1, μ_2, and μ_3? (Select one.)

(a) $\mu_1 = \mu_2$, and μ_3 differs from μ_1 and μ_2.

(b) $\mu_1 = \mu_3$, and μ_2 differs from μ_1 and μ_3.

(c) $\mu_2 = \mu_3$, and μ_1 differs from μ_2 and μ_3.

(d) All three μ's are different from one another.

12.22 Four chemicals were used to combat plant lice on sugar beets. Each chemical was used in one plot. From each plot of many sugar beet plants, 25 leaves were collected and the number of plant lice on each was recorded. The data are provided along with some numerical summaries.

Chemical 1	Chemical 2	Chemical 3	Chemical 4
12	13	26	13
17	24	14	10
6	4	2	10
8	6	7	13
18	10	18	3
4	18	13	10
21	10	21	34
15	5	22	12
25	18	12	2
2	10	22	17
20	19	20	12
11	16	5	11
17	16	23	14
14	20	27	25
17	18	29	14
31	5	13	18
23	16	13	23
4	16	17	9
28	23	19	32
26	24	16	32
18	33	16	34
18	9	19	29
30	18	25	20
21	27	31	25
33	24	16	24

Descriptives

LICE

	N	Mean	Std. Deviation
Chemical 1	25	12.00	6.45
Chemical 2	25	14.96	7.44
Chemical 3	25	18.36	7.00
Chemical 4	25	24.00	6.74

(a) Construct the ANOVA table and test at the 5% significance level whether there is a difference in the mean number of plant lice between the four chemicals.

(b) What are the assumptions needed for the analysis in (a) to be valid? Create appropriate plots for assessing the validity of the ANOVA assumptions and comment.

(c) The output for performing the follow-up multiple comparisons analysis using the Bonferroni method is as follows. Give an interpretation of this output.

Multiple Comparisons

Dependent Variable: LICE
Bonferroni

(I) chemical	(J) chemical	Mean Difference (I−J)	Std. Error	Sig.	95% Confidence Interval Lower Bound	Upper Bound
1	2	−2.96	1.96	0.801	−8.23	2.31
	3	−6.36*	1.96	0.010	−11.63	−1.09
	4	−12.00*	1.96	0.000	−17.27	−6.73
2	1	2.96	1.96	0.801	−2.31	8.23
	3	−3.40	1.96	0.513	−8.67	1.87
	4	−9.04*	1.96	0.000	−14.31	−3.77
3	1	6.36*	1.96	0.010	1.09	11.63
	2	3.40	1.96	0.513	−1.87	8.67
	4	−5.64*	1.96	0.029	−10.91	−0.37
4	1	12.00*	1.96	0.000	6.73	17.27
	2	9.04*	1.96	0.000	3.77	14.31
	3	5.64*	1.96	0.029	0.37	10.91

* The mean difference is significant at the 0.05 level.

12.23 A survey was conducted on homes in a region that were sold during the past two years. A random sample of 42 homes located in County 1 revealed a mean length of time on the market of 6.2 weeks with a standard deviation of 2.2 weeks. For the random sample of 34 homes located in County 2, the mean was 3.9 weeks with a standard deviation of 1.6 weeks. For the random sample of 47 homes located in County 3, the mean was 3.1 weeks with a standard deviation of 1.8 weeks.

(a) Construct the ANOVA table and test at the 5% significance level whether there is a difference in the mean length of time on the market between the three counties.

(b) The following are the Bonferroni 95% confidence intervals for the three pairwise comparisons:

Confidence interval for $\mu_1 - \mu_2$: $2.3 \pm 1.07 \Rightarrow (1.23, 3.37)$.

Confidence interval for $\mu_1 - \mu_3$: $3.1 \pm 0.98 \Rightarrow (2.12, 4.08)$.

Confidence interval for $\mu_2 - \mu_3$: $0.8 \pm 1.04 \Rightarrow (-0.24, 1.84)$.

Give an interpretation of these confidence intervals.

12.7   Two-way ANOVA and Interaction*

The previous sections have focused on the **one-way ANOVA** procedure, where "one-way" refers to having only one explanatory variable (or factor) and one quantitative response variable. This section presents a brief overview of **two-way ANOVA**, which examines the effect of two explanatory variables (or factors) on the mean response. An agriculturalist may wish to learn about the effect of the type of fertilizer used and the amount of fertilizer used on the yield of tomatoes. A marketing experiment is conducted to study the how the color of the packaging and the shape of the packaging for a hand soap affects the attractiveness rating of the product by a consumer. In each of these examples, the researcher is interested in the *individual effect* of each explanatory variable on the mean response and also in the *combined effect* of the two explanatory variables on the mean response.

The individual effect of each factor on the response is called a **main effect**. If one of the factors does not have an effect on the response, we say there is no main effect due to that factor. Besides assessing the main effects of each factor on the response, an interesting feature in two-way analyses is the possibility of **interaction** between the two factors. We say there is interaction between two factors if the effect of one factor on the mean response depends on the specific level of the other factor. The interpretation of the factor main effects can be more difficult when interaction is present. An example of a statement that describes interaction is "Fertilizer II gave a higher yield when used in small quantities, while Fertilizer I performed best when applied in larger amounts."

❖ EXAMPLE 12.10   Growing Tomatoes _____

Let's consider the agriculturalist that wished to learn about the effect of the type of fertilizer used and the amount of fertilizer used on the yield of tomatoes. An experiment is conducted involving two types of fertilizer and three amounts of fertilizer being applied. Standard-sized plots are randomly allocated to a particular treatment combination of type and amount. The following table provides the mean tomato yield (in bushels per standard plot) for each possible treatment combination and for each factor level:

	Amount A (small)	Amount B (medium)	Amount C (large)	Overall mean
Fertilizer I	12	17	16	15
Fertilizer II	20	13	12	15
Overall mean	16	15	14	15

The response variable for this experiment is tomato yield. The two factors are type of fertilizer used and the amount of fertilizer used. Type of fertilizer has two levels while amount of fertilizer has three levels, for a total of six treatments or factor level combinations.

If we looked at the results for just type of fertilizer alone, ignoring amount of fertilizer (i.e., the row means), there does not appear to be an effect due to fertilizer type (the mean yields were both 15). If we looked at the results for just amount of fertilizer alone, ignoring type of fertilizer (i.e., the column means), there does not appear to be much of an effect due to fertilizer amount (the mean yields were very similar and around 15). Descriptively, the best amount appears to be Amount A (small). So if Fertilizer I were the least expensive, we might use it in the smallest amount. The mean yield for this treatment combination however only 12. The optimal combination would be Fertilizer II in small amounts. The effect that describes these data is called *interaction*.

---

*This section is optional.

The **profile plot** of this tomato yield data is provided. The response variable (mean tomato yield) is on the vertical axis, and the levels of one of the factors (fertilizer amount) are placed on the horizontal axis. The mean responses are then plotted and lines are used to connect the responses for the levels of the second factor (fertilizer type).

In this graph, we see that the profiles (for the two types of fertilizer) are not parallel, and, in fact, they even cross. When the profiles are not parallel, we say there is **evidence of interaction** between these two factors. If there is no evidence of interaction (i.e., parallel profiles), we say the factors appear to be **additive**.

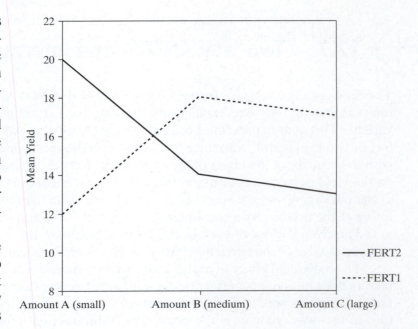

 **12.6 Growing Corn**

The table that follows provides the results of a second experiment for assessing the effect of the type of fertilizer used and the amount of fertilizer used on the yield of corn. The mean corn yields (in bushels per standard plot) for each treatment combination and each factor level are reported.

	Amount A (small)	Amount B (medium)	Amount C (large)	Overall mean
Fertilizer I	16	20	14	16.7
Fertilizer II	8	12	6	8.7
Overall mean	12	16	10	12.7

(a) Does there appear to be a type of fertilizer effect? Explain.

(b) Does there appear to be an amount of fertilizer effect? Explain.

(c) Does there appear to be an interaction effect between these two factors? Support your answer by sketching the profile plot for these data.

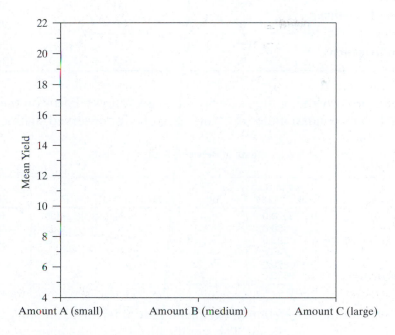

Many of the ideas presented for one-way ANOVA extend to the two-way ANOVA. We again assume that the data are approximately normal and that the various treatment groups have the same population standard deviation, but possible different means. We will have an estimate of the population standard deviation that will play a role in the tests performed. We will again use *F*-statistics for significance tests, but instead of one *F*-test, we will now have three. One *F*-test is used to assess the significance of the *interaction* between the two factors. The other two *F*-tests are used to assess the *main effect* of each of the two factors. The results of a two-way ANOVA are summarized in an ANOVA table based on splitting the total variation and total degrees of freedom among the two main effects and the inter-action. The computational details behind a two-way analysis are complex and will not be presented in this text. Many statistical packages can perform a two-way analysis and pro-grams are available for TI graphing calculators to perform such analysis (see the ■ TI Quick Steps for more details).

Since the presence of interaction can influence how the factor results are interpreted, the test for interaction is examined first. A profile plot will also be helpful in interpreting the results of the tests. We turn to an example that focuses on interpretation rather than computation.

❖ EXAMPLE 12.11 Growing Barley

Three levels of fertilizer were used in a field experiment, with and without irrigation. The six treatment combinations were assigned at random to 12 plots, with 2 plots receiving each treatment combination. Barley yields were measured in bushels per acre. The table that fol-lows provides the observed barley yields for each possible treatment combination. A two-way ANOVA is to be performed to assess the effects of amount of fertilizer and irrigation use on barley yield. All tests are to be conducted using a 5% significance level.

	Low Level Fertilizer	Medium Level Fertilizer	High Level Fertilizer
Without Irrigation	135 165	90 66	75 93
With Irrigation	125 95	127 105	120 136

The SPSS output for performing the two-way ANOVA follows: Focus on the *p*-values that are reported in the last column under the "Sig." heading for observed significance level.

Tests of Between-Subjects Effects

Dependent Variable: YIELD

Source	Type III Sum of Squares	df	Mean Square	F	Sig.
AMTFERT	2328.0	2	1164.0	4.06	0.077
IRRIG	588.0	1	588.0	2.05	0.202
AMTFERT * IRRIG	4392.0	2	2196.0	7.66	0.022
Error	1720.0	6	286.7		
Total	156880.0	12			
Corrected Total	9028.0	11			

The significance test of the *interaction* between amount of fertilizer and irrigation usage has a *p*-value of 0.022, which is significant. The significance test of the *main effect* for irrigation usage has a *p*-value of 0.202, indicating a nonsignificant result. For the main effect of amount of fertilizer, the *p*-value is smaller, but still not significant at a 5% level. Although irrigation usage and amount of fertilizer individually do not significantly effect barley yield, there is evidence of an interaction between these two factors based on the data. The effect of irrigation usage on barley yield depends on the amount of fertilizer used.

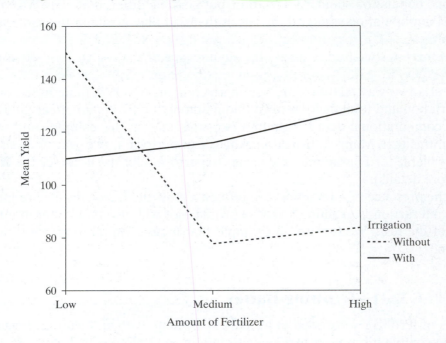

A profile plot portrays this visually. If no irrigation is used, the low level of fertilizer appear to perform well, but the higher levels of fertilizer appear to perform very poorly. If irrigation is used, the yields do not seem to vary as much across the levels of fertilizer, with higher levels performing just slightly better. ❖

✳ Exercises

12.24 If you believe two factors may influence a response, why must you study both factors at all levels of each of the other factor, rather than studying one factor at a time (holding the second factor constant)?

12.25 Give an example of two factors each at two levels that show no evidence of interaction. Fill in the mean responses in the following table:

		Factor A	
		Level 1	**Level 2**
Factor B	*Level 1*		
	Level 2		

12.26 Make a profile plot of the mean responses from Exercise 12.25. Explain how a profile plot can be used to assess whether there is interaction between two factors.

12.27 In Chapter 3, we examined types of studies. Exercise 3.42 provided results of an experiment involving two factors—type of fertilizer and amount of water. The crop yields, in pounds, from equal-sized plots of wheat are provided in the following table:

	Water		
Fertilizer	*A*	*B*	*C*
I	30, 25, 29	32, 34, 29	25, 22, 23
II	31, 37, 33	31, 32, 30	28, 24, 25
III	33, 29, 24	35, 41, 31	30, 29, 29
IV	31, 30, 32	33, 30, 31	26, 29, 25

(a) Compute the average yield for each treatment combination and construct a profile plot.

(b) Based on the profile plot, does there appear to be evidence of interaction between the two factors? Explain.

REAL DATA

12.28 In testing food products for palatability, General Foods employed a seven-point scale from −3 (terrible) to +3 (excellent), with zero representing "average." The company's standard method for testing palatability was to conduct a taste test with 50 people, 25 men and 25 women. The experiment reported here involved the effects on palatability of a coarse screen versus a fine screen and of a low concentration versus a high concentration of a liquid component. Four groups of 50 consumers each were recruited from local churches and club groups. Persons were assigned randomly to the four treatment groups as they were recruited. The experiment was replicated four times, so that there were 16 groups of 50 consumers each in the entire experiment. (*Source:* E. Street and M.G. Carroll, "Preliminary Evaluation of a Food Product," *Statistics: A Guide to the Unknown* (Judith M. Tanur, ed.) Holden-Day, San Francisco, 1972, 220–238.) The following table provides the total palatability score for the four replications:

	Coarse Screen	**Fine Screen**
Low Liquid	35	104
	39	129
	77	97
	16	84
High Liquid	24	65
	21	94
	39	86
	60	64

(a) Compute the average total score across the four replications for each treatment combination and construct a profile plot.

(b) Based on the profile plot, does there appear to be evidence of interaction between the two factors? Explain.

REAL DATA

12.29 Mr. John McKinley, President of Texaco, made a statement to the Air and Water Pollution Subcommittee of the Senate Public Works Committee on June 26, 1973. He cited the Octel filter, developed by Associated Octel Company, as effective in reducing pollution. However, questions had been raised about the effects of pollution filters on aspects of vehicle performance, including noise levels. He referred to data as evidence that the Octel filter was at least as good as a standard silencer in controlling vehicle noise levels. The data consisted of noise-level readings (in decibels) and the factors were type of filter (two types—standard silencer and Octel filter) and vehicle size (three sizes—small, medium, and large). SPSS was used to produce the following two-way ANOVA. (*Source:* A.Y. Lewin and M.F. Shakun, *Policy Sciences: Methodology and Cases*, Pergamon Press, 1976, p. 313)

Tests of Between-Subjects Effects

Dependent Variable: NOISE

Source	Type III Sum of Squares	df	Mean Square	F	Sig.
SIZE	26051.39	2	13025.69	199.12	0.000
TYPE	1056.25	1	1056.25	16.15	0.000
SIZE * TYPE	804.17	2	402.08	6.15	0.006
Error	1962.50	30	65.42		
Total	23657575.00	36			
Corrected Total	29874.31	35			

(a) Focusing on the *p*-values that are reported in the last column, comment on the results of the three *F*-tests.

(b) The profile plot follows. What is the evidence that there may be an interaction between type of filter and size of car? Describe the pattern of this possible interaction.

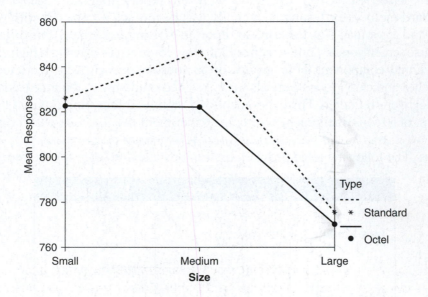

Chapter Summary

In this chapter, we extended the idea of making comparisons of the means of populations — this time with a technique that allows the comparison of the means from two or more normal populations. The technique for testing the equality of several population means is called the **analysis of variance (ANOVA)** because the test statistic is based on the ratio of two different measures of variation. If this ratio is large, there is evidence that the population means are not equal. Computing this test statistic, called an **F-statistic**, involves a number of calculations that are better left for the computer or calculator to come up with. This new test statistic brings with it a new distribution called the **F-distribution**, which is indexed by a set of degrees of freedom.

This chapter focused on the methods involved in **one-way ANOVA**, which examines the effect of one explanatory variable on the mean response. If the one-way ANOVA *F*-test results in rejecting the null hypothesis, we can only conclude that at least one of the population means appears to be different. There are other questions of interest, such as Which means are different? and How different are they? The follow-up analysis involves making **multiple comparisons**. We discussed the need for methods that allow many comparisons to be made while still having one overall (family) significance level apply to the set of comparisons. A summary of the steps involved in a one-way ANOVA is presented next:

F-Test for Comparing Many Treatments

1. **State the hypotheses** to be tested and **set the significance level** α for the test. A common significance level used is 0.05.

   H_0: $\mu_1 = \mu_2 = \ldots = \mu_I$.
   H_1: The population means are not all equal.

2. **Assess whether the assumptions underlying the test seem valid.** Do the data for each group appear to come from an approximately normal distribution? Does the equal population standard deviation assumption seem reasonable based on comparing the sample standard deviations? Can the data for each group be assumed to be a random sample from the larger population of interest? Can the samples be assumed to be independent of each other?

3. **Compute the observed F-statistic and corresponding p-value.** Standard computer output and many calculators provide this value and the corresponding *p*-value

$$F = \frac{\text{Variation BETWEEN the sample means}}{\text{Natural Variation WITHIN the samples}} = \frac{MSB}{MSW}.$$

   Note that the $\sqrt{MSW} = \hat{\sigma} = s$ is the unbiased estimator of σ.

4. **Determine whether the result is statistically significant** by comparing the *p*-value to the significance level. If the *p*-value is less than or equal to α, we reject the null hypothesis and say the results are statistically significant at that level α.

5. **Write the conclusion** using the language of the problem. If the null hypothesis was rejected, do a follow-up **multiple comparisons** procedure for assessing which population means are different and how different they are.

We ended the chapter with a brief overview of **two-way ANOVA**, which examines the effect of two explanatory variables (or factors) on the mean response. In two-way ANOVA, the researcher is interested in the *individual effect* of each explanatory variable on the mean response, called the **main effects**, and also in the *combined effect* of the two explanatory variables on the mean response. An interesting feature in two-way analyses is the possibility of **interaction** between the two factors. We say there is interaction between two factors if the effect of one factor on the mean response depends on the specific level of the other factor. In two-way ANOVA, an *F*-test for the interaction effect is conducted first as the interpretation of the main effects can be more difficult when interaction is present. A **profile plot** is a useful tool for visualizing potential interaction between the two factors.

Key Terms

Be sure you can describe, in your own words, and give an example of each of the following key terms from this chapter:

analysis of variance (ANOVA)

mean square between (MSB)

mean square within (MSW)

sum of squares between (SSB)

sum of squares within (SSW)

Bonferroni method

Tukey method

one-way ANOVA

F-distribution

variability between sample means

variability within sample means

point estimator of the common population standard deviation $= \hat{\sigma} = s$

multiple comparisons

two-way ANOVA

main effect

interaction effect

profile plot

conservative degrees of freedom

Exercises

12.30 Three fuel additives were compared using an ANOVA technique. Similar cars were randomly assigned to receive one of the three additives, driven 100 highway miles, and the miles per gallon (mpg) were recorded. A summary of the results is provided.

	Additive 1 (0 mL)	Additive 2 (5 mL)	Additive 3 (10 mL)
Sample size	11	11	11
Sample mean	19.8	22.9	31.5
Sample standard deviation	1.94	3.59	3.62

```
- - - - - O N E W A Y - - - - -
  Variable  MPG   By Variable  ADDITIVE   Fuel Additive
                            Analysis of Variance
                         Sum of          Mean         F       F
      Source       D.F.  Squares        Squares      Ratio   Prob.
Between Groups       2   799.2727      399.6364     40.3303  .0000
Within Groups       30   297.2727        9.9091
Total               32  1096.5455
```

(a) State the hypotheses for assessing whether the additives are equally effective with respect to the average mpg.

(b) Based on the preceding results, would you reject the null hypothesis at the 5% significance level?

12.31 A psychologist wished to compare three methods for reducing stress levels in university students. A certain test was used to measure the degree of stress. A high score on the test indicated more stress. Students with high scores were used in the experiment. Eleven students were randomly assigned to each of the three methods. All treatments were continued for a one-semester period. Each student was given the test at the end of the semester. The data were used to assess whether or not there are differences among mean scores for the three methods.

```
Dep var:  SCORES   N: 33   Multiple R:  .365   Squared multiple R:  .133
                          Analysis of Variance
   Source    Sum-of-squares   DF  Mean-square   F-ratio      P
   METHODS        14.970       2     7.485       2.308      0.117
   Error          97.273      30     3.242
```

(a) Test the hypothesis that the mean scores for the three methods are equal at the 10% level.

(b) One of the assumptions in one-way ANOVA is that the random samples are drawn from distributions with the same variance. What is our estimate of this variance?

12.32 An agricultural experiment was conducted to study the effects of two herbicides on crop yield. Both herbicides were applied to 11 plots each, and 11 more plots were used as a control without using any herbicide. The data on yields are summarized.

	Sample Mean	Sample Variance
Herbicide 1	90.2	4.5
Herbicide 2	89.3	3.8
Control	85.0	3.4

(a) Use these data to complete the one-way ANOVA table.

Source of Variation	Sum of Squares	Degrees of Freedom	Mean Square	F
Between (groups)				
Within (error)				
Total	271.5			

(b) We wish to assess whether there are any differences in the mean yields for the three treatments. State the null hypothesis and the alternative hypotheses.

(c) Perform the test in (b). Use a significance level of 0.05.

12.33 We want to compare three brands of film for the color intensity of pictures. The larger the value, the more intense the color. Using the same camera, pictures were taken for each brand of film. The data follow:

Brand 1	34, 36, 33, 32, 39, 30, 30, 29, 32, 34, 28, 31, 29, 32, 33
Brand 2	40, 38, 40, 39, 30, 31, 33, 32, 31, 32, 35, 35, 37, 33
Brand 3	29, 30, 31, 27, 32, 31, 33, 32, 28, 31, 33, 36, 31, 34

(a) We wish to test the hypothesis that the mean color brightness is the same for the three brands of film. State the null hypothesis and the alternative hypotheses.

An ANOVA table for these data was computed with the following results:

Source	SS	DF	MS	F
Between (groups)				
Within (error)	351.5			
Total	441.1			

(b) Complete the ANOVA table shown.

(c) Give the point estimate of the common population standard deviation.

(d) Give the p-value for the test in (a).

12.34 The mean phosphorus content of leaves of three types of trees, represented by μ_1, μ_2, and μ_3, is to be compared. Five observations were obtained on each of the three types. The following summary statistics were calculated from these observations:

Sample mean phosphorus content Type 1: 0.776.
Sample mean phosphorus content Type 2: 0.708.
Sample mean phosphorus content Type 3: 0.460.

$$SSB = 0.27664 \text{ and } SST = 0.37444.$$

(a) We wish to test the hypothesis that the mean phosphorus contents are the same for the three types of trees. State the null hypothesis and the alternative hypothesis.

(b) Complete the ANOVA table.

Source	SS	DF	MS	F
Between (groups)				
Within (error)				
Total				

(c) Perform the test in (a). Use a significance level of 0.05.

12.35 Bakery products are displayed at three levels (bottom, middle, top) in supermarkets. The following monthly sales (in $100s) have been recorded in several supermarkets adopting one of three levels.

Bottom	58.2	53.7	55.6		
Middle	73.0	78.1	76.5	82.0	78.4
Top	52.5	49.8	56.0	51.9	53.3

(a) Set up the null and alternative hypotheses to test whether there are significant differences in the mean sales of the products displayed on the three levels.

(b) Obtain the MSW and MSB and carry out the F-test at a 5% level of significance for (a).

(c) State carefully the assumptions necessary for carrying out the ANOVA.

TI Quick Steps

Finding Areas under an F-Distribution Using the TI

The TI has a function that is built into the calculator for finding areas under any F-distribution. This function is located under the DISTR menu and allows you to specify the set of degrees of freedom. An ANOVA F-test resulted in a test statistic value of $F_{OBS} = 5.5$. The distribution for the statistic under H_0 is an F-distribution with 3 and 20 degrees of freedom. We wish to compute the p-value for the test. We need to find the area under the $F(3, 20)$ distribution to the right of 5.5. The steps are shown.

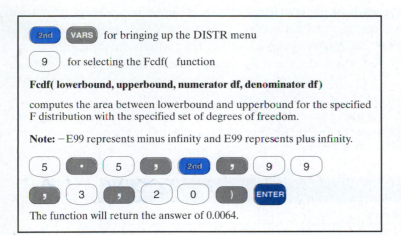

Performing a One-way ANOVA F-test Using the TI

The TI has an ANOVA option under the TESTS menu located under the STATS menu that allows you to perform a one-way ANOVA F-test. The data must first be stored in the editor with each sample entered as a separate list. The input to the ANOVA option is to provide the lists, separated with a comma, that contain the data. The F-test statistic, the *p*-value, the sum of squares, the mean squares, the degrees of freedom, and the estimated common population standard deviation are reported. The steps to produce the results are as follows:

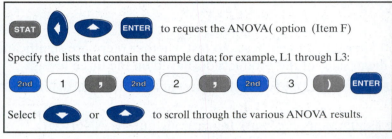

With the ad data from Example 12.5 entered into L1, L2, and L3, the resulting screen of results should look like

Note: The term *Factor* represents the *between* source of variation. The term *Error* represents the *within* source of variation. So we have $MSB = 303.68588$ and $MSW = 120.370357$.

Performing a One-Way ANOVA Using a Program for the TI

The Texas Instruments Web site www.education.ti.com provides much information about the various graphing calculators. The educator section of this Web site provides links to a program archive that contains many programs for various TI graphing calculators, which can be viewed and downloaded to a computer or to a graphing calculator. Programs are also available on disk by calling 1-800-TI-CARES. One of the programs available for the TI is called A1ANOVA. This program is quite lengthy, but can be used to perform a one-way ANOVA, a randomized block design, and a two-factor factorial experiment. Another advantage of this program is that it can analyze the original data (which must be entered as a matrix) or use summary measures (the sample sizes, sample means and sample standard deviations) as input. The next example provides the steps for performing a one-way ANOVA for comparing three groups using the summary measures as the input.

Consider Example 12.7, in which three drugs for the treatment of headache pain were compared. The summary measures for the three sets of data are given here.

	Drug 1	Drug 2	Drug 3
Sample size	5	7	7
Sample mean	8.22	9.3	6.8
Sample standard deviation	1.654	1.617	1.600

The steps for running the A1ANOVA program (once you have it saved in your TI) are as follows:

1. Press PRGM and hightlight the program named A1ANOVA and press enter, so the name prgmA1ANOVA appears in your window. Press ENTER to start running the A1ANOVA program and you should see the shown menu in your window.

```
ANOVA USES [D][E]
1:ONE-WAY ANOVA
2:RAN BLOCK DESI
3:2WAY FACTORIAL
4:QUIT
```

2. Select 1:ONE-WAY ANOVA and a screen will appear that lets you know there are two options for inputting the data. Press ENTER and select the 2nd option in this list, which states that the sample means and standard deviations will be the input.

```
DATA INPUT WITH
1:DATA MAT [D]
2:X1,Sx1,n1,X2,..
3:QUIT
```

3. Various questions will appear in your window and you answer them by typing in the numerical response and pressing ENTER. The first few answers are shown in the window at the right.

```
HOW MANY LEVELS
?3
LEVEL           1

MEAN=?8.22
S.D.=?1.654
SIZE=?5
```

4. After you enter the sample mean, sample standard deviation, and sample size for the last group, press ENTER and the ANOVA table will appear in your window.

```
     DF    SS
FAC 2  21.981473
ERR 16 41.990998

   F=4.19
   P=0.034
  SP=1.620011535
```

5. If you press ENTER one more time, the 95% confidence interval for each population mean is provided. This is a confidence interval for the mean using the pooled standard error estimate and provides information about each population mean individually. These intervals can be examined to see where the differences appear to be if the null hypothesis of equality of the means is rejected.

The program code for the A1ANOVA follows on the next page.

A1ANOVA Program

```
Menu("ANOVA USES[D][E]","ONE-WAY ANOVA",P,"RAN BLOCK DESIGN",Q,"2WAY FACTORIAL",R,
"QUIT",S)
Lbl S:Stop
Lbl P
Disp "1. USE A N*2 MAT":Disp "[D],OBSERVATIONS":Disp "IN COL 1, FACTOR":Disp
"LEVEL IN COL 2-":Disp "INTEGERS 1,2..N."
Disp "2. ENTER MEAN,SD":Disp "AND SAMP SIZES."
Pause :ClrHome
Lbl 5:ClrHome
Menu("DATA INPUT WITH","DATA MAT [D]",2,"X̄1,Sx1,n1,X̄2..",1,"QUIT",3)
Lbl 3:ClrHome
Stop
Lbl 1:ClrHome
9→F:Disp "HOW MANY LEVELS"
Input N
N→dim(L₄:N→dim(L₅:N→dim(L₆:0→G
0→T
For(I,1,N)
Disp "LEVEL",I:Input "MEAN=?",V:V→L₄(I)
Input "S.D.=?",V:V→L₅(I)
Input "SIZE=?",V:V→L₆(I)
L₆(I)+G→G
L₄(I)*L₆(I)+T→T:End
T/G→M:0→R:0→S
For(I,1,N)
S+(M-L₄(I))²L₆(I)→S:R+(L₅(I))²*(L₆(I)-1)→R:End
S/(N-1)→D
R/(G-N)→E
Goto 4
Lbl 2:ClrHome
1→F:dim([D]:Ans(1)→G:2→N
For(I,1,G)
If [D](I,2)>N
[D](I,2)→N:End
N→dim(L₄:N→dim(L₅:N→dim(L₆:Fill(0,L₄):Fill(0,L₅):Fill(0,L₆):0→T:0→S:{N,4}→dim([E]
For(I,1,G)
T+[D](I,1)→T
S+([D](I,1))²→S:[D](I,2)→B
L₅(B)+1→L₅(B)
L₄(B)+[D](I,1)→L₄(B):L₆(B)+([D](I,1))²→L₆(B):End
S-T²/G→P:0→A
For(I,1,N)
A+(L₄(I))²/L₅(I)→A:End
A-T²/G→S:P-S→R
S/(N-1)→D
R/(G-N)→E
Lbl 4
ClrHome
Disp "     DF   SS"
Output(2,1,"FAC":Output(2,5,N-1):Output(2,8,S)
Output(3,1,"ERR"):Output(3,5,(G-N)):Output(3,8,R)
Output(5,3,"F="):Output(5,5,round((D/E),2))
Output(6,3,"P="):Fix 3
Output(6,5,round(Fcdf(D/E,E99,N-1,G-N),3):Float
Output(7,2,"SP="):Output(7,5,√(E):Pause
ClrHome
If F=9:Goto V
Disp "[LEV N MEAN SD]"
For(I,1,N)
I→[E](I,1)
L₅(I)→[E](I,2)
L₄(I)/L₅(I)→M
```

(continued)

```
M→[E](I,3)
(L₆(I)-(L₄(I))²/L₅(I))/(L₅(I)-1)→V
√(V→[E](I,4)
End:Pause  [E]
Lbl  V:ClrHome
G-N→C
1.96(1+2/(1+8*C))→W
C*(e^(W*W/C)-1)→C:√(C)→C
C*√(E→W
ClrList  L₂,L₃
If  9=F:Then
For(I,1,N,1)
W/√((L₆(I))→Z
L₄(I)-Z→L₂(I)
L₄(I)+Z→L₃(I)
End:Else
For(I,1,N,1)
W/√((L₅(I))→Z
L₄(I)/L₅(I)-Z→L₂(I)
L₄(I)/L₅(I)+Z→L₃(I):End:End
(max(L₃)-min(L₂))/10→W
min(L₂)-W→Xmin
max(L₃)+W→Xmax
0→Xscl:0→Yscl
1→Ymin:192→Ymax:120/(N+1)→H
PlotsOff  :FnOff  :ClrDraw
Text(1,1,"0.95  CI.S  -LEVEL  1  AT  TOP")
H+40→Y
For(I,N,1,-1)
Line(L₂(I),Y,L₃(I),Y)
Y+H→Y:End
Trace:Pause
ClrHome:Stop
Lbl  Q:ClrHome
Disp  "DATA  IN  N*3  MAT":Disp  "[D].OBSERVATIONS"
Disp  "COL  1,COL  2(A)+":Disp  "COL  3(B)  CONTAIN"
Disp  "FACTOR  LEVEL  AND":Disp  "BLOCK-INTEGERS"
Disp   "1,2..."
Goto  9
Lbl  R:ClrHome
Disp  "EQUAL  REPLICATES":Disp  "DATA  IN  N*3  MAT ":Disp  "[D],1ST  COL-DATA":Disp  "2ND
COL-A  LEVELS":Disp  "3RD  COL-B  LEVELS":Disp  "LEVELS-INTEGERS":Disp  "STARTING  WITH
1."
Lbl  9
Pause  :ClrHome
Menu("CONT  OR  QUIT","CONTINUE",A,"QUIT",B)
Lbl  B:ClrHome:Stop
Lbl  A
dim([D]:Ans(1)→R:2→K:1→L
For(I,1,R)
If  [D](I,3)>K
[D](I,3)→K
If  [D](I,2)>L
[D](I,2)→L:End
K→dim(L₆:K→dim(L₅:Fill(0,L₆):Fill(0,L₅):L→dim(L₄:L→dim(L₃:Fill(0,L₄):Fill(0,L₃):
0→T:0→S
For(I,1,R)
T+[D](I,1)→T
S+([D](I,1))²→S:[D](I,3)→C
[D](I,2)→D
L₅(C)+1→L₅(C)
L₃(D)+1→L₃(D)
```

(continued)

```
L₆(C)+[D](I,1)→L₆(C):L₄(D)+[D](I,1)→L₄(D):End
S-T²/R→G:0→A
For(I,1,K)
A+(L₆(I))²/L₅(I)→A:End
0→B:For(I,1,L)
B+(L₄(I))²/L₃(I)→B:End
A-T²/R→U
B-T²/R→V
K*L→M:M→dim(L₃
Fill(0,L₃):1→I
For(J,1,K)
For(E,1,L)
For(F,1,R)
If     (([D](F,3)=J)*([D](F,2)=E))
L₃(I)+[D](F,1)→L₃(I):End
1+I→I
End:End
0→Q
For(I,1,M)
Q+(L₃(I))²→Q
End
Q/(R/M)-U-V-T²/R→P:G-U-V-P→E
ClrHome
Disp "    DF    SS"
Output(2,1,"A":Output(2,4,L-1):Output(2,7,U)
Output(3,1,"B":Output(3,4,K-1):Output(3,7,V)
If    (R/M)-1:Then:P→E:Output(4,1,"ER"):(K-1)*(L-1)→Z:Goto    E:End
Output(4,1,"AB")
Lbl  E
Output(4,4,(K-1)*(L-1)):Output(4,7,P):Output(5,1,"
If  (R/M)=1
Goto  C
Output(5,1,"ER":Output(5,4,(R-K*L)):Output(5,7,E):(R-K*L)→Z
Lbl  C
Output(6,4,"F(A)=")
U/(L-1)/(E/Z)→F:Output(6,9,round(F,2)
Output(7,7,"P="):Fix    3
Output(7,9,round(Fcdf(F,ε99,L-1,Z),3):Float
Output(8,4,"F(B)=")
(U/(K-1))/(E/Z)→F:Output(8,9,round(F,2):Pause    :ClrHome
Output(1,1,"B        P="):Fix  3
Output(1,9,round(Fcdf(F,ε99,K-1,Z),3):Float
If  (R/M)=1
Goto  D
Output(2,3,"F(AB)=")
P/((K-1)*(L-1))/(E/(R-K*L))→F
Output(2,9,round(F,2))
Output(3,7,"P="):Fix    3
Output(3,9,round(Fcdf(F,ε99,L-1,R-K*L),3):Float
Lbl  D
√((E/Z)→S
Output(5,3,"S=":Output(5,5,S)
Return
```

90th percentiles of the *F*-distribution

F(numerator df, denominator df)

0.90

0.10

$F_{.90}$ *F*

Table V: *F-Distribution*
90th Percentiles of the *F*-distribution

	Degrees of Freedom for Numerator							
	1	**2**	**3**	**4**	**5**	**6**	**7**	**8**
1	39.86	49.50	53.59	55.83	57.24	58.20	58.91	59.44
2	8.53	9.00	9.16	9.24	9.29	9.33	9.35	9.37
3	5.54	5.46	5.39	5.34	5.31	5.28	5.27	5.25
4	4.54	4.32	4.19	4.11	4.05	4.01	3.98	3.95
5	4.06	3.78	3.62	3.52	3.45	3.40	3.37	3.34
6	3.78	3.46	3.29	3.18	3.11	3.05	3.01	2.98
7	3.59	3.26	3.07	2.96	2.88	2.83	2.78	2.75
8	3.46	3.11	2.92	2.81	2.73	2.67	2.62	2.59
9	3.36	3.01	2.81	2.69	2.61	2.55	2.51	2.47
10	3.29	2.92	2.73	2.61	2.52	2.46	2.41	2.38
11	3.23	2.86	2.66	2.54	2.45	2.39	2.34	2.30
12	3.18	2.81	2.61	2.48	2.39	2.33	2.28	2.24
13	3.14	2.76	2.56	2.43	2.35	2.28	2.23	2.20
14	3.10	2.73	2.52	2.39	2.31	2.24	2.19	2.15
15	3.07	2.70	2.49	2.36	2.27	2.21	2.16	2.12
16	3.05	2.67	2.46	2.33	2.24	2.18	2.13	2.09
17	3.03	2.64	2.44	2.31	2.22	2.15	2.10	2.06
18	3.01	2.62	2.42	2.29	2.20	2.13	2.08	2.04
19	2.99	2.61	2.40	2.27	2.18	2.11	2.06	2.02
20	2.97	2.59	2.38	2.25	2.16	2.09	2.04	2.00
21	2.96	2.57	2.36	2.23	2.14	2.08	2.02	1.98
22	2.95	2.56	2.35	2.22	2.13	2.06	2.01	1.97
23	2.94	2.55	2.34	2.21	2.11	2.05	1.99	1.95
24	2.93	2.54	2.33	2.19	2.10	2.04	1.98	1.94
25	2.92	2.53	2.32	2.18	2.09	2.02	1.97	1.93
30	2.88	2.49	2.28	2.14	2.05	1.98	1.93	1.88
40	2.84	2.44	2.23	2.09	2.00	1.93	1.87	1.83
60	2.79	2.39	2.18	2.04	1.95	1.87	1.82	1.77
120	2.75	2.35	2.13	1.99	1.90	1.82	1.77	1.72

Degrees of Freedom for Denominator

95th percentiles of the *F*-distribution

F(numerator df, denominator df)

Table V: *F-Distribution*
95th Percentiles of the *F*-distribution

	Degrees of Freedom for Numerator							
	1	**2**	**3**	**4**	**5**	**6**	**7**	**8**
1	161.45	199.50	215.71	224.58	230.16	233.99	236.77	238.88
2	18.51	19.00	19.16	19.25	19.30	19.33	19.35	19.37
3	10.13	9.55	9.28	9.12	9.01	8.94	8.89	8.85
4	7.71	6.94	6.59	6.39	6.26	6.16	6.09	6.04
5	6.61	5.79	5.41	5.19	5.05	4.95	4.88	4.82
6	5.99	5.14	4.76	4.53	4.39	4.28	4.21	4.15
7	5.59	4.74	4.35	4.12	3.97	3.87	3.79	3.73
8	5.32	4.46	4.07	3.84	3.69	3.58	3.50	3.44
9	5.12	4.26	3.86	3.63	3.48	3.37	3.29	3.23
10	4.96	4.10	3.71	3.48	3.33	3.22	3.14	3.07
11	4.84	3.98	3.59	3.36	3.20	3.09	3.01	2.95
12	4.75	3.89	3.49	3.26	3.11	3.00	2.91	2.85
13	4.67	3.81	3.41	3.18	3.03	2.92	2.83	2.77
14	4.60	3.74	3.34	3.11	2.96	2.85	2.76	2.70
15	4.54	3.68	3.29	3.06	2.90	2.79	2.71	2.64
16	4.49	3.63	3.24	3.01	2.85	2.74	2.66	2.59
17	4.45	3.59	3.20	2.96	2.81	2.70	2.61	2.55
18	4.41	3.55	3.16	2.93	2.77	2.66	2.58	2.51
19	4.38	3.52	3.13	2.90	2.74	2.63	2.54	2.48
20	4.35	3.49	3.10	2.87	2.71	2.60	2.51	2.45
21	4.32	3.47	3.07	2.84	2.68	2.57	2.49	2.42
22	4.30	3.44	3.05	2.82	2.66	2.55	2.46	2.40
23	4.28	3.42	3.03	2.80	2.64	2.53	2.44	2.37
24	4.26	3.40	3.01	2.78	2.62	2.51	2.42	2.36
25	4.24	3.39	2.99	2.76	2.60	2.49	2.40	2.34
30	4.17	3.32	2.92	2.69	2.53	2.42	2.33	2.27
40	4.08	3.23	2.84	2.61	2.45	2.34	2.25	2.18
60	4.00	3.15	2.76	2.53	2.37	2.25	2.17	2.10
120	3.92	3.07	2.68	2.45	2.29	2.18	2.09	2.02

Degrees of Freedom for Denominator

99th percentiles of the *F*-distribution

F(numerator df, denominator df)

0.99   0.01

$F_{.99}$ F

Table V: *F-Distribution*
99th Percentiles of the *F*-distribution

	Degrees of Freedom for Numerator							
	1	2	3	4	5	6	7	8
1	4052.18	4999.50	5403.35	5624.58	5763.65	5858.99	5928.36	5981.07
2	98.50	99.00	99.17	99.25	99.30	99.33	99.36	99.37
3	34.12	30.82	29.46	28.71	28.24	27.91	27.67	27.49
4	21.20	18.00	16.69	15.98	15.52	15.21	14.98	14.80
5	16.26	13.27	12.06	11.39	10.97	10.67	10.46	10.29
6	13.75	10.92	9.78	9.15	8.75	8.47	8.26	8.10
7	12.25	9.55	8.45	7.85	7.46	7.19	6.99	6.84
8	11.26	8.65	7.59	7.01	6.63	6.37	6.18	6.03
9	10.56	8.02	6.99	6.42	6.06	5.80	5.61	5.47
10	10.04	7.56	6.55	5.99	5.64	5.39	5.20	5.06
11	9.65	7.21	6.22	5.67	5.32	5.07	4.89	4.74
12	9.33	6.93	5.95	5.41	5.06	4.82	4.64	4.50
13	9.07	6.70	5.74	5.21	4.86	4.62	4.44	4.30
14	8.86	6.51	5.56	5.04	4.69	4.46	4.28	4.14
15	8.68	6.36	5.42	4.89	4.56	4.32	4.14	4.00
16	8.53	6.23	5.29	4.77	4.44	4.20	4.03	3.89
17	8.40	6.11	5.18	4.67	4.34	4.10	3.93	3.79
18	8.29	6.01	5.09	4.58	4.25	4.01	3.84	3.71
19	8.18	5.93	5.01	4.50	4.17	3.94	3.77	3.63
20	8.10	5.85	4.94	4.43	4.10	3.87	3.70	3.56
21	8.02	5.78	4.87	4.37	4.04	3.81	3.64	3.51
22	7.95	5.72	4.82	4.31	3.99	3.76	3.59	3.45
23	7.88	5.66	4.76	4.26	3.94	3.71	3.54	3.41
24	7.82	5.61	4.72	4.22	3.90	3.67	3.50	3.36
25	7.77	5.57	4.68	4.18	3.85	3.63	3.46	3.32
30	7.56	5.39	4.51	4.02	3.70	3.47	3.30	3.17
40	7.31	5.18	4.31	3.83	3.51	3.29	3.12	2.99
60	7.08	4.98	4.13	3.65	3.34	3.12	2.95	2.82
120	6.85	4.79	3.95	3.48	3.17	2.96	2.79	2.66

Degrees of Freedom for Denominator

Chapter 13

IS THERE A RELATIONSHIP BETWEEN TWO OR MORE QUANTITATIVE VARIABLES?

13.1 Introduction

Recall from Chapter 3 that many statistical studies are conducted to try to detect relationships between variables—that is, to attempt to show that some explanatory variable "causes" the values of some response variable to occur.

- Does blood pressure predict life expectancy?
- Do SAT scores predict college performance?
- Do midterm scores predict final scores?

In this chapter, we will discuss methods for displaying and describing the relationship between two or more quantitative variables. The

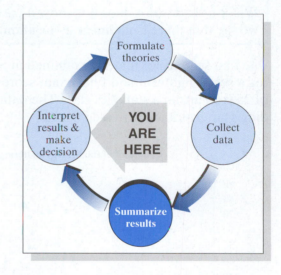

data used to study the relationship between two variables are called **bivariate data**. Bivariate data are obtained by measuring both variables on the same individual or unit. We might record both midterm score and final exam score for a sample of students. These data will help us study the association between the two variables. We will be able to see if certain midterm test scores tend to occur more often with certain final exam scores than with other final exam scores.

Recall that for **quantitative** variables (i.e., height, length, and time) we examine the relationship using a graph called a scatterplot. In this chapter, we will see if we can summarize the relationship with a "simple" model. We will discuss linear regression, which is a method for developing an equation of a line that predicts the value of one quantitative variable from the value of the other quantitative variable. We will also discuss correlation, which measures the strength and direction of the linear relationship between two quantitative variables. We will discuss the various inference techniques that are done in regression, including interval estimation and hypothesis testing. In the last section, we will extend this idea to two or more quantitative explanatory variables, called multiple regression.

13.2 Displaying the Relationship

Let's begin by recalling some of the language that is used in the study of relationships and the graph used to display the relationship called a scatterplot. These ideas were first introduced in Section 4.4.6.

> *Definitions:* Studies are often conducted to attempt to show that some explanatory variable "causes" the values of some response variable to occur. The **response** or **dependent** variable is the response of interest, the variable we want to predict, and is usually denoted by y. The **explanatory** or **independent** variable attempts to explain the response and is usually denoted by x.
>
> A **scatterplot** shows the relationship between two quantitative variables x and y. The values of the x variable are marked on the horizontal axis, and the values of the y variable are marked on the vertical axis. Each pair of observations (x_i, y_i), is represented as a point in the plot.
>
> Two variables are said to be **positively associated** if, as x increases, the values of y tend to increase. Two variables are said to be **negatively associated** if, as x increases, the values of y tend to decrease.
>
> When a scatterplot does not show a particular direction, neither positive, nor negative, we say that there is **no linear association.**

The first scatterplot that we examined in Section 4.4.6 was for a set of midterm and final exam scores from a sample of 25 students in a statistics class. The data and corresponding scatterplot are provided again.

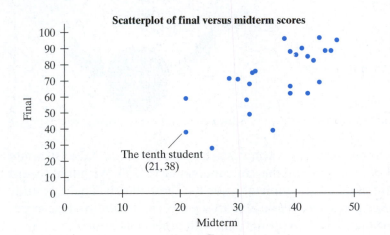

Scatterplot of final versus midterm scores

The tenth student (21, 38)

Student Number	x Midterm Score	y Final Score
1	39	62
2	44	69
3	32	68
4	40	86
5	45	88.5
6	46	88.5
7	33	76
8	39	66.5
9	32.5	75
10	**21**	**38**
11	30	71
12	39	88
13	44	96.5
14	28.5	71.5
15	38	96
16	43	82.5
17	42	85
18	25.5	28
19	47	95
20	36	39
21	31.5	58
22	32	49
23	42	62
24	21	59
25	41	90

The association between these two scores is positive. Higher midterm scores generally tend to go with higher final exam scores. But there are some exceptions. Overall, the relationship is approximately linear and of moderate strength.

In our first "Let's do it!" we practice making a scatterplot and describing the relationship that we see.

 **13.1** **Selling Beer**

The data that follow give the media expenditures (in millions of dollars) and the shipments of bottles (in millions) for 10 major brands of beer. These data were published on October 20, 1997, in *Superbrands 1998*. Make a scatterplot to examine the relationship between x = expenditures and y = shipments. Comment on the relationship with respect to form, direction, strength, and any departures or unusual values.

Brand	x = Media Expenditure (Millions of $)	y = Shipments (Millions of bottles)
Budweiser	120.0	36.3
Bud Light	68.7	20.7
Miller Lite	100.1	15.9
Coors Light	76.6	13.2
Busch	8.7	8.1
Natural Light	0.1	7.1
Miller Genuine Draft	21.5	5.6
Miller High Life	1.4	4.4
Busch Light	5.3	4.3
Milwaukee's Best	1.7	4.3

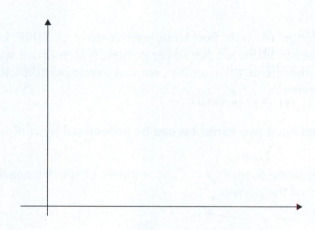

Comments:

Before we turn to developing a model to describe the relationship that is portrayed in a scatterplot, we take a look at three cautionary notes to keep in mind throughout this chapter.

Notes of Caution

1. **An observed relationship between two variables does not imply that there is some causal link between the two variables.**

   For example, consider the following scatterplot of IQ score versus shoe size:

   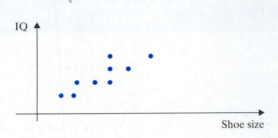

   As people age, their shoe size increases, and so does their IQ. Although there is a positive association, there is no causal link between the two variables shoe size and IQ.

   Most studies attempt to show that some explanatory variable "causes" the values of the response to occur. While we can never positively determine whether or not there is a distinct cause-and-effect relationship, we can assess whether there appears to be a significant relationship.

2. **A relationship between two variables can be influenced by confounding variables.**

   Consider the following scatterplot of the number of sport magazines read in a month versus the height of the person:

   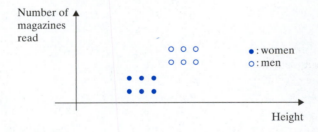

   Overall, there appears to be a positive association between height and number of magazines. However, for each gender, there does not appear to be an association. Gender is a confounding variable, and aggregating the data across gender can result in misleading conclusions. Any study, especially an observational study, has the potential to be wrongly interpreted because of confounding variables.

3. Unusual data points (outliers) can mislead the association, especially if the data set is small.

Consider the following scatterplot of the percentage of people who speak English versus population size:

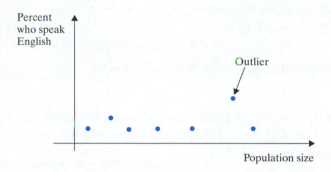

The eight points in the scatterplot represent eight countries from Central and South America selected at random. The outlier is Mexico City.

13.3   Modeling a Linear Relationship

13.3.1   Simple Linear Regression

We saw from the scatterplot that the underlying form of the relationship between midterm and final exam scores is approximately linear. So it would make sense to come up with an equation of a straight line to model that relationship. The equation of this line could serve as a nice, compact summary of the relationship between midterm and final exam scores. This equation could be used for predicting a future student's final exam score based on the midterm score. But which equation, which line, do we use? If two different people were asked to eyeball a best-fitting line to our scatter of points, we would probably end up with two slightly different equations of a line. Two possible lines are shown.

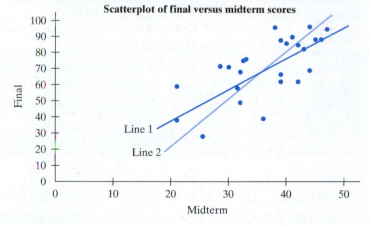

So the question remains as to how to find a best-fitting line.
The method we will use for finding this line is called **least squares regression**, the resulting line is called the **least squares regression line**, and the equation that describes this line is called the **least squares regression equation**. Although beyond the scope of this text, the basic ideas of regression analysis can be extended to handle nonlinear relationships. Situations in which the response depends on two or more explanatory variables are discussed in Section 13.11.

Let's step back a minute to the basic ideas of a straight line. We want to find a compact description for the straight-line pattern evident in our scatterplot—that is, an equation of a line $y = a + bx$, where y is the response variable, x is the explanatory variable, b is the **slope** of the line, and a is the **y-intercept**.

Equation of a Line

$y = a + bx$ where

b = **slope**—the amount y changes when x is increased by 1 unit.

a = **y-intercept**—the value of y when x is set equal to zero.

Our goal is to find a straight line that fits the data, a line that comes as close as possible to all of the points. Consider the data on midterm and final exam scores and the corresponding picture.

The observation for Student 18 corresponds to the point ($x = 25.5$, $y = 28$). It is the circled point in the picture. Thus, at $x = 25.5$, we observed a response $y = 28$. However, if we were to use the line given as our model, we would predict the response to be approximately 48. This vertical difference, the observed value of y minus the predicted value of y, represents our "error" in prediction. The **predicted value of y** is often represented by $\hat{y}$, read "y-hat," so we can distinguish between the response value predicted from the line and the observed value of y. We would want to make these so-called errors as small as possible. These vertical deviations of the points from the line are called the **residuals**. For each point in the scatterplot, there is a corresponding residual. We will study residuals in more detail in Section 13.4.

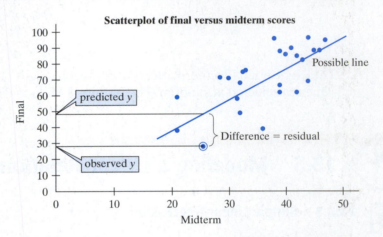

> *Definition:*   A **residual** is the difference between the observed response, y_i, and the predicted response, $\hat{y}_i$, using the regression line. Each pair of observations (x_i, y_i)—that is, each point on the scatterplot—produces a residual.
>
> $$\text{The } i^{\text{th}} \text{ residual} = e_i = y_i - \hat{y}_i = y_i - (a + bx_i).$$

We would like to find the line that would make all of the residuals as small as possible. Notice that some of the residuals will be positive and some will be negative. Thus, we can't simply try to make the sum of the residuals small, since the positive and negative residuals will cancel out in the sum. One approach is to consider the squared vertical distances, as we did when we computed the standard deviation as a measure of variability. An observation with a residual of -2 would have a "squared error" of $(-2)^2 = 4$. An observation with a residual of +2 would have the same "squared error" of 4. A best-fitting line could be defined as the line that minimizes the sum of the squared vertical deviations of the data points from the line. This method is called the **method of least squares**, and the resulting line is called the **least squares regression line** or the estimated regression line.

> *Definition:* The **least squares regression line**, given by $\hat{y} = a + bx$, is the line that makes the sum of the squared vertical deviations of the data points from the line as small as possible. Performing the regression is often stated as *regress y on x*.

We will discuss later how to compute the actual values for a and b. For now, we will focus on how to interpret the values of a and b and how to use the estimated regression line for making predictions. The least squares regression line can be used for **prediction** by substituting your value of x into the equation and computing the resulting predicted value, $\hat{y}$. If two variables have a strong relationship, we have more confidence in the ability to predict the value of one variable from the value of the other one. When the relationship is linear, the least squares regression line can be used for prediction.

Consider the data on the midterm and final exam scores for the 25 students. The least squares regression line for regressing final exam scores, y, on midterm exam scores, x, is given by $\hat{y} = 7.5 + 1.75x$. This line has been added to the scatterplot as shown. The estimated slope of $b = 1.75$ tells us that for a one point increase in the midterm score, we would *expect*, on average, an increase of 1.75 points

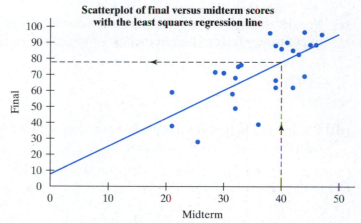

Scatterplot of final versus midterm scores with the least squares regression line

for the final exam score. The estimated y-intercept of $a = 7.5$ tells us that if someone were to score 0 points on the midterm, we would predict that he or she would get 7.5 points on the final exam. Suppose that a new student scores 40 points on the midterm. Based on our model, what would be his or her predicted final exam score? Plug the value of $x = 40$ into our estimated equation. The predicted final exam score is $\hat{y} = 7.5 + 1.75(40) = 77.5$ points.

If you find the midterm score $x = 40$ on the accompanying scatterplot, and go up to the regression line and then over to the final exam score axis, you can check that the predicted final exam score is approximately 77.5. This is a prediction. It certainly does not imply that this particular student will score exactly 77.5 points. In fact, Student 4 scored 40 points on the midterm, but received 86 points on the final. As we can see in the scatterplot, the observed values of y do vary around the estimated regression line. In general, when the points lie close to the estimated regression line, indicating a strong linear association, we are more confident in our predictions.

Let's do it! 13.2 Childhood Growth

The growth of children from early childhood through adolescence generally follows a linear pattern. Data on the heights of female Americans during childhood, from four to nine years old, were compiled and the least squares regression line was obtained as $\hat{y} = 80 + 6x$, where $\hat{y}$ is the predicted height in centimeters and x is age in years. Note that 1 inch is equal to 2.54 centimeters.

(a) Interpret the value of the estimated slope $b = 6$.

(b) Would interpretation of the value of the estimated y-intercept, $a = 80$, make sense here? If yes, interpret it. If no, explain why not.

(c) Predict the height for a female American who is eight years old. Give your answer first in centimeters and then in inches.

(d) Predict the height for a female American who is 25 years old. Give your answer first in centimeters and then in inches.

(e) Why do you think your answer to (d) was so inaccurate?

One of the dangers to be aware of when predicting is called **extrapolation**. Extrapolation is when you use your fitted regression line to predict outside the range of the original data. There is no guarantee that the relationship will extend beyond the range for which we have data. However, using the equation for a small extrapolation is sometimes performed. In (d) of the previous "Let's do it! 13.2", the height of a 25-year-old was predicted using a model developed for heights of four- to nine-year-olds. Our prediction is very inaccurate because the straight line that models heights for children cannot be used to model heights for adults.

❖ EXAMPLE 13.1 Another Type of Extrapolation

Before you calculate a regression line and use it for prediction, you should always look at the scatterplot first. Suppose that the scatterplot of your data is as shown.

You would like to predict the value of y for when $x = 50$. Even though a value of $x = 50$ is inside the range of the original data, it falls in the gap. No data on the x-variable between 38 and 70 were obtained. The relationship between the two variables in this middle range may not be linear. You should take some data in this range first.

Another possible explanation for the appearance of the first scatterplot is that there are two somewhat distinct subgroups in the sample. One set of the observations could be for, say, the females in the sample, while the other set of observations was for males.

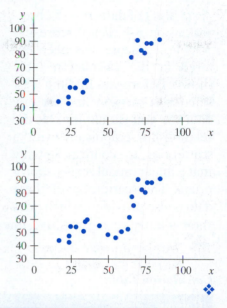

❖

> **Definition:** **Extrapolation** is using the regression line to predict the value of a response corresponding to a value of x that is outside the range of the data used to determine the regression line. Extrapolation can lead to unreliable predictions.

  ### 13.3.2 Calculating the Least Squares Regression Line

Recall that the least squares regression line, given by $\hat{y} = a + bx$, is the line that minimizes the sum of the squared vertical deviations of the data points from the line. The actual procedure to find the a and b involves taking derivatives, setting the derivatives equal to zero, and solving for a and b. The resulting expressions for a and b are given.

The Least Squares Regression Line

The least squares regression line is given by $\hat{y} = a + bx$, where

$$\text{slope} = b = \frac{\sum(x_i - \bar{x})(y_i - \bar{y})}{\sum(x_i - \bar{x})^2} = \frac{n(\sum x_i y_i) - (\sum x_i)(\sum y_i)}{n(\sum x_i^2) - (\sum x_i)^2}$$

$$y\text{-intercept} = a = \bar{y} - b\bar{x}.$$

The sums in the preceding expressions run over all of the observed values. Sometimes, the subscripts on the x and y variables are omitted for easier reading. The equation for the estimated slope is expressed in two equivalent ways. The right-hand-side expression is often easier to use if you compute these values by hand. As you can see, you need to first find the slope of the estimated regression line, b, and then use it to determine the estimated y-intercept, a.

Often, the calculation of these values for a and b is done with the help of a calculator or a computer. For details on finding these values with a TI graphing calculator, see Example 13.3 and the ■ TI Quick Steps at the end of this chapter. We will also see where these values are located in the regression output for various computer packages. However, if you have to find these values for a set of data by hand, we recommend the use of a calculation table, as demonstrated in the next example. We will do it together for a small set of data on x = Test 1 score and y = Test 2 score.

❖ EXAMPLE 13.2 Test 1 versus Test 2: By Hand

A small set of data, $n = 5$ observations on $x =$ Test 1 score and $y =$ Test 2 score, has been entered in the calculation table shown. Before you find the least squares regression line for any data set, you should look at the relationship graphically with a scatterplot to confirm initially that a linear model seems appropriate. The scatterplot of these data is also provided, which does show a fairly strong, positive, linear association.

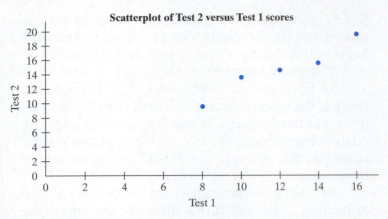

Scatterplot of Test 2 versus Test 1 scores

Calculation Table

Column	1 x_i	2 y_i	3 x_i^2	4 $x_i y_i$	5 y_i^2
	8	9	64	72	81
	10	13			
	12	14			
	14	15			
	16	19			
Total					

First you fill in the values for Columns 3, 4, and 5 by doing exactly what the column heading says to do. For example, for Column 3, we are to square each value of x from Column 1 and write it down in the same row under Column 3 ($x_i = 8 \rightarrow x_i^2 = 64$). Once you have filled in all of the column entries, you total or sum the column entries. By summing up each column, you are effectively putting a summation sign, Σ, in front of the respective column heading as a label for that total value. When you sum up the x_i^2-values in Column 3, you are finding Σx_i^2.

Completed Calculation Table

Column	1 x_i	2 y_i	3 x_i^2	4 $x_i y_i$	5 y_i^2
	8	9	64	72	81
	10	13	100	130	169
	12	14	144	168	196
	14	15	196	210	225
	16	19	256	304	361
Total	$\Sigma x_i = 60$	$\Sigma y_i = 70$	$\Sigma x_i^2 = 760$	$\Sigma x_i y_i = 884$	$\Sigma y_i^2 = 1032$

For computing the coefficients a and b, you only need the totals for the first four columns. Column 5 will become useful in an upcoming section when we introduce the correlation coefficient. So let's take the column totals and put them in the appropriate places in the expression for the estimated slope coefficient b:

$$b = \frac{n(\sum x_i y_i) - (\sum x_i)(\sum y_i)}{n(\sum x_i^2) - (\sum x_i)^2} = \frac{5(884) - (60)(70)}{5(760) - (60)^2} = \frac{220}{200} = 1.1.$$

Once we have the value for the estimated slope, we can find the estimated y-intercept:

$$a = \bar{y} - b\bar{x} = \frac{70}{5} - (1.1)\frac{60}{5} = 0.8.$$

The least squares equation for regressing Test 2 score on Test 1 score is given by $\hat{y} = 0.8 + 1.1x$. The slope of the line is $b = 1.1$. This means that Test 2 scores are expected to go up by 1.1 points on average for each additional point scored on Test 1. A student who scored 15 points on Test 1 is predicted to score $\hat{y} = 0.8 + 1.1(15) = 17.3$ points on Test 2. ❖

❖ EXAMPLE 13.3   Test 1 versus Test 2: Using the TI Calculator _____

To obtain the least squares regression line using the TI graphing calculator, we would first need to enter the data.

L1	L2
8	9
10	13
12	14
14	15
16	19

Enter the values of the quantitative variable x = Test 1 into L1 and enter the corresponding values of the quantitative variable y = Test 2 into L2. To get the least squared regression equation, we use the following sequence of buttons:

Your output screen should provide the least squares regression equation as

$$y = a + bx,$$

with the y-intercept of $a = 0.8$ and the slope of $b = 1.1$.

Caution   There are two linear regression options: LinReg($ax + b$) and LinReg($a + bx$). We request the latter option, which uses b to represent the slope. ❖

 **13.3** **Oil-Change Data**

The following table presents data on x = the number of oil changes per year and y = the cost of repairs for a random sample of 10 cars of a certain make and model, from a given region:

Number of oil changes per year (x)	3	5	2	3	1	4	6	4	3	0
Cost of repairs ($) ($y$)	300	300	500	400	700	400	100	250	450	600

(a) Make a scatterplot of the points as a check for linearity and outliers. Comment on your plot.

(b) Find the least squares regression line for regressing cost on number of oil changes. Describe what the estimated y-intercept and estimated slope represent.

(c) Use your least squares regression line to predict the cost of car repairs for a car that had four oil changes.

THINK ABOUT IT

If you exchange x for y, will you get the same regression line? The equation $\hat{y} = a + bx$ would be obtained from the regression of y on x. The equation $\hat{x} = c + dy$ would be obtained from the regression of x on y. Try it for the data in "Let's do it! 13.3" on oil changes.

Computers make finding the least squares regression line easy, especially for very large data sets. However, we are responsible for

- telling the computer which is the explanatory variable, x, and which is the response variable, y;

- examining a scatterplot first to check that a linear model is appropriate and check for outliers;

- interpreting the results of the regression analysis.

When you request a regression analysis from the computer, you typically get a lot of output by default. The standard regression output using the regression analysis option within Microsoft EXCEL for the regression of final exam score, y, on midterm exam score, x, is as follows:

```
Line
1   Regression Statistics

2   Multiple R          0.6911
3   R Square            0.4776
4   Adjusted R Square   0.4549
5   Standard Error      14.021
6   Observations        25

7   Analysis of Variance
8                   df   Sum of Squares  Mean Square  F        Significance F
9   Regression       1   4134.138        4134.138     21.029   0.0001
10  Residual        23   4521.602         196.591
11  Total           24   8655.740

12              Coefficients  Standard Error  t Statistic  p-value    Lower 95%  Upper 95%
13  Intercept   7.521 ← a     14.235          0.528        0.6021     -21.926    36.969
14  Midterm     1.754 ← b      0.3826         4.586        0.0001       0.963     2.546
```

Do not let this output overwhelm you. There is a lot more information presented here than we will need for our purposes. In Line 6, we see that there were $n = 25$ pairs of observations used in this regression analysis, corresponding to the scores for the 25 students. The equation for the least squares regression line is obtained from the numbers under the heading "Coefficients" in Lines 12 through 14. The y-intercept is $a = 7.521$ or approximately 7.5. The slope, the coefficient for x (midterm exam score), is $b = 1.754$ or approximately 1.75. So the least squares regression line for regressing final exam scores, y, on midterm exam scores, x, is given by $\hat{y} = 7.5 + 1.75x$. If you will use the equation for prediction, do not round the values for a and b for more accurate results.

Exercises

13.1 Suppose that 100 recent residential sales of houses in a city are used to fit a least squares regression line relating the sale price, y (in dollars), to the amount of living space, x (in hundreds of square feet). Houses in the sample range from 1500 square feet to 3500 square feet. The resulting least squares equation is

$$\hat{y} = -30{,}000 + 7000x.$$

(a) Can you estimate the mean sale price of a 3000-square-foot house? If yes, give the estimate. If no, explain why not.

(b) Can you estimate the mean sale price of a 5000-square-foot house? If yes, give the estimate. If no, explain why not.

13.2 Data were gathered to estimate the regression line for a model where a gymnast's score (y) depends on the gymnast's physical condition as measured by his or her weight (x). The estimated regression line is given by $\hat{y} = 1190 - 1.9x$. Complete the following sentence: Every additional pound that a gymnast weighs drops the predicted score by about _____ points, on average.

13.3 During World War II, there was a scarcity of foods rich in fat (meat, butter, eggs, etc.) in Norway (and other countries). Along with a decline in the consumption of fat, a decline in the death rate from atherosclerosis was also observed. (Atherosclerosis is the formation of fatty deposits on the walls of the arteries, which can lead to heart disease or stroke.) The following table gives the consumption of fat in kilograms per year per person (x) and the number of deaths from atherosclerosis per 100,000 people in a year (y) for Norway from 1938 to 1947:

Year	1938	1939	1940	1941	1942	1943	1944	1945	1946	1947
Rate (x)	14.4	16.0	11.6	11.0	10.0	9.6	9.2	10.4	11.4	12.5
Deaths (y)	29.1	29.7	29.2	26.0	24.0	23.1	23.0	23.1	25.2	26.1

(a) Produce a scatterplot. Circle the observation corresponding to the year 1940.

(b) What is the equation of the least squares regression line of y on x?

(c) If the consumption of fat were 13 kilograms per year per person, what would be the estimated annual death rate (for Norway) due to atherosclerosis?

13.4 The following data represent trends in cigarette consumption per capita (in hundreds) and lung cancer mortality (per 100,000) for Canadian males:

Cigarette Consumption (x)	11.8	12.5	15.7	19.2	21.9	23.3
Mortality Rate (y)	10.4	16.5	22.9	26.6	33.8	42.8

(a) Construct a scatterplot of the data and briefly describe the overall pattern. (Sketch the scatterplot from your graphing calculator window, use the computer, or sketch by hand.)

(b) Give the equation of the least squares regression line of y = mortality rate on x = cigarette consumption.

(c) Interpret the slope of the regression line. (Be specific.)

(d) Use the least squares regression equation to predict the lung cancer mortality rate when the cigarette consumption per capita is 2000.

13.5 The accompanying table reports percentages of on-time arrivals and on-time departures at 22 major U.S. airports for the fourth quarter of 1991.

(a) Make a scatterplot of these data and briefly describe the overall pattern.

(b) Give the equation of the least squares regression line.

Airport	% On-Time Arrivals	% On-Time Departures
Boston	75.4	83.3
Seattle–Tacoma	75.6	86.6
Los Angeles	76.0	83.9
Chicago	77.0	80.8
Denver	77.0	83.1
Dallas	77.1	84.3
Miami	77.8	88.6
San Francisco	79.6	85.7
Newark	80.1	87.0
Atlanta	81.1	88.1
Minneapolis	81.1	83.8
St. Louis	81.2	85.7
Philadelphia	81.5	86.1
San Diego	82.5	86.5
New York (LaGuardia)	83.1	88.9
Orlando	84.1	91.3
Washington	84.9	89.9
Dulles International	85.8	88.8
Las Vegas	86.0	89.0
Charlotte	86.4	87.6
Detroit	87.9	89.8
Raleigh	90.3	92.6

13.6 Milk samples were obtained from 14 Holstein–Friesian cows, and each was analyzed to determine uric-acid concentration (y), measured in mol/L. In addition to acid concentration, the total milk production (x), measured in kg/day, was recorded for each cow. The data were entered into a computer, and the following regression output was obtained:

```
* * * *  M U L T I P L E   R E G R E S S I O N   * * * *
Equation Number 1    Dependent Variable.. ACIDCONC
Variable(s) Entered on Step Number    1..  MILKPROD


Multiple R            .88789
R Square              .78835
Adjusted R Square     .77071
Standard Error       21.48174


Analysis of Variance
                   DF      Sum of Squares        Mean Square
Regression          1         20625.84845        20625.84845
Residual           12          5537.58012          461.46501
F =      44.69645        Signif F =  .0000


----------------- Variables in the Equation -----------------
Variable            B          SE B         Beta        T    Sig T
MILKPROD       -5.202654     .778195     -.887889    -6.686  .0000
(Constant)    321.241328   23.712311                 13.547  .0000
```

What is the equation of the least squares regression line?

13.4 Residual Analysis

For each data point—that is, for each pair of observations (x,y)—the corresponding residual is the difference between the observed response y and the predicted response $\hat{y}$ for that x. When looking at data, we stated that you should examine the overall pattern and then look for any deviations from that overall pattern. We have focused on the case in which the overall pattern is a linear relationship. The deviations from that overall pattern are the residuals.

Consider the midterm and final exam scores for the 25 students in a statistics class. The original data, the predicted responses, and corresponding residuals are given in the accompanying table. Let's check the first line in the listing. Student 1 scored 39 points on the midterm and 62 points on the final. Using the least squares regression equation, the predicted final exam score is

$$\hat{y} = 7.5 + 1.75(39) = 75.75 \text{ points.}$$

Student Number	x Midterm Score	y Final Score	Predicted $\hat{y}$	Residual
1	39	62	75.94	−13.94
2	44	69	84.71	−15.71
3	32	68	63.66	4.34
4	40	86	77.70	8.30
5	45	88.5	86.47	2.03
6	46	88.5	88.22	0.28
7	33	76	65.41	10.59
8	39	66.5	75.94	−9.44
9	32.5	75	64.54	10.46
10	21	38	44.36	−6.36
11	30	71	60.15	10.85
12	39	88	75.94	12.06
13	44	96.5	84.71	11.79
14	28.5	71.5	57.52	13.98
15	38	96	74.19	21.81
16	43	82.5	82.96	−0.46
17	42	85	81.20	3.80
18	25.5	28	52.26	−24.26
19	47	95	89.98	5.02
20	36	39	70.68	−31.68
21	31.5	58	62.78	−4.78
22	32	49	63.66	−14.66
23	42	62	81.20	−19.20
24	21	59	44.36	14.64
25	41	90	79.45	10.55

The 75.94 in the listing is a bit more accurate, since we rounded off the y-intercept and slope coefficients to report the regression equation. Using the more precise y-intercept and slope, we find that the predicted final exam score is

$$\hat{y} = 7.521396 + 1.754348(39) = 75.94.$$

The residual for the first observation is then

$$e_1 = \text{observed response} - \text{predicted response}$$
$$= y_1 - \hat{y}_1 = 62 - (75.94) = -13.94.$$

You can verify the residual for another student.

❖ EXAMPLE 13.4 Finding Residuals

A sociologist was hired by a university to investigate the relationship between the number of unauthorized days that employees are absent per year and the number of years the employee has worked at the university. The data and corresponding scatterplot are as follows:

x = Number of years working at the university	y = Number of absent days
1	8
3	6
4	7
6	7
8	6
10	4
12	5
14	4
18	2

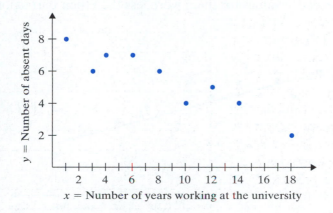

The least squares regression line for these data is given by

$$\hat{y} = 8.09 - 0.313x.$$

The following table provides data, the predicted values, and corresponding residuals:

x	y	$\hat{y}$	residuals = $y - \hat{y} = e$
1	8	7.777	0.223
3	6	7.151	−1.151
4	7	6.838	0.162
6	7	6.212	0.788
8	6	5.586	0.414
10	4	4.960	−0.960
12	5	4.334	0.666
14	4	3.708	0.292
18	2	2.456	−0.456

The TI can be used to obtain the residuals. With the *x* data in L1 and the *y* data in L2, the steps for obtaining and storing the residuals in L4 are as follows:

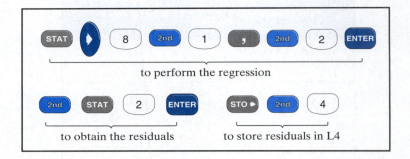

Note that one property of the residuals is that they sum up to zero. If we sum the residuals in this example we obtain −0.022, which is close but not exactly zero due to round-off errors.

Also notice that some residuals are positive and some residuals are negative Does this make sense? Let's look at the scatterplot with the regression line superimposed on the plot. The three colored points that lie below the regression line produced the three negative residuals because the observed *y* values for them were less than their corresponding $\hat{y}$ values.

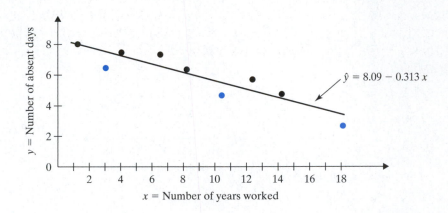

Since residuals measure how far the observations are from the regression line, they are often used to assess the *fit* of the regression line to the data. We might display these vertical deviations graphically using a **residual plot**. By plotting the residuals against the explanatory variable *x*, we effectively magnify the deviations (that is, change the *y*-axis from response to vertical deviations), which allows for a better and closer examination of the deviations. Since the residuals from the least squares regression line always sum to zero (or very close to zero due to round-off error), a horizontal reference line at the value zero is included on a residual plot. We should see an **unstructured horizontal band of points centered around zero** such as in Figure A, if the linear model is appropriate. Often the calculation and plotting of the residuals is done with the help of a calculator or computer. For details on making a residual plot with a TI graphing calculator, see Example 13.4 and the ■ TI Quick Steps at the end of this chapter.

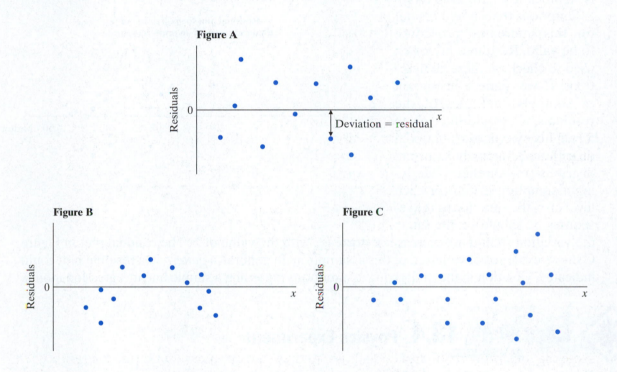

In Figure B, the residuals show a slight curved pattern. There appears to be some curvature in the data that is evident in the residuals, but perhaps not so evident on the larger scaled scatterplot of *y* versus *x*. This implies that the linear model may not be appropriate. Perhaps a model that also includes a quadratic explanatory variable, x^2, may provide a better fit to the data. In Figure C, the residuals show a slight increase in variation with the larger values of *x*. A trumpet-shaped residual plot in which the residuals fan out tells us that the fit for values of *x* at one end will not be as good as the fit for values of *x* at the other end. For Figure C, the predicted responses will be less precise for the larger values of *x*.

The accompanying figures show the scatterplot and corresponding residual plot for our midterm and final exam example. Since the residual plot shows a random scatter of points centered around zero, we conclude it appears that a linear regression model is appropriate.

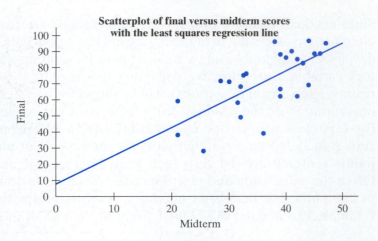

In the next section, we will discuss the statistical test for assessing if the linear relationship is significant. There are certain assumptions that should be met for this inferential procedure to be valid. Residuals are often used to check on these assumptions. If we made a histogram or stem-and-leaf plot of the residuals, we would like to see a roughly symmetric, unimodal distribution—actually a normal distribution. Another underlying assumption is that for each level of x, the variation in the responses, y, should be the same;

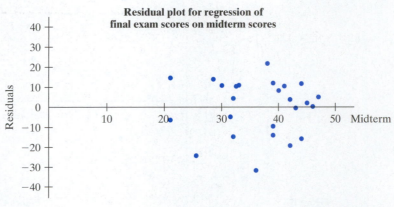

the variation should not increase or decrease with the value of x. The residual plot in Figure C shows evidence of violation of this assumption. In general, a *pattern* in a residual plot could indicate a violation in the underlying assumptions regarding a simple linear regression model.

L e t's d o it! 13.4 Physics Experiment

Students in a physics class are studying free fall to determine the relationship between the distance an object has fallen and the amount of time since release. The scatterplot and residual plot for the data are shown.

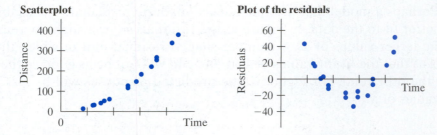

Based on these plots, does the linear regression model seem appropriate? Explain.

❖ EXAMPLE 13.5 **Laws of Motion**

When Galileo was formulating the laws of motion, he considered the motion of a body starting from rest and falling just under gravity. He first thought that the velocity of a falling body under that condition was proportional to the distance it had fallen. From the following experimental data, can you accept Galileo's original hypothesis?

x = distance(ft)	y = velocity(ft/sec)
0	0
1	8
2	11.3
3	13.9
4	16

The following is a scatterplot of these data:

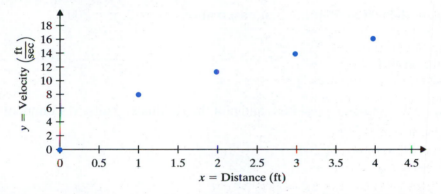

The least squares regression line is $\hat{y} = 2.26 - 3.79x$. When we superimpose the regression line onto the scatterplot, we produce the following picture:

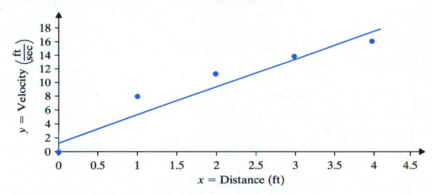

The corresponding residuals are provided in the next table:

x = distance	y = velocity	residual
0	0	−2.26
1	8	1.95
2	11.3	1.46
3	13.9	0.27
4	16	−1.42

Note: The sum of the residuals is zero. The residual values that lie below the regression line are negative. The residual values that lie above the regression line are positive. ❖

Let's do it! 13.5 Oil-Change Data (revisited)

Refer back to the oil-change example in "Let's do it! 13.3."

Number of oil changes per year (x)	3	5	2	3	1	4	6	4	3	0
Cost of repairs ($) ($y$)	300	300	500	400	700	400	100	250	450	600

(a) Calculate the least squares regression line:_____

(b) Calculate the residual for the first observations ($x = 3$, $y = 300$).

(c) Calculate the remaining residuals and plot the residuals against x. Comment on the plot.

Does the linear model seem appropriate? Explain.

13.5   Influential Points and Outliers

So far, we have primarily focused on the overall pattern of a scatterplot. Sometimes, there can be individual points that lie outside the pattern. There are two main types of unusual points in regression—an **outlier** and an **influential point**. More formally, an **outlier** is an observation that is far from the predicting line, and so produces a large, positive or negative, residual. A point is said to be **influential** in the regression setting if removing it markedly changes the position of the regression line. A point that is at an extreme position on the *x*-variable is likely to have a strong influence on the position of the regression line. There are various diagnostic procedures that measure the degree of influence for each observation. One such measure is called Cook's *D* statistic. Although discussion of such procedures go beyond the scope of this text, they are generally available in most statistical computer packages. The following two sets of data show us that influential points do not have to be outliers and that outliers are not necessarily influential.

Data Set I   Consider the following set of data and corresponding scatterplot:

x	y	
1	1	
1	2	
2	1.5	
2.5	2.5	
3	3	
3.5	3	
4	3.5	
4	4	
4.5	4	
5	5	
5	6	
5.5	6	
2	6	← **Point A**

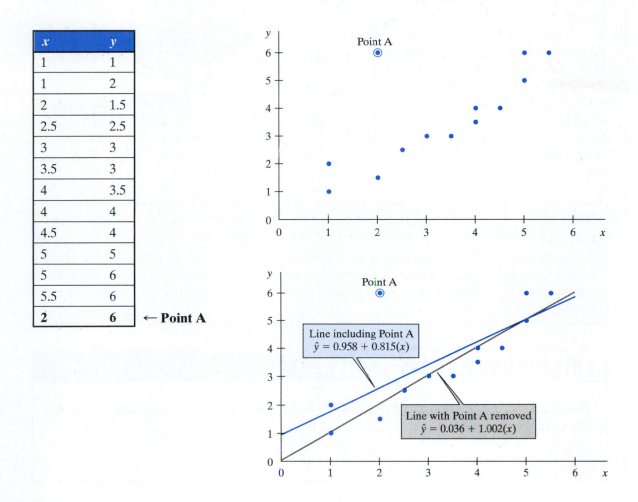

Point A produces a very large, positive, residual and appears to be an outlier. However, it is not very influential since removing it does not change the estimated regression line very much. Thus, Point A is an *outlier, but is not influential.*

Data Set II Consider the following set of data and corresponding scatterplot:

x	y
1	3
1.5	2
2	3
2	4
2.5	1
2.5	2
3	1
3	2
3	3
3.5	2
4	1
7	**7**	← **Point B**

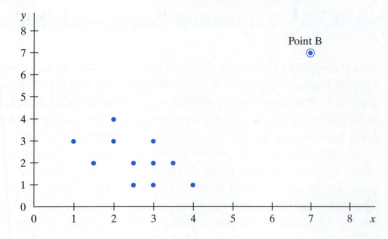

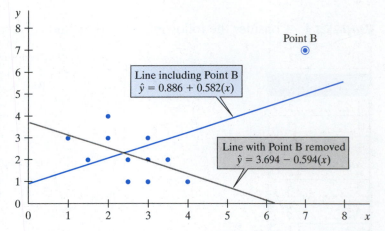

Point B does not produce the largest residual. The point at $x = 4$ and $y = 1$ generates the largest residual. However, Point B is very influential since removing it markedly changes the estimated regression line. The slope goes from being positive to being negative. Thus, Point B is *influential, but is not an outlier.*

Definitions: An **outlier** in regression is an observation with a residual that is unusually large (positive or negative) as compared to the other residuals.

An **influential point** in regression is an observation that has a great deal of influence in determining the regression equation. Removing such a point would markedly change the position of the regression line. Observations that are somewhat extreme for the value of x are often influential.

THINK ABOUT IT

Observations that are somewhat extreme for the value of x are often influential. Are they necessarily outliers?

Let's do it! 13.6 Influential or Outlier?

Consider the two sets of data and corresponding scatterplots shown. Two points are marked with the labels C and D. Some possible descriptions are also provided. Which description best matches which point? You may wish to compute the estimated regression line and residual including the point to assess if it is an outlier, and then remove the point to assess how influential it is.

Data Set III

x	y
0	0
0	1
1	1
0.5	2
1	2
1.5	2
1.5	3
2	3
2	4
2.5	4
3	5
3	6
3.5	5.5
5	**3**	← **Point C**

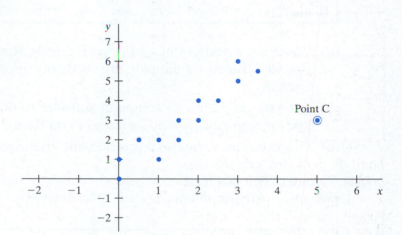

Data Set IV

x	y
1	3
1.5	2
2	3
2	4
2.5	1
2.5	**2**	← **Point D**
3	1
3	2
3	3
3.5	2
4	1

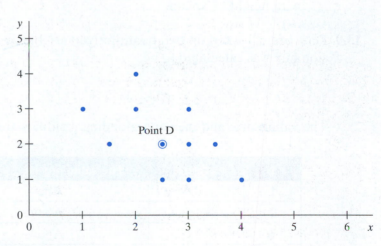

Descriptions

1. Outlier, but not influential
2. Outlier and influential
3. Influential, but not an outlier
4. Neither influential nor an outlier

Point C = _____   Point D = _____

✳ Exercises

13.7 The following are the golf scores of 12 members of a women's golf team in two rounds of tournament play (a golf score is the number of strokes required to complete the course, so low scores are better):

Player	1	2	3	4	5	6	7	8	9	10	11	12
Round 1 (x)	79	91	88	83	106	101	81	85	96	88	91	90
Round 2 (y)	81	89	90	86	90	105	78	81	89	90	96	93

(a) Construct a scatterplot of the scores from Round 2 versus the scores from Round 1. Based on the plot, are there any apparent influential observations? Explain.

(b) Use your calculator or computer software to find the least squares regression line for regressing Round 2 scores (y) on Round 1 scores (x).

(c) Graph this least squares regression line on the scatterplot in (a).

13.8 Consider the following data, where x = height in inches and y = average weight in pounds for American females aged 30 to 39 years:

Height (x)	58	60	62	64	66	68	70	72
Average weight (y)	113	118	124	131	137	145	153	164

The least squares regression line is given by: $\hat{y} = -96.9 + 3.58x$.

(a) Provide a sketch of the corresponding residual plot.

(b) Based on the residual plot, does it appear that average weight increases linearly with height? Explain.

13.9 Data are collected on two quantitative variables x and y. The least squares regression line was calculated to be

$$\hat{y} = -2 + 7x.$$

The values of x and the corresponding residuals are given in the following table:

x	residual	The value of y is _____ the regression line.		
1	2.0	Above	Below	On
2	−1.5	Above	Below	On
3	−1.0	Above	Below	On
4	3.0	Above	Below	On
5	−2.5	Above	Below	On
6	0.0	Above	Below	On

For each observed x value, circle the appropriate description for the corresponding observed y value.

13.10 A high **leverage point** is a point that has a very high or very low value of the independent variable (i.e., an outlier in the x values). A high leverage point that at the same time is an outlier of the dependent variable is known as a highly influential point. The data that follow were collected from 24 mortgage applications. The variable x is the yearly income (in thousands) of home purchasers and the variable y is the sales price of the house (in thousands).

obs	x	y
1	33.5	110
2	34	100
3	35.9	116
4	36	110
5	39	125
6	39	119.9
7	40.5	130.6
8	40.9	120.8
9	42.5	129.9
10	33	99.9
11	32	105
12	31.9	101

obs	x	y
13	31.5	99.5
14	31	102.5
15	30	93.5
16	29.2	96.5
17	28.5	94
18	25	84.9
19	70	185
20	65	110
21	54.6	170
22	50	150.7
23	45	140
24	44	135.5

(a) Construct the scatterplot for this data.

(b) Are there any high leverage points? If so, which observation(s)?

(c) Report the least squares regression line both with and without the influential points (if any).

(d) Did the equation of the line change greatly or not much?

13.6   Statistically Significant Relationship?

Researchers must rely on data from only a sample in order to assess whether a relationship exists between two variables. Since the values for the coefficients a and b are based on information from a sample, these values are sample statistics. These statistics are used as estimates of the corresponding parameters in the population. The slope and y-intercept for the population are (unknown) parameters. Even though a relationship may be apparent in the sample, it is possible that it will not extend to the population. Researchers use statistics to assess the significance of an observed relationship by measuring the chance that a relationship as strong or stronger would be observed, assuming that there really is no relationship in the population. This chance is a p-value and is used for deciding between two hypotheses.

To better understand the actual hypotheses that this *p*-value corresponds to, think about the following questions:

T**HINK** A**BOUT** I**T** Slope of Zero

Consider the equation of a line for relating two variables: $\hat{y} = a + bx$.

Suppose that the *y*-intercept, *a*, is equal to 10 and slope, *b*, is equal to zero.

What would be the value of the response, $\hat{y}$, if *x* were equal to 2?

What would be the value of the response, $\hat{y}$, if *x* were equal to 12?

What would be the value of the response, $\hat{y}$, for any value of *x*?

What would it mean if the slope for the regression line for a population were equal to zero?

A question of interest in regression is whether or not the estimated slope, based on a sample, is "significantly" different from zero. If in the population the slope were equal to zero, the response variable *y* would not depend (linearly) on the explanatory variable *x* at all. The so-called explanatory variable would not be very *explanatory* with respect to helping to predict the response variable *y* in this case. If the estimated slope is significantly different from 0, we say that the observed relationship is statistically significant. The hypotheses of interest in linear regression are as follows:

H_0: The slope of the linear regression line using all of the population values is equal to zero.

H_1: The slope of the linear regression line using all of the population values is *not* equal to zero.

This is one of the primary inferences made in linear regression, assessing whether there is a significant linear relationship, with a nonzero slope, between the two variables. Recall that the *p*-value is a value between 0 and 1 that measures how likely the observed result or something even more extreme, is if the null hypothesis H_0 is true. A small *p*-value indicates that the observed relationship, or a relationship that is even stronger, is unlikely if the null hypothesis is true. The smaller the *p*-value, the more support for the alternative hypothesis H_1.

Let's take a look at the output for the midterm and final exam scores data to see where this *p*-value is reported.

```
Line
1    Regression Statistics

2    Multiple R           0.6911
3    R Square             0.4776
4    Adjusted R Square    0.4549
5    Standard Error       14.021
6    Observations         25

7    Analysis of Variance
8                      df   Sum of Squares  Mean Square  F        Significance F
9    Regression        1    4134.138        4134.138     21.029   0.0001 ← p-value
10   Residual          23   4521.602         196.591
11   Total             24   8655.740

12                Coefficients   Standard Error   t Statistic   p-value           Lower 95%   Upper 95%
13   Intercept    7.521 ← a       14.235          0.528        0.6021            -21.926      36.969
14   Midterm      1.754 ← b        0.3826         4.586        0.0001 ← p-value    0.963       2.546
```

Lines 12 through 14 are used to assess the significance of the respective coefficients. Line 14 presents information about the slope. The *p*-value in Line 14 of 0.0001 is used to assess whether the estimated slope is statistically significantly different from zero. Since this *p*-value is very small, we have strong evidence for supporting H_1. There appears to be a significant positive linear association between midterm and final exam scores.

Line 13 presents information about the *y*-intercept. The *p*-value in Line 13 is used to test whether the *y*-intercept for the population linear regression is equal to zero. In many cases, it is not important to have the linear relationship go through the origin, so this is not often a test of interest. We will discuss later the details behind the analysis of variance output presented in Lines 7 through 11. We will note that the test performed in Line 9 is the same test for the slope performed in Line 14 in the case of the linear regression model with one explanatory variable. The *p*-value in Line 9, located under the heading of Significance F, is the same as in Line 14. In Chapter 12, we saw one type of analysis of variance and discussed the corresponding *F*-test. The Multiple R entry in Line 2 will be discussed in Section 13.7 on correlation.

❖ EXAMPLE 13.6   Service Time

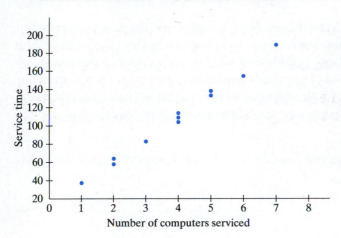

A computer-repair technician recorded data on the number of computers serviced and the amount of time to complete the service for 11 randomly selected service visits. The number of computers serviced, *x*, ranged from 1 to 7, while the time to complete the service, *y*, ranged from about 30 minutes to nearly 3.5 hours. The scatterplot of the data shows a strong, positive linear relationship between these two variables.

Using a statistical computing package called SPSS, the following regression output was obtained:

```
        * * * *  M U L T I P L E   R E G R E S S I O N * * * *
    Equation Number 1    Dependent Variable..    TIME
    Variable(s) Entered on Step Number    1..    NUMBERPC
    Multiple R            .99643
    R Square              .99287
    Adjusted R Square     .99207
    Standard Errorv       4.02524


    Analysis of Variance
                    DF        Sum of Squares      Mean Square
    Regression       1           20296.35861      20296.35861
    Residual         9             145.82320         16.20258
    F =     1252.66228      Signif F =  .0000


    ----------------- Variables in the Equation ------------------
    Variable         B              SE B       Beta      T        Sig T
 *  NUMBERPC      24.834254 ← b    .701672    .996427  35.393    .0000 ← p-value
 ** (Constant)    10.193370 ← a   2.999410             3.398    .0079
```

(a) Obtain the estimated linear regression equation $\hat{y} = a + bx$.

Note: The column with the heading "B" contains the values for b and a.

SPSS reports the estimated slope first, in the * row, and the y-intercept information in the ** row.

$$\hat{y} = 10.19 + 24.83x$$

(b) Is there evidence of a significant (nonzero) linear relationship between the number of computers serviced and the service time? Explain.

Note: SPSS uses the header "Sig" for presenting the p-value of a test.

Since the p-value for testing the hypotheses H_0 (the slope of the linear regression equation using all the population values is equal to zero) versus H_1 (the slope of the linear regression equation using all the population values is *not* equal to zero) is nearly zero (p-value < 0.00005), we conclude that there does appear to be a significant (nonzero) linear relationship between the number of computers serviced and the service time. The number of computers serviced is a significant linear predictor of service time.

(c) Predict the number of minutes required for service when it is reported that five computers are down.

$$\hat{y} = 10.19 + 24.83(5) = 134.3 \, \text{minutes}$$ ❖

Let's do it! 13.7 Size of Homes and Selling Price

There are many factors that affect the selling price of a home. The total dwelling size and the assessed value are just two factors. Data were gathered on homes in a Milwaukee, Wisconsin, neighborhood. A scatterplot revealed a linear relationship between the total dwelling size of a home in 100 square feet and its selling price in dollars. The following is the regression output for the least squares regression of selling price on total dwelling size.

```
Dep var: PRICE      N:   20    Multiple R:  .913    Squared multiple R:  .834
Adjusted squared multiple R:   .825    Standard error of estimate: 3377.192


Variable    Coefficient    Std error    Std coef Tolerance    T      P(2 tail)
CONSTANT    11947.010      4748.133      0.000    .             2.516    0.022
SIZE         2749.622       288.980      0.913    .100E+01      9.515    0.000


                     Analysis of Variance
   Source    Sum-of-squares    DF   Mean-square    F-ratio      P
Regression    .103257E+10      1    .103257E+10    90.533     0.000
  Residual    .205298E+09     18    .114054E+08
```

(a) How many homes were included in this study?

(b) Obtain the least squares regression line for predicting selling price from the size of the home.

(c) Is there evidence of a significant (nonzero) linear relationship between price and size? Explain.

(d) The total dwelling size for another home in this neighborhood is 1620 square feet. Use the least squares regression equation to estimate the selling price of this home.

✳ Exercises

13.11 You suspect that the number of grievances filed by workers at an assembly plant is related to the local unemployment rate. From a scatterplot of the data, it seems appropriate to fit a linear model for predicting the number of grievances from the unemployment rate (in percent). The regression output is given:

```
Dep var: GRIEV   N:    7    Multiple R: .951    Squared multiple R:  .904
Adjusted squared multiple R: .885      Standard error of estimate:  4.316


   Variable    Coefficient    Std error    Std coef  Tolerance    T      P(2 tail)
CONSTANT          137.083       12.639        0.000      .        10.846    0.000
UNEMPLO           -10.877        1.584       -0.951   .100E+01    -6.869    0.001


                              Analysis of Variance
    Source    Sum-of-squares    DF   Mean-square    F-ratio      P
Regression        1166.472       1     1166.472      7.073     0.001
  Residual         494.728       3      164.909
```

(a) Is the association between number of grievances (*y*) and unemployment rate (*x*) positive or negative? Explain how the regression equation can tell you the answer to this question.

(b) The local unemployment rate was 8% last month. Predict the number of grievances for that month.

(c) Is there evidence of a significant (nonzero) linear relationship between the unemployment rate and the number of grievances? Use a 5% significance level.

13.12 Milk samples were obtained from 14 Holstein–Friesian cows, and each was analyzed to determine uric-acid concentration (*y*), measured in mol/L. In addition to acid concentration the total milk production (*x*), measured in kg/day, was recorded for each cow. The data were entered into a computer, and the following regression output was obtained:

```
        * * * *   M U L T I P L E    R E G R E S S I O N    * * * *
Equation Number 1    Dependent Variable.. ACIDCONC
Variable(s) Entered on Step Number    1..  MILKPROD


Multiple R          .88789
R Square            .78835
Adjusted R Square   .77071
Standard Error    21.48174


Analysis of Variance
                    DF       Sum of Squares      Mean Square
Regression           1         20625.84845      20625.84845
Residual            12          5537.58012        461.46501
F =      44.69645       Signif F =  .0000


------------------ Variables in the Equation ------------------
Variable             B        SE B        Beta        T    Sig T
MILKPROD      -5.202654    .778195    -.887889    -6.686   .0000
(Constant)   321.241328  23.712311                13.547   .0000
```

(a) Give the equation of the least squares regression line.

(b) Using the variable names and units, explain what the value of the slope represents.

(c) Is there evidence of a significant (nonzero) linear relationship between these two variables? Explain your answer, including a description of the information in the preceding output that allows you to answer this question.

13.7 Correlation: How Strong Is the Linear Relationship?

Sometimes a researcher is not concerned with quantifying the relationship between two variables with an equation but is only interested in measuring the extent to which the two variables are related. A scatterplot displays the form, direction, and strength of the relationship between the two quantitative variables. Interpretation of the strength based on how tightly clustered the points are about the underlying form in the scatterplot is somewhat subjective and can even be deceiving. A change in the scaling of the axes or zooming in with the graphing calculator can change your perception of strength, as shown in the following figures:

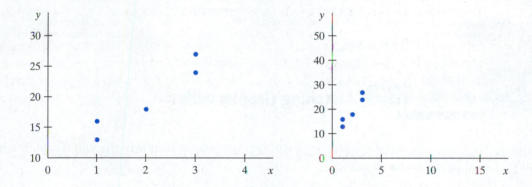

In this section, we turn to a numerical measure of the strength of the linear relationship that can add to our visual perception of the strength, called the sample **correlation coefficient**, denoted by r.

> *Definition:* The sample **correlation coefficient r** measures the strength of the **linear relationship** between two quantitative variables. It describes the direction of the linear association and indicates how closely the points in a scatterplot are to the least squares regression line.

Here are some of the features of a correlation coefficient.

1. Range	$-1 \le r \le +1$. Because of the way it is defined, values for the correlation coefficient are always between -1 and $+1$.
2. Sign	The sign of the correlation coefficient indicates direction of association—negative $[-1, 0)$ or positive $(0, +1]$.

3. Magnitude

The magnitude of the correlation coefficient indicates the strength of the linear association. If the data follow a straight line, $r = +1$ (if the slope is positive) or $r = -1$ (if the slope is negative), indicating a perfect linear association. If $r = 0$, then there is no linear association.

4. Measures Strength

The correlation coefficient only measures the strength of the *linear* association. It is important to look at the scatterplot first to examine the type of association that is present.

5. Unitless

The correlation coefficient is computed using standard scores of the two variables. It has no unit of measure, and the absolute value of r will not change if the units of measurement for x or y are changed. The correlation between x and y is the same as the correlation between y and x.

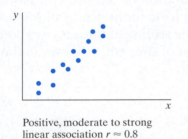

Positive, moderate to strong
linear association $r \approx 0.8$

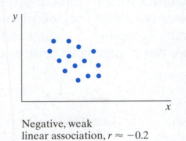

Negative, weak
linear association, $r \approx -0.2$

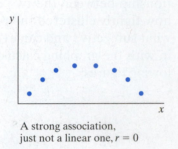

A strong association,
just not a linear one, $r = 0$

Let's do it! 13.8 Matching Graphs with *r*

Assuming that the following graphs are displayed on approximately the same scales, which value of r best matches each graph?

$r = 0$

$r = +1$

$r = -1$

$r = 0.6$

$r = -0.2$

$r = -0.8$

$r = 0.1$

Graph A _____

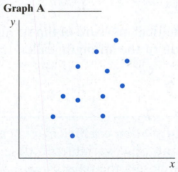

Graph B _____

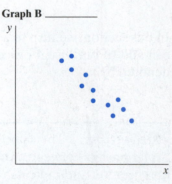

Graph C _____

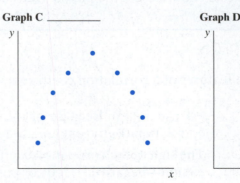

Graph D _____

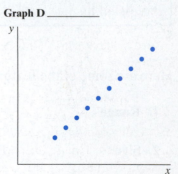

Let's do it! 13.9 Is There an Error?

Identify and explain the error in each of the following situations. If there is no error, write *no error*.

(a) Our study shows that the correlation between the number of hours a child spends watching television per week and the child's reading level is $r = -1.04$.

(b) We found that the correlation between a voter's political party and income is $r = 0.38$.

Given that the scatterplot shows a linear relationship is plausible, a strong correlation, a value of r close to $+1$ or -1, tells us that the two variables are closely associated in a linear way. A positive correlation implies that they increase together, while a negative correlation implies that as one variable increases the other tends to decrease. We cannot stress enough the importance of examining the scatterplot along with the reporting of a correlation. Correlations can be misleading. Any study dealing with relationships must be on the lookout for possible confounding variables. Even a very strong correlation does not imply that changes in the explanatory variable will actually cause changes in the response variable. Outliers can also have quite an impact on correlations. The next examples illustrate these cautionary notes regarding correlation.

❖ EXAMPLE 13.7 Watch Out for Third Variables

Consider the accompanying scatterplot. The correlation is positive, indicating that an increase in the number of TV hours watched by children corresponds to an increase in the reading level of the children. What is going on here?

Details of the data tell you that the children were of different ages, ranging from first grade through third grade.

The lower scatterplot at the right shows you the data for the three separate grades, denoted by the different plotting symbols. If you examine the data for each grade separately, you see that the correlation coefficient within each grade is negative, not positive. The variable *grade* is a third variable that is *confounded* with the response, reading level.

The comparison between the two variables is complicated because the subjects are at different grade levels. *Aggregating the results* across the three grades is not appropriate. This is an example of Simpson's paradox, which we will see again in Chapter 14.

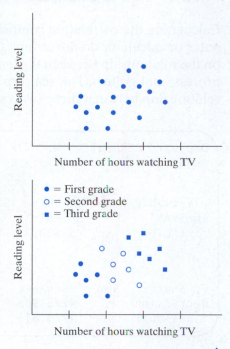

❖

❖ **EXAMPLE 13.8** **Correlation Does Not Imply Causation!** _____

Consider two variables: x = height of a person and y = salary of a person. A scatterplot of data on these two variables shows a positive association, with a positive value for the correlation coefficient r. Does your height cause you to earn a certain salary level? Of course not. Upon examining the data further, you find that the ages of the people sampled range from 12 years old to 30 years old. Obviously, age influenced both height and salary. This positive association is due to **common response**—the observed association is explained by a third variable that is common to both variables. This common response creates an association even though there is no direct causal link between the two variables. ❖

❖ **EXAMPLE 13.9** **Correlation Is Not Resistant to Unusual Points** _____

Consider the accompanying data and scatterplot.

x	2	3	4	5	6	7	8	9	10	*15*
y	3	4	3	6	5	7	7	9	8	6

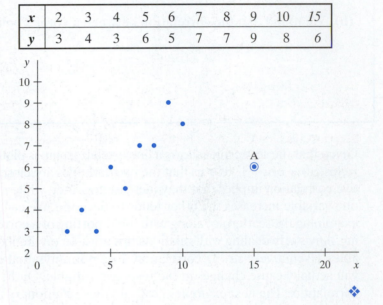

The point A in the scatterplot ($x = 15$, $y = 6$) is an observation that stands apart from the remaining points.

The correlation coefficient for these data **with** point A is $r = 0.648$. If we remove the observation and recompute the correlation coefficient, we have $r = 0.927$.

Did r change? How did it change? The correlation coefficient can change dramatically when an outlier or an influential point is removed. ❖

 13.7.1 **How to Calculate r**

Calculating the correlation coefficient can be quite cumbersome, so we often have a computer or calculator do the arithmetic for us. Consider a study by the PGA tour that focused on the relationship between the length of putt (in feet) and the percentage of putts made by professional golfers. The scatterplot was examined and indicated an approximate linear relationship between these two variables. The computer output is provided:

```
Dep var: PERCENT    N:   18    Multiple R:   .840    Squared multiple R:   .706
Adjusted squared multiple R:   .688      Standard error of estimate:    10.547


   Variable  Coefficient    Std error    Std coef Tolerance     T      P(2 tail)
  CONSTANT        96.681       6.289         0.000  .          15.373    0.000
    LENGTH        -5.970       0.963        -0.840  .100E+01   -6.201    0.000

                             Analysis of Variance
     Source   Sum-of-squares    DF   Mean-square      F-ratio        P
  Regression       4276.908      1      4276.908        38.448      0.000
    Residual       1779.815     16       111.238
```

In the first line of output, you have the dependent variable y = percent, the number of observations $n = 18$, and the "Multiple R: .840," which is the absolute value of the correlation coefficient r. The sign of r can be determined by looking at the sign of the slope, which here is -5.970. Hence the correlation between length of putt and percentage of putts made is $r = -0.84$. The least squares regression line is given by $\hat{y} = 96.68 - 5.97x$, where x is length. The p-value for testing if the slope is significantly different from zero is found under the heading P in the analysis of variance of the P(2 tail) value in the row corresponding to the variable length. We conclude that the estimated slope is significantly different from zero since the p-value is very small, less than 0.0005.

There are a number of equivalent formulas for expressing the correlation coefficient. Which one to use depends on what you have already computed from the data.

$$r = \frac{1}{(n-1)} \sum \left(\frac{x_i - \bar{x}}{s_X} \right) \left(\frac{y_i - \bar{y}}{s_Y} \right) = \frac{n(\sum x_i y_i) - (\sum x_i)(\sum y_i)}{\sqrt{n(\sum x_i^2) - (\sum x_i)^2} \sqrt{n(\sum y_i^2) - (\sum y_i)^2}}$$

As usual, the sums in the preceding expressions run over all of the observed values. Sometimes the subscripts on the x and y variables are omitted for easier reading. Often the calculation of the correlation is done with the help of a calculator or computer. For details on finding the correlation with a TI graphing calculator, see Example 13.11 and the ■ TI Quick Steps at the end of this chapter. We will also see where these values are located in the regression output for various computer packages. The right-hand-side expression is often easier to use if you have to compute these values by hand or have already worked through the calculation table used in getting the least squares regression line.

❖ **EXAMPLE 13.10 Getting the Correlation for Test 1 versus Test 2 (by hand)** _____

We already have computed the summation quantities in Example 13.2 that are needed for finding r. The completed calculation table is shown.

Completed Calculation Table

Column	1 x_i	2 y_i	3 x_i^2	4 $x_i y_i$	5 y_i^2
	8	9	64	72	81
	10	13	100	130	169
	12	14	144	168	196
	14	15	196	210	225
	16	19	256	304	361
Total	$\sum x_i = 60$	$\sum y_i = 70$	$\sum x_i^2 = 760$	$\sum x_i y_i = 884$	$\sum y_i^2 = 1032$

Using these values, we can compute the correlation coefficient:

$$r = \frac{n(\sum x_i y_i) - (\sum x_i)(\sum y_i)}{\sqrt{n(\sum x_i^2) - (\sum x_i)^2} \sqrt{n(\sum y_i^2) - (\sum y_i)^2}}$$

$$= \frac{5(884) - (60)(70)}{\sqrt{5(760) - (60)^2} \sqrt{5(1032) - (70)^2}} = 0.965$$

The large positive corre-
lation coefficient and the
accompanying scatterplot
indicate a strong, positive,
linear association between
Test 1 and Test 2 scores.

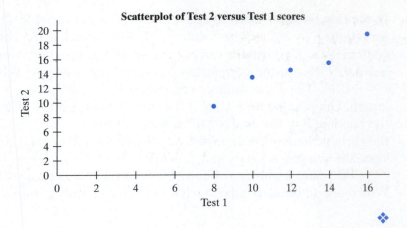

Scatterplot of Test 2 versus Test 1 scores

❖

❖ EXAMPLE 13.11 Birth Rates—Obtaining the Correlation Using the TI

We gathered data from 1970 for 12 nations on the percentage of women aged 14 or older
who were economically active and the crude birth rate. (We define the crude birth rate as
the number of births in a year per 1000 population size.) We are interested in the relationship of the crude birth rate (y) and the percentage of women who were economically active
(x), as given in the following table:

Nation	x	y
Algeria	2	48
Argentina	19	21
Denmark	34	14
E. Germany	40	11
Guatemala	8	41
India	12	37
Ireland	20	22
Jamaica	20	31
Japan	37	19
Phillipines	19	42
USA	30	15
Soviet Union	46	18

The corresponding scatterplot is

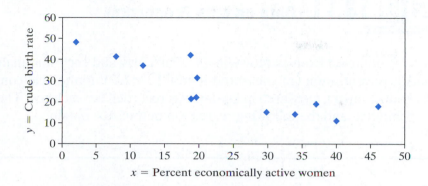

Since the scatterplot shows a negative, approximate linear association, performing a simple linear regression seems valid. The least squares regression line for these data is given by

$$\hat{y} = 45.7 - 0.8x.$$

The estimated slope of -0.8 can be interpreted as follows: For every 1% increase in the percentage of women economically active, the crude birth rate is expected to decrease by about 0.8.

The correlation coefficient for these data is $r = -0.86$, showing a somewhat strong negative linear association. To get the regression line and the correlation coefficient using the TI, we first need to turn on the diagnostic option. If the x data are in L1 and the y data are in L2, then the steps are as follows:

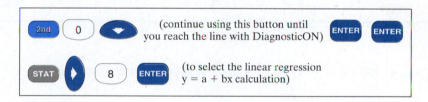

Let's do it! 13.10 Oil-Change Data (revisited)

Refer back to the oil-change exercise in "Let's do it! 13.3"

(a) Based on your scatterplot, give an estimate for the value of the correlation coefficient.

Estimate of $r =$ _____.

(b) Use the data and compute the correlation coefficient and interpret it.

Let's do it! 13.11 Data on Milk Production

Milk samples were obtained from 14 Holstein–Friesian cows, and each was analyzed to determine uric-acid concentration (y), measured in mol/L. In addition to acid concentration, the total milk production (x), measured in kg/day, was recorded for each cow. The data were entered into a computer, and the following regression output was obtained:

```
* * * *   MULTIPLE  REGRESSION   * * * *
Equation Number 1    Dependent Variable.. ACIDCONC
Variable(s) Entered on Step Number    1.. MILKPROD


Multiple R            .88789
R Square              .78835
Adjusted R Square     .77071
Standard Error      21.48174


Analysis of Variance
                  DF        Sum of Squares        Mean Square
Regression         1          20625.84845        20625.84845
Residual          12           5537.58012          461.46501
F =      44.69645      Signif F =  .0000


------------------ Variables in the Equation ------------------
Variable          B         SE B       Beta        T   Sig T
MILKPROD      -5.202654    .778195  -.887889    -6.686   .0000
(Constant)   321.241328  23.712311              13.547   .0000
```

(a) What is the equation of the least squared regression line?

(b) What is the correlation between x and y? $r =$ _____.

(c) We are interested in whether the linear relationship is significant. To find out, we test the following hypotheses:

> H_0: The slope of the linear regression line using all the population values equals zero.

> H_1: The slope of the linear regression line using all the population values does not equal zero.

Circle the p-value in the output that is used to test these hypotheses.

At the level of significance of 0.05, we would (circle one)

> accept H_0.   reject H_0.   can't tell.

 **13.12** **Three Data Sets**

Shown in the accompanying table are three regression data sets—that is, three sets of (x, y) data.

Analyze the three data sets. Examine the scatterplot for each pair, and find the corresponding regression line and correlation coefficient.

Data Set I		Data Set II		Data Set III	
x	y	x	y	x	y
8	10.04	8	11.14	8	9.46
6	8.95	6	10.14	6	8.77
11	9.58	11	10.74	11	14.74
7	10.81	7	10.77	7	9.11
9	10.33	9	11.26	9	9.81
12	11.96	12	10.10	12	10.84
4	9.24	4	8.13	4	8.08
2	6.26	2	5.10	2	7.39
10	12.84	10	11.13	10	10.15
5	6.82	5	9.26	5	8.42
3	7.68	3	6.74	3	7.73

Data Set I

Scatterplot   Regression Line   Correlation

Data Set II

Scatterplot   Regression Line   Correlation

Data Set III

Scatterplot   Regression Line   Correlation

Is the linear regression model appropriate for all three data sets? Explain why or why not.

Let's stop for a moment and summarize the underlying ideas in performing a linear regression analysis. Some notes of caution are also listed.

Regression Analysis Steps

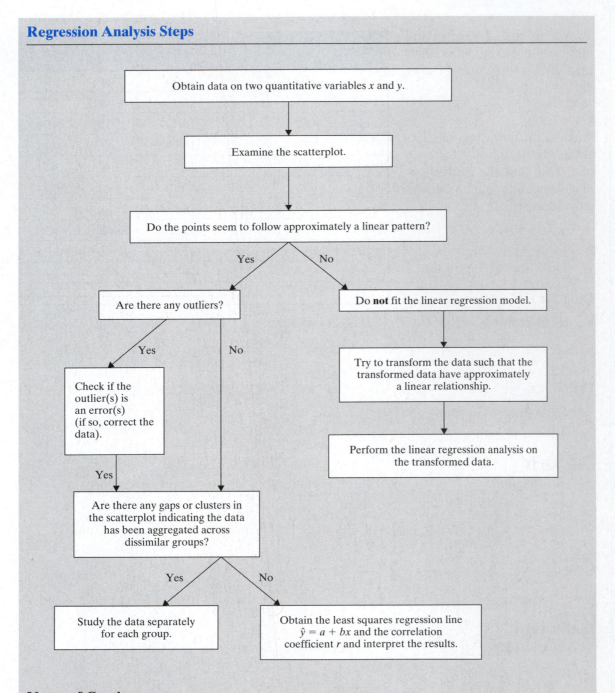

Notes of Caution

Do not use the correlation and linear regression model if . . .

- the relationship is explained better by a different curve or pattern other than a linear function.
- outliers are present.
- the data set appears to be an aggregate of two or more nonhomogeneous sets of data.

THINK ABOUT IT

Suppose that you have a set of data on x and y and you calculate the correlation coefficient r. Would the value of r change if you removed one of the data points? If so, would it decrease or increase? Does it depend on which point is removed?

Let's do it! 13.13 How Does the Correlation Change?

Consider the following data on X and Y.

x	2	3	4	5	5	6
y	1	2	2	3	4	5

(a) Determine the correlation between X and Y.

$r_{XY} = $ _____

(b) Transform your X data to $M = -2(X) + 5$. Find the correlation between M and Y.

$r_{MY} = $ _____

(c) Transform your Y data to $N = -5(Y) - 1$. Find the correlation between X and N.

$r_{XN} = $ _____

(d) Transform your X data to $L = 7(X) + 6$. Find the correlation between L and Y.

$r_{LY} = $ _____

(e) Transform your Y data to $T = (1/4)(Y) - \pi$. Find the correlation between X and T.

$r_{XT} = $ _____

(f) Find the following additional correlations:

$r_{LT} = $ _____

$r_{MN} = $ _____

(g) Write up a brief conclusion as to what you have observed about how the correlation coefficient changes with linear transformations of X and Y.

How Does the Correlation Change?

1. The correlation between the variables X and Y is the same as the correlation between Y and X, $r_{XY} = r_{YX}$.

2. Consider the following general linear transformation of X and of Y: $L = c + dX$, $M = e + fY$. The letters $c, d, e,$ and f represent the actual numerical values that define the transformation. We have the following results for the correlation between M, the transformed Y, and L, the transformed X:

$$r_{XY} = r_{LM} \quad \text{if } d > 0 \text{ and } f > 0 \text{ or if } d < 0 \text{ and } f < 0.$$
$$r_{XY} = -r_{LM} \quad \text{if } d > 0 \text{ and } f < 0 \text{ or if } d < 0 \text{ and } f > 0.$$

13.7.2 Relationship between *r* and the Slope

The correlation coefficient r and the slope of the least squares regression line b are related by the formula

$$b = r\left(\frac{s_Y}{s_X}\right).$$

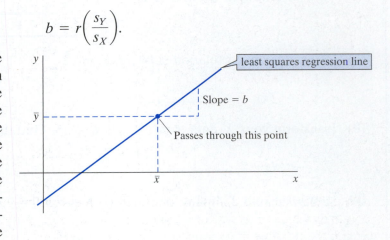

Observe that r and b have the same sign and are related through the standard deviations of the x and y data. If we combine the preceding expression for the slope with our expression for the y-intercept $a = \bar{y} - b\bar{x}$, we see that we only need the sample means, sample standard deviations, and the correlation coefficient to be able to derive the least squares regression line.

Note: The least squares regression line *always* passes through the point $(\bar{x}, \bar{y})$.

Let's do it! 13.14 Standard Score on the Midterm

The scores on the midterm and final exam for 500 students were obtained. The possible values for each exam are between 0 and 100. The least squares regression line for predicting the final exam score from the midterm exam score was obtained for these data. Suppose that the correlation coefficient is 0.5, $r = 0.5$.

Susan, a student in this class, received a midterm score that was one standard deviation above the average midterm score. Suppose that the average and standard deviation for the midterm scores were 80 and 10, respectively. Also, suppose that the average and standard deviation for the final exam scores were 60 and 20, respectively. Predict Susan's final exam score.

How Does the Slope Change?*

1. The slope of the least squares regression of Y on X will not necessarily be the same as the slope of the least squares regression of X on Y.

$$\text{Slope for regression of } Y \text{ on } X: b_{Y \text{ on } X} = r\left(\frac{s_Y}{s_X}\right)$$

$$\text{Slope for regression of } X \text{ on } Y: b_{X \text{ on } Y} = r\left(\frac{s_X}{s_Y}\right)$$

2. Consider the following general linear transformations of X and $Y: L = c + dX$ and $M = e + fY$. The letters $c, d, e,$ and f represent the actual numerical values that define the transformation. For the slope for the regression of M, the transformed Y, on L, the transformed X, we have

$$M = a^* + b^*L, \text{ where } b^* = r_{LM}\left(\frac{s_M}{s_L}\right) = r_{LM}\left(\frac{|f|s_Y}{|d|s_X}\right)$$

and where the correlation between the variables L and M is given by $r_{LM} = r_{XY}$ or $r_{LM} = -r_{XY}$, depending on the signs of the values of d and e.

Let's do it! 13.15 Vending Machines

A vending-machine company operates coffee vending machines in office buildings. The company wants to study the relationship between the number of cups of coffee sold per day, y, and the number of persons working in each building, x. Data were collected by the company for 10 locations. After assessing that the relationship is generally linear, the company computed the least squares regression equation and summary information given:

$$\hat{y} = 4 + 1.79x \qquad r = 0.975 \qquad n = 10$$

$$\bar{x} = 14 \quad s_X = 7.483 \qquad \bar{y} = 29 \qquad s_Y = 13.703$$

(a) At one of the office buildings there were 30 cups of coffee sold and 15 people working at this location. What is the residual for this observation?

---

*This summary regarding the slope is optional.

(b) Suppose that at one of the locations, the number of occupants was two standard deviations of x above the average number of occupants. What is the estimated number of cups of coffee sold per day at this location?

(c) The company's vending machines charge $0.80 for a cup of coffee. What is the correlation between coffee sales (in dollars) and the number of people working in each building?

(d) What is the slope of the least squares regression of coffee sales (in dollars) on the number of people working in each building?*

 Exercises

13.13 Assuming that we have a linear relationship, which of the following correlations shows the strongest linear relationship?

(a) $r = +0.55$

(b) $r = +0.33$

(c) $r = -0.77$

(d) $r = -0.344$

13.14 Two variables have a perfect positive correlation ($+1$). You obtain a z-score on the x variable for one subject of $+2$. Which of the following would be the subject's z-score on the y variable?

(a) 0

(b) 2

(c) -2

(d) not enough information to answer the question.

13.15 Assuming that we have a linear relationship, a large negative correlation shows that:

(a) as one variable increases, the other variable tends to increase.

(b) as one variable decreases, the other variable tends to decrease.

(c) as one variable increase, the other variable tends to decrease.

(d) as one variable increases, the other variable tends to stay the same.

13.16 A researcher obtains a correlation coefficient of exactly zero between variables x and y. Based on this finding, the researcher concludes that there is no relationship between x and y. What are some factors that could cause a correlation of zero when in fact there very well may be a relationship?

---

*Part (d) pertains to how a slope changes and is optional.

13.17 (a) Draw a scatterplot between two quantitative variables x and y that includes an outlier that makes the value of r, the correlation coefficient, misleading.

(b) Draw a scatterplot between two quantitative variables x and y for which a curve (different from a linear function) would fit the data better.

(c) Draw a scatterplot that shows the aggregate of two different data sets, for which the aggregated data have a negative linear association, while for each of the two data sets there is no linear association.

13.18 For each of the following statements about the correlation, select the best answer:

(a) A study found that there is a strong positive correlation between a person's height and annual income. The most reasonable explanation is

(i) employers like to pay more to tall people, and there is a linear relationship between an employee's height and income.

(ii) employers like to pay more to tall people, but there may not be a linear relationship between an employee's height and income.

(iii) the positive correlation is due to measurement error.

(iv) the observed association between the height and annual income is due to common response.

(b) Which of the following pairs of variables can be studied using the correlation?

(i) hair color of wife and hair color of husband.

(ii) ethnic background and annual income.

(iii) height of wife and height of husband.

(iv) religion and family size.

13.19 *Visualizing r.* Without actually computing the correlation, give an estimate of the value of r for each of the following sets of data (you may wish to sketch out a scatterplot first):

(a)

x	y
1	5.2
2	7.6
3	10.0
4	12.4
5	14.8

$r =$ ____

(b)

x	y
−4	16
−3	9
−2	4
−1	1
0	0
1	1
2	4
3	9
4	16

$r =$ ____

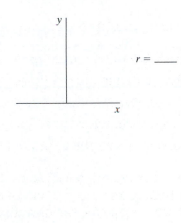

13.20 Influential or outlier? The article "Crop Improvement for Tropical and Subtropical Australia: Designing Plants for Difficult Climates" (*Field Crops Research*, 1991) gave the accompanying data on x = crop duration in days for soybeans and y = crop yield in tons/hectare.

x	92	92	96	100	102	102	106	106	121	143
y	1.7	2.3	1.9	2.0	1.5	1.7	1.6	1.8	1.0	0.3

(a) Construct a scatterplot of the data.

(b) The largest and last value of x in the sample greatly exceeds the remaining ones. The equation of the least squares regression line with all 10 points is given by $\hat{y} = 5.207 - 0.034x$. Give the equation of the least squares line with the last point excluded. Does removal of this observation greatly affect the equation of the line? Explain.

(c) Observations that, if removed, greatly affect the equation of the line are called (select one) influential. outliers.

(d) Next let's examine what effect, if any, the deletion of the last point has on the value of r. With all 10 observations, $r = -0.94$. Give the value of r with the last point excluded.

13.21 Want to be a weather forecaster? How does the technique of predicting tomorrow's high temperature to be the same as today's high temperature perform? Data for $n = 35$ United States cities on y = TODAY__A (today's actual high temperature in degrees Fahrenheit) and x = YESTER__A (yesterday's actual high temperature in degrees Fahrenheit) have been analyzed to produce the following output:

```
Dep var: TODAY_A      N:   35    Multiple R:  .908    Squared multiple R:  .824
Adjusted squared multiple R:  .819     Standard error of estimate:       6.754


  Variable    Coefficient    Std error    Std coef Tolerance    T    P(2 tail)
CONSTANT          6.863         4.269        0.000    .         1.608    0.117
YESTER_A          0.862         0.069        0.908   .100E+01  12.425    0.000


                          Analysis of Variance
  Source    Sum-of-squares    DF  Mean-square     F-ratio      P
Regression      7041.677       1     7041.677     154.372    0.000
  Residual      1505295       33       45.615
```

(a) What is the value of the correlation coefficient between today's high temperature and yesterday's high temperature?

(b) What would you predict that today's high temperature would be for a randomly chosen city in the United States if you knew that yesterday's high temperature for that city was 30 degrees Fahrenheit?

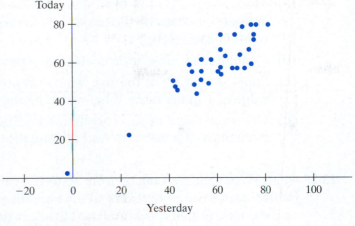

(c) If the data had been recorded in degrees Celsius rather than in degrees Fahrenheit, how would this change the value of the correlation coefficient? Note that Celsius = $\left(\frac{5}{9}\right)$ (Fahrenheit − 32).

13.22 Consider the summaries of the average summer temperature and the corresponding average winter temperature for the 50 United States.

```
TOTAL OBSERVATIONS:       50                    PEARSON CORRELATION MATRIX

                      SUMMER   WINTER                        SUMMER   WINTER
N OF CASES                50       50            SUMMER       1.000
MINIMUM               55.500   10.000            WINTER       0.574    1.000
MAXIMUM               90.000   72.500
MEAN                  75.240   33.730
STANDARD DEV           5.747   12.995
```

The scatterplot of average summer temperature versus average winter temperature indicated that the assumption of a linear relationship is plausible.

(a) What is the correlation between summer and winter temperatures?

(b) Give the least squares regression equation for regressing y = summer temperature on x = winter temperature.

13.23 Give the correlation coefficient if the data follow exactly the relationship $y = 2 - 3x$.

13.24 When adults are asked their weight, the reported weight tends to be less than their actual weight as measured by a scale. But there is a strong relationship between reported weight and measured weight, because heavy people usually report higher weights than do light people. Here are the measured weights x (in pounds) and the reported weights y (in pounds) for five women.

Measured Weight (x)	112	123	178	141	135
Reported Weight (y)	110	120	165	125	129

(a) Make a scatterplot of these data. Which observation do you think will have the greatest influence on the position of the regression line and the value of the correlation coefficient?

(b) Compute the correlation coefficient r between x and y.

(c) Suppose that all of the five women reported a weight 5 pounds less than the value of y in the table. What would be the value of r?

(d) Use your value of the correlation coefficient r and the means and standard deviations of x and y to find the equation of the least squares regression line.

13.25 A group of children ranging from 10 to 12 years of age was given a verbal test. Each subject was asked a question and the time (in seconds) before the subject responded and the total number of words used in answering were recorded. The purpose of the study was to look at the effect of *the silence interval before response* on *the number of words used.*

(a) What is the explanatory variable?

(b) What is the response variable?

The following computer output gives the result of regressing the number of words used on the silence interval before response based on a sample of eight children:

SUMMARY OUTPUT

Regression Statistics	
Multiple R	0.194
R Square	0.038
Adj R square	−0.123
Standard Error	15.6502
Observations	8

ANOVA

	Df	SS	MS	F	Significance F
Regression	1	57.297	57.297	0.234	0.646
Residual	6	1469.578	244.930		
Total	7	1526.875			

	Coefficients	Standard Error	t Stat	P-value	Lower 95%	Upper 95%
Intercept	70.837	16.878	4.197	0.006	29.538	112.137
X Variable 1	−0.282	0.582	−0.484	0.646	−1.707	1.144

(c) Find the equation of the linear regression line for regressing *the number of words used* on *the silence interval before response.*

(d) Give the value of the correlation coefficient between the number of words used and the silence interval before response.

(e) If the silence interval before response had been recorded in minutes (instead of seconds), the value of the correlation coefficient would (select one)

    increase.   *decrease.*   *stay the same.*

(f) Give the *p*-value for testing whether or not there is a significant (nonzero) linear relationship between these two variables. Using a 10% significance level, is there evidence of a significant (nonzero) linear relationship between these two variables?

13.26 In placing a weekly order, a concessionaire who provides services at a baseball stadium must know what size of crowd is expected during the coming week in order to know how much food to order. Since advance ticket sales (*in thousands*) give an indication of expected attendance, food needs (*in thousands*) might be predicted on the basis of advanced sales. Data from 7 previous weeks of games are given in the table.

Advance ticket sales in thousands (*x*)	54.0	48.1	28.8	62.4	64.4	59.5	42.3
Hot dogs purchased in thousands (*y*)	39.1	35.9	20.8	42.4	46.0	40.7	29.9

(a) Sketch the scatterplot for regressing *y* on *x*. Be sure to label your axes.

(b) Give the equation of the least squares regression line for regressing *y* on *x*.

(c) Draw (or superimpose) the regression line in (b) on the scatterplot in (a).

(d) Complete the following sentence to give a correct interpretation of the estimated slope of this regression line (be sure you include correct units):

For an additional _____ tickets sold in advance, we would expect an increase in the number of hot dogs purchased to be about _____.

(e) Should you predict purchase of hotdogs for an advanced ticket sales of $58,000? If yes, give your prediction. If no, explain why not.

(f) Sketch the residual plot based on this regression analysis. Be sure to label your axes and circle on your plot the residual for the last observation when the advance ticket sales were $42,300.

(g) Based on the residual plot, does the linear model seem appropriate? Explain.

13.27 The following table gives the midterm exam score, *x*, and the final exam score, *y*, for a sample of 10 students in a class:

Student	1	2	3	4	5	6	7	8	9	10
Midterm Score	55	40	73	81	90	40	83	75	65	51
Final Score	58	42	70	85	87	41	87	70	67	55

(a) Obtain the scatterplot of *y* versus *x* with the fitted regression line. Label your axes clearly and give some values on each axes.

(b) Obtain the least squares regression line of *y* on *x*.

(c) What is the predicted final score for a student who scored 70 on the midterm?

(d) Obtain the residual corresponding to Student 3.

13.28 The number of car accidents at a main intersection in Ann Arbor, Michigan, is recorded for each year from 1990–1999. The following table gives information on the total number of cars in Ann Arbor, *x* (in thousands), and the number of car accidents at the main intersection, *y*, for each year:

Year	1990	1991	1992	1993	1994	1995	1996	1997	1998	1999
x	3.5	3.8	3.9	3.9	3.8	4	4.1	4.3	4.4	4.4
y	20	23	24	27	30	28	32	32	35	35

(a) Obtain the least squares regression line of y on x.

(b) What is the predicted number of car accidents at this intersection if the number of cars in Ann Arbor is 3800?

(c) Obtain the residual corresponding to the year 1993.

(d) Obtain the residual plot for the data given in the table. Label your axes clearly and provide some values for each axes.

(e) Comment on the fit of the regression line on the data based on the residual plot.

13.29 The PJH&D Company is in the process of deciding whether to purchase a maintenance contract for its new word-processing system. The company feels that maintenance expense should be related to usage, and it has collected the following information on weekly usage (hours) and annual maintenance expense:

Observation #	1	2	3	4	5	6	7	8	9	10
x = Weekly Usage (hours)	13	10	20	28	32	17	24	31	40	38
y = Annual Maintenance Expense (in $100s)	17	22	30	37	47	30.5	32.5	39	51.5	40

(a) Sketch the scatterplot for regressing y on x. Be sure to label your axes and circle on your plot Observation #4.

(b) Find the equation of the least squares regression line for regression y on x.

(c) Draw (or superimpose) the regression line in (b) on the scatterplot in (a).

(d) Suppose that the weekly usage for a company is 21 hours. Predict their annual maintenance expense. Show your work and include your *units*.

(e) Using the new word-processing system incurs other expenses besides maintenance costs. Suppose that weekly utility costs, C, in dollars, is related to weekly usage by the following expression: $C = 3X - 0.80$. What would be the correlation between weekly utility cost, C, and annual maintenance cost, Y?

(f) Sketch the residual plot on the basis of this regression analysis. Be sure to label your axes and circle on your plot the residual for Observation #4.

(g) Based on the residual plot, does the linear model seem appropriate? Explain.

13.30 A sociologist was hired by a city hospital to study the effect of *distance to work* (miles between an employee's home on work) on the *number of days absent* (number of unauthorized days the employee is absent per year).

(a) What is the explanatory variable?

(b) What is the response variable?

The following computer output gives the result of regressing the number of days absent on distance to work based on a sample of 10 employees:

SUMMARY OUTPUT

Regression Statistics

Multiple R	0.843
R Square	0.711
Adj R square	0.675
Standard Error	1.289
Observations	10

ANOVA

	df	SS	MS	F	Significance F
Regression	1	32.699	32.699	19.668	0.0022
Residual	8	13.301	1.663		
Total	9	46			

	Coefficients	Standard Error	t Stat	P-value	Lower 95%	Upper 95%
Intercept	8.0978	0.8088	10.0119	0.0000	6.2327	9.9630
X Variable 1	−0.3442	0.0776	−4.4348	0.0022	−0.5232	−0.1652

(c) Find the equation of the linear regression line for regressing number of days absent on distance to work. (Round all numbers to two decimal places.)

(d) Give the correlation coefficient between number of days absent and distance to work.

(e) Using a 5% significance level, is there evidence of a significant (nonzero) linear relationship between these two variables? Explain.

(f) Suppose that the hospital had a total of 2000 employees, with 1500 classified as supporting staff and 500 classified as medical staff. The sample of 10 employees was obtained by taking a simple random sample of 8 supporting-staff employees and a simple random sample of 2 medical-staff employees.

   (i) What sampling method was used to obtain the 10 employees used in this study?

   (ii) Using your TI with a seed value of 109 or Row 8, Column 1 of the random number table, give the resulting labels (in order selected) for the 8 supporting-staff employees selected to be in this study.

13.31 A student wonders whether people of similar heights tend to date each other. She measures herself, her dormitory roommate, and the women in the adjoining rooms; then she measures the next man each woman dates. Here are the data (heights are recorded in inches).

Height of Woman (x)	66	64	66	65	70	65
Height of Man (y)	72	68	70	68	71	65

(a) Make a scatterplot of these data.

(b) Based on the scatterplot, do you expect the correlation to be positive or negative? Near ±1 or not?

(c) Compute the correlation r between the heights of the men and women.

(d) Does the correlation help answer the question of whether women tend to date men taller than themselves? Explain.

(e) How would r change if all the men were 12 inches shorter than the heights given in the table?

(f) If every woman dated a man exactly 3 inches taller than she, what would be the correlation between male and female heights?

(g) Compute the slope of the regression line of male height on female height.

(h) Suppose that the heights of the men were measured in centimeters (cm) rather than in inches, but that the heights of the women remained in inches. (There are 2.54 cm to an inch.)

   (i) Now what would be the correlation between male and female heights?

   (ii) What would be the new slope of the regression line of male height on female height?

13.32 A random sample of seven households was obtained, and information on their income and food expenditures for the past month was collected. The data (in hundreds of dollars) are given.

Income ($100s) ($x$)	22	32	16	37	12	27	17
Food Expend. ($100s) ($y$)	7	8	5	10	4	6	6

(a) Make a scatterplot of the data, find the least squares regression line, and superimpose it on your scatterplot.

(b) Should you predict the food expenditure for a household with an income of $5200? Explain.

(c) How much would food expenditure change for a $100 increase in income?

(d) Find the residual for the third observation ($x = 16$, $y = 5$).

(e) Is the following statement true or false? "The correlation coefficient between income and food expenditure is approximately $0.93." Explain.

13.8   Regression Effect[*]

- Will a special reading program improve the reading levels for students with extremely low reading levels?

- Will carrying a "good luck" charm improve the batting average of baseball players experiencing an early season slump?

- Will an extra review section for students whose Exam 1 scores were the lowest lead to higher scores on Exam 2?

Although these interventions may have some effect, another possible explanation, due to nonrandom sampling of the subjects, is called the **regression effect**.

> *Definition:*   In nearly all test–retest situations, the lowest group on the first test will on average show some improvement on the second test—and the highest group on the first test will on average perform worse. This is the **regression effect**.

---

[*]This is an interesting, but optional, topic.

This regression effect can also be described in terms of a "skill" and "luck" model for test scores. We can think of an individual's test score as being equal to the actual skill level of that individual plus some contribution due to luck. A common model for the luck component is to assume that luck averages out to zero.

Test 1 = Actual Skill Level + Luck for Test 1

Test 2 = Actual Skill Level + Luck for Test 2

A very high score on Test 1 is generally due to a high skill level and some good luck. Since the luck component, also referred to as *chance* or *random variation*, has an expected value of zero, we would *expect* a high score on Test 2, due to the common high skill level, but not quite as high as Test 1.

   We present the idea of the regression effect in this chapter because it arises in situations when the two measures are associated. In this case, it is possible to express the expected response (expected value of y) at a given value of x in terms of this correlation. Starting with the least squares regression equation, and doing a bit of algebra, we can reexpress the least squares regression line as follows:

$$\frac{\text{expected } Y (\text{for given } x) - \text{average of } Y}{\text{standard deviation of } Y} = (r)\frac{x - \text{average of } X}{\text{standard deviation of } X}.$$

Since the correlation is generally a fraction, $-1 < r < 1$, the quantity on the left side is smaller, in absolute value, than the quantity to the right of the correlation. For example, if the value of x is three standard deviations above the average of X and the correlation is 0.6, then the expected value of Y for this value of x is only $1.8 (=(0.6)3)$ standard deviations above the average of Y—also extreme, but not as extreme as the value of x.

❖ EXAMPLE 13.12   Test Scores

Consider two test scores you obtain in a course. Suppose that you score three standard deviations higher than the average on the first test. What do you expect your score to be on the subsequent test? Your score on the first test is substantially higher than the average—an indication of your effort, ability, and perhaps some good luck. Your effort and ability may still provide you with an above-average performance on the second test, but the contribution from *luck* may not be the same.

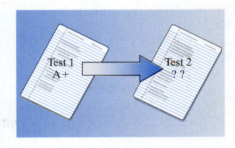

   The presence of a common factor contributing to both test scores implies the correlation between the test scores is positive. And the regression effect says we *expect* a lower performance on the second test, relative to the standards on that test—but this is expected, on average, and does not apply to every individual. If we wanted to learn whether or not positive reinforcement works, we must use this "treatment" on a random sample of students—not just those whose performance was exceptionally high on the first test.   ❖

❖ EXAMPLE 13.13   Alzheimer's Disease

Alzheimer's is a disease that generally fluctuates quite a bit from day to day—random fluctuations. When testing various drug treatments, it is very difficult to separate what portion of the "improvement" is due to the regression effect, what portion is due to the random fluctuations, and what portion is attributed, if any, to the actual treatment.

   What is it called when the effects of various factors cannot be separated? The factors are said to be confounded.   ❖

Be Careful: Two Issues Regarding the Regression Effect

1. It is a statement about what we would expect on average—it says nothing about variance. The variation in the measurements is not decreasing.

2. It is a statement about what we would expect on average—it says nothing about what will occur for an individual.

Let's do it! 13.16 Heights of Mothers and Daughters

In a study on the relationship between the heights of mothers and daughters, a regression model was proposed with x, the explanatory variable, the height of the mother, and y, the response variable, the height of the daughter. Two hundred mother–daughter pairs were selected and their heights were measured in centimeters. The following statistics were calculated:

$$\bar{x} = 150, \quad s_X = 10, \quad \bar{y} = 150, \quad s_Y = 10, \quad r = 0.8.$$

(a) What are the units of measurement for $\bar{y}$, s_Y, and r?

(b) If we measure heights in feet instead of centimeters, will the value of the correlation change?

(c) Determine the estimated regression line.

(d) Jane is 140 cm tall. What is her standardized height—that is, what is her z-score? Give the predicted height of her daughter, first in centimeters, and then also in standardized units.

(e) Repeat (d) for Mary of height 150 and Betty of height 160.

(f) The predicted heights of the three daughters have been pulled toward the mean. This is why the regression effect is also referred to as *regression toward the mean*. Do you think the variance of heights of all daughters is less than the variance of the heights of all mothers? Explain.

Exercises

13.33 You have a new drug that you claim should help reduce high blood pressure. You put an ad in the paper asking for volunteers for your blood-pressure study. You conduct an experiment and get a significant result and conclude that your new drug significantly reduces blood pressure. Should you be surprised? Think about what kind of people you would expect to volunteer for such a study.

13.34 Explain how the regression effect can bias the results of an experiment.

 # 13.9   The Squared Correlation r^2— What Does It Tell Us?

Recall that r, the correlation coefficient, gives us the strength and the direction of the linear relationship between two quantitative variables x and y. Since $-1 \le r \le 1$, the square of the correlation, r^2, will be $0 \le r^2 \le 1$. We also learned how the slope, b, of the least squares regression line relates to the correlation coefficient r, namely,

$$r = b\left(\frac{s_x}{s_y}\right).$$

So why do we need to compute the value of r^2? In regression, the total sum of squares, SST, is the sum of the squared difference between the observed y values and the average of the y values, $\overline{y}$:

$$\text{SST} = \sum_{i=1}^{n}(y_i - \overline{y})^2 \quad \text{where } n = \text{the sample size.}$$

The residual sum of squares, SSE, is the sum of the squared differences between the observed y values and the predicted values, $\hat{y}$:

$$\text{SSE} = \sum_{i=1}^{n}(y_i - \hat{y}_i)^2.$$

These sums of squares form a regression identity that is summarized in the following box:

The Regression Identity

$$\text{SST} = \text{SSE} + \text{SSR}$$

where

$$\text{SSR} = \text{Regression Sum of Squares}$$
$$= \Sigma(\hat{y} - \overline{y})^2$$

$$\text{SSE} = \text{Residual (or Error) Sum of Squares}$$
$$= \Sigma(y - \hat{y})^2$$

$$\text{SST} = \text{Total Sum of Squares}$$
$$= \Sigma(y - \overline{y})^2$$

It can be shown that the square of the correlation coefficient is equal to

$$r^2 = \frac{\text{SST} - \text{SSE}}{\text{SST}} = \frac{\text{SSR}}{\text{SST}}.$$

Note that when $r \neq 0$, we will have SSE $<$ SST. Thus, it is better to use the regression line to predict a value y for a particular new value of x than it would be to use the average of all of the y values, namely, $\bar{y}$, as the predicted value. The quantity r^2 is generally denoted in computer output as R^2, and is often reported as a percent. For example, an $r^2 = 0.75$ means that about 75% of the variation in the response variable y can be explained by the linear relationship between x and y.

Coefficient of Determination

$$r^2 = \frac{\text{SST} - \text{SSE}}{\text{SST}} = \frac{\text{SSR}}{\text{SST}}$$

represents the proportion of the total variation in the y values that can be explained by the linear relationship between x and y.

❖ EXAMPLE 13.14   Coefficient of Determination for Test Scores _____

Let us revisit Example 13.3 on Test 1 versus Test 2 scores. The data and various calculations are summarized in the following table:

x	y	$\hat{y}$	$\bar{y}$	$(y - \bar{y})^2$	$(y - \hat{y})^2$
8	9	9.6	14	$(-5)^2 = 25$	$(-0.6)^2 = 0.26$
10	13	11.8	14	$(-1)^2 = 1$	$(1.2)^2 = 1.44$
12	14	14	14	$0^2 = 0$	$0^2 = 0$
14	15	16.2	14	$1^2 = 1$	$(1.2)^2 = 1.44$
16	19	18.4	14	$5^2 = 25$	$(0.6)^2 = 0.36$
				SST $= 52$	SSE $= 3.6$

The value of r^2 is thus $r^2 = \frac{52 - 3.6}{52} = 0.9308$. This large percentage of about 93% indicates that a substantial amount of the variation in the Test 2 scores can be explained by the regression line $\hat{y} = 0.8 + 1.1\ x$ and therefore the Test 1 score is very useful for predicting the score on Test 2.   ❖

Values of r^2 that are close to zero indicate that the regression equation will not be very useful for making predictions. Values of r^2 that are close to one indicate that the regression equation will be useful for making predictions.

13.10   Inference in Simple Linear Regression

In a **simple linear regression model** there is a single quantitative independent variable, x, that is used to predict the value of a quantitative response variable, y. The relationship between the two variables is assumed to be a linear function. In many cases, the assumption of linearity might not be reasonable for a wide range of values of the explanatory (or independent) variable x. This first plot shows a nonlinear relationship over the range of $x = 0$ to $x = 3$.

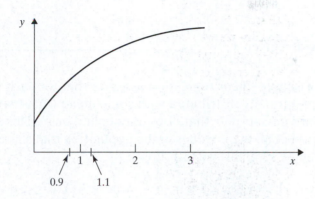

However, if we look only at the values near $x = 1$ (see next plot), a straight-line model would work very well.

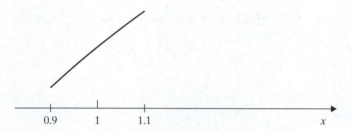

13.10.1   The Simple Linear Regression Model

We begin our more formal inference discussion by distinguishing between the **estimated regression line**, which describes the sample data, from the **true regression line**, which describes the population from which the sample was selected.

   In the previous sections, we represented the estimated regression line based on the sample data by $\hat{y} = a + bx$. This linear equation is found using the data (x_i, y_i), $i = 1, \ldots, n$ with the property that it has the smallest sum of squared differences between the observed values y_i and the corresponding predicted values $\hat{y}_i$. The difference $e_i = y_i - \hat{y}_i$ is the residual for the ith observation. For the ith observation, we can write the identity

$$y_i = \hat{y}_i + e_i = a + b\,x_i + e_i.$$

The true regression line for the population will be written as $E(Y) = \alpha + \beta x$, where $E(Y)$ is the mean of all the population values of the response variable y that share the same value of the explanatory variable x. The α represents the true y-intercept of the true linear function using all values of the population and the β represents the slope of the true linear function using all the values of the population. The accompanying picture displays these ideas for the population:

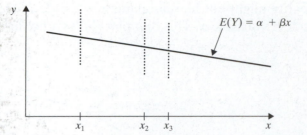

$E(Y) = $ mean response at a given x.

The statistical model for simple linear regression assumes that for each value of x the values of the response y are normally distributed with some mean (that may depend on x in a linear way) and a standard deviation σ that does not depend on x. This standard deviation σ is the standard deviation of all the y values in the population that share the same value of x. These assumptions can be summarized as follows:

$$\text{For each } x, Y \sim N(E(Y), \sigma), \qquad \text{where } E(Y) = \alpha + \beta x.$$

We can visualize the model with the following picture:

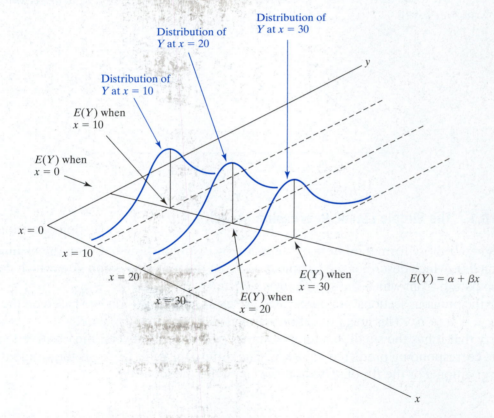

The basic parts of a linear regression analysis are

- estimate the unknown parameters (α, β, σ) based on some data.
- use the estimated equation for making predictions.
- assess if the linear relationship is statistically significant.
- provide interval estimates (confidence intervals) for our predictions.
- understand and check the assumptions of our model.
- measure the strength of the linear relationship with the correlation.

The sample data provide point estimators of the corresponding population parameters. So, the quantities a, b, and $\hat{y}$ are statistics and the quantities α, β and $E(Y)$ are parameters.

Regression Parameters Estimates

The estimate of the mean response $E(Y)$ is given by

$$\widehat{E(Y)} = \hat{y} = a + bx.$$

The estimate of the *y*-intercept α is given by

$$\hat{\alpha} = a.$$

The estimate of the slope β is given by

$$\hat{\beta} = b.$$

The estimate for the standard deviation σ is given by

$$\hat{\sigma} = s = \sqrt{\frac{\text{SSE}}{n-2}},$$

where SSE = sum of squared residuals

$$= \Sigma(y_i - \hat{y}_i)^2 = \Sigma e_i^2.$$

Finally, the sample correlation coefficient $r = \hat{\rho}$ is a point estimate of the population correlation ρ.

Let's turn to an example in which these estimates are computed.

❖ **EXAMPLE 13.15** **Test 1 versus Test 2 Revisited Using the TI Calculator**

Consider the data on Test 1 and Test 2 scores from Example 13.3. With the Test 1 data entered into L1 and the Test 2 data entered into L2, the following steps are needed to perform a regression analysis:

Test 1	Test 2
8	9
10	13
12	14
14	15
16	19

For bringing up the Linear Regression tests
Be sure to match Xlist with L1 and Ylist with L2

To put Y_1 in the RegEQ line you enter the following steps:

Your input screen should look like this:

```
LinRegTTest
Xlist:L1
Ylist:L2
Freq :1
β & ρ:≠0 <0 >0
RegEQ:Y1
Calculate
```

Selecting the Calculate option and pressing ENTER should provide an output screen that looks like this:

```
LinRegTTest
y=a+bx
β ≠ 0 and ρ ≠ 0
t=6.350852961
p=.0078978382
df=3
a=.8
b=1.1
s=1.095445115
r²=.9307692308
r=.9647638212
```

From this output screen, we have the following estimates:

$$\hat{\alpha} = a = 0.8,$$

$$\hat{\beta} = b = 1.1,$$

$$\hat{\sigma} = s = 1.0954,$$

$$\hat{\rho} = r = 0.96476,$$

and $\hat{y} = 0.8 + 1.1x$ is the estimated regression line. ❖

 ### 13.10.2   Testing Hypotheses about the Linear Relationship

Recall that the simple linear regression model is given by $E(Y) = \alpha + \beta x$. If $\beta = 0$, then the variables x and y are not linearly related, and the expected response is just a constant $E(Y) = \alpha$.

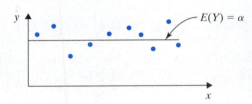

In this case, for any value of the variable x, we would have the **same** value α for the expected response $E(Y)$. Knowing the value of x would not help to predict the value of y. In Section 13.5, we discussed the idea of assessing the significance of the linear relationship. We also examined the p-value in computer output for testing about the value of the population slope β. More formally, the test of interest can be expressed as

H_0: $\beta = 0$ (the slope of the regression line using all the values in the population is zero).

H_1: $\beta \neq 0$.

The alternative hypothesis is a two-sided test, though a one-sided alternative hypothesis test does exist, where H_1: $\beta < 0$ or H_1: $\beta > 0$.

There are a number of ways to test this hypothesis. One way is through a t-test statistic. Recall that the general form for a t-test statistic is

$$t = \frac{\text{point estimate} - \text{hypothesized value}}{\text{standard error of the point estimate}}.$$

Our point estimate for β is b, and we have the hypothesized value of zero. It can be shown that the standard error for b is

$$SE(b) = \frac{s}{\sqrt{\sum (x - \bar{x})^2}}.$$

The t-test statistic for testing hypotheses about the population slope is given by

$$T = \frac{b - 0}{\text{SE}(b)}, \text{ which has a } t\text{-distribution with } n - 2 \text{ degrees of freedom.}$$

❖ EXAMPLE 13.16 Test 1 versus Test 2 Revisited Using the TI Calculator

Recall the Test 1 versus Test 2 data from Example 13.3. We would like to assess if Test 1 scores are a significant linear predictor for Test 2 scores, that is, $H_0: \beta = 0$ versus $H_1: \beta \neq 0$, using a 5% level of significance.

x	$(x - \bar{x})^2$
8	16
10	4
12	0
14	4
16	16
$\bar{x} = 12$	Sum $= 40$

Recall the estimated linear regression model: $\hat{y} = 0.8 + 1.1x$. So our estimated slope is $b = 1.1$, our estimated population standard deviation is $s = 1.0954$, and, from the calculation table, we have

$$\sqrt{\Sigma(x - \bar{x})^2} = \sqrt{40}.$$

Thus, we have

$$\text{SE}(b) = \frac{1.0954}{\sqrt{40}} = 0.1732,$$

and our observed test statistic is

$$t = \frac{b - 0}{\text{SE}(b)} = \frac{1.1 - 0}{0.1732} = 6.35.$$

To calculate the p-value using the TI, we have the following steps:

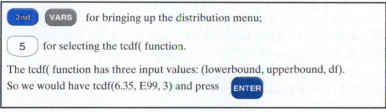

2nd VARS for bringing up the distribution menu;

5 for selecting the tcdf(function.

The tcdf(function has three input values: (lowerbound, upperbound, df). So we would have tcdf(6.35, E99, 3) and press ENTER

Your screen should look like this:

```
tcdf(6.35,E99,3)
        0.0039504221
```

Since the test is a two-sided test, the p-value $= 2(0.00395) = 0.0079$. Furthermore, since the P-value is less than $\alpha = 0.05$, we reject the null hypothesis $H_0: \beta = 0$ at the 0.05 level of significance. We conclude that the Test 1 score is a useful linear predictor of the Test 2 score. ❖

We can also test the two-sided hypothesis about the population slope at the $\alpha = 0.05$ level of significance by constructing the 95% confidence interval for the true slope. The formula for the interval is as follows:

Confidence Interval for the Slope

A confidence interval for β is given by

$$b \pm t^* \, [SE(b)],$$

where t^* is an appropriate percentile of a t-distribution with $n - 2$ degrees of freedom (from Table IV) and

$$SE(b) = \frac{s}{\sqrt{\sum (x - \overline{x})^2}}.$$

THINK ABOUT IT

Suppose that the t-test rejected the null hypothesis H_0: $\beta = 0$ in favor of the alternative hypothesis H_1: $\beta \neq 0$ at the 5% significance level. Would the corresponding 95% confidence interval for the true slope β contain the value zero?

Select one: Yes   No

❖ **EXAMPLE 13.17** **Test 1 versus Test 2 Revisited**

Consider the data on Test 1 and Test 2 scores from Example 13.3. We have an estimated slope $b = 1.1$, an estimated standard deviation $s = 1.0954$, and $\sum (x - \overline{x})^2 = 40$. We wish to compute the 95% confidence interval for β.

The 95% confidence interval for β is given by

$$b \pm t^* \, [SE(b)],$$

where $t^* = 3.182$ can be found from Table IV on page 617 using a t-distribution with $n - 2 = 5 - 2 = 3$ degrees of freedom and a 95% confidence level.

So our confidence interval is given by

$$\left[1.1 \pm 3.182 \left(\frac{1.0954}{\sqrt{40}} \right) \right] \;\Rightarrow\; [1.1 \pm 3.182(.1732)] \;\Rightarrow\; (0.548, 1.6511).$$

Interpretation: We estimate with 95% confidence that the score for Test 2 will increase between 0.548 and 1.6511 points for a one point increase in the Test 1 score, on average. ❖

A third way to test the significance of a linear relationship is with an F-test. The F-test, provided the assumptions hold, is a ratio of two estimates of the population variance, σ^2:

$$F = \frac{\text{MSR}}{\text{MSE}} = \frac{\text{mean square for regression}}{\text{mean square for error (or residuals)}}$$

$$= \frac{\text{only a good estimator of } \sigma^2 \text{ if } H_0: \beta = 0 \text{ is true; otherwise it tends to be too big}}{\text{a generally good estimator of } \sigma^2}$$

So, we would reject the null hypothesis $H_0: \beta = 0$ for large values of the F-test statistic. Note that the direction of extreme is to the right. This is similar to the F-test we discussed in Chapter 12 on one-way ANOVA. The F-test is provided in standard regression computer output. The SPSS output for Test 1 versus Test 2 scores regression analysis is as follows:

```
        * * * * MULTIPLE REGRESSION * * * *
Equation Number 1 Dependent Variable.. Test2score
Variable(s) Entered on Step Number 1.. Test1score


Multiple R                   .9647
R Square                     .93076
Adjusted R Square            .908
Standard Error               1.0954


Analysis of Variance
                    DF      Sum of Squares   Mean Square    F-test     Signif F
Regression          1            48.4           48.4         40.33      0.008
Residual            3             3.6            1.2


----------------- Variables in the Equation -----------------------------
Variable         B          SE B        T        Sig T
Test1score       1.1        .1732       6.351     0.008
(constant)       0.8        2.135       0.375     0.733
```

The observed F-test statistic value is 40.33 and the corresponding p-value is very small at 0.008. Notice the p-value is the same as for the t-test in Example 13.16. Once again, we conclude that the Test 1 score is a useful linear predictor of the Test 2 score.

❖ **EXAMPLE 13.18** **Interpreting Computer Output** _____

We would like to study the relationship between two quantitative variables: $x = $ Quiz 1 and $y = $ Quiz 2. Significance will be assessed at the 5% level. The data shown in the following table have already been entered into SPSS:

Quiz 1	Quiz 2
0	6
2	5
4	8
6	7
8	9

The following regression output was obtained:

```
        * * * * MULTIPLE REGRESSION * * * *
Equation Number 1 Dependent Variable.. QUIZ2
Variable(s) Entered on Step Number 1.. QUIZ1


Multiple R              .800
R Square                .640
Adjusted R Square       .520
Standard Error


Analysis of Variance
                 DF      Sum of Squares   Mean Square    F-test    Signif F
Regression        1      6,400            6.400          5.333     0.104
Residual          3      3.600            1.200


------------------- Variables in the Equation ----------------------------
Variable        B        SE B      Beta           T        Sig T
QUIZ1           0.4       .173      .8.351         2.309     .104
(constant)      5.4       .849                     6.364     .008
```

From the output, we see that most of the computations are already given to us.

- The equation of the estimated linear regression line is $\hat{y} = 5.4 + 0.4x$.
- The 95% confidence interval for b is given by

$$b \pm t^*(\text{SE}(b)) \Rightarrow 0.4 \pm 2.31(0.173) \Rightarrow (-0.15, 0.95).$$

- The observed t-test statistic is $t = 2.309$, which has a two-sided p-value of 0.104. At the 5% level, there is no significant linear relationship between Quiz 1 scores and Quiz 2 scores.
- The observed F-test statistic is $F = 5.333$, and its p-value is also 0.104. Again, we cannot reject H_0: $\beta = 0$ at the significance level $\alpha = 0.05$. ❖

Let's do it! 13.17 **Height and Weight**

The National Center for Health Statistics publishes data on heights and weights. We obtained the following data from 10 randomly selected males 18–24 years of age:

x = Height (inches)	69	71	70	67	75	66	71	71	67	65
y = Weight (pounds)	151	159	163	153	198	126	163	185	133	175

(a) Draw a scatterplot. Comment on the relationship between height and weight.

(b) Obtain the estimated regression line. Give an interpretation of b, the slope of the estimated regression line.

(c) Find the correlation r and r^2. Give an interpretation of r^2.

(d) Test H_0: $\beta = 0$ versus H_1: $\beta \neq 0$ at the 10% significance level.

(e) Estimate σ, the standard deviation of the values of y in the population that share the same x value.

13.10.3   Predicting for an Individual versus Estimating the Mean

What is the difference between $\hat{y}$ and $\widehat{E(Y)}$ for a given value of x? Consider the quiz data in Example 13.17. The predicted Quiz 2 score for a particular student who scored 6 on Quiz 1 can be found as

$$\hat{y} = 5.4 + 0.4(6) = 7.8 \text{ points.}$$

If we want to estimate the mean Quiz 2 score for all the students in the population who scored a 6 on Quiz 1, we would also use the estimated regression line and obtain $\widehat{E(Y|x = 6)} = 5.4 + 0.4(6) = 7.8$ points. So our point estimate for predicting a future individual observation and for estimating the mean response are found using the same least squares regression equation. What about their standard errors? With the standard errors we could produce an interval estimate or perhaps do a test. Let's step back a moment to the results in Chapter 8 on the sampling distribution of a mean. If we consider the population of individuals, the standard deviation for the population of individual values is just σ. If we consider the population of means, the standard deviation for the population of mean value is $\frac{\sigma}{\sqrt{n}}$. We see that the standard deviation of the individual values is larger, so a prediction interval for an individual response will be wider than a confidence interval for a mean response. Here are the actual formulas.

Prediction Interval for an Individual Response

The prediction interval for an individual response y for a given value of the explanatory variable $x = x_i$ is

$$\hat{y} \pm t^* s \sqrt{1 + \frac{1}{n} + \frac{(x_i - \bar{x})^2}{\sum_j (x_j - \bar{x})^2}},$$

where t^* is an appropriate percentile of the t-distribution with $n - 2$ degrees of freedom.

Note: The prediction interval will be narrower if the value of x_i is closer to $\bar{x}$. If n is also large, the prediction interval will be approximately equal to

$$\hat{y} \pm t^*(s).$$

Confidence Interval for a Mean Response

The confidence interval for the mean response $E(Y)$ for a given value of the explanatory variable $x = x_i$ is

$$\hat{y} \pm t^* s \sqrt{\frac{1}{n} + \frac{(x_i - \bar{x})^2}{\sum_j (x_j - \bar{x})^2}},$$

where t^* is an appropriate percentile of the t-distribution with $n - 2$ degrees of freedom.

Note: The confidence interval will be narrower if the value of x_i is closer to $\bar{x}$.

❖ **EXAMPLE 13.19**   **Quiz 1 versus Quiz 2 revisited** _____

Recall our data set on Quiz 1 and Quiz 2 for five students from Example 13.18. The data are provided in the following table, along with some of the results found earlier in Example 13.18:

Quiz 1	Quiz 2
0	6
2	5
4	8
6	7
8	9

$\hat{y} = 5.4 + 0.4x$

$s = 1.095$

$\bar{x} = 4$ $\qquad \sum(x - \bar{x})^2 = 40$

We wish to compute the 95% prediction interval for an individual student who scores 6 on Quiz 1. Therefore, we have

$$\hat{y} = 5.4 + 0.4(6) = 7.8,$$

$t^* = 3.182$ (from Table IV for a 95% confidence level with three degrees of freedom), and

$$SE(\hat{y}) = s\sqrt{1 + \frac{1}{n} + \frac{(x_i - \bar{x})^2}{\sum_j (x_j - \bar{x})^2}} = 1.095\sqrt{1 + \frac{1}{5} + \frac{(6-4)^2}{40}} = 1.248.$$

So the 95% prediction interval is given by

$$7.8 \pm (3.182)(1.248) \Rightarrow 7.8 \pm 3.97 \Rightarrow (3.83, 11.77).$$

Also,

$$\widehat{E(Y|x = 6)} = 5.4 + 0.4(6) = 7.8,$$

$t^* = 3.182$, and

$$SE[(\widehat{E(Y|x = 6)})] = s\sqrt{\frac{1}{n} + \frac{(x_i - \bar{x})^2}{\sum_j (x_j - \bar{x})^2}} = 1.095\sqrt{\frac{1}{5} + \frac{(6-4)^2}{40}} = 0.599.$$

So the 95% confidence interval for the mean Quiz 2 score of all the students who scored 6 points on Quiz 1 is

$$7.8 \pm (3.182)(0.599) \Rightarrow 7.8 \pm 1.9 \Rightarrow (5.89, 9.71).$$

As you can see, the confidence interval for the mean response is narrower than the prediction interval for an individual response. ❖

Let's do it! 13.18 Weekly Gross Revenue versus Amount of Dollars Spent on Advertisement

The following data give the weekly gross revenue in $1000s ($y$) for a Show Time movie theater, given the amount of dollars spent on television advertisement in $1000s ($x$):

x	y
5	99
2	90
4	95
2.5	92
3	95
3.5	94
3	93
3	94

(a) Make a scatterplot of the data.

(b) Calculate the estimated linear regression line using the data.

(c) Calculate the 95% confidence interval for β.

(d) Interpret the 95% confidence level in (c).

(e) Calculate the 95% prediction interval for the weekly gross revenue for an individual movie theater when $x = 4.5$.

(f) Calculate the 95% confidence interval for the mean weekly gross revenue for all movie theaters when $x = 4.5$—that is, for $E(Y|x = 4.5)$.

13.10.4   Checking Assumptions for Simple Linear Regression

First, you should always examine the scatterplot of y versus x to assess if a linear model seems reasonable. Next, let's recall the simple linear regression model for the response. The responses Y are assumed to be normally distributed with a mean $E(Y) = \alpha + \beta x$ and a standard deviation of σ. Earlier we expressed these assumptions with the following statement:

$$\text{For each } x, Y \text{ is } N\big(E(Y),\ \sigma\big) \quad \text{where } E(Y) = \alpha + \beta x.$$

We could use the following model to describe the components of a given response Y:

$$Y = \alpha + \beta x + \varepsilon.$$

The ε's are the **true error terms** (that we do not observe). These are the difference of the response y and the true mean (for a given x). These errors are to have a normal distribution with a mean of zero and a standard deviation of σ. We can't see these true errors, but we do have their "estimates"—the **residuals**. The residuals are the difference between the observed response y and the estimated mean (for a given x). So we examine the residuals to check the assumptions regarding the error terms. The residuals should look like they are a random sample from a normal population with mean zero and standard deviation σ.

How Would You Examine the Residuals to Check for . . .

. . . random sample?	Plot the residuals versus x. The assumption of a random sample would be reasonable if the plot shows a random scatter of points in a horizontal band with no apparent pattern. 

Lack of independence often occurs when the data are collected over time. For the independence part of a random sample to be reasonable, the plot of the residuals versus time should not exhibit a pattern.

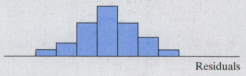

. . . normality?   The histogram of the residuals should look like a normal distribution with no outliers.

. . . common standard deviation (that does not depend on x)?   The residual plot (plot of residuals versus x) should show roughly a horizontal band of approximately the same width.

The following residual plot shows evidence that the variability in the responses tends to increase as x increases:

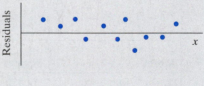

Let's turn to one last full regression problem, which will include checking of the assumptions.

❖ EXAMPLE 13.20 Effect of Drug on Reaction Time_____

An experiment was conducted involving 12 subjects to study the effect of the percentage of a certain drug in the bloodstream on the length of time (in seconds) it takes to react to a stimulus. The data are as follows:

Amount of Drug (%)	1	1.5	2	2.5	3	3.5	4	4.5	5	5.5	6	6.5
Reaction Time (seconds)	1.0	0.8	1.8	1.4	2.1	1.8	2.2	3.0	2.75	3.0	4.1	4.9

The scatterplot shows a positive, approximately linear association between the amount of drug and the reaction time:

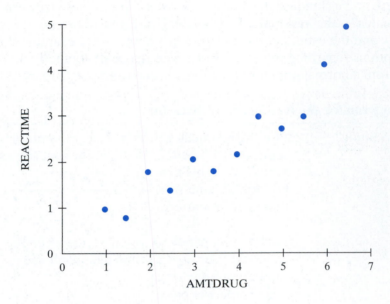

The SPSS regression output is given in the following table:

Model Summary

Model	R	R Square	Adjusted R Square	Std. Error of the Estimate
1	.939	.882	.871	.4385

a. Predictors: (Constant), AMTDRUG
b. Dependent Variable: REACTIME

From this output, we have $s = \hat{\sigma} = 0.4385$, $r^2 = 0.882$, and $|r| = 0.939$. An analysis of variance gives the following table:

ANOVA

Model		Sum of Squares	df	Mean Square	F	Sig.
1	Regression	14.430	1	14.430	75.048	.000
	Residual	1.923	10	.192		
	Total	16.352	11			

a. Predictors: (Constant), AMTDRUG
b. Dependent Variable: REACTIME

So the p-value ~ 0, degrees of freedom $= n - 2 = 12 - 2 = 10$, and $s^2 = \hat{\sigma}^2 = $ MSE $= 0.192$. The coefficients output is as follows:

Coefficients

Model		Unstandardized Coefficients		Standardized Coefficients	t	Sig.
		B	Std. Error	Beta		
1	(Constant)	2.174E-02	.303		.072	.944
	AMTDRUG	.635	.073	.939	8.663	.000

a. Dependent Variable: REACTIME

From the coefficients output, we have

$$a = 0.02174, \quad b = 0.635, \quad SE(b) = 0.073, \quad \text{and} \quad p\text{-value} \sim 0.$$

(a) Which variable is the dependent variable and which is the independent variable?

The dependent variable is $y =$ reaction time,
and the independent variable is $x =$ amount of drug.

(b) What is the equation of the least squares regression line?

$$\hat{y} = 0.02174 + 0.635x.$$

(c) Predict the reaction time for a subject with 6% of the drug in the bloodstream.

$$\hat{y} = 0.02174 + 0.635(6) = 3.83 \text{ seconds.}$$

(d) In the SPSS output, in the row labeled AMTDRUG, there is a t-value of 8.663. The "Sig" value in the same row is 0.000. This is a p-value. For what null and alternative hypotheses is this the appropriate p-value? State your decision for this test using a 1% significance level.

H_0: $\beta = 0$ (the slope of the regression line using all the values in the population is zero).

H_1: $\beta \neq 0$.

Since the p-value $\approx 0 < 0.01 = \alpha$, we would reject the null hypothesis.
There appears to be a significant nonzero linear relationship between $x =$ amount of drug and $y =$ reaction time.

(e) In the SPSS output, in the row labeled AMTDRUG, there is an SE value of 0.073. Explain what this standard error means in terms of repetitions of the experiment.

$SE(b) = 0.073$. In repeated samples of size $n = 12$, we would estimate the average distance of the possible b values from the true population slope β to be about 0.073.

(f) What proportion of variation in reaction times can be explained by the linear relationship between reaction time and amount of drug?

$$r^2 = 0.882 \Rightarrow 88.2\%.$$

(g) In the simple linear regression model, we assume that the y-values are normally distributed, with means that vary linearly with x and with same standard deviation σ. What is the estimate of σ here?

$$s = \hat{\sigma} = 0.4385.$$

(h) The following is a plot of the residuals against x:

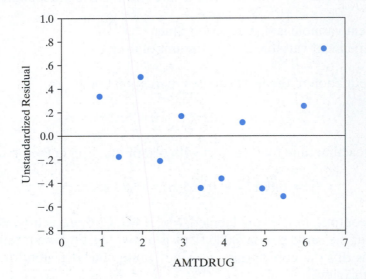

Based on this plot, do the assumptions for a linear regression model seem reasonable?

Since the residual plot shows a random scatter of points around zero with no apparent pattern, the linear regression model seems reasonable. ❖

Let's do it! 13.19 Numerical Literacy

Data were obtained on the numerical literacy score on a college entrance exam and the calculus final exam score at the end of the first year of college for a random sample of 24 students. These data were used to perform a regression of calculus final exam score on numerical literacy score. The SPSS output is as follows:

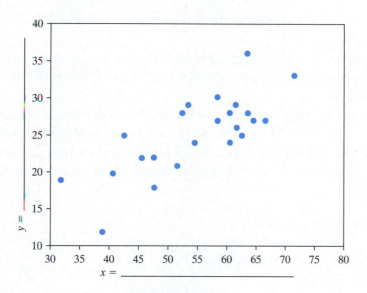

Model Summary

Model	R	R Square	Adjusted R Square	Std. Error of the Estimate
1	.761[a]	.580	.561	3.3822

a. Predictors: (Constant), Numerical Literacy

ANOVA[b]

Model		Sum of Squares	df	Mean Square	F	Sig.
1	Regression	347.290	1	347.290	30.359	.00002[a]
	Residual	251.669	22	11.439		
	Total	598.958	23			

a. Predictors: (Constant), Numerical Literacy
b. Dependent Variable: Calc Final

Coefficients[a]

Model		Unstandardized Coefficients		Standardized Coefficients	t	Sig.
		B	Std. Error	Beta		
1	(Constant)	4.148	3.899		1.064	.29887
	Numerical Literacy	.380	.069	.761	5.510	.00002

a. Dependent Variable: Calc Final

We also have $\bar{x} = 55.71$ and $\sum(x - \bar{x})^2 = 2410.96$.

(a) The axes on the preceding scatterplot are missing the variable names. Write in the correct variable names for x and y.

(b) Give the equation of the least squares regression line for regressing calculus final exam score on numerical literacy score.

(c) At a 5% significance level, is there a significant (nonzero) linear relationship between numerical literacy score and calculus final exam score? Give the *p*-value.

p-value = _____ .   Circle one: *Yes*   *No*

(d) The value of r^2 is 0.58. This implies that (*clearly circle your answer*)

- The correlation between numerical literacy score and calculus final exam score is 0.58.
- 58% of the variability observed in the calculus final exam scores can be explained by the linear relationship between numerical literacy score and calculus final exam score.
- 58% of the variability observed in the numerical literacy scores can be explained by the relationship between numerical literacy score and calculus final exam score.
- 58% of the variability observed in the numerical literacy scores can be explained by the linear relationship between numerical literacy score and calculus final exam score.

(e) Based on the accompanying residual plot, select the correct comment regarding the fit of the regression line. *Clearly circle your answer*.

- The fit is good because the residuals form a pattern around the *x*-axis.
- The fit is good because the residuals indicate a departure from the linear regression line.
- The fit is good because there are no influential points in the residual plot.

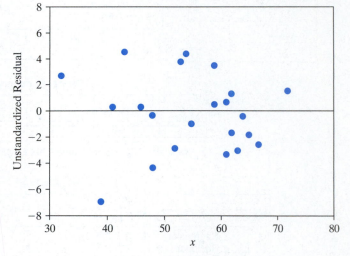

- The fit is good because the residuals form an unstructured band around the *x*-axis.

(f) A student beginning his first year of college had a numerical literacy score of 45 on the entrance exam. Give a 95% prediction interval for the calculus final exam score for this student.

(g) Give a 95% confidence interval for the mean calculus final exam score for all students with a numerical literacy score of 45 on the entrance exam.

✳ Exercises

13.35 Research on digestion requires accurate measurements of blood flow through the lining of the stomach. A promising way to make such measurements easily is to inject mildly radioactive microscopic spheres into the blood stream. The spheres lodge in tiny blood flow to be measured from outside the body. Medical researchers compared blood flow in the stomachs of dogs, measured by use of microspheres, with simultaneous measurements taken using a catheter inserted into a vein. The data, in milliliters of blood per minute, are provided along with the scatterplot and SPSS regression output.

Spheres (x)	Vein (y)
4.0	3.3
4.7	8.3
6.3	4.4
8.2	9.3
12.0	10.7
15.9	16.4
17.4	15.4
18.1	17.6
20.2	21
23.9	21.7

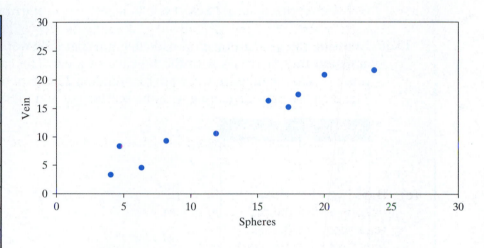

Model Summary

Model	R	R Square	Adjusted R Square	Std. Error of the Estimate
1	.967	.936	.928	1.7566

ANOVA

Model		Sum of Squares	df	Mean Square	F	Sig.
1	Regression	360.570	1	360.570	116.849	.000
	Residual	24.686	8	3.086		
	Total	385.256	9			

Coefficients

Model		Unstandardized Coefficients		Standardized Coefficients		
		B	Std. Error	Beta	t	Sig.
1	(Constant)	1.031	1.224		.843	.424
	SPHERES	.902	.083	.967	10.810	.000

We also have $\bar{x} = 13.07$ and $\Sigma(x - \bar{x})^2 = 443.2$.

(a) What is the correlation between the variables vein blood flow rate and spheres blood flow rate? Also, interpret the value of $R^2 = 0.936$.

(b) Give the estimated regression line for predicting vein blood flow rate from spheres blood flow rate.

(c) Construct the 95% confidence interval for the slope.

(d) We wish to assess whether the two methods for measuring blood flow give similar results. Using a 5% significance level, can we conclude that the slope is significantly different from unity? State the hypotheses to be tested. Using your result from (c), state your decision and explain how you reached it.

(e) Predict the vein blood flow rate for a particular dog whose sphere blood flow rate is 12 ml/minute.

(f) Give a 95% prediction interval for the vein blood flow rate for a particular dog whose sphere blood flow rate is 12 ml/minute.

(g) How would a 95% confidence interval for the mean vein blood flow rate for all dogs whose sphere blood flow rate is 12 ml/minute compare with the interval you reported in (g)? Select one: *wider* *narrower* *same width*.

13.36 Emission rates of carbon monoxide in grams per mile were taken to see whether the emission rate increases with the mileage of a car. The carbon monoxide emission rate (y) and the mileage (x) were recorded on 22 cars of the same make and model. The data are provided along with the scatterplot and SPSS regression output.

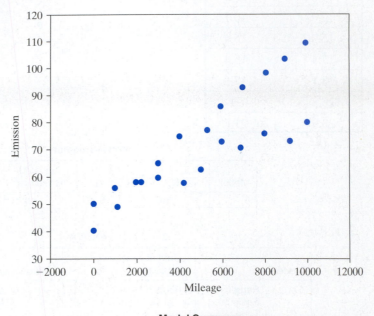

Mileage	Emission
0	50
0	40
1000	56
1100	49
2000	58
2200	58
3000	60
3000	65
4000	75
4200	58
5000	63
5300	77
6000	73
6000	86
6900	71
7000	93
8000	76
8100	98
9000	103
9200	73
10000	80
10000	109

Model Summary

Model	R	R Square	Adjusted R Square	Std. Error of the Estimate
1	.856	.733	.719	9.6112

ANOVA

Model		Sum of Squares	df	Mean Square	F	Sig.
1	Regression	5059.814	1	5059.814	54.775	.00000
	Residual	1847.504	20	92.375		
	Total	6907.318	21			

Coefficients

Model		Unstandardized Coefficients		Standardized Coefficients	t	Sig.
		B	Std. Error	Beta		
1	(Constant)	47.190	3.861		12.222	.00000
	MILEAGE	.00480	.001	.856	7.401	.00000

We also have $\bar{x} = 5045.45$ and $\sum(x - \bar{x})^2 = 219{,}594{,}545.5$.

(a) Give the estimated regression line for predicting the carbon monoxide emission rate from the mileage on the car's odometer.

(b) Compute the residual for the first observation ($x = 0$, $y = 50$).

(c) Using a 1% significance level, can we conclude that the carbon monoxide emission rate *increases* with the car's mileage? State the hypotheses to be tested, the observed value of the test statistic, the corresponding p-value, and your decision.

(d) Predict the carbon monoxide emission rate of *a particular car* with an odometer reading of 7,500 miles.

(e) Give a 95% prediction interval for the carbon monoxide emission rate of *a particular car* with an odometer reading of 12,000 miles.

(f) The R^2 value is 0.733. Interpret this number in terms of the percent of variation of the carbon monoxide emission rate.

13.11 Multiple Regression

The previous sections have focused on the analysis involved in a simple linear regression problem. The model in simple linear regression stated the mean of the response variable as a linear function of a single explanatory variable. In multiple regression, the mean of the response variable is a function of two or more explanatory variables. This technique is used quite often in practice as there are generally many possible explanatory variables that may be useful for predicting the value of a response variable.

In Section 13.2, we examined the relationship between midterm exam scores and final exam scores. There certainly are some other possible factors that may be related to final exam scores, such as number of classes missed during the term or total homework points. The corresponding multiple regression model in this case would look like this:

Final Score = β_0 + β_1(Midterm Score) + β_2(Number of Missed Classes) + β_3 (Homework Score), where the parameters β_0, β_1, β_2, and β_3 would be estimated from the data.

The descriptive and inferential tools that we studied in the previous sections are also useful in the multiple regression setting.

- We will use the **method of least squares** to determine the estimates of the population parameter coefficients in the regression model. The estimates are often expressed as b_0, b_1, b_2, etc. The formulas for these least squares estimates is complicated and we will let the computer or calculator do the computations.

- We will have **residuals**, defined again as the difference between the observed and predicted response, that should look like they came from a normal distribution, with a mean of zero and an unknown standard deviation σ (that does not depend on the values of the explanatory variables).

- The **population standard deviation** σ of the responses about their mean, for a given set of x-values, will be estimated from the root of the mean square error **MSE**. The estimate is denoted by s. The MSE is again an average of the squared residuals and can be found in the standard ANOVA table output.

- The **degrees of freedom** associated with the estimate s is $n - p - 1 = n - (p + 1)$, where n is the sample size and p is the number of explanatory variables in the model. Note that in the simple regression model, the value of $p = 1$, so the degrees of freedom are $n - 2$.

- We will have an **ANOVA table** with an **F-test** for assessing the overall significance of the regression model. The null hypothesis is that all of the regression coefficients $\beta_1, \beta_2, \ldots, \beta_n$ for the explanatory variables are equal to zero. This null hypothesis states that none of the explanatory variables are useful linear predictors of the response variable. The alternative hypothesis is that at least one of these variables is linearly related to the response.

- We will have an R^2 quantity, the squared multiple correlation, which provides the proportion of the variability in the response that is explained by the explanatory variables in the multiple linear regression.

- We can obtain a **confidence interval** or **perform a t-test** for each of the regression coefficients $\beta_0, \beta_1, \beta_2$, etc., as we did in simple linear regression. These inferences provide information about the significance of each explanatory variable in predicting the value of the response variable.

In Example 13.21, we will show a multiple regression analysis using data on the midterm and final exam scores along with the number of classes missed. We will rely on the computer output for providing the results and focus on their interpretation. Multiple regression can also be performed using the TI graphing calculator and the program A2MULTREG. Details for obtaining and running this program are provided in the TI Quick Steps at the end of this chapter.

❖ EXAMPLE 13.21   Predicting Final Exam Scores

In Section 13.2, we examined the relationship between midterm and final exam scores. We found the association to be positive and approximately linear. The linear relationship was statistically significant—that is, midterm scores were a significant linear predictor of final exam scores. Another variable that may provide some information about the final exam score is the number of classes that the student missed over the course of a term. This second explanatory variable will be added to our model for predicting final exam scores and the significance of the multiple linear regression model will be assessed using a 5% significance level. We have already examined the data on midterm scores, so we will start here with a quick look at the scatterplot of missed classes versus final exam scores that is shown.

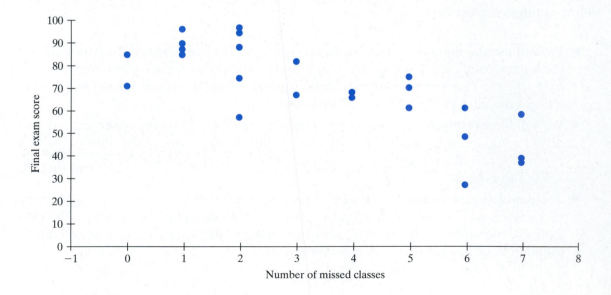

As we may have guessed, there seems to be a negative, somewhat linear, relationship between number of classes missed and final-exam score. Thus, we might expect that the coefficient for this variable in the multiple regression model to be negative. In general, it is important to examine each of the explanatory variables to understand something about each variable before they are incorporated into a larger, more complicated model.

We now turn to the multiple regression model. In this case the model can be expressed generally as

$$\text{final exam score} = \beta_0 + \beta_1(\text{midterm score}) + \beta_2(\text{number of missed classes}).$$

The data for the $n = 25$ students for these three variables were entered into SPSS and produced the following multiple linear regression output:

Variables Entered/Removed[b]

Model	Variables Entered	Variables Removed	Method
1	MISSED, MIDTERM[a]	.	Enter

a. All requested variables entered.
b. Dependent Variable: FINAL

Model Summary[b]

Model	R	R Square	Adjusted R Square	Std. Error of the Estimate
1	.857[a]	.734	.710	10.2315

a. Predictors: (Constant), MISSED, MIDTERM
b. Dependent Variable: FINAL

ANOVA[b]

Model		Sum of Squares	df	Mean Square	F	Sig.
1	Regression	6352.682	2	3176.341	30.342	.000[a]
	Residual	2303.058	22	104.684		
	Total	8655.740	24			

a. Predictors: (Constant), MISSED, MIDTERM
b. Dependent Variable: FINAL

Coefficients[a]

Model		Unstandardized Coefficients B	Std. Error	Standardized Coefficients Beta	t	Sig.
1	(Constant)	51.949	14.179		3.664	.001
	MIDTERM	.977	.326	.385	2.994	.007
	MISSED	−4.839	1.051	−.592	−4.604	.000

a. Dependent Variable: Final

This output contains some information about the model, the corresponding ANOVA table, and about the coefficients. The first box shows us that two explanatory variables were entered into the model—namely, midterm score and number of missed classes—and that the response variable is final exam score.

The model summary box provides the value of $R^2 = 0.734$, which implies that about 73.4% of the variability in the final exam scores can be explained by the explanatory variables of midterm score and number of classes missed in the multiple linear regression model. The standard error of estimate of 10.2315 is our estimate of the population standard deviation σ.

The ANOVA F-test statistic is 30.342, with a p-value of practically zero. We would reject the null hypothesis that $\beta_1 = \beta_2 = 0$ and conclude that at least one of the two regression coefficients is different from zero in the population regression equation. The square root of the MSE would also provide the estimated population standard deviation.

From the coefficients box, we find the parameter estimates and thus our estimated regression line of:

prediced final score = 51.949 + 0.977(midterm score) − 4.839(number of missed classes).

Also in the coefficients box, we have the t-test statistics for testing whether a particular regression coefficient is equal to zero. These tests assess the significance of each explanatory variable assuming that all of the other explanatory variables are included in the model. The p-values can be found in the last column. All of these p-values are for the two-sided alternative. Both midterm score and number of missed classes have a very small p-value and are significant at the 5% level.

As in the simple linear regression analysis, we should examine the residuals to assess whether the underlying model is appropriate. We could make a residual plot for each of the explanatory variables in the model and also plot the residuals against the predicted values. We would hope to see approximately a horizontal band of points around zero with no apparent pattern. A histogram could also be examined of the residuals to assess whether the normal model for the error terms seems reasonable. From the following histogram of the standardized residuals, the normality assumption seems reasonable.

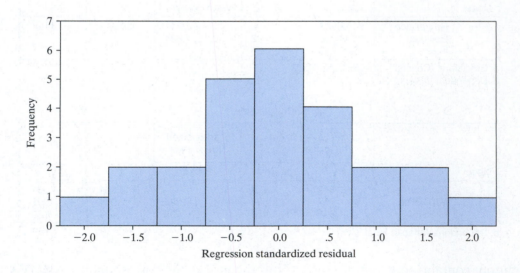

 **13.20** **Predicting Mental Capacity**

Are the size and weight of your brain an indicator of your mental capacity? A study was conducted in which magnetic resonance imaging (MRI) was used to determine the size of the brain of the subjects. (SOURCE: Willerman, L., Schultz, R., Rutledge, J. N., and Bigler, E. (1991), "In Vivo Brain Size and Intelligence," *Intelligence*, 15, 223–228.) Information about body size (height and weight) was also included. One response variable was full scale IQ score based on the four Wechsler subtests. The three explanatory variables are body weight (in pounds), body height (in inches), and MRI count (total pixel count from the 18 MRI scans performed on each subject). Our multiple linear regression model can be expressed as

$$IQ = \beta_0 + \beta_1(\text{MRI count}) + \beta_2(\text{body weight}) + \beta_3(\text{body height}).$$

The data for 20 female subjects were entered into SPSS, and the following output was produced:

Model Summary[b]

Model	R	R Square	Adjusted R Square	Std. Error of the Estimate
1	.360[a]	.130	−.034	24.08

a. Predictors: (Constant), MRICOUNT, HEIGHT, WEIGHT
b. Dependent Variable: IQ

ANOVA[b]

Model		Sum of Squares	df	Mean Square	F	Sig.
1	Regression	1382.376	3	460.792	.795	.515[a]
	Residual	9277.424	16	579.839		
	Total	10659.800	19			

a. Predictors: (Constant), MRICOUNT, HEIGHT, WEIGHT
b. Dependent Variable: IQ

Coefficients[a]

Model		Unstandardized Coefficients		Standardized Coefficients	t	Sig.
		B	Std. Error	Beta		
1	(Constant)	35.775	190.361		.188	.853
	WEIGHT	−.131	.432	−.094	−.303	.766
	HEIGHT	−.720	2.914	−.070	−.247	.808
	MRICOUNT	1.638E-04	.000	.387	1.479	.159

a. Dependent Variable: IQ

(a) Give the estimated multiple regression line for predicting IQ from weight, height and MRI count.

(b) Give the value of R^2 and interpret it in terms of this regression problem.

(c) Give the estimate of the population standard deviation σ.

(d) Examine the *t*-tests in the coefficients output. For each test, state the hypotheses being tested, the resulting *p*-value, and your conclusion at a 5% level.

✳ Exercises

**REAL
DATA**

13.37 Are the size and weight of your brain an indicator of your mental capacity? A study was conducted by Willerman, et al. (1991), in which magnetic resonance imaging (MRI) was used to determine the size of the brain of the subjects. Information about body size (height and weight) was also included. One response variable was full scale IQ score based on the four Wechsler subtests. The three explanatory variables are body weight (in pounds), body height (in inches), and MRI count (total pixel count from the 18 MRI scans performed on each subject). Our multiple linear regression model can be expressed as

$$IQ = \beta_0 + \beta_1(\text{MRI count}) + \beta_2(\text{body weight}) + \beta_3(\text{body height.})$$

The data for 20 male subjects were entered into SPSS and the following output was produced:

Model Summary

Model	R	R Square	Adjusted R Square	Std. Error of the Estimate
1	.730[a]	.533	.433	18.46

a. Predictors: (Constant), MRICOUNT, WEIGHT, HEIGHT

ANOVA[b]

Model		Sum of Squares	df	Mean Square	F	Sig.
1	Regression	5438.796	3	1812.932	5.319	.012[a]
	Residual	4771.481	14	340.820		
	Total	10210.278	17			

a. Predictors: (Constant), MRICOUNT, WEIGHT, HEIGHT
b. Dependent Variable: IQ

Coefficients[a]

Model		Unstandardized Coefficients B	Unstandardized Coefficients Std. Error	Standardized Coefficients Beta	t	Sig.
1	(Constant)	183.994	109.411		1.682	.115
	WEIGHT	4.222E-02	.253	.035	.167	.870
	HEIGHT	−3.909	1.335	−.607	−2.928	.011
	MRICOUNT	2.148E-04	.000	.491	2.651	.019

a. Dependent Variable: IQ

(a) Give the estimated multiple regression line for predicting IQ from weight, height, and MRI count for males.

(b) Give the value of R^2 and interpret it in terms of this regression problem.

(c) Give the estimate of the population standard deviation σ.

(d) Examine the t-tests in the coefficients output. For each test, state the hypotheses being tested, the resulting p-value, and your conclusion.

13.38 In Example 13.21, we examined the multiple regression model for predicting final exam scores from midterm scores and the number of classes missed for that term. We found that both midterm score and number of missed classes have a very small p-value and were significant at the 5% level. Another variable that may provide some information about the final exam score is the homework score for the student at the end of the term. This additional explanatory variable will be added to our model for predicting final exam scores and the significance of the multiple linear regression model will be assessed using a 5% significance level.

Variables Entered/Removed[b]

Model	Variables Entered	Variables Removed	Method
1	HOMEWORK, MIDTERM, MISSED[a]	.	Enter

a. All requested variables entered.
b. Dependent Variable: FINAL

Model Summary[b]

Model	R	R Square	Adjusted R Square	Std. Error of the Estimate
1	.857[a]	.735	.697	10.4476

a. Predictors: (Constant), HOMEWORK, MIDTERM, MISSED

ANOVA[b]

Model		Sum of Squares	df	Mean Square	F	Sig.
1	Regression	6363.553	3	2121.184	19.433	.000[a]
	Residual	2292.187	21	109.152		
	Total	8655.740	24			

a. Predictors: (Constant), HOMEWORK, MIDTERM, MISSED
b. Dependent Variable: FINAL

Coefficients[a]

Model		Unstandardized Coefficients		Standardized Coefficients		
		B	Std. Error	Beta	t	Sig.
1	(Constant)	59.515	28.004		2.125	.046
	MIDTERM	.948	.345	.374	2.747	.012
	MISSED	−5.108	1.371	−.625	−3.725	.001
	HOMEWORK	−.680	2.154	−.046	−.316	.755

a. Dependent Variable: FINAL

(a) Give the estimated multiple regression line for predicting final exam score from midterm score, number of missed classes, and homework score.

(b) Give the value of R^2 and interpret it in terms of this regression problem.

(c) Give the estimate of the population standard deviation σ.

(d) Examine the *t*-tests in the coefficients output. For each test, state the hypotheses being tested, the resulting *p*-value, and your conclusion. Note: The homework scores were generally very good scores overall, so it does not do as well at explaining final exam scores, especially after accounting for midterm scores and number of classes missed already in the model.

Chapter Summary

In this chapter, we focused on displaying, describing, and modeling the relationship between two quantitative variables. We have data on two different quantitative variables that are measured on the same unit. Sometimes, one of the variables can be considered the explanatory variable and the other the response variable. Other times we simply wish to study the relationship between the two quantitative variables.

We examine the relationship graphically with a scatterplot. When the data suggest a linear relationship between the two variables, we use the data to find a best fitting line, the least squares regression line. This equation can be used for making predictions. We must be careful that we predict responses for values of x that are in the range of the data collected. We have a measure of the strength of the linear relationship between the two variables with the correlation coefficient. We can also assess whether the slope for the regression line for the population is significantly different from the value of zero, which would support using the explanatory variable x as a linear predictor for the response variable y. In the last section, we extended the idea of simple linear regression to handling more than one explanatory variable called multiple regression.

Key Terms

Be sure you can describe, in your own words, and give an example of each of the following key terms from this chapter:

scatterplot	strength of a relationship	correlation coefficient
response variable	simple linear regression	regression effect
explanatory variable	method of least squares	least squares regression line
dependent variable	slope	prediction interval
independent variable	y-intercept	regression model
positive association	prediction	confidence interval for the
negative association	residuals	mean response
direction of a relationship	influential points	linear transformation
form of a relationship	regression coefficients	multiple regression
F-test for slope	t-test for slope	

Exercises

13.39 A research group reports the correlation coefficient for two variables, height and gender, to be 0.75. What can you conclude from this information?

13.40 Suppose that 100 recent residential home sales in a city are used to fit a least squares regression line relating the selling price, y (in dollars), to the amount of living space, x (in hundreds of square feet). Homes in the sample range from 1500 square feet to 3500 square feet. The resulting least squares equation is $\hat{y} = -30{,}000 + 7000x$, and the correlation coefficient is $r = 0.76$. Suppose that the amount of living space were measured in square yards instead of square feet. What would the correlation between x and y be now? (Recall that 1 yard = 3 feet.)

13.41 A company that markets and repairs computers collects some data on the length of their service calls in minutes, y, and the number of components that needed to be repaired, x. The least squares regression line for these data was $\hat{y} = 4.162 + 15.509x$, and the value of the correlation coefficient was 0.993.

    (a) Describe the nature of the association between x and y. Specify whether it is strong or weak and whether it is positive or negative.

    (b) Predict the length of the service call when two components need to be repaired.

(c) In the data that were collected to develop the preceding equation, one particular service call, which required the repair of two components, took 29 minutes. Calculate the residual for that call.

(d) The company charges $50 for each service call plus $1 for each minute that the service call lasts. What is the numerical value of the correlation between x and the dollar cost of the call?

(e) What is the statistical term for a service call that lasted much longer than the preceding regression equation would predict?

13.42 The data that follow were collected in an investigation of the relationship between the time that a sugar cube takes to dissolve in a glass of water and the temperature of the water. The investigator took a given quantity of water, measured its temperature (in °C), added a sugar cube, and stirring gently, measured the time until the sugar was completely dissolved (in seconds). The results of nine trials were as follows:

Temperature (x)	24	45	54	80	67	16	31	35	19
Time (y)	65	39	22	13	19	80	50	46	72

The following sums were computed from the data:

$$\Sigma x_i = 371, \quad \Sigma y_i = 406, \quad \Sigma x_i^2 = 19{,}209, \quad \Sigma x_i y_i = 12{,}624, \quad \Sigma y_i^2 = 22{,}960.$$

(a) Give the correlation r between the time for the sugar cube to dissolve and the temperature of the water.

(b) What would the correlation be if the data were recorded in minutes and °F rather than seconds and °C? [Recall that $°F = \left(\frac{9}{5}\right)°C + 32$.]

(c) Produce a plot of the residuals versus temperature. Based on this plot, does the linear model relating time and temperature seem appropriate? Explain.

13.43 The shoe sizes (in cm) and scores on a standardized math exam are shown for 12 children in grades 2, 4, and 6. A researcher would like to examine the relationship between shoe size and score on the standardized math exam. In particular, she would like to develop a model to predict the math exam score from a child's shoe size.

Shoe size (x)	21	25	21	24	16	16	17	18	20	17	21	21
Exam score (y)	65	74	69	77	46	52	53	48	65	53	67	55

(a) Sketch the corresponding scatterplot. Be sure to mark all important features.

(b) Give the least squares regression line for regressing exam score on shoe size. Sketch this regression line on your scatterplot.

(c) Interpret the value of the slope coefficient of the regression line; that is, explain precisely what the value signifies about the relationship between shoe size and exam score.

(d) If a child's shoe size is 23 cm, what do you predict he will score on the math exam?

(e) Give a confounding variable that would account for the apparent association between shoe size and math exam score.

13.44 In a study of the shelf life of a packaged cereal, the number of days on the shelf (x) and the percent of moisture content in the package (y) were recorded. The data and corresponding least squares regression line for relating y = percent moisture content and x = shelf time are provided.

x	0	6	10	16	24	27	34
y	2.8	2.9	3.4	3.5	4.0	3.9	4.3

Note: $r = 0.979$ and $\hat{y} = 2.79 + 0.045x$.

(a) Predict the percent of moisture content for a cereal package that has been on the shelf for three *weeks*.

(b) Sketch the residual plot.

(c) Based on the residual plot, does it appear that the percentage of moisture content increases linearly with the number of days on the shelf?

13.45 A statistical computer package was used to examine the relationship between the ranking (x) of a male tennis player (lower value implies a higher rank, 1 = best) and corresponding dollar earnings for a year (y).

```
Dep var:EARNINGS       N:    32    Multiple R:   .765    Squared multiple R:   .586
Adjusted squared multiple R:   .572      Standard error of estimate:    95798.163


  Variable    Coefficient    Std error    Std coef Tolerance    T     P(2 tail)
CONSTANT      375337.492     34679.516      0.000   .           10.823   0.000
RANKING       -11940.924      1834.151     -0.765   .100E+01    -6.510   0.000


                            Analysis of Variance
   Source    Sum-of-squares    DF   Mean-square    F-ratio     P
Regression    .388974E+12       1   .388974E+12     42.384   0.000
  Residual    .275319E+12      30   .917729E+10
```

(a) Predict the earnings for a male tennis player with a ranking of 10.

(b) What is the value of the correlation between ranking and earnings?

(c) Suppose that the $10,000 fee for participating in tournaments was subtracted off of each of the player's earnings, so we have $y^* = y - 10{,}000$.

   (i)   What would be the correlation between ranking x and y^*?

   (ii) What would be the value of the slope for regressing y^* on x?

13.46 A car dealer specializing in Corvettes enlarged his facilities and offered a number of models for sale during the open house. From his data on price (in $1000s) and age (in years) of Corvettes, he generated the following EXCEL output:

```
EXCEL REGRESSION OUTPUT
--------------------
Regression Statistics
--------------------
Multiple R        0.890
R Square          0.791
Adjusted R        0.775
Square
Standard Error    4.343
Observations       15
--------------------

ANOVA
------------------------------------------------------------------------
                df        SS         MS         F       Significance F
------------------------------------------------------------------------
Regression       1     930.626    930.626    49.342        0.00001
Residual        13     245.190     18.861
Total           14    1175.816
------------------------------------------------------------------------
            Coefficients  Standard Error  t Stat  p-value  Lower 95%  Upper 95%
------------------------------------------------------------------------
Intercept       33.726        2.588      13.030   0.00000   28.135     39.318
X Variable 1    -1.862        0.265      -7.024   0.00001   -2.435     -1.289
------------------------------------------------------------------------
```

(a) How many observations were taken?

(b) Report the least-squares regression line.

(c) Predict the value of a car that is 20 years old. Comment on your prediction.

(d) Calculate the residual for the observation $(x = 5, y = 20)$.

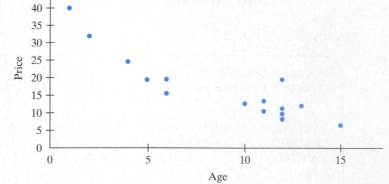

(e) Is there evidence of a significant (nonzero) linear relationship between age and price? Explain.

13.47 One of the factors thought to contribute to the incidence of skin cancer is the ultraviolet (UV) radiation from the sun. The amount of UV radiation a person (in the United States) receives depends on the person's latitude north. The following table gives the rates of malignant skin cancer (melanoma) and the degrees latitude north for nine areas throughout the United States:

Degrees latitude north (x)	32.8	33.9	34.1	37.9	40.0	40.8	41.7	42.2	45.0
Melanoma rate (per 100,000) (y)	9.0	5.9	6.6	5.8	5.5	3.0	3.4	3.1	3.8
Residuals	1.52	−1.14	−0.36	0.35	0.89	−1.29	−0.53	−0.63	1.19

(a) What is the correlation between melanoma rate and degrees latitude north?

(b) What is the equation of the least squares regression line for regressing melanoma rate on degrees latitude north?

(c) Ann Arbor, Michigan, is located 42.2 degrees latitude north; what melanoma rate would you predict for Ann Arbor?

(d) The population of Ann Arbor is 120,000. How many cases of melanoma would you expect to get in Ann Arbor?

(e) Anchorage (Alaska) is located 61.1 degrees latitude north. *Should* you predict its melanoma rate? If yes, give the prediction. If no, explain why not.

(f) If the X data were transformed to $M = -(X) + 10$,

   (i)   What is the correlation between M and Y?

   (ii) What is the slope for regressing Y on M?

(g) The corresponding residuals for these data are also provided in the preceding table. Provide a sketch of the residual plot.

(h) Based on this residual plot, does the linear model seem appropriate? Explain your answer.

13.48 Suppose that we have gathered the following data to measure the effect of TV viewing on reading scores in elementary school children:

Grade	x Number Hours TV/Week	y Reading Score
2	4	605
2	5	510
2	8	560
2	12	435
2	20	405
4	16	755
4	20	810
4	22	635
4	22	615
4	25	540
4	30	640

(a) Enter the data into your calculator as follows:

    L1: TV hours for second graders;

    L2: reading scores for second graders;

    L3: TV hours for fourth graders;

    L4: reading scores for fourth graders;

    L5: TV hours for all children;

    L6: reading scores for all children.

(b) Obtain the least squares regression line for reading score (y) versus number of hours of TV watching (x) for all elementary school children. (Use L5 and L6.) Interpret the values of the slope and the intercept.

(c) Obtain the least squares regression line for reading score (y) versus number of hours of TV watching (x) for second graders only. (Use L1 and L2.) Interpret the values of the slope and the intercept.

(d) Do (c) for the fourth graders only. (Use L3 and L4.)

(e) Explain why the slope is positive for the regression line with all children and negative for the grades individually.

■ *TI Hint*: You can plot all the data on the same scatterplot with different plot marks for second and fourth graders as follows: Turn *both Plot 1 and Plot 2 on*, plot L1 and L2 in Plot 1 with one plotting mark (say, the box symbol), and plot L3 and L4 in Plot 2 using a different plotting mark (say, the + symbol).

13.49 Determine whether the following statements are true or false (a true answer is always true):

(a) The correlation coefficient between income and gender is $r = 0.7$.

(b) When the correlation coefficient is zero, it means that the two variables are not related in any way.

13.50 An economist was interested in the production costs for companies supplying a particular chemical for use in fertilizers. The explanatory variable x is the number of tons of chemical produced per year (in thousands). The response variable y is the production costs per ton (in dollars). Data on these two variables were used to produce the regression output that follows.

(a) Give the equation of the least squares regression line for regressing cost on number of tons produced.

(b) Interpret the value of the slope coefficient of the regression line—that is, explain precisely what the value signifies about the relationship between number of tons produced and cost.

(c) Is there evidence of a significant (nonzero) linear relationship between these two variables? Use a 5% significance level. Include in your explanation what information in the computer output allows you to answer this question.

```
* * * *   M U L T I P L E   R E G R E S S I O N   * * * *
Equation Number 1 Dependent Variable.. COST cost per ton ($)
Variable(s) Entered on Step 1.. TONS  Number of tons (1000s)


Multiple R          .88620
R Square            .78536
Adjusted R Square   .74243
Standard Error     4.49862


Analysis of Variance
                    DF       Sum of Squares      Mean Square
Regression           1          370.24071         370.24071
Residual             5          101.18786          20.23757
F =      18.29472    Signif F =  .0079


----------------- Variables in the Equation -----------------
Variable          B           SE B        Beta        T   Sig T
TONS        -6.619885     1.547703    -.886205    -4.277  .0079
(Constant)  65.775279     6.100889                10.781  .0001
```

13.51 The following table provides data on the average teacher salary (x) in thousands of dollars and the pupil-to-staff ratio (y) for Western suburbs of Boston in 1992:

Suburb	x Average Teacher Salary	y Pupil-to-staff Ratio
(1) Arlington	40	13.4
(2) Belmont	44	14.4
(3) Brookline	35	11.7
(4) Cambridge	42	11.7
(5) Dedham	37	13.0
(6) Lexington	48	13.9
(7) Newton	56	17.8
(8) Somerville	40	14.3
(9) Waltham	38	13.9
(10) Watertown	41	13.1

(a) Make the scatterplot for these data. Be sure to label your axes.

(b) Give the least squares regression line for regressing ratio (y) on salary (x) and superimpose the line on the scatterplot.

(c) Calculate the residual for the fourth observation.

(d) Verify that the point $(\bar{x}, \bar{y})$ falls on the least squares regression line.

(e) Is there any observation that would be called an influential point? If so, clearly circle that observation in the scatterplot and select the reason why it is influential from the following list:

(i) It produces a large residual.

(ii) It has a large y-value.

(iii) Removing it would markedly change the regression line.

13.52 The following table lists distances (the explanatory variable, in miles) and cheapest airline fares (the response variable, in dollars) to certain destinations for passengers flying out of Baltimore, Maryland (for January 8, 1995):

Observation	Destination	Distance (x)	Airfare (y)
1	Atlanta	576	178
2	Boston	370	138
3	Chicago	612	94
4	Dallas–Fort Worth	1216	278
5	Detroit	409	158
6	Denver	1502	258
7	Miami	946	198
8	New Orleans	998	188
9	New York	189	98
10	Orlando	787	179
11	Pittsburgh	210	138
12	St. Louis	737	98

(a) Make a scatterplot of these data. Be sure to label your axes.

(b) Is airfare linearly associated with distance? If so, is the association positive or negative?

(c) Would you characterize the association as weak, moderate, or strong? Explain.

(d) Give the least squares regression line for regressing airfare (y) on distance (x).

(e) Looking at the data, Atlanta is 576 miles from Baltimore. What airfare would the regression line have predicted for Atlanta?

(f) The actual airfare to Atlanta at that time was $178. Determine the numerical value of the residual for this first observation.

13.53 Milk samples were obtained from 14 Holstein–Friesian cows, and each was analyzed to determine uric-acid concentration (y), measured in mol/L. In addition to acid concentration, the total milk production (x), measured in kg/day, was recorded for each cow. The data were entered into a computer and the following regression output was obtained:

```
* * * *   M U L T I P L E   R E G R E S S I O N   * * * *
Equation Number 1    Dependent Variable.. ACIDCONC
Variable(s) Entered on Step Number   1..  MILKPROD


Multiple R          .88789
R Square            .78835
Adjusted R Square   .77071
Standard Error    21.48174


Analysis of Variance
                   DF        Sum of Squares       Mean Square
Regression          1            20625.84845      20625.84845
Residual           12             5537.58012        461.46501
F =      44.69645       Signif F =  .0000


------------------ Variables in the Equation ------------------
Variable              B          SE B        Beta        T    Sig T
MILKPROD       -5.202654      .778195    -.887889    -6.686   .0000
(Constant)    321.241328    23.712311              13.547   .0000
```

(a) Give the equation of the least squares regression line.

(b) Using the variable names and units, explain what the value of the slope represents.

(c) Is there evidence of a significant (nonzero) linear relationship between these two variables? Explain your answer, including a description of the information in the preceding output that allows you to answer this question.

13.54 Consider the following scatterplot for a hypothetical set of exam scores:

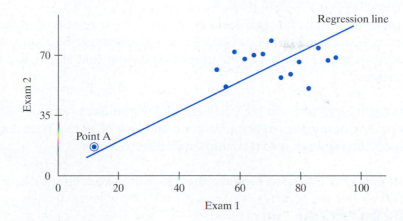

The observation marked Point A is called (select one)

(a) an outlier with respect to the regression line.

(b) an influential point.

(c) a residual.

13.55 Data on y = life expectancy and x = the number of people per television set for a sample of 22 countries were used to produce the following regression output (the values of x ranged from 1 to 250):

```
Dependent Variable..   LIFEEXP   Life Expectancy
Independent Variable.. PERTV     Number of People per Television Set


Multiple R           .80381
R Square             .64611
Adjusted R Square    .62842
Standard Error      6.62249


Analysis of Variance
                   DF        Sum of Squares        Mean Square
Regression          1           1601.44262         1601.44262
Residual           20            877.14829           43.85741
F =      36.51475       Signif F =   .0000


----------------- Variables in the Equation -----------------
Variable           B          SE B          Beta          T    Sig T
PERTV        -.115611       .019132      -.803810      -6.043   .0000
(Constant)   70.717499     1.733859                    40.786   .0000
```

(a) Write down the least squares regression line.

(b) Predict the life expectancy for a country that has 100 people per television set.

(c) What is the correlation coefficient between life expectancy and the number of people per television set?

(d) Select your answer: The point $(\bar{x}, \bar{y})$ [always sometimes never] falls on the least squares regression line.

(e) Is there evidence of a significant (nonzero) linear relationship between life expectancy and the number of people per television set? Explain your answer and describe the information in the preceding output that allows you to answer this question.

REAL DATA

13.56 Listed are the ages of a sample of 24 couples taken from marriage licenses filed in Cumberland County, Pennsylvania in June and July of 1993. These data will be used to perform the regression of husbands' age on wives' age.

The data have been used to produce the following two scatterplots:

Couple	Age of Wife	Age of Husband
1	22	25
2	32	25
3	50	51
4	25	25
5	33	38
6	27	30
7	45	60
8	47	54
9	30	31
10	44	54
11	23	23
12	39	34
13	24	25
14	22	23
15	16	19
16	73	71
17	27	26
18	36	31
19	24	26
20	60	62
21	26	29
22	23	31
23	28	29
24	36	35

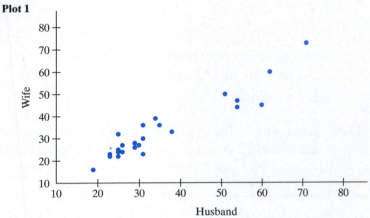

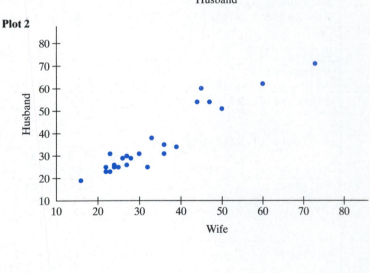

(a) Which scatterplot is the appropriate plot if we are regressing husbands' age on wives' age?

(b) Give the least squares regression line for regressing husbands' age on wives' age.

(c) Give the correlation coefficient.

(d) Interpret the value of the slope coefficient of the regression line. That is, explain precisely what the value signifies about the relationship between husbands' ages and wives' ages.

(e) Compute the residual for Couple 17.

13.57 To understand the pattern of variation in plant species in the Mediterranean grassland, data were collected on y = density of plant species (the number of species per 0.04 square meter) and x = altitude (in thousands of meters) from 12 experimental plots. The data are provided along with the scatterplot and SPSS regression output. Note that an altitude of 800 meters would correspond to an $x = 0.80$.

Altitude	Density
0.64	15
0.64	20
0.86	16
0.86	18.5
0.89	13.5
0.89	16
1.22	11
1.22	19.5
1.45	11.5
1.45	12
1.72	8
1.72	8.5

Model Summary

Model	R	R Square	Adjusted R Square	Std. Error of the Estimate
1	.782	.611	.572	2.6539

ANOVA

Model		Sum of Squares	df	Mean Square	F	Sig.
1	Regression	110.629	1	110.629	15.707	.003
	Residual	70.434	10	7.043		
	Total	181.062	11			

Coefficients

Model		Unstandardized Coefficients B	Std. Error	Standardized Coefficients Beta	t	Sig.
1	(Constant)	23.354	2.452		9.526	.000
	Altitude (thousands of meters)	−8.168	2.061	−.782	−3.963	.003

We also have $\bar{x} = 1.13$ and $\Sigma(x - \bar{x})^2 = 1.66$.

(a) What is the direction of the association between density of plant species and altitude based on the scatterplot?

(b) Give the equation of the least squares regression line.

(c) Compute the residual for the first observation ($x = 0.64$, $y = 15$).

(d) Construct the 95% confidence interval for the slope β and use the confidence interval in to test H_0: $\beta = 0$ versus H_1: $\beta \neq 0$ at the 5% level.

(e) Estimate the mean plant species density at an altitude of 750 meters.

(f) What is the standard error of your estimate in (e)?

(g) What percent of variability in plant density can be explained by its linear relationship with altitude?

13.58 Physical characteristics of sharks were studied by marine researchers. Data on x = body length (feet) and y = jaw width (inches) for 12 sharks were obtained. Because it is difficult to measure jaw width in living sharks, researchers would like to estimate jaw width from body length, which is more easily measured. A scatterplot did support a linear model.

Model Summary

Model	R	R Square	Adjusted R Square	Std. Error of the Estimate
1	.883	.779	.757	1.488

ANOVA

Model		Sum of Squares	df	Mean Square	F	Sig.
1	Regression	78.07	1	78.07	35.28	.00014
	Residual	22.13	10	2.21		
	Total	100.20	11			

Coefficients

Model		Unstandardized Coefficients		Standardized Coefficients	t	Sig.
		B	Std. Error	Beta		
1	(Constant)	−3.03	3.25		−.94	.37180
	LENGTH	1.20	.20	.88	5.94	.00014

We also have $\bar{x} = 16$ and $\Sigma(x - \bar{x})^2 = 54.3$.

(a) Find the equation of the estimated regression line for predicting the jaw width from the body length.

(b) The researcher wants to assess whether there is a significant linear relationship between body length and jaw width. Write the hypotheses to be tested.

(c) Obtain the p-value for the test in (b) from the given output.

(d) Do you accept or reject the null hypothesis in (b)? Use $\alpha = 0.05$.

(e) Find the value of the coefficient of determination and explain its meaning.

(f) Construct a 95% prediction interval for the jaw width for a particular shark whose body length is 15 feet.

(g) Construct a 95% confidence interval for the mean jaw width for all sharks whose body length is 15 feet.

TI Quick Steps

Simple Linear Regression

The TI graphing calculator can be used to make a scatterplot, find the least squares regression line, and superimpose the line on the scatterplot. The steps are given using the data from Example 13.3 on Test 1 and Test 2 scores.

Entering the Paired Observations—Data in L1 and L2

Be sure you have all previous data cleared out!

Then press

and you are ready to start.

$x = L1$	$y = L2$
8	9
10	13
12	14
14	15
16	19

You can either enter point by point as:

⟨8⟩ **ENTER** ▶ ⟨9⟩ **ENTER** ◀ etc.

or you can enter all the values of *x* first:

⟨8⟩ **ENTER** ⟨1⟩ ⟨0⟩ **ENTER** • • •

⟨1⟩ ⟨6⟩ **ENTER**

▶ to enter all the values of *y* in the correct order in L2

Your screen should look like this:

```
L1      L2      L3      2
8       9       ------
10      13
12      14
14      15
16      19
------  ======
L2(6) =
```

Obtaining a Scatterplot of the Data in $X = $ L1 and $Y = $ L2

2nd ⟨**Y=**⟩ **ENTER**   For Plot 1
(STAT PLOT)

ENTER   To turn on the plot we want the "on" to be blinking

▼   To highlight the scatterplot icon

ENTER   Check that your Xlist and Ylist contain the correct data L1 for Xlist and L2 for Ylist

ZOOM ⟨9⟩   To "see" the plot in the window

Be sure all graphs have been cleared out! If the plot suggests that a linear regression line will fit the points, we next want to get the estimated least squares regression line and superimpose this line onto the scatterplot.

Your screen should look like this:

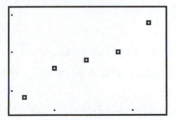

Obtaining the Least Squares Regression Line of $Y = L2$ on $X = L1$

To obtain both the equation of the line and the correlation coefficient, we first need to turn ON the DIAGNOSTIC.

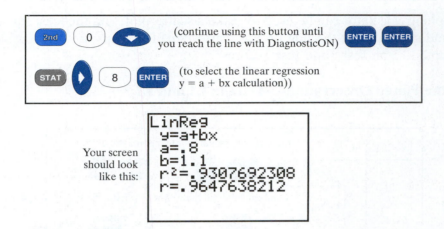

Your screen should look like this:

```
LinReg
 y=a+bx
 a=.8
 b=1.1
 r²=.9307692308
 r=.9647638212
```

Caution: There are two linear regression options: LinReg $(ax + b)$ and LinReg $(a + bx)$. We are requesting the latter option, which has b representing the slope.

Note that the value of the correlation coefficient r and r^2 is also reported with this output.

Superimposing the Regression Line on the Scatterplot

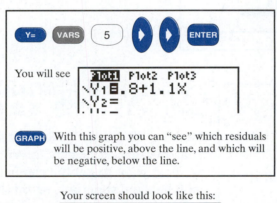

You will see

```
Plot1 Plot2 Plot3
\Y1=.8+1.1X
\Y2=
```

GRAPH   With this graph you can "see" which residuals will be positive, above the line, and which will be negative, below the line.

Your screen should look like this:

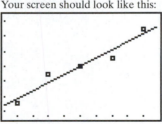

Obtaining the Residuals

The TI can be used to obtain the residuals. With the *x* data in L1 and the *y* data in L2, the steps for obtaining and storing the residuals in L4 are as follows:

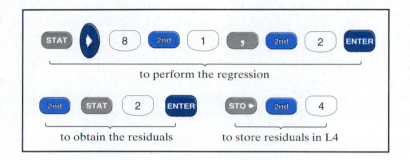

to perform the regression

to obtain the residuals   to store residuals in L4

Now L4 contains all the residuals. You can verify that they add up to zero.

Your screen should look like this:

L2	L3	L4	4
9	9.6	⁻.6	
13	11.8	1.2	
14	14	0	
15	16.2	⁻1.2	
19	18.4	.6	
------	------		

L4(1)= ⁻.6

How Good Is the Regression Line?
The Residual Plot

Once you have your residual plot, ask, "Do you see a random scatter of points in an approximate horizontal band around zero, or is there an apparent pattern in the residuals that may indicate that the linear regression may not be appropriate?"

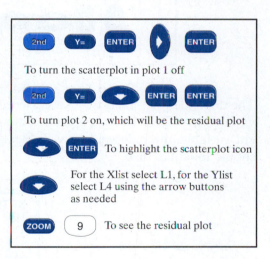

To turn the scatterplot in plot 1 off

To turn plot 2 on, which will be the residual plot

To highlight the scatterplot icon

For the Xlist select L1, for the Ylist select L4 using the arrow buttons as needed

To see the residual plot

Your screen should look like this:

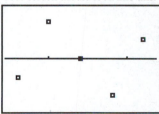

Regression Analysis Using the A2MULTREG Program

The Texas Instruments Web site, **www.education.ti.com**, provides a lot of information about the various graphing calculators. The education section of this Web site provides links to a program archive that contains many programs for various TI graphing calculators, which can be viewed and downloaded to a computer or a graphing calculator. Programs are also available on disk by calling 1-800-TI-CARES. One of the programs available for the TI-83 Plus is called A2MULREG.

This program is quite lengthy, but can be used to obtain the ANOVA table for simple or multiple regression. This program does require that the data be entered into a matrix. An example is given to demonstrate the steps for performing a simple linear regression analysis; however, the steps can be modified to perform a multiple regression analysis.

Recall the data for Test 1 and Test 2 scores from Example 13.3.

Test 1	Test 2
8	9
10	13
11	14
14	15
16	19

To use the A2MULREG program, we need to enter the data as a matrix.
The following steps will open Matrix D and set up its size to be five rows by two columns:

The dependent variable y = Test 2 scores must be entered in Column 1 and the independent variable x = Test 1 scores must be entered in Column 2. If there is more than one independent variable, the independent variables are entered in Columns 2, 3, etc. Your input screen should look like this:

```
MATRIX[D] 5×2
[9     8    ]
[13    10   ]
[14    12   ]
[15    14   ]
[19    16   ]
```

Press PRGM , highlight the program named <u>A2MULREG</u>, and press enter so the name <u>prgmA2MULREG</u> appears in your window. Press ENTER to start running the A2MULREG program, and you should see the following menu in your window:

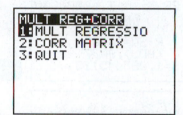

```
DATA IN MAT [D]
COL Y,X1,X2,..XN
Y MUST BE IN THE
1ST COL OF [D].

[A],[B],[C],[D],
[E]+[F] USED.
```

To select multiple regression, you should press ENTER .

Your screen window should look like this:

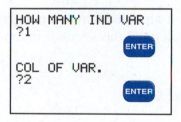

```
MULT REG+CORR
1:MULT REGRESSIO
2:CORR MATRIX
3:QUIT
```

Press 1 to select multiple regression.

Various questions will appear in your window, and you answer them by typing in the numerical response and then pressing ENTER .

The first few answers are shown in the following window:

```
HOW MANY IND VAR
?1
                  ENTER

COL OF VAR.
?2
                  ENTER
```

After you have entered the number of independent variables, in this case just one, and the column that contains the independent variable(s), Column 2, you press ENTER. Since you are performing a simple linear regression, you will be asked whether you want to make a scatterplot of your data. Your window will look like this:

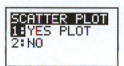

```
SCATTER PLOT
1:YES PLOT
2:NO
```

You select Yes by pressing ENTER.

Your window will look like this:

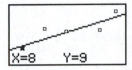

Pressing ENTER one more time will produce the ANOVA table for regression. Your output window will look like this:

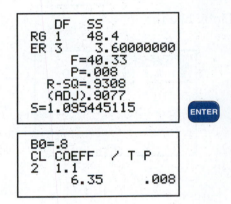

So we see that our *F*-test statistic is $F = 40.33$, the *p*-value is 0.008, and $r^2 = 0.9304$; hence, the correlation coefficient is $r = 0.9647$, the estimated standard deviation $\hat{\sigma} = s = 1.095$, and the *y*-intercept $a = BO = 0.8$. The estimated slope is found under the "COEFF" heading $b = 1.1$, and the *t*-test for the slope is $t = 6.35$ with a two-sided *p*-value of 0.008.

To obtain a prediction interval for an individual in a population whose value of *x* is 11 and a confidence interval for the mean of all the *y* values in the population whose value of *x* is 11, press ENTER.

Your screen window will look like this:

```
MAIN MENU
1:CONF+PRI INTER
2:RESIDUALS
3:NEW MODEL
4:QUIT
```

To select number 1, the confidence interval and predicted interval, press ENTER.

Your screen window will look like this:

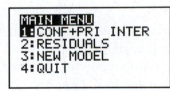

Enter **3.182** for the *t** value for this example, and press ENTER.

The new screen will look like this:

```
┌──────────────────────────┐
│  X FOR COL               │
│                    2     │
│  ?                       │
│                          │
└──────────────────────────┘
```

Enter the value of x— for our example it is a Test 1 score of 11—and press ENTER .

The output screen with both the confidence interval and the predicted interval will look like this:

```
┌──────────────────────────┐
│  C.I. FROM/TO            │
│        11.2465843        │
│        14.5534157        │
│  P.I. FROM/TO            │
│        9.042030031       │
│        16.75796997       │
│  YHAT =12.9              │
└──────────────────────────┘
```

So we see that for $x = 11$, the 95% confidence interval for mean Test 2 score, $E(Y)$, is (11.2464, 14.5540), the 95% predicted interval for an individual Test 2 score, y, is (9.0415, 16.7584), which is wider, and that the predicted Test 2 score is $\hat{y} = 12.9$ points.

Program A2MULREG.83p

```
ClrHome
Disp "DATA IN MAT [D]":Disp "COL
Y,X1,X2...XN":Disp "Y MUST BE IN THE":Disp
"1ST COL OF [D]."
Disp "     "
Disp "[A],[B],[C],[D]."
Disp "[E]+[F] USED."
Pause :ClrHome
dim([D])→L₆:L₆(1)→N:L₆(2)→M
Menu("MULT REG+CORR","MULT
REGRESSION",1,"CORR MATRIX",X,"QUIT",Y)
Lbl 1
0→Z:0→dim(∟YVAL:0→dim(∟YHAT:0→dim(∟RES:0→dim(
∟SRES:0→dim(∟LEVER:0→dim(∟COOKD
ClrHome:FnOff
Disp "HOW MANY IND VAR":Input
D:D+1→P:P→dim(L₅
M→L₅(P)
{N,P}→dim([E]
Fill(1,[E])
For(J,1,D)
Disp "COL. OF VAR.",J
Input H:H→L₅(J):End
If D≠1:Goto 3
Menu("SCATTER PLOT","YES PLOT",2,"NO",3)
Lbl 2
ClrList L₁,L₂
0→Xscl:0→Yscl
For(I,1,N)
[D](I,H)→L₁(I)
[D](I,1)→L₂(I)
End:PlotsOff
FnOn 6
LinReg(a+bx) L₁,L₂:"a+bX"→Y₆
PlotsOn 3:ExprOff:Plot3(Scatter,L₁,L₂,□)
ZoomStat:Trace
ExprOn
Lbl 3
For(J,1,D)
L₅(J)→C:J+1→K
For(I,1,N)
[D](I,C)→[E](I,K):End:End
{P,P}→dim([C]
([E]ᵀ*[E])⁻¹→[C]:{N,1}→dim([B]:0→Y:For(I,1,N)
[D](I,1)→[B](I,1):Y+[B](I,1)→Y
End
{P,1}→dim([A]
[C]*([E]ᵀ*[B])→[A]:0→A
For(I,1,N)
A+([B](I,1))²→A:End:0→B
[E]ᵀ*[B]→[F]
For(I,1,P)
B+[A](I,1)*[F](1,1)→B:End
A-B→E:N-P→F
√((E/F))→S:0→A
For(I,1,N)
A+([B](I,1))²→A:End:A-Y²/N→T
```

```
T-E→R:R/T→Q
(R/D)/(E/F)→V
1-(E/F)/(T/(N-1))→J
ClrHome
Output(1,1,"   DF  SS")
Output(2,1,"RG"):Output(2,4,D):Output(2,7,R)
Output(3,1,"ER"):Output(3,4,F):Output(3,7,E):
Output(4,1,"     F="):Output(4,7,round(V,2)
Output(5,5,"P=":Fix
3:Output(5,7,round(Fcdf(V,ε99,D,F),3):Float:O
utput(6,2,"R-SQ="):Output(6,7,round(Q,4))
Output(7,2,"(ADJ)"):Output(7,7,round(J,4))
Output(8,1,"S="):Output(8,3,S)
Pause :ClrHome
Output(1,1,"B0="):[A](1,1)→H:Output(1,4,H)
1→A:0→B:D→C:0→E:Lbl 4:1→J
If C>3:Then:1→E:3+B→B:Else:B+C→B:End
Output(2,1,"CL COEFF / T  P")
For(I,A,B):I+1→K:Output(J+2,1,L₅(I))
[A](K,1)→H:Output(J+2,3,H)
[A](K,1)/(S*√([C](K,K))→H:Output(J+3,5,round(
H,2):Fix 3
Output(J+3,12,round(2*tcdf(abs(H),ε99,F),3)
Float:2+J→J
End
If E=0:Goto 5
Pause :ClrHome
3+A→A:C-3→C:0→E:Goto 4
Lbl 5:Pause
Lbl A
Menu("MAIN MENU","CONF+PRI
INTERV",C,"RESIDUALS",R,"NEW
MODEL",1,"QUIT",Z)
Lbl C:ClrHome
Disp "FOR C.I. OR P.I.":Disp "  D.F.
ERR.=",F:Input "T*=?",T
D→dim(L₆
{P,1}→dim([F]
Fill(1,[F])
Lbl D:ClrHome
For(I,1,D)
Disp "X FOR COL",L₅(I):Input V
I+1→K:V→L₆(I)
V→[F](K,1):End
[F]ᵀ*[A]
Ans(1,1)→H
[F]ᵀ*[C]*[F]
Ans(1,1)→Q
T*S*√(Q→E
ClrHome
Disp "C.I. FROM/TO",(H-E):Disp (H+E)
T*S*√((1+Q)→E
Disp "P.I. FROM/TO",(H-E):Disp
(H+E):Output(7,1,"YHAT="):Output(7,6,H):Pause
Menu("AGAIN OR RETURN","AGAIN",D,"MAIN
MENU",A,"QUIT",Z)
Lbl R
```

(continued)

```
1→Z
PlotsOff :PlotsOn 3:ClrHome
FnOff :AxesOn
0→Xscl:0→Yscl
For(I,1,N)
[E]*[A]→[F]
[F](I,1)→ιYHAT(I):[D](I,1)→ιYVAL(I):ιYVAL(I)-
ιYHAT(I)→ιRES(I)
End
{P,1}→dim([F]
For(I,1,N)
For(J,1,P)
[E](I,J)→[F](J,1):End
[C]*[F]→[B]
[F]ᵀ[B]
Ans(1,1)→ιLEVER(I):End
For(I,1,N)
ιRES(I)/(S*√((1-
ιLEVER(I))→ιSRES(I):ιRES(I)² *ιLEVER(I)/(P*S² *
(1-ιLEVER(I))² )→ιCOOKD(I):End
SetUpEditor
ιYVAL,ιYHAT,ιRES,ιSRES,ιLEVER,ιCOOKD
Lbl U:ιSRES→L₄
Menu("RES OR STAND RES","RESIDUAL
PLOT",S,"STAND.RES.PLOT",T,"DURBIN WATSON
D",W,"MAIN MENU",A,"QUIT",Z)
Lbl S:ClrHome
ιRES→L₄:Goto V
Lbl T:ClrHome
Lbl V
Menu("PLOT OF RESID ","VS YHAT.",6,"VS AN IND
VAR.",7,"VS ROW NUMBER.",8,"PREVIOUS
MENU",U,"MAIN MENU",A,"QUIT",Z)
Lbl 7:ClrHome
Input "WHAT COL?",L:0→dim(L₆
For(I,1,N)
[D](I,L)→L₆(I)
End:Goto 9
Lbl 8:ClrHome
For(I,1,N):I→L₆(I):End
Goto 9
Lbl 6:ClrHome
For(I,1,N):ιYHAT(I)→L₆(I):End
Lbl 9:ExprOff
Plot3(Scatter,L₆,L₄,□):ZoomStat:Trace:ExprOn
Goto U
Lbl W
ClrHome:0→R
For(I,1,N)
(ιRES(I))² +R→R
End:0→T
For(I,1,N-1)
(ιRES(I+1)-ιRES(I))² +T→T:End
T/R→W
Disp "D-W D=",W:Disp "N=",N
Disp "K=",D
Pause :Goto U
Lbl X
ClrHome
```

```
{1,N}→dim([A])
Fill(1,[A])
(1/N)*[A]*[D]→[B]:{N,M→dim([E]
For(J,1,M)
For(I,1,N)
[D](I,J)-[B](1,J)→[E](I,J)
End:End
[E]ᵀ[E]→[A]
{M,M}→dim([B])
Fill(0,[B])
Disp "CORR MATRIX":For(I,1,M)
(1/√([A](I,I)))→[B](I,I):End
round([B]*[A]*[B],3)→[C]
Pause [C]
ClrHome:Goto Y
Lbl Z
If Z=0:Goto Y
ClrHome
Disp "UNDER STAT EDIT":Disp "Y IN ιYVAL,THEN"
Disp "ιYHAT,ιRES,ιSRES":Disp "ιLEVER,ιCOOKD"
Disp "FOR LEVERAGE AND":Disp "COOK DISTANCE"
Lbl Y
Return
```

Chapter 14

ANALYSIS OF COUNT DATA

14.1 Introduction

This chapter is about the *analysis of count data*—that is, data on **qualitative variables**. A lot of data reported in newspapers and magazines are qualitative or categorical. Recall from Chapter 4 that qualitative variables are those that classify the observations into **categories**. Some variables are qualitative by nature—such as gender, race, religion, or occupation. Some variables are made qualitative by how they are measured—such as weight of a package classified as light (under 5 pounds), medium (between 5 and 20 pounds), or heavy (over 20 pounds). In Section 4.3, we used pie charts and bar charts for displaying the distribution of a single

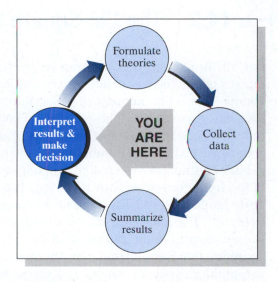

qualitative variable. We saw that the association between two qualitative variables can be displayed in the form of a conditional distribution of one variable given the other variable. We were also introduced to a technique for assessing whether the relationship between two qualitative variables is significant, called a chi-square test of association.

In this chapter, we will examine techniques for making decisions when our data are composed of counts. We will discuss three tests. If we have qualitative data on just one variable, a **test of goodness of fit** is used to assess if the qualitative data "fit" or are consistent with a particular discrete model for the percentages in each category. The null hypothesis would state the hypothesized discrete model. A **test of homogeneity** is used to assess if two or more populations are homogeneous or alike with respect to the distribution for some categorical variable. The null hypothesis is that the distributions are the same across the two or more populations. A **test of independence** or **association** determines whether two qualitative variables are related or not for a given population. The null hypothesis is that the two variables are independent, that there is no apparent association.

The **main idea** underlying all three tests can be summarized as follows:

Big Idea for Chi-Square Tests

1. The data consist of *observed counts*—that is, how many of the items or subjects fall into each category.

2. We will compute *expected counts under* H_0—that is, the counts that we would expect to see for each category if the corresponding null hypothesis were true.

3. We will *compare the observed and expected counts to each other via a test statistic* that will be a measure of how close the observed counts are to the expected counts under H_0. So if this "distance" is large, we have some support for rejecting H_0.

The test statistic that is computed for all three tests is called a **chi-square test statistic**. In the next section, we present some background about this statistic and its distribution that will allow us to perform the various tests. In the remaining sections, the details of how to perform each test will be motivated and derived through examples.

14.2 The Chi-Square Statistic

For all three tests, the data can be presented in a frequency table. The frequency table provides the **observed counts** (that is, the number of observations that fall into each category or cell). These observed counts are sometimes denoted by the letter O. The sum of all the observed counts will be the total number of observations n. Data on gender and number of spoken languages are presented

		Gender	
		Female	Male
Speak two or more languages?	Yes	8	4
	No	2	7

$n = 21$

in the frequency table shown. Among the 21 people, 8 were female who could speak two or more languages, and 7 were male who could not speak two or more languages. The cell entries of 8, 4, 2, and 7 are the observed counts. For each category we will determine the expected count, often represented by the letter E. The **expected counts** are computed assuming that a stated null hypothesis is true. We compare the observed counts with the corresponding expected counts and assess whether they are in agreement or not. This is accomplished by computing a **chi-square test statistic** that measures the distance between the observed and expected counts across all cells. We cannot simply average the distances across all cells, since some will be positive and some negative, leading to some cancellations. Instead we compute the squared distances, standardized by dividing by the expected count to put the size of the numerator into perspective.

$$\text{Chi-square test statistic} = X^2 = \sum_{\text{all cells}} \frac{(O - E)^2}{E}.$$

Definitions: The **observed counts** are the data, the number of observations that fall into each category or cell.

The **expected counts** are the number of observations that would be expected to fall into each category or cell if the null hypothesis being tested were true.

The **chi-square test statistic** measures the distance between the observed and expected counts across all cells and is computed as

$$X^2 = \sum_{\text{all cells}} \frac{(O - E)^2}{E}.$$

THINK ABOUT IT

What values could the X^2 statistic take on? Could you get an X^2 statistic that is negative?

If the observed counts are very close to the expected counts under H_0, then the X^2 test statistic value should be small. Larger values of X^2 indicate disagreement between the observed and expected counts. Since only larger values support the alternative hypothesis, all chi-square tests are *one-sided tests* with the direction of extreme being *to the right*. The *p*-value will be the probability of observing a X^2 test statistic value as large or larger than the observed X^2 value, assuming that the null hypothesis is true.

To compute the *p*-value we need to know the distribution of X^2 under the null hypothesis. If H_0 were true, and if we repeatedly took a simple random sample(s) of the same size(s) from the same population(s), recorded the observed counts, computed the corresponding X^2 value, and then made a histogram of the resulting X^2 values, the histogram would be approximated by a distribution called a **chi-square distribution**. This approximation holds best when all of the expected counts are at least 5. Similar to the *t*-distribution, there are many chi-square distributions, one for each degree of freedom, written $\chi^2(\text{df})$. The appropriate value for the degrees of freedom will be presented with each specific test in the upcoming sections. Various chi-square distributions are sketched at the right.

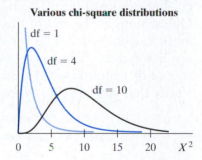

Various chi-square distributions

Note: We use the Greek symbol for chi, χ^2, when referring to the chi-square distribution. We use the English letter X^2 when referring to the general test statistic, and X^2_{OBS} represents the observed test statistic value.

An interesting aspect about a chi-square distribution, which could be shown mathematically is that its expected value or mean is equal to the degrees of freedom and its variance is twice the degrees of freedom. So if the X^2 test statistic has a chi-square distribution with four degrees of freedom, then the mean or expected value for X^2 is 4, and the standard deviation is $\sqrt{8} \approx 2.83$.

Properties of the Chi-Square Distribution $\chi^2(df)$

- The distribution is not symmetric and is skewed to the right.
- The values are nonnegative.
- There is a different chi-square distribution for different degrees of freedom.
- The mean of the chi-square distribution is equal to its degrees of freedom (df) and is located to the right of the mode.
- The variance of the chi-square distribution is 2(df).

THINK ABOUT IT

If under H_0, the test statistic X^2 follows approximately a chi-square distribution with four degrees of freedom, then we would expect the value of X^2 to be about 4, give or take about 2.83, if H_0 is true. Recall that we reject H_0 for large values of X^2. What decision would you make if you observed a test statistic value of $X^2_{\text{OBS}} = 2$? Explain.

Many calculators have the ability to compute areas under a chi-square distribution. The TI graphing calculator can draw a chi-square distribution, can compute areas under a chi-square distribution, and has an option under the STAT TESTS menu that allows you to perform chi-square tests of homogeneity and independence. For details on finding areas for chi-square distributions with a TI graphing calculator, see the next example and the ■ TI Quick Steps at the end of this chapter. Table VI, presented at the end of this chapter, summarizes some percentiles for various chi-square distributions. Since chi-square distributions are not symmetrical, lower and upper percentiles are given. The 95th percentile of a chi-square distribution with three degrees of freedom is 7.82 and is denoted by $\chi^2_{0.95}(3) = 7.82$. Verify that you can find this value using Table VI.

Chi-square distribution with three degrees of freedom

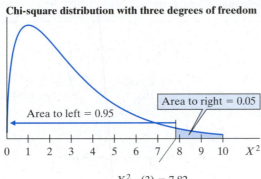

$$X^2_{0.95}(3) = 7.82$$

❖ EXAMPLE 14.1 Working with the Chi-Square Distribution

The objective of the study in Example 4.5 was to assess whether having a pet increased the length of survival for coronary heart disease patients. From a descriptive standpoint there seemed to be an advantage to having a pet. Approximately 94.3% of the patients with a pet survived for one year, while only 71.8% of those without a pet survived for one year. Is this difference of 22.5% significant? Is there a significant relationship between pet status and survival status? A chi-square test of independence was performed to assess if having a pet is associated with a higher survival rate. The details of this test will be discussed in Section 13.5. In this example, we will focus on working with the chi-square distribution to find the *p*-value.

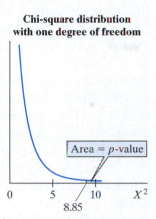

Chi-square distribution with one degree of freedom

Area = *p*-value

Suppose that the observed chi-square test statistic value was $X^2_{OBS} = 8.85$. We want to measure the chance of getting a value of 8.85 or larger under the null hypothesis of no association. The distribution for X^2 under H_0 is a chi-square distribution with one degree of freedom. The picture portrays this chance as the area under the chi-square distribution to the right of $X^2_{OBS} = 8.85$.

Using the TI

The TI has a function that is built into the calculator for finding areas under any chi-square distribution. This function allows you to specify the degrees of freedom and is located under the DISTR menu. The TI steps to compute the area to the right of 8.85 under the $\chi^2(1)$ distribution are shown.

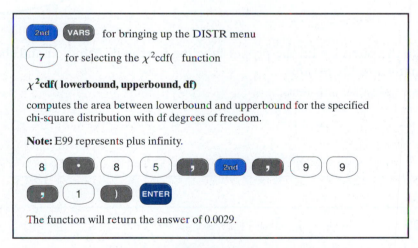

2nd VARS for bringing up the DISTR menu

7 for selecting the χ^2cdf(function

χ^2**cdf(lowerbound, upperbound, df)**

computes the area between lowerbound and upperbound for the specified chi-square distribution with df degrees of freedom.

Note: E99 represents plus infinity.

8 . 8 5 , 2nd , 9 9

, 1) ENTER

The function will return the answer of 0.0029.

Your screen should look like this

```
X²cdf(8.85,E99,1
)
        .0029308889
```

Using Table VI

Table VI provides percentiles for various chi-square distributions. The rows designate the degrees of freedom and the columns specify which percentile is being reported. Recall that the 90th percentile of a distribution would be the value (on the horizontal axis) such that the area to the left of that value is 0.90, or equivalently, the area to the right of that value is 0.10. Our observed chi-square statistic is 8.85. Looking in the first row of Table VI (for a chi-square distribution with one degree of freedom), we see that the largest value is only 7.879 and is the 99.5[th] percentile. Our test statistic value is even larger than the 99.5[th] percentile. Since the X^2_{OBS} value of 8.85 is larger than the 99.5[th] percentile, using only Table VI, we would report that the *p*-value is even smaller than 0.005.

Since the *p*-value for the test of no association is so small, we would reject H_0. The data support that there is a statistically significant relationship between pet status and survival status. ❖

14.1 The Chi-Square Distribution

A chi-square test resulted in an observed test statistic value of $X^2_{OBS} = 5.7$. This test is based on two degrees of freedom. You are asked to find the *p*-value for this test. Recall that all chi-square tests are *one-sided tests* with the direction of extreme being *to the right*. So the *p*-value will be the probability of observing a X^2 test statistic value as large or larger than the observed X^2 value, assuming that the null hypothesis is true.

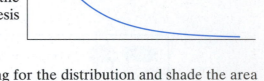

(a) Complete the picture with the proper labeling for the distribution and shade the area corresponding to the *p*-value.

(b) Find the *p*-value for this test.

✳ Exercises

14.1 Using Table VI, find the numerical value for each of the following percentiles.

(a) The 90th percentile for a chi-square distribution with four degrees of freedom, $\chi^2_{.90}(4)$.

(b) $\chi^2_{.95}(4)$.

(c) $\chi^2_{.99}(4)$.

(d) $\chi^2_{.10}(4)$.

(e) $\chi^2_{.90}(6)$.

(f) $\chi^2_{.90}(12)$.

14.2 For each of the scenarios that follow, report the *p*-value for the chi-square test. If you use the χ^2cdf(function on the TI, you can report the exact *p*-value. If you use Table VI, you can report bounds for the *p*-value.

    (a) The observed X^2 statistic value is 3.2 and the null distribution is the chi-square distribution with one degree of freedom.

    (b) The observed X^2 statistic value is 1.7 and the null distribution is the chi-square distribution with two degrees of freedom.

    (c) The observed X^2 statistic value is 16.5 and the null distribution is the chi-square distribution with five degrees of freedom.

14.3 For each of the scenarios that follow, report the *p*-value for the chi-square test. If you use the χ^2cdf(function on the TI, you can report the exact *p*-value. If you use Table VI, you can report bounds for the *p*-value.

    (a) The observed X^2 statistic value is 0.8 and the null distribution is the chi-square distribution with one degree of freedom.

    (b) The observed X^2 statistic value is 8.2 and the null distribution is the chi-square distribution with four degrees of freedom.

    (c) The observed X^2 statistic value is 31.4 and the null distribution is the chi-square distribution with ten degrees of freedom.

 14.3   Test of Goodness of Fit

For a test of goodness of fit, we assume that we have one random sample from one population. The response is a qualitative variable with K possible categories or cells. The data consist of how many observations fall into each of the K cells. The test assesses whether the data "fit" a particular distribution. The null hypothesis would state the hypothesized distribution. The alternative hypothesis is that the distribution is not as given in H_0 and is often not formally stated. The hypotheses may be written as sentences or stated mathematically.

Suppose that you roll a die 120 times and record how many times each of the values 1 through 6 occur. The number of categories is $K = 6$. Your question may be, "Is the die fair?" If the die is fair, what would you expect to see? The statement that the die is fair is the same as saying, "The observed outcomes 'fit' with the expected distribution, namely, that the probability of each of the six values is 1/6." The null hypothesis could be expressed as H_0: The die is fair, or H_0: $p_1 = \frac{1}{6}, p_2 = \frac{1}{6}, p_3 = \frac{1}{6}, p_4 = \frac{1}{6}, p_5 = \frac{1}{6}, p_6 = \frac{1}{6}$, where p_i is the probability that a random toss of the die results in the ith outcome. The alternative hypothesis would be that the die is not fair.

Now suppose that there are four lab sections for a statistics course. They are scheduled to meet weekly on Wednesdays at various times of the day. Data were collected on the number of students who selected each of the $K = 4$ sections. Your question may be, "Are the sections equally preferred or do students have a preference for certain sections?" If the sections are equally preferred, the number of students would be equally distributed across the four sections, with approximately 25% in each lab. The null hypothesis could be expressed as H_0: There is no preference (equally distributed), or H_0: $p_1 = 0.25, p_2 = 0.25, p_3 = 0.25, p_4 = 0.25$, where p_i is the probability that a randomly selected student will select section i. The alternative hypothesis would be that there was a preference (not equally distributed).

We will illustrate the concept of testing for goodness of fit through the following scenario regarding the contribution of working women to the household income.

❖ **EXAMPLE 14.2** **Women Are Key Earners**

The idea of what it means to be a family's provider was updated based on a new study entitled "Women: The New Providers," which showed that 55% of working women contribute half or more of their household's income. In the study, 26% of the working women surveyed said they provided about half of their family's income, 11% said they provided more than half, and 18% said they were the household's sole earner (i.e., they provide all of their family's income). Two-thirds of the 1502 women (or 1001 women) in the survey had jobs outside of the home and thus were considered "working women." The figures for the 1001 working women (as reported in the *Ann Arbor News*, June 1995, article entitled "Women Are Key Earners") are summarized in the following table:

| | **Proportion of Family's Income Provided** | | | |
| | *Less than* | | *More than* | |
	Half	*Half*	*Half (not all)*	*All*
Observed Counts	451	260	110	180
(percent)	(45%)	(26%)	(11%)	(18%)

Suppose that, from previous studies and government figures, it was accepted that 49% of the working women contributed half or more of their household's income, 24% provided about half, 10% more than half (but not all), and 15% all. This is the hypothesized discrete model shown at the right.

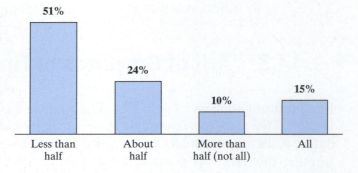

Question: Does the recent survey data support that the distribution has changed significantly from the previous model?

We have a response variable, "proportion of family's income provided by women," that is actually continuous (could be any value from 0 to 100%), but that (in our case) has been broken down into four categories: *Less than half, Half, More than half (but not all),* and *All.* The **number of categories** (or cells) is denoted by K, and in our case we have $K = 4$. We have **observed counts** or frequencies for each category, which are denoted by O_i = observed count in the ith category. For our example we have $O_1 = 451$, $O_2 = 260$, $O_3 = 110$, and $O_4 = 180$. The total of these observed counts is our sample size $n = 1001$.

If we let p_i be the proportion of working women in the ith category, then our question of interest corresponds to the null hypothesis: H_0: $p_1 = 0.51$, $p_2 = 0.24$, $p_3 = 0.10$, $p_4 = 0.15$. If this null hypothesis is true, how many working women in a sample of size $n = 1001$ would we expect to . . .

provide *less than half* of the family's income?	51% of 1001 or **510.5**
provide about *half* of the family's income?	24% of 1001 or **240.2**
provide *more than half (but not all)* of the family's income?	10% of 1001 or **100.1**
provide *all* of the family's income?	15% of 1001 or **150.2**

These are the **expected counts** under the null hypothesis and denoted by

$$E_1 = 510.5, E_2 = 240.2, E_3 = 100.1, \text{ and } E_4 = 150.2.$$

If H_0 were true, we would expect 510.5 women to provide less than half of the family's income. However, we observed 451 women. Is this too big of a difference? Should we reject H_0? The following table provides both the observed and expected counts under the hypothesized model:

| | Proportion of Family's Income Provided | | | |
	Less than Half	Half	More than Half (not all)	All
Observed Counts	451	260	110	180
Expected Counts	510.5	240.2	100.1	150.2

Our chi-square test statistic provides a measure of the distance between the observed and expected counts across all cells. The observed measure of the distance is

$$X^2_{\text{OBS}} = \sum_{\text{all cells}} \frac{(O - E)^2}{E}$$

$$= \frac{(451 - 510.5)^2}{510.5} + \frac{(260 - 240.2)^2}{240.2} + \frac{(110 - 100.1)^2}{100.1} + \frac{(180 - 150.2)^2}{150.2}$$

$$= 15.5.$$

Recall that larger values of X^2 support the alternative hypothesis. Is a value of 15.5 large enough to reject H_0? To answer this, we need to know what the distribution of X^2 is under the null hypothesis. For the test of goodness of fit, the degrees of freedom are $K - 1$, where K is the number of categories or cells. For our women's income example we have df $= 4 - 1 = 3$.

Once we have computed the test statistic based on the data, we can find the corresponding p-value. The chi-square test statistic value is $X^2_{\text{OBS}} = 15.5$. We want to measure the chance of getting a value of 15.5 or larger under the null hypothesis. The model for X^2 under H_0 is a chi-square distribution with three degrees of freedom. The following picture portrays this chance as the area under the chi-square distribution to the right of the observed value of 15.5:

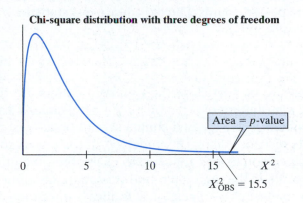

Chi-square distribution with three degrees of freedom

Area = p-value

$X^2_{\text{OBS}} = 15.5$

Using TI

Using the χ^2cdf(distribution function on the TI, with $X^2_{\text{OBS}} = 15.5$ and three degrees of freedom, we find that the p-value is X^2cdf(15.5, E99, 3) = 0.0014.

Using Table VI

Since a chi-square statistic of 15.5 is larger than the 99.5th percentile for a $\chi^2(3)$ distribution (which is 12.838), the p-value is even smaller than 0.005.

Since the p-value is so small, we would reject H_0. It appears that the distribution has changed significantly from the previous model for proportion of family's income provided by working women. ❖

THINK ABOUT IT   Meaning of Degrees of Freedom

If there were $K = 4$ values, all between zero and 1, that must add up to 1, then you would only need to know $K - 1 = 3$ of them. With the constraint that they must add to 1, you could figure out the fourth value. For example, if you have that $p_1 + p_2 + p_3 + p_4 = 1$ and that $p_1 = 0.2$, $p_2 = 0.3$, $p_3 = 0.2$, then p_4 must be equal to _____.

Summary of the Test of Goodness of Fit

- The data are assumed to be a **random sample of size n from a population**. The response is a **qualitative variable with K possible categories** or cells. The number of observations that fall into each category, called the **observed counts O_i ($i = 1, 2, \ldots K$)**, are recorded.

- We want to test hypotheses about whether the data "fit" a particular discrete model for the percentages in each category. The **null hypothesis would state the hypothesized discrete model** often in terms of p_i ($i = 1, 2, \ldots K$), where p_i is the probability that a random observation falls in the ith category. The probabilities specified in the null hypothesis must sum to one. The alternative hypothesis is that the distribution is not as given in H_0. The **significance level α** is selected.

- The number of observations that we would expect to see fall into each category if the null hypothesis were true are computed. These **expected counts** are computed as $E_i = np_i$, where p_i is specified in the null hypothesis. The expected counts will sum to the sample size n and do not need to be rounded to the nearest whole number because they are what we expect, on average.

- We base our decision about the "fit" of the hypothesized model by comparing the observed and expected counts. The **chi-square test statistic**,

$$X^2 = \sum_{\text{all cells}} \frac{(O - E)^2}{E},$$

provides a measure of how close the observed counts are to the counts expected under H_0. The distribution of X^2 under H_0 is a chi-square distribution with **$K - 1$ degrees of freedom**. The expected counts should be at least 5 to ensure that this is the appropriate distribution for X^2 under H_0. If the observed counts and the corresponding expected counts are not close to each other, then the X^2 test statistic can be large and will be in the right tail of the chi-square distribution. Thus the **direction of extreme** for the test of goodness of fit is always one-sided and "**to the right**."

- The **p-value** is the chance, assuming that H_0 is true, of getting an X^2 value as large or larger than the observed value. This chance corresponds to the area under a $\chi^2(K - 1)$ distribution to the right of the observed value. The p-value is compared with the significance level α, and a decision is made. A conclusion is written, using the language of the problem.

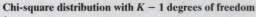

Chi-square distribution with $K - 1$ degrees of freedom

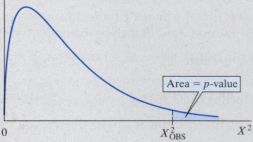

Area = p-value

0 X^2_{OBS} X^2

 **14.2** Clock Just Keeps on Ticking

The *New York Times* article entitled "Another Day Older, and the Clock Just Keeps On Ticking" (March 26, 1995) presented the model given below for responses to the question "Looking ahead to retirement, what do you expect to be your major source of income—Social Security, an employer-sponsored pension plan or your own retirement savings?"

1. Social Security	22%
2. Employer-sponsored pension plan	30%
3. Savings	39%
4. Other	5%
5. Don't know–no answer	4%

A survey of 315 adult Americans from Hawaii gave the following responses to the income-expectations question:

Response	1	2	3	4	5	
Observed Counts	60	103	111	20	21	Total $O = n = 315$
Expected Counts						Total $E = n = 315$

Do the data indicate that adult Americans from Hawaii have a different distribution of major sources of retirement income? State the null hypothesis. (**Hint:** p_1 is the proportion of adults from Hawaii stating that their major source of income will be Social Security and based on the model being tested, $p_1 = 0.22$.)

$$H_0: \ p_1 = 0.22, \ p_2 = \underline{\qquad}, \ p_3 = \underline{\qquad}, \ p_4 = \underline{\qquad}, \ p_5 = \underline{\qquad}.$$

Compute the expected counts and enter them in the previous table.
Compute the observed test statistic.

$$X^2_{\text{OBS}} = \sum_{\text{all cells}} \frac{(O - E)^2}{E} =$$

Find the *p*-value.

State your decision and conclusion using a 5% significance level.

Let's do it! 14.3 Student Preferences

Students who register for an introductory chemistry course must also sign up for a laboratory section. There are eight sections available, scheduled to meet at various times and taught by various instructors. These eight sections are open to all students. Data were gathered on the number of students who selected each of the eight sections, for a total of 152 students.

(a) State the null hypothesis for testing if the data support that each section was equally likely to be chosen.

H_0: _____

H_1: Not all probabilities specified in H_0 are correct.

(b) Suppose that these data are statistically significant at the 5% level, indicating that there does seem to be a preference for certain sections. What might be some possible explanations for the preference(s)? How would you go about assessing what the preference(s) are?

Remember that conclusions must be worded carefully to avoid implying statements that you cannot support.

 ## Exercises

14.4 The manager of a local coffee shop wants to learn about the distribution of the types of coffee ordered. Type of coffee is generally classified as latte, espresso, or cappuccino. A random sample of 120 customers was obtained, and the number of customers ordering each of the three types of coffee was recorded. The observed counts are as follows:

Type of Coffee	1 = Latte	2 = Espresso	3 = Cappuccino
Observed # of Customers	32	37	51

(a) Give the appropriate null hypothesis to test whether the three types of coffee are ordered equally often.

(b) If the three types of coffee are ordered equally often, how many lattes would you expect to be ordered in a random sample of 120 customers?

(c) The observed test statistic is $X^2 = 4.85$. Using a significance level of 0.05, test the hypotheses in (a). Show all work.

14.5 An experiment is conducted to try to determine the most effective over-the-counter pain relief medication. The five best-selling brands of pain relief are tested on $n = 200$ randomly selected subjects. For each subject, the "best" pain relief medicine is recorded, based on various criteria. The results are presented below. You are asked to test the claim that the five brands ($K = 5$) of pain relief medicine are equally effective.

	Brand of Pain Relief				
	A	**B**	**C**	**D**	**E**
Observed Counts	38	44	56	37	25

(a) State the null hypothesis.

(b) Compute the expected counts.

(c) Compute the observed test statistic.

(d) Find the p-value.

(e) State your decision and conclusion using a 5% significance level.

14.6 A genetics experiment is conducted on garden peas. Two pea plants yielded 440 plants, 320 with green seeds and 120 with yellow seeds. The researcher hypothesizes that in the parent plants, the allele (a group of possible mutational forms of a gene) for the green is dominant to the allele for the yellow and that the parent plants were both heterozygous (inherited different alleles) for this trait. Based on Mendel's laws of genetics, the predicted ratio of offspring from the crossbreeding under this hypothesis would be 3 to 1—namely, 75% green and 25% yellow. Perform a test of goodness of fit to assess if the observed data are consistent with those expected under Mendel's law. Use a 5% significance level and show all details.

14.7 According to a professor of genetics, a crossing of red and white peas should produce offspring that are 25% red, 50% pink, and 25% white. The experiment (conducted by a graduate student) produced the following results in 144 crossings. Do the data provide sufficient evidence to support the professor's theory using a 1% significance level?

Color	**White Peas**	**Pink Peas**	**Red Peas**
Observed Counts	45	69	30

(a) State the appropriate null hypothesis to be tested.

(b) Compute the test statistic, find the p-value, and state your decision and corresponding conclusion.

14.8 A very expensive dice has been made such that the pips on each face are emeralds. It has been hypothesized that the weight of the emeralds will cause the "five" or "six" faces to fall downward more often than expected. In other words, the "one" and "two" faces are thought to fall upward more frequently than if the dice were fair. The hypothesized model (H_0) is that the "one" and "two" will each occur with probability $\frac{1}{5}$, the "three" and "four" will each occur with probability $\frac{1}{6}$, and the "five" and "six" will each occur with probability $\frac{2}{15}$. To test this model, a study is conducted by rolling the die $n = 120$ times and recording the number of times each of the six possible outcomes occurred ($K = 6$). The data are provided. Test the hypothesized model using a 10% significance level.

Outcome	1	2	3	4	5	6
Observed Counts	26	25	18	19	17	15

14.9 Air bags in cars have become an increasingly important safety feature to consumers. It has been stated that 45% of nondomestic cars are equipped with a driver-side air bag, 15% with both a driver-side and passenger-side air bag, and 40% with no air bags. A random sample of 48 domestic cars was obtained, and the air-bag status for each was recorded. The results are as follows:

	0 Air Bags	1 Air Bag	2 Air Bags	
Number of Domestic Cars	16	23	9	Total = 48

Are the domestic car air-bag data consistent with the distribution of air bags for nondomestic cars? Test the hypothesis H_0: $p_0 = 0.40$, $p_1 = 0.45$, $p_2 = 0.15$, where p_i is the proportion of domestic cars equipped with i air bags, $i = 0, 1, 2$. Report your observed test statistic, p-value, and decision at a 1% significance level.

REAL
DATA

14.10 The EPIC-MRA-Mitchell Research poll found that 33% of respondents believed that same-sex marriages should be allowed, while 63% disagreed and 4% were undecided. Details of the survey are shown at the right. A task force would like to assess if such attitudes have significantly changed over the past year. For last year, only 13% strongly agreed ($= 1$), 11% somewhat agreed ($= 2$), 17% somewhat disagreed ($= 3$), 56% strongly disagreed ($= 4$), and 3% were undecided ($= 5$). Test the hypothesis that there has been no change in attitude: Compute the expected counts, the observed test statistic, and the p-value. State your decision and conclusion using a 5% significance level.

Poll Results

The national poll question about same-sex marriages and the results:

"If two people love each other, they should be able to get married even if they are of the same sex."

- 19% — Strongly agree
- 14% — Somewhat agree
- 13% — Somewhat disagree
- 50% — Strongly disagree
- 4% — Undecided/don't know

The telephone survey of $n = 1000$ registered voters was done June 21–26 and has a margin of error of three percentage points either way.

REAL
DATA

14.11 **What Americans Do Out of Doors.** Consider the accompanying display of the results of a survey of 1993 adults interviewed in person April 16–23, 1994. (SOURCE: *The New York Times*, April 2, 1995, p. F11.)

(a) How many of the sampled adults actively participated in fishing last year?

(b) How many of the sampled adults actively participated in motorboating last year?

(c) Could there have been some sampled adults who actively participated in both fishing and motorboating last year? If so, how many? If not, why not?

(d) Could this information, as it is provided, be used to perform a goodness of fit test to assess whether it follows a specified distribution of activities participated in last year? Explain.

(e) Where did the 1993 sample size come from? And how does it relate to the statement that the survey has a margin of (sampling) error of plus or minus two percentage points?

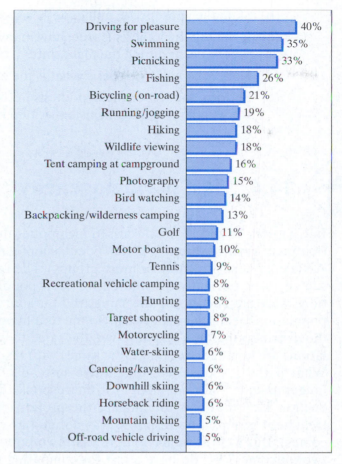

(f) Would you feel comfortable using these data to predict what proportion of Americans will actively participate in downhill skiing next year? How about to predict what proportion of Florida residents will actively participate in downhill skiing next year? or New Hampshire residents?

(g) What further information or questions do you have about how these data were obtained?

14.12 In order to study the effect of shift time on work-related accidents, a union collected data on work-related accidents in a particular manufacturing plant. Suppose that the same number of workers work in each shift. The number of accidents was recorded as follows:

Time of Accident	Observed Number of Accidents
Morning Shift	40
Evening Shift	45
Night Shift	75

(a) The researcher wishes to test the hypothesis that the probability of an accident is the same for the morning and evening shifts while the probability of an accident on the night shift is twice as likely as the morning or evening shifts.

(b) State the appropriate null hypothesis to be tested.

(c) If the null hypothesis is true, how many accidents in the sample would you expect to occur during the night shift?

(d) Under the null hypothesis, what is the expected value for the test statistic?

(e) Perform the test of the hypothesis in (b). Use a 5% significance level. Show all work and include a conclusion written in terms of the problem.

 # 14.4   Test of Homogeneity

In Chapters 11 and 12, we discussed techniques that allowed us to compare two or more treatment groups—the two independent samples *t*-test (for two groups) and the one-way ANOVA *F*-test (for three or more groups). The response was a quantitative variable assumed to be normally distributed for each population, the population standard deviations were assumed to be the same, and the null hypothesis was the statement that the population means were all equal. So the null hypothesis was essentially a statement that the distribution of the response was the same for all populations—that is, that the populations are homogenous (alike, the same) with respect to the distribution of the response. What if the response was qualitative instead of quantitative? We certainly could no longer assume a normal model for the response. The two independent samples *z*-test in Section 11.5 allows us to compare the proportion of successes for two populations or treatment groups. The response was defined as either success or failure (just two outcomes). But what if we have more than two populations or the response has more than two outcomes? We do have a test for comparing two or more populations when the response is qualitative (with two or more possible outcomes)—called the **chi-square test of homogeneity**.

For a test of homogeneity, we assume that we have *C* independent random samples, one from each of the *C* populations under study. The response is a qualitative variable with *R* possible categories. For each sample, we record the number of observations that fall into each of the *R* categories. These data are summarized as counts in the form of a two-way table. For consistency, we will have the columns refer to the different populations and the rows refer to the categories. The test assesses whether the *C* populations are homogeneous with respect to the discrete distribution for the response. The null hypothesis would state that the populations are homogeneous with respect to the distribution of the response. The alternative hypothesis is that the populations are not homogeneous and is often not formally stated. We will focus on expressing the hypotheses as sentences rather than stating them mathematically.

Two teaching methods are available for training bank tellers. We would like to assess if the quality of learning is the same for the two teaching methods by comparing performance ratings for tellers trained

| | | Training Method | |
		Method I	*Method II*
Job Performance	*Below Average*	8	10
	Average	21	25
	Above Average	21	15

under the two methods. Of the 100 incoming trainees available for the study, half are randomly assigned to Method I and the remaining trainees are taught with Method II. One month after the training is completed, the job performance of the 100 tellers was recorded as below average, average, or above average. We have *C* = 2 populations to be compared.

We assume that the data represent two independent random samples from these two populations. The response, job performance, has $R = 3$ categories, so the data can be summarized with a two-way frequency table. The null hypothesis would be that the job performance distribution is the same for the two methods.

An employee of an electronics store wanted to learn whether or not the repair records differed for the three primary color television manufacturers. She obtained random samples from the sales records for each of the three manufacturers and recorded whether or not the television had required service during the first one-year period. We have $C = 3$ populations to be compared. We assume that the data represent three independent random samples from these three populations. The response, service status, has $R = 2$ categories. The null hypothesis would be that the proportion of televisions requiring service is the same for the three manufacturers.

		Manufacturer		
		Brand A	**Brand B**	**Brand C**
Service Status	*Required Service*	12	10	5
	Did Not Require Service	50	38	27

We will illustrate the concept of testing for homogeneity through the scenario that follows regarding the incidence of insomnia for various allergy treatments.

❖ EXAMPLE 14.3 Allergy Relief

REAL DATA

A number of drugs have been recently approved by the Food and Drug Administration (FDA) for the treatment of seasonal allergy relief. Two such drugs are called Seldane-D (Marion Merrell Dow Inc.) and Claritin-D (Schering Corporation). In the brief summary (Schering Corporation, 1995) for Claritin-D, some details are provided regarding adverse experiences. The summary states that in controlled clinical trials (experiments) using the recommended dose of one tablet every 12 hours, the incidence of reported adverse events was similar to those reported with placebo, with the exception of insomnia and dry mouth, which were statistically significant at a 5% significance level.

THINK ABOUT IT

What is meant by *controlled* clinical trials? Recall Chapter 3 principles for designing an experiment.

If tests for comparing the incidence of various adverse experiences were performed using a 5% significance level, then the *p*-value for comparing the incidence rates of *dry mouth* was (circle one)

    greater than 0.05   less than or equal to 0.05   cannot tell

Data on the incidence of insomnia are summarized in the following table:

Observed Counts

		Treatment Group				
		Claritin-D	*Loratadine*	*Pseudoephedrine*	*Placebo*	
Insomnia	*Yes*	164	22	104	28	318
	No	859	521	444	894	2718
		1023	543	548	922	3036

In the table, we have four treatment groups that we want to compare—the new drug (Claritin-D), two standard drugs (loratadine and pseudoephedrine), and a placebo drug group. The data on insomnia incidence are displayed in a two-way contingency table. The response variable is categorical, namely, "Yes" or "No." We have four independent random samples of sizes $n_1 = 1023$, $n_2 = 543$, $n_3 = 548$, and $n_4 = 922$, one from each population.

An adverse event, such as insomnia, is an outcome that is not desirable. Researchers will often report the proportion of individuals in a treatment group that experience the adverse event, otherwise referred to as the **risk** of that event. Very often, the risk is reported as a percent rather than a proportion. The risk of insomnia for the four treatment groups in our study is as follows:

▪ The proportion of Claritin-D patients reported having as experienced insomnia is $\frac{164}{1023} = 0.160$ or 16.0%.

▪ The proportion of Loratadine patients reported having as experienced insomnia is $\frac{22}{543} = 0.041$ or 4.1%.

▪ The proportion of pseudoephedrine patients reported having as experienced insomnia is $\frac{104}{548} = 0.190$ or 19.0%.

▪ The proportion of placebo patients reported having as experienced insomnia is $\frac{28}{922} = 0.030$ or 3.0%.

How do these various risks compare? Is the risk of insomnia the same for the four treatment groups? We wish to assess whether the insomnia risk is the same across the four populations, or whether differences exist. That is, we wish to test whether the four populations are homogenous with respect to insomnia risk:

H_0:   The insomnia incidence distributions for the four populations are the same or **homogeneous**—namely, $p_1 = p_2 = p_3 = p_4 = p$, where p_i is the insomnia rate for the ith population.

From the totals, which have been added to the contingency table, we see that, overall, 318 out of 3036 patients, or approximately 10.5%, did experience insomnia. We refer to this overall rate as $p = 0.10474$. If the insomnia incidence rate is the same for the four populations, then approximately 10.5% (actually 10.474%) of the patients from each group are expected to report having experienced insomnia. Therefore, under the null hypothesis, we have

▪ The number of Claritin-D patients *expected* to report having experienced insomnia = $n_1 p = (1023)(0.10474) = 107.15$.

▪ The number of Loratadine patients *expected* to report having experienced insomnia = $n_2 p = (543)(0.10474) = 56.87$.

- The number of pseudoephedrine patients *expected* to report having experienced insomnia $= n_3 p = (548)(0.10474) = 57.40$.
- The number of placebo patients *expected* to report having experienced insomnia $= n_4 p = (922)(0.10474) = 96.57$.

Expected Counts

		Claritin-D	Loratadine	Pseudoephedrine	Placebo	
				Treatment Group		
Insomnia	*Yes*	107.15	56.87	57.40	96.57	318
	No	915.85	486.13	490.60	825.43	2718
		1023	543	548	922	3036

Let's examine how we actually computed an expected count. Since 318 out of 3036 patients (10.474%) had said "yes," and there were 1023 Claritin-D patients, then we would expect about 10.474% of these 1023 patients to say "yes" if the distributions are the same.

Expected number of Claritin-D patients reporting "yes"

$$= (1023)\left(\frac{318}{3036}\right) = \frac{(\text{total Claritin-D})(\text{total yes})}{\text{overall total}} = 107.15$$

Thus, the expected count for a cell in the two-way contingency table is obtained using what is called the cross-product rule.

$$\textbf{Expected count} = \frac{(\textbf{column total})(\textbf{row total})}{\textbf{overall total}}$$

How do these expected counts compare to the observed counts? Does it appear that the incidence of insomnia is the same for the four treatment groups? Again, we turn to our measure of distance between the observed counts and the counts that we would expect to see if the four populations did have the same insomnia incidence rate. We have

$$\textbf{Chi-square test statistic} = X^2 = \sum_{\text{all cells}} \frac{(O - E)^2}{E},$$

where O is the observed count for a given cell and E is the expected count under the null hypothesis. If the observed counts are very close to the expected counts under H_0, then the X^2 test statistic value should be small. Larger values of X^2 indicate disagreement between the observed and expected counts. Suppose that H_0 were true and we repeatedly took random samples of the same sizes from the same populations, recorded the observed counts for the cells, and computed the corresponding X^2 value. If we made a histogram of the resulting X^2 values, this distribution would be approximately a chi-square distribution with degrees of freedom given by

$$df = (\text{number of rows} - 1)(\text{number of columns} - 1).$$

In our 2-by-4 contingency table, knowing only one of the two row percentages (number of rows $-$ 1) and three of the four column percentages (number of columns $-$ 1), we could determine the remaining column and row percentage, since the rows and the columns must each add to 100%. Thus, the degrees of freedom are $(1)(3) = 3$.

The observed measure of the distance between the observed and expected counts for our insomnia data is

$$X^2_{\text{OBS}} = \frac{(164 - 107.15)^2}{107.15} + \frac{(22 - 56.87)^2}{56.87} + \frac{(104 - 57.40)^2}{57.40} + \frac{(28 - 96.57)^2}{96.57}$$

$$+ \frac{(859 - 915.85)^2}{915.85} + \frac{(521 - 486.13)^2}{486.13} + \frac{(444 - 490.60)^2}{490.60} + \frac{(894 - 825.43)^2}{825.43}$$

$$= 154.2.$$

What do you think such a large X^2 value means? Recall that the mean of a chi-square variable with three degrees of freedom is equal to the degrees of freedom, so the mean is 3. The variance will be $2(3) = 6$, so the standard deviation is $\sqrt{6} \approx 2.45$. If we standardize the observed value of our test statistic, we see that 154.2 is about $\frac{(154.2 - 3)}{(2.45)} = 61.7 \approx 62$ standard deviations above the expected value for X^2 if the distributions were the same. Thus, we have strong evidence that the insomnia rates are not the same for the four populations. In comparing Claritin-D to the placebo, the rates are 16% versus 3%, or a **relative risk** of $\frac{0.16}{0.03} = 5.33$. The risk of experiencing insomnia is 533% higher for Claritin-D patients as compared to placebo patients.

Statistical output using SPSS for the homogeneity test based on these data is shown. The frequency table is first presented followed by two chi-square test statistics. Our measure of distance is called the *Pearson chi-square statistic*, named after the English statistician Karl Pearson (1857–1936), who invented the chi-square statistic in 1900. The observed chi-square test statistic is reported as 154.22408, and the corresponding *p*-value is very small (significance = 0.00000, which implies that the *p*-value < 0.000005). This extremely small *p*-value confirms the significant difference. Thus, H_0 is rejected at the 5% significance level.

```
INSOMNIA Insomnia Indicator  by  TRTMENT  Treatment group
            TRTMENT                        Page 1 of 1
          Count |
                |Claritin Loratadi Pseudoe  Placebo    Row
                |    1.00|    2.00|    3.00|    4.00|  Total
INSOMNIA  -------+--------+--------+--------+--------+
          1.00  |  164   |   22   |  104   |   28   |  318
    yes         |        |        |        |        |  10.5
          -------+--------+--------+--------+--------+
          2.00  |  859   |  521   |  444   |  894   |  2718
    no          |        |        |        |        |  89.5
          -------+--------+--------+--------+--------+
         Column   1023      543      548      922     3036
          Total    33.7     17.9     18.1     30.4    100.0
```

Chi-Square	Value	DF	Significance
Pearson	154.22408	3	.00000
Likelihood Ratio	168.27079	3	.00000

The following statistical output was produced with the use of the *Chisquare Test* option in MINITAB. Both the observed and expected counts are presented in the table output. The observed chi-square test statistic value and the degrees of freedom are computed. The *p*-value is not provided with this output, but can be obtained with MINITAB using the chi-square probability distribution commands.

```
MTB > Chi Square C1-C4.
Expected counts are printed below observed counts
             C1          C2          C3          C4        Total
    1       164          22         104          28          318
         107.15       56.88       57.40       96.57
    2       859         521         444         894         2718
         915.85      486.12      490.60      825.43
Total      1023         543         548         922         3036
ChiSq = 30.160 + 21.385 + 37.834 + 48.691 +
         3.529 +  2.502 +  4.426 +  5.697 =
154.224
df = 3
```

Definitions: The **risk** of an outcome for a category is the proportion of items in that category having that outcome. It is common to report the risk as a percent rather than as a proportion.

The **relative risk** of an outcome for two categories of an explanatory variable is the ratio of the risks for each category. The relative risk is unity if the two risks are the same. The relative risk is greater than unity if the numerator risk (in the top) is higher than the denominator risk (in the bottom). The relative risk is less than unity if the numerator risk is lower than the denominator risk. The denominator risk is often the risk for the placebo or standard level of the explanatory variable and is referred to as the **baseline risk.**

Note: Risk data are often presented in the news, but without enough information to truly understand the risk. In Section 3.6, we discussed some questions to ask in order to judge the information being presented about a study.

Summary of the Test of Homogeneity

- The data consist of **C independent random samples**, one from each of the C populations. The response is a **qualitative variable with R possible categories**. The number of observations in the ith sample that fall into the jth category, called the **observed count O_{ij}** ($i = 1, 2, \ldots, C$; $j = 1, 2, \ldots, R$), is recorded.

- We want to test the hypothesis that the C populations are homogeneous with respect to the discrete distribution for the response. The **null hypothesis would state that populations are homogeneous** and the alternative hypothesis is that populations are not homogeneous. The **significance level α** is selected.

- The number of observations that we would expect to see fall into each category if the null hypothesis were true are computed. These **expected counts** are computed using the cross-product rule as

$$E_{ij} = \frac{(i\text{th column total})\,(j\text{th row total})}{\text{overall total}}.$$

- We base our decision about the homogeneity of the C populations by comparing the observed and expected counts. The **chi-square test statistic**,

$$X^2 = \sum_{\text{all cells}} \frac{(O - E)^2}{E},$$

provides a measure of how close the observed counts are to the counts expected under H_0. The distribution of X^2 under H_0 is a chi-square distribution with **$(R - 1)(C - 1)$ degrees of freedom**. The expected counts should be at least 5 to ensure that this is the appropriate distribution for X^2 under H_0. If the observed counts and the corresponding expected counts are not close to each other, then the X^2 test statistic can be large and will be in the right tail of the chi-square distribution. Thus, the **direction of extreme** for the test of homogeneity is always one-sided and "**to the right**."

- The **p-value** is the chance, assuming that H_0 is true, of getting a X^2 value as large or larger than the observed value. This chance corresponds to the area under a $\chi^2((R - 1)(C - 1))$ distribution to the right of the observed X^2 value. The p-value is compared with the significance level α, and a decision is made. A conclusion is written using the language of the problem.

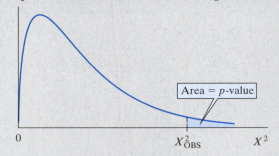

Chi-square distribution with $(R - 1)(C - 1)$ degrees of freedom

Area = p-value

0 X^2_{OBS} X^2

TI Note: The TI has a χ^2-test option under the TESTS menu located under the STAT menu that allows you to perform a test of homogeneity. The table of observed counts must first be stored as a matrix of counts. Details for creating this matrix and performing the test are provided in the ■ TI Quick Steps at the end of this chapter.

Let's do it! 14.4 Preventing Pregnancies

Three oral contraceptive drugs for women are tested. Of the 300 women available for the study, 100 were randomly assigned to each drug group. The responses were classified as pregnant (Row 1) or not pregnant (Row 2). You are asked to assess if the three drugs (C1, C2, and C3) are equally effective at preventing pregnancy.

```
MTB > ChiSquare C1-C3.
Expected counts are printed below observed counts
               C1         C2         C3      Total
    1           5         12         16         33
            11.00      11.00      11.00

    2          95         88         84        267
            89.00      89.00      89.00

Total         100        100        100        300

ChiSq = 3.273 +   0.091 +   2.273 +
        0.404 +   0.011 +   0.281 =   6.333

df = 2
```

(a) If the drugs are equally effective, how many women receiving Drug 1 (C1) would you expect to be pregnant?

(b) What is the distribution of the test statistic under the null hypothesis that the three drugs are equally effective? (*Hint:* Which distribution will you use to find the *p*-value?)

(c) Based on the results, are the drugs equally effective? Use a 1% significance level. Explain.

Let's do it! 14.5 The Role of Randomization in a Clinical Study

A study report synopsis submitted by a pharmaceutical company to the FDA regarding the results of a double-blind study of *New Drug* in patients with chronic bronchitis stated that "there were no statistically significant differences among the three treatment groups with respect to gender (M, F), age (in years), weight (in kg), smoking habit (non, ex, smoker), alcohol consumption (daily, weekly, monthly, rarely, never), and special diet (yes, no) (called *Demographic* characteristics)." Data on smoking habit are as follows:

		Treatment Group		
		Group 1	*Group 2*	*Group 3*
Smoking Habit	**Nonsmoker**	38 ( )	50 ( )	37 ( )
	Ex-smoker	52 ( )	48 ( )	48 ( )
	Smoker	102 ( )	102 ( )	108 ( )

Does it appear that the randomization did a good job in controlling for this potential confounding variable? That is, does the distribution for smoking habit appear to be the same for the three treatment groups?

(a) State the hypotheses to be tested in terms of the populations and response variable.

H_0:

H_1:

(b) Compute the row and column totals and then find the expected counts and record them in the preceding table in parentheses.

(c) Compute the observed test statistic: X^2_{OBS} =

(d) What is the degrees of freedom for this homogeneity test? df =

(e) What is the corresponding *p*-value?

(f) Using a 5% significance level, what is your decision?   reject H_0 accept H_0

Based on the data, it appears that

Exercises

14.13 A total of 210 emphysema patients entering a clinic over a one-year period were treated with one of two drugs, either the standard drug or an experimental compound, for a period of one week. After this period, each patient's condition was rated as greatly improved, improved, or no change.

		Patient's Condition			
		No Change	Improved	Greatly Improved	
Therapy	Standard	20	35	45	100
	Experimental	15	45	50	110
		35	80	95	210

(a) If the distribution for patient condition is the same for the standard drug as for the experimental drug, how many standard drug patients would be expected to show no change?

(b) The observed test statistic for assessing if the distribution for patient condition is the same for the standard drug as for the experimental drug is $X^2_{OBS} = 6.51$. At the 1% significance level, does there appear to be a significant difference in the distribution of patient condition for the two drugs? Explain.

14.14 As part of pharmaceutical testing for drowsiness as a side effect of a drug, 200 patients were randomly assigned to one of two groups of 100 patients each: One group received the drug, and the other group received a placebo. The number of patients in each group who fell asleep in the first hour after receiving the treatment was recorded. Results are shown as follows:

```
SLEEP  Sleep  Status  by  TRTGROUP  Treatment  Group
                    TRTGROUP        Page 1 of 1
            Count   |
                    |drug      placebo
                    |                    Row
                    |    1.00|    2.00|  Total
SLEEP     ----------+--------+--------+
            1.00    |   36   |   20   |     56
         sleep      |        |        |   28.0
                    +--------+--------+
            2.00    |   64   |   80   |    144
         no sleep   |        |        |   72.0
                    +--------+--------+
            Column      100      100      200
            Total      50.0     50.0    100.0


    Chi-Square                 Value          DF      Significance
    ---------------            ----------     ----    ------------
    Pearson                    6.34921         1          .01174
    Continuity Correction      5.58036         1          .01816
    Likelihood Ratio           6.41720         1          .01130
```

(a) It was stated that the study was conducted in a double-blind fashion. Explain what *double-blind* means.

(b) State the hypotheses for performing a test of homogeneity to compare the incidence of drowsiness for the two treatment groups.

(c) If the incidence of drowsiness is the same for the two groups, how many placebo patients would you expect to fall asleep in the first hour after receiving the treatment?

(d) Using a 5% significance level, assess whether the incidence of drowsiness is the same for the two groups. Give the test statistic value, *p*-value, decision, and corresponding conclusion.

14.15 Is there a relationship between whether a woman is in a union and her salary? We hear a lot about how much a woman makes compared to a man, but it is possible that there are also differences among women. A random sample of 30 unionized women and 40 nonunionized women was taken and their salary (low, medium and high) was recorded to see if the distribution of salary is the same for union members versus nonunion members. The data and SPSS output are summarized in the following tables:

Union Status* Salary Level Crosstabulation

Count

		Salary Level			
		High	Low	Medium	Total
Union	Nonunion	2	16	22	40
Status	Union	12	4	14	30
Total		14	20	36	70

Chi-Square Tests

	Value	df	Asymp. Sig. (2-sided)
Pearson Chi-Square	14.998[a]	2	.001
Likelihood Ratio	15.994	2	.000
N of Valid Cases	70		

a. 0 cells (.0%) have expected count less than 5. The minimum expected count is 6.00.

(a) State the hypotheses for assessing whether there the distribution of salary is the same for union members and nonunion members.

(b) Assuming the distribution of salary is the same for union members and nonunion members, how many union members would you expect to earn a low salary?

(c) Assuming the distribution of salary is the same for union members and nonunion members, what is the distribution of the test statistic?

(d) Using a significance level of 0.01, perform the chi-square test for assessing if there the distribution of salary is the same for union members and nonunion members. State the observed test statistic value, the *p*-value, your decision, and your conclusion in words.

14.16 A certain job in a car assembly plant involves a great deal of stress. A study was conducted to assess whether there is any difference in how men versus women adjust to the stress. A random sample of 25 men and 25 women employed in this job were surveyed. The results are given as follows:

		Adjustment Status		
		Well Adjusted	*Not Well Adjusted*	
Gender	*Men*	18	7	25
	Women	13	12	25

At the 1% significance level, test whether the adjustment status is the same for men versus women.

REAL DATA

14.17 A recent survey of randomly selected eighth, tenth, and twelfth graders in the Brighton Area Public Schools reveals some disturbing results (SOURCE: *The Ann Arbor News*, "Drug Use Is Up Sharply among Brighton High School Kids," June 7, 1995). Part of the results are presented in the following table:

		Grade		
		Eighth Graders	*Tenth Graders*	*Twelfth Graders*
Used Alcohol in the Last 30 Days?	*Yes*	34%	48%	68%
	No	66%	52%	32%

We wish to assess if the alcohol usage rate is the same across the three populations—that is, to test if the three populations are homogeneous with respect to incidence of alcohol usage.

(a) State the appropriate null and alternative hypotheses.

(b) Give an 5% level decision rule.

(c) The data in the table are reported in percentages. What do you need to know to be able to get the observed counts?

(d) Assume that the results are based on random samples of 100 students in each of the three grade levels, for a total of 300 students surveyed. Perform the test and state your conclusion.

 14.5 Test of Independence

Many studies, whether observational or experimental, are conducted to try to detect relationships between two (or more) variables. The researcher is interested in determining whether the category that an individual falls in for one variable depends on the category they are in for the other variable. When the two variables are both qualitative, a **chi-square test for association** can be used to assess whether the relationship between the two variables is significant. This chi-square test is often called the **test of independence** because the null hypothesis will be the statement that the two variables are indeed independent for the population under study. We were first introduced to the ideas of this test of independence back in Section 4.3.3. In this section, we will provide the details for performing the test of independence and discover that the method is identical to the test of homogeneity—the only differences being how the data are obtained and how the hypotheses are stated.

For a test of independence, we have just one random sample from one population under study. For each unit in the sample, two responses (both qualitative) are recorded. One response has *R* possible categories and the other has *C* possible categories. The responses are presented in a two-way frequency table. The test assesses whether the two variables are independent for the population. The null hypothesis would state that there is no relationship or association between the two variables for the population—the variables are independent. The alternative hypothesis is that there is a relationship between the two variables for the population. We will again focus on expressing the hypotheses as sentences rather than stating them mathematically.

A State Highway Patrol Department would like to assess whether the cause of an accident is related to the outcome of the accident. The Department decides to focus on accidents that occur along the major toll road that crosses the state. A random sample of 250 accident reports over the past six months was obtained. The accidents were cross-classified by primary cause of the accident (speeding, reckless driving, drinking, other) and by the outcome of the accident (death, no death). We have one population, all accidents occurring along the major toll road. We have measured two responses: cause, with four categories; and outcome, with two categories. These data are summarized with a two-way frequency table. The null hypothesis would be that there is no relationship between the cause of accidents and the outcome.

		Outcome	
		Death	**No Death**
Cause	*Speeding*	21	41
	Reckless	10	30
	Drinking	39	71
	Other	5	33

A statewide survey of employed adults was conducted to assess the relationship between union membership and attitude toward spending on state welfare programs. The state agency obtained a random sample of 400 employed adults who were questioned regarding union membership and attitude toward decreased state spending on welfare programs. Employees were classified by whether they were union or nonunion and by whether they supported, were neutral, or were opposed to decreased spending on state welfare programs. We have one population, all employed adults in the state. We have measured two responses: union status, with two categories, and attitude, with three categories. These data are summarized with a two-way frequency table. The null hypothesis would be that there is no relationship between union membership and attitude toward decreased spending on state welfare programs.

		Union Status	
		Union	**Nonunion**
Attitude	*Support*	112	94
	Neutral	36	72
	Oppose	28	58

We will illustrate the concept of testing for independence through the following scenario regarding the relationship between blood pressure and heart attacks.

❖ **EXAMPLE 14.4** **High Blood Pressure Linked to Heart Attacks?**____

Many studies have suggested that there is a link between high blood pressure and heart attacks. In one study, white male subjects aged 35 to 64 were classified according to whether their systolic blood pressure was low (less than 140 millimeters of mercury) or high (140 or higher) and then followed for five years to determine whether or not they suffered from a heart attack during the five years. The data are summarized in the following table:

		Heart Attack?	
		Yes	**No**
Blood Pressure	*Low*	21	2655
	High	55	3283

Do these data support the hypothesis that heart attack status is independent of or dependent on blood pressure level? We wish to test these hypotheses:

H_0: Heart attack status is independent of blood pressure level for the given population.

H_1: Heart attack status is dependent on blood pressure level for the given population.

The proportion of low-blood-pressure subjects who had a heart attack is $\frac{21}{2676} = 0.0078$ or 0.78%. The proportion of high-blood-pressure subjects who had a heart attack is $\frac{55}{3338} = 0.0165$ or 1.65%.

Overall, 76 of the 6014 subjects, or 1.26%, had a heart attack. If heart attack status is independent of blood pressure level, then we would expect about 1.26% of low-blood-pressure subjects to have a heart attack and about 1.26% of the high-blood-pressure subjects to have a heart attack. The expected number of low-blood-pressure subjects (out of the 2676 subjects) with a heart attack under H_0 is

$$0.0126(2676) = \frac{(76)(2676)}{6014} = 33.82.$$

This value is obtained using the same cross-product rule for finding expected counts as in the test of homogeneity.

$$\text{Expected count} = \frac{(\text{column total})(\text{row total})}{\text{overall total}}$$

Note: In Section 14.4, when we addressed whether or not the distribution of insomnia incidence was the same for the four treatment groups, we were actually trying to assess whether or not insomnia incidence was independent of treatment group. In other words, ***the test of homogeneity and the test of independence are carried out exactly the same way***.

We have found one expected count for the first cell: 33.82. How about the other expected counts? Their values can be found simply by subtraction, since both the row and column totals must stay the same. The following table presents the observed and expected outcomes for all cells:

		Heart Attack?	
		Yes	*No*
Blood Pressure	*Low*	21 (33.82)	2655 (2642.18)
	High	55 (42.18)	3283 (3295.82)

Are the observed and corresponding expected counts different enough to reject H_0 and support the hypothesis that there is an association between blood pressure level and heart attack status? Let's measure the distance between the observed counts and those expected under H_0 using our X^2 test statistic. We obtain

$$X^2_{\text{OBS}} = \frac{(21-33.82)^2}{33.82} + \frac{(2655-2642.18)^2}{2642.18} + \frac{(55-42.18)^2}{42.18} + \frac{(3283-3295.82)^2}{3295.82} \approx 8.87.$$

For a two-by-two frequency table, we have $(2 - 1)(2 - 1) = 1$ degree of freedom. Notice that we only needed to compute one expected count. The other three expected counts were constrained, since the row and column totals had to be preserved.

What is the *p*-value for this test?

The *p*-value is the probability of getting a value of X^2 of 8.87 or larger, under the null hypothesis—that is, assuming that X^2 has a chi-square distribution with one degree of freedom. Recall that being too big is the only direction of extreme that makes sense for the chi-square test statistic. The *p*-value found with a calculator or computer is 0.0029, which is quite small. Our observed result is very unusual. It would occur (or something more extreme) only about 0.29% of the time if H_0 were true. For any significance level of 0.0029 or higher, we would reject the null hypothesis. We conclude that there is sufficient evidence that blood pressure level and heart attack status are associated, with a higher proportion of high blood pressure subjects having heart attacks.

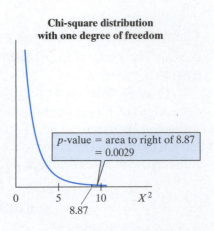

The corresponding output using MINITAB is shown below. As the output states, the expected counts are printed below the observed counts in the frequency table. The observed test statistic value is 8.865 with one degree of freedom. The *p*-value is not provided with this output but can be obtained with MINITAB using the chi-square probability distribution commands. The *p*-value for this test is 0.0029.

```
MTB > ChiSquare  C1  C2.
Expected counts  are  printed below observed counts
             C1          C2       Total
    1        21        2655        2676
           33.82     2642.18
    2        55        3283        3338
           42.18     3295.82
Total        76        5938        6014
Chi Sq = 4.860 +    0.062 +  3.894 + 0.050 = 8.865
df = 1
```

❖

Summary of the Test of Independence

- The data consist of **one random sample from one population. Two qualitative responses are measured for each unit** in the sample. The number of units that fall into the *i*th category of the first variable and the *j*th category of the second variable is called the **observed count** O_{ij} $(i = 1, 2, \ldots, C; \quad j = 1, 2, \ldots, R)$.

- We want to test the hypothesis that the two variables are independent for the population. The **null hypothesis would state that there is no association between the two variables** (the two variables are independent), and the alternative hypothesis is that there is a relationship between the two variables. The **significance level** α is selected.

- The number of observations that we would expect to see fall into each category if the null hypothesis were true is computed. Using the cross-product rule, we compute these **expected counts** as

$$E_{ij} = \frac{(i\text{th column total})\,(j\text{th row total})}{\text{overall total}}.$$

- We base our decision about the relationship between the two variables by comparing the observed and expected counts. The **chi-square test statistic**

$$X^2 = \sum_{\text{all cells}} \frac{(O - E)^2}{E},$$

provides a measure of how close the observed counts are to the counts expected under H_0. The distribution of X^2 under H_0 is a chi-square distribution with $(R - 1)(C - 1)$ **degrees of freedom**. The expected counts should be at least 5 to ensure that this is the appropriate distribution for X^2 under H_0. If the observed counts and the corresponding expected counts are not close to each other, then the X^2 test statistic can be large and will be in the right tail of the chi-square distribution. Thus the **direction of extreme** for the test of independence is always one-sided and "**to the right**."

- The **p-value** is the chance, assuming that H_0 is true, of getting an X^2 value as large or larger than the observed value. This chance corresponds to the area under a $\chi^2((R - 1)(C - 1))$ distribution to the right of the observed X^2 value. The p-value is compared with the significance level α, and a decision is made. A conclusion is written using the language of the problem.

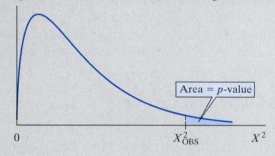

Chi-square distribution with $(R - 1)(C - 1)$ degrees of freedom

TI Note: The TI has a χ^2-test option under the TESTS menu located under the STAT menu that allows you to perform a test of independence. The table of observed counts must first be stored as a matrix of counts. Details for creating this matrix and performing the test are provided in the ■TI Quick Steps at the end of this chapter.

Note: The test of homogeneity and the test of independence are carried out exactly the same way. The primary difference is in how the data are obtained. In the **homogeneity test** we have independent random samples from two or more populations and we measured one qualitative response. The sample sizes were known in advance and were either the row or column totals. In the **test of independence**, we have just one random sample from one population of interest. We measure two qualitative responses, and we assess if these two variables are associated. Only the overall total, the sample size, is known before the measurements are made.

Let's do it! 14.6 Hodgkin's Disease and Tonsillectomies

A study investigated whether any relationship exists between Hodgkin's disease and tonsillectomies. The counts shown are based on a random sample of 85 patients suffering from Hodgkin's disease and who had a sibling of the same gender who was free of the disease and whose age was within five years of the patient's age.

(a) What is the appropriate null hypothesis of interest?

H_0:

(b) Carry out a test at the 5% significance level and state your conclusion.

		Sibling	
		No Tonsillectomy	*Tonsillectomy*
Patient	*No Tonsillectomy*	37	7
	Tonsillectomy	15	26

In Chapter 3, we learned that any study has the potential to be misinterpreted because of confounding variables. In Example 13.7, we saw that the presence of a third variable can enhance or mask a relationship between two quantitative variables. A relationship can appear to exist if the third variable is not considered, but is shown not to exist between the two variables if the third variable is taken into account, or vice versa. This type of reversal of conclusion is referred to as **Simpson's paradox**. Our next hypothetical example shows that a lurking variable can change or even reverse the apparent relationship between two categorical variables.

❖ EXAMPLE 14.5 Height and Income _____

A study of the relationship between height and income of executives was conducted. Executives were classified by their height as either short (under 5'10'') or tall (5'10'' or taller) and according to their income level with categories defined as low or high. The data are as follows:

		Income		
		Low	*High*	
Height	*Short*	1000	500	1500
	Tall	600	825	1425
		1600	1325	2925

Based on these data, the

- proportion of short individuals with low income $= 1000/1500 = 66.7\%$.
- proportion of tall individuals with low income $= 600/1425 = 42.1\%$.
- proportion of short individuals with high income $= 500/1500 = 33.3\%$.
- proportion of tall individuals with high income $= 825/1425 = 57.9\%$.

In general, taller executives tend to earn a higher income. There does seem to be a relationship between height and level of income. In fact, performing a chi-square test of independence results in an extremely large test statistic of $X^2_{\text{OBS}} = 177.9$ and a corresponding p-value that is nearly zero. We would certainly reject the null hypothesis of independence and conclude that there is very strong evidence of an association between height and income.

Chi-Square Tests

	Value	df	Asymp. Sig.
Pearson Chi-Square	177.911[b]	1	.000
Continuity Correction[a]	176.921	1	.000
Likelihood Ratio	179.682	1	.000
N of Valid Cases	2925		

a. Computed only for a 2×2 table
b. 0 cells (.0%) have expected count less than 5. The minimum expected count is 645.51.

Do you think the short executives are mostly men or women? Who is more likely to be in the tall executive group? What happens if we examine the same data broken down by gender? The tables are presented next, along with the results of a chi-square test of independence for each subgroup:

		Women Income		
		Low	High	
Height	*Short*	800	100	900
	Tall	200	25	225
		1000	125	

		Men Income		
		Low	High	
Height	*Short*	200	400	600
	Tall	400	800	1200
		600	1200	

Chi-Square Tests

	Value	df	Asymp. Sig.
Pearson Chi-Square	.000[b]	1	1.000
Continuity Correction[a]	.000	1	1.000
Likelihood Ratio	.000	1	1.000
N of Valid Cases	1125		

a. Computed only for a 2×2 table
b. 0 cells (.0%) have expected count less than 5. The minimum expected count is 25.00.

Chi-Square Tests

	Value	df	Asymp. Sig.
Pearson Chi-Square	.000[b]	1	1.000
Continuity Correction[a]	.000	1	1.000
Likelihood Ratio	.000	1	1.000
N of Valid Cases	1800		

a. Computed only for a 2×2 table
b. 0 cells (.0%) have expected count less than 5. The minimum expected count is 200.00.

When the data are analyzed separately for each subgroup, there is no evidence of any association between height and income. The chi-square test statistics are both in fact the smallest they could be, a value of zero (which implies what about the observed counts and the expected counts?). The proportion of short women with a low income is the same as the proportion of tall women with a low income (namely, 88.9%. Can you verify this?). The proportion of short men with a low income is the same as the proportion of tall men with a low income (namely, 33%). This is an example of **Simpson's paradox**, otherwise known as **aggregation bias**, with the lurking variable being gender. The reversal occurs because of the higher proportion of short women executives who generally have a lower income (perhaps due to entering the workforce later than most of the male executives). *Observed associations can be misleading when there are lurking variables.* ❖

Let's do it! 14.7 Hiring Based on Race?

Suppose that data accumulated from the records of hiring results for two municipal departments of a certain city over the last two years provided the following results:

Department of Public Works	Hired	Not Hired
Race = White	140	60
Race = Nonwhite	32	8

Department of Law Enforcement	Hired	Not Hired
Race = White	6	54
Race = Nonwhite	50	150

(a) Compute the row totals for each department. Answer the following questions:

> *For Public Works:*
> > What proportion of *white* applicants were hired?
> > What proportion of *nonwhite* applicants were hired?
>
> *For Law Enforcement:*
> > What proportion of *white* applicants were hired?
> > What proportion of *nonwhite* applicants were hired?

(b) Which group, white or nonwhite, appears to be hired at a lower rate?

(c) Complete the following table by combining (aggregating) the above data into one table:

	Hired	Not Hired
Race = White		
Race = Nonwhite		

(d) For this aggregated table, which group, white or nonwhite, appears to be hired at a lower rate?

(e) What accounts for the reversal in the direction of hiring rates?

Of the Public Works 240 applicants, _____% were nonwhite.

Of the Law Enforcement 260 applicants, _____% were nonwhite.

The reversal occurs because
the Public Works hiring rates were generally (circle one)   *higher*   *lower*
compared to the Law Enforcement rates, and the Public Works applicants were primarily
(circle one)   *white*   *nonwhite*.

The two departments were not homogeneous with respect to hiring rates and the distribution of race within each applicant pool was not comparable.

Definition: **Simpson's paradox (or aggregation bias)** occurs when some overall conclusion concerning a set of items fails to hold within some subgroups of those items. Comparisons based on the aggregated results can be misleading.

Hand and Eye Dominance

People are often classified as right handed or left handed. People can also be classified as right eyed or left eyed, in the sense that either the right or the left eye will be dominant. Eye dominance can be assessed by first focusing on an object across the room with both eyes looking through a small opening (e.g., use your two hands to form a small circle to look through). Then, close one eye at a time; when the object appears to move (out of the circle), then the eye that is closed is your dominant eye. How is hand dominance related to eye dominance? Or are these two traits independent?

		Hand Dominance	
		Right	*Left*
Eye Dominance	*Right*	Cell 1	Cell 2
	Left	Cell 3	Cell 4

Before you collect some data, think about the following questions:

(a) If a large proportion of those sampled fell into Cells 2 and 3, what might that indicate?

(b) If a large proportion of those sampled fell into Cells 1 and 4, what might that indicate?

(c) Collect data and record your results. Does there appear to be a strong association between hand and eye dominance? Give supporting details.

(d) A theory has been proposed that the relationship between hand and eye dominance may be different for females versus males. That is to say, gender may be a *confounding* variable. Recall that a three-way contingency table can be presented as several two-way tables side by side. When gathering your data for (c), did you record any other information, such as gender, age, race, and so on? Such demographic data can be useful for examining the responses of interest for various subgroups. Separate the data from (c) into two tables, one for females and one for males. Assess the association between hand and eye dominance by gender. Write a brief conclusion and provide supporting evidence.

		Females Hand Dominance	
		Right	*Left*
Eye Dominance	*Right*	Cell 1	Cell 2
	Left	Cell 3	Cell 4

		Males Hand Dominance	
		Right	*Left*
Eye Dominance	*Right*	Cell 1	Cell 2
	Left	Cell 3	Cell 4

Exercises

14.18 A State Highway Patrol Department would like to assess if the cause of an accident is related to the outcome of the accident. The Department decides to focus on accidents that occur along the major toll road that crosses the state. A random sample of 250 accident reports over the past six months were obtained. The accidents were cross-classified by primary cause of the accident (speeding, reckless driving, drinking, other) and by the outcome of the accident (death, no death). These data are summarized in the following table:

		Outcome	
		Death	*No Death*
Cause	*Speeding*	21	41
	Reckless	10	30
	Drinking	39	71
	Other	5	33

(a) State the appropriate null hypothesis for assessing whether there is a relationship between the cause of accidents and the outcome.

(b) Construct a table of expected counts.

(c) Compute the chi-square test statistic.

(d) At a 1% level of significance, give your decision and conclusion.

14.19 A geneticist wanted to learn whether the gene that could cause someone to have a Morton's toe (second toe bigger than the fat first toe) was related to the gene that allowed people to roll their tongues. A sample of 100 people was obtained, and for each person their toe status (BIGTOE: 1 = Morton's toe, 2 = not Morton's toe) and tongue status (TONGUE: 1 = can roll, 2 = can't roll) were recorded. The data were entered into SPSS, and the following output was produced:

```
BIGTOE    by    TONGUE
                      TONGUE          Page 1 of 1
              Count |
              Exp Val |
                      |                Row
                      |   1.00|   2.00|  Total
BIGTOE        --------+-------+--------+
              1.00  |   36  |    14 |     50
                    | 30.0  |  20.0 |  50.0%
                    +-------+--------+
              2.00  |   24  |    26 |     50
                    | 30.0  |  20.0 |  50.0%
                    +-------+--------+
            Column       60      40     100
             Total    60.0%   40.0%  100.0%
        Chi-Square        Value              DF      Significance

-------------------------------------       ----    -------------------
Pearson                 6.00000              1            .01431
Likelihood Ratio        6.07230              1            .01373
```

(a) State the appropriate null hypothesis for assessing if the two genes are related.

(b) If the two characteristics are independent, what is the expected number of people with Morton's toe who can roll their tongue?

(c) Using a 0.05 significance level, does it appear that there is an association between big-toe status and roll-tongue status? Give the test statistic value, p-value, decision, and what your decision means in terms of the problem.

14.20 A random sample of 100 adults, all of whom agreed to participate in a study on weight and hypertension, was obtained. The results shown will be used to assess if there is a relationship between weight status (classified as overweight, normal weight, and underweight) and hypertensive status (Yes, No).

		Over	Weight Normal	Under	
Hypertensive?	*Yes Hypertensive*	15	15	10	40
	Not Hypertensive	10	35	15	60
		25	50	25	100

(a) If hypertensive status and weight status are independent, how many subjects would you expect to be classified as overweight and hypertensive?

(b) Briefly explain why we reject H_0 when the chi-square statistic is too large.

(c) Using a 5% significance level, does there appear to be an association between hypertensive status and weight status? (Give the test statistic value, p-value, and decision.)

(d) Based on these data, what can we conclude about the relationship between hypertensive status and weight status?

14.21 A survey was conducted by taking a random sample of 400 people who were questioned regarding union membership and attitude toward decreased national spending on social welfare programs. The count data and test results are presented in the following SPSS output:

```
MEMBER      union membership by ATTITUDE    Spending on Welfare Attitude
                       ATTITUDE                 Page   1 of 1
            Count  |
                   |Support Neutral   Opposed
                   |                             Row
                   |    1.00|   2.00|    3.00| Total
MEMBER      -------+--------+-------+---------+
            1.00 |   112  |   36  |   28  |  176
      Union      |        |       |       |  44.0
            -------+--------+-------+---------+
            2.00 |    94  |   72  |   58  |  224
      Nonunion   |        |       |       |  56.0
            -------+--------+-------+---------+
            Column    206     108     86     400
             Total    51.5    27.0    21.5   100.0
         Chi-Square              Value          DF        Significance
---------------------         ----------       ----      ------------
Pearson                        18.54498          2           .00009
Likelihood Ratio               18.72207          2           .00009
```

(a) What is the null hypothesis of interest here?

(b) If H_0 is true, how many people would you expect to be classified as union and opposed?

(c) Using a 1% significance level, test the hypothesis you specified in (a). Give the test statistic value, p-value, decision, and conclusion stated in terms of variables union membership and attitude.

(d) Suppose that, instead of taking a random sample of 400 people, the investigators had taken a random sample of 176 union members and a random sample of 224 nonunion members and had then classified each person according to their attitude about spending.

   (i)   What null hypothesis would be appropriate then?

   (ii)   What would your conclusion be using a test with significance level of 0.01?

14.22 As part of a standard course evaluation, students were asked to rate the course as poor, good, or excellent overall. They were also asked to indicate whether the course was a required course or taken as an elective. The dean of the college is interested in assessing if the course rating is related to the reason for taking the course. The data were entered into SPSS and the following output was produced:

Reason for Taking Course* Overall Course Rating Crosstabulation

Count

		Overall Course Rating			Total
		excellent	good	poor	
Reason for Taking Course	elective	20	25	4	49
	required	12	29	10	51
Total		32	54	14	100

Chi-Square Tests

	Value	df	Asymp. Sig. (2-sided)
Pearson Chi-Square	4.830[a]	2	.089
Likelihood Ratio	4.935	2	.085
Linear-by-Linear Association	4.776	1	.029
N of Valid Cases	100		

a. 0 cells (.0%) have expected count less than 5. The minimum expected count is 6.86.

(a) If the null hypothesis were true, what is the expected number of students who took the course as a requirement and rate the course as poor overall?

(b) If there is no relationship between course ratings and reason for taking the course, what is the distribution of the test statistic?

(c) Using a 5% level of significance, give the test statistic value, the p-value, your decision, and your conclusion.

14.23 A university offers only two degree programs. A group, claiming the university discriminated against women in the admissions process, presented the data shown in the table in their support.

	Men	Women
Admitted	90	60
Denied	110	140

(a) Based on these data, what proportion of men was admitted? What proportion of women was admitted? Does there appear to be an association between gender and admission status? Support your answer by performing the chi-square test of independence using a 5% significance level.

The university's response was that these data do show an association, but not as the result of discrimination. The admission data were provided for each of the two programs.

	Engineering	
	Men	*Women*
Admitted	80	20
Denied	80	20

	English	
	Men	*Women*
Admitted	10	40
Denied	30	120

(b) Based on these data, what proportion of men was admitted to the Engineering program? What proportion of women was admitted to the Engineering program? What proportion of men was admitted to the English program? What proportion of women was admitted to the English program? Does there appear to be an association between gender and admission status with each program? Support your answer by performing a chi-square test of independence for each program. Use a 5% significance level.

(c) What is the statistical term for such a reversal in the association conclusion?

14.24 A school newspaper reports the following: John and Peter finished their four years of swimming competition with impressive results.

- Overall, John won 60% of his competitions while Peter won 50% of his competitions. They each competed 500 times over the four years.
- John has won 70% of his backstroke competitions and 30% of his butterfly-stroke competitions.
- Peter has won an impressive 75% of his backstroke competitions and 45% of his butterfly-stroke competitions.

Is there a mistake in this reporting? How can Peter be so successful in both the backstroke and the butterfly stroke, yet overall he is less successful as compared to John? Do you need to call the newspaper to report a mistake?

Consider the following data:

	John's Results	
	# Competitions	*# Won*
Backstroke	375	263
Butterfly stroke	125	37
	500	300

	Peter's Results	
	# Competitions	*# Won*
Backstroke	83	62
Butterfly stroke	417	188
	500	250

(a) Verify that John won 60% of all his competitions and that Peter won 50% of all his competitions.

(b) Verify that John has won 70% of his backstroke competitions and 30% of his butterfly-stroke competitions.

(c) Verify that Peter has won 75% of his backstroke competitions and 45% of his butterfly-stroke competitions.

(d) Complete the following:

Both John and Peter do better at the (circle one)   backstroke   butterfly stroke. (That is, they won the higher percentage of such competitions.)

Of John's 500 competitions, _____ % were backstroke competitions.

Of Peter's 500 competitions, _____ % were backstroke competitions.

Although Peter was more successful than John at both the backstroke and the butterfly-stroke competitions, John was more successful overall since a higher percentage of his competitions were with his better stroke, the backstroke.

The comparison between the two swimmers is complicated because the swimmers swam a different percentage of competitions of the two types of strokes. *Aggregating the results* across the two types of strokes is not appropriate because of this imbalance. This is an example of **Simpson's paradox**.

14.25 A comparison between the promotion rates of male and female employees of a certain company revealed the following information:

	Promoted	*Not Promoted*
Males	120	880
Females	260	740

(a) Overall, which gender, males or females, had a higher promotion rate? Give numerical support to your answer.

(b) The numbers given reflect the overall activity within the company. There are various job categories or levels within the company. Could females have a lower promotion rate than males at every job level? Explain.

14.26 The average salaries for two types of employees are listed for two companies.

	Support Staff			**Technical Staff**	
	Average Salary	*Number Employed*		*Average Salary*	*Number Employed*
Company A	$22,000	300	*Company A*	$52,000	700
Company B	$24,000	800	*Company B*	$56,000	200

(a) What is the overall average salary for the employees

(i)   at Company A?

(ii)   at Company B?

(b) Which company has the highest average salary

(i)   for support staff?

(ii)   for technical staff?

(c) This is an example of Simpson's paradox or aggregation bias. Explain why the overall average salary for Company A is higher than the overall average salary at Company B, yet the results in (b) contradict this.

Chapter Summary

Our world is inundated with statistics. This daily barrage is not a matter of mere counting but of inference and decision in the face of uncertainty. Money markets, family life, and medical discoveries are all subject to tests of significance and data analysis. This all began in 1900 when Karl Pearson published his chi-square test of goodness of fit. This test provided a formula for measuring how well a theoretical hypothesis fits some data—how well theory and observations correspond. The chi-square tests can be used for hypotheses and data for which the observations fall into discrete categories. Much of the information that we encounter each day is categorical in nature. Is the proportion of phone calls made over various holidays equally distributed? Is there an association between high-school grade and the usage of various types of drugs for the past month? Is the distribution of academic ranking for various departments at a college the same for males as for females?

In this chapter, we have studied these chi-square testing techniques that allow us to analyze such count data. We can assess how well a particular discrete model fits based on the observed counts from a random sample from the population of interest (test of goodness of fit). Likewise, we can assess if populations are the same with respect to the distribution for a discrete characteristic (test of homogeneity). Finally, we can assess if two characteristics are associated for a given population (test of independence).

Although we have learned three different tests, the underlying concepts behind them are the same. Our data consist of observed counts (how many sampled items fell into various categories). These data are presented in a frequency table (one-way for goodness of fit, two-way for homogeneity, and two-way for independence). We determine what we would expect the counts to be under the null hypothesis. The distance between these expected counts (under the null hypothesis) and the observed counts is measured using a test statistic X^2. A chi-square distribution (with a certain number of degrees of freedom) is the approximate distribution for the possible values for this test statistic under the assumption that the corresponding null hypothesis is true. You can safely use the approximate chi-square distribution if the expected counts are at least 5. Knowing this distribution shows us what to expect under H_0 and allows us to assess if our observed results are consistent with H_0 or are unusual under H_0.

If our results are unusual under H_0, such that we reject the null hypothesis, we can do some follow-up descriptive analyses that might help us better understand why we rejected H_0. We could look at the *contribution* to the test statistic made by each category or cell, denoted by $\frac{(\text{observed} - \text{expected})^2}{\text{expected}}$. The cells with the largest values have contributed the most to the result being declared statistically significant. For the two-way frequency tables, we can examine various percentages and make a bar chart (recall Section 4.3.3 for details). These informal analyses can help describe the nature of the association in a test of independence.

The various components of the three chi-square tests are summarized in the following table:

	Goodness of Fit	**Homogeneity**	**Independence**
H_0	a particular discrete model for X $p_i\,(i = 1, 2, \ldots K)$	the distribution of X is the same for C populations	the two discrete characteristics, X and Y, are independent
Expected Count	$E_i = np_i$	$\dfrac{(\text{row total})(\text{column total})}{\text{overall total}}$	$\dfrac{(\text{row total})(\text{column total})}{\text{overall total}}$
Test Statistic	$X^2 = \displaystyle\sum_{\text{all cells}} \dfrac{(O - E)^2}{E}$	(Same test statistic for all three tests)	
Degrees of Freedom	$K - 1$	$(R - 1)(C - 1)$	$(R - 1)(C - 1)$

We ended this chapter with some examples that show it can be risky to combine information across subgroups that are not homogeneous. A comparison between two variables that holds for each subgroup can be changed or even reversed when the data are aggregated across the subgroups. This paradox is known as Simpson's paradox.

Key Terms

Be sure you can describe, in your own words, and give an example of each of the following key terms from this chapter:

categorical/count data
goodness of fit of a
   discrete model
independence of two
   categorical variables

homogeneity of two or
   more populations
degrees of freedom
risk
expected counts
chi-square distribution

chi-square test statistic
association
independence
relative risk
Simpson's paradox
aggregation bias

Exercises

14.27 Mendel's theory of genes indicate that offspring carry traits of their parents in certain frequencies. Two genes, A and B, are to be considered. If a hybrid AB is crossed with another hybrid, then 25% of the resulting offspring should be AA, 50% should be AB, and 25% should be BB. An experiment is performed resulting in 188 offspring from hybrid parents.

   (a) State the hypothesis for assessing Mendel's theory of the frequencies of offspring's genes.

   (b) If Mendel's theory is correct, how many of the offspring in the experiment would you expect to be BB?

   (c) The observed test statistic for these data is $X^2_{\text{OBS}} = 3.2$. Do these data appear to support Mendel's theory? Explain.

REAL DATA

14.28 M&M's were first produced in 1941, in six colors (brown, yellow, red, violet, green, orange). In 1949, tan replaced violet. In early 1995, Mars, Inc., concluded that people don't like tan and they conducted a survey to pick a new color. The winning color was royal blue. In the *Detroit Free Press* (March 31, 1995) an article entitled "No Sweet Secret: M&M's Go Blue," contained the following information about M&M's:

> The candy company makes 300,000 pieces a day. In the plain bags, 30 percent should be brown. You also get 20 percent each of red and yellow; 10 percent each of green, orange and now, blue.

We wish to assess whether the color distribution actually holds. We will take a random sample of 10 plain M&M bags of such candies.

(a) Let Y = color for plain M&M's. Write out the appropriate null hypothesis for assessing the distribution for Y using a test of goodness of fit.

(b) State the corresponding 10% level decision rule for testing the null hypothesis in (a).

(c) In our random sample of 10 plain M&M bags, what proportion of the M&M's are expected to be red under the null hypothesis model? What do you need to know in order to be able to get the expected number of red M&M's?

14.29 A particular paperback book is published in a choice of three different covers. A certain bookstore keeps copies of each cover on its racks. To test the hypothesis that sales of this book are equally divided among the three choices, a random sample of 120 purchases is obtained. The sales results are as follows:

Cover Type	1	2	3
Observed # Books Purchased	31	47	42

(a) Give the appropriate null hypothesis to test that the three cover types have the same purchase rate.

(b) Compute the observed test statistic. Using a significance level of 0.10, test the hypotheses in (a). Show all work.

14.30 The model for the number of children in a family in Green City 10 years ago stated that 10% of the families had no children, 65% had one or two children, 20% had three or four children, and 5% had five or more children. A study was conducted to determine whether this 10-year-old model still holds for families in Green City. A random sample of 450 families in Green City were classified with respect to the number of children. The results are shown in the following table:

	Number of Children				
	None	One or Two	Three or Four	Five or More	
Observed Counts	53	313	69	15	450
Expected Counts					

(a) What is the appropriate null hypothesis of interest here?

(b) Carry out the appropriate test at the 5% significance level.

    (i) Compute all expected counts and write them in the table.

    (ii) Compute the appropriate test statistic.

    (iii) Find the *p*-value.

    (iv) State your decision and conclusion in terms of the fit of the 10-year-old model.

REAL DATA

14.31 The University of California's Berkeley campus was one of the first leading universities to employ affirmative action in its admissions policies. The enrollment for 1994 undergraduates, as reported in the *New York Times*, June 4, 1995, article entitled "Success on Diversity Faces Strong Attack in California," has *whites* accounting for only 32% of the student body, *Asian-Americans* accounting for 39%, with 13.8% *Hispanic*, 5.5% *black*, and the remaining 9.7% classified as *other* (for other groups or those who did not identify their race). The model for racial and ethnic background of undergraduates at UC Berkeley in 1984 was 61% *whites*, 24% *Asian-Americans*, 6% *Hispanic*, 5% *black*, and 4% *other*. Do the 1994 data support that the distribution has changed significantly?

(a) State the null hypothesis.

(b) The data for 1994 are reported in percentages.

    What do you need to know in order to be able to get the observed counts?

(c) Consider the following excerpt from the article:

> Enrollment of Blacks, which peaked at 1,647 or 7.7% of undergraduates, in 1989, currently stands at 1,127, or 5.5 percent. Enrollment of Hispanics, which peaked at 3,158, or 15.1 percent of undergraduates, in 1991, is now 2,828, or 13.8 percent.

    Based on this new information, try to determine the observed counts for each category. What is the total number of observations for 1994?

(d) Based on your total number of observations in (c), do you think the data will be statistically significant? Explain.

(e) Assess whether or not the distribution for racial and ethnic background has changed from the 1984 model. Show supporting details, and write a brief conclusion.

14.32 Many industrial air pollutants adversely affect plants. For example, sulfur dioxide causes leaf damage. In a study of the effect of a sulfur dioxide in the air on three types of garden vegetables, 40 plants of each type are exposed to a given concentration of sulfur dioxide under controlled greenhouse conditions. The observed counts are given in the accompanying table.

		Leaf Damage		
		Severe	*Not Severe or None*	
	Lettuce	32 ( )	8 ( )	40
Plant Type	*Spinach*	28 ( )	12 ( )	40
	Tomato	21 ( )	19 ( )	40
		81	39	120

(a) What is the appropriate null hypothesis of interest here?

(b) Carry out a test at the 10% significance level. Compute all six expected counts and write them in the parentheses in the table. Compute the appropriate test statistic. State your decision and conclusion.

14.33 The quality-assurance unit of a large pharmaceutical company was engaged in comparing two new formulations of a drug product in tablet form to assess whether they have similar potency acceptance rates. A random sample of 100 tablets was obtained from a pilot batch for each formulation. The number of tablets classified as acceptable with regard to potency is shown for each formulation.

```
FORMULAT  Drug Formulation  by  POTENCY  Potency Check
                POTENCY            Page 1 of 1
         Count |              Not
               | Accept     Accept      Row
               | 1.00|      2.00|   Total
FORMULAT    -------+---------+----------+
        1.00 |    84    |    16    |    100
  Formulation 1 |        |          |    50.0
            -------+---------+----------+
        2.00 |    96    |     4    |    100
  Formulation 2 |        |          |    50.0
            -------+---------+----------+
          Column      180        20       200
           Total      90.0      10.0     100.0
      Chi-square                  Value        DF    Significance
      ------------                -------       ----  ------------

Pearson                          8.00000        1      .00468
Continuity Correction            6.72222        1      .00952
Likelihood Ratio                 8.51038        1      .00353
```

(a) What is the appropriate null hypothesis of interest here?

(b) Using a 1% significance level, give the value of the chi-square test statistic, the p-value, your decision, and the corresponding conclusion for testing the null hypothesis in (a).

14.34 The presence of air bags in an automobile is often a key factor in the decision to purchase a particular model. Automobiles can be classified into six categories, roughly corresponding to size. The chi-square computer output based on these data is provided.

```
Table of     TYPE    (row) by   AIR BAGS    (columns)
Frequencies
              1.000    2.000     3.000     TOTAL
            ---------------------------------------
    1.000      22        10         6         38
    2.000       6        16         6         28
    3.000       6         9         4         19
    4.000       8        16         7         31
    5.000       5        10         8         23
    6.000      12         6         4         22
            ---------------------------------------
Total         59        67        35        161
Test statistic                     Value       DF      Prob
    Pearson chi-square            19.221       10      .038
    Likelihood ratio chi-square   18.973       10      .041
```

Note: TYPE: 1 = small, 2 = sporty, 3 = compact, 4 = midsize, 5 = large, 6 = van

AIR BAGS: 1 = none, 2 = driver only, 3 = driver and passenger side

(a) What is the appropriate chi-square test to assess if there is a relationship between the type of automobile and the level of air-bag implementation that is standard equipment? Explain.

(b) If the appropriate null hypothesis is true, how many small cars would you expect to have both driver and passenger air bags as standard equipment?

(c) Report the observed value of the test statistic, the corresponding *p*-value, and your decision, using a 5% significance level, and interpret your decision in terms of the problem.

14.35 A study was conducted to examine the role of sports as a social determinant for elementary school children. The subjects were students in Grades 4 to 6 from school districts in Michigan. The researchers stratified their sample, selecting students from urban, suburban, and rural school districts. Students were asked to indicate whether good grades, athletic ability, or popularity was most important to them as a personal goal. The questionnaire also asked for demographic information, such as gender, grade level, age, and race. (SOURCE: M. A. Chase and G. M. Dummer, "The Role of Sports as a Social Determinant for Children," *Research Quarterly for Exercise and Sport*, 63, 418–424, 1992. Data set available through the Statlib Data and Story Library, DASL, available on the Internet at http://lib.stat.cmu.edu/DASL/) One of the questions to be addressed is whether there is an association between the type of school area and the students' choice of good grades, athletic ability, or popularity as most important. A two-way table dividing the students' goal responses by type of district is presented in the following table:

REAL
DATA

		District Type		
		Rural	*Suburban*	*Urban*
	Good Grades	57	87	24
Goal	**Popularity**	50	42	6
	Athletic Ability	42	22	5
		149	151	35	335

(a) Compute the percentage of responses for each column and enter them in the next table. For example, since 57 of the 149 rural district children response was "good grades," the percentage is (57/149)100 = 38.3%.

		District Type		
		Rural	*Suburban*	*Urban*
	Good Grades	38.3%		
Goal	**Popularity**			
	Athletic Ability			
		100%	100%	100%

(b) Complete this sentence: The table in (a) displays the conditional distribution of _____ given _____.

(c) From the table in (a), does it appear that the emphasis on good grades increases as the school areas become more urban, while the emphasis on popularity decreases?

(d) Assess whether the association is significant. Use a significance level of 0.05. Show all details.

14.36 (Continuation of Exercise 14.35) In the previous exercise, you expressed the data as percentages by type of district. That is, you gave the conditional distribution of personal goal choice given district type.

(a) Give the overall percentages for personal goal choice by completing the following sentence:

Overall, _____% of the students considered good grades as a personal goal, _____% of the students selected popularity, and _____% of the students selected athletic ability.

If the distribution of personal goal is to be homogeneous for the three district types, we would expect the entries in the first row of this conditional distribution table to be approximately equal to the overall rate of _____%. The extent to which the row entries deviate from the overall row rate represents the nonhomogeneity between district types.

A plot of these percentages (made with SPSS) helps summarize the pattern in the conditional table. The horizontal axis displays the district types, and the vertical axis displays the goal percentages. Such a plot, sometimes referred to as a **profile plot**, is as follows:

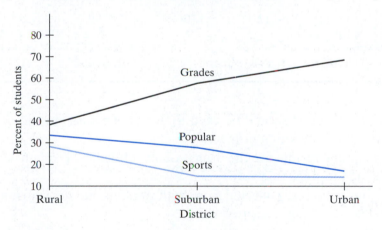

(b) Enhance this plot by adding the overall pecentages that you computed in part (a) as horizontal reference lines.

This plot provides evidence as to why the test resulted in the conclusion that the distribution of personal goal differs among the three district types.

14.37 Recall the data from "Let's do it! 14.5." Here we were examining whether or not randomization did a good job in controlling for the potential lurking variable of smoking status.

| | | **Treatment Group** | | |
		Group 1	*Group 2*	*Group 3*
Smoking Habit	*Nonsmoker*	38	50	37
	Ex-smoker	52	48	48
	Smoker	102	102	108

(a) Give the conditional distribution of smoking status given treatment group.

(b) Make a profile plot of the percentages to summarize the pattern in the conditional table. (See Exercise 14.36 for an example and definition of a profile plot.)

(c) Give the overall percentages for smoking status by completing this sentence: Overall, _____ % of the subjects were nonsmokers, _____ % of the subjects were ex-smokers, and _____ % subjects were smokers.

(d) Enhance this plot by adding the overall percentages that you computed in (c) as horizontal reference lines.

(e) Does this plot support your decision as to whether or not randomization did a good job in controlling for the potential confounding variable of smoking status? Explain.

14.38 Two oral contraceptive drugs for women are tested. Of the 2200 women available for the study, 1100 were randomly assigned to each drug group. The responses were classified as pregnant (RESPONSE = 1) or not pregnant (RESPONSE = 2).

```
Table of RESPONSE      (row) by    DRUG    (columns)
Frequencies    1.000        2.000       TOTAL
               ----------------------------
       1.000     595          905         1500
       2.000     505          195          700
               ----------------------------
    Total       1100         1100         2200
Test statistic                       Value   DF    Prob
Pearson chi-square                  201.352   1   0.000
Likelihood ratio chi-square         206.698   1   0.000
```

(a) What type of test should be used to determine whether the drugs are equally effective at preventing pregnancy?

(b) Based on the results, are the drugs equally effective? Explain. If not, which drug has a higher pregnancy rate?

The previous data were collected from both smokers and nonsmokers. The responses for these two subgroups are provided.

```
Smokers
Table of RESPONSE (row) by  DRUG   (columns)
Frequencies  1.000      2.000        TOTAL
             ------------------------
      1.000    95         900          995
      2.000     5         100          105
             ------------------------
Total         100        1000         1100

Test statistic              Value   DF    Prob
Pearson chi-square           2.632   1   0.105
Likelihood ratio chi-square  3.085   1   0.079
```

```
Nonsmokers
Table of RESPONSE (row) by DRUG (columns)
Frequencies  1.000      2.000     TOTAL
             ------------------
     1.000    500          5        505
     2.000    500         95        595
             ------------------
Total        1000        100       1100
Test statistic                 Value    DF    Prob
Pearson chi-square             74.133    1   0.000
Likelihood ratio chi-square    91.555    1   0.000
```

(c) Does the conclusion obtained in (b) hold true for each of these subpopulations at the 5% level? for smokers? for nonsmokers?

(d) Briefly explain why the aggregated results are misleading.

TI Quick Steps

Finding Areas Under a Chi-Square Distribution Using the TI

The TI has a function that is built into the calculator for finding areas under any chi-square distribution. This function allows you to specify the degrees of freedom and is located under the DISTR menu. A chi-square test resulted in an observed test statistic value of 15.5. The distribution for X^2 under H_0 is a chi-square distribution with three degrees of freedom. We wish to compute the p-value for the test. We need to find the area under the $\chi^2(3)$ distribution to the right of 15.5. The TI steps are provided.

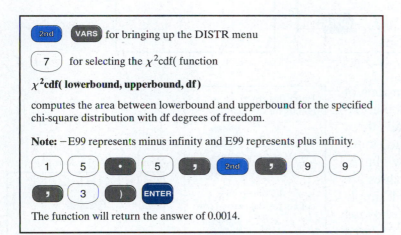

[2nd] [VARS] for bringing up the DISTR menu

[7] for selecting the χ^2cdf(function

χ^2**cdf(lowerbound, upperbound, df)**

computes the area between lowerbound and upperbound for the specified chi-square distribution with df degrees of freedom.

Note: −E99 represents minus infinity and E99 represents plus infinity.

[1] [5] [•] [5] [,] [2nd] [,] [9] [9]

[,] [3] [)] [ENTER]

The function will return the answer of 0.0014.

Your screen should look like

```
χ²cdf(15.5,ε99,3
)
         .0014355859
```

Performing a Chi-Square Test of Independence or Homogeneity Using the TI

The TI has a χ^2-Test option under the TESTS menu located under the STAT menu that allows you to perform a test of independence or a test of homogeneity. The table of observed counts must first be stored as a matrix of counts. A matrix is a two-way array, similar to a two-way frequency table. The TI has 10 matrix variables, [**A**] through [**J**]. The χ^2-Test option can compute the expected counts and store them in another matrix, provide the value of the X^2 test statistic and the p-value, and report the degrees of freedom. To demonstrate the steps, we consider the test of independence to assess if having a pet is associated with a higher survival rate. The table is as follows:

		Pet Status	
		No Pet	*Pet*
Survival Status	*Alive*	28	50
	Dead	11	3

Putting Observed Counts in a Matrix

Before you can put the observed counts into a matrix, you must define the size or dimension of the matrix, specifying first the number of rows followed by the number of columns. The following steps show how to define Matrix [**A**] to have two rows and two columns.

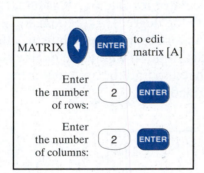

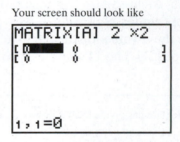

After you have set the dimensions of the Matrix [**A**], you can view the matrix (up to seven rows and three columns displayed at a time) and enter the observed counts. In a new matrix, all values are zero. If some values are not zero, it is fine since you will simply enter values over them. The following steps show how to enter the observed counts from the preceding table:

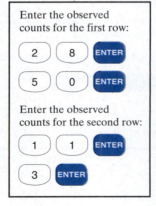

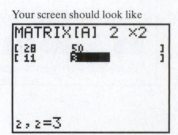

Performing the Chi-Square Test

To perform the chi-square test, you will need to specify two matrices. The matrix name that has the observed counts, in this case [**A**], and the matrix name to which you want the expected counts to be stored; we will use [**E**]. You then have the option to have the test statistic and *p*-value reported (Calculate) or to have the chi-square distribution under the null hypothesis displayed with the *p*-value area shaded and the test statistic and *p*-value reported (Draw). The steps to produce the calculated results are as follows:

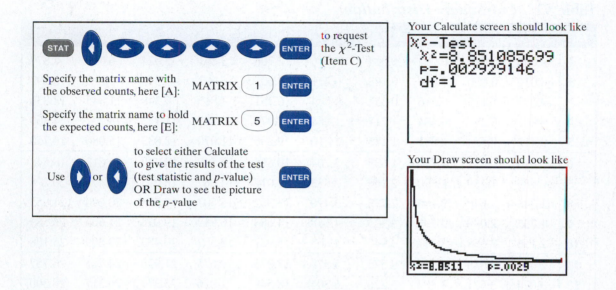

If you press MATRIX 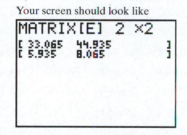, the corresponding table of expected counts is displayed in your window:

Your screen should look like

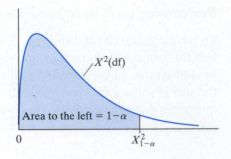

Area to the left = $1-\alpha$

$X^2_{1-\alpha}$

Table VI *Chi-Square Distribution*

DF	$X^2_{.005}$	$X^2_{.01}$	$X^2_{.025}$	$X^2_{.05}$	$X^2_{.10}$	$X^2_{.90}$	$X^2_{.95}$	$X^2_{.975}$	$X^2_{.99}$	$X^2_{.995}$
1	0.000	0.000	0.001	0.004	0.016	2.706	3.841	5.024	6.635	7.879
2	0.010	0.020	0.051	0.103	0.211	4.605	5.991	7.378	9.210	10.597
3	0.072	0.115	0.216	0.352	0.584	6.251	7.815	9.348	11.345	12.838
4	0.207	0.297	0.484	1.711	1.064	7.779	9.488	11.143	13.277	14.860
5	0.412	0.554	0.831	1.145	1.610	9.236	11.070	12.833	15.086	16.750
6	0.676	0.872	1.237	1.635	2.204	10.645	12.592	14.449	16.812	18.548
7	0.989	1.239	1.690	2.167	2.833	12.017	14.067	16.013	18.475	20.278
8	1.344	1.646	2.180	2.733	3.490	13.362	15.507	17.535	20.090	21.955
9	1.735	2.088	2.700	3.325	4.168	14.684	16.919	19.023	21.666	23.589
10	2.156	2.558	3.247	3.940	4.865	15.987	18.307	20.483	23.209	25.188
11	2.603	3.053	3.816	4.575	5.578	17.275	19.675	21.920	24.725	26.757
12	3.074	3.571	4.404	5.226	6.304	18.549	21.026	23.337	26.217	28.300
13	3.565	4.107	5.009	5.892	7.042	19.812	22.362	24.736	27.688	29.819
14	4.075	4.660	5.629	6.571	7.790	21.064	23.685	26.119	29.141	31.319
15	4.601	5.229	6.262	7.261	8.547	22.307	24.996	27.488	30.578	32.801
16	5.142	5.812	6.908	7.962	9.312	23.542	26.296	28.845	32.000	34.267
18	6.265	7.015	8.231	9.390	10.865	25.989	28.869	31.526	34.805	37.156
20	7.434	8.260	9.591	10.851	12.443	28.412	31.410	34.170	37.566	39.997
24	9.886	10.856	12.401	13.848	15.659	33.196	36.415	39.364	42.980	45.559
30	13.787	14.953	16.791	18.493	20.599	40.256	43.773	46.979	50.892	53.672
40	20.707	22.164	24.443	26.509	29.051	51.805	55.758	59.342	63.691	66.766
60	35.534	37.485	40.482	43.188	46.459	74.397	79.082	83.298	88.379	91.952
120	83.852	86.923	91.573	95.705	100.624	140.233	146.567	152.211	158.950	163.648

WHAT IF THE ASSUMPTIONS DON'T HOLD?

15.1 Introduction

In Chapter 9, we outlined the steps that are followed for all hypothesis testing procedures:

- State the population(s) and corresponding parameter(s) of interest.
- State the competing theories—that is, the null and alternative hypotheses.
- State the significance level for the test.
- ***Collect and examine the data and assess whether the assumptions are valid***.
- Compute a test statistic using the data and determine the *p*-value.
- Make a decision and state a conclusion using a well-written sentence.

So far, the tests we have performed have covered two main situations:

1. The response is **quantitative** and can be modeled with a **normal** density. Then the inferential procedures, such as *t*-tests and the *F*-tests associated with analysis of variance, were appropriate. The normality assumption was less crucial if our sample sizes were large enough.
2. The response is **categorical**, modeled with a discrete function. Then the data were analyzed with chi-square tests or large-sample *z*-tests.

What if the response is quantitative and modeled with some density function, but we cannot assume a particular ***parametric*** model, such as the normal distribution? For this reason, the alternative tests that we are about to discuss are sometimes referred to as **nonparametric** tests. Perhaps the observed data are skewed, or some extreme outliers are present. As we learned in Chapter 5, the mean is not the best measure of center in such cases. The median is a better choice, resistant to skewness and extreme observations. Thus, the statistical tests discussed in this chapter are used to analyze hypotheses about the **median** instead of the mean. These tests will be based on the sign of the observation or the ranked observations. Section 15.2 answers the following question: What if we have a small random sample from a nonnormal population? A one-sample t-test should not be used if either skewness or outliers are present. In this case, the **sign test** be used to test hypotheses about the median of the population under study. In Section 15.3, we return to the two independent samples design for comparing two nonnormal populations because the two-sample t-test of Chapter 11 is no longer appropriate. The **Wilcoxon rank sum test** can be used to test the hypothesis that the medians of two populations are equal. Section 15.4 addresses the analysis of paired data when the paired t-test of Chapter 11 (Section 11.3) is no longer appropriate. The **Wilcoxon signed rank test** can be used to test the hypothesis that the median of the differences is equal to zero. Finally, in Section 15.5, we present the ideas for the **Kruskal–Wallis test**, which is used to compare the medians of more than two populations when normality for the response is not reasonable. We will focus on the use of these nonparametric tests when the normal assumption for the quantitative response is not satisfied and the sample sizes are small. These nonparametric tests can also be used when the response variable is not quantitative, but the outcomes can be ranked in order of magnitude (such qualitative data that can be ranked is sometimes called *ordinal* data).

 ## 15.2 One Sample—Sign Test

The sign test is analagous to the one-sample t-test for a population mean from Chapter 10, Section 10.2, but dropping the normality assumption. We assume that we have one random sample from one population. We measure one continuous response, which has some density function as its model, not necessarily normal. The median is generally used as a measure of center instead of the mean when the shape of the density function is skewed or perhaps is not known. Since the median is also the 50th percentile, we will denote the median of a population by $\pi_{0.5}$. So we are interested in testing the hypothesis

$$H_0: \pi_{0.5} = \pi_{0.5}^0$$

against one of the alternative hypotheses,

$$H_1: \pi_{0.5} \neq \pi_{0.5}^0 \quad \text{or} \quad H_1: \pi_{0.5} > \pi_{0.5}^0 \quad \text{or} \quad H_1: \pi_{0.5} < \pi_{0.5}^0,$$

where $\pi_{0.5}^0$ is the hypothesized value of the median.

The data are a random sample of size n. In theory, the values of our random variables should all be different from the hypothesized median (the probability of a continuous random variable being equal to any one value is zero, there is no area under a density curve at a single point). However, due to rounding, a few observations may be equal to the median. If this is the case, those observations are not used. Thus, we have the effective sample size n = the number of observations that are **not equal** to $\pi_{0.5}^0$. The test will be conducted by counting the number of observations that exceed the hypothesized median, and the **test statistic** is denoted by R = the number of observations which exceed $\pi_{0.5}^0$.

Why is this a reasonable test statistic? The statistic R will tend to be large when many of the values in the sample are larger than $\pi_{0.5}^0$, the hypothesized median—that is, R will tend to be large when the sample is drawn from a population whose median is larger than $\pi_{0.5}^0$. So, large values of R would support $H_1: \pi_{0.5} > \pi_{0.5}^0$. Similarly, a small value of R suggests that many of the values in the sample are smaller than $\pi_{0.5}^0$, the hypothesized median—that is, R will tend to be small when the sample is drawn from a population whose median is smaller than $\pi_{0.5}^0$. So, small values of R would support $H_1: \pi_{0.5} < \pi_{0.5}^0$. So the direction of extreme will match with the direction in the alternative hypothesis.

To conduct our test, we need to compute the **p-value**. To compute the p-value, we need to know the distribution of the test statistic under the null hypothesis.

THINK ABOUT IT **The Meaning of the Median**

We have a sample of size n and we define a success outcome if the observation exceeds the hypothesized median. If the null hypothesis is true (that is, the hypothesized median is the true median), then the probability of a success (getting a value larger than the median) is just 0.5 by definition.

Our test statistic is *counting the number of successes* in a random sample of size n, with success probability p. So our test statistic will have a _____ distribution when the null hypothesis is true.

Hint: Look back to Section 7.5.2.

If H_0 is true, then the probability that the random variable of interest exceeds the hypothesized median is equal to 1/2. The test statistic R, which is counting the number of "successes" (the number of observations which exceed $\pi_{.5}^0$) in n trials, has a **binomial distribution** with n and $p = 1/2$.

- The possible values of $R = 0, 1, 2, 3, \ldots, n$.
- The mean (average or expected value) of $R = np = n/2$.
- The variance of $R = np(1 - p) = n/4$.
- The standard deviation of $R = \sqrt{n}/2$.

So we can compute the p-value, the probability of observing a value for R as extreme or more extreme than that observed, under the assumption that H_0 is true, by using the binomial distribution:

$$P(R = r | H_0 \text{ true}) = \binom{n}{r}(1/2)^r(1 - 1/2)^{n-r} = \binom{n}{r}(1/2)^n.$$

Let's turn to an example to put the sign test concepts to work.

❖ **EXAMPLE 15.1** **Salaries**

Nationwide, the median starting salary for graduates of a certain field is $27,000. The department chair at a local university claims that the starting salaries for their graduates tend to be higher. Distributions for salaries are typically skewed to the right. The starting salaries for a random sample of $n = 11$ graduates from this department were obtained and are as follows:

27,500	25,500	24,000	*29,500*	*31,000*	*28,500*
29,000	*34,000*	26,500	*28,000*	*28,000*	

Since the normality assumption is not reasonable and our sample size is small, we turn to the sign test for conducting the comparison.

State the hypotheses $H_0: \pi_{0.5} = 27{,}000$ versus $H_1: \pi_{0.5} > 27{,}000$.

State the significance level for the test $\alpha = 0.05$.

Compute the value of the test statistic

Note that, since none of the observations are equal to 27,000, we have $n = 11$. There are eight observations (shown in bold teal italics) that exceed 27,000, so our observed test statistic is $R = 8$.

Report the p-value for the test

$$
\begin{aligned}
p\text{-value} &= P(R \geq 8 | H_0 \text{ true}) \\
&= P(R = 8) + P(R = 9) + P(R = 10) + P(R = 11) \\
&= \binom{11}{8}(1/2)^{11} + \binom{11}{9}(1/2)^{11} + \binom{11}{10}(1/2)^{11} + \binom{11}{11}(1/2)^{11} \\
&= 0.1134
\end{aligned}
$$

Recall that the TI has the binompdf(and binomcdf(functions that could be used to compute this p-value with the calculator. (See the TI Quick Steps at the end of Chapter 7.) For this example, the p-value would be computed as $1 - \text{binomcdf}(11, .5, 7) = 0.1133$.

What is your decision and conclusion?

Since our p-value is larger than the significance level of 0.05, we cannot reject H_0, and we conclude that the department chair's claim of higher salaries is not supported by the data.   ❖

Remarks About Our First Sign Test

- Instead of the actual salaries, the data could have been given to you as

+	−	−	+	+	+
+	+	−	+	+	

  where "+" means the salary was above 27,000 and "−" means the salary was below 27,000.

  Then R = number of positive signs—hence the name "sign test."

- If one of the 11 observations had been equal to 27,000, then $n = 10$ would have been used in calculating the p-value.

The use of the sign test for testing hypotheses about a population median is summarized in the next box.

Sign Test

Assumptions	The sample is selected randomly from a continuous population whose model may not neccesarily be normal.
Hypotheses	H_0: $\pi_{0.5} = \pi_{0.5}^0$ versus
	H_1: $\pi_{0.5} > \pi_{0.5}^0$ or H_1: $\pi_{0.5} < \pi_{0.5}^0$ or H_1: $\pi_{0.5} \neq \pi_{0.5}^0$,
	where $\pi_{0.5}^0$ is the hypothesized value of the median.
Data and Test Statistic	n = the number of observations that are **not equal** to $\pi_{0.5}^0$ R = the number of observations that exceed $\pi_{0.5}^0$
p-value	For a *one-sided to the right* test, the p-value = $P(R \geq R_{\mathrm{OBS}})$.
	For a *one-sided to the left* test, the p-value = $P(R \leq R_{\mathrm{OBS}})$.
	For a *two-sided* test, compute both the number of observations that exceed $\pi_{0.5}^0$ and the number of observations less than $\pi_{0.5}^0$; call R_{OBS} = the larger of these two numbers.

$$\text{Then, the } p\text{-value} = 2P(R \geq R_{\mathrm{OBS}}).$$

The p-value probabilities are computed using a binomial distribution with n and $p = 1/2$.

Notes:

- To perform the test, we need only to know the value of R and do not need to know the actual values of the responses.

- If the salaries had a normal distribution, then we would have performed a one-sample t-test.

- Just as the **paired t-test** is actually a one-sample t-test on the differences, this ***sign test could also be applied to differences if we have paired data***, with typically a hypothesized median of zero.

Let's do it! 15.4 Test Scores

A real estate appraiser wants to verify the market value for homes on the east side of the city that are very similar in size and style. The appraiser wants to test the popular belief that the median sales price is $37.80 per square foot for such homes. He will use a significance level of 0.01. The model for sales has typically been skewed, so the assumption of a normal distribution as a model for sales is not appropriate. Using the sign test, test the appropriate hypotheses about the median sales price.

(a) State the appropriate null and alternative hypotheses about the population median sales price for such homes.

(b) Suppose that the random sample of six sales was selected. The sampled sales prices per square foot are

   $35.00 $38.10 $30.30 $37.20 $29.80 $35.40

Does it appear that the popular perception of the market value is valid for this neighborhood? Give the value of the test statistic and the *p*-value.

(c) Based on your *p*-value in (b), are the results statistically significant at a 1% significance level?

❋ Exercises

15.1 Consider the following sample of 10 observations:

   23.9   19.5   11.4   18.0   14.9   15.1   22.4   17.9   20.1   31.6

Use a sign test with a significance level of 0.05 to test each of the following hypotheses:

(a) $H_0: \pi_{0.5} = 16$ versus $H_1: \pi_{0.5} > 16$.

(b) $H_0: \pi_{0.5} = 16$ versus $H_1: \pi_{0.5} < 16$.

(c) $H_0: \pi_{0.5} = 16$ versus $H_1: \pi_{0.5} \neq 16$.

(d) What assumption about the data is required for the preceding tests to be valid?

15.2 The numbers of packages handled by a shipping office on 12 randomly selected days are as follows:

$$1158 \quad 991 \quad 1462 \quad 1311 \quad 1093 \quad 1538$$
$$1216 \quad 1117 \quad 1361 \quad 1389 \quad 1602 \quad 1710$$

At the 0.05 significance level, test the claim that the median number of packages handled per day is greater than 1100.

15.3 For the past month of April, the median round-trip-ticket cost for an airline with a discount fare was $258. A simple random sample of 15 round-trip discount fares during the month of May provided the following data:

$$310 \quad 260 \quad 265 \quad 255 \quad 300 \quad 310 \quad 230 \quad 250$$
$$265 \quad 280 \quad 290 \quad 240 \quad 285 \quad 250 \quad 260$$

We wish to assess whether there is sufficient evidence that the median round-trip-ticket cost for discount fares has increased in May. Test at the 5% level of significance. State the appropriate null and alternative hypotheses, provide the observed test statistic value, *p*-value, and decision, and report whether the results are statistically significant.

15.4 An investigator wants to know about the median size of bank passbook savings accounts. He has a hunch that the median account is far less than $5000. He also knows that the distribution of such account sizes is far from being symmetric. He selects a sample of 20 accounts and determine their sizes. The results are as follows:

$$500.12 \quad 183.62 \quad 246.89 \quad 1000.00 \quad 683.40 \quad 10{,}563.00 \quad 2465.60$$
$$10.12 \quad 296.30 \quad 843.10 \quad 4683.00 \quad 562.00 \quad 24.10 \quad 25.96$$
$$183.40 \quad 627.10 \quad 483.70 \quad 400.00 \quad 925.80 \quad 3672.89$$

Is this sufficient evidence at a 1% level to suggest that median account is less than $5000? State the null and alternative hypotheses and show all details.

15.3 Two Independent Samples—Rank Sum Test

Suppose that we have two independent samples obtained under two different conditions. The appropriate test to perform depends on the assumptions we can make regarding the underlying distribution of the data.

❖ EXAMPLE 15.2 Which Test? _____

Suppose that we want to compare two treatments for the common cold. A sample of people is provided with a **treatment**, vitamin C, while a second sample of people is provided with a **placebo**, a similar pill, but without an active ingredient.

Case 1

If we record the outcome as the number of days until a person suffers from their next cold, then the procedure we use depends on how we model this variable. When the waiting time until the next cold is modeled as a ***normal*** variable, the **two independent samples *t*-test** would be used.

Case 2

If we recorded the outcome as the number of people in the sample with a cold, then we could arrange the data in a table with two rows representing treatment group and two columns indicating whether a person had a cold. The data in the contingency table would provide the observed number of people in a treatment group who had a cold. The analysis would be based on the **chi-square test of homogeneity** to compare the two treatment groups.

		Treatment Group	
		Vit C	Placebo
Have a Cold?	Yes		
	No		

Case 3

When the waiting times until a person is infected are recorded, and these waiting times are modeled with some ***unknown density***, an alternative procedure, the **rank sum test**, may be used. ❖

For the **(Wilcoxon) rank sum test**, we assume that each set of data is a random sample, the two samples are independent, and both populations have continuous distributions of the same shape. The last assumption requires that the densities differ only with respect to location and not otherwise. This assumption is not essential, but the test will have limited power without this assumption. We are interested in testing the null hypothesis

$$H_0: f = g$$

against one of the alternative hypotheses,

$$H_1: f \neq g \quad \text{or} \quad H_1: f > g \quad \text{or} \quad H_1: f < g,$$

where f represents the density for the first population and g represents the density for the second population. The one-sided alternative hypotheses are shown in picture form in the case of the densities being skewed to the right:

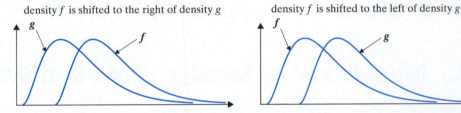

Assuming the shapes of densities f and g are the same, these hypotheses are often expressed in terms of the medians of the two distributions. So the null hypothesis could be written as

$$H_0: \pi_{0.5}(f) = \pi_{0.5}(g).$$

Since we assumed that the densities differed only in location and not in shape, acceptance of equal medians would imply that we thought the two populations have the same density model.

We have two samples, of sizes n_1 and n_2, from the two populations. For consistency, we will give the label of "first sample" to the sample with the fewest observations; thus, $n_1 \leq n_2$. We rank all the $(n_1 + n_2)$ observations from smallest to largest. The smallest observation is assigned a rank of 1 and the largest observation would receive a rank of $(n_1 + n_2)$. Theoretically, there shouldn't be any ties because we assumed the data are modeled with a continuous density function. However, in application, some rounding of the measurements would lead to a few ties, and so we assign to each tie the *average rank of the tied items.* For instance, if the third and fourth smallest observations were equal, since they would have received the ranks of 3 and 4, we assign to each the average rank of 3.5. The *test statistic*, labeled W, is the sum of the ranks in the first (smaller) sample.

Why is this a reasonable test statistic? The statistic W will tend to be large when the values in the first sample are generally larger than those in the second sample. That is, W will tend to be large when the first sample is drawn from a population that is centered at a larger value than the population from which the second sample is drawn. (A large W would support $f > g$.) Similarly, a small value of W suggests that the center of the first population is smaller than the center of the second population. (A small W would support $f < g$.)

Our next task is to decide whether the observed result is surprising under the null hypothesis of no difference. This requires a frame of reference. What do we expect our statistic W to equal when there is no difference between the two populations? What can we say about the standard deviation and the percentiles of the distribution of W under the null hypothesis?

 ### 15.3.1 Frame of Reference

What is the **smallest possible value** for W? The **minimum value** for our test statistic occurs when the observations in the first sample are all smaller than those found in the second sample:

$$\min W = 1 + 2 + \cdots + n_1 = \frac{n_1(n_1 + 1)}{2}.$$

THINK ABOUT IT **Why This Formula for the Minimum?**

Here is one way you can derive the formula for the sum of the first n natural numbers:

$$\text{Sum}_n = 1 + 2 + 3 + \cdots + n.$$

The calculation is made easy if we write this sum out backward:

$$\text{Sum}_n = 1 + 2 + 3 + \cdots + (n - 1) + n$$
$$\text{Sum}_n = n + (n - 1) + (n - 2) + \cdots + 2 + 1.$$

Adding these two expressions up by columns, we get

$$2\text{Sum}_n = (n + 1) + (n + 1) + (n + 1) + \cdots + (n + 1).$$

Since there are n terms, each equal to $(n + 1)$, the formula for the sum is

$$\text{Sum}_n = \frac{n(n + 1)}{2}.$$

The sum of the numbers $1, 2, 3, \ldots, 100$ would be given by

$$\text{Sum}_{100} = \frac{100(100 + 1)}{2} = 5,050.$$

Continuing with our frame of reference we ask, "What is the **largest possible value** for W?" The **maximum value** occurs when all of the observations in the first sample are larger than all of the observations in the second sample. The maximum value of the test statistic would then be computed as follows:

$$\max W = (n_2 + 1) + (n_2 + 2) + \cdots + (n_2 + n_1) = n_1 n_2 + \frac{n_1(n_1 + 1)}{2}.$$

What is the **expected value** for W, assuming that H_0 is true? When we rank $(n_1 + n_2)$ numbers, the average rank is the sum of the numbers, $1, 2, \ldots, n_1 + n_2$, divided by the number of these $(n_1 + n_2)$ numbers. So the average rank is given by

$$\text{Average rank} = \frac{(n_1 + n_2)(n_1 + n_2 + 1)}{2(n_1 + n_2)} = \frac{(n_1 + n_2 + 1)}{2}.$$

If there isn't a difference between the two populations—that is, H_0 is true—the first sample should have ranks that average out to this value. Since there are n_1 observations in this sample, we have the **expected value** of our test statistic under the null hypothesis:

$$E[W | \text{no difference}] = \mu_W = \frac{n_1(n_1 + n_2 + 1)}{2}.$$

A similar calculation can be used to compute the **standard deviation** of the test statistic when there is no difference between the treatments. The result of that calculation yields

$$\text{standard deviation}[W | \text{no difference}] = \sigma_W = \sqrt{\frac{n_1 n_2(n_1 + n_2 + 1)}{12}}.$$

To conduct our test, we need to compute a p-value. If we were testing $H_0: f = g$ against the alternative hypothesis $H_1: f > g$, then larger values of W would be considered more extreme and would support H_1. So the p-value would be the chance of getting a W as large or larger than the one observed, under the null distribution. If we were testing $H_0: f = g$ against the alternative hypothesis $H_1: f < g$, then smaller values of W would be considered more extreme and would support H_1. So our p-value would be the chance of getting a W as small or smaller than the one observed under the null distribution. If we were testing $H_0: f = g$ against the alternative $H_1: f \neq g$, then values of W that are either too big or too small would be considered more extreme and would support H_1. We would obtain the p-value by doubling a one-sided probability calculation. For samples with 10 or fewer observations in each of the two groups, the value of this probability can be found in Table VII. A portion of the table is shown for the case in which the sample sizes are $n_1 = 4$ and $n_2 = 5$:

Smaller Sample Size = 4 Larger Sample Size = 5		
W_{lower}	Tail Probability	W_{upper}
10	**0.008**	**30**
11	0.016	29
12	0.032	28
⋮	⋮	⋮
15	0.143	25

The table has three columns. The first line entry tells us that $P[W \leq 10] = P[W \geq 30] = 0.008$.

Let's put the Wilcoxon rank sum test to work through an example.

❖ **EXAMPLE 15.3** **Two Teaching Methods** _____

To compare two teaching methods, nine students were randomly assigned to one of two classrooms. Following a period of instruction, all were tested and their scores recorded:

Class I	55	43	71	66	
Class II	67	76	83	78	91

Since test scores on such an exam are typically skewed to the left and the sample sizes are small ($n_1 = 4$, $n_2 = 5$), we turn to the rank sum test.

State the hypotheses H_0: $\pi_{0.5}$(Class I) $= \pi_{0.5}$(Class II).

H_1: $\pi_{0.5}$(Class I) $\neq \pi_{0.5}$(Class II).

State the significance level $\alpha = 0.05$

Compute the value of the test statistic

All nine scores are ranked from smallest to largest. The ranks are recorded in the following table:

Class I	2	1	5	3	
Class II	4	6	8	7	9

The test statistic is the sum of the ranks of students in Class I, because there are fewer students in that class.

$$W = 2 + 1 + 5 + 3 = 11.$$

There are four students in Class I and five in Class II. If the two classes are not different, then the expected value and standard deviation of the statistic can be computed from the previous frame of reference formulas. The expected value is

$$\mu_W = \frac{n_1(n_1 + n_2 + 1)}{2} = \frac{4(4 + 5 + 1)}{2} = 20.$$

The standard deviation can also be found through substitution to be

$$\sigma_W = \sqrt{\frac{n_1 n_2(n_1 + n_2 + 1)}{12}} = \sqrt{\frac{4(5)(4 + 5 + 1)}{12}} = 4.08.$$

Report the *p*-value for the test. To decide whether our observed test statistic result of 11 is surprising under the hypothesis of no difference, we need to calculate the *p*-value. For samples with 10 or fewer observations in each of the two treatments, the *p*-value can be found in Table VII. A portion of the table is shown for the case in which the sample sizes are $n_1 = 4$ and $n_2 = 5$:

W_{lower}	Smaller Sample Size = 4 Larger Sample Size = 5	
	Tail Probability	W_{upper}
10	0.008	30
11	**0.016**	**29**
12	0.032	28
13	0.056	27
14	0.095	26
15	0.143	25

For our example, we had $W_{\mathrm{OBS}} = 11$ and we were doing a two-sided test. Our *p*-value for this test would be *p*-value = $2P[W \leq 11 | H_0 \text{ is true}] = 2(0.016) = 0.032$. At a 5% significance level, we would conclude that there is a significant difference between the two methods in terms of median test score. ❖

A summary of the rank test for testing hypotheses about two population medians is provided below:

Wilcoxon Rank Sum Test

Assumptions The two samples are random and independent. The models for the two populations are continuous and of the same shape.

Hypotheses H_0: $\pi_{0.5}(population\ 1) = \pi_{0.5}(population\ 2)$ versus

H_1: $\pi_{0.5}(population\ 1) > \pi_{0.5}(population\ 2)$ or

H_1: $\pi_{0.5}(population\ 1) < \pi_{0.5}(population\ 2)$ or

H_1: $\pi_{0.5}(population\ 1) \neq \pi_{0.5}(population\ 2)$,

where $\pi_{0.5}$ represents the medians for the populations.

Procedure Label the sample with the smallest sample size Group 1 with sample size n_1; that is, let $n_1 \leq n_2$. Rank all $(n_1 + n_2)$ observations (if any ties, assign the average of the ranks).

Test Statistic W = sum of the ranks in the first (smallest) sample.

p-value For a *one-sided to the right* test, the *p*-value = $P[W \geq W_{\mathrm{OBS}} | H_0 \text{ is true}]$.

For a *one-sided to the left* test, the *p*-value = $P[W \leq W_{\mathrm{OBS}} | H_0 \text{ is true}]$.

For a *two-sided* test, the *p*-value is obtained by doubling a one-sided probability calculation.

Notes:

■ Although tied observations can be assigned ranks by averaging, the number of ties should be small relative to the number of observations to ensure the validity of the test.

■ If the two populations had a normal distribution with equal population standard deviation, then we would have performed a two independent samples *t*-test for comparing the two population means.

 **15.2** **Starting Salaries**

Random samples of starting salaries for math majors from two colleges are given in the following table, in $1000s:

College 1	30	35	29	37.5	32	40		
College 2	28.5	38	30.5	26	37	29	33	32

Is there sufficient evidence to say that the median starting salary for College 1 math majors is higher than the median starting salary for College 2 math majors?

(a) State the appropriate null and alternative hypotheses.

(b) Rank all the data, and calculate the value of the test statistic W.

(c) What is the smallest possible value of W? What is the largest possible value?

(d) If the null hypothesis is true, what are the mean and standard deviation of W?

(e) Use Table VII to calculate the p-value, and state your conclusion.

 Exercises

15.5 In the small city of Brownburgh, there are two competing real estate brokers. Broker G claims in its advertising that the homes it sells spend a shorter time on the market as compared to homes listed through Broker K. To investigate this claim, you select a random sample of real estate sales from each broker, measuring the length of time (in days) each home was on the market. You come up with the following data:

Broker G	14	28	29	34	53	54	62	71	113
Broker K	47	63	81	87	115	139			

The variable *time on the market* is believed to have a right-skewed distribution. You decide to use a ranking procedure to test the claim of Broker G.

(a) Clearly state the null and alternative hypotheses.

(b) Order the observations and assign ranks. What is the value of the observed test statistic? Show all work.

(c) Assuming that the null hypothesis is true, what would we expect the value of the test statistic to be?

(d) What is the standard deviation of the test statistic? Round your answer to two decimal places.

(e) How many standard deviations is the observed value from (b) above or below its expected value? (Fill in the blank and select the correct answer.)
_____standard deviations ***above*** ***below*** the expected value.

(f) Using the observed test statistic from (b), find the *p*-value for this test and give your decision at the 5% level.

(g) Interpret what the decision you made in (f) means. Write out your conclusion. Include the words *broker, claim*, and *time on the market*.

15.6 Continuation of Exercise 15.5. Now the variable *time on the market* is believed to have a normal distribution. You use a *t*-test to test the claim of Broker G.

(a) Clearly state the hypotheses you wish to test. Make sure to clearly identify all parameters.

(b) Some summary measures are provided: $\bar{x}_G = 50.89$, $s_G^2 = 881.11$, $\bar{x}_K = 88.67$, $s_K^2 = 1136.67$. What is the observed value of the test statistic?

(c) The distribution of the above test statistic relies on the assumption that the population distributions are normal. In addition to that assumption, what other assumption do we have to make about the two populations of real estate sales?

(d) Assuming the null hypothesis is true, what is the distribution of the test statistic?

(e) Find the *p*-value for this test, and give your decision at the 5% level and your conclusion.

15.7 A razor-blade manufacturer advertises that its twin-blade disposable razor "gets you lots more shaves" than any single-blade disposable razor on the market. A rival company that has been very successful selling their single-blade razors plans to test this claim. Independent random samples of six single-blade razors and five twin-blade razors are taken. The number of shaves that each razor gets before indicating a new blade is needed is recorded. The results are as follows:

Twin-Blade Razor		Single-Blade Razor	
8	15	10	13
17	11	6	7
9		5	4

Suppose that the distribution of the number of shaves per blade for both types of razor is highly right skewed.

(a) Sketch an example of a right-skewed distribution.

(b) State the appropriate hypotheses to test if the twin-blade razors get more shaves per blade than the single-blade razors.

(c) Calculate the appropriate test statistic and find its corresponding *p*-value.

(d) What is your decision at the 5% significance level?

(e) Suppose that we could assume that the distribution of the number of shaves per blade for both types of razor is normal (and not right skewed).

    (i) What additional assumption would we need to make in order to be able to perform a *t*-test?

    (ii) What would be the distribution of the *t*-test statistic if the mean number of shaves for the two populations of razors were equal?

15.8 Professor Anderson wants to compare selling time of residential homes for two cities. She takes independent random samples of 10 recently sold homes from each city and records the selling time (in days).

Part I: Assuming the distribution of selling time for each city is **normal** with equal population standard deviation,

    (a) what type of statistical test will Professor Anderson use?

    (b) what is the sampling distribution of the test statistic under the null hypothesis of no difference?

Part II: Assuming the distribution of selling time for each city is **skewed to the right**,

    (c) what type of statistical test will Professor Anderson use?

    (d) what is the expected value of the test statistic under the null hypothesis of no difference?

    (e) if the observed test statistic is 134, give the *p*-value for assessing whether there is any difference between the distributions (a two-sided test).

15.9 In a dog show with 16 contestants, 6 were trained at Training School A and 10 were trained at Training School B. The overall scores for the dogs are as follows:

School A	21	15	16	24	22	19				
School B	10	18	21	19	17	16	14	15	12	20

Is there a significant difference between the median scores for the two schools at the level 0.10?

15.4 Paired Samples—Signed Rank Test

The design we consider next involves paired data. If you recall from Section 11.3, in a paired design, analysis is on the differences between the measurements. When the differences are a random sample of normally distributed random variables, the **paired *t*-test** would be used. If, however, the differences had a density that wasn't suitably modeled with a normal density, a Wilcoxon **signed rank test** could be used.

    The assumptions for the (Wilcoxon) signed rank test are that differences can be considered a random sample and the distribution for the differences is continuous. We denote the median of this distribution by $\pi_{0.5}(D)$. We are interested in testing the null hypothesis

$$H_0: \pi_{0.5}(D) = 0$$

against one of the following alternative hypotheses:

$$H_1: \pi_{0.5}(D) \neq 0 \quad \text{or} \quad H_1: \pi_{0.5}(D) > 0 \quad \text{or} \quad H_1: \pi_{0.5}(D) < 0.$$

Since we have a random sample from a continuous distribution, the differences that we see should, in principle, all be different from zero. Due to rounding, however, we may see some that are equal to zero, and if we do, these differences are not used. Let n = the number of differences that are not equal to zero.

Since in a paired design the differences are analyzed, if you have not already done so, you first need to compute the values of the differences. Next, discard any zero differences; that is, keep the n nonzero differences. Then take the absolute value of the differences; in other words, drop the sign of the differences. The n absolute differences are ranked from smallest to largest. The smallest difference is assigned a rank equal to 1, the next smallest absolute difference has a rank of 2, and so on. If there are any ties, we assign the average rank. The *test statistic*, labeled W^+, is found by adding up the ranks of the positive differences.

Why is this a reasonable test statistic? The statistic W^+ will tend to be large when more of the differences are positive; that is, W^+ will tend to support $\pi_{0.5}(D) > 0$. Similarly, a small value of W^+ suggests more of the differences were negative, which would support $\pi_{0.5}(D) < 0$.

Our next task is to decide whether the observed result is surprising under the null hypothesis of no difference. This requires a frame of reference. What do we expect our statistic W^+ to equal when there is no difference between the two methods? What can we say about the standard deviation and the percentiles of the distribution of W^+ under the null hypothesis?

15.4.1 Frame of Reference

The **minimum value** for our test statistic occurs when none of the observed differences are positive (i.e., they are all negative). If they are all negative, there are no ranks to add up, so

$$\min W^+ = 0.$$

The **maximum value** occurs when all of the observed differences are positive. If they are all positive, we will add up the ranks from 1 to n, so

$$\max W^+ = \frac{n(n+1)}{2}.$$

The **expected value** of our test statistic under the null hypothesis is halfway between the minimum and the maximum:

$$E[W^+|\text{no difference}] = \mu_{W^+} = \frac{n(n+1)}{4}.$$

A similar calculation can be used to compute the **standard deviation** of the test statistic when there is no difference between the treatments. The result of that calculation yields

$$\text{standard deviation}[W^+|\text{no difference}] = \sigma_{W^+} = \sqrt{\frac{n(n+1)(2n+1)}{24}}.$$

To conduct the test, we will need to compute a p-value. If we were testing H_0: $\pi_{0.5}(D) = 0$ against the alternative hypothesis H_1: $\pi_{0.5}(D) > 0$, then larger values of W^+ would be considered more extreme and would support H_1. So the p-value would be the chance of getting a W^+ as large or larger than the one observed, under the null distribution. If we were testing H_0: $\pi_{0.5}(D) = 0$ against the alternative hypothesis H_1: $\pi_{0.5}(D) < 0$, then smaller values of W^+ would be considered more extreme and would support H_1. So the p-value would be the chance of getting a W^+ as small or smaller than the one observed, under the null distribution. If we were testing H_0: $\pi_{0.5}(D) = 0$ against the alternative hypothesis H_1: $\pi_{0.5}(D) \neq 0$, then values of W^+ that are either too big or too small would be considered more extreme and would support H_1. We would obtain the p-value by doubling a one-sided probability calculation. For samples with 10 or fewer nonzero differences, the value of the p-value can be found in Table VIII. A portion of the table is shown next:

W^+_{lower}	$n = 11$ Tail Probability	W^+_{upper}
7	**0.009**	59
8	0.012	58
9	0.016	57
10	0.021	56
⋮	⋮	⋮
18	0.103	48

The table has three columns. The first line in the $n = 11$ entry tells us that $P[W^+ \leq 7] = P[W^+ \geq 59] = 0.009$.

Let's put this information about the signed rank test to work in our next example.

❖ **EXAMPLE 15.4** **Weight Program** _____

The before-diet and after-diet weights (in pounds) of a sample of 12 people in a special diet program are given, along with their differences, computed as (before less after):

Before	155	168	172	180	174	142	135	196	200	149	165	139
After	152	170	166	180	160	143	132	180	189	147	168	138
diff = B − A	3	−2	6	0	14	−1	3	16	11	2	−3	1
absolute diff	3	2	6		14	1	3	16	11	2	3	1
Rank	6	3.5	8		10	1.5	6	11	9	3.5	6	1.5

We wish to assess whether the program was effective at the 5% significance level; that is, were the weights significantly reduced. In terms of the median of the differences, we wish to test

$$H_0: \pi_{0.5}(D) = 0 \quad \text{versus} \quad H_1: \pi_{0.5}(D) > 0.$$

There are 12 differences; however, one of the differences is equal to zero, so we effectively have a sample size of $n = 11$. The absolute value of the differences is computed in the fourth line of the table, with the zero difference having been discarded. These absolute differences have been ranked in the last row of the table. There are a number of tied observations, which required assigning average ranks. For example, the first two smallest observations were both equal to one they would have received the ranks of 1 and 2, so they both are assigned the average rank of 1.5. Similarly, for the next smallest value of two, both are assigned the average of the ranks of 3 and 4, or 3.5. Check out the rest of the ranks. The value of our test statistic will be

$$W^+ = \text{the sum of the ranks of the positive differences;}$$

$$W^+_{\text{OBS}} = 6 + 8 + 10 + 6 + 11 + 9 + 3.5 + 1.5 = 55.$$

If there was no difference in the weights before versus after, then the expected value and standard deviation of the test statistic can be computed from the previous formulas. The expected value is

$$\mu_{W^+} = \frac{n(n+1)}{4} = \frac{11(11+1)}{4} = 33.$$

The standard deviation can also be found through substitution to be

$$\sigma_{W^+} = \sqrt{\frac{n(n+1)(2n+1)}{24}} = \sqrt{\frac{11(11+1)(22+1)}{24}} = 11.25.$$

To decide whether our observed test statistic result of 55 is surprising under the hypothesis of no difference, we need to calculate the *p*-value. For samples with 10 or fewer nonzero differences, the value of this probability can be found in Table VIII. A portion of the table is shown for $n = 11$ nonzero differences:

W^+_{lower}	$n = 11$ Tail Probability	W^+_{upper}
7	0.009	59
8	0.012	58
9	0.016	57
10	0.021	56
11	**0.027**	**55**
12	0.034	54
13	0.042	53
14	0.051	52
15	0.062	51
16	0.074	50
17	0.087	49
18	0.103	48

For our example, we had $W^+_{\text{OBS}} = 55$ and a one-sided to the right alternative of $H_1: \pi_{0.5}(D) > 0$. Our *p*-value for this test would be *p*-value $= P[W^+ \geq 55 | H_0 \text{ is true}] = 0.027$. At a 5% significance level, we would conclude that the weight program does appear to be effective at reducing weight. ❖

A summary of the signed rank test for testing hypotheses about the population median difference for a paired design is given next.

Wilcoxon Signed Rank Test

Assumptions   The sample of differences is a random sample from a population of differences. The model for the differences is continuous.

Hypotheses $H_0: \pi_{0.5}(D) = 0$ versus
$H_1: \pi_{0.5}(D) \neq 0$ or $H_1: \pi_{0.5}(D) > 0$ or $H_1: \pi_{0.5}(D) < 0$.

Procedure

(1) Compute the differences.

(2) Discard any differences equal to zero, so n = the number of nonzero differences.

(3) Take the absolute value of the differences (write them without their sign).

(4) Rank the n absolute differences (if ties, assign the average rank).

Test Statistic W^+ = sum of the ranks of the positive differences

p-value   For a *one-sided to the right* test, the p-value = $P[W^+ \geq W^+_{OBS} | H_0$ is true].

For a *one-sided to the left* test, the p-value = $P[W^+ \leq W^+_{OBS} | H_0$ is true].

For a *two-sided* test, the p-value is obtained by doubling a one-sided probability calculation.

Notes:

- Although tied (absolute) differences can be assigned ranks by averaging, the number of ties should be small relative to the number of observations to ensure the validity of the test.

- If the population of differences had a normal distribution, we would have performed a paired t-test.

Let's do it! 15.3 Bonus Plan

The data that follow represent the number of computer units sold per week by a random sample of 8 salespersons before and after a bonus plan was introduced. The differences, defined as (after less before), are provided.

Before	54	25	80	76	63	82	94	72
After	53	30	79	79	70	90	92	81
Difference	−1	5	−1	3	7	8	−2	9

(a) A paired *t*-test was conducted (see computer output that follows) using a 5% level. Does it appear that the bonus plan significantly ***increased*** sales (on average)? State the hypotheses, the *p*-value, decision, and answer to the preceding question.

```
Paired Samples t-test on AFTER vs BEFORE with 8 cases.
     Mean difference =      3.500
     SD difference =      4.408
     DF =      7
     T =      2.246
     Prob =      .060
```

(b) Suppose that the assumption of normality for the differences were no longer plausible, so the signed rank procedure is to be used. Again the significance level is set to 0.05. Compute the signed rank test statistic, and state your decision and conclusion.

❋ Exercises

15.10 In studying the difference of difficulty between two tests, 15 students were selected at random and asked to take Tests A and B. Their scores were recorded as follows:

Student	1	2	3	4	5	6	7	8	9	10	11	12	13	14	15
Test A	87	92	66	72	75	96	85	82	69	70	91	80	52	84	72
Test B	96	91	80	85	91	95	89	94	78	68	97	92	75	92	95

(a) Name the nonparametric test one could use to test whether Tests A and B have different degrees of difficulty, and perform the test at a 5% level.

(b) Name the appropriate parametric test. Do not perform the test.

15.11 Shown here are data on the number of baggage-related complaints per 1000 passengers for 10 airlines during two consecutive months:

Airline	December Complaints	January Complaints
American	8.9	8.0
Delta	8.2	7.9
Continental	7.9	8.2
Eastern	7.5	7.8
Northwest	9.6	6.5
Pan American	5.0	5.1
Piedmont	12.3	11.0
TWA	11.2	10.9
United	7.7	7.4
USAir	8.6	7.9

Use a 5% significance level and the Wilcoxon signed rank test (show all steps) to determine whether the data indicate the number of baggage-related complaints for the airline industry has decreased over the two months studied. What is your conclusion?

15.12 A manufacturing firm wants to determine whether a difference between task completion times exists for two production methods. A sample of 15 workers was selected, and each worker completed a production task using both production methods. The production method that each worker used first was selected randomly. Thus, each worker in the sample provided a pair of observations, the task completion time under Method 1 and under Method 2. A signed rank test based on these 15 (nonzero) differences (Method 1 − Method 2) gave an observed test statistic of $W^+ = 94$. Is this sufficient evidence to suggest that there is a *difference* between task completion times for the two methods? Use a significance level of 0.05. State the null and alternative hypotheses and show all details.

15.13 As a test of a new insect repellent, the number of mosquito bites received during a period of time, with and without the repellent, for 12 people was recorded. Assume that the difference in number of bites (without repellent − with repellent) is **not** necessarily a normal distribution. The researcher decides to use a signed rank test to analyze the results.

(a) Formulate appropriate null and alternative hypotheses to assess whether the repellent is effective at reducing the number of bites received when used.

(b) The data resulted in an observed test statistic of 11. Carry out the test to see whether the repellent is effective at 5% level of significance. Show all details clearly.   (***Note:*** Assume that none of the differences were equal to zero.)

15.5   Kruskal–Wallis Test

In this chapter, we have studied the nonparametric sign and rank tests that are analogous to the one-sample t-test, the paired t-test, and the two-independent-samples t-test. There is a nonparametric procedure that can be used in the case of more than two populations under study. Here, we will mention this test by name only, explain the main idea, and look at computer output for performing the test.

With $I = 2$ or more independent random samples from continuous distributions (with the same shape, but possibly different locations), we could perform the **Kruskal–Wallis** rank procedure. This procedure, as in the rank sum test, begins by ranking all observations and adding up the ranks for each sample. The sum of the ranks for each sample is used to compute a test statistic that measures the variation among the average ranks shown for the *I* groups, much like the *F*-test statistic measures the variation among the sample means in ANOVA. Under the null hypothesis that the distributions for the populations are all the same, the Kruskal–Wallis test statistic has approximately a chi-squared distribution with *I* − 1 degrees of freedom. Our last example provides the scenario and computer output for a Kruskal–Wallis test.

❖ EXAMPLE 15.5 Weight Program

An experimenter was interested in studying the effect of sleep deprivation on the ability to detect moving objects on a radar screen. A total of 20 subjects was available for the study; five subjects were randomly assigned to each of the four treatment groups (Group 1 = 4 hours without sleep; Group 2 = 12 hours without sleep; Group 3 = 20 hours without sleep; and Group 4 = 28 hours without sleep). After being deprived of sleep, each subject was given a test; and the number of times they failed to spot a moving object during the set test period was recorded. The results are as follows:

	4 Hours	12 Hours	20 Hours	28 Hours
	36	38	46	76
	21	45	74	66
	20	46	67	62
	26	22	61	44
	21	40	59	61
Median:	21	40	61	62

Past data have indicated that the normality assumption for the distribution of the response is not valid. We turn to the nonparametric method, the Kruskal–Wallis test, for assessing whether the median number of failures is the same for the four sleep-deprivation treatments. A significance level of 0.01 is to be used. The SPSS output is given next:

Ranks

	GROUP	N	Mean Rank
FAILURES	1	5	3.4
	2	5	8.1
	3	5	15.2
	4	5	15.3
	Total	20	

Test Statistics[a,b]

	FAILURES
Chi-Square	14.50
df	3
Asymp. Sig.	.002

a. Kruskal–Wallis Test
b. Grouping Variable: GROUP

Looking at the sample medians, we do not see much of a difference between the 20- and 28-hour groups, but a larger difference is evident between the 4- and 12-hour groups and both the 12- and 20- and 28-hour groups. Indeed, the 20- and 28-hour groups had the largest average ranks. The Kruskal–Wallis test statistic value was 14.50. Using the approximate chi-square distribution with three degrees of freedom, we find that the *p*-value is 0.002. The data are statistically significant at the 1% level. There is strong evidence that the distributions of the number of failures for the four sleep-deprivation treatments are not all equal. ❖

Chapter Summary

In many of our previous tests having to do with population means (one-sample t, paired t, two-independent-samples t, ANOVA, and regression), we assumed that the continuous data came from normal population(s). What if the normality assumption does not seem to be plausible? The tests are somewhat robust to departures from normality—that is, they still work reasonably well; but if the departures are strong, we need to look for alternative methods.

In this chapter, we examined some alternatives. Their general name, **nonparametric**, implies that they do not require a particular parametric form for the distribution. The assumptions are stated in terms of some density function(s) and are not necessarily normal. The hypotheses for these nonparametric tests are generally stated in terms of the median of the population(s) instead of the mean.

The one-sample test is called the **sign test**. It, along with the **paired sign test** for paired observations, is based on a statistic that follows a binomial distribution. These two tests are actually a type of a more general binomial test.

The other tests presented here were based on the rankings of the data, rather than the actual values and are called **rank tests**. The nonparametric rank test for paired data is called the **signed rank test**. The signed rank test is more powerful than the paired sign test. The test for two independent samples is the **rank sum test**. For more than two populations, we have the **Kruskal–Wallis test** for comparing the distributions.

In all of these tests, we have concentrated on the case of small sample sizes, providing tables for the rank tests to determine p-values. If the sample sizes are large, we could go back to the parametric procedures, as they work reasonably well, even if there is some skewness.

Key Terms

Be sure you can describe, in your own words, and give an example of each of the following key terms from this chapter:

nonparametric	median	Kruskal–Wallis test
sign test	paired sign test	
signed rank test	rank sum test	

Exercises

15.14 A field experiment was conducted to assess the effect of two methods for reducing car speed on a particular section of road. Method 1 involved using parked police cars (with no actual police officer in the car) and Method 2 involved active patrolling by a police car. For each method, the speeds of eight randomly selected cars driving through the section were recorded and are given here:

Method 1	69	73	80	71	82	79	74	78
Method 2	72	68	74	70	76	71	67	75

Assume that the speeds of cars under each method are values of a random sample that can be modeled with a density, but the assumption of normality is *not* appropriate.

(a) State the appropriate hypotheses for a test aimed at determining whether the distribution of speeds under Method 1 are significantly higher than the speeds under Method 2.

(b) What is the expected value of the test statistic when the null hypothesis is true?

(c) Calculate the observed test statistic, give the *p*-value for your test in (a) and state your decision, using a 10% significance level. Show all details clearly.

(d) Suppose that you could assume the two independent random samples each come from a normal population with common population standard deviation. What test statistic would you use to assess whether the speeds under Method 1 are significantly higher than the speeds under Method 2? (Just give the formula for the test statistic; do not compute the test-statistic value.)

15.15 The data that follow are a subset of data from Ijzermans (1970). He examined a random sample of stainless steel and recorded the percentages of chromium as follows (SOURCE: *Corrosion Science,* vol. 10, 1970):

Steel	1	2	3	4	5	6	7	8
% Cr	17.4	17.9	17.6	18.1	17.6	18.9	16.9	17.5

Test the hypothesis that the median of chromium present is 18% versus the alternative hypothesis that the median is not 18%. Use a 5% level of significance.

15.16 Jorge would like to know whether or not he should market his new improved hot salsa. Eighteen tasters were asked to taste both Jorge's original hot salsa and the new improved Jorge's hot salsa. Of the 18 tasters, 13 preferred the new improved hot salsa. Two tasters could not tell the difference, and three tasters preferred the old salsa. Is this sufficient evidence to suggest that a majority of the tasters preferred the new improved hot salsa? Use a significance level of 5% and show all details.

Hint: Consider the differences (new − original) in the pairs of ratings by each observer. A positive difference means that the observer preferred the new salsa, a negative difference means that the observer preferred the original salsa, and a difference of zero means that the observer could not tell the difference. Test the hypotheses $H_0: \pi_{0.5}(D) = 0$ against $H_1: \pi_{0.5}(D) > 0$, using a sign test and zero as the hypothesized median value. Here, R is the number of differences that exceed zero—that is, the number of raters who preferred the new topping. What is the value of n? (Be careful!)

15.17 A two-independent-samples *t*-test was used to decide whether two methods of teaching reading yielded different average scores. The sample means differed by three, and the standard error of the estimate of the difference in means was 1.5.

(a) What is the value of the *t* statistic for testing if the two methods yield the same mean score?

(b) If there were 10 students in each of the two groups, what can you say about the *p*-value of the test?

(c) Suppose that the assumption of normality of the test scores was suspect and a nonparametric test was used instead of the two-independent-samples *t*-test. Name the nonparametric test you would use.

(d) Continuation of (c). If the hypothesis that the two teaching methods have the same median is true, what is the expected value of the nonparametric test statistic?

15.18 A researcher randomly divided 18 children into two groups of 9 each. One group was given training in creative problem solving and the other was not. All children were given a series of problems and asked to generate possible solutions. The number of solutions generated by each subject is shown in the following table:

Group 1 Training	12	16	19	8	10	13	9	15	14
Group 2 No Training	15	10	11	6	5	5	9	8	11

We are interested in determining whether these data provide evidence in favor of the hypothesis that training leads to *more* solutions. Do *not* assume that the observations are drawn from normal populations.

(a) What nonparametric test is appropriate here?

(b) What are the appropriate null and alternative hypotheses?

(c) What is the observed value of the test statistic?

(d) What is the p-value of the test?

(e) Using a test with significance level of 0.05, what is your conclusion?

(f) If you were told that the observations were obtained from random samples from normal populations with equal population standard deviation, what test would you perform?

15.19 Air France offers two different package tours from the United States to France. One, which includes airfare and five nights in a luxurious hotel resort on the beach, is advertised at $1345. The other package, for an extended weekend to Paris, which includes airfare and three nights in a nice hotel downtown Paris, is advertised at $1287. Both package deals are offered throughout the year. The number of bookings for each package is recorded every day. A random sample of bookings from seven days resulted in the bookings per day shown. Do you believe that one package is more popular than the other?

Day	1	2	3	4	5	6	7
Paris Weekend	20	28	24	24	30	20	28
Beach Package	23	20	17	19	30	16	30

Assume that no normal assumption can be made and that the observations form pairs of booking numbers for each day. Air France wishes to test the following hypotheses:

H_0: The median of the difference in the number of bookings per day is zero.

H_1: The Paris weekend is more popular, resulting in a median of the difference greater than zero.

(a) A rank procedure is to be used. Which one?

(b) Calculate the appropriate test statistic and its corresponding p-value. Show all work.

(c) What is your decision at the 10% significance level?

(d) What does your decision in (c) suggest about the popularity of the two package deals to France?

15.20 A restaurant manager wishes to compare mean daily sales between two of his restaurants located several miles apart in the same city. For a one-week period, the following sales data are recorded:

Day	Sunday	Monday	Tuesday	Wednesday	Thursday	Friday	Saturday
Restaurant 1	1916	991	1112	1212	1461	2017	2406
Restaurant 2	1771	904	996	1258	1385	1867	2192
Difference	145	87	116	−46	76	150	214

SPSS was used to provide the following output:

Paired Samples Test

	Paired Differences					
	Mean	Std. Deviation	Std. Error Mean	t	df	Sig. (2-tailed)
Pair 1 REST1- REST2	106.00	81.23	30.70	3.45	6	.014

(a) Explain why these data can be considered matched pairs.

(b) Clearly state the hypotheses to test whether the mean daily sales for Restaurant 1 is higher. Make sure to identify all parameters.

(c) Finish the following sentence to interpret the standard error as an approximate average distance: The standard error of 30.70 from the preceding output means that if we were to take repeated samples (of the same size, from the same population), we would estimate _____.

(d) What assumption would we have to make about the distribution of the difference in sales between the two restaurants in order for the preceding *t*-test to be valid?

(e) Suppose that we cannot make this assumption from (d), but we know that the distribution is quite skewed. What test could we use to establish whether there is equality in mean daily sales between the two restaurants?

(f) Calculate the appropriate test statistic for testing H_0: $\pi_{0.5}(D) = 0$ versus H_1: $\pi_{0.5}(D) > 0$.

(g) Find the corresponding *p*-value, and give your decision and conclusion at the 5% level.

Table VII *Tail Probabilities for the Rank Sum Statistic under H_0*

Tail Probability $= P[W \leq W_{lower}] = P[W \geq W_{upper}]$

Smaller Sample Size = 2 Larger Sample Size = 3		
W_{lower}	Tail Probability	W_{upper}
2	0.000	10
3	0.100	9
4	0.200	8

Smaller Sample Size = 2 Larger Sample Size = 4		
W_{lower}	Tail Probability	W_{upper}
2	0.000	12
3	0.067	11
4	0.133	10

Smaller Sample Size = 2 Larger Sample Size = 5		
W_{lower}	Tail Probability	W_{upper}
2	0.000	14
3	0.048	13
4	0.095	12
5	0.190	11

Smaller Sample Size = 2 Larger Sample Size = 6		
W_{lower}	Tail Probability	W_{upper}
2	0.000	16
3	0.036	15
4	0.071	14
5	0.143	13

Smaller Sample Size = 2 Larger Sample Size = 7		
W_{lower}	Tail Probability	W_{upper}
2	0.000	18
3	0.028	17
4	0.056	16
5	0.111	15

Smaller Sample Size = 2 Larger Sample Size = 8		
W_{lower}	Tail Probability	W_{upper}
2	0.000	20
3	0.022	19
4	0.044	18
5	0.089	17
6	0.133	16

Smaller Sample Size = 2 Larger Sample Size = 9		
W_{lower}	Tail Probability	W_{upper}
2	0.000	22
3	0.018	21
4	0.036	20
5	0.073	19
6	0.109	18

Smaller Sample Size = 2 Larger Sample Size = 10		
W_{lower}	Tail Probability	W_{upper}
3	0.015	23
4	0.030	22
5	0.061	21
6	0.091	20
7	0.136	19

Table VII   *Tail Probabilities for the Rank Sum Statistic under H_0* *(continued)*

Tail Probability $= P[W \le W_{lower}] = P[W \ge W_{upper}]$

Smaller Sample Size = 3 Larger Sample Size = 3				Smaller Sample Size = 3 Larger Sample Size = 4				Smaller Sample Size = 3 Larger Sample Size = 5		
W_{lower}	Tail Probability	W_{upper}		W_{lower}	Tail Probability	W_{upper}		W_{lower}	Tail Probability	W_{upper}
5	0.000	16		5	0.000	19		5	0.000	22
6	0.050	15		6	0.029	18		6	0.018	21
7	0.100	14		7	0.057	17		7	0.036	20
8	0.200	13		8	0.114	16		8	0.071	19
								9	0.125	18

Smaller Sample Size = 3 Larger Sample Size = 6				Smaller Sample Size = 3 Larger Sample Size = 7				Smaller Sample Size = 3 Larger Sample Size = 8		
W_{lower}	Tail Probability	W_{upper}		W_{lower}	Tail Probability	W_{upper}		W_{lower}	Tail Probability	W_{upper}
5	0.000	25		5	0.000	28		5	0.000	31
6	0.012	24		6	0.008	27		6	0.006	30
7	0.024	23		7	0.017	26		7	0.012	29
8	0.048	22		8	0.033	25		8	0.024	28
9	0.083	21		9	0.058	24		9	0.042	27
10	0.131	20		10	0.092	23		10	0.067	26
				11	0.133	22		11	0.097	25
								12	0.139	24

Smaller Sample Size = 3 Larger Sample Size = 9				Smaller Sample Size = 3 Larger Sample Size = 10		
W_{lower}	Tail Probability	W_{upper}		W_{lower}	Tail Probability	W_{upper}
7	0.009	32		7	0.007	35
8	0.018	31		8	0.014	34
9	0.032	30		9	0.024	33
10	0.050	29		10	0.038	32
11	0.073	28		11	0.056	31
12	0.105	27		12	0.080	30
				13	0.108	29

Table VII   *Tail Probabilities for the Rank Sum Statistic under H_0 (continued)*

Tail Probability $= P[W \leq W_{lower}] = P[W \geq W_{upper}]$

Smaller Sample Size = 4 Larger Sample Size = 4			Smaller Sample Size = 4 Larger Sample Size = 5			Smaller Sample Size = 4 Larger Sample Size = 6		
W_{lower}	Tail Probability	W_{upper}	W_{lower}	Tail Probability	W_{upper}	W_{lower}	Tail Probability	W_{upper}
9	0.000	27	9	0.000	31	11	0.010	33
10	0.014	26	10	0.008	30	12	0.019	32
11	0.029	25	11	0.016	29	13	0.033	31
12	0.057	24	12	0.032	28	14	0.057	30
13	0.100	23	13	0.056	27	15	0.086	29
14	0.171	22	14	0.095	26	16	0.129	28
			15	0.143	25			

Smaller Sample Size = 4 Larger Sample Size = 7			Smaller Sample Size = 4 Larger Sample Size = 8			Smaller Sample Size = 4 Larger Sample Size = 9		
W_{lower}	Tail Probability	W_{upper}	W_{lower}	Tail Probability	W_{upper}	W_{lower}	Tail Probability	W_{upper}
11	0.006	37	12	0.008	40	13	0.010	43
12	0.012	36	13	0.014	39	14	0.017	42
13	0.021	35	14	0.024	38	15	0.025	41
14	0.036	34	15	0.036	37	16	0.038	40
15	0.055	33	16	0.055	36	17	0.053	39
16	0.082	32	17	0.077	35	18	0.074	38
17	0.115	31	18	0.107	34	19	0.099	37
						20	0.130	36

Smaller Sample Size = 4 Larger Sample Size = 10		
W_{lower}	Tail Probability	W_{upper}
13	0.007	47
14	0.012	46
15	0.018	45
16	0.027	44
17	0.038	43
18	0.053	42
19	0.071	41
20	0.094	40
21	0.120	39

Table VII   *Tail Probabilities for the Rank Sum Statistic under H_0 (continued)*

Tail Probability $= P[W \leq W_{lower}] = P[W \geq W_{upper}]$

Smaller Sample Size = 5 Larger Sample Size = 5			Smaller Sample Size = 5 Larger Sample Size = 6			Smaller Sample Size = 5 Larger Sample Size = 7		
W_{lower}	Tail Probability	W_{upper}	W_{lower}	Tail Probability	W_{upper}	W_{upper}	Tail Probability	W_{upper}
16	0.008	39	17	0.009	43	18	0.009	47
17	0.016	38	18	0.015	42	19	0.015	46
18	0.028	37	19	0.026	41	20	0.024	45
19	0.048	36	20	0.041	40	21	0.037	44
20	0.075	35	21	0.063	39	22	0.053	43
21	0.111	34	22	0.089	38	23	0.074	42
			23	0.123	37	24	0.101	41

Smaller Sample Size = 5 Larger Sample Size = 8			Smaller Sample Size = 5 Larger Sample Size = 9			Smaller Sample Size = 5 Larger Sample Size = 10		
W_{lower}	Tail Probability	W_{upper}	W_{lower}	Tail Probability	W_{upper}	W_{lower}	Tail Probability	W_{upper}
19	0.009	51	20	0.009	55	21	0.010	59
20	0.015	50	21	0.014	54	22	0.014	58
21	0.023	49	22	0.021	53	23	0.020	57
22	0.033	48	23	0.030	52	24	0.028	56
23	0.047	47	24	0.041	51	25	0.038	55
24	0.064	46	25	0.056	50	26	0.050	54
25	0.085	45	26	0.073	49	27	0.065	53
26	0.111	44	27	0.095	48	28	0.082	52
			28	0.120	47	29	0.103	51

Table VII   *Tail Probabilities for the Rank Sum Statistic under H_0* *(continued)*

Tail Probability $= P[W \leq W_{lower}] = P[W \geq W_{upper}]$

Smaller Sample Size = 6 Larger Sample Size = 6			Smaller Sample Size = 6 Larger Sample Size = 7			Smaller Sample Size = 6 Larger Sample Size = 8		
W_{lower}	Tail Probability	W_{upper}	W_{lower}	Tail Probability	W_{upper}	W_{lower}	Tail Probability	W_{upper}
24	0.008	54	25	0.007	59	27	0.010	63
25	0.013	53	26	0.011	58	28	0.015	62
26	0.021	52	27	0.017	57	29	0.021	61
27	0.032	51	28	0.026	56	30	0.030	60
28	0.047	50	29	0.037	55	31	0.041	59
29	0.066	49	30	0.051	54	32	0.054	58
30	0.090	48	31	0.069	53	33	0.071	57
31	0.120	47	32	0.090	52	34	0.091	56
			33	0.117	51	35	0.114	55

Smaller Sample Size = 6 Larger Sample Size = 9			Smaller Sample Size = 6 Larger Sample Size = 10		
W_{lower}	Tail Probability	W_{upper}	W_{lower}	Tail Probability	W_{upper}
28	0.009	68	29	0.008	73
29	0.013	67	30	0.011	72
30	0.018	66	31	0.016	71
31	0.025	65	32	0.021	70
32	0.033	34	33	0.028	69
33	0.044	63	34	0.036	68
34	0.057	62	35	0.047	67
35	0.072	61	36	0.059	66
36	0.091	60	37	0.074	65
37	0.112	59	38	0.090	64
			39	0.110	63

Table VII *Tail Probabilities for the Rank Sum Statistic under H_0 (continued)*

Tail Probability $= P[W \leq W_{lower}] = P[W \geq W_{upper}]$

Smaller Sample Size = 7 Larger Sample Size = 7			Smaller Sample Size = 7 Larger Sample Size = 8			Smaller Sample Size = 7 Larger Sample Size = 9		
W_{lower}	Tail Probability	W_{upper}	W_{lower}	Tail Probability	W_{upper}	W_{lower}	Tail Probability	W_{upper}
34	0.009	71	36	0.010	76	37	0.008	82
35	0.013	70	37	0.014	75	38	0.011	81
36	0.019	69	38	0.020	74	39	0.016	80
37	0.027	68	39	0.027	73	40	0.021	79
38	0.036	67	40	0.036	72	41	0.027	78
39	0.049	66	41	0.047	71	42	0.036	77
40	0.064	65	42	0.060	70	43	0.045	76
41	0.082	64	43	0.076	69	44	0.057	75
42	0.104	63	44	0.095	68	45	0.071	74
			45	0.116	67	46	0.087	73
						47	0.105	72

Smaller Sample Size = 7 Larger Sample Size = 10		
W_{lower}	Tail Probability	W_{upper}
39	0.009	87
40	0.012	86
41	0.017	85
42	0.022	84
43	0.028	83
44	0.035	82
45	0.044	81
46	0.054	80
47	0.067	79
48	0.081	78
49	0.097	77
50	0.115	76

Table VII   *Tail Probabilities for the Rank Sum Statistic under H_0* (continued)

Tail Probability $= P[W \leq W_{lower}] = P[W \geq W_{upper}]$

Smaller Sample Size = 8 Larger Sample Size = 8				Smaller Sample Size = 8 Larger Sample Size = 9				Smaller Sample Size = 8 Larger Sample Size = 10		
W_{lower}	Tail Probability	W_{upper}		W_{lower}	Tail Probability	W_{upper}		W_{lower}	Tail Probability	W_{upper}
46	0.010	90		48	0.010	96		50	0.010	102
47	0.014	89		49	0.014	95		51	0.013	101
48	0.019	88		50	0.018	94		52	0.017	100
49	0.025	87		51	0.023	93		53	0.022	99
50	0.032	86		52	0.030	92		54	0.027	98
51	0.041	85		53	0.037	91		55	0.034	97
52	0.052	84		54	0.046	90		56	0.042	96
53	0.065	83		55	0.057	89		57	0.051	95
54	0.080	82		56	0.069	88		58	0.061	94
55	0.097	81		57	0.084	87		59	0.073	93
56	0.117	80		58	0.100	86		60	0.086	92
								61	0.102	91

Smaller Sample Size = 9 Larger Sample Size = 9				Smaller Sample Size = 9 Larger Sample Size = 10				Smaller Sample Size = 10 Larger Sample Size = 10		
W_{lower}	Tail Probability	W_{upper}		W_{lower}	Tail Probability	W_{upper}		W_{lower}	Tail Probability	W_{upper}
59	0.009	112		61	0.009	119		74	0.009	136
60	0.012	111		62	0.011	118		75	0.012	135
61	0.016	110		63	0.014	117		76	0.014	134
62	0.020	109		64	0.017	116		77	0.018	133
63	0.025	108		65	0.022	115		78	0.022	132
64	0.031	107		66	0.027	114		79	0.026	131
65	0.039	106		67	0.033	113		80	0.032	130
66	0.047	105		68	0.039	112		81	0.038	129
67	0.057	104		69	0.047	111		82	0.045	128
68	0.068	103		70	0.056	110		83	0.053	127
69	0.081	102		71	0.067	109		84	0.062	126
70	0.095	101		72	0.078	108		85	0.072	125
71	0.111	100		73	0.091	107		86	0.083	124
				74	0.106	106		87	0.095	123
								88	0.109	122

Source: Adapted from C. Kraft and C. van Eeden, *A Nonparametric Introduction to Statistics*, Macmillan, New York, 1968.

Table VIII   *Tail Probabilities for the Rank Sum Statistic under H_0*

Tail Probability $= P[W^+ \leq W^+_{lower}] = P[W^+ \geq W^+_{upper}]$

W^+_{lower}	$n = 3$ Tail Probability	W^+_{upper}
	0.000	7
0	0.125	6
1	0.250	5

W^+_{lower}	$n = 4$ Tail Probability	W^+_{upper}
	0.000	11
0	0.062	10
1	0.125	9
2	0.188	8

W^+_{lower}	$n = 5$ Tail Probability	W^+_{upper}
	0.000	16
0	0.031	15
1	0.062	14
2	0.094	13
3	0.156	12

W^+_{lower}	$n = 6$ Tail Probability	W^+_{upper}
	0.000	22
0	0.016	21
1	0.031	20
2	0.047	19
3	0.078	18
4	0.109	17

W^+_{lower}	$n = 7$ Tail Probability	W^+_{upper}
0	0.008	28
1	0.016	27
2	0.023	26
3	0.039	25
4	0.055	24
5	0.078	23
6	0.109	22

W^+_{lower}	$n = 8$ Tail Probability	W^+_{upper}
1	0.008	35
2	0.012	34
3	0.020	33
4	0.027	32
5	0.039	31
6	0.055	30
7	0.074	29
8	0.098	28
9	0.125	27

W^+_{lower}	$n = 9$ Tail Probability	W^+_{upper}
3	0.010	42
4	0.014	41
5	0.020	40
6	0.027	39
7	0.037	38
8	0.049	37
9	0.064	36
10	0.082	35
11	0.102	34

W^+_{lower}	$n = 10$ Tail Probability	W^+_{upper}
5	0.010	50
6	0.014	49
7	0.019	48
8	0.024	47
9	0.032	46
10	0.042	45
11	0.053	44
12	0.065	43
13	0.080	42
14	0.097	41
15	0.116	40

W^+_{lower}	$n = 11$ Tail Probability	W^+_{upper}
7	0.009	59
8	0.012	58
9	0.016	57
10	0.021	56
11	0.027	55
12	0.034	54
13	0.042	53
14	0.051	52
15	0.062	51
16	0.074	50
17	0.087	49
18	0.103	48

Table VIII   *Tail Probabilities for the Rank Sum Statistic under H_0 (continued)*

$$\text{Tail Probability } = P[W^+ \leq W^+_{lower}] = P[W^+ \geq W^+_{upper}]$$

W^+_{lower}	$n = 12$ Tail Probability	W^+_{upper}
10	0.010	68
11	0.013	67
12	0.017	66
13	0.021	65
14	0.026	64
15	0.032	63
16	0.039	62
17	0.046	61
18	0.055	60
19	0.065	59
20	0.076	58
21	0.088	57
22	0.102	56

W^+_{lower}	$n = 13$ Tail Probability	W^+_{upper}
12	0.009	79
13	0.011	78
14	0.013	77
15	0.016	76
16	0.020	75
17	0.024	74
18	0.029	73
19	0.034	72
20	0.040	71
21	0.047	70
22	0.055	69
23	0.064	68
24	0.073	67
25	0.084	66
26	0.095	65
27	0.108	64

W^+_{lower}	$n = 14$ Tail Probability	W^+_{upper}
16	0.010	89
17	0.012	88
18	0.015	87
19	0.018	86
20	0.021	85
21	0.025	84
22	0.029	83
23	0.034	82
24	0.039	81
25	0.045	80
26	0.052	79
27	0.059	78
28	0.068	77
29	0.077	76
30	0.086	75
31	0.097	74
32	0.108	73

W^+_{lower}	$n = 15$ Tail Probability	W^+_{upper}
19	0.009	101
20	0.011	100
21	0.013	99
22	0.015	98
23	0.018	97
24	0.021	96
25	0.024	95
26	0.028	94
27	0.032	93
28	0.036	92
29	0.042	91
30	0.047	90
31	0.053	89
32	0.060	88
33	0.068	87
34	0.076	86
35	0.084	85
36	0.094	84
37	0.104	83

Source: Adapted from C. Kraft and C. van Eeden, *A Nonparametric Introduction to Statistics*, Macmillan, New York, 1968.

APPENDIX

Table I: *Random Numbers*

Column Row	1–5	6–10	11–15	16–20	21–25	26–30	31–35	36–40	41–45	46–50	51–55	56–60	61–65	66–70
1	10480	15011	01536	02011	81647	91646	69179	14194	62590	36207	20969	99570	91291	90700
2	22368	46573	25595	85393	30995	89198	27982	53402	93965	34095	52666	19174	39615	99505
3	24130	48360	22527	97265	76393	64809	15179	24830	49340	32081	30680	19655	63348	58629
4	42167	93093	06243	61680	07856	16376	39440	53537	71341	57004	00849	74917	97758	16379
5	37570	39975	81837	16656	06121	91782	60468	81305	49684	60672	14110	06927	01263	54613
6	77921	06907	11008	42751	27756	53498	18602	70659	90655	15053	21916	81825	44394	42880
7	99562	72905	56420	69994	98872	31016	71194	18738	44013	48840	63213	21069	10634	12952
8	96301	91977	05463	07972	18876	20922	94595	56869	69014	60045	18425	84903	42508	32307
9	89579	14342	63661	10281	17453	18103	57740	84378	25331	12566	58678	44947	05585	56941
10	85475	36857	53342	53988	53060	59533	38867	62300	08158	17983	16439	11458	18593	64952
11	28918	69578	88231	33276	70997	79936	56865	05859	90106	31595	01547	85590	91610	78188
12	63553	40961	48235	03427	49626	69445	18663	72695	52180	20847	12234	90511	33703	90322
13	09429	93969	52636	92737	88974	33488	36320	17617	30015	08272	84115	27156	30613	74952
14	10365	61129	87529	85689	48237	52267	67689	93394	01511	26358	85104	20285	29975	89868
15	07119	97336	71048	08178	77233	13916	47564	81056	97735	85977	29372	74461	28551	90707
16	51085	12765	51821	51259	77452	16308	60756	92144	49442	53900	70960	63990	75601	40719
17	02368	21382	52404	60268	89368	19885	55322	44819	01188	65255	64835	44919	05944	55157
18	01011	54092	33362	94904	31273	04146	18594	29852	71585	85030	51132	01915	92747	64951
19	52162	53916	46369	58586	23216	14513	83149	98736	23495	64350	94738	17752	35156	35749
20	07056	97628	33787	09998	42698	06691	76988	13602	51851	46104	88916	19509	25625	58104
21	48663	91245	85828	14346	09172	30168	90229	04734	59193	22178	30421	61666	99904	32812
22	54164	58492	22421	74103	47070	25306	76468	26384	58151	06646	21524	15227	96909	44592
23	32639	32363	05597	24200	13363	38005	94342	28728	35806	06912	17012	64161	18296	22851
24	29334	27001	87637	87308	58731	00256	45834	15398	46557	41135	10367	07684	36188	18510
25	02488	33062	28834	07351	19731	92420	60952	61280	50001	67658	32586	86679	50720	94953
26	81525	72295	04839	96423	24878	82651	66566	14778	76797	14780	13300	87074	79666	95725
27	29676	20591	68086	26432	46901	20849	89768	81536	86645	12659	92259	57102	80428	25280
28	00742	57392	39064	66432	84673	40027	32832	61362	98947	96067	64760	64584	96096	98253
29	05366	04213	25669	26422	44407	44048	37937	63904	45766	66134	75470	66520	34693	90449
30	91921	26418	64117	94305	26766	25940	39972	22209	71500	64568	91402	42416	07844	69618
31	00582	04711	87917	77341	42206	35126	74087	99547	81817	42607	43808	76655	62028	76630
32	00725	69884	62797	56170	86324	88072	76222	36086	84637	93161	76038	65855	77919	88006
33	69011	65795	95876	55293	18988	27354	26575	08625	40801	59920	29841	80150	12777	48501
34	25976	57948	29888	88604	67917	48708	18912	82271	65424	69774	33611	54262	85963	03547
35	09763	83473	73577	12908	30883	18317	28290	35797	05998	41688	34952	37888	38917	88050

Table I: *Random Numbers (continued)*

Column Row	1-5	6-10	11-15	16-20	21-25	26-30	31-35	36-40	41-45	46-50	51-55	56-60	61-65	66-70
36	91576	42595	27958	30134	04024	86385	29880	99730	55536	84855	29080	09250	79656	73211
37	17955	56349	90999	49127	20044	59931	06115	20542	18059	02008	73708	83517	36103	42791
38	46503	18584	18845	49618	02304	51038	20655	58727	28168	15475	56942	53389	20562	87338
39	92157	89634	94824	78171	84610	82834	09922	25417	44137	48413	25555	21246	35509	20468
40	14577	62765	35605	81263	39667	47358	56873	56307	61607	49518	89656	20103	77490	18062
41	98427	07523	33362	64270	01638	92477	66969	98420	04880	45585	46565	04102	46880	45709
42	34914	63976	88720	82765	34476	17032	87589	40836	32427	70002	70663	88863	77775	69348
43	70060	28277	39475	46473	23219	53416	94970	25832	69975	94884	19661	72828	00102	66794
44	53976	54914	06990	67245	68350	82948	11398	42878	80287	88267	47363	46634	06541	97809
45	76072	29515	40980	07391	58745	25774	22987	80059	39911	96189	41151	14222	60697	59583
46	90725	52210	83974	29992	65831	38857	50490	83765	55657	14361	31720	57375	56228	41546
47	64364	67412	33339	31926	14883	24413	59744	92351	97473	89286	35931	04110	23726	51900
48	08962	00358	31662	25388	61642	34072	81249	35648	56891	69352	48373	45578	78547	81788
49	95012	68379	93526	70765	10592	04542	76463	54328	02349	17247	28865	14777	62730	92277
50	15664	10493	20492	38391	91132	21999	59516	81652	27195	48223	46751	22923	32261	85653
51	16408	81899	04153	53381	79401	21438	83035	92350	36693	31238	59649	91754	72772	02338
52	18629	81953	05520	91962	04739	13092	97662	24822	94730	06496	35090	04822	86774	98289
53	73115	35101	47498	87637	99016	71060	88824	71013	18735	20286	23153	72924	35165	43040
54	57491	16703	23167	49323	45021	33132	12544	41035	80780	45393	44812	12512	98931	91202
55	30405	83946	23792	14422	15059	45799	22716	19792	09983	74353	68668	30429	70735	25499
56	16631	35006	85900	98275	32388	52390	16815	69290	82732	38480	73817	32523	41961	44437
57	96773	20206	42559	78985	05300	22164	24369	54224	35083	19687	11052	91491	60383	19746
58	38935	64202	14349	82674	66523	44133	00697	35552	35970	19124	63318	29686	03387	59846
59	31624	76384	17403	53363	44167	64486	64758	75366	76554	31601	12614	33072	60332	92325
60	78919	19474	23632	27889	47914	02584	37680	20801	72152	39339	34806	08930	85001	87820
61	03931	33309	57047	74211	63445	17361	62825	39908	05607	91284	68833	25570	38818	46920
62	74426	33278	43972	10110	89917	15665	52872	73823	73144	88662	88970	74492	51805	99378
63	09066	00903	20795	95452	92648	45454	09552	88815	16553	51125	79375	97596	16296	66092
64	42238	12426	87025	14267	20979	04508	64535	31355	86064	29472	47689	05974	52468	16834
65	16153	08002	26504	41744	81959	65642	74240	56302	00033	67107	77510	70625	28725	34191
66	21457	40742	29820	96783	29400	21840	15035	34537	33310	06116	95240	15957	16572	06004
67	21581	57802	02050	89728	17937	37621	47075	42080	97403	48626	68995	43805	33386	21597
68	55612	78095	83197	33732	05810	24813	86902	60397	16489	03264	88525	42786	05269	92532
69	44657	66999	99324	51281	84463	60563	79312	93454	68876	25471	93911	25650	12682	73572
70	91340	84979	46949	81973	37949	61023	43997	15263	80644	43942	89203	71795	99533	50501

Table I: *Random Numbers (continued)*

Column Row	1–5	6–10	11–15	16–20	21–25	26–30	31–35	36–40	41–45	46–50	51–55	56–60	61–65	66–70
71	91227	21199	31935	27022	84067	05462	35216	14486	29891	68607	41867	14951	91696	85065
72	50001	38140	66321	19924	72163	09538	12151	06878	91903	18749	34405	56087	82790	70925
73	65390	05224	72958	28609	81406	39147	25549	48542	42627	45233	57202	94617	23772	07896
74	27504	96131	83944	41575	10573	08619	64482	73923	36152	05184	94142	25299	84387	34925
75	37169	94851	39117	89632	00959	16487	65536	49071	39782	17095	02330	74301	00275	48280
76	11508	70225	51111	38351	19444	66499	71945	05422	13442	78675	84081	66938	93654	59894
77	37449	30362	06694	54690	04052	53115	62757	95348	78662	11163	81651	50245	34971	52924
78	46515	70331	85922	38329	57015	15765	97161	17869	45349	61796	66345	81073	49106	79860
79	30986	81223	42416	58353	21532	30502	32305	86482	05174	07901	54339	58861	74818	46942
80	63798	64995	46583	09785	44160	78128	83991	42865	92520	83531	80377	35909	81250	54238
81	82486	84846	99254	67632	43218	50076	21361	64816	51202	88124	41870	52689	51275	83556
82	21885	32906	92431	09060	64297	51674	64126	62570	26123	05155	59194	52799	28225	85762
83	60336	98782	07408	53458	13564	59089	26445	29789	85205	41001	12535	12133	14645	23541
84	43937	46891	24010	25560	86355	33941	25786	54990	71899	15475	95434	98227	21824	19585
85	97656	63175	89303	16275	07100	92063	21942	18611	47348	20203	18534	03862	78095	50136
86	03299	01221	05418	38982	55758	92237	26759	86367	21216	98442	08303	56613	91511	75928
87	79626	06486	03574	17668	07785	76020	79924	25651	83325	88428	85076	72811	22717	50585
88	85636	68335	47539	03129	65651	11977	02510	26113	99447	68645	34327	15152	55230	93448
89	18039	14367	64337	06177	12143	46609	32989	74014	64708	00533	35398	58408	13261	47908
90	08362	15656	60627	36478	65648	16764	53412	09013	07832	41574	17639	82163	60859	75567
91	79556	29068	04142	16268	15387	12856	66227	38358	22478	73373	88732	09443	82558	05250
92	92608	82674	27072	32534	17075	27698	98204	63863	11951	34648	88022	56148	34925	57031
93	23982	25835	40055	67006	12293	02753	14827	23235	35071	99704	37543	11601	35503	85171
94	09915	96306	05908	97901	28395	14186	00821	80703	70426	75647	76310	88717	37890	40129
95	59037	33300	26695	62247	69927	76123	50842	43834	86654	70959	79725	93872	28117	19233
96	42488	78077	69882	61657	34136	79180	97526	43092	04098	73571	80799	76536	71255	64239
97	46764	86273	63003	93017	31204	36692	40202	35275	57306	55543	53203	18098	47625	88684
98	03237	45430	55417	63282	90816	17349	88298	90183	36600	78406	06216	95787	42579	90730
99	86591	81482	52667	61582	14972	90053	89534	76036	49199	43716	97548	04379	46370	28672
100	38534	01715	94964	87288	65680	43772	39560	12918	86537	62738	19636	51132	25739	56947

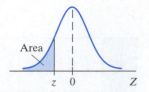

Table II: *Areas Under the Standard Normal Distribution*

z	.00	.01	.02	.03	.04	.05	.06	.07	.08	.09
−3.4	0.0003	0.0003	0.0003	0.0003	0.0003	0.0003	0.0003	0.0003	0.0003	0.0002
−3.3	0.0005	0.0005	0.0005	0.0004	0.0004	0.0004	0.0004	0.0004	0.0004	0.0003
−3.2	0.0007	0.0007	0.0006	0.0006	0.0006	0.0006	0.0006	0.0005	0.0005	0.0005
−3.1	0.0010	0.0009	0.0009	0.0009	0.0008	0.0008	0.0008	0.0008	0.0007	0.0007
−3.0	0.0013	0.0013	0.0013	0.0012	0.0012	0.0011	0.0011	0.0011	0.0010	0.0010
−2.9	0.0019	0.0018	0.0017	0.0017	0.0016	0.0016	0.0015	0.0015	0.0014	0.0014
−2.8	0.0026	0.0025	0.0024	0.0023	0.0023	0.0022	0.0021	0.0021	0.0020	0.0019
−2.7	0.0035	0.0034	0.0033	0.0032	0.0031	0.0030	0.0029	0.0028	0.0027	0.0026
−2.6	0.0047	0.0045	0.0044	0.0043	0.0041	0.0040	0.0039	0.0038	0.0037	0.0036
−2.5	0.0062	0.0060	0.0059	0.0057	0.0055	0.0054	0.0052	0.0051	0.0049	0.0048
−2.4	0.0082	0.0080	0.0078	0.0075	0.0073	0.0071	0.0069	0.0068	0.0066	0.0064
−2.3	0.0107	0.0104	0.0102	0.0099	0.0096	0.0094	0.0091	0.0089	0.0087	0.0084
−2.2	0.0139	0.0136	0.0132	0.0129	0.0125	0.0122	0.0119	0.0116	0.0113	0.0110
−2.1	0.0179	0.0174	0.0170	0.0166	0.0162	0.0158	0.0154	0.0150	0.0146	0.0143
−2.0	0.0228	0.0222	0.0217	0.0212	0.0207	0.0202	0.0197	0.0192	0.0188	0.0183
−1.9	0.0287	0.0281	0.0274	0.0268	0.0262	0.0256	0.0250	0.0244	0.0239	0.0233
−1.8	0.0359	0.0351	0.0344	0.0336	0.0329	0.0322	0.0314	0.0307	0.0301	0.0294
−1.7	0.0446	0.0436	0.0427	0.0418	0.0409	0.0401	0.0392	0.0384	0.0375	0.0367
−1.6	0.0548	0.0537	0.0526	0.0516	0.0505	0.0495	0.0485	0.0475	0.0465	0.0455
−1.5	0.0668	0.0655	0.0643	0.0630	0.0618	0.0606	0.0594	0.0582	0.0571	0.0559
−1.4	0.0808	0.0793	0.0778	0.0764	0.0749	0.0735	0.0721	0.0708	0.0694	0.0681
−1.3	0.0968	0.0951	0.0934	0.0918	0.0901	0.0885	0.0869	0.0853	0.0838	0.0823
−1.2	0.1151	0.1131	0.1112	0.1093	0.1075	0.1056	0.1038	0.1020	0.1003	0.0985
−1.1	0.1357	0.1335	0.1314	0.1292	0.1271	0.1251	0.1230	0.1210	0.1190	0.1170
−1.0	0.1587	0.1562	0.1539	0.1515	0.1492	0.1469	0.1446	0.1423	0.1401	0.1379
−0.9	0.1841	0.1814	0.1788	0.1762	0.1736	0.1711	0.1685	0.1660	0.1635	0.1611
−0.8	0.2119	0.2090	0.2061	0.2033	0.2005	0.1977	0.1949	0.1922	0.1894	0.1867
−0.7	0.2420	0.2389	0.2358	0.2327	0.2296	0.2266	0.2236	0.2206	0.2177	0.2148
−0.6	0.2743	0.2709	0.2676	0.2643	0.2611	0.2578	0.2546	0.2514	0.2483	0.2451
−0.5	0.3085	0.3050	0.3015	0.2981	0.2946	0.2912	0.2877	0.2843	0.2810	0.2776
−0.4	0.3446	0.3409	0.3372	0.3336	0.3300	0.3264	0.3228	0.3192	0.3156	0.3121
−0.3	0.3821	0.3783	0.3745	0.3707	0.3669	0.3632	0.3594	0.3557	0.3520	0.3483
−0.2	0.4207	0.4168	0.4129	0.4090	0.4052	0.4013	0.3974	0.3936	0.3897	0.3859
−0.1	0.4602	0.4562	0.4522	0.4483	0.4443	0.4404	0.4364	0.4325	0.4286	0.4247
−0.0	0.5000	0.4960	0.4920	0.4880	0.4840	0.4801	0.4761	0.4721	0.4681	0.4641

SOURCE: *Walpole/Myers/Myers*, Sixth Edition.

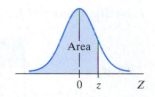

Table II: *Areas Under the Standard Normal Distribution (continued)*

z	.00	.01	.02	.03	.04	.05	.06	.07	.08	.09
0.0	0.5000	0.5040	0.5080	0.5120	0.5160	0.5199	0.5239	0.5279	0.5319	0.5359
0.1	0.5398	0.5438	0.5478	0.5517	0.5557	0.5596	0.5636	0.5675	0.5714	0.5753
0.2	0.5793	0.5832	0.5871	0.5910	0.5948	0.5987	0.6026	0.6064	0.6103	0.6141
0.3	0.6179	0.6217	0.6255	0.6293	0.6331	0.6368	0.6406	0.6443	0.6480	0.6517
0.4	0.6554	0.6591	0.6628	0.6664	0.6700	0.6736	0.6772	0.6808	0.6844	0.6879
0.5	0.6915	0.6950	0.6985	0.7019	0.7054	0.7088	0.7123	0.7157	0.7190	0.7224
0.6	0.7257	0.7291	0.7324	0.7357	0.7389	0.7422	0.7454	0.7486	0.7517	0.7549
0.7	0.7580	0.7611	0.7642	0.7673	0.7704	0.7734	0.7764	0.7794	0.7823	0.7852
0.8	0.7881	0.7910	0.7939	0.7967	0.7995	0.8023	0.8051	0.8078	0.8106	0.8133
0.9	0.8159	0.8186	0.8212	0.8238	0.8264	0.8289	0.8315	0.8340	0.8365	0.8389
1.0	0.8413	0.8438	0.8461	0.8485	0.8508	0.8531	0.8554	0.8577	0.8599	0.8621
1.1	0.8643	0.8665	0.8686	0.8708	0.8729	0.8749	0.8770	0.8790	0.8810	0.8830
1.2	0.8849	0.8869	0.8888	0.8907	0.8925	0.8944	0.8962	0.8980	0.8997	0.9015
1.3	0.9032	0.9049	0.9066	0.9082	0.9099	0.9115	0.9131	0.9147	0.9162	0.9177
1.4	0.9192	0.9207	0.9222	0.9236	0.9251	0.9265	0.9278	0.9292	0.9306	0.9319
1.5	0.9332	0.9345	0.9357	0.9370	0.9382	0.9394	0.9406	0.9418	0.9429	0.9441
1.6	0.9452	0.9463	0.9474	0.9484	0.9495	0.9505	0.9515	0.9525	0.9535	0.9545
1.7	0.9554	0.9564	0.9573	0.9582	0.9591	0.9599	0.9608	0.9616	0.9625	0.9633
1.8	0.9641	0.9649	0.9656	0.9664	0.9671	0.9678	0.9686	0.9693	0.9699	0.9706
1.9	0.9713	0.9719	0.9726	0.9732	0.9738	0.9744	0.9750	0.9756	0.9761	0.9767
2.0	0.9772	0.9778	0.9783	0.9788	0.9793	0.9798	0.9803	0.9808	0.9812	0.9817
2.1	0.9821	0.9826	0.9830	0.9834	0.9838	0.9842	0.9846	0.9850	0.9854	0.9857
2.2	0.9861	0.9864	0.9868	0.9871	0.9875	0.9878	0.9881	0.9884	0.9887	0.9890
2.3	0.9893	0.9896	0.9898	0.9901	0.9904	0.9906	0.9909	0.9911	0.9913	0.9916
2.4	0.9918	0.9920	0.9922	0.9925	0.9927	0.9929	0.9931	0.9932	0.9934	0.9936
2.5	0.9938	0.9940	0.9941	0.9943	0.9945	0.9946	0.9948	0.9949	0.9951	0.9952
2.6	0.9953	0.9955	0.9956	0.9957	0.9959	0.9960	0.9961	0.9962	0.9963	0.9964
2.7	0.9965	0.9966	0.9967	0.9968	0.9969	0.9970	0.9971	0.9972	0.9973	0.9974
2.8	0.9974	0.9975	0.9976	0.9977	0.9977	0.9978	0.9979	0.9979	0.9980	0.9981
2.9	0.9981	0.9982	0.9982	0.9983	0.9984	0.9984	0.9985	0.9985	0.9986	0.9986
3.0	0.9987	0.9987	0.9987	0.9988	0.9988	0.9989	0.9989	0.9989	0.9990	0.9990
3.1	0.9990	0.9991	0.9991	0.9991	0.9992	0.9992	0.9992	0.9992	0.9993	0.9993
3.2	0.9993	0.9993	0.9994	0.9994	0.9994	0.9994	0.9994	0.9995	0.9995	0.9995
3.3	0.9995	0.9995	0.9995	0.9996	0.9996	0.9996	0.9996	0.9996	0.9996	0.9997
3.4	0.9997	0.9997	0.9997	0.9997	0.9997	0.9997	0.9997	0.9997	0.9997	0.9998

Source: *Walpole/Myers/Myers*, Sixth Edition.

Table III: *Selected Percentiles of the Standard Normal Distribution*

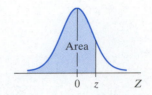

z	Area
−4.265	0.00001
−3.719	0.0001
−3.090	0.001
−2.576	0.005
−2.326	0.01
−2.054	0.02
−1.960	0.025
−1.881	0.03
−1.751	0.04
−1.645	0.05
−1.555	0.06
−1.476	0.07
−1.405	0.08
−1.341	0.09
−1.282	0.10
−1.036	0.15
−0.842	0.20
−0.674	0.25
−0.524	0.30
−0.385	0.35
−0.253	0.40
−0.126	0.45
0	0.50
0.126	0.55
0.253	0.60
0.385	0.65
0.524	0.70
0.674	0.75
0.842	0.80
1.036	0.85
1.282	0.90
1.341	0.91
1.405	0.92
1.476	0.93
1.555	0.94
1.645	0.95
1.751	0.96
1.881	0.97
1.960	0.975
2.054	0.98
2.326	0.99
2.576	0.995
3.090	0.999
3.719	0.9999
4.265	0.99999

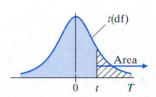

Table IV: *Upper Percentiles of t-distributions*

df	0.40	0.30	0.20	0.10	0.05	0.025	0.01	0.005
Area to the Right								
1	0.325	0.727	1.376	3.078	6.314	12.706	31.821	63.657
2	0.289	0.617	1.061	1.886	2.920	4.303	6.965	9.925
3	0.277	0.584	0.978	1.638	2.353	3.182	4.541	5.841
4	0.271	0.569	0.941	1.533	2.132	2.776	3.747	4.604
5	0.267	0.559	0.920	1.476	2.015	2.571	3.365	4.032
6	0.265	0.553	0.906	1.440	1.943	2.447	3.143	3.707
7	0.263	0.549	0.896	1.415	1.895	2.365	2.998	3.499
8	0.262	0.546	0.889	1.397	1.860	2.306	2.896	3.355
9	0.261	0.543	0.883	1.383	1.833	2.262	2.821	3.250
10	0.260	0.542	0.879	1.372	1.812	2.228	2.764	3.169
11	0.260	0.540	0.876	1.363	1.796	2.201	2.718	3.106
12	0.259	0.539	0.873	1.356	1.782	2.179	2.681	3.055
13	0.259	0.538	0.870	1.350	1.771	2.160	2.650	3.012
14	0.258	0.537	0.868	1.345	1.761	2.145	2.624	2.977
15	0.258	0.536	0.866	1.341	1.753	2.131	2.602	2.947
16	0.258	0.535	0.865	1.337	1.746	2.120	2.583	2.921
17	0.257	0.534	0.863	1.333	1.740	2.110	2.567	2.898
18	0.257	0.534	0.862	1.330	1.734	2.101	2.552	2.878
19	0.257	0.533	0.861	1.328	1.729	2.093	2.539	2.861
20	0.257	0.533	0.860	1.325	1.725	2.086	2.528	2.845
21	0.257	0.532	0.859	1.323	1.721	2.080	2.518	2.831
22	0.256	0.532	0.858	1.321	1.717	2.074	2.508	2.819
23	0.256	0.532	0.858	1.319	1.714	2.069	2.500	2.807
24	0.256	0.531	0.857	1.318	1.711	2.064	2.492	2.797
25	0.256	0.531	0.856	1.316	1.708	2.060	2.485	2.787
26	0.256	0.531	0.856	1.315	1.706	2.056	2.479	2.779
27	0.256	0.531	0.855	1.314	1.703	2.052	2.473	2.771
28	0.256	0.530	0.855	1.313	1.701	2.048	2.467	2.763
29	0.256	0.530	0.854	1.311	1.699	2.045	2.462	2.756
30	0.256	0.530	0.854	1.310	1.697	2.042	2.457	2.750
40	0.255	0.529	0.851	1.303	1.684	2.021	2.423	2.704
60	0.254	0.527	0.848	1.296	1.671	2.000	2.390	2.660
120	0.254	0.526	0.845	1.289	1.658	1.980	2.358	2.617
z*	0.253	0.524	0.842	1.282	1.645	1.960	2.326	2.576
	20%	40%	60%	80%	90%	95%	98%	99%
				Confidence Level				

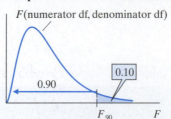

90th percentiles of the F-distribution

Table V: *F-Distribution*
90th Percentiles of the F-distribution

		Degrees of Freedom for Numerator						
	1	**2**	**3**	**4**	**5**	**6**	**7**	**8**
1	39.86	49.50	53.59	55.83	57.24	58.20	58.91	59.44
2	8.53	9.00	9.16	9.24	9.29	9.33	9.35	9.37
3	5.54	5.46	5.39	5.34	5.31	5.28	5.27	5.25
4	4.54	4.32	4.19	4.11	4.05	4.01	3.98	3.95
5	4.06	3.78	3.62	3.52	3.45	3.40	3.37	3.34
6	3.78	3.46	3.29	3.18	3.11	3.05	3.01	2.98
7	3.59	3.26	3.07	2.96	2.88	2.83	2.78	2.75
8	3.46	3.11	2.92	2.81	2.73	2.67	2.62	2.59
9	3.36	3.01	2.81	2.69	2.61	2.55	2.51	2.47
10	3.29	2.92	2.73	2.61	2.52	2.46	2.41	2.38
11	3.23	2.86	2.66	2.54	2.45	2.39	2.34	2.30
12	3.18	2.81	2.61	2.48	2.39	2.33	2.28	2.24
13	3.14	2.76	2.56	2.43	2.35	2.28	2.23	2.20
14	3.10	2.73	2.52	2.39	2.31	2.24	2.19	2.15
15	3.07	2.70	2.49	2.36	2.27	2.21	2.16	2.12
16	3.05	2.67	2.46	2.33	2.24	2.18	2.13	2.09
17	3.03	2.64	2.44	2.31	2.22	2.15	2.10	2.06
18	3.01	2.62	2.42	2.29	2.20	2.13	2.08	2.04
19	2.99	2.61	2.40	2.27	2.18	2.11	2.06	2.02
20	2.97	2.59	2.38	2.25	2.16	2.09	2.04	2.00
21	2.96	2.57	2.36	2.23	2.14	2.08	2.02	1.98
22	2.95	2.56	2.35	2.22	2.13	2.06	2.01	1.97
23	2.94	2.55	2.34	2.21	2.11	2.05	1.99	1.95
24	2.93	2.54	2.33	2.19	2.10	2.04	1.98	1.94
25	2.92	2.53	2.32	2.18	2.09	2.02	1.97	1.93
30	2.88	2.49	2.28	2.14	2.05	1.98	1.93	1.88
40	2.84	2.44	2.23	2.09	2.00	1.93	1.87	1.83
60	2.79	2.39	2.18	2.04	1.95	1.87	1.82	1.77
120	2.75	2.35	2.13	1.99	1.90	1.82	1.77	1.72

Degrees of Freedom for Denominator

95th percentiles of the *F*-distribution

F(numerator df, denominator df)

0.95

0.05

$F_{.95}$ *F*

Table V: *F-Distribution*
95th Percentiles of the F-distribution

	Degrees of Freedom for Numerator							
	1	2	3	4	5	6	7	8
1	161.45	199.50	215.71	224.58	230.16	233.99	236.77	238.88
2	18.51	19.00	19.16	19.25	19.30	19.33	19.35	19.37
3	10.13	9.55	9.28	9.12	9.01	8.94	8.89	8.85
4	7.71	6.94	6.59	6.39	6.26	6.16	6.09	6.04
5	6.61	5.79	5.41	5.19	5.05	4.95	4.88	4.82
6	5.99	5.14	4.76	4.53	4.39	4.28	4.21	4.15
7	5.59	4.74	4.35	4.12	3.97	3.87	3.79	3.73
8	5.32	4.46	4.07	3.84	3.69	3.58	3.50	3.44
9	5.12	4.26	3.86	3.63	3.48	3.37	3.29	3.23
10	4.96	4.10	3.71	3.48	3.33	3.22	3.14	3.07
11	4.84	3.98	3.59	3.36	3.20	3.09	3.01	2.95
12	4.75	3.89	3.49	3.26	3.11	3.00	2.91	2.85
13	4.67	3.81	3.41	3.18	3.03	2.92	2.83	2.77
14	4.60	3.74	3.34	3.11	2.96	2.85	2.76	2.70
15	4.54	3.68	3.29	3.06	2.90	2.79	2.71	2.64
16	4.49	3.63	3.24	3.01	2.85	2.74	2.66	2.59
17	4.45	3.59	3.20	2.96	2.81	2.70	2.61	2.55
18	4.41	3.55	3.16	2.93	2.77	2.66	2.58	2.51
19	4.38	3.52	3.13	2.90	2.74	2.63	2.54	2.48
20	4.35	3.49	3.10	2.87	2.71	2.60	2.51	2.45
21	4.32	3.47	3.07	2.84	2.68	2.57	2.49	2.42
22	4.30	3.44	3.05	2.82	2.66	2.55	2.46	2.40
23	4.28	3.42	3.03	2.80	2.64	2.53	2.44	2.37
24	4.26	3.40	3.01	2.78	2.62	2.51	2.42	2.36
25	4.24	3.39	2.99	2.76	2.60	2.49	2.40	2.34
30	4.17	3.32	2.92	2.69	2.53	2.42	2.33	2.27
40	4.08	3.23	2.84	2.61	2.45	2.34	2.25	2.18
60	4.00	3.15	2.76	2.53	2.37	2.25	2.17	2.10
120	3.92	3.07	2.68	2.45	2.29	2.18	2.09	2.02

Degrees of Freedom for Denominator

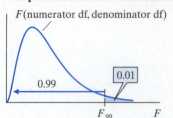

99th percentiles of the F-distribution

Table V: *F-Distribution*
99th Percentiles of the F-distribution

	Degrees of Freedom for Numerator							
	1	**2**	**3**	**4**	**5**	**6**	**7**	**8**
1	4052.18	4999.50	5403.35	5624.58	5763.65	5858.99	5928.36	5981.07
2	98.50	99.00	99.17	99.25	99.30	99.33	99.36	99.37
3	34.12	30.82	29.46	28.71	28.24	27.91	27.67	27.49
4	21.20	18.00	16.69	15.98	15.52	15.21	14.98	14.80
5	16.26	13.27	12.06	11.39	10.97	10.67	10.46	10.29
6	13.75	10.92	9.78	9.15	8.75	8.47	8.26	8.10
7	12.25	9.55	8.45	7.85	7.46	7.19	6.99	6.84
8	11.26	8.65	7.59	7.01	6.63	6.37	6.18	6.03
9	10.56	8.02	6.99	6.42	6.06	5.80	5.61	5.47
10	10.04	7.56	6.55	5.99	5.64	5.39	5.20	5.06
11	9.65	7.21	6.22	5.67	5.32	5.07	4.89	4.74
12	9.33	6.93	5.95	5.41	5.06	4.82	4.64	4.50
13	9.07	6.70	5.74	5.21	4.86	4.62	4.44	4.30
14	8.86	6.51	5.56	5.04	4.69	4.46	4.28	4.14
15	8.68	6.36	5.42	4.89	4.56	4.32	4.14	4.00
16	8.53	6.23	5.29	4.77	4.44	4.20	4.03	3.89
17	8.40	6.11	5.18	4.67	4.34	4.10	3.93	3.79
18	8.29	6.01	5.09	4.58	4.25	4.01	3.84	3.71
19	8.18	5.93	5.01	4.50	4.17	3.94	3.77	3.63
20	8.10	5.85	4.94	4.43	4.10	3.87	3.70	3.56
21	8.02	5.78	4.87	4.37	4.04	3.81	3.64	3.51
22	7.95	5.72	4.82	4.31	3.99	3.76	3.59	3.45
23	7.88	5.66	4.76	4.26	3.94	3.71	3.54	3.41
24	7.82	5.61	4.72	4.22	3.90	3.67	3.50	3.36
25	7.77	5.57	4.68	4.18	3.85	3.63	3.46	3.32
30	7.56	5.39	4.51	4.02	3.70	3.47	3.30	3.17
40	7.31	5.18	4.31	3.83	3.51	3.29	3.12	2.99
60	7.08	4.98	4.13	3.65	3.34	3.12	2.95	2.82
120	6.85	4.79	3.95	3.48	3.17	2.96	2.79	2.66

Degrees of Freedom for Denominator

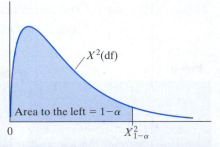

Area to the left = $1-\alpha$

Table VI: *Chi-Square Distribution*

DF	$X^2_{.005}$	$X^2_{.01}$	$X^2_{.025}$	$X^2_{.05}$	$X^2_{.10}$	$X^2_{.90}$	$X^2_{.95}$	$X^2_{.975}$	$X^2_{.99}$	$X^2_{.995}$
1	0.000	0.000	0.001	0.004	0.016	2.706	3.841	5.024	6.635	7.879
2	0.010	0.020	0.051	0.103	0.211	4.605	5.991	7.378	9.210	10.597
3	0.072	0.115	0.216	0.352	0.584	6.251	7.815	9.348	11.345	12.838
4	0.207	0.297	0.484	1.711	1.064	7.779	9.488	11.143	13.277	14.860
5	0.412	0.554	0.831	1.145	1.610	9.236	11.070	12.833	15.086	16.750
6	0.676	0.872	1.237	1.635	2.204	10.645	12.592	14.449	16.812	18.548
7	0.989	1.239	1.690	2.167	2.833	12.017	14.067	16.013	18.475	20.278
8	1.344	1.646	2.180	2.733	3.490	13.362	15.507	17.535	20.090	21.955
9	1.735	2.088	2.700	3.325	4.168	14.684	16.919	19.023	21.666	23.589
10	2.156	2.558	3.247	3.940	4.865	15.987	18.307	20.483	23.209	25.188
11	2.603	3.053	3.816	4.575	5.578	17.275	19.675	21.920	24.725	26.757
12	3.074	3.571	4.404	5.226	6.304	18.549	21.026	23.337	26.217	28.300
13	3.565	4.107	5.009	5.892	7.042	19.812	22.362	24.736	27.688	29.819
14	4.075	4.660	5.629	6.571	7.790	21.064	23.685	26.119	29.141	31.319
15	4.601	5.229	6.262	7.261	8.547	22.307	24.996	27.488	30.578	32.801
16	5.142	5.812	6.908	7.962	9.312	23.542	26.296	28.845	32.000	34.267
18	6.265	7.015	8.231	9.390	10.865	25.989	28.869	31.526	34.805	37.156
20	7.434	8.260	9.591	10.851	12.443	28.412	31.410	34.170	37.566	39.997
24	9.886	10.856	12.401	13.848	15.659	33.196	36.415	39.364	42.980	45.559
30	13.787	14.953	16.791	18.493	20.599	40.256	43.773	46.979	50.892	53.672
40	20.707	22.164	24.443	26.509	29.051	51.805	55.758	59.342	63.691	66.766
60	35.534	37.485	40.482	43.188	46.459	74.397	79.082	83.298	88.379	91.952
120	83.852	86.923	91.573	95.705	100.624	140.233	146.567	152.211	158.950	163.648

Table VII: *Tail Probabilities for the Rank Sum Statistic under H_0*

Tail Probability $= P[W \leq W_{lower}] = P[W \geq W_{upper}]$

Smaller Sample Size = 2		
Larger Sample Size = 3		
W_{lower}	**Tail Probability**	W_{upper}
2	0.000	10
3	0.100	9
4	0.200	8

Smaller Sample Size = 2		
Larger Sample Size = 4		
W_{lower}	**Tail Probability**	W_{upper}
2	0.000	12
3	0.067	11
4	0.133	10

Smaller Sample Size = 2		
Larger Sample Size = 5		
W_{lower}	**Tail Probability**	W_{upper}
2	0.000	14
3	0.048	13
4	0.095	12
5	0.190	11

Smaller Sample Size = 2		
Larger Sample Size = 6		
W_{lower}	**Tail Probability**	W_{upper}
2	0.000	16
3	0.036	15
4	0.071	14
5	0.143	13

Smaller Sample Size = 2		
Larger Sample Size = 7		
W_{lower}	**Tail Probability**	W_{upper}
2	0.000	18
3	0.028	17
4	0.056	16
5	0.111	15

Smaller Sample Size = 2		
Larger Sample Size = 8		
W_{lower}	**Tail Probability**	W_{upper}
2	0.000	20
3	0.022	19
4	0.044	18
5	0.089	17
6	0.133	16

Smaller Sample Size = 2		
Larger Sample Size = 9		
W_{lower}	**Tail Probability**	W_{upper}
2	0.000	22
3	0.018	21
4	0.036	20
5	0.073	19
6	0.109	18

Smaller Sample Size = 2		
Larger Sample Size = 10		
W_{lower}	**Tail Probability**	W_{upper}
3	0.015	23
4	0.030	22
5	0.061	21
6	0.091	20
7	0.136	19

Table VII: *Tail Probabilities for the Rank Sum Statistic under H_0 (continued)*

Tail Probability $= P[W \leq W_{lower}] = P[W \geq W_{upper}]$

W_{lower}	Tail Probability	W_{upper}
Smaller Sample Size = 3	**Larger Sample Size = 3**	
5	0.000	16
6	0.050	15
7	0.100	14
8	0.200	13

W_{lower}	Tail Probability	W_{upper}
Smaller Sample Size = 3	**Larger Sample Size = 4**	
5	0.000	19
6	0.029	18
7	0.057	17
8	0.114	16

W_{lower}	Tail Probability	W_{upper}
Smaller Sample Size = 3	**Larger Sample Size = 5**	
5	0.000	22
6	0.018	21
7	0.036	20
8	0.071	19
9	0.125	18

W_{lower}	Tail Probability	W_{upper}
Smaller Sample Size = 3	**Larger Sample Size = 6**	
5	0.000	25
6	0.012	24
7	0.024	23
8	0.048	22
9	0.083	21
10	0.131	20

W_{lower}	Tail Probability	W_{upper}
Smaller Sample Size = 3	**Larger Sample Size = 7**	
5	0.000	28
6	0.008	27
7	0.017	26
8	0.033	25
9	0.058	24
10	0.092	23
11	0.133	22

W_{lower}	Tail Probability	W_{upper}
Smaller Sample Size = 3	**Larger Sample Size = 8**	
5	0.000	31
6	0.006	30
7	0.012	29
8	0.024	28
9	0.042	27
10	0.067	26
11	0.097	25
12	0.139	24

W_{lower}	Tail Probability	W_{upper}
Smaller Sample Size = 3	**Larger Sample Size = 9**	
7	0.009	32
8	0.018	31
9	0.032	30
10	0.050	29
11	0.073	28
12	0.105	27

W_{lower}	Tail Probability	W_{upper}
Smaller Sample Size = 3	**Larger Sample Size = 10**	
7	0.007	35
8	0.014	34
9	0.024	33
10	0.038	32
11	0.056	31
12	0.080	30
13	0.108	29

Table VII: *Tail Probabilities for the Rank Sum Statistic under H_0 (continued)*

Tail Probability $= P[W \leq W_{lower}] = P[W \geq W_{upper}]$

Smaller Sample Size = 4 Larger Sample Size = 4		
W_{lower}	Tail Probability	W_{upper}
9	0.000	27
10	0.014	26
11	0.029	25
12	0.057	24
13	0.100	23
14	0.171	22

Smaller Sample Size = 4 Larger Sample Size = 5		
W_{lower}	Tail Probability	W_{upper}
9	0.000	31
10	0.008	30
11	0.016	29
12	0.032	28
13	0.056	27
14	0.095	26
15	0.143	25

Smaller Sample Size = 4 Larger Sample Size = 6		
W_{lower}	Tail Probability	W_{upper}
11	0.010	33
12	0.019	32
13	0.033	31
14	0.057	30
15	0.086	29
16	0.129	28

Smaller Sample Size = 4 Larger Sample Size = 7		
W_{lower}	Tail Probability	W_{upper}
11	0.006	37
12	0.012	36
13	0.021	35
14	0.036	34
15	0.055	33
16	0.082	32
17	0.115	31

Smaller Sample Size = 4 Larger Sample Size = 8		
W_{lower}	Tail Probability	W_{upper}
12	0.008	40
13	0.014	39
14	0.024	38
15	0.036	37
16	0.055	36
17	0.077	35
18	0.107	34

Smaller Sample Size = 4 Larger Sample Size = 9		
W_{lower}	Tail Probability	W_{upper}
13	0.010	43
14	0.017	42
15	0.025	41
16	0.038	40
17	0.053	39
18	0.074	38
19	0.099	37
20	0.130	36

Smaller Sample Size = 4 Larger Sample Size = 10		
W_{lower}	Tail Probability	W_{upper}
13	0.007	47
14	0.012	46
15	0.018	45
16	0.027	44
17	0.038	43
18	0.053	42
19	0.071	41
20	0.094	40
21	0.120	39

Table VII: *Tail Probabilities for the Rank Sum Statistic under H_0 (continued)*

Tail Probability $= P[W \leq W_{lower}] = P[W \geq W_{upper}]$

Smaller Sample Size = 5 Larger Sample Size = 5			Smaller Sample Size = 5 Larger Sample Size = 6			Smaller Sample Size = 5 Larger Sample Size = 7		
W_{lower}	Tail Probability	W_{upper}	W_{lower}	Tail Probability	W_{upper}	W_{upper}	Tail Probability	W_{upper}
16	0.008	39	17	0.009	43	18	0.009	47
17	0.016	38	18	0.015	42	19	0.015	46
18	0.028	37	19	0.026	41	20	0.024	45
19	0.048	36	20	0.041	40	21	0.037	44
20	0.075	35	21	0.063	39	22	0.053	43
21	0.111	34	22	0.089	38	23	0.074	42
			23	0.123	37	24	0.101	41

Smaller Sample Size = 5 Larger Sample Size = 8			Smaller Sample Size = 5 Larger Sample Size = 9			Smaller Sample Size = 5 Larger Sample Size = 10		
W_{lower}	Tail Probability	W_{upper}	W_{lower}	Tail Probability	W_{upper}	W_{lower}	Tail Probability	W_{upper}
19	0.009	51	20	0.009	55	21	0.010	59
20	0.015	50	21	0.014	54	22	0.014	58
21	0.023	49	22	0.021	53	23	0.020	57
22	0.033	48	23	0.030	52	24	0.028	56
23	0.047	47	24	0.041	51	25	0.038	55
24	0.064	46	25	0.056	50	26	0.050	54
25	0.085	45	26	0.073	49	27	0.065	53
26	0.111	44	27	0.095	48	28	0.082	52
			28	0.120	47	29	0.103	51

Table VII: *Tail Probabilities for the Rank Sum Statistic under H_0* (continued)

Tail Probability $= P[W \leq W_{lower}] = P[W \geq W_{upper}]$

Smaller Sample Size = 6 Larger Sample Size = 6		
W_{lower}	Tail Probability	W_{upper}
24	0.008	54
25	0.013	53
26	0.021	52
27	0.032	51
28	0.047	50
29	0.066	49
30	0.090	48
31	0.120	47

Smaller Sample Size = 6 Larger Sample Size = 7		
W_{lower}	Tail Probability	W_{upper}
25	0.007	59
26	0.011	58
27	0.017	57
28	0.026	56
29	0.037	55
30	0.051	54
31	0.069	53
32	0.090	52
33	0.117	51

Smaller Sample Size = 6 Larger Sample Size = 8		
W_{lower}	Tail Probability	W_{upper}
27	0.010	63
28	0.015	62
29	0.021	61
30	0.030	60
31	0.041	59
32	0.054	58
33	0.071	57
34	0.091	56
35	0.114	55

Smaller Sample Size = 6 Larger Sample Size = 9		
W_{lower}	Tail Probability	W_{upper}
28	0.009	68
29	0.013	67
30	0.018	66
31	0.025	65
32	0.033	34
33	0.044	63
34	0.057	62
35	0.072	61
36	0.091	60
37	0.112	59

Smaller Sample Size = 6 Larger Sample Size = 10		
W_{lower}	Tail Probability	W_{upper}
29	0.008	73
30	0.011	72
31	0.016	71
32	0.021	70
33	0.028	69
34	0.036	68
35	0.047	67
36	0.059	66
37	0.074	65
38	0.090	64
39	0.110	63

Table VII: *Tail Probabilities for the Rank Sum Statistic under H_0 (continued)*

Tail Probability $= P[W \leq W_{lower}] = P[W \geq W_{upper}]$

Smaller Sample Size = 7 Larger Sample Size = 7			Smaller Sample Size = 7 Larger Sample Size = 8			Smaller Sample Size = 7 Larger Sample Size = 9		
W_{lower}	Tail Probability	W_{upper}	W_{lower}	Tail Probability	W_{upper}	W_{lower}	Tail Probability	W_{upper}
34	0.009	71	36	0.010	76	37	0.008	82
35	0.013	70	37	0.014	75	38	0.011	81
36	0.019	69	38	0.020	74	39	0.016	80
37	0.027	68	39	0.027	73	40	0.021	79
38	0.036	67	40	0.036	72	41	0.027	78
39	0.049	66	41	0.047	71	42	0.036	77
40	0.064	65	42	0.060	70	43	0.045	76
41	0.082	64	43	0.076	69	44	0.057	75
42	0.104	63	44	0.095	68	45	0.071	74
			45	0.116	67	46	0.087	73
						47	0.105	72

Smaller Sample Size = 7 Larger Sample Size = 10		
W_{lower}	Tail Probability	W_{upper}
39	0.009	87
40	0.012	86
41	0.017	85
42	0.022	84
43	0.028	83
44	0.035	82
45	0.044	81
46	0.054	80
47	0.067	79
48	0.081	78
49	0.097	77
50	0.115	76

Table VII: *Tail Probabilities for the Rank Sum Statistic under H_0 (continued)*
Tail Probability $= P[W \leq W_{lower}] = P[W \geq W_{upper}]$

Smaller Sample Size = 8 Larger Sample Size = 8			Smaller Sample Size = 8 Larger Sample Size = 9			Smaller Sample Size = 8 Larger Sample Size = 10		
W_{lower}	Tail Probability	W_{upper}	W_{lower}	Tail Probability	W_{upper}	W_{lower}	Tail Probability	W_{upper}
46	0.010	90	48	0.010	96	50	0.010	102
47	0.014	89	49	0.014	95	51	0.013	101
48	0.019	88	50	0.018	94	52	0.017	100
49	0.025	87	51	0.023	93	53	0.022	99
50	0.032	86	52	0.030	92	54	0.027	98
51	0.041	85	53	0.037	91	55	0.034	97
52	0.052	84	54	0.046	90	56	0.042	96
53	0.065	83	55	0.057	89	57	0.051	95
54	0.080	82	56	0.069	88	58	0.061	94
55	0.097	81	57	0.084	87	59	0.073	93
56	0.117	80	58	0.100	86	60	0.086	92
						61	0.102	91

Smaller Sample Size = 9 Larger Sample Size = 9			Smaller Sample Size = 9 Larger Sample Size = 10			Smaller Sample Size = 10 Larger Sample Size = 10		
W_{lower}	Tail Probability	W_{upper}	W_{lower}	Tail Probability	W_{upper}	W_{lower}	Tail Probability	W_{upper}
59	0.009	112	61	0.009	119	74	0.009	136
60	0.012	111	62	0.011	118	75	0.012	135
61	0.016	110	63	0.014	117	76	0.014	134
62	0.020	109	64	0.017	116	77	0.018	133
63	0.025	108	65	0.022	115	78	0.022	132
64	0.031	107	66	0.027	114	79	0.026	131
65	0.039	106	67	0.033	113	80	0.032	130
66	0.047	105	68	0.039	112	81	0.038	129
67	0.057	104	69	0.047	111	82	0.045	128
68	0.068	103	70	0.056	110	83	0.053	127
69	0.081	102	71	0.067	109	84	0.062	126
70	0.095	101	72	0.078	108	85	0.072	125
71	0.111	100	73	0.091	107	86	0.083	124
			74	0.106	106	87	0.095	123
						88	0.109	122

Source: Adapted from C. Kraft and C. van Eeden, *A Nonparametric Introduction to Statistics,* Macmillan, New York, 1968.

Table VIII: *Tail Probabilities for the Signed Rank Statistic under* H_0

Tail Probability $= P[W^+ \leq W^+_{lower}] = P[W^+ \geq W^+_{upper}]$

W^+_{lower}	$n = 3$ Tail Probability	W^+_{upper}	W^+_{lower}	$n = 4$ Tail Probability	W^+_{upper}	W^+_{lower}	$n = 5$ Tail Probability	W^+_{upper}
	0.000	7		0.000	11		0.000	16
0	0.125	6	0	0.062	10	0	0.031	15
1	0.250	5	1	0.125	9	1	0.062	14
			2	0.188	8	2	0.094	13
						3	0.156	12

W^+_{lower}	$n = 6$ Tail Probability	W^+_{upper}	W^+_{lower}	$n = 7$ Tail Probability	W^+_{upper}	W^+_{lower}	$n = 8$ Tail Probability	W^+_{upper}
	0.000	22	0	0.008	28	1	0.008	35
0	0.016	21	1	0.016	27	2	0.012	34
1	0.031	20	2	0.023	26	3	0.020	33
2	0.047	19	3	0.039	25	4	0.027	32
3	0.078	18	4	0.055	24	5	0.039	31
4	0.109	17	5	0.078	23	6	0.055	30
			6	0.109	22	7	0.074	29
						8	0.098	28
						9	0.125	27

W^+_{lower}	$n = 9$ Tail Probability	W^+_{upper}	W^+_{lower}	$n = 10$ Tail Probability	W^+_{upper}	W^+_{lower}	$n = 11$ Tail Probability	W^+_{upper}
3	0.010	42	5	0.010	50	7	0.009	59
4	0.014	41	6	0.014	49	8	0.012	58
5	0.020	40	7	0.019	48	9	0.016	57
6	0.027	39	8	0.024	47	10	0.021	56
7	0.037	38	9	0.032	46	11	0.027	55
8	0.049	37	10	0.042	45	12	0.034	54
9	0.064	36	11	0.053	44	13	0.042	53
10	0.082	35	12	0.065	43	14	0.051	52
11	0.102	34	13	0.080	42	15	0.062	51
			14	0.097	41	16	0.074	50
			15	0.116	40	17	0.087	49
						18	0.103	48

Table VIII: *Tail Probabilities for the Signed Rank Statistic under H_0 (continued)*

Tail Probability $= P[W^+ \leq W^+_{lower}] = P[W^+ \geq W^+_{upper}]$

W^+_{lower}	n = 12 Tail Probability	W^+_{upper}	W^+_{lower}	n = 13 Tail Probability	W^+_{upper}	W^+_{lower}	n = 14 Tail Probability	W^+_{upper}
10	0.010	68	12	0.009	79	16	0.010	89
11	0.013	67	13	0.011	78	17	0.012	88
12	0.017	66	14	0.013	77	18	0.015	87
13	0.021	65	15	0.016	76	19	0.018	86
14	0.026	64	16	0.020	75	20	0.021	85
15	0.032	63	17	0.024	74	21	0.025	84
16	0.039	62	18	0.029	73	22	0.029	83
17	0.046	61	19	0.034	72	23	0.034	82
18	0.055	60	20	0.040	71	24	0.039	81
19	0.065	59	21	0.047	70	25	0.045	80
20	0.076	58	22	0.055	69	26	0.052	79
21	0.088	57	23	0.064	68	27	0.059	78
22	0.102	56	24	0.073	67	28	0.068	77
			25	0.084	66	29	0.077	76
			26	0.095	65	30	0.086	75
			27	0.108	64	31	0.097	74
						32	0.108	73

W^+_{lower}	n = 15 Tail Probability	W^+_{upper}
19	0.009	101
20	0.011	100
21	0.013	99
22	0.015	98
23	0.018	97
24	0.021	96
25	0.024	95
26	0.028	94
27	0.032	93
28	0.036	92
29	0.042	91
30	0.047	90
31	0.053	89
32	0.060	88
33	0.068	87
34	0.076	86
35	0.084	85
36	0.094	84
37	0.104	83

Source: Adapted from C. Kraft and C. van Eeden, *A Nonparametric Introduction to Statistics*, Macmillan, New York, 1968.

ANSWERS TO ODD-NUMBERED EXERCISES

Chapter 1

1.1 In hypothesis testing, the purpose is to determine whether there is sufficient evidence with which to reject the null hypothesis (H_0), which generally reflects the prevailing viewpoint. The alternative hypothesis (H_1) is often what someone sets out to try to prove.

1.3 H_0: The 5-year survival rate for those using the vaccine is equal to 10%.

H_1: The 5-year survival rate for those using the vaccine is greater than 10%.

1.5 (a) A Type I error may be more serious as one might accidentally shoot a loaded gun.

(b) A Type I error may be more serious as we might approach a dog that could bite.

(c) A Type II error may be more serious as we might waste the time and gas to drive to the mall expecting it open when it is closed.

(d) A Type I error may be more serious as we might ruin our watch if it gets wet.

1.7 (a) H_0: The average tomato yield for Brand A fertilizer is the same as the tomato average yield for the more expensive Brand B fertilizer.

H_1: The average tomato yield for the more expensive Brand B fertilizer is greater than the average tomato yield for the Brand A fertilizer.

(b) Type I error: Spend more money on Brand B when it really is not better than Brand A in terms of the average tomato yield.

Type II error: Continue to use Brand A when Brand B results in a higher tomato yield on average.

1.9 The owner would conclude the patrons are older and the owner would spend the time and money to remodel, when the crowd is actually not older. The owner would have spent money unnecessarily and the remodeling may not appeal to some of the patrons, but in general, it is not a serious error.

1.11 (a) The null hypothesis was accepted.

(b) No, a complaint was not registered.

(c) Yes, a Type II error may have been made. The cans are thought to be containing the stated sodium content when actually they contain higher amounts of sodium on average.

1.13 If α is decreased then β will increase, so the possible value is 0.30.

1.15 (a) $\alpha = 2/30 = 0.067$ and $\beta = 20/30 = 0.667$.

(b) Decision Rule 2: Reject H_0 if the selected voucher is $\leq \$2$ or $\geq \$9$. The significance level α is $6/30 = 0.20$ and the level of β is $12/30 = 0.40$. Enlarging the rejection region resulted in increasing the level of α from 0.067 to 0.20 while decreasing the level of β from 0.667 to 0.40.

1.17 No, we need a decision rule that states when we reject or accept H_0.

1.19 (a) False: $\alpha + \beta$ does not need to equal 1. The value of α is calculated under H_0 while the value of β is calculated under H_1.

(b) False: A Type II error is the chance of accepting H_0 when H_1 is true.

(c) True.

(d) False: H_0 is rejected if the sample shows evidence against it.

(e) False: The sample size does not influence the alternative hypothesis. The alternative hypothesis H_1 can be one-sided no matter what the size of the sample is.

1.21 (a) H_0: The shown box is Box A.

H_1: The shown box is Box B.

(b) One-sided to the left.

(c) Reject H_0 if the selected token is $5 or less.

(d) $\alpha = 2/25 = 0.08$.

(e) $\beta = 11/25 = 0.44$.

(f) Reject H_0.

1.23 (a) False.

(b) False.

(c) True.

1.25   (a)   The frequency plots are provided below:

Bag **X**

```
                          X
                          X
              X           X           X
              X           X           X
              X           X           X
              X           X           X
              X           X           X
              X           X           X
   X          X           X           X           X
 ──────────────────────────────────────────────────
  Blue       Brown      Yellow      Green        Red
```

Bag **Y**

```
   X                                              X
   X                                              X
   X                                              X
   X                                              X
   X                                              X
   X                                              X
   X          X                       X           X
   X          X                       X           X
   X          X           X           X           X
 ──────────────────────────────────────────────────
  Blue       Brown      Yellow      Green        Red
```

(b)   No, the response being recorded is the color, which has no particular ordering for the outcomes. So it is not appropriate to discuss a direction of extreme in this case.

1.27   The p-value should be small in order to reject the null hypothesis H_0. A small p-value indicates that the observed data or data even more extreme is very unlikely or unusual if the null hypothesis is true. In general, we reject H_0 if p-value $\leq \alpha$, the significance level.

1.29   (a)   Frequency plots for the two completing hypotheses.

H_0:

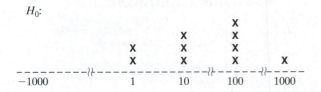

```
                                    X
                          X         X
              X           X         X
              X           X         X         X
 ─ ─ ─ ─ ─ ─ ─〉〉─ ─ ─ ─ ─ ─ ─ ─ ─ ─〉〉─ ─ ─ ─〉〉─ ─ ─ ─
  −1000      1          10       100      1000
```

H_1:

```
              X
              X
              X         X
   X          X         X         X
 ─ ─ ─ ─ ─ ─ ─〉〉─ ─ ─ ─ ─ ─ ─ ─ ─ ─〉〉─ ─ ─ ─〉〉─ ─ ─ ─
  −1000      1          10       100      1000
```

(b) $\alpha = 2/10 = 0.20$

(c) $\beta = 3/8 = 0.375$

(d)   No, we did not actually observe a voucher.

1.31   (a)   One-sided to the left.

(b) $\alpha = 1/6 = 0.1667$ and $\beta = 7/10 = 0.70$

(c) p-value $= 2/6 = 0.333$. We cannot reject H_0 and conclude that selected die appears to be Die A.

1.33   (a) H_0: The shown bag is Bag A.

H_1: The shown bag is Bag B.

(b)   Two-sided.

(c)   (i) p-value $= 4/40 = 0.10$.

(ii)   Yes, since the p-value is $\leq \alpha$.

(iii)   No, since the p-value is $> \alpha$.

(d)   (i)   The p-value is 1.

(ii)   No, since the p-value is $> \alpha$.

(iii)   No, since the p-value is $> \alpha$.

1.35   (a)   All new model 100-watt light bulbs produced at Claude's plant.

(b) H_0: The population of all new model 100-watt light bulbs (produced at Claude's plant) has an average lifetime equal to 40 hours. H_1: The population of all new model 100-watt light bulbs (produced at Claude's plant) has an average lifetime greater than 40 hours.

(c)   10%

(d)   Any value between 0 and 0.10.

(e)   Yes, if the p-value is less than or equal to 0.10, then the p-value is also less than or equal to 0.15 so it is significant at the 0.15 level. However, if we only know that the p-value is less than or equal to 0.10, we cannot be sure whether the p-value is also less than or equal to 0.05. Without further information we cannot determine if the data would also be significant at the 0.05 level.

1.37   (a) H_0

(b) p-value > 0.10

(c)   Type II error

(d)   One-sided to the right

1.39   (a)   Type I error

(b)   We decide that the average cost is higher than $350 while it is really not. Maybe you decide that the cost is too high and decide to attend a different college, while in reality you could have attended this college after all.

(c) p-value ≤ 0.10

(d)   Yes.

1.41   (a)   Study A: 0.001, Study B: 0.11, Study C: 0.03

(b)   Study B

(c)   Type I error

(d)   Study A: one-sided to the right, Study B: two-sided, Study C: one-sided to the left

1.43   (a) and (b)   See the chart.

	Null Hypothesis	Alternative Hypothesis	p-value
Study A	The true proportion of females is equal to 0.60.	The true proportion of females is not equal to 0.60.	0.08
Study B	The average time to relief for all Treatment I users is equal to the average time to relief for all Treatment II users.	The average time to relief for all Treatment I users is less than the average time to relief for all Treatment II users.	0.005
Study C	The true average income of adults who work two jobs is equal to $70,000.	The true average income of adults who work two jobs is greater than $70,000.	0.20

(c)   Study C

(d)   Type I error

1.45 (a) One-sided to the left.
(b) See graph for α (under H_0) and β (under H_1).
(c) See graph for p-value (under H_0).
(d) Since the p-value is larger than α, the result is not statistically significant.

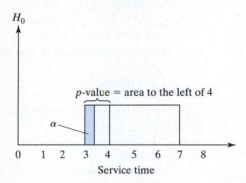

H_0

p-value = area to the left of 4

α

Service time

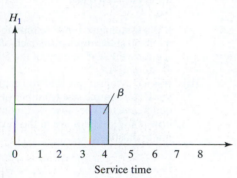

H_1

β

Service time

1.47 (a) True.
(b) False.
1.49 (c) To be statistically significant at the 5% level means the p-value is less than or equal to 0.05. However, we do not know if the p-value is less than or equal to 0.01 or if it is between 0.01 and 0.05, so the answer is "sometimes yes" (if the p-value is also ≤ 0.01) and "sometimes no" (if the p-value is > 0.01).
1.51 (a) You observed a yellow ball, which is the most extreme result that you could get. So the p-value is the chance of observing a yellow ball under the null hypothesis, which is $1/5 = 0.20$. Since the p-value is larger than 10%, the result is not statistically significant.
(b) The data consist of selecting two balls with replacement (and the order is not important).

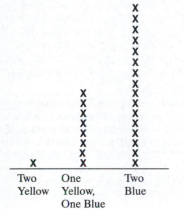

We observed one yellow and one blue ball. Results that are even more extreme would be observing two yellow balls. So the p-value is $(1 + 8)/25 = 0.36$. Since the p-value is larger than 10%, the result is not statistically significant.

1.53 Statement (a) since the effect is small so we would need a larger sample size to detect it.
1.55 (a) (i) 0.0368
(ii) 45
(iii) 0.289
(iv) There is a 28% chance we decide the bag shown is bag A when it really is bag B.
(b) (i) 0.1737
(ii) Reject H_0
(iii) 0.2789
(iv) Accept H_0
(v) $35
1.57 (a) 190
(b) 1140

Chapter 2

2.1 35% is a parameter, 28% is a statistic.
2.3 Answer is (c), statistic is to a sample.
2.5 (a) Using the calculator with a seed value of 291 the selected persons are 39 (79 years old), 3 (75 years old), and 24 (70 years old). The average age is 74.67 years.
(b) 22 (36 years old), 34 (89 years old), and 29 (89 years old). The average age is 71.33 years, different from the mean calculated in (a).
(c) 69.19 years.
(d) 74.67 and 71.33 are statistics, while the mean in (c) is a parameter.
2.7 (a) The population consists of (the planned vote for) the 100 U.S. Senators.
(b) $N = 100$.
(c) The sample consists of (the planned vote for) the ten selected U.S. Senators.
(d) $n = 10$.
2.9 (a) The value of 34 is a statistic.
(b) Response bias.
2.11 Response bias.
2.13 Nonresponse bias.
2.15 Nonresponse bias.
2.17 (a) Yes.
(b) Without replacement.
2.19 (a) With the calculator the selected ID numbers are: 179, 2274, 3327. With the random number table the selected ID numbers are: 2398, 2258, 3540.
(b) A statistic.
2.21 (a) H_0: The proportion of dissatisfied customers equals 0.10.

H_1: The proportion of dissatisfied customers is less than 0.10.
(b) With the calculator the selected customers are: 34318, 15553, 8461, 614. With the random table the selected customers are: 15409, 23336, 29490, 43127.
(c) Type II error.
(d) 0.21.
(e) No, the p-value is > 0.05.
(f) (ii) Statistic.

2.23 (a) Reject H_0.
 (b) Yes.
 (c) No.
 (d) Type I error.
 (e) $\$1_1$ and $\$1_2$ $\$1_1$ and $\$5_1$ $\$1_1$ and $\$5_2$
  $\$1_2$ and $\$5_1$ $\$1_2$ and $\$5_2$ $\$5_1$ and $\$5_2$.
 (f) $\$1_1$ and $\$1_2$ $\$1_1$ and $\$1_3$ $\$1_1$ and $\$1_4$
  $\$1_2$ and $\$1_3$ $\$1_2$ and $\$1_4$ $\$1_3$ and $\$1_4$.
 (g) p-value = 1/6.

2.25 (a) 0.50.
 (b) 0.10.
 (c) 0.10. This is not a simple random sample.

2.27 Stratified random sampling.

2.29 (a) Stratified random sampling.
 (b) No.
 (c) Selection bias.

2.31 $(0.20)(16) + (0.50)(43) + (0.30)(71) =$
 46 years.

2.33 (a) All former university graduate students.
 (b) Stratified random sampling.
 (c) False.

2.35 (a) Stratified random sampling.
 (b) Selection bias.

2.37 (a) Stratified random sampling.
 (b) 0.
 (c) Book A and Book C.
 (d) Book A and Book D.
 (e) 1/4.

2.39 (a) Stratified random sampling.
 (b) High: 20 clients, Moderate: 125 clients, Low: 45 clients.
 (c) With the calculator the selected clients are: 163, 2196, 214, 2462, 740. With the random number table the selected clients are: 1887, 1209, 2294, 954, 1869.

2.41 (a) With the calculator, the label of the first student selected is 3. With the random number table, the label of the first student selected is 3.
 (b) The students in the sample are those with ID numbers 3, (3 + 4 =) 7, (7 + 4 =) 11, and (11 + 4 =) 15, for a total of 4 students.

2.43 (a) Systematic sampling (1 in 30).
 (b) Stratified random sampling. Yes, 25 freshmen.

2.45 (a) (i) $0.02(500) + 0.03(1200) + 0.05(18000) = 10 + 36 + 900 = 946.$
  (ii) Stratified Random Sampling.
  (iii) With the calculator the selected labels are 432, 232, 304, 412, 372. With the random number table, we might assign the first High category driver the labels 001 and 501. We would assign the second High category driver 002 and 502. This assignment pattern would continue until the 500th High category driver who would be assigned the labels 000 and 500. Reading off labels from row 60, column 1, we have: 789, 191, 947, 423, and 632. This would correspond to selecting the 289th, 191st, 447th, 423rd, and 532nd High category drivers in the list of 500 High category drivers.

 (b) (i) With the calculator or the random number table, the first selected label is 15. Thus the selected labels will be 15, 35, 55, 75, 95, and so on.
  (ii) Since the 500 High category drivers divide evenly into groups of 20 (500/20 = 25), there will be a total of 25 High category drivers in the systematic 1-in-20 sample.

2.47 False, the chance depends on the number of clusters.

2.49 (a) Cluster sampling.
 (b) No.
 (c) Poor design together with bad luck.

2.51 (a) Cluster sampling.
 (b) 12 students.
 (c) 36 students.
 (d) We do not know the exact value; something between 0 and 36.
 (e) No.

2.53 (a) It is a systematic 1 in 45 sampling resulting in 50 students (1 from each of the 50 sections, the 33rd in each of the 50 lists of 45 students).
 (b) It is a cluster sample and you cannot know the sample size (number of students selected) because we do not know how many students are in the various major clusters.

2.55 (a) False.
 (b) False.
 (c) True.

2.57 (a) (i) Convenience sampling.
  (ii) Yes, a selection bias.
  (iii) The calculated average is expected to be higher than the true average, as all of the books in the sample have already been checked out at least once, and may include some of the more popular books.
 (b) Cluster sampling.
 (c) (i) Stratified random sampling.
  (ii) Overall estimate:
  $$\left(\tfrac{400}{1200}\right)(20) + \left(\tfrac{200}{1200}\right)(15) + \left(\tfrac{600}{1200}\right)(10) =$$
  142 times checked out.
 (d) For each of the three categories of books the following stages are followed.
  Stage 1: Divide the books into clusters according to the last digit of the call number (0 through 9). Take a simple random sample of 3 digits from the list of 0, 1, 2, 3, 4, 5, 6, 7, 8, and 9. The clusters of books (in that category) with call numbers ending with those selected digits are selected.
  Stage 2: Within each of the selected clusters of books from Stage 2, select a simple random sample of 7 books. Note that with this multistage sampling plan, we will have a total of 3 categories × 3 clusters × 7 books = 63 books.

2.59   (a)   With the calculator or the random number table, the selected region is 3 = Southwest.

(b)   Stratified random sampling.

(c)   (i)   1-in-10 systematic sampling.

(ii)   0.10.

(iii)   With the calculator, the first five selected cans are 7, (7 + 10 =) 17, (17 + 10 =) 27, (27 + 10 =) 37, and (27 + 10 =) 47. With the random number table we might label the first can 1, the second can 2, ..., and the 10th can 0. Then the first five selected cans are 7, (7 + 10 =) 17, (17 + 10 =) 27, (27 + 10 =) 37, and (27 + 10 =) 47.

(iv) $\frac{125}{10} = 12\frac{5}{10} \Rightarrow$ 12 or 13 cans. However, there will not be a 7th can to select in the last group. Thus, the total number of cans in the sample will be 12.

(d)   (i) H_0.

(ii)   Two possible values are 0.12 and 0.15.

(iii)   Yes, a Type II error.

Chapter 3

3.1   (a)   Explanatory variable: Amount of coffee consumed.
Response variable: Exam performance.

(b)   Explanatory variable: Hours of counseling per week.
Response variable: Grade point average.

(c)   Explanatory variable: Type of driver (levels are good or bad).
Response variable: Reaction time on a driving test.

3.3   (a)   Observational study.

(b)   Blood cholesterol level.

(c)   Frequency of egg consumption.

(d)   Diet, age, amount of exercise, and history of high cholesterol are a few possible confounding variables.

3.5   (a)   Observational study.

(b)   Homelessness.

(c)   Some are: separation status (whether or not they were separated from their parents), poverty status (whether or not they were raised in poverty), family problem status (whether or not they have family problems, abuse status (whether or not they were sexually or physically abused).

(d)   All homeless people in Los Angeles.

(e)   The homeless people in Los Angeles that were surveyed.

(f)   Statistic.

3.7   (a) H_1.

(b)   *p*-value was less or equal to 0.05.

(c)   (iii) Observational study, Prospective.

3.9   (a)   Observational study (retrospective).

(b)   Response variable is kidney stones status (whether or not the subject develops kidney stones); Explanatory variable is amount of calcium in diet.

(c)   A diet high in calcium lowers the risk of developing kidney stones in men and women.

3.11   (a)   Observational study (retrospective).

(b)   Response variable is Alzheimer's disease status, explanatory variable is linguistic ability.

(c)   The chance of low linguistic ability given that a person has Alzheimer's.

3.13   (a)   Durability.

(b)   Dye color and type of cloth.

(c)   20.

(d) $20 \times 6 = 120$.

3.15   (a)   Batches of feed stock.

(b)   Yield of the process.

(c)   Temperature (2 levels) and Stirring Rate (3 levels).

(d)   6 treatments.

(e)   24 units.

(f)   Design layout table:

		Factor 1: Temperature	
		50	**70**
	60	4 batches	4 batches
Factor 2: Stirring Rate	100	4 batches	4 batches
	140	4 batches	4 batches

3.17   (a)   Weight.

(b)   Fiber content and carbohydrate content.

(c)   Fiber levels (3) Low, Medium, and High. Carbohydrate levels (2) Low and Medium.

(d) $3 \times 2 \times 20 = 120$.

3.19   The rats are labeled 1 through 8. Use a calculator with seed = 1209 or a random number table with row = 24, column = 6 to select the 4 rats to receive the treatment. The other 4 rats will not receive the treatment. At the end of the week the effectiveness of the vaccine against the virus will be measured.

3.21   (a)   Experiment.

(b)   Cause of death.

(c)   Estrogen therapy with two levels: receive the treatment, do not receive the treatment.

(d)   Placebo.

H_0: Women who receive estrogen therapy do not have a lower risk of dying from a heart disease than women who do not receive the estrogen therapy.

H_1: Women who receive estrogen therapy have a lower risk of dying from a heart disease than women who do not receive the estrogen therapy.

(e)   *p*-value was less than or equal to 0.05.

(f)   Yes, Type 1 error.

3.23   Answer is d.

3.25   (a)   With the calculator the selected rats were: 14, 13, 5, 16, 9, 3, 4, 2, 7, 10. With the random number table the selected rats were: 4, 13, 8, 5, 12, 15, 2, 6, 9, 20.

(b)   Have another researcher make and record the measurements.

3.27 Observational study. No active treatment was imposed.

3.29 (a) Observational study.
(b) Attitude towards mathematics.
(c) Gender.
(d) Cluster sampling.
(e) Selection bias.

3.31 (a) It is not clear from the article.
(b) It may not be appropriate to extend these results to the general male population, nor to the female population.
(c) Confounding variables.

3.33 (a) Temperature (3 levels) and baking time (2 levels).
(b) Taste.
(c) 6 treatments.
(d) 36 batches.

3.35 Lack of blinding of the subjects and of the experimenter(s) or evaluator(s).

3.37 (a) (iii) confounding variable.
(b) (i) False
(ii) False
(iii) True
(iv) False

3.39 The placebo effect is a phenomenon in which receiving medical attention, even administration of an inert drug, improves the condition of the subjects.

3.41 (a) Experiment.
(b) Proportion of juice added to the drink.
(c) Rating of the juice on a scale of 1 to 100.
(d) (iii) Confounding variable.
(e) (ii) Single-blinded.
(f) With the calculator, all five volunteers received 5% juice.
   With the random number table, the first four volunteers received a 10% juice and the last one received 5% juice (using 0–4 to represent 5%, using 5–9 to represent 10%).
(g) False.

3.43 (a) Experiment.
(b) Intestinal discomfort measured on a scale (0–5).
(c) Explanatory variable is type of milk. The treatments are treated milk (has a lactose enzyme) and ordinary milk.
(d) The population under study is people who say they are lactose intolerant. The sample is not random because they are volunteers.

3.45 (a) Observational study.
(b) We don't know whether the patients in both groups were in the same physical condition.
(c) Assign to the 100 patients numbers 1–100. With the calculator or with the random number table select at random the first 50 patients. Those patients will receive treatment 1, the other patients will receive treatment 2.
(d) No. The treatment is an operation.

3.47 (a) Experiment.
(b) Using the calculator. With seed = 23 the selected men were: BJ(3), AB(14), JP(4), ZB(16), TN(20), CF(17), JW(7), TD(19), MK(2), SK(5).
   Using the calculator. With seed = 41 the selected women were: NL(12), BG(2), MM(5), JB(4), AK(10), KS(6), KB(18), SL(11), JG(19), NI(17).
   With the random number table. Men(row 12, column 1): RM(9), TG(18), MK(2), VN(8), WB(12), SK(5), BH(11), BJ(3), RM(9), JD(1).
   With the random number table. Women(row 22, column 1): CI(16), NI(17), MA(3), SM(7), BG(2), KS(6), EG(15), AK(10), MM(5), CJ(13).
(c) Confounding variable.

3.49 (a) Using the calculator with seed = 45, the first five subjects selected are: 3351, 3140, 860, 703, and 5488.
   With the table of random numbers (row 10, column 5) the first five subjects selected are: 5368, 5753, 3425, 3988, and 5306.
(b) The placebo effect is a phenomenon in which receiving medical attention, even administration of an inert drug, improves the condition of the subjects.
(c) 3.5% of 4058 = 142.
(d) Statistic.
(e) H_0.
(f) p-value > 0.05.
(g) Type II error.

3.51 (a) Observational study.
(b) Explanatory variable: Amount of wine. Response variable: IQ score.
(c) Amount of wine was likely self-reported and may result in a response bias.
(d) Who paid for the study? Who were the subjects on the study? How many subjects were selected?

3.53 (a) Experiment.
(b) Explanatory variable: Behavioral therapy status.
   Response variable: Insomnia status.
(c) The number of people in each treatment group is unknown. We only know that they were 75 in total.
(d) How were the results analyzed? How were the 75 subjects selected? Was the placebo treatment truly a placebo?

3.55 Answers will vary.

Chapter 4

4.1 (a) Qualitative
(b) Quantitative-continuous
(c) Qualitative
(d) Quantitative-discrete
(e) Qualitative
(f) Quantitative-continuous

4.3 (iv) Qualitative

4.5 A more useful measure is the proportion or percent of students who failed the midterm.

4.7

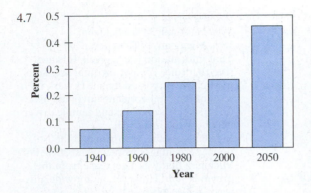

4.9 (a) The sales are represented by the areas of the bars, not just the height of the bars.

(b)

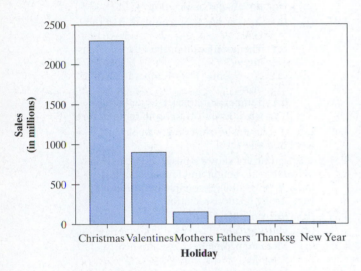

4.11 (a) $0.20(620) = 124$

(b) No, background is a categorical variable. It is not appropriate to discuss shape in a bar graph since the order of the categories is arbitrary.

4.13 (a) The conditional distribution of years of experience given gender is:

Years of Experience:		1–6	7–12	13 +
Gender	*Male*	25.7%	37.9%	36.4%
	Female	42.3%	35.1%	22.6%

(b) Male coaches generally have more years of coaching experience than females. There is a high percent of relatively *new* female coaches.

4.15 (a) Men: 65.08%; Women: 39.83%

(b) The conditional distribution is:

	Firearms	Drugs	Hanging	Gases	Other
Men	65.1%	5.9%	15.1%	6.9%	7.0%
Women	39.8%	27.0%	13.4%	8.6%	11.1%

(d) Men who commit suicide overwhelmingly choose firearms while women are more evenly split between firearms and drugs.

4.17 (a) 0.72

(b) 0.48

(c) Yes, since the *p*-value of 0.01431 is less than 0.05.

4.19

```
              X
        X  X  X
        X  X  X
     X  X  X  X  X  X
     X  X  X  X  X  X              X
     0  1  2  3  4  5  6  7  8   Goals scored
```

4.21 Answers will vary. Stem-and-leaf plot:

4.23 **Note:** 83|4 represents 83.4

```
83 | 4
84 |
85 |
86 | 7
87 | 5 7
88 | 0 4 5 8
89 | 1
90 | 5
```

4.25 (a)

```
Student B              Student A
              4 | 5 8
              5 | 2
              6 | 9
     8 6 5 4 3 | 7 | 6 9
     7 4 2 1 0 | 8 | 0 6
              9 | 2 4
```

(b) Student A's scores range from 45 to 94. Student A did receive the highest score of the two students. However, Student A also received the lowest score of the two students. The scores for Student B ranged from 73 to 87. Student B performed very consistently in the 70's and 80's.

4.27 (a) 781 hours

(b) If you wanted the brand that gives the longest lasting light bulb.

(c) If you wanted the brand that lasts longest on average.

4.29 (a) **Note:** 80|8 represents an on-time departure rate of 80.8%.

```
% On-Time     6 4 | 75 |         % On-Time
Arrival         0 | 76 |         Departures
            8 1 0 0 | 77 |
                    | 78 |
                  6 | 79 |
                  1 | 80 | 8
            5 2 1 1 | 81 |
                  5 | 82 |
                  1 | 83 | 1 3 8 9
                9 1 | 84 | 3
                  8 | 85 | 7 7
                4 0 | 86 | 1 5 6
                  9 | 87 | 0 6
                    | 88 | 1 6 8 9
                    | 89 | 0 8 9
                  3 | 90 |
                    | 91 | 3
                    | 92 | 6
```

(b) The arrivals range from 75.4% to 90.3% with a gap on the high side between Detroit (87.9%) and Raleigh (90.3%). The departures range from 80.8% to 92.6%. There is less variation among the departures as compared to the arrivals. The departures distribution is somewhat bimodal with a gap on the low side between Chicago (80.8%) and Denver (83.1%).

(c) Airports perform better overall with respect to on-time departures. The center value for on-time departures (86.8%) is higher than the center value for on-time arrivals (81.15%). The range for the on-time departures is smaller than for the on-time arrivals. With the exception of Chicago, all of the % on-time departures are higher than the center value for the % on-time arrivals.

4.31 (a) $3 + 6 + 8 + 10 + 5 = 32$.
(b) $9/32 = 0.2813$

4.33 (a) Skewed to the right
(b) 16%
(c) 2 people
(d) One customer spent between 350 and 400 seconds but we don't know the actual time.

4.35 Back-to-back histograms

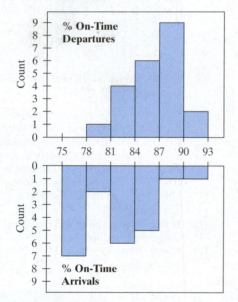

4.37 (a) For 1925: somewhat uniform with an extreme modal class of 100+. For 1960: skewed to the right with the most frequent price class of 20–25. For 1994: skewed to the right with the most frequent price class of 10–15.
(b) A general shift from many stocks priced 100+ to more stocks priced between 5–40.
(c) Yes. A lot of stock splits would lead to more stocks at lower prices.

4.39 Time Series I matches Description (1) since we would expect the proportion of babies born that are girls to fluctuate around 0.50. Time Series II matches Description (2) since cumulating the number of batteries that fail would increase over time.

4.41 (b) An increasing trend.

4.43 (a) 25% National, 30% Michigan
(b) 18% Michigan, 17% National
(c) That information is not given. It is not true that since 25% of the fatal accidents in Michigan involved female drivers, then 75% of the fatal accidents involved male drivers because a male and female driver may both die in a crash.
(d) In 1983, both the Michigan and the National percentages were around 20%. For Michigan that was a decrease from the previous year. However, nationwide, this was a slight increase from 1982.
(e) More women are working consequently, more women are driving.

4.45 (a) Plot C
(b) Negative linear association

4.47 (i) (a) X = Year; Y = Population of the U.S.
(c) The population of the U.S. is increasing.
(ii) (a) X = Time; Y = Temperature of a drink
(c) Temperature increases and after a while levels off to room temperature.

4.49 Let X = hours of exercise per week, and Y = level of cholesterol.

4.51 The scatterplot shows a positive association between male height and female height. The linearity might be reasonable but it is not a very strong linear association.

4.53 (b) Yes, positive
(c) Fairly strong

4.55 (a) Median income decreased to nearly $36,000.
(b) Median income increased from $35,000 to nearly $40,000.
(c) Median income decreased back to approximately $37,000.
(d) Predictions would be difficult without further information to help explain the changes.

4.57 (a) Is this an average SAT score? What data were used to get this average? Is the spending an average amount spent? Across all schools? Just some schools?
(b) The range is 400 to 1600. With the updated graph the increase in spending and decrease in SAT would be less dramatic.

4.59 (a) quantitative continuous
(b) quantitative continuous
(c) quantitative continuous
(d) quantitative discrete
(e) qualitative
(f) quantitative discrete
(g) quantitative continuous

4.61 (a) 8196 students
(b) $\frac{88}{251} \Rightarrow 35.06\%$.

4.63 Answer is: (d) The conditional distribution of Blood Type given Rh Factor.

4.65 (a) Observational study
(b) Color blindness (with responses being yes or no)

(c) Gender (with levels being male or female)
(d) Statistic (2% is a sample percentage)
(e) Stratified with female = stratum I and male = stratum II
(f) (i) Yes; (ii) Can't tell; (iii) No
(g) The answer is (ii).

4.67 (a) Time is continuous, but since it was recorded to the nearest 5 minute interval it is discrete.
(b) The distribution is skewed to the right with an outlier at 105 minutes. The waiting times range from 5 minutes to 105 minutes. The median is 32.5 minutes.
(c) Since the distribution of waiting times is skewed, the mean waiting time is not an appropriate measure of the typical waiting time. The median is resistant to outlier and skewed distributions. Therefore, the median (32.5 minutes) is an appropriate typical waiting.
(d) It does not say how that day was selected or whether the 30 patients were all of the patients on that day or a sample.

4.69 (a) Amount of Radiation Emitted (mR/hr).
(b) No. We only know that the range falls between 0.1 and 0.9. We don't know the exact upper and lower limits of the range.
(c) There are 10 out of 20 stores for 50%.
(d) Since 0.5/0.08 = 6.25, the maximum we could use is 6 small television sets.

4.71 (b) A decreasing trend.

4.73 (b) Observations 1 through 13 seem to be varying around the target value of 20.000 cm. Observations #16 and #17 are extremely low.
(c) What happened after observation #13? Did the operator fall asleep? Did the machine settings get bumped?

4.75 (b) Yes, the relationship does appear to be linear.
(c) The relationship appears to be positive.
(d) There are no unusual values or outliers.

4.77 (a) A person born in 1946 would be 19 years old in 1965 and a person born in 1964 would be 1 year old (or younger) in 1965.
(b) Some of the 1,867,000 people being tracked over time have died.
(c) No, there heights of the bars are not given, nor any axis with values to be able to approximate the heights.
(d) The distribution is skewed to the right for both 1965 and 1995, with the skewness being stronger in 1965. In projected 2025 the distribution is nearly uniform.

4.79 (a) Assaulted more often between 10 PM and 2 AM and less often between 6 and 8 AM.
(b) Answers will vary.
(c) How was assault defined? Were these data based on all cities, counties, rural areas?

Chapter 5

5.1 (c) Mean

5.3 The overall average for the 40 students would be:
$$\frac{25(82) + 15(74)}{40} = 79.$$

5.5 The median may have been preferred if the distribution of prices was skewed to the right.

5.7 (a) Mean:
$$\frac{x_1 + x_2 + x_3 + x_4 + x_5 + x_6 + x_7 + x_8 + x_9 + x_{10} + x_{11}}{11}$$
$$= \frac{350 + 32}{11} = 34.727$$
(b) We cannot find the new median.

5.9 (a) Yes, I agree. The mode is the value occurring most frequently.
(b) For the values: 1, 1, 1, 2, 20; the median = 1 = minimum and the mean = 5. So 4 out of 5 or 80% of the values are below the mean

5.11 (a) (i) $s = 9.016$
(ii) $s = 0$
(iii) $s = 2.669$
(b) The standard deviation is so large due to one very large outlier, namely the value of 27.
(c) (i) range = $27 - 1 = 26$
(ii) range = $7 - 7 = 0$
(iii) range = $11 - 3 = 8$
The range is misleading for data set (i). If the outlier of 27 were removed, the range would be 8.

5.13 (a) $670,000
(b) Skewed to the right
(c) The mean.
(d) It is the largest value that is not considered a potential outlier.

5.15 (a) True
(b) True
(c) False
(d) True
(e) False

5.17 Part (b)

5.19 (a) Min = 28, Q1 = 32, Median = 41, Q3 = 48, Max = 50.5
(b) 15 years old
(c) 31 years old
(d) IQR = $48 - 32 = 16$ years
(e) $1.5(16) = 24$; lower fence = $32 - 24 = 8$; Since 20 is > 8, the observation would not be an outlier.

5.21 (a) equal to
(b) larger than
(c) Larger than, because the mean will be in the middle so there are many large deviations. Whereas for plot 2, the mean is drawn to the left where most of the observations are, and for those observations the deviations are small.
(d) Plot 1: range = 4, mean = 2, standard deviation = 1.46; Plot 2: range = 4, mean = 1.33, standard deviation = 1.29.

5.23 Mean in Celsius = $\frac{5}{9}(28 - 32) = -2.22$, Standard deviation in Celsius = $\left|\frac{5}{9}\right|(10) = 5.55$

5.25 (a) 3.43 °F, 953.2 °F.
(b) Min = 949, Q1 = 950, Median = 953.5, Q3 = 955, Max = 959

(c) It could increase as much as you want. It will never change the median.

(d) We can see that there is an upward trend, the temperature seems to increase slowly.

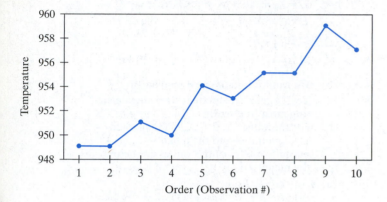

(e) 1.91 °C, 511.78 °C.

(f) Min = 509.44, Q1 = 510, Median = 511.94, Q3 = 512.78, Max = 515

5.27 (a) Quantitative standard score = −0.41
Analytical standard score = 0.70
Verbal standard score = 1.91

(b) Verbal comprehension

5.29 Lab score 2 = $\frac{[(100)(64) - (70)(52)]}{30}$ = 92

5.31 (b) The mean would be higher than the median.

5.33 Yes. For the values 1, 3, 20, the mean = 8 and the standard deviation = 10.44.

5.35 (a) These data appear to be skewed to the left with an outlier at zero.

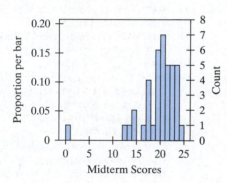

(b) The five-number summary would be preferable since the extreme outlier of zero would influence the mean. Min = 0, Q1 = 17.5, Median = 20, Q3 = 22, Max = 24

(c) With the outlier removed, the mean and standard deviation may now be appropriate measures to summarize these data. Mean = 19.564, Standard deviation = 2.927

5.37 (a) Negative means that the after is larger than the before; blood pressure increased.

(b) sample mean = 5; sample standard deviation = 8.74.

5.39 (a) 60 bottles

(b) 52 bottles

(c) 64.05, give or take about 3.15.

(d) Min = 58, Q1 = 62, Median = 64, Q3 = 66.5, Max = 70

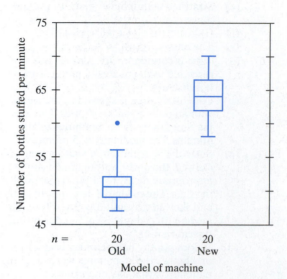

5.41 (a) False: Q1 and the median of 1, 2, 2, 2, 2, 2, 3, 4, 5 are both 2, whereas Q3 is 3.5.

(b) False: If there are 90 boxes that weigh 4 lbs, and 10 boxes that weigh 14 lbs, the total average is [(90)(4) + (10)(14)]/100 = 5 lbs, but there are 90 boxes that weigh less than 5 lbs and only 10 that weigh more than 5 lbs.

(c) False: Both of the following data sets have a mean of 3 and a standard deviation of 1. However, their histograms are not the same. Data set 1: 1, 3, 3, 3, 3, 3, 3, 3, 5; Data set 2: 2, 3, 4.

5.43 (e) none of the above

5.45 (a) Yes

(b) Yes

(c) 4

5.47 (a) Five-number summary: Min = 16.55, Q1 = 26.95, Median = 45.585, Q3 = 51.90, Max = 69.85

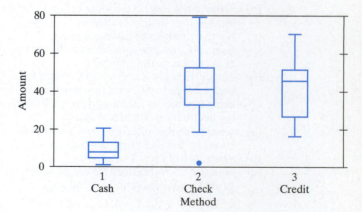

(b) Mean = 41.65, Standard Deviation, s = 14.06

(c) Cash: $\frac{38}{100}$, Check: $\frac{40}{100}$, Credit Card: $\frac{22}{100}$

(d) $1

(e) Yes

5.49   The 20 represents 2 standard deviations.

5.51   (b)   (ii) increasing trend
       (c)   Min = 2.4 mm, Q1 = 3.5 mm, Median = 4.3 mm, Q3 = 4.7 mm, Max = 5.3 mm.
       (d)   IQR = 4.7 − 3.5 = 1.2 mm.
       (e)   (iii) the new median = the old median.
       (f)   (ii) will be larger than the old mean.

5.53   (a)   (ii) a prospective observational study
       (b)   Min = 14, Q1 = 24, Median = 27.5, Q3 = 32.5, Max = 44.
       (c)   (i)   False
             (ii) Can't tell
             (iii) Can't tell
             (iv) True

5.55   (a)   Station 2 with a range of 20 − 7 = 13.
       (b)   Can't tell.
       (c)   (i) $\bar{x}$ = 9.67, s = 3.14
             (ii) Min = 3, Q1 = 9, Median = 10.5, Q3 = 11, Max = 15
             (iii) IQR = 11 − 9 = 2; 1.5(2) = 3; Lower fence = 9 − 3 = 6; Upper fence = 11 + 3 = 14. So any observation below 6 or above 14 is a possible outlier, namely, 3, 5, and 15.
             (iv) Modified boxplot is provided.

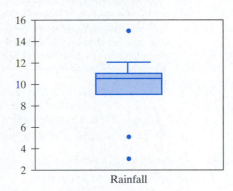

       (v) $\bar{x}$ = 10.67, s = 3.14

Chapter 6

6.1   They are all the same.

6.3   (b)   the mean of the distribution becomes zero and the standard deviation becomes 1.

6.5   (a)   (680 − 500)/100 = 1.8.
       (b)   (27 − 18)/6 = 1.5.
       (c)   Eleanor did.

6.7   (a)   0.9998
       (b)   1
       (c)   0.7995
       (d)   0.7795
       (e)   0.0104
       (f)   0.0321
       (g)   0.4155

6.9   (a)   normalcdf(−E99, 70, 63, 5) = 91.92%.
       (b)   normalcdf(−E99, 70, 67, 4) = 77.34%.

6.11   (a) μ = 7.
       (b)   68%.

6.13   (b)   Brand A: 0.7486; Brand B: 0.8413. So a higher proportion of Brand B fuses last longer than 980 days as compared to Brand A.

6.15   invNorm(0.05, 30, 5) = 21.78 The maximum duration is 21.78 minutes.

6.17   (a)   0.6915.
       (b)   2.8 years.

6.19   (b)   Both give the same. Johnson's = normalcdf(90, E99, 80, 5) = 0.0227. Anderson's = normalcdf(90, E99, 70, 10) = 0.0227.
       (c)   Johnson's = normalcdf(−E99, 50, 80, 5) = 9.9 E−10. Anderson's = normalcdf(−E99, 50, 70, 10) = 0.02275. Anderson gives a higher proportion of E's.
       (d)   invNorm(.90, 69, 9) = 80.53.

6.21   (a)   0.0047.
       (b)   56.7 lesions.

6.23   (b)   (ii) p-value = normalcdf(4.8, E99, 4.6, 0.1) = 0.02275.
       (c)   We cannot reject H_0 because the p-value is 0.022 > significance level = α = 0.01.

6.25   (a) α = normalcdf(−E99, 8, 15, 3) = 0.0098.
       (b) β = normalcdf(8, E99, 10, 3) = 0.7475.
       (c)   p-value = normalcdf(−E99, 8.5, 15, 3) = 0.01513.

6.27   (a)   1.28 standard deviations below the mean for A/B cutoff and 1.28 standard deviations above the mean for B/C cutoff.
       (b)   Mean = (50 + 250)/2 = 150 and (50 − 150)/σ = −1.38, so σ = 78.

6.29   (a) $\bar{x}$ = 255.4 grams, s = 2.769 grams.
       (b)   The distribution appears to be approximately symmetric, bell-shaped, unimodal.
       (c)   The percentages are 71.7%, 95%, and 100%.
       (d)   These data appear to resemble a normal distribution approximately.
       (e)   The product is being filled within the required range approximately 90% of the time.

6.31   (a)   The proportion is approximately zero. This could be considered the p-value since it is the chance of the data or more extreme data being observed if the null hypothesis is true.
       (b)   Proportion (X < 9) = practically zero. We reject H_0 since the p-value is nearly zero.

6.33   (a)   0.67.
       (b)   0.33.

6.35   (a)

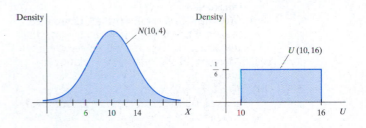

       (b)   (i)   False.
             (ii) True.

6.37 (b) 3/10 or 30%.
    (c) (i) one-sided to the right.
        (ii) p-value = 1/10 = 0.10
        (iii) Reject H_0 since p-value < α = 0.15.

6.39 (b) (ii) $\alpha = (3.2 - 3)/4 = 0.05$
        (iv) $\beta = (4 - 3.2)/4 = 0.20$
    (c) (ii) p-value = $(4.1 - 3)/4 = 0.275$.
    (d) No, since the p-value = $0.275 > \alpha = 0.05$.

6.41 (a) 25%.
    (b) More than 1 pound.
    (c) More than 1 pound.

6.43 0.20.

6.45 (a)

Y	Brown	Red	Yellow	Green	Orange	Blue
Proportion	0.30	0.20	0.20	0.10	0.10	0.10

    (b) No, the variable Y is qualitative.
    (c) 0.20.

6.47 Proportion$(X < 36) = 0.0013$.

6.49 (a) Proportion $(64.5 \le X \le 72) = 0.3590$.
    (b) $1000(0.359) = 359$ women.
    (c) 67.7 inches.

6.51 (a) Proportion$(X > 31.5) = 0.3085$ or 30.85%.
    (b) 34.93 pounds.

6.53 (a) 557
    (b) Proportion $(X < 507) = 0.6736$.

6.55 Mean = 9.75, Standard deviation = 0.375.

6.57 (a) Proportion$(X > 50) = 0.0013$.
    (b) 0.0015.
    (c) 48.58.

6.59 (a) (i) α = normalcdf(13.2, E99, 10, 2) = 0.055
        (ii) β = normalcdf(−E99, 13.2, 12, 2) = 0.7257.
    (b) p-value = normalcdf(11.2, E99, 10, 2) = 0.274.

6.61 (a) p-value = Proportion$(X \ge 62.5) = 0.062$.
    (b) We reject H_0 at the 10% significance level, since the p-value is < 0.10.

6.63 Proportion$(X > 20) = 0.2810$.

6.65 (a) invNorm(0.95, 15, 3) = 19.93 minutes.
    (b) Mean of $Y = 1.2(15) - 4 = 14$ minutes. Standard deviation of $Y = 1.2(3) = 3.6$ minutes.
    (c) invNorm(0.95, 14, 3.6) = 19.92 minutes.

6.67 (a)

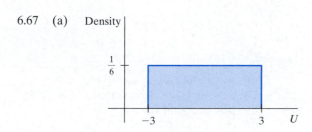

(b) IQR = $1.5 - (-1.5) = 3$.

(c)

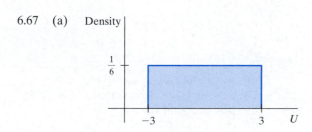

6.69 (a) 0.06.
    (b) 0.61.
    (c) Skewed to the right.
    (d) The median will be between 1 and 2. The mean will be larger than the median.

6.71 (a) Proportion$(X > 2)$ = normalcdf(2, E99, 1, 2) = 0.3085 or 30.85%.
    (b) Proportion$(T = 1)$ = Proportion$(T = 4) = 0.05$; Proportion$(T > 2) = 0.65$.
    (c) Proportion$(L > 2) = 2/3$.

6.73 (a) Proportion$(X > 3) = 0.3085$.
    (b) 0.5.
    (c) Proportion$(X > 3) = 1/8$.
    (d) 17/25.

6.75 (a) More than 6.
    (b) The proportion of the observations to the right of 6 is much larger than the proportion of observations to the left of 6. Since the median divides the total area under the curve in two equal parts, the median must be larger than 6.
    (c) 0.5.

6.77 (a) 1/6.
    (b) 49%.
    (c) p-value = 0.5 (1/6) = 1/12 = 0.083.
    (d) Since the p-value is < 0.10, we reject H_0.

6.79 (b) Proportion$(X < 300$ under $H_1) = 0.01$.
    (c) (i) one-sided to the right.
    (d) $\alpha = (800 - 700)/600 = 1/6$ or 16.7%.
    (e) $\beta = 0.5$.
    (f) p-value = $(800 - 400)/600 = 2/3 = 0.67$.
    (g) No, p-value > α.

Chapter 7

7.1 (a) Personal probability.
    (b) Relative-frequency approach.

7.3 The long-run relative frequency approach may not be appropriate because not all events are repeatable.

7.5 Answers will vary.

7.7 (a) Mrs. Smith is right.
    (b) Answers will vary.

7.9 (a) With the calculator assign 1 as the winning number and 2, 3, 4, and 5 as the losing numbers. With the random number table we can assign 01–20 as the winning numbers and the other remaining two-digit numbers to be the losing numbers.
    (b) With the calculator we get the numbers:

4, 4, 2, 4, 3, 1, 4, 4, 1, 1,   1, 2, 3, 5, 2, 5, 5, 5, 2, 1, 5,
5, 2, 1, 1, 2, 2, 5, 4, 4, 4,   5, 4, 5, 3, 2, 1, 4, 4, 1, 3,
2, 1, 1, 4, 1, 5, 1, 3, 1, 5.

The first five numbers obtained with the random number table are:

91, 57, 64, 25, 95, and all represent losses.

(c) With the calculator we have $14/50 = 0.28$.

7.11 (a) $S = \{$EEE, EEG, EGE, GEE, EGG, GEG, GG$\}$.

(b) $A = \{$GEG, EGE$\}$.

7.13 (a) $S = \{$F, P$\}$.

(b) $S = \{$FFFF, FFFP, FFPF, FFPP, FPFF, FPFP, FPPF, FPPP, PFFF, PFFP, PFPF, PFPP, PPFF, PPFP, PPPF, PPPP$\}$.

(c) $S = \{0, 1, 2, 3, 4, 5 \ 6, 7, 8, 9, 10, 11, 12, 13, 14, 15, 16, 17, 18, 19, 20\}$.

7.15 (a) No, since you can have a king of hearts.

(b) Yes, since you cannot have a card that is both a heart and a spade.

(c) No, since you can have a king of spades.

7.17 (a) Error: $0.19 + 0.38 + 0.29 + 0.15 = 1.01 > 1$.

(b) Error: $0.49 + 0.52 = 0.92 < 1$.

(c) Error: $-0.25 < 0$.

(d) Error: Probability is 0 since the events "heart" and "black" are disjoint.

7.19 (a) 0.50.

(b) 0.70.

(c) 0.50.

7.21 (a) 0.40.

(b) 0.30.

(c) 0.25.

7.23 (a) $P(D) = 110/1160$.

(b) $P(D|O) = 20/470$.

(c) $P(O \text{ and } D) = 20/1160$
$P(O) = 470/1160 \quad P(D) = 110/1160$.
The two events O and D are *not* independent because $P(O \text{ and } D) \neq P(O)P(D)$.

7.25 (a) $P(\text{Hypertension}) = P(\text{H}) = 87/180$.

(b) $P(\text{H}|\text{Heavy smoker}) = 30/49$.

(c) No, since $P(\text{"Hypertension" and "Heavy smoker"}) \neq P(\text{"Hypertension"}) \, P(\text{"Heavy smoker"})$.

$30/180 \neq (87/180)(49/180)$.

7.27 (b) 0.18.

7.29 0.9676.

7.31 $(0.6)(0.6)(0.6) = 0.216$.

7.33 (a) If A and B are independent events then $P(A \text{ and } B) = (0.4)(0.2) = 0.08$.

(b) If A and B are mutually disjoint events then $P(A \text{ or } B) = P(A) + P(B) = 0.6$.

7.35 (a) False.

(b) True.

7.37 (a) No.

(b) Yes.

7.39 (a) 1/3.

(b) 0.7143.

7.41 (a) 0.68.

(b) 0.8235.

7.43 (a) 0.034.

(b) 0.01156.

(c) $(0.1)(0.05)/0.034 = 0.147$.

7.45 (a) 0.0215.

(b) $(0.7)(0.02)/0.0215 = 0.651$.

7.47 (a) 0.2.

(b) 0.1.

(c) 0.4.

(d) 0.25.

7.49 (a) $S = \{$T, WT, WWT$\}$.

(b) $X = 1, 2, \text{ or } 3$.

(c) $S = \{$T, WT, WWT, WWWT, ...$\}$ so $X = 1, 2, 3, 4, ...$

7.51 (a) $S = P(X = 4) = 0.05$.

(b) 0.95.

(c) $E(X) = 1.9$.

7.53 (b) $E(X) = 1.5$, standard deviation =

7.55 (a) The expected value for Y is one.

(b) The variance of Y is zero.

7.57 (a) $P(X - 4) = 1/6$.

(b) $P(\text{at least six cars}) = 0.833$.

(c) $E(X) = 6.833$.

(d) $E(Y) = 2(6.833) - 1 = 12.677$.

7.59 (b) $32 = 1 + 5 + 10 + 10 + 5 + 1$.

7.61 (a) Yes. Each day you win or you don't win, and the probability of winning remains the same all the time, and the outcomes are independent to each other.

(b) No. The responses are not independent.

7.63 (a) H_0: $p = 0.01$ versus H_1: $p > 0.01$.

(b) p-value $= 0.1655$.

(c) If we were to take many, many samples of size 18 from this population, we can expect to see 16.55% of them to have 1 or more audited tax returns, if indeed 1% of tax returns are audited.

(d) Accept the null hypothesis.

7.65 (b) $E(X) = 5$.

(c) 3/10.

(d) (i) to the right
(ii) p-value $= 1/10$
(iii) We reject H_0, p-value $= \alpha$.

7.67 (a) $P(T < 18) = 0.1151$.

(b) $(200)(0.1151) = $ approximately 23.

7.69 (a) $P(1050 < X < 1250)$ is the probability that a randomly selected applicant has an SAT score between 1050 and 1250.
$P(1050 < X < 1250) = 0.7745$.

(b) (i) 97.5%.
(ii) About 1013.

7.71 $\{$DDD, DDN, DND, NDD, NND, NDN, DNN, $NNN\}$.

7.73 (a) $0.12 + 0.10 - 0.17 = 0.05$.

(b) 0.078.

7.75 (a) $P(\text{you win}) = 0.10$.

(b) $P(\text{you win}) = 0.41$.

7.77 (a) 0.60.

(b) 0.35.

(c) 0.18.

(d) 0.30.

(e) 0.60.

(f) No. $0.18 \neq (0.60)(0.35)$.

7.79 (a) 0.30.

(b) 5/11.

(c) No. The events can occur at the same time.

(d) No. $P(\text{Nonsmoker and 0 visit}) = 0.25 \neq P(\text{Nonsmoker}) \, P(\text{0 visit}) = (11/20)(0.3)$.

7.81 (a) Not independent.

(b) Not mutually exclusive.

7.83 The answer is (a).

7.85 (a) Not independent.
    (b) Not mutually exclusive.
    (c) 0.40.

7.87 0.84.

7.89 0.042.

7.91 (a) False.
    (b) True.
    (c) False.
    (d) False.

7.93 (a) $P(X = 4) = 0.05$.
    (b) 0.85.
    (c) 0.25.
    (d) $E(X) = 1.65$.

7.95 (a) 0.80.
    (b) $E(X) = 2.4$.

7.97 (b) Mode = 4.
    (c) $E(X) = 4.58$.
    (d) Less than.
    (e) 0.70.
    (f) 0.75.
    (g) Response bias.

7.99 (a) 0.60.
    (b) $E(X) = 3.19$.

7.101 0.8704.

7.103 (a) 0.37580.
    (b) 0.00569.

7.105 (a) $P(X > 592) = 0.69$.
    (b) 0.478.
    (c) They are not independent events.

Chapter 8

8.1 The sampling distribution of a statistic is the distribution of the values of the statistic in all possible samples of the same size from the same population. We often generate the empirical sampling distribution using some form of simulation.

8.3 The answer is (b).

8.5 (a) Histogram C.
    (b) Histogram B.

8.7 (a) Yes.
    (b) Yes.
    (c) No. $np < 5$.

8.9 0.02275.

8.11 The answer is (iii).

8.13 (a) Approximately normal with mean of 0.85 and standard deviation of 0.0357.
    (b) The sample proportion is 90/100 or 0.90.
    (c) p-value = 0.0808.
    (d) Since the p-value is more than 0.05, the new method is not significantly more accurate at the 5% significance level.

8.15 (a) The true standard deviations are given by:
    $n = 50$, $p = 0.1$ standard deviation = 0.0424.
    $n = 50$, $p = 0.3$ standard deviation = 0.0648.
    $n = 50$, $p = 0.7$ standard deviation = 0.0648.
    $n = 50$, $p = 0.5$ standard deviation = 0.0707.
    $n = 100$, $p = 0.5$ standard deviation = 0.050.

    (b) Answers will vary.
    (c) Answers will vary.

8.17 False. The central limit theorem states that for a large sample size the distribution of $\overline{X}$ is approximately normal; it will not be exactly normal. The larger the sample size, the better the approximation.

8.19 (a) Since X is $N(1250, 150)$, the probability is 0.4719.
    (b) Since $\overline{X}$ is $N(1250, 150/6)$, the probability is 0.9772.
    (c) The probability in (b) is higher since the distribution of $\overline{X}$ is more concentrated around 1250 (i.e., has a smaller standard deviation) than that of X.

8.21 (a) 0.0098.
    (b) $\overline{X}$ is $N(170, 30/2)$.
    (c) $P(\overline{X} > 190) = 0.09121$.

8.23 (a) Fail to operate implies that the total weight exceeds 5000 or equivalently that the average or mean weight exceeds 555.6. The probability is 0.1498.

8.25 (a) Histogram C.
    (b) Histogram F.

8.27 (a) 0.61.
    (b) $E(X) = 1.2$.
    (c) $\overline{X}$ is approximately $N(0.61, 0.9/10)$.
    (d) $P(\overline{X} < 1) = 0.0131$.

8.29 Note that the total number of tickets desired being less than or equal to 220 is equivalent to having the sample mean number of tickets desired for 100 students being less than or equal to 2.2. The sample mean has an $N(2.1, 0.2)$ distribution and the probability is given by $P(\overline{X} < 2.2) = 0.6915$.

8.31 (a) False.
    (b) True.
    (c) False.
    (d) False.

8.33 (a) Histogram B.
    (b) No.

8.35 0.0207.

8.37 (a) The three remaining possible samples are: (9,1), (9,5), (9,9). The sample means for all nine possible samples are: 1, 3, 5, 3, 5, 7, 6, 7, 9.
    (b) 3/9.
    (c) 5.

8.41 (a) $1 - (0.39 + 0.24 + 0.14 + 0.09 + 0.05 + 0.03) = 0.06$.
    (b) Histogram C.

8.43 (a) 0.4078.
    (b) 0.0808.

8.45 (a) (iii) Can't tell.
    (b) $\overline{X}$ is approximately $N(-3, 4/20)$.
    (c) $P(\overline{X} < -2.8) = 0.8413$.

8.47 No.

8.49 Pablo is correct because as the size of each sample used increases, the sampling distribution has less variability. Eduardo is also correct because the value of the parameter p does influence the variability for the sampling distribution of a sample proportion.

Chapter 9

9.1 (a) $H_0: p = 0.50$ where p is the proportion of Democrats in Wayne County.

$H_1: p > 0.50$.

(b) $H_0: p_1 = p_2$ where p_1 is the proportion of male births in Wayne County and p_2 is the proportion of male births in Oakland County.

$H_1: p_1 \neq p_2$

9.3 (iv) is equal to α.

9.5 (ii) p-value = 0.005.

9.7 (a) The population of pregnant women who work with a computer 1–20 hours per week.

(b) The test statistic value $z = 1.4772$. We cannot reject H_0 because the p-value = $0.0698 > 0.01$.

(c) No, the results are not statistically significant at the level 0.01.

9.9 (a) $\hat{p} = 0.7289$.

(b) The p-value is 0.0930. We cannot reject the null hypothesis.

9.11 $H_0: p = 0.2$ versus $H_1: p > 0.2; \hat{p} = 0.72$. The p-value is nearly 0. Reject H_0.

9.13 $H_0: p = 0.20$.

$H_1: p < 0.20$.

The test statistic value $= z = -1.5811$, p-value = 0.0568. We cannot reject H_0 at the level 0.05. The results are not statistically significant at 0.05.

9.15 (a) It may be a sample of typical working mothers with young children.

(b) $H_0: p = 0.5$ versus $H_1: p \neq 0.50$.

(c) No. We do not know how many of the 1200 had children under six.

9.17 Reject H_0 if $Z \geq 2.576$ or $Z \leq -2.576$.

9.19 (a) $\hat{p} = 17/40 = 0.425$.

(b) The 99% confidence interval for the population proportion is $(0.22367, 0.62633)$.

(c) The margin of error $= E = 0.20133$.

9.21 (a) The 95% confidence interval for the population proportion is $(0.66784, 0.73216)$.

(b) p represents the true proportion of all operating vehicles that are equipped with air bags.

9.23 (a) 633.

(b) A statistic.

(c) $\hat{p} = 0.63$.

(d) Standard error of the sample proportion = 0.01524.

(e) $(0.60062, 0.66033)$.

(f) The margin of error $= E = 0.02986 \approx 0.03$. Yes, it is close to the stated value.

9.25 $(0.0151, 0.1099)$.

9.27 A confidence interval is constructed for estimating a population parameter.

9.29 (a) $\hat{p}$ is approximately $N\left(p, \sqrt{\frac{p(1-p)}{n}}\right)$ where $n = 4000$.

(b) No.

(c) Yes.

(d) $\hat{p} = 0.53$, $(0.51, 0.55)$.

9.31 (a) The interval provides plausible values for the true population proportion of abused children with above average anxiety level, based on this sample of 60 such children.

(b) If we repeat this procedure many times, we would expect the true proportion p to lie inside about 95% of the confidence intervals produced in this manner.

9.33 At least 1504.

9.35 (a) (i)–(iv) false. (v) true.

(b) $n \geq 601$

9.37 (a) A statistic.

(b) $(0.79199, 0.92801)$.

(c) $E = 0.06801$.

(d) $n \geq 601$.

(e) $H_0: p = 0.87$ versus $H_1: p < 0.87$.

(f) The test statistic $= z = -0.2974$. The p-value = 0.3831. You cannot reject H_0 at the 0.05 level.

9.39 $H_0: p = 2/3$ versus $H_1: p \neq 2/3$.

9.41 (a) $H_0: p = 0.72$ versus $H_1: p > 0.72$. One-sided to the right.

(b) $H_0: p = 0.90$ versus $H_1: p < 0.90$. One-sided to the left.

(c) $H_0: p = 0.60$ versus $H_1: p \neq 0.60$. Two-sided.

9.43 (a) No, since we do not know the number of pregnant women.

(b) $(0.1344, 0.1656)$.

(c) Approximately 100.

9.45 (a) The results are statistically significant at the 5% level.

(b) Sometimes.

9.47 (a) $(0.491, 0.549)$.

(b) The margin of error is 0.029, that is, approximately 2.9%.

9.49 (a) $H_0: p = 0.50$ versus $H_1: p > 0.50$.

(b) $z = 1.7244$; p-value = 0.0423.

(c) Yes, the p-value is smaller than 0.05.

9.51 (a) $H_0: p = 0.60$ versus $H_1: p > 0.60$.

(b) The test statistic $= z = 1.4434$. The p-value = 0.0745. The results are not statistically significant. We cannot reject H_0 at the 0.05 level.

(c) No. The sample size should be large.

9.53 (a) $(0.44126, 0.51874)$.

(b) The margin of error $= E = 0.03874$.

(c) The margin of error will increase to 0.06066.

9.55 Answers will vary.

Chapter 10

10.1 (a) $H_0: \mu = 33°$ F versus $H_1: \mu < 33°$ F.

(b) $H_0: \mu = 26$ years versus $H_1: \mu > 26$ years.

(c) $H_0: \mu = 200$ versus $H_1: \mu \neq 200$.

10.3 (a) $H_0: \mu = 5000$ versus $H_1: \mu < 5000$.

(b) $z = -2.5$.

(c) p-value = 0.0062.

(d) Yes, since the p-value < 0.01.

10.5 H_0: $\mu = 60$ versus H_1: $\mu < 60$, the test statistic $= z = -2.5$, p-value $= 0.0062$, so we reject H_0.

10.7 (a) tcdf(2.6, E99, 9) $= 0.014369$.
 (b) tcdf($-$E99, -1.4, 49) $= 0.0839$.
 (c) 2 tcdf($-$E99, -2, 14) $= 2(0.03264) = 0.06528$.

10.9 (a) H_0: $\mu = 16$ versus H_1: $\mu \neq 16$.
 (b) Assume the distribution of the amount of syrup is normal.
 (c) The company would turn off the equipment to investigate and lose money.
 (d) The company either loses money or customers would be dissatisfied.
 (e) $t = -1.28$, p-value $= 0.2167$. Do not reject the null hypothesis.
 (f) Our advice to the company is to let the machine continue to run.

10.11 (a) $t = 1.46$; p-value $= 0.0858$.
 (b) Since the p-value is less than 0.10, we reject the null hypothesis.
 (c) Yes, a Type 1 error. The probability of having committed an error is either 0 or 1.

10.13 (a) H_0: $\mu = 12$ versus H_1: $\mu > 12$.
 (b) $t = 1.3765$, p-value $= 0.1136$.
 (c) Assume the data are a random sample and concentration has a normal distribution.
 (d) If the mean concentration were 12 mg/kg, we would see a test statistic of 1.376 or even larger about 11.4% of the time in repeated sampling.
 (e) Do not reject the null hypothesis since the p-value is $> \alpha$.

10.15 (a) $t = 0.475$, p-value $= 0.323$.
 (b) Since p-value is $> \alpha$, we cannot reject the null hypothesis.

10.17 Yes, we have $t = 7.67$ and a p-value of nearly 0. We reject the null hypothesis at the 0.01 level.

10.19 (a) $\bar{x} = 13.5$.
 (b) $(13.094, 13.906)$
 (c) If we repeat this method over and over, yielding many 95% confidence intervals for μ, we expect 95% of these intervals to contain μ.

10.21 (a) $(434.99, 675.41)$.
 (b) The margin of error $= E = 120.21$.
 (c) If we repeat this method over and over, yielding many 95% confidence intervals for μ, we expect 95% of these intervals to contain μ.

10.23 $\bar{x} = 49.2$, the margin of error $= 5$.

10.25 Student #4 is correct, the other students are incorrect.

10.27 $(-3.282, 7.502)$.

10.29 (a) $(75.9, 84.1)$.
 (b) $E = 4.1$
 (c) No. 1 cannot reject the null hypothesis because 82 is inside the confidence interval.
 (d) Narrower.

10.31 (a) Yes.
 (b) Can't tell.
 (c) Can't tell.

(d) No.
(e) Yes.

10.33 The answer is (b).

10.35 (a) H_0: $\mu = 17.1$ versus H_1: $\mu < 17.1$.
 (b) $t = -2.3159$; p-value $= 0.00229$.
 (c) Since the p-value is less than 0.10, we reject the null hypothesis at the 0.10 level.

10.37 (a) 0.9332.
 (b) (i) $1 - (0.9332)^4 = 0.242$.
  (ii) 0.9987.
 (c) H_0: $\mu = 270$ versus H_1: $\mu < 270$.
 (d) The test statistic value $= t = -1.3333$; the p-value $= 0.1019$.
 (e) Since the p-value > 0.05, we cannot reject the null hypothesis.
 (f) There is not sufficient evidence to say that such pregnancies are shorter than 270 days on average.

10.39 (a) Stratified random sampling.
 (b) Using the calculator with a seed value of 331 the selected male student are:

  35, 8, 4, 38, 23, 17, 36, 13, 6, 24.

  Using the random number table (Row 18, Column 1) the selected male students are:

  1, 15, 40, 33, 36, 29, 4, 31, 27, 30.

 (c) (i) $\bar{x} = 7.18$; $s = 1.08$.
  (ii) Min $= 4.2$; $Q_1 = 6.6$; Median $= 7.1$; $Q_3 = 7.9$; Max $= 9.2$.
 (d) H_0: $\mu = 7.5$ versus H_1: $\mu < 7.5$. The test statistic value $= t = -1.3254$; p-value $= 0.1004$.
 (e) Since the p-value > 0.05, we cannot reject H_0. The data are not statistically significant at the 0.05 level.
 (f) $(40/80)(6.82) + (40/80)\,7.18 = 7$.

10.41 (a) Min $= 230$; $Q_1 = 250$; Median $= 265$; $Q_3 = 290$; Max $= 310$.
 (b) H_0: $\mu = \$258$ versus H_1: $\mu > \$258$. The test statistic value $= t = 1.8752$; p-value $= 0.0409$.
 (c) Since the p-value < 0.05, we reject H_0. The data are statistically significant at the 0.05 level.

10.43 (a) $(625.3, 658.7)$.
 (b) (i) False.
 (c) (ii) False.

10.45 (a) $(11.397, 11.603)$.
 (b) 0.10303.

10.47 (a) Apartments in the suburb.
 (b) Living area of the apartment in square feet.
 (c) 1325 square feet.
 (d) $(1316.67, 1333.33)$.
 (e) No.
 (f) Increase the confidence level or decrease the sample size.

10.49 (a) False.
 (b) False.
 (c) False.
 (d) True.
 (e) Accept H_0.

10.51 (a) No.
 (b) 7.4
 (c) Accept H_0.
 (d) Accept H_0.

Chapter 11

11.1 Paired—Each female twin is directly linked to her twin brother.

11.3 Paired—Each woman is directly related (linked) to her husband.

11.5 (a) One example: Ronald & Kyle, Kerri & Sonya, Lee & Pablo, and Emily & Sara.

(b) One example: Ronald & Kyle (20), Kerri & Sonya (18), Emily & Lee, Pablo & Sara.

(c) Let Ronald = 1, Lee = 2, Kyle = 3, Pablo = 4; and Kerri = 1, Emily = 2, Sara = 3, Sonya = 4. Using the TI: (seed = 18): 2, 2 (skip), 1, so the males are: Ronald & Lee, Kyle & Pablo; (seed = 33): 3, 1, so the females are: Kerri & Sara, Emily & Sonya. Using the random number table: (Row 8, Column 1): 3, 1, so the males are: Ronald & Kyle, Lee & Pablo; (Row 22, Column 6): 2, 3, so the females are: Emily & Sara, Kerri & Sonya.

11.7 (a) $H_1: \mu_D > 0$

(b) The data are a random sample from a normal population with unknown population standard deviation.

(c) Mean = 2, Standard deviation = 3.464, $t = 1.5275$, p-value = 0.0887, accept H_0.

(d) The correct answer is (i).

11.9 (a) Let $\mu_D = \mu_1 - \mu_2$. The hypotheses are: $H_0: \mu_D = 0$ vs. $H_1: \mu_D \neq 0$.

(b) The data are a random sample from a normal population with unknown population standard deviation.

(c) There is only one pair (the third pair) for which the highest weight is on scale 2.

(d) Excluding the third pair, we have $(-0.258, 5.458)$.

(e) (ii) The p-value would be greater than 0.05, since 0 is in the interval.

11.11 (a) The observed sample mean is 1.45 standard errors above the hypothesized mean of 0.

(b) 0.095

(c) No, because $p > \alpha$.

(d) No, a confidence interval can only be used for a two-sided test.

11.13 (a) $H_0: \mu_D = 0$ versus $H_1: \mu_D > 0$ where D = second score − first score.

(b) This is a paired t-test so we need to assume that the differences are a random sample from a normal population.

(c) The standard error of the mean difference = $s_D/\sqrt{n} = 41.4/\sqrt{4} = 20.7$. We estimate the average distance of the possible sample mean differences from μ_D to be about 20.7 points.

(d) We have $\bar{d} = 100/4 = 25$. So the observed test statistic is $t = 25/20.7 = 1.208$. Using Table IV: $0.15 < p$-value < 0.20. Using the TI: p-value = 0.157. Decision: accept H_0. We conclude that there is not sufficient evidence to say students improve their score the second time they take the SAT.

(e) (ii) about 5% of the time.

11.15 (a) The differences are: 7, 6, −1, 5, 6, 1.

(b) $\bar{d} = 4$

(c) $H_0: \mu_D = 0$ versus $H_1: \mu_D > 0$ where D = Before − After.

(d) The population of differences is assumed to be normal.

(e) $t = \dfrac{\bar{d} - 0}{s_D/\sqrt{n}} = \dfrac{4 - 0}{3.225/\sqrt{6}} = 3.038$

(f) Using Table IV: $0.01 < p$-value < 0.02. Using the TI: p-value = 0.0144.

(g) Yes

(h) Type I Error

11.17 (a) Common (or equal) population standard deviations.

(b) Yes, the sample standard deviations are similar.

(c) $s_p = \sqrt{\dfrac{(16 - 1)(90)^2 + (16 - 1)(100)^2}{16 + 16 - 2}} = 95.1315$, so the 90% confidence interval is given by $(600 - 550) \pm (1.697)(95.1315)\sqrt{\frac{1}{16} + \frac{1}{16}} \Rightarrow (-7.08, 107.08)$. Using the TI would yield: $(-7.086, 107.09)$.

11.19 (a) $(-0.6385, -0.0615)$

(b) The interval provides a range of plausible values for $\mu_1 - \mu_2$ at a 95% confidence level. If we repeat this procedure over and over, yielding many 95% confidence intervals for $\mu_1 - \mu_2$, we would expect that approximately 95% of these intervals would contain the true parameter value $\mu_1 - \mu_2$.

(c) $H_0: \mu_1 = \mu_2$ versus $H_1: \mu_1 \neq \mu_2$

11.21 (a) 303

(b) We do not know; it depends on the sample sizes and the sample standard deviations.

11.23 (a) $H_0: \mu_1 = \mu_2$ versus $H_1: \mu_1 \neq \mu_2$

(b) $t = -2.579$, p-value = 0.0229

(c) Reject the null hypothesis.

(d) (i) Approximately 95% of the intervals produced with this method are expected to contain $\mu_1 - \mu_2$.

(ii) We would accept the null hypothesis since 9 is in the interval.

11.25 (a) $H_0: \mu_1 = \mu_2$ versus $H_1: \mu_1 > \mu_2$, where 1 = students with a C or higher and 2 = students with a D or an E.

(b) We have $\bar{x}_1 = 4.6$, $\bar{x}_2 = 2.25$, $s_1 = 3.4$, $s_2 = 1.3$, $s_p = 2.7$, $t = 1.841$, and a p-value = 0.0421. We would reject H_0 and conclude that students who earn a C or higher do appear to spend more hours per week outside of class on course work on average as compared to students who receive a D or an E.

(c) Each sample is a random sample from a normal population. The two population standard deviations are assumed to be equal. The two samples are assumed to be independent. The assumption of equal population standard deviation is somewhat suspect, based on boxplots or comparing the sample standard deviations. However, the sample sizes are nearly the same. If we do not assume equal population standard deviations, the test statistic would be $t = 2.01$ and the p-value would be 0.034, so the same conclusion would be reached. Normality seems reasonable.

11.27 (a) $H_0: \mu_1 = \mu_2$ versus $H_1: \mu_1 > \mu_2$
 (b) $t = 1.839$, p-value $= 0.073/2 = 0.0365$; reject H_0 and conclude that the Strat method does appear to produce higher scores on average as compared to the Basal method.
 (c) Same as in part (b).
 (d) Similar. Note the two sample standard deviations are very close.

11.29 (a) $H_0: p_1 = p_2$ versus $H_1: p_1 > p_2$
 (b) $\hat{p}_1 = \frac{34}{40} = 0.85$, $\hat{p}_2 = \frac{19}{35} = 0.543$, $\hat{p} = 0.7067$
 (c) $z = 2.915$ with a p-value of 0.0018, reject the null hypothesis.

11.31 (a) $H_0: p_1 = p_2$ versus $H_1: p_1 < p_2$, where $1 = 1992$ and $2 = 1983$
 (b) $x_1 = (1250)(0.46) = 575$, $x_2 = (1251)(0.53) \approx 663$, $\hat{p} = 0.495$, $z = -3.50$, p-value $= 0.00023$, reject H_0.

11.33 $n_1 = n_2 = \frac{1}{2}\left(\frac{1.96}{0.01}\right)^2 = 19{,}208$

11.35 (a) Observational study
 (b) $1 =$ Parents of infants who died of SIDS $2 =$ Parents of healthy infants
 (c) There is not much detail given. The sample size from each group was 200. Parents of 200 *similar* healthy infants were selected, indicating a possibility of a paired design.

11.37 (a) The sampled differences represent a random sample from a normal population.
 (b) $\frac{0.06}{2} = 0.03$, reject the null hypothesis.

11.39 (a) Sample mean
 (b) Scenario 1 = independent samples; Scenario 2 = paired samples
 (c) Scenario 2
 (d) $H_0: \mu_1 - \mu_2 = 0$ versus $H_1: \mu_1 - \mu_2 < 0$, $t = -0.673$, p-value $= 0.2547$, accept H_0
 (e) $(-1.22, 3.22)$

11.41 (a) Independent samples design
 (b) Paired design

11.43 (a) Treatment group $\hat{p}_1 = 0.0273$, Placebo group $\hat{p}_2 = 0.04137$
 (b) $H_0: p_1 = p_2$ versus $H_1: p_1 \neq p_2$, $z = -2.47$, p-value $= 0.0135$, fail to reject H_0
 (c) $\frac{(0.0273 - 0.04137)}{0.04137} = -0.34$, a 34% reduction

11.45 (a) 9.74
 (b) $H_0: \mu_D = 0$ (where $D =$ Male $-$ Female) and $H_1: \mu_D > 0$
 (c) $t = 2.216$; p-value $= 0.030/2 = 0.015$; reject H_0 and conclude that the male sports do seem to dominate for private schools.

11.47 (a) $H_0: \mu_1 - \mu_2 = 0$ versus $H_1: \mu_1 - \mu_2 > 0$
 (b) $s_p = \sqrt{10{,}039{,}260.33} = 3168.48$
 (c) $t = 1.748$; p-value $= 0.097/2 = 0.0485$; reject H_0.
 (d) It appears that they do.
 (e) No, if we repeated this process many times, we would expect 95% of the confi-

dence intervals generated to contain the true difference in the population means, $\mu_1 - \mu_2$.
 (f) $t = 1.748$; p-value $= .049$; reject H_0; $(-500.52, 5454.52)$. The conclusions are essentially the same.

Chapter 12

12.1 We test the equality of the means by comparing two estimates of the common population variance σ^2. Both will be unbiased estimators if the means of the populations are equal. If the null hypothesis is not true, then the value of MSB/MSW, i.e., the F-ratio tends to be inflated, because the MSB overestimates σ^2.

12.3 (a) Under H_0, all have the same normal distribution, with common mean and common population variance. Under H_1, all have a normal distribution with at least one mean different from the others, but the variances are the same.
 (b) Under H_0, the F-test statistic has an $F(2,30)$ distribution.

12.5 (a) 2.74
 (b) 2.99
 (c) 5.64
 (d) 7.95
 (e) 2.33
 (f) 1.94

12.7 (a) 4 methods
 (b) $n = 64$ observations
 (c) p-value $= 0.0000003$, so the result is statistically significant.

12.9 (a) Experiment
 (b) Number of hours without sleep (or sleep deprivation level).
 (c) Number of failures (to detect a moving object).
 (d) Using the TI: Group 1 = 9, 20, 4, 12, 13; Group 2 = 3, 15, 16, 2, 6; Group 3 = 17, 1, 7, 11, 10; leaving the remaining in Group 4 = 5, 8, 14, 18, 19.
 Using the table we let subject 1 have labels 01, 21, 41, 61, 81 and subject 2 have labels 02, 22, 42, 62, 82, and so on through subject 20 with labels 20, 40, 60, 80, 00. Group 1 = 5, 7, 13, 8, 17; Group 2 = 14, 19, 6, 15, 18; Group 3 = 16, 3, 20, 1, 11; leaving the remaining in Group 4 = 2, 4, 9, 10, 12.
 (e) $H_0: \mu_1 = \mu_2 = \mu_3 = \mu_4$ H_1: At least one population mean is different.

12.11 (a) $H_0: \mu_1 = \mu_2 = \mu_3$ versus H_1: At least one μ_i is different; p-value $= 0.012$, reject H_0.
 (b) 282 units
 (c) By the side-by-side boxplot, the variances seem the same.

12.13 (a)

Source	SS	DF	MS	F
Between	104,855	3	34951.6	9.915
Within	70,500	20	3525	
Total	175,355	23		

(b) MSW = 3525
(c) $H_0: \mu_1 = \mu_2 = \mu_3 = \mu_4$ H_1: At least one population mean is different.
(d) *p*-value = 0.0003
(e) Reject H_0, it appears that not all mean physical fitness scores are the same.

12.15 (a) Accept H_0 since the *p*-value = 0.117 is larger than 0.10.
(b) MSW = 3.242
(c) A large *F*-ratio means that MSB is over-estimating the common population variance. This happens when the population means are not all the same.

12.17 $H_0: \mu_1 = \mu_2 = \mu_3$ versus H_1: At least one population mean is different, where 1 = Group A, 2 = Group B, and 3 = Group C.

Descriptives

MINUTES

	N	Mean	Std. Deviation
Group A	6	17.000	1.789
Group B	6	14.000	1.414
Group C	6	12.667	1.751
Total	18	14.556	2.431

ANOVA

MINUTES

	Sum of Squares	df	Mean Square	F	Sig
Between Groups	59.111	2	29.556	10.726	.001
Within Groups	41.333	15	2.756		
Total	100.444	17			

Since the *p*-value is less than 0.01, we would reject H_0 and conclude that at least one population mean time is different. The three diets do not appear to be the same with respect to the mean time.

12.19 Overall Type I error $= 1 - (1 - \alpha)^c$

c	P(Type I Error)
5	.226
10	.401
25	.723
50	.923
100	.994

12.21 The answer is (c).

12.23 (a) $H_0: \mu_1 = \mu_2 = \mu_3$ H_1: At least one μ_i is different.

Source	SS	DF	MS	F
Between	223.96	2	111.98	31.11
Within	431.96	120	3.6	
Total	655.92	122		

p-value $= P(F \geq 31.11) \approx 0.000$, where *F* has (2,120) degrees of freedom. With this very small *p*-value, we would reject

H_0 and conclude that the mean time on the market for at least one county is different from the others.

(b) We conclude that there is a significant difference between the mean time on market for County 1 and each of the other two counties (County 2 and County 3).

12.25

		Factor A	
		Level 1	Level 2
Factor B	Level 1	20	30
	Level 2	50	60

12.27 (a)

Fertilizer		Water	
	A	B	C
I	28	31.67	23.33
II	33.67	31	25.67
III	28.67	35.67	29.33
IV	31	31.33	26.67

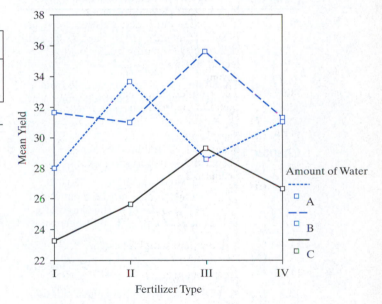

(b) Since the profiles for the three amounts of water are not parallel, there is evidence of interaction. Overall water amount C performs the worst. Amount A works best with Fertilizer II, and Amount B works best with Fertilizer I.

12.29 (a) The significance test for interaction is significant with a *p*-value of 0.006. The significance tests for main effects are also significant with both *p*-values less than 0.0005.
(b) Although the profiles are not parallel, they do not cross. The largest difference between the two types of filters appears to occur for medium sized cars, while there is not much of a difference between the filter types for small or large sized cars.

12.31 (a) We accept H_0 since the p-value of 0.117 is greater than 0.10.
    (b) MSE = 3.242.

12.33 (a) $H_0: \mu_1 = \mu_2 = \mu_3$ H_1: at least one μ_i is different.
    (b)

Source	SS	DF	MS	F
Between	89.6	2	44.8	5.1
Within	351.5	40	8.79	
Total	441.1	42		

    (c) 2.965
    (d) p-value = 0.0106

12.35 (a) $H_0: \mu_1 = \mu_2 = \mu_3$ H_1: at least one μ_i is different.
    (b)

	Sample Size	Sample Mean	Sample Standard Deviation
Bottom	3	55.83	2.259
Middle	5	77.60	3.264
Top	5	52.70	2.255

    The overall mean is 63.0, $n = 13$, $I = 3$, SSB = 1750.477, MSB = 875.238, SSW = 73.161, MSW = 7.316, $F = 119.63$, p-value ≈ 0, reject H_0.
    (c) Each sample is a random sample from a normal population. The three samples are independent. The population variances are equal.

Chapter 13

13.1 (a) $\hat{y} = -30{,}000 + 7{,}000(30) = 180{,}000$.
    (b) $\hat{y} = -30{,}000 + 7{,}000(50) = 320{,}000$. The estimation may not be meaningful since 5000 is outside of the range of x values used to calculate the regression equation.

13.3 (b) $\hat{y} = 13.487 + 1.0649x$.
    (c) $\hat{y} = 13.487 + 1.0649(13) = 27.33$ deaths per 100,000 people.

13.5 (a) A positive approximately linear relationship.
    (b) $\hat{y} = 42.989 + 0.539x$.

13.7 (a) (106, 90) could be influential since it is not consistent with the linear pattern and has a large x-value.
    (b) $\hat{y} = 32.007 + 0.634x$.

13.9 Above, Below, Below, Above, Below, On.

13.11 (a) Negative. The slope of the regression equation $b = -10.877$.
    (b) The regression equation is $\hat{y} = 137.083 - 10.877x$. The predicted number of grievances is $\hat{y} = 137.083 - 10.877(8) = 50.067$.
    (c) Yes. The p-value = 0.001 < 0.05. The data are statistically significant at the 0.05 level.

13.13 (c) $r = -0.77$.

13.15 (c) As one variable increases, the other variable tends to decrease.

13.17 Answers will vary.

13.19 (a) 1
    (b) 0

13.21 (a) $r = 0.908$.
    (b) $\hat{y} = 6.863 + 0.862(30) = 32.72$.
    (c) Unchanged.

13.23 $r = -1$.

13.25 (a) Silence interval before response.
    (b) Number of words used in the response.
    (c) $\hat{y} = 70.837 - 0.282x$
    (d) $r = -0.194$.
    (e) Stay the same.
    (f) Since the p-value = 0.646 > 0.10, we cannot reject the null hypotheses. There is no significant (nonzero) linear relationship between the two variables.

13.27 (b) $\hat{y} = 4.8595 + 0.9394x$.
    (c) $\hat{y} = 4.8595 + 0.9394(70) = 70.6175$.
    (d) residual = e = 70 - 73.436 = -3.436.

13.29 (b) $\hat{y} = 10.528 + 0.953x$.
    (d) $\hat{y} = 10.528 + 0.953(21) = \$3{,}055$ in annual maintenance expenses.
    (e) $r = 0.9252$.
    (g) Yes, the residual plot shows a random scatter in a horizontal band around 0.

13.31 (b) Expect the correlation to be positive, but not close to 1.
    (c) $r = 0.565$.
    (d) No, a truism of statistics is "correlation, not causation."
    (e) Unchanged.
    (f) $r = 1$.
    (g) 0.682
    (h) (i) Unchanged.
        (ii) 1.73

13.33 This may be a consequence of the regression effect. The volunteers will probably have higher blood pressure on average than normal people.

13.35 (a) $r = 0.967$. R Square = 0.936 means that 93.6% of the total variation in vein blood flow rate can be explained by the linear regression between sphere blood flow rate and vein blood flow rate.
    (b) $\hat{y} = 1.031 + 0.902x$.
    (c) $[0.902 \pm 2.306(0.083)] \Rightarrow$ $[0.902 \pm 0.1924] \Rightarrow [0.711, 1.093]$.
    (d) $H_0: \beta = 1$ versus $H_1: \beta \neq 1$. Since 1 is inside the confidence interval we cannot reject H_0 at the 0.05 level.
    (e) $\hat{y} = 1.031 + 0.902(12) = 11.855$.
    (f)

$$\left[11.855 \pm 2.306(1.7566)\sqrt{1 + 1/10 + \frac{(12 - 13.07)^2}{443.2}}\right] \Rightarrow [7.602, 16.108].$$

    (g) Narrower.

13.37 (a) IQ = 183.994 + 0.04222 (Weight) − 3.909 (Height) + 0.0002148 (MRIcount).
    (b) R Square = 0.533. About 53.3% of the variability in IQ can be explained by the explanatory variables weight, height, and MRI count in the multiple linear regression model.

(c) The estimate of the population standard deviation is 18.46.

(d) Each *t*-test statistic is for testing whether the particular regression coefficient is equal to zero. The *p*-values for height and MRI count are < 0.05, so they are statistically significant at the 0.05 level. Weight, however, is not statistically significant at the 0.05 level.

13.39 We need both variables to be quantitative. Gender is qualitative.

13.41 (a) There is a strong positive linear relationship since $r = 0.987$ is close to 1.
(b) $\hat{y} = 4.162 + 15.509(2) = 35.18$.
(c) Residual $= e = 29 - 35.18 = -6.18$.
(d) Unchanged.
(e) Outlier.

13.43 (b) $\hat{y} = -1.781 + 3.145x$.
(c) For a one shoe size increase, we would expect an exam score increase of 3.145 points.
(d) $\hat{y} = -1.781 + 3.145(45) = 70.554$ points.
(e) Age (or grade level).

13.45 (a) Predicted earnings $\hat{y} = 375337.492 - 11940.924(10) = \$255,982.25$.
(b) $r = -0.765$.
(c) (i) Unchanged.
(ii) Unchanged.

13.47 (a) $r = -0.857$.
(b) $\hat{y} = 20.56 - 0.399x$.
(c) $\hat{y} = 20.56 - 0.399(42.2) = 3.72$.
(d) $120,000 = (1.2)(100,000)$ so we expect to have $(1.2)(3.72) = 4.46$, that is, 4 or 5 cases of melanoma.
(e) No, 61.1 is outside the range of the *x*-values used to calculate the regression equation.
(f) (i) $r = 0.857$.
(ii) $\hat{y} = 16.57 + 0.399x$.
(h) Yes, the residuals form a random scatter around the zero line.

13.49 (a) $\hat{y} = 65.775279 - 6.619885x$.
(b) If the chemical production increases by 1 thousand ton per year, the production costs per ton are expected to decrease by \$6.62.
(c) Yes. The *p*-value $= 0.0079 < 0.05$.

13.51 (b) $\hat{y} = 4.08 + 0.2289x$.
(c) Residual $= e = 11.7 - 13.697 = -1.997$.
(d) $(42.1, 13.72); \hat{y} = 4.08 + 0.2289(42.1) = 13.72$.
(e) Yes. $(56, 17.8)$ (iii) Removing this observation would markedly change the regression line.

13.53 (b) Yes, positive.
(c) Fairly strong.
(d) $\hat{y} = 83.27 + 0.117x$.
(e) Predicted airfare $\hat{y} = 83.27 + 0.117(576) = \$150.66$.
(f) Residual $= e = 178 - 150.66 = 27.34$.

13.55 The answer is (ii).

13.57 (a) Negative.
(b) $\hat{y} = 23.354 - 8.168x$.
(c) Residual $= e = 15 - 18.126 = -3.126$.
(d) $[-8.1675 \pm 2.228(2.061)] \Rightarrow [-30.0494, 13.714]$
(e) $\hat{y} = 23.354 - 8.168(0.75) = 17.229$.

(f) 12.6539.
(g) R Square $= 0.611 = 61.1\%$.

Chapter 14

14.1 (a) 7.779
(b) 9.488
(c) 13.277
(d) 1.064
(e) 10.645
(f) 18.549

14.3 (a) With the TI: *p*-value $= 0.3711$; Using Table VI; $0.10 < p\text{-value} < 0.90$.
(b) With the TI: *p*-value $= 0.0845$; Using Table VI: $0.05 < p\text{-value} < 0.10$.
(c) With the TI: *p*-value $= 0.0005$; Using Table VI: *p*-value < 0.005.

14.5 (a) H_0: $p_A = p_B = p_C = p_D = p_E = 0.20$.
(b) The five expected counts are all $(200)(0.20) = 40$.
(c) $X^2_{OBS} = 12.75$
(d) *p*-value $= 0.0126$
(e) Reject H_0, conclude the five pain relief medicines are not all equally effective.

14.7 (a) H_0: $p_1 = 0.25$, $p_2 = 0.50$, $p_3 = 0.25$
(b) $E_1 = E_3 = (0.25)(144) = 26$ and $E_2 = (0.50)(144) = 72$; $X^2 = 3.375$; df $= 2$; $0.15 < p\text{-value} < 0.20$ (or *p*-value $= 0.185$ using the TI). So we accept H_0 and conclude that the professor's theory is supported by the data.

14.9 $X^2_{OBS} = 1.074$; *p*-value is 0.5845, accept H_0.

14.11 (a) About 26% of 1993 or 518.
(b) About 10% of 1993 or 200.
(c) Yes, at most 200.
(d) No, the categories are not disjoint.
(f) No, we do not know anything about how the sample was obtained.
(g) Is it a random sample? Where was the survey conducted?

14.13 (a) 16.67
(b) *p*-value $= 0.0386$, accept H_0, conclude the distribution for patient condition is the same.

14.15 (a) H_0: The distribution of salary for union members is the same as the distribution of salary for nonunion members.

H_1: The distribution of salary for union members is not the same as the distribution of salary for nonunion members.
(b) $E = \frac{20(30)}{70} = 8.57$
(c) $X^2(2)$ distribution
(d) Test statistic: 14.998; *p*-value $= 0.001$; Reject H_0 and conclude the distribution of salary does not appear to be the same for union members and nonunion members.

14.17 (a) H_0: The distribution of alcohol usage is the same for the 3 populations; H_1: The distribution of alcohol usage is not the same for the 3 populations.
(b) Reject H_0 if $X^2 > c$ where $c = X^2_{.95}(2) = 5.99$, or if the *p*-value is less than or equal to 0.05.
(c) You need to know the sample sizes.

(d) $X^2_{OBS} = 23.36$, p-value $= 0.0000085$, reject H_0, conclude that the distribution of alcohol usage does not appear to be the same for the 3 populations.

14.19 (a) H_0: Big toe status and roll tongue status are independent.
(b) 30
(c) $X^2_{OBS} = 6.00$, p-value $= 0.01431$, reject H_0 and conclude that there does appear to be an association between big toe status and roll tongue status.

14.21 (a) H_0: Union membership and attitude are independent.
(b) 37.84
(c) $X^2_{OBS} = 18.54498$, p-value $= 0.00009$, reject H_0, conclude that it appears that membership and attitude are associated.
(d) (i) H_0: The distribution of attitude is the same for the two populations.
(ii) Reject H_0.

14.23 (a) Proportion of men admitted: $90/200 = 0.45$ or 45%. Proportion of women admitted: $60/200 = 0.30$ or 30%. There appears to be an association between gender and admission status. The chi-square test of independence would give a test statistic of 9.6 and a p-value of 0.0019 (or using Table VI: p-value < 0.005.) So there is strong evidence of an association between the two variables.
(b) Both departments admitted equally women and men applicants (50% in the engineering department, but only 25% in the English department). There does not appear to be any association between gender and admission status for either program. In fact the test statistic value for both programs would be 0 and the p-value $= 1$.
(c) Simpson's paradox or aggregation bias.

14.25 (a) The promotion rate for females of $300/1000$ (or 30%) is higher than the promotion rate for males of $120/1000$ (or 12%). A chi-square test homogeneity would yield an extremely large test statistic value of 63.68 and a p-value that is nearly 0. There is very strong support to say the incidence of promotion is not the same for male and female employees.
(b) Yes, as shown in the following possible example where females have a lower promotion rate at both job levels, but a higher proportion of females are in the job level the higher promotion rates overall.

Job Level 1	#	% promoted	# promoted
Males	100	30%	30
Females	900	28%	252

Job Level 2	#	% promoted	# promoted
Males	900	10%	90
Females	100	8%	8

14.27 (a) H_0: $p_1 = 0.25$, $p_2 = 0.50$, $p_3 = 0.25$
(b) 47
(c) p-value $= 0.202$, accept H_0, which supports Mendel's theory.

14.29 (a) H_0: $p_1 = \frac{1}{3}$, $p_2 = \frac{1}{3}$, $p_3 = \frac{1}{3}$ or $p_1 = p_2 = p_3 = \frac{1}{3}$
(b) $E_1 = E_2 = E_3 = (120)\left(\frac{1}{3}\right) = 40$

$$X^2_{OBS} = 3.35$$

Using the TI: p-value $= 0.1873$.
Using Table VI: $0.15 < p$-value < 0.20.
So we accept H_0.

14.31 (a) H_0: $p_W = 0.61$, $p_A = 0.24$, $p_H = 0.06$, $p_B = 0.05$, $p_O = 0.04$
(b) We need to know the sample size.
(c) Approximately $n = \frac{1127}{0.055} = 20{,}490$.
(d) The table of observed and expected counts is:

	White	Asian American	Hispanic	Black	Other
Observed	6557	7991	2828	1127	1987
Expected	12,498.9	4917.6	1229.4	1024.5	819.6

(e) $X^2_{OBS} = 8497$, p-value ≈ 0, reject H_0, conclude that the 1984 model does not hold.

14.33 (a) H_0: The potency acceptance rates are the same for the 2 formulations.
(b) $X^2_{OBS} = 8$, p-value $= 0.00468$, we reject H_0, conclude that there appears to be a significant difference in the rates, with formulation 2 having the higher acceptance rate.

14.35 (a) The completed table is given by:

		District Type		
		Rural	Suburban	Urban
	Good Grades	38.3%	57.6%	68.6%
Goal	Popularity	33.6%	27.8%	17.1%
	Athletic Ability	28.2%	14.6%	14.3%

(b) The conditional distribution of goal given district type.
(c) Yes
(d) $X^2_{OBS} = 18.56$, p-value $= 0.00096$, we reject H_0.

14.37 (a) The conditional distribution of smoking status given treatment group is:

		Treatment Group		
		Group 1	Group 2	Group 3
	Nonsmoker	19.8%	25.0%	19.2%
Smoking Habit	Ex-Smoker	27.1%	24.0%	24.9%
	Smoker	53.1%	51.0%	55.9%

(c) 21.4% nonsmokers, 25.3% ex-smokers, 53.5% smokers.

(e) Yes

Chapter 15

15.1 (a) $R_{OBS} = 7$; p-value $= 0.1719 > 0.05$, accept H_0.

(b) $R_{OBS} = 7$; p-value $= 0.9453 > 0.05$, accept H_0.

(c) $R_{OBS} = 7$; p-value $= 2(0.1719) = 0.3438 > 0.05$, accept H_0.

(d) The sample is selected randomly from a continuous population.

15.3 $H_0: \pi_{0.5} = 258$ versus $H_1: \pi_{0.5} > 258$

$R_{OBS} = 10$; p-value $= 0.15093 > 0.05$ so accept H_0. The results are not statistically significant at the 5% level.

15.5 (a) $H_0: \pi_{0.5} (pop\ K) = \pi_{.5} (pop\ G)$ versus $H_1: \pi_{0.5} (pop\ K) > \pi_{.5} (pop\ G)$

(b) $W = 5 + 9 + 11 + 12 + 14 + 15 = 66$

(c) 48

(d) 8.49

(e) 2.12 standard deviations above

(f) 0.018

(g) The claim made by broker G, that its homes spend less time on the market, is supported by the data.

15.7 (b) $H_0: \pi_{0.5} (pop\ 1) = \pi_{0.5} (pop\ 2)$ versus $H_1: \pi_{0.5} (pop\ 1) > \pi_{0.5} (pop\ 2)$ where 1 = twin and 2 = single.

(c) $W = 5 + 6 + 8 + 10 + 11 = 40$; p-value $= 0.041$

(d) Reject H_0.

(e) (i) Equal population standard deviations $\sigma_1 = \sigma_2$

(ii) $t(9)$ distribution

15.9 $H_0: \pi_{0.5} (pop\ A) = \pi_{0.5} (pop\ B)$ versus $H_1: \pi_{0.5} (pop\ A) \neq \pi_{0.5} (pop\ B)$.

$W = 66$; p-value $= 2(0.059) = 0.118$

Accept H_0 and conclude that there is not a significant difference between the median scores for the two schools at the 10% level.

15.11 $H_0: \pi_{0.50} (D) = 0$ versus $H_1: \pi_{0.50} (D) > 0$ where D = December − January

$W^+ = 46$ and the p-value $= 0.032$ so we reject H_0. There is sufficient evidence to say the median number of baggage-related complaints has decreased overall.

15.13 (a) $H_0: \pi_{0.50}(D) = 0$ versus $H_1: \pi_{0.50}(D) \neq 0$ where D = without − with

(b) $W^+ = 11$ and the p-value $= 0.013$ so we reject H_0.

15.15 $H_0: \pi_{0.5} = 18$ versus $H_1: \pi_{0.5} \neq 18$

$R_{OBS} = 2$; p-value $= 0.2892 > 0.05$ so we accept H_0. The results are not statistically significant at the 5% level. The data do not refute the claim that the median percentage of chromium present is 18%.

15.17 (a) $t = 3/1.5 = 2$

(b) $df = 18$, p-value is 0.0608 or from Table IV: $0.05 < p\text{-value} < 0.10$.

(c) Wilcoxon rank sum test.

(d) $\mu_W = \frac{10(21)}{2} = 105$

15.19 (a) Signed rank test

(b) $W^+ = 3 + 4 + 5 + 6 = 18$ and the p-value $= 0.078$

(c) Reject H_0

(d) The Paris weekend seems to be more popular than the beach package.

INDEX